Thin Film
Solar Cells

Thin Film Solar Cells

Kasturi Lal Chopra
Indian Institute of Technology
New Delhi, India

and

Suhit Ranjan Das
National Research Council
Ottawa, Ontario, Canada

PLENUM PRESS ● NEW YORK AND LONDON

Library of Congress Cataloging in Publication Data

Chopra, Kasturi L., 1933-
Thin film solar cells.

Bibliography: p.
Includes index.
1. Solar cells. 2. Thin films. I. Das, Suhit Ranjan, 1931- . II. Title.
TK2960.C48 1983 621.31'244 82-2126
ISBN 0-306-41141-5

A Division of Plenum Publishing Corporation
233 Spring Street, New York, N.Y. 10013

Printed in the United States of America

Dedicated to

OUR PARENTS
who have been a source of inspiration to us
in our endeavor to explore
the frontiers of science and technology
in the service of mankind

Preface

> *"You, O Sun, are the eye of the world*
> *You are the soul of all embodied beings*
> *You are the source of all creatures*
> *You are the discipline of all engaged in work"*
>
> — Translated from Mahabharata
> 3rd Century BC

Today, energy is the lifeline and status symbol of "civilized" societies. All nations have therefore embarked upon Research and Development programs of varying magnitudes to explore and effectively utilize renewable sources of energy. Albeit a low-grade energy with large temporal and spatial variations, solar energy is abundant, cheap, clean, and renewable, and thus presents a very attractive alternative source. The direct conversion of solar energy to electricity (photovoltaic effect) via devices called solar cells has already become an established frontier area of science and technology. Born out of necessity for remote area applications, the first commercially manufactured solar cells — single-crystal silicon and thin film CdS/Cu_2S — were available well over 20 years ago. Indeed, all space vehicles today are powered by silicon solar cells. But large-scale terrestrial applications of solar cells still await major breakthroughs in terms of discovering new and radical concepts in solar cell device structures, utilizing relatively more abundant, cheap, and even exotic materials, and inventing simpler and less energy intensive fabrication processes.

No doubt, this extraordinary challenge in R/D has led to a virtual explosion of activities in the field of photovoltaics in the last several years. Such new devices as barrier layer, inversion layer, cascade junction, tandem junction, spectrum splitting and shifting, and photoelectrochemical cells have emerged. Exotic materials of the type a-Si : H(F), a-SiC : H, Zn_3P_2, $CuInSe_2$, quaternaries and penternaries, and graded multicomponent semiconductors are very attractive options today. New materials technologies such as ribbon, LASS, TESS, RTR, spray pyrolysis, magnetron sputtering, ARE, chemical solution growth, and screen printing are much talked about. The theoretical understanding of the photovoltaic process has advanced considerably as a result of the development of more refined and realistic models of various types of junctions and the associated solid state physics, particularly that applicable to polycrystalline materials. Nevertheless, the R/D explosion has not yet reached its culmination.

A characteristic feature of good R/D activity is turbulent evolution of the field through critical published literature, seminars, conferences, reviews, and books. If scientific attention is a measure of the importance of a field, photovoltaics is among the top few frontiers today which continue to be hotly debated in conferences and in reviews all over the world. In the area of published literature, H. J. Hovel's book *Solar Cells* (1975) has served as a useful textbook for several years. A recent textbook, *Solar Cell Device Physics* by S. J. Fonash (1981), has brought the subject of the physics of junctions up to date. D. Pulfrey's book *Photovoltaic Power Generation* (1978) contains basic theory of junctions and emphasizes systems aspects of photovoltaic conversion. *Solar Energy Conversion: The Solar Cell* by Richard C. Neville (1978) provides good coverage of the related solid state physics, device design, and systems. An indispensable handbook for the systems engineer is provided by H. Rauschenbach in his book *Solar Cell Arrays* (1980). However, none of these books has addressed the problem from the *universally* accepted point of view that viable devices for terrestrial applications must necessarily be Thin Film Solar Cells. Such aspects as thin film materials; the associated preparation, measurement, and analysis techniques; and device technology are not discussed, or are treated perfunctorily at best in these books. The importance of these aspects and the fact that we in the Thin Film Laboratory have been engaged in very extensive R/D activities with thin film solar cells for over a decade have inspired us to undertake the job of writing a comprehensive book on the subject.

This book, consisting of 12 chapters, begins with a scientific, technological, and economic justification of "why thin film solar cells?" in Chapter 1. This is followed by a detailed description of the electron transport and optical processes in monocrystalline, polycrystalline, and amorphous semiconductors, and in metal films in Chapter 2. Different types of electronic junctions and the associated physics are presented in Chapter 3. Chapter 4 discusses measurement techniques for the analysis of junctions. A comprehensive review of the major deposition techniques of interest to the field of thin film solar cells forms Chapter 5. The significant physical properties of thin film materials for solar cell applications are discussed in Chapter 6. Chapters 7 through 10 are devoted to a description of the fabrication and critical discussion/analysis of the performance of solar cells based on copper sulfide, polycrystalline silicon, new and emerging materials, and amorphous silicon. An attractive alternative to an all solid state solar cell is a photoelectrochemical cell, which is the theme of Chapter 11. The last chapter deals with novel and futuristic concepts that have been proposed and experimented with for obtaining high-efficiency solar cells. Finally, six appendixes deal with the solar spectrum, antireflection coatings, grid design, solar cell arrays, concentrators, and degradation and encapsulation of solar cells. In all, the book is illustrated with over 200 figures and contains 42 tables of accumulated data.

In a rapidly changing field, strong and prejudiced views, errors, and omissions are inevitable. This book is no exception and we take full responsibility for it. We do, however, appeal to our readers to be critical and to communicate its shortcomings to us. As emphasized in Chapter 1, an unambiguous choice of a viable thin film solar cell material has yet to emerge. And, thus, some or even a substantial part of what is presented here as technology may become obsolete very soon. Indeed, we would be very disappointed if this did not happen!

Despite the rapidly evolving nature of the field, this book represents the first major attempt to expose graduate students and R/D scientists and engineers to a comprehensive treatment of many facets of materials, technologies, and solid state physics of thin film solar cell devices. We earnestly hope that this book, aside from serving as a text and research-cum-reference volume, will inspire the readers toward much awaited innovations in the field to make thin film solar cells viable enough to serve the energy-hungry societies of tomorrow.

Acknowledgments

No book of this size can be written without consulting and looting a vast number of published papers, reviews, and books. We are grateful to numerous authors and publishing companies for permitting us to reproduce figures and data in tables. We thank all the members of the Thin Film Laboratory for their generous help and assistance in various forms in the preparation of the manuscript. In particular, we are grateful to R. C. Kainthla for his help in writing Chapter 11 and to D. K. Pandya, V. Dutta, I. J. Kaur, E. Shanthi (Iyer), Sarita Thakoor, P. K. Gupta, and Rajesh Mehta for collecting material for the appendixes. R. C. Budhani, Bharat Bhushan, S. Harshvardhan, S. Major, O. S. Panwar, T. V. Rao, Bodhraj, Madhu Banerjee, Satyendra Kumar, M. Rajeshwari, Jagriti Singh, K. Chidambaram, G. B. Reddy, Harminder Singh, and Mangal Singh assisted with figures, tables, and references. We are indebted to D. F. Williams and A. Banerjee for reading the manuscript and offering their critical comments.

We gratefully acknowledge the support of Mr. N. S. Gupta for drawing the figures and of V. N. Sharma, S. D. Malik, H. S. Sawhney, Keshav Giri, Susan Farrell, Nicole Paquette, and Kim Burke for typing the manuscript. We thank the Indian Institute of Technology, New Delhi, for extending its facilities and providing financial support.

Finally, the book owes much to the exemplary patience and moral support of our neglected families, in particular our wives.

K. L. Chopra and S. R. Das

Contents

Chapter 6. Properties of Thin Films for Solar Cells 275

Chapter 7. Cu_2S Based Solar Cells 349

1

Why Thin Film Solar Cells?

1.1. *Introduction*

The 1970s have brought the end to thoughts of unlimited cheap energy. Today, energy consumption per capita is synonymous with the standard of living of a nation, as can be seen vividly in Figure 1.1. Whether or not man should continue an energy-intensive-based lifestyle may be open to debate, but there is little doubt that the present world rate of energy consumption is alarming in view of the rapid depletion of existing conventional resources.

The situation can perhaps be put into perspective most vividly by considering the reserves and current use of the most well-known major energy resource — petroleum. The known petroleum reserves in the world can provide about 70 Quads (1 Quad $\sim 10^{18}$ joules), and the cumulative production is about 1.6 Quads. Even if it is assumed that there will be no increase in consumption (a major assumption in view of the increasing world population and the trend toward higher technologies), it is apparent mankind is heading toward an energy crisis that can be averted only through conservation coupled with a renewable energy technological revolution. Action to develop new sources of energy, however, must take place immediately if these new sources are to be in place when they are required. This urgency is due to the fact that the growth and decay of the market penetration of conventional energy sources, according to Marchetti,[1] follow a logistic behavior of the type

$$F/(1-F) = \exp A(t-T)$$

where F is the market fraction, t is the time, T is the date on which $F = 0.5$, and A is a rate constant. T and A vary from country to country. Figure 1.2 shows that on a global scale it would require a century to displace 50% and 20 to 21 years to displace 5% of other sources of energy

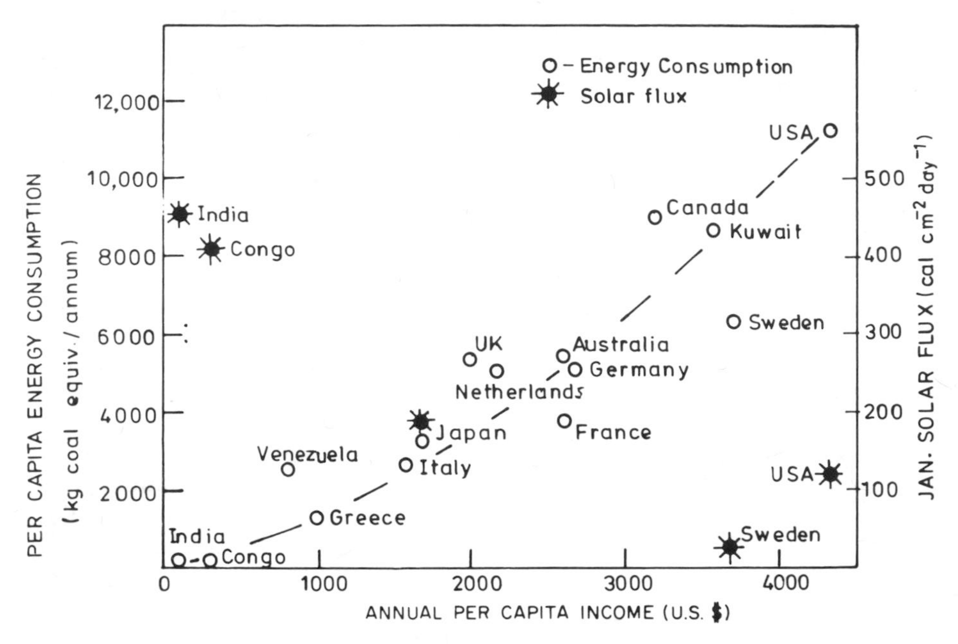

Figure 1.1. Energy consumption as a function of per capita income for several countries (data from Parikh[26]). Also shown are the data points for solar flux for some of the countries (data from Meinel and Meinel[27]).

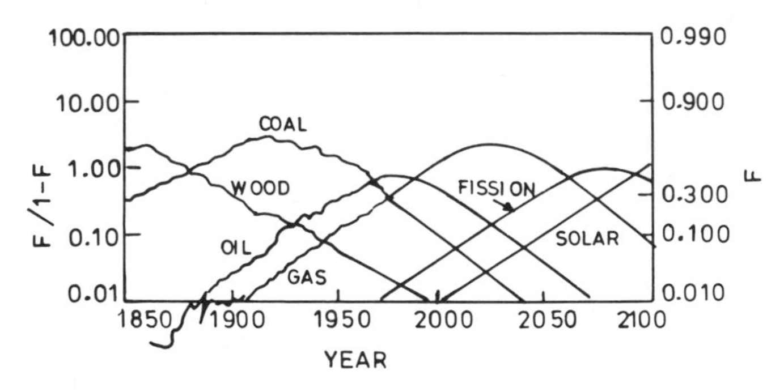

Figure 1.2. Energy market penetration history and projection for the world (from Marchetti[1]) for various energy sources. F is the market fraction defined in the text.

by solar technologies (other than hydropower). The numbers are half for the United States. Note that even the fission power scenario is no better.

Renewable energy sources that can be developed involve the harnessing of more or less continuous natural energy flows (such as sunlight, ocean currents, waves, falling water, and wind) or natural energy stocks whose replenishment is far greater than projected human use (ocean thermal gradients, etc.). Of this wide range of energy sources, sunlight or solar energy is surely one of the most attractive. Solar energy is diffused, intermittent, spatially and temporally variable, and ubiquitous and diverse in its manifestations. And, quoting Denis Hayes,[2] "no country uses as much energy as is contained in the sunlight that strikes just its buildings." Ironically, the poorer and less developed nations of the world have more of this natural gift (see Figure 1.1). The magnitude of solar energy as a potential energy source is seen when one recognizes that the total solar energy falling on the face of the earth is approximately 7.45×10^{17} kWh annually (~ 3200 Q) as compared with the present world consumption (in the form of electricity and other forms) of about 0.5×10^{14} kWh annually.

The major problem facing the technologist is thus how can he use solar energy to provide preferably high-grade energy in a useful form which is economically competitive, recognizing the added constraint that the materials he uses must also be readily available and abundant.

Though not the only method, the direct conversion of solar energy to electricity by a photovoltaic technology has a number of technological and social advantages over other energy technologies. For example, in addition to the positive aspects of using sunlight, photovoltaic systems are quiet, require little maintenance, have no critical size and indeed size can be matched to load with little loss in efficiency, can be physically located near the load, and are environmentally benign in operation. It is thus one of the most attractive future technologies.

1.2. *Solar Energy Conversion*

The solar spectra in the form of energy density E_λ and photon flux N_{ph} are shown in Figure 1.3a, b for different air mass conditions (see Appendix A). The AM0 spectrum can be approximated by the spectrum of a blackbody radiating at 5900 K. The energy density extends from 0.3 to 2.5 μm and peaks at a wavelength of 0.5 μm.

The cheap but low-grade solar energy may be converted to other forms of higher-grade energy through one of several methods, such as photothermal, photochemical, photoelectrochemical, photobiochemical, and photovoltaic. Among these, as noted above, the cleanest and most direct and efficient mode of conversion to electrical power is with the help of photovoltaic (PV) or solar cell devices.

To appreciate the scale of application of such devices, let us examine the electric power requirements of the Unites States and India. We assume a worldwide figure of 0.6 kW m^{-2} solar flux averaged over an entire year and a 10% conversion efficiency PV system. To produce 1% of the total

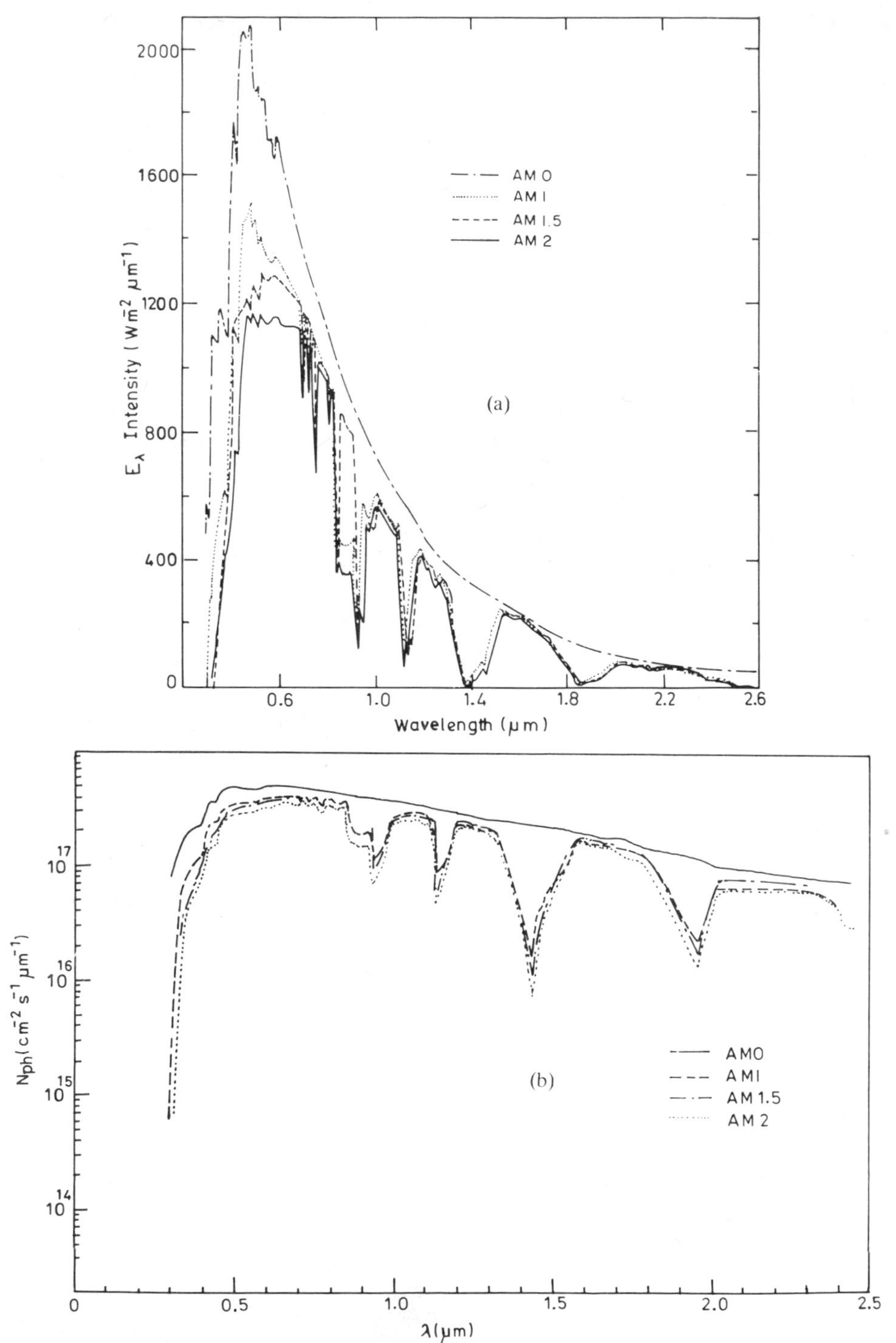

Figure 1.3. Solar spectra for different atmospheric conditions in the form (a) energy density E_λ and (b) photon flux N_{ph}.

electric power requirement of the United States and India today would need PV panels of 100 and 2 MW_p per year, respectively, where W_p is a peak watt, the power produced by a solar cell at the maximum power point of the load curve. If the solar cell is exposed to AM1 solar illumination at normal incidence (1 kW m^{-2}) and it has a 10% conversion efficiency, the 1 m^2 of cell area would deliver 100 W_p. In the year 2000, the numbers of PV panel requirements would escalate by a factor of at least 20. A 100-MW_p PV plant would require a surface area coverage of 200 km^2. The present-day worldwide production of PV panels (primarily single crystal silicon) is about 5 MW_p. To increase the PV power generation capacity at a uniform rate of 1% every 10 years would require that a PV program deliver 2000 MW_p per year, which is more than 1000 times greater than the present entire world capacity. These realities led Ehrenreich and Martin[3] to remark that "the naive impatience of some solar advocates to deploy solar cells immediately and on a large scale appears to be associated with an inadequate appreciation of the technological problem." The situation for underdeveloped countries is fortunately not so serious in view of their smaller demands and thus modest goals of 1 to 5% of their needs by PV systems by the year 2000 are technically feasible.

In all countries, however, there are even now specific tasks for which small PV power systems seem ideal, e.g., as stand-alone systems providing power for repeater stations for telecommunications and for pumps to provide water for irrigation and human use. Indeed, optimized PV systems are expected to be in the 1 to 10 MW_p capacity serving, in the first instance, rural, underdeveloped, and remote areas.

1.3. *Efficient Conversion*

Basically, photovoltaic conversion occurs through three separate processes: (1) the absorption of light to create electron–hole pairs in an appropriate semiconductor, (2) collection and separation of these carriers by an internal electric field, and then, (3) distribution to an external load. Absorption and carrier generation occurs most for photons of energy greater than the bandgap of the material. Obviously, for greater carrier generation and, consequently, higher photocurrent, the bandgap energy should be small. However, the open-circuit voltage that is available is determined by the bandgap and is, for optimally designed devices, about half the bandgap (see Chapter 3). Thus, for high-efficiency solar cells with high photocurrents and high open-circuit voltage, the semiconductor bandgap has to be matched with the solar spectrum, and it can be shown that the optimum bandgap is in the range of 1.1 to 1.5 eV. Also the material should have a high value of absorbance to ensure capture of all available photons. Since high absorbance is often found in limited wavelength regions for any one material, if we wish to use the full solar spectrum it is essential to have a series of solar cells of varying and/or graded bandgaps arranged in cascade or acting in tandem or constructed in an integrated, multilayer, tandem structure. Alternatively, the solar spectrum is split into different spectral regions by filter mirrors and each of the split beams is directed to individual solar cells with response and bandgap matched to that particular region. These novel concepts are discussed in Chapter 12. By appropriate combinations of as many as 36 cells, theoretical efficiencies as high as 72% are expected.[4] In Figure 1.4 we show the fractions of solar spectrum utilized and the conversion efficiencies expected for different cascade combinations employing cells of 1, 2, 3, and 36 bandgaps.

These are not strictly academic exercises in calculations. Indeed, tandem solar cells with efficiencies approaching 27% have been achieved. It is to be noted that if we could utilize the full spectrum, then we would expect a Carnot cycle efficiency of $\eta = (T_1 - T_2)/T_1 = 95\%$, taking the sun's temperature to be about 6000 K.[4]

Efficient conversion is also possible with a hybrid system[5] wherein silicon or germanium PV systems deliver electricity and a heat-extracting fluid extracts the thermal energy from the heated panels. The overall efficiency of the integrated photovoltaic/photothermal system could be quite high.

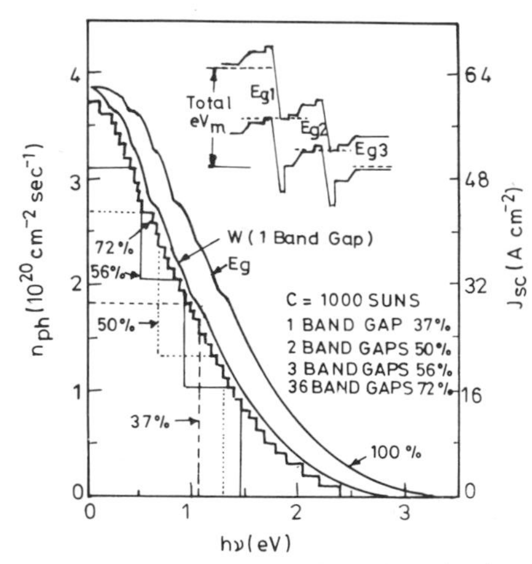

Figure 1.4. Graphical representation of the efficiencies of solar cells of 1, 2, 3, and 36 bandgaps. The step heights are equal to the photon flux absorbed by each energy gap and the step widths (measured from the origin) are equal to the maximum energy per absorbed photon delivered to the load. The efficiency of each cell is given by the ratio of the area enclosed by the steps and the area under the outer curve, labeled 100%. The outer curve is E_g vs. n_{ph} (the absorbed photon flux) and the area under the outer curve is the solar power per unit area. The inner curve is W (the work per absorbed photon) vs. n_{ph} (after Henry[4]).

1.4. *Materials Requirements for Efficient Conversion*

It is clear that high conversion efficiency is possible by employing multiple/tandem/cascade junctions, or by integrating photovoltaic/ photothermal systems. This, however, imposes a serious constraint on the choice of suitable PV materials. If we examine the available materials (see Chapter 3) with an appropriate bandgap, absorption coefficient α, minority carrier lifetime τ, or minority carrier diffusion length L_D, and surface recombination velocity S_r, the choice is limited to such materials as Si, GaAs, CdTe, Cu_2S, Zn_3P_2, InP, and $CuInSe_2$. Except for Si, all other materials, by virtue of their electrical and optical properties must necessarily be in thin film form. However, it is not sufficient for the PV material to have the desired optical and electrical properties; it is also necessary that it be readily available in abundant supply in the world. If in the long term we wish to produce 10% of the total world requirement of electricity, solar cells of 10% efficiency will require about 10^7 kg per year of the materials. The availability and present world production of some of the useful materials are listed in Table 1.1. It is evident that the choice is limited to such materials as InP, GaAs, Cu_2S, Zn_3P_2, and $CuInSe_2$ in thin film form. Although Si is more fortunately placed in this regard vis-a-vis other materials, it is possible that, owing to other technical and economic reasons, Si will also be used in thin film form for the very large-scale application envisaged above.

Thus, thin film materials and their associated technologies offer an extremely attractive approach toward the production of efficient cost-effective PV cells. The technological aspects of thin film PV cells can be put into perspective when we consider the mechanisms governing the formation of thin films in general.

1.5. *What Are Thin Film Materials?*

A thin material created *ab initio* by an atom/molecule/ion/cluster of species-by-atom/molecule/ion/cluster of species condensation process is defined as a "thin film." Thin materials may also be formed from a liquid or a paste, in which case it is called a "thick film." It is not the thickness that is important in defining a film, but rather the way it is created with the consequential effects on its microstructure and properties.

Depending on how the atoms/molecules/ions/clusters of species are created for the condensation process, the methods for depositing thin films are termed physical vapor deposition (PVD), chemical vapor deposition (CVD), electrochemical deposition (ECD), or mixtures of PVD and CVD (hybrid). These methods are described in Chapter 5.

Table 1.1. Production, Cost, and Concentration in Earth's Crust of Elements Appearing in Solar Cell Materials[24,25]

Element	World production (kg/year)	Cost ($/kg)	Concentration (at.%)
Se	1.3×10^{6}	33.0	—
S	3.4×10^{10}	0.05	—
Sb	6.7×10^{7}	4.0	—
As	—	0.44	—
Mg	2.5×10^{8}	2.2	1.84
Bi	3.0×10^{6}	13.0	—
Sn	1.8×10^{8}	13.0	2.0×10^{-4}
P	1.2×10^{11}	0.03	0.11
In	4.7×10^{4}	300.0	—
Cu	6.8×10^{9}	1.3	5.5×10^{-3}
Co	2.0×10^{7}	14.0	2.5×10^{-3}
Pb	3.7×10^{9}	0.7	1.3×10^{-3}
Zn	4.3×10^{9}	0.8	7.0×10^{-3}
Pt	8.3×10^{4}	6400.0	1.0×10^{-6}
Cd	1.8×10^{7}	6.0	2.0×10^{-5}
Te	2.6×10^{5}	44.0	—
Ti	2.6×10^{7}	6.5	0.44
Ni	6.8×10^{8}	4.4	7.5×10^{-3}
Fe	2.0×10^{11}	0.07	1.92
Hg	8.3×10^{6}	3.8	8.0×10^{-6}
Ag	1.2×10^{7}	144.0	7.0×10^{-6}
Au	—	—	4.0×10^{-7}
Ga	1.0×10^{4}	550.0	—
Ge	2.0×10^{4}	300.0	—
Si	5.0×10^{8}	1.0	21.22
Al	1.4×10^{10}	1.2	6.47
Cr	9.0×10^{9}	0.06	1.0×10^{-2}
Mo	—	—	1.5×10^{-4}
W	—	—	1.5×10^{-4}

Vapor atoms impinging on a substrate lose their kinetic energy and are absorbed on the surface as ad-atoms. The movement of these ad-atoms depends on numerous conditions, in particular the energy of the vapor atoms, the rate of impingement, the absorption and desorption activation energies, the topography and chemical nature of the substrate, and lastly, the substrate temperature.[6-9] An ad-atom is not stable by itself. As dimers, trimers, and multimers are formed by random collision processes on the substrate, their stability increases simply because of the increasing number of bonds between the ad-atoms which overcome the disruptive surface energy. As a critical size of these monomers is reached and a "nucleation" barrier crossed, the ad-atom cluster becomes stable and is chemically absorbed. The size of the critical nucleus in most cases is of atomic dimensions.

The growth of a thin film can take place by one of three modes: (1) layer-by-layer, which occurs if either on one extreme the ad-atoms have little mobility (as in amorphous deposits), or under the extreme conditions of very low supersaturation, single crystal substrate, and ultrahigh vacuum deposition; (2) Stranski–Kaitchev mode, in which case the film grows just as in the layer-by-layer mode and then converts itself into three-dimensional nuclei; and (3) three-dimensional growth of the discrete nuclei. The last mode is most common for oriented thin films. If the size of nucleation centers is small and ad-atom mobility is large, the growth yields a platelet type of growth which resembles two-dimensional growth.

For our purpose, it suffices to say that growth takes place at nucleation sites both laterally (in the plane of the film) and perpendicular to the film. The number of grains and grain size is more or less determined by the density of nucleation centers unless large-scale coalescence and/or recrystallization leads to sintering and increase of grain sizes. The growth in the perpendicular direction takes place in a columnar fashion and thus is anisotropic with a grain size perpendicular to the substrate determined by the film thickness and, of course, by recrystallization/coalescence processes. Generally, the lateral grain size will be about the film thickness for small thicknesses or a fraction of the thickness in case of very thick films. This mode of growth, consisting of nucleation and ad-atom mobility dominated coalescence, has the following consequences:

1. Different technologies for large-area coatings, ranging from the very simple spray process to the highly sophisticated molecular beam epitaxy (MBE) exhibit similar stages of growth. The film is first discontinuous, next a network, and finally continuous. The critical thickness t_c, at which the film becomes continuous, depends markedly on nucleation sites and substrate temperature T_s.

2. Nucleation centers can be modified. For example, by depositing SiO on glass the nucleation centers are increased and t_c is lowered. Thus a layer-by-layer growth is possible. Also, by depositing at low substrate temperatures, which kills ad-atom mobility, a layer-by-layer growth is possible.

3. Mismatch of randomly formed nuclei results in a large variety of structural defects: point (vacancies), line (dislocations), and planar (stacking faults). These defects are obviously connected with the density of nucleation centers and hence the grain size. With grain sizes of about 100 Å, one may expect frozen-in vacancies of about 10^{-2} at.% in thin (~ 1000 Å) films. Similarly, a dislocation density of about 10^{12} lines cm^{-2} is commonly observed. In oriented or epitaxial films, dislocation densities are lower — about 10^4–10^6 lines cm^{-2}.

4. Owing to the granular structure and mismatch of grains, large stresses are developed, both compressive and tensile, of the order of 10^{10} dynes cm^{-2}, approaching the fracture strength of most materials.

5. The effective surface area of a film depends on its microstructure. If the film has a porous or columnar nature (as obtained by oblique deposition), the area increases with film thickness. For growth at high T_s with large grains and for epitaxial layers, the effective area approaches the geometrical area.

6. The smaller the grain size (micropolycrystalline), the larger the changes in lattice constant expected.

7. Solubility conditions are not restrictive in the vapor phase. Therefore, co-deposition of vapors results in alloys and compounds over an extended range of solubility. a-Si : H is a good example.[10]

8. Since the free energies of different structures are very similar, the additional energies due to surface, electrostatic, and strain energies, etc., result in the easy formation of metastable structures — primarily polymorphs which are usually obtained at high temperatures and/or high pressures. β-Ta is an example of such a metastable structure.[6]

9. Co-deposition and/or multiple deposition allows a spatial and thickness gradient of composition and thus films with spatially variable properties.

10. By using appropriate substrates and deposition conditions, the resulting film microstructure can be varied from amorphous at one extreme to epitaxial at the other. Thus amorphous, micropolycrystalline, oriented, and epitaxial deposits are possible in films of the same material, as in the case of Si.

11. Surface roughness depends on mode of growth. Layer-by-layer (low ad-atom mobility) growth yields atomistically smooth films. Oriented growth, where grains develop orientation, generally yields rather rough surfaces, as for example in CdS.[11] However, W or Pt oriented films are very smooth.[6]

1.6. *Role of Thin Films in Solar Cells*

We have already mentioned that from the point of view of availability of raw materials, a PV technology must necessarily be based on the thin film form of suitable photovoltaic materials. Thin film devices would typically be about 5 to 50 μm thick, in contrast to bulk devices which are about 150 to 250 μm thick. It should be pointed out that the ultimate lower limit of the cost of bulk devices is defined by the cost of the wafer itself and thus cost lowering below the price of wafers is not possible. Hence, even for Si, a thin film technology needs to be developed to meet the cost goals.

This conclusion is adequately illustrated in Figure 1.5, which depicts the module/array price history of PV technology.[12]

An equally important though often not equally appreciated aspect of thin films in relation to photovoltaic technology is the unique properties of thin films and the very large variety of deposition techniques available with the following outstanding features:

1. Thin films can be deposited over large areas in any predetermined shape and structure. The fabrication of large-area devices restricts the number of interconnections to a bare minimum. Thus array configurations can be made simpler and high packing density can be achieved. Integrated interconnections are possible and a single-step contact, interconnection, and encapsulation process has already been demonstrated in the fabrication of Cu_2S/CdS backwall cells.

2. A large variety of deposition processes are available and a single process or a combination of processes can be selected to fabricate large-area, interconnected photovoltaic panels on an automated, continuous production basis. Several schemes[13] have been proposed for continuous production of Cu_2S/CdS solar cells. Photon Power is producing 60 cm × 50 cm panels, with 60 solar cells in each panel, series connected in an interdigited pattern, on a continuous roller plant by spray deposition and chemiplating techniques.

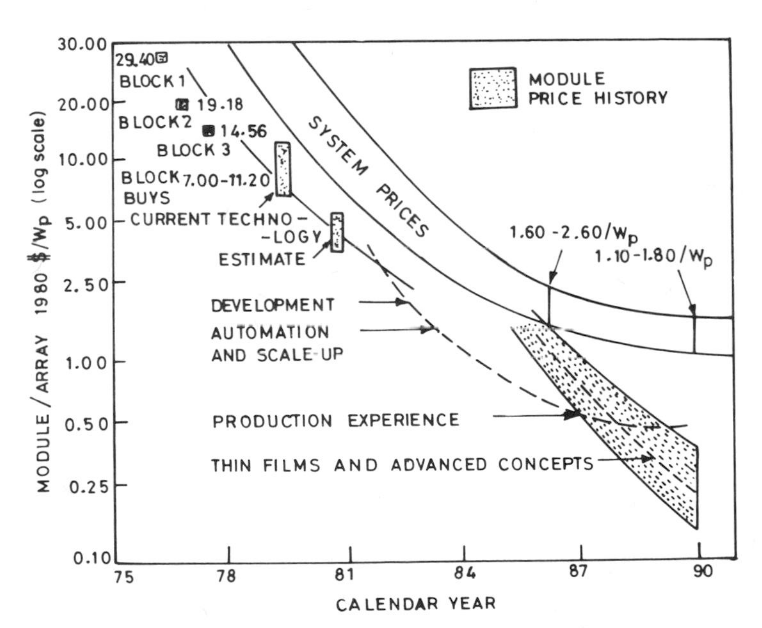

Figure 1.5. Photovoltaic module/array price history and projections (after Deb and Wallace[12]).

The deposition techniques which have been utilized for thin film solar cell fabrication include evaporation, sputtering, spray pyrolysis, chemical deposition, chemical vapor deposition, electrodeposition, glow discharge decomposition, ion-exchange reactions, electrophoresis, and screen printing. The spray pyrolysis process and the chemical deposition techniques are particularly attractive for large-area devices. These two processes have been developed to coat substrates of dimensions of the order of $1\ m \times 1\ m$ reproducibly and uniformly. In-process doping profiles or multilayer structures can be obtained by suitably modifying solution composition during film growth.

Single-step device fabrication is possible with thin film deposition techniques. Thin film silicon solar cells have been fabricated by CVD in a single deposition cycle.[14] $CuInSe_2/CdS$ high-efficiency devices have also been fabricated in a single vacuum evaporation cycle using a three-source technique.[15] Multilayer CdS/PbS stacks have been prepared by chemical deposition techniques for making selective surfaces.[16] The same method can be employed to fabricate multilayer PV devices. An all-spray Cu_2S/CdS solar cell has been fabricated in a single roller deposition process.[17] Cu_2S/CdS heterojunction devices can also be fabricated in a single pumpdown by a sputtering process using a multicathode arrangement. a-Si devices of the $p/i/n$ type have been fabricated in a single run by glow discharge decomposition of silane.[10] *In situ* change of film composition and doping is effected by changing the gas mixtures at the appropriate time. With suitable equipment, all these processes can be programmed and controlled by microprocessors to fabricate reproducibly large-area devices with complicated, multilayer, interdigited structures.

3. Control of the doping profile makes it possible to obtain a desired profile in the growing film and thus to incorporate a drift field for efficient carrier collection. Such gradient doping has been employed in a multilayer n^+-CdS : In/n-CdS/p-$CuInSe_2$/p^+-$CuInSe_2$: Cu thin film device[15] to obtain high photocurrent values (39 mA cm^{-2}) and high collection efficiencies. Gradient doping has also been employed in thin film Si solar cells[14] to provide a back surface field. Alternatively, gradient composition materials with variable bandgaps have been deposited to provide a minority carrier reflecting field at the top surface, as in the case of GaAlAs/GaAs devices.[18] Since the solid solubility conditions in thin films are considerably relaxed owing to the atom (molecule, ion)-by-atom (molecule, ion) deposition process, a large variety of solid solutions have been prepared by several deposition techniques. The spray pyrolysis and chemical deposition techniques have also been extended to deposit ternary alloys of CdZnS, CdSSe, PbHgSe, PbHgS, and PbCdS with gradient composition and variable bandgaps. A large number of quaternary and penternary alloy films have been prepared by spraying. In this context we wish to emphasize

the applicability and suitability of thin film techniques for fabricating multilayer, internally connected ITSC (integrated tandem solar cell) devices. The operation of these devices requires that the photocurrent in all the series-connected cells be matched and, moreover, that the spectral response of each cell be matched to the portion of the solar spectrum incident upon it. This obviously requires that each cell have a suitable bandgap and as these cells are constructed in a monolithic structure, the lattice parameters should be well matched. To obtain the desired bandgap and lattice structures, it is necessary to synthesize multicomponent alloys of variable and well-controlled composition. Further, heterojunction devices also require that the electron affinities be matched. Finally, the absorber layer thickness, the composition gradient, and the doping profiles have to be finely tuned and critically controlled. These conditions can be met only by thin film techniques.

4. The microstructural features of the absorber layer sensitively influence the PV performance of a solar cell and, in some cases, specific microstructures (textured surface, columnar growth, corrugated back surface) may be necessary to obtain the desired performance. A wide variety of microstructures (ranging from epitaxial growth to amorphous structures) and consequently properties can be obtained by simply varying the deposition conditions during the growth of the films.

Thin films are usually polycrystalline with a large number of grains ranging from 0.1 to 10 μm, depending on the deposition process and conditions. These grain boundaries provide recombination surfaces for minority carriers and thus degrade the performance of a device. Further, the grain boundaries provide easy diffusion paths for mobile ions and atoms. This can cause deleterious effects in device operation owing to interdiffusion of certain atomic species or by diffusion of a particular element from one surface to another, creating shorting paths. Techniques to overcome these problems are currently being investigated in several laboratories.

The electrical activity of grain boundaries can be neutralized by hydrogenation (passivation),[19] by selective anodization of the grain boundaries, rendering them ineffective,[20] or by heavy doping with appropriate dopants to provide a directed field away from the grain boundary surfaces.[14] Diffusion through grain boundaries can be minimized or eliminated by providing multilayered diffusion barriers. This concept is utilized in spray-deposited Cu_2S/CdS/CdS : Al_2O_3 thin film solar cells, where a gradient composition multilayer CdS/CdS : Al_2O_3 stack prevents the diffusion of Cu from the top surface, both during fabrication as well as during subsequent operation, thus lending greater stability to the structure.[21]

A serious problem with thin film devices, particularly Cu_2S/CdS cells, has been the structural, microstructural, and stoichiometry changes during operation, which lead to instability and, finally, degradation. Doping with impurities has been tried in our laboratory with some success to stabilize the structure and phase of the Cu_2S films and, thereby, arrest degradation.

1.7. *Progress of Thin Film Solar Cells*

It is to be noted that only recently have all thin film Cu_2S/ZnCdS and $CuInSe_2$/ZnCdS solar cells achieved efficiencies exceeding 10%, but on small areas. However, research and development efforts on thin film solar cells have gained impetus only in the last couple of years and since then a steady increase in the conversion efficiencies has been reported, as illustrated in Figure 1.6. One of the most heartening features is the progress in a-Si : H solar cells, which have now exhibited efficiencies approaching 10%. The trend for a-Si cells over the last 2 to 3 years shows an upward growth. Similarly, Cu_2S/CdS cells and modified Cu_2S/ZnCdS cells also show an upward trend over the last couple of years. Composite, gradient-doped CdS : Al_2O_3 spray-deposited cells have achieved greater stability and are steadily improving in performance. Among other materials, $CuInSe_2$ cells have improved considerably in the last year. Large-area

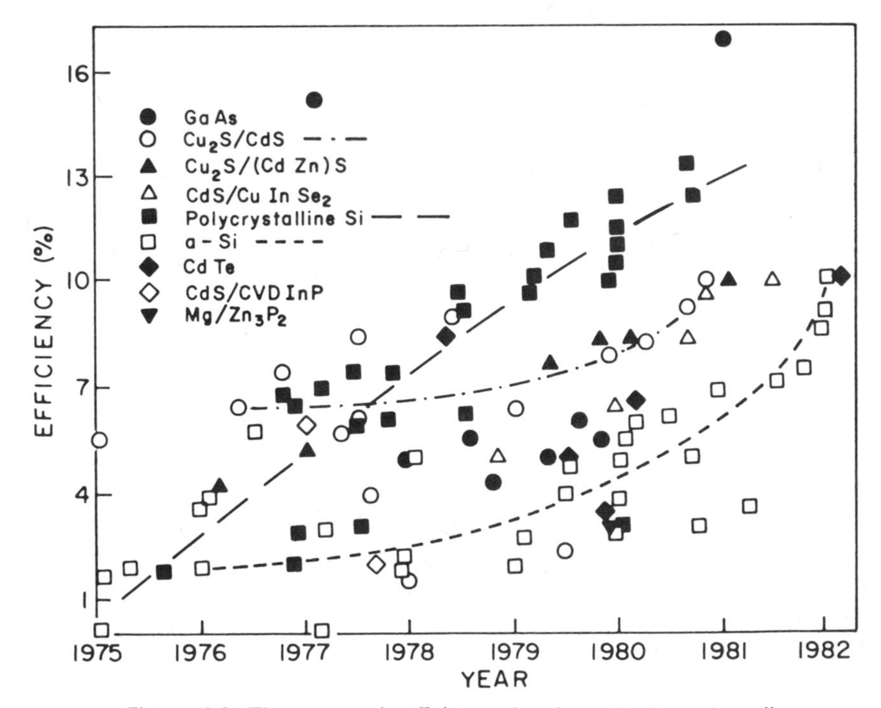

Figure 1.6. The progress in efficiency of various thin film solar cells.

Si and GaAs solar cells have also exhibited efficiencies from 8 to 9% and recrystallized and/or epitaxial Si and GaAs solar cells have achieved efficiencies in the range of 10 to 17%, respectively.

1.8. *Production of Thin Film Solar Cells*

Historically, the first thin film solar cell pilot plant production was undertaken by Clevite Corporation in the United States and S.A.T. in France. However, owing to poor market potential, the effort was given up. At present, thin film solar cell modules, primarily Cu_2S/CdS, are being produced on a pilot plant scale at several places including: SES, Inc. (USA), Stuttgart (Germany), IRD (UK), NUKEM (Germany), St. Gobain (France), and Photon Power (USA). The last-named firms are producing arrays and modules of Cu_2S/CdS by the spray process. Large efforts on Cu_2S/CdS cells have also been undertaken at IEC, Delaware, I.I.T., Delhi, and the C.N.R.S. Laboratories, France. At I.I.T., Delhi, thin film Cu_2S/CdS modules are being produced by both evaporated and spray processes. The typical flow charts of the two processes, as followed in the authors' laboratory, are shown in Figures 1.7a and b.

Both the evaporated and spray processes are viable and show promise of success. The scale of production today at different places is 10 kW. Plants of from 1 to 5 MW have been proposed and are entirely reasonable. We must emphasize that all these processes based on thin film technology are cheaper, simpler, and take less input in terms of materials and energy. It should also be pointed out that problems associated with Cd washing, handling, and recovery have been solved. In the case of $SnO_x : F$, used as the base contact for the spray-deposited cells, a substantial fraction of the sprayed material is recovered in a closed-cycle process. Thus the loss of material is minimal. Moreover, although Cd is known to be toxic, it should be emphasized that CdS is entirely safe and stable up to very high temperatures. With adequate precautions in the handling of Cd, no problems are anticipated.

In comparison to thin film Cu_2S/CdS cells, the production of bulk Si (primarily single crystal) solar cells is in the range of 5 MW internationally. For large-scale applications, Si production capability does not exist and the trend today is toward ribbon technology. However, single crystal ribbon technology is energy-intensive and costly and requires a large materials input. Although the scope for immediate applications exists, the long-range potential is low. On the other hand, thin film a-Si solar cells offer a very promising future. At present, Sanyo, Fuji, and Sharp of Japan are producing 2 to 3 MW/year of such cells for low-power electronic applications.

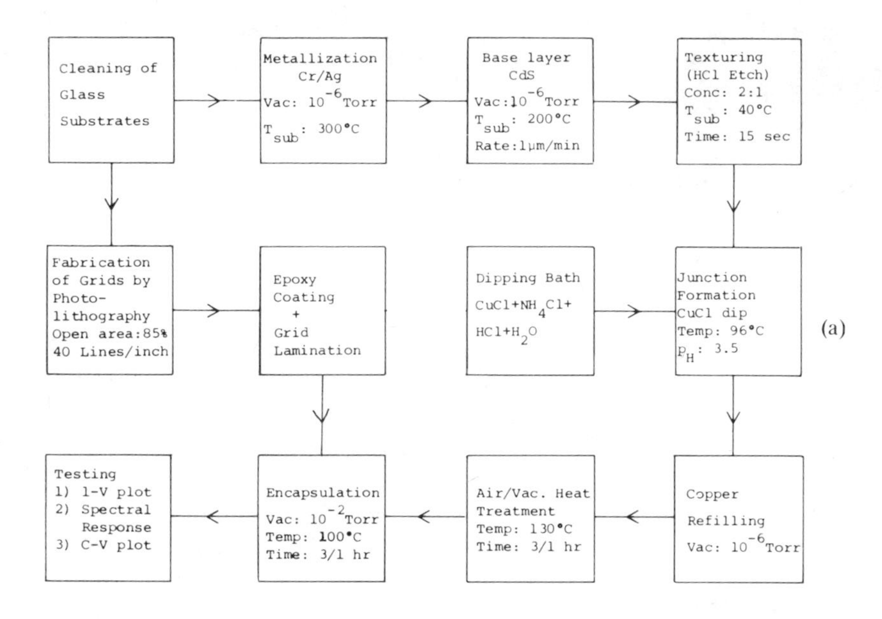

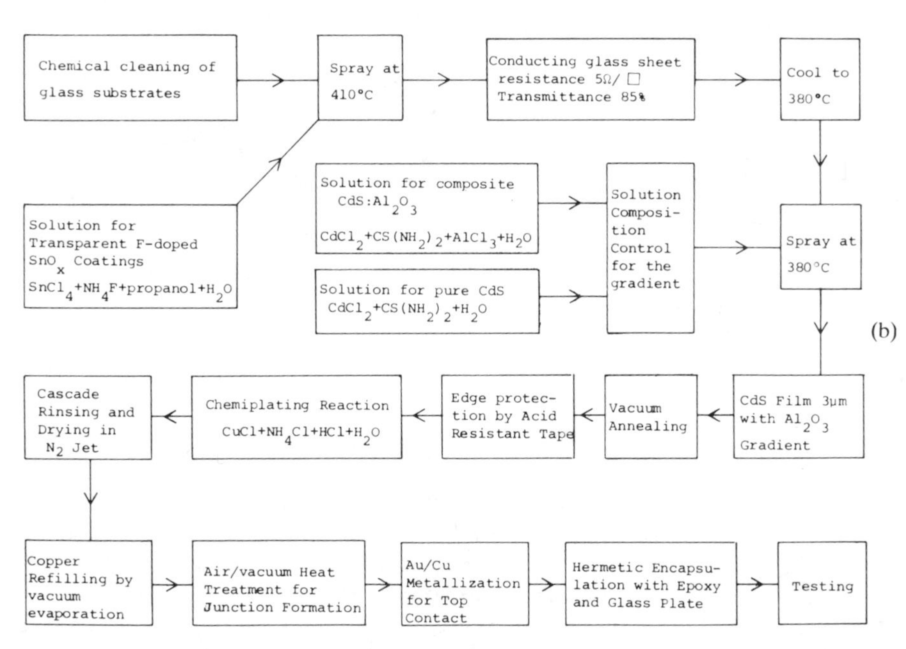

Figure 1.7. Flow chart for the production of Cu_2S/CdS thin film solar cells by (a) evaporated process and (b) spray process.

To conclude this section, we quote Ehrenreich, "The potential payoff would be immense if a truly inexpensive technology based on a thin film system were to be developed."

1.9. *Conclusions*

It is evident that for the future well-being of nations, a supply of energy based on a renewable source which is economically and environmentally acceptable has to be developed. Direct conversion of solar energy to electric power by thin film PV devices appears to be the ideal solution. The major considerations of a thin film PV energy technology can be summarized as follows:

1. Efficiencies greater than or equal to 10% on large areas appear to be feasible. Higher efficiencies and better stabilities are expected.
2. Cell performance is structure and deposition parameter sensitive. Better understanding of the structure sensitive properties, particularly at the interface, is necessary.
3. Cascade, tandem, variable bandgap, gradient-doped, multilayer, integrated junction cells have been demonstrated.
4. New and exotic materials must be investigated.
5. R/D is required on interface physics, grain boundary passivation, grain boundary diffusion, composite thin film structures, gradient bandgap materials and properties, degradation processes, and life tests.
6. Material economy, energy economy, small payback period, high conversion efficiency cells, simple production processes, tailor-made materials, and simple module/panel fabrication make thin film solar cells the only viable system.
7. There is today no clear choice among the existing thin film solar cells and associated production technologies.

It would not be improper to say that although it was discovered in 1955, it is only in the last 5 years that we know what a "thin film solar cell" is — structurally and electronically. And the more we understand it, the more we realize how complex the problem of controlling all the parameters is. But it should be possible. We should not forget that according to Henry Kelly,[22] "2 billion dollars in Federal funds [went into] the Clinch River breeder reactor design to produce subsidized electricity with a capital cost of $\$5/W_p$. A substantially smaller outlay would almost certainly produce PV arrays at $\$1/W_p$ or less installed in remote areas."

The principal conclusions of the APS Study Group on Solar Photovoltaic Energy Conversion emphasizes that an intensive, imaginative, well-funded, and well-managed R/D program offers the greatest hope for the long-term success of PV technology as a way of effectively utilizing our solar income.

Finally, we quote Frank von Hippel and Robert H. Williams from the *Bulletin of the Atomic Scientist* (1977): "The best approach may be to 'let a thousand flowers bloom' and then to cultivate the most promising varieties. We have a society where hundreds of thousands of citizens are skilled in science and technology. It may take a solar technological revolution to remind us once again that the true strength of such a society is only revealed when its members are given the opportunity to show what they can do."

2

Basic Physical Processes in Solar Cell Materials

2.1. *Introduction*

The basic physical processes underlying the operation of a solar cell consist of various optical interactions and generation/recombination and transport of carriers in different parts of the cell. We present in this chapter a brief theoretical description of these processes in semiconductor, insulator, and metal films, primarily to serve as a reference. For more detailed analyses, the reader is referred to standard text books[1–11] on the subject. The solid state physics of junctions is discussed in Chapter 3.

2.2. *Semiconductor Statistics*

We begin with the semiconductor statistics which form the basis of transport calculations in such materials. It is necessary, at this point, to consider the band structure of semiconductors, i.e., the energy–momentum (E–k) relationship of the electrons in the semiconductor, which is fundamental to an understanding of the electrical and optical properties. An electron in a crystal is affected by the potential from all the atoms of the crystal, and, hence, the allowed energies of electrons in crystals are very different from those in an isolated atom. The simplest form of energy bands in a crystal consists of a parabolic conduction band with a minimum at $k = 0$, a parabolic valence band with maximum at $k = 0$, and both bands with spherical energy surfaces. This type of energy band is shown in Figure 2.1a and the simplified energy band diagram is depicted in Figure 2.1b. Most real crystals have much more complex band structures, which have been studied theoretically using a variety of computer methods.[4,5] However, the main source of information about band structure is analysis of

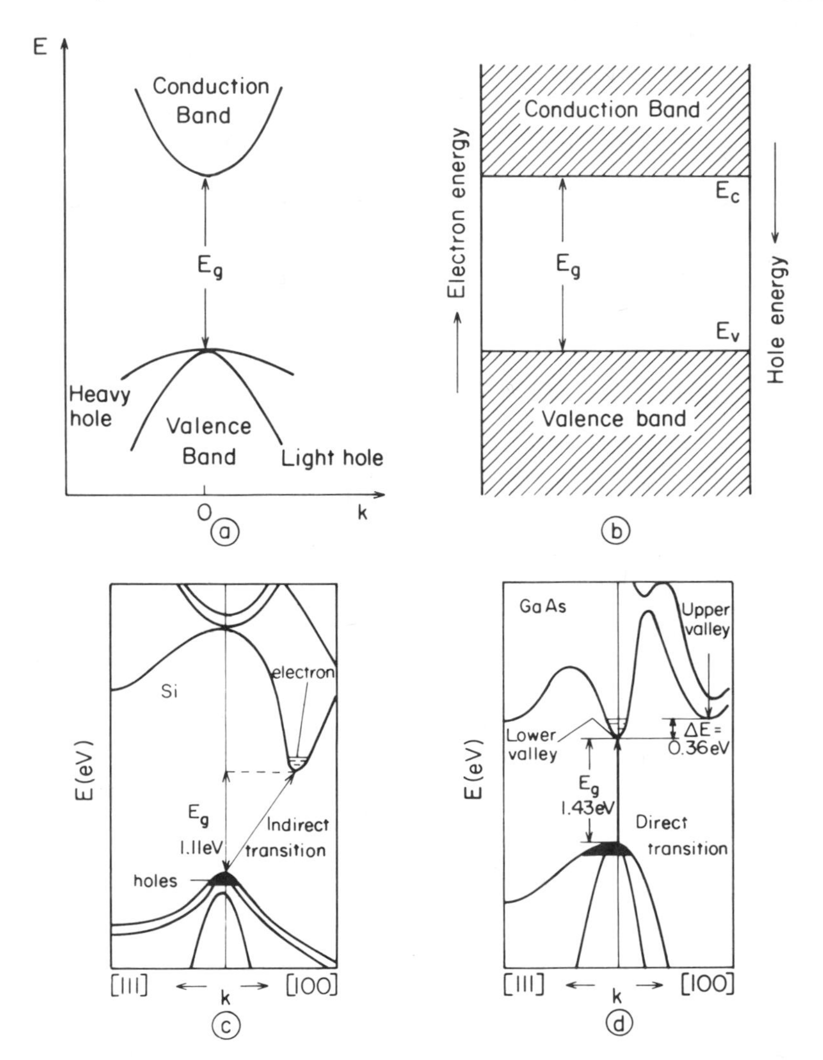

Figure 2.1. (a) The simplest form of band structure in a solid with parabolic conduction and valence bands with extrema at $k = 0$. (b) Simplified energy band diagram . (c) and (d) Band structures in Si and GaAs, respectively. The indirect (in Si) and direct (in GaAs) optical transitions are also represented schematically.

experimental data. Obviously, the band structure in a solid is determined primarily by the crystal structure. The crystallographic data for various semiconductors used in solar cells is given in Table 2.1.

The calculations of band structures in Ge and Si[12–15] have served as a basis for calculating the band pattern in III-V and II-VI compounds. The energy band structures in Si and GaAs are depicted in Figure 2.1c, d. The

Table 2.1. Bulk Structural and Optical Properties of Various Solar Cell Materials

Semi-conductor	Crystal structure	Lattice constant (Å) *a*	*b*	*c*	E_g (eV)	Type	dE_g/dT ($eV\,K^{-1}$)	T_c ($10^{-6}\,°C^{-1}$)
Si	Cubic	5.4301			1.11	I	-4×10^{-4}	2.3
Ge	Cubic	5.6576			0.66	I	-4.5×10^{-4}	5.8
GaAs	Cubic	5.6534			1.43	D	-5.0×10^{-4}	5.9
CdTe	Cubic	6.481			1.44	D	-5.4×10^{-4}	5.9
CdSe	Hexagonal	4.299		7.010	1.7	D	-4.6×10^{-4}	4.8
	Sphalerite	6.084						
CdS	Cubic	5.818		—	2.42	D	-5.2×10^{-4}	4.0
	Hexagonal	4.136		6.713				
Cu_2S	Hexagonal	3.961		6.722	1.2	D		
	Orthorhombic	11.849	27.330	13.497				
Cu_2Se	Cubic	5.852			1.4	I		
Zn_3P_2	Tetragonal	8.097		11.45	~1.6	D		
InP	Cubic	5.869			1.27	D	-4.7×10^{-4}	4.5
$CuInSe_2$	Tetragonal	5.782		11.620	1.01	D	-1.5×10^{-4}	
$CuInS_2$	Tetragonal	5.51		11.05	1.5	D		
$CuInTe_2$	Tetragonal	6.179		12.36	0.9	I	-3.2×10^{-4}	
SnO_2	Tetragonal	4.737		3.185	3.8	D	-2.0×10^{-4}	4.0
In_2O_3	Cubic	10.118			2.8	D	-7.7×10^{-4}	10.2
ZnO	Hexagonal	3.249		5.205	3.3	D	-9.5×10^{-4}	7.2
	Cubic	4.58						
Cd_2SnO_4	Orthorhombic	5.568	9.887	3.902	~2.9			
$Zn_{0.43}Cd_{0.57}S$	Hexagonal	4.00		6.52	2.92	D		

T_c — coefficient of thermal expansion, E_g — optical gap, *I* — indirect, *D* — direct.

important differences from the simple band structure discussed above are obvious. Energy extrema for electrons (in the conduction band) and holes (in the valence band) in Si are at different points in k-space. In III-V compounds like GaAs, the energy extrema may occur both at different points and at one point. It is to be noted that, in general, the valence bands are made up of two or more branches of the E–k curve, each of which has its own type of holes. The constant energy surfaces can be either deformed spheres or ellipsoids.

The semiconductor statistics are derived on the basis of the simple picture of parabolic conduction and valence bands. In the intrinsic case, the number of occupied conduction band levels n is given by

$$n = \int_{E_c}^{E_{top}} N(E)F(E)dE \tag{2.1}$$

where E_c is the energy at the bottom of the conduction band, E_{top} is the energy at the top, and $N(E)$ is the density of states. For low carrier densities and temperatures, $N(E)$ is expressed as

$$N(E) = M_c[2(E - E_c)]^{1/2}(m_{de})^{3/2}(1/\pi^2)(1/\hbar^3) \tag{2.2}$$

where M_c is the number of equivalent minima in the conduction band and m_{de} is the density-of-state effective mass for electrons[9] given by

$$m_{de} = (m_1^* m_2^* m_3^*)^{1/3} \tag{2.3}$$

where m_1^*, m_2^*, and m_3^* are the electron effective masses along the principal axes of the ellipsoid energy surface. The electron effective mass is defined by

$$m^* \equiv \hbar^2/(\partial^2 E/\partial k^2) \tag{2.4}$$

From experimental data we obtain a single effective mass for GaAs and two effective masses for Ge and Si, viz., m_l^* along the symmetry axes and m_t^* transverse to the symmetry axes. The term $F(E)$ in equation (2.1) is the Fermi–Dirac distribution function:

$$F(E) = \frac{1}{1 + \exp[(E - E_f)/kT]} \tag{2.5}$$

where k is the Boltzmann constant, T is the absolute temperature, and E_f the Fermi energy. Thus, n is evaluated to be

$$n = N_c \frac{2}{\pi^{1/2}} F_{1/2}\left(\frac{E_c - E_f}{kT}\right) \tag{2.6}$$

where N_c is the effective density of states in the conduction band, expressed by the relation

$$N_c \equiv 2\left(\frac{2\pi m_{de} kT}{h^2}\right)^{3/2} M_c \tag{2.7}$$

and $F_{1/2}(\eta_f)$ is the Fermi–Dirac integral[16] with $\eta_f = (E_c - E_f)/kT$. For nondegenerate semiconductors, i.e., E_f several kT below E_c, $F_{1/2}(\eta_f)$ approaches $[\pi^{1/2}\exp(-\eta_f)]/2$ and equation (2.6) becomes

$$n = N_c \exp\left(-\frac{E_c - E_f}{kT}\right) \tag{2.8}$$

In a similar manner, we have for the hole density near the top of the valence band the expression

$$p = N_v \exp\left(-\frac{E_f - E_v}{kT}\right) \tag{2.9}$$

for nondegenerate conditions. The effective density of states in the valence band is given by

$$N_v \equiv 2\left(\frac{2\pi m_{\rm dh}kT}{h^2}\right)^{3/2} \tag{2.10}$$

where $m_{\rm dh}$ is the density-of-state effective mass of the holes in the valence band[9]:

$$m_{\rm dh} = (m_{\rm lh}^{*3/2} + m_{\rm hh}^{*3/2})^{2/3} \tag{2.11}$$

where $m_{\rm lh}^*$ is the light-hole effective mass and $m_{\rm hh}^*$ is the heavy-hole effective mass. The effective masses of electrons and holes in different semiconductors at 300 K are given in Table 2.2.

At finite temperatures, thermal excitation of electrons from the valence band to the conduction band leads to an equal number of electrons in the conduction band and holes left behind in the valence band, i.e., $n = p = n_i$. This thermal excitation is balanced by recombination of electrons in the conduction band with holes in the valence band. Recombination phenomena are discussed in greater detail later in this chapter.

The Fermi level for an intrinsic semiconductor is obtained by equating equations (2.8) and (2.9):

$$E_f = E_i = \frac{E_c + E_v}{2} + \frac{kT}{2}\ln\frac{N_v}{N_c} = \frac{E_c + E_v}{2} + \frac{3kT}{4}\ln\left(\frac{m_{\rm dh}}{m_{\rm de}}\right) \tag{2.12}$$

At $T = 0$, the Fermi level lies in the middle of the forbidden gap. As the temperature increases, the Fermi level moves toward the band in which the effective mass for the density of states is less.

The intrinsic carrier concentration n_i is given by

$$n_i^2 = np = N_c N_v \exp(-E_g/kT) \tag{2.13}$$

where $E_g = E_c - E_v$. Thus n_i can be written as

$$\begin{aligned} n_i &= (N_c N_v)^{1/2} \exp(-E_g/2kT) \\ &= 4.9 \times 10^{15}\left(\frac{m_{\rm de}m_{\rm dh}}{m_0^2}\right)^{3/4} T^{3/2} \exp(-E_g/2kT) \end{aligned} \tag{2.14}$$

where m_0 is the free-electron mass.

Doping a semiconductor with either donors or acceptors introduces impurity energy levels. A donor level is termed neutral if it is filled by an

Table 2.2. Electrical Properties of Several Bulk Semiconductors[a]

Semiconductor	χ (eV)	N_c (cm^{-3})	m^*/m_0 Electron	m^*/m_0 Hole	μ(cm^2 V^{-1} s^{-1}) μ_n	μ_p	L_D(μm) L_n	L_p	τ(μs) τ_n	τ_p	ε	Typical dopants p-type	n-type
Si	4.05	2.8×10^{19}	$m_l^* = 0.97$, $m_t^* = 0.19$	$m_{lh}^* = 0.16$, $m_{hh}^* = 0.5$	1350	480	10	6	8.5	10.5	12.0	B, Al, Ga In	P, As, Sb Bi
Ge	4.0	1.04×10^{19}	$m_l^* = 1.6$, $m_t^* = 0.082$	$m_{lh}^* = 0.04$, $m_{hh}^* = 0.3$	3900	1900					16	B, Al, Ga In	P, As, Sb
GaAs	4.07	4.7×10^{17}	$m_l^* = 0.068$	$m_{lh}^* = 0.087$, $m_{hh}^* = 0.475$	8000	300	6		50	110	11.5	Zn, Cd, Ge Si	Si, Sn, Ge Se, Te
CdTe	4.3	10^{15}	0.14	0.35	700	65	2.5	3.3		0.054	9.6	Li, Sb, P N	I
CdSe	4.95	3.6×10^{17}	0.13	0.45	600						10	Cu	Cl, Br, I
CdS	4.5	10^{19}	0.35	0.07	340						9.0–10.3	Cu, Ag	Cl, Br, I Al, Ga, In
Cu_2S	4.3		0.205	0.7		24							
InP	4.4	$(0.4–4) \times 10^{17}$	0.07	0.4	4500	100					12.1	Zn, Cd	Se, Te
$CuInSe_2$	4.15				320	10							
SnO_2	4.8–4.9	9.5×10^{19}	0.104		2.0 (350 °C)								Sb, F
In_2O_3	4.3–4.4	7.8×10^{19}										Sn	
Cd_2SnO_4	4.55										8.5		H, Li, Zn Al, In
ZnO	4.2	1×10^{18} (300 °C)	0.38	1.8	190								

[a] χ is the electron affinity, N_c the effective density of states at the conduction band edge, m^*/m_0 the effective mass, μ the mobility (at 300 K), L_D the diffusion length of minority carriers, τ the lifetime of minority carrier, and ε the static dielectric constant.

electron and positive if it is empty. An acceptor level is neutral if empty and negative if filled by an electron. The impurity energy levels can be calculated on the basis of the hydrogen atom model. The ionization energy for the hydrogen atom is

$$E_H = m_0 q^4 / 32\pi^2 \varepsilon_0^2 \hbar^2 = 13.6 \text{ eV} \tag{2.15}$$

where q is the electronic charge and ε_0 is the permittivity of free space. The ionization energy E_d of the donor can be obtained by replacing m_0 by the conductivity effective mass of electrons, m_{ce},[9] and ε_0 by the permittivity of the semiconductor, ε_s, where m_{ce} is given by

$$m_{ce} = 3\left(\frac{1}{m_1^*} + \frac{1}{m_2^*} + \frac{1}{m_3^*}\right)^{-1} \tag{2.16}$$

E_d can be written as (from equation 2.15)

$$E_d = (\varepsilon_0/\varepsilon_s)^2 (m_{ce}/m_0) E_H \tag{2.17}$$

Similarly, the ionization level for acceptors E_a can be calculated. Ionization energies for different donors and acceptors have been calculated for several semiconductors. Some typical dopants for various semiconductors are also listed in Table 2.2.

In Figure 2.2 we show schematically the simplified band diagram, the

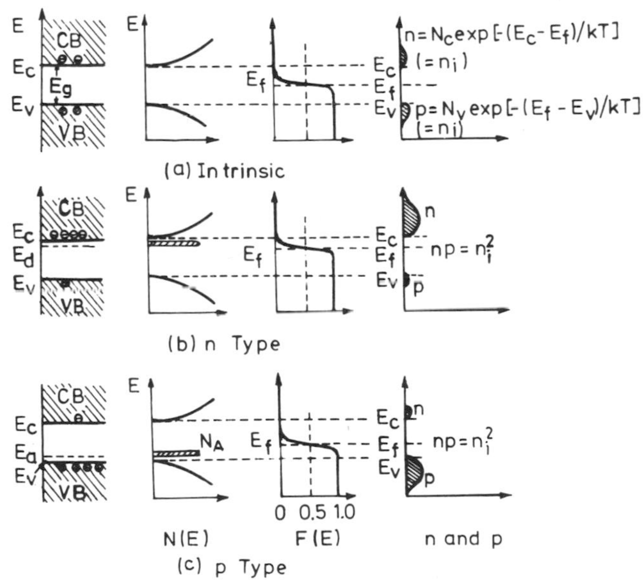

Figure 2.2. Schematic representation of band diagram, density of states $N(E)$, Fermi–Dirac distribution $F(E)$, and the carrier concentration n or p for an intrinsic, n-type, and p-type semiconductor (after Sze[9]).

density of states, the Fermi–Dirac distribution function, and the carrier concentrations for intrinsic n-type and p-type semiconductors. As noted above, the Fermi level in an intrinsic semiconductor, given by equation (2.12), lies very close to the middle of the bandgap. With the introduction of impurity atoms, the Fermi level adjusts itself to preserve charge neutrality. Suppose N_D (cm^{-3}) donor impurities are added. For charge neutrality

$$n = N_D^+ + p \tag{2.18}$$

where n is the electron density in the conduction band, p is the hole density in the valence band, and N_D^+ is the number of ionized donors given by

$$N_D^+ = N_D\left[1 - \frac{1}{1+(1/g)\exp[(E_d - E_f)/kT]}\right] \tag{2.19}$$

where g is the ground state degeneracy of the donor impurity level and is equal to 2 because a donor level can accept one electron of either spin or can have no electron. Similarly for acceptor impurity concentration N_A (cm^{-3}), the ionized acceptor density is expressed by

$$N_A^- = \frac{N_A}{1+(1/g)\exp[(E_a - E_f)/kT]} \tag{2.20}$$

For acceptor levels, the factor g is equal to 4.[9]

To determine the Fermi level, we rewrite equation (2.18) as

$$N_c \exp\left(-\frac{E_c - E_f}{kT}\right) = N_D\left[\frac{1}{1+2\exp[(E_f - E_d)/kT]}\right] + N_v \exp\left(\frac{E_v - E_f}{kT}\right) \tag{2.21}$$

Thus for a given set of N_c, N_D, N_v, E_c, E_d, E_v, and T, the Fermi level E_f can be uniquely determined. For low temperatures, the Fermi level rises toward the donor level (for an n-type semiconductor) and the donor level is partially filled with electrons. The electron density is given by

$$n \simeq \left(\frac{N_D - N_A}{2N_A}\right) N_c \exp(-E_d'/kT) \tag{2.22}$$

for $N_A \gg \frac{1}{2}N_c \exp(-E_d'/kT)$, where $E_d' \equiv (E_c - E_d)$, or

$$n \simeq \frac{1}{2^{3/2}}(N_D N_c)^{1/2} \exp(-E_d'/2kT) \tag{2.23}$$

for $N_D \gg \frac{1}{2}N_c \exp(-E_d'/kT) \gg N_A$. At high temperatures, $n \approx p \gg N_D$ and the intrinsic range is observed. At very low temperatures, most impurities are frozen out and the carrier concentration is given by equation (2.22) or (2.23), depending on the compensation conditions. In the intermediate temperature range, the electron density remains essentially constant.

It should be noted that even when impurity atoms are added, the np product is still given by n_i^2 and is independent of the added impurities. At elevated temperatures, when most of the donors and acceptors are ionized, the neutrality condition can be approximated by

$$n + N_A = p + N_D \tag{2.24}$$

From equations (2.13) and (2.24), the concentrations of electrons n_{n0} and holes p_{n0} in an n-type semiconductor is obtained, viz.,

$$\begin{aligned} n_{n0} &= \tfrac{1}{2}\{(N_D - N_A) + [(N_D - N_A)^2 + 4n_i^2]^{1/2}\} \\ &\approx N_D \qquad \text{if } |N_D - N_A| \gg n_i \text{ and } N_D \gg N_A \end{aligned} \tag{2.25}$$

$$p_{n0} = n_i^2/n_{n0} \simeq n_i^2/N_D \tag{2.26}$$

and

$$E_c - E_f = kT \ln(N_c/N_D)$$

From equation (2.12),

$$E_f - E_i = kT \ln(n_{n0}/n_i) \tag{2.27}$$

Similarly, for a p-type semiconductor, the hole p_{p0} and electron n_{p0} concentrations are given by:

$$\begin{aligned} p_{p0} &= \tfrac{1}{2}\{(N_A - N_D) + [(N_A - N_D)^2 + 4n_i^2]^{1/2}\} \\ &\approx N_A \qquad \text{if } |N_A - N_D| \gg n_i \text{ and } N_A \gg N_D \end{aligned} \tag{2.28}$$

$$n_{p0} = n_i^2/p_{p0} \simeq n_i^2/N_A \tag{2.29}$$

and

$$E_f - E_v = kT \ln(N_v/N_A)$$

or

$$E_i - E_f = kT \ln(p_{p0}/n_i) \tag{2.30}$$

The subscripts n and p refer to the type of semiconductors and 0 refers to the thermal equilibrium condition.

Thus far we have been discussing the nondegenerate case. For degenerate semiconductors, i.e., E_f lying inside an energy band about or more than $5kT$ away from the energy extremum, the electron concentration is independent of temperature and is given by the relation

$$n = \frac{8\pi}{3}\left(\frac{2m_{\mathrm{de}}}{h^2}\right)^{3/2}(E_f - E_c)^{3/2} \tag{2.31}$$

The hole concentration when the Fermi level lies more than $5kT$ below E_v is given by the temperature-independent expression

$$p = \frac{8\pi}{3}\left(\frac{2m_{\mathrm{dh}}}{h^2}\right)^{3/2}(E_v - E_f)^{3/2} \tag{2.32}$$

It may be noted here that in an n-type material where $n_{n0} > p_{n0}$, electrons are called the majority carriers and holes the minority carriers. The reverse is true for p-type semiconductors.

The preceding discussion applies to the bulk of the semiconductor. At the surface, because of the interruption of the periodic lattice structure of the crystal, localized states exist within the forbidden gap.[17–21] These surface states can be classified into two categories[9,22,23]: (i) fast surface states associated with the semiconductor surface itself and with a transition time of the order of 10^{-6} s for charge transfer with the conduction or valence band and (ii) slow surface states associated with an oxide layer or other absorbed species on the semiconductor surface and with a transition time of the order of seconds for charge exchange. The density of fast states is about 10^{15} cm^{-2}, there being, in general, one surface state per surface atom. The density of slow states is believed to be greater than that of fast states. Tamm[17] and Shockley[18] invoked different potential models to describe the surface states. Accordingly, we have Tamm states or Shockley states. However, of greater importance are the characteristics of surface states. A brief description follows.

A surface state can act as a donor state if it is neutral when filled and becomes positive when it gives up an electron. An acceptor surface state is neutral when empty and can become negative by accepting an electron. To formulate surface state statistics, we first consider the simplest case in which each center at the surface can capture or release only one electron, thereby introducing a single allowed energy level E_t in the forbidden gap. These are called the single-charge surface states. We assume that the density N_t of the centers is sufficiently low to neglect mutual interaction and that all centers are identical in characteristics. Defining n_t and p_t as the equilibrium densities of occupied and unoccupied centers, respectively, and E_t^f as an effective energy level to allow for the multiplicity of the surface states, and noting that N_t, p_t, and n_t refer to densities per unit area, we have the following relations:

$$p_t = N_t - n_t \tag{2.33}$$

At the surface,

$$(E_t^f - E_f)/kT = [(E_t^f - E_i)/kT] - u_s \tag{2.34}$$

where u_s is the surface potential. We now have

$$n_t/N_t = F_n(E_t^f) \tag{2.35}$$

$$p_t/N_t = F_p(E_t^f) \tag{2.36}$$

where the Fermi–Dirac distribution functions $F_n(E_t^f)$ and $F_p(E_t^f)$ are given by

$$F_n(E_t^f) = \frac{1}{1 + \exp[(E_t^f - E_i)/kT - u_s]} \tag{2.37}$$

$$F_p(E_t^f) = \frac{1}{1 + \exp[u_s - (E_t^f - E_i)/kT]} \tag{2.38}$$

From the above equations, when $u_s = (E_t^f - E_i)/kT$, i.e, when the Fermi level passes through E_t^f, then $n_t = p_t = N_t/2$.

In general, for several surface states at energy levels E_{t_j} and of densities N_{t_j}, the density of electrons captured in all surface states will be given by the sum $\sum_j N_{t_j} F_n(E_t^f)$.

In addition to the simple surface states considered above, it is possible to have complex surface states which are capable of capturing more than one electron and for each condition of charge, these centers can exist in several quantum states (ground and excited states) with different energies. The density of centers charged j times and corresponding to the mth state, M_{jm}, has been given by Shockley and Last[24] to be

$$M_{jm} \propto N_t \exp[(jE_f - E_{jm})/kT] \tag{2.39}$$

where E_{jm} is the corresponding energy level and N_t is the total density of the centers. From normalization requirements we have

$$\sum_{jm} M_{jm} = N_t \tag{2.40}$$

Thus, summing equation (2.39) over all m to obtain the density of centers charged j times (M_j) yields

$$M_j = \sum_m M_{jm} \propto N_t Z_j \exp(jE_f/kT) \tag{2.41}$$

where

$$Z_j \equiv \sum_m \exp(-E_{jm}/kT) \tag{2.42}$$

We have already mentioned the surface potential u_s. We now consider the origin and magnitude of this potential. Taking an n-type semiconductor at 0 K with a number of acceptor surface states and assuming that the donor levels in the bulk n-type semiconductor are initially all occupied by electrons at 0 K, we get the situation as depicted in Figure 2.3a. A region of fixed positive space charge is now created in the bulk of the semiconductor by the flow of electrons from bulk donor states into empty acceptor states. The donor states become ionized and positively charged and the acceptor surface states become negatively charged. Thus an electric field E is set up in the space charge region which repels further electrons from the surface and equilibrium is attained. The situation at this stage is depicted in Figure 2.3b. We note that all the surface states are assumed to be filled. It should be noted here that at temperatures above 0 K, where some of the donors are already ionized, a similar situation holds and electrons from the conduction band states near the surface fill the surface acceptor levels.

Corresponding to the electric field E, there exists a gradient of electrostatic potential $V(x)$ in the space charge region. The electrostatic potential V is higher in the bulk of the semiconductor than it is at the surface by an amount u_s. The energy bands in the n-type semiconductor are thus bent upward at the surface. The net result of the presence of acceptor surface states on the n-type semiconductor surface is to produce a negative surface charge, a positive space-charge region, and an upward bending of the energy bands at the surface by an amount eu_s.

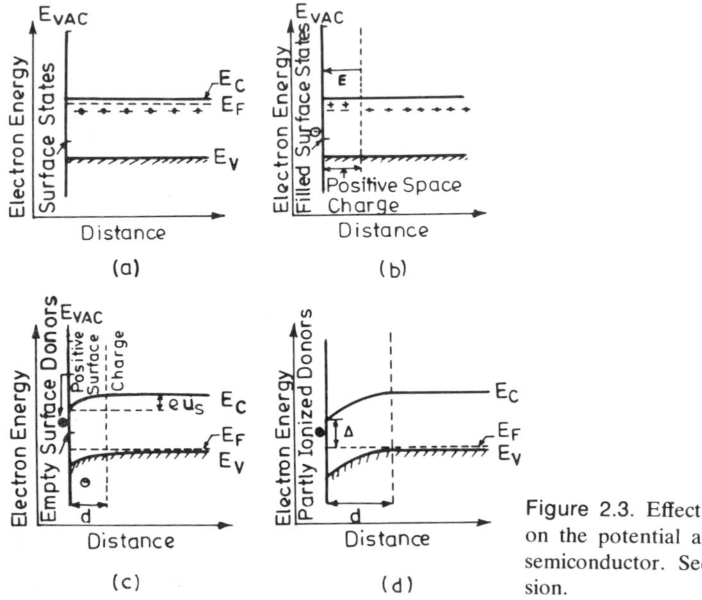

Figure 2.3. Effect of surface states on the potential at the surface of a semiconductor. See text for discussion.

In the case of a p-type semiconductor with donor surface states, we have, at equilibrium, all the surface donors ionized, as a result of having given up their electrons to acceptors in the bulk. As a consequence, a positive space charge and a region of negative space charge in the bulk exist with the latter being due to fixed ionized acceptors. The energy bands are bent downward at the surface, and this is shown in Figure 2.3c.

In the two cases treated above, since we have assumed all the surface states to be ionized, the Fermi level position at the surface is determined by the Fermi level in the bulk. The amount eu_s of band bending can then be pictured as a raising or lowering of the conduction and valence band edges at the surface relative to the fixed Fermi level. However, if the density of surface states is large such that not all of them are ionized then the situation is different. Consider a p-type semiconductor with donor surface states at an energy level Δ below the conduction band edge. Electrons can flow from the surface donors into the p-type semiconductor bulk only until the Fermi level at the surface coincides with the surface donor energy. This leaves some of the surface donors still un-ionized and occupied by electrons. Thus the Fermi level is fixed or pinned at the surface at the donor energy, i.e., at an energy Δ below the conduction band edge. This condition is depicted in Figure 2.3d. The magnitude of u_s for the condition in Figure 2.3b can be calculated to be

$$u_s = 2\pi e n_s^2 / \varepsilon N_d \tag{2.43}$$

where N_d is the bulk donor density, n_s is the surface acceptor density, and ε is the dielectric constant of the semiconductor. A typical value of u_s for Si (at 300 K) containing 10^{16} acceptors (cm^{-3}) is 0.9 V.[25]

2.3. *Transport Parameters*

In this section we discuss the transport of charge carriers through a semiconductor due to an applied electric field.

2.3.1. *Crystalline Semiconductors*

At low electric fields, the drift velocity v_d is proportional to the electric field strength E. Thus,

$$v_d = \mu E \tag{2.44}$$

where the proportionality constant μ is defined as the mobility of the carrier. The mobility is significantly affected by various scattering mechanisms, each of which is operative in a particular temperature range. The scattering can be due to: (i) impurity ions, (ii) thermal lattice vibrations or phonons, (iii) impurity atoms, (iv) vacancies and point defects, (v) dislocations, (vi) grain boundaries, cleavage planes, and crystal surfaces, and (vii) charge carriers.

The mobility of the charge carrier can be expressed in terms of the relaxation time τ as[26]

$$\mu = e\langle\tau\rangle/m^* = \mu(T) \tag{2.45}$$

where the relaxation time is defined as equal to the time the nonequilibrium state persists, on the average, after the fields responsible for it have been switched off. The relaxation time is characteristic of the collision process which restores the system to equilibrium. The temperature variation of mobility may result from the temperature dependence of the effective mass. However, we shall neglect effective mass variation and confine ourselves to the temperature dependence of the averaged relaxation time, which can be written as

$$\langle\tau\rangle = \frac{\int_0^\infty x^{3/2}\exp(-x)\tau(x)dx}{\int_0^\infty x^{3/2}\exp(-x)dx} \tag{2.46}$$

where $x = E/kT$. If the relaxation time has a power dependence on the energy of the form

$$\tau(E) = \tau_0 E^p = \tau_0(kT)^p x^p \tag{2.47}$$

where p is the power and τ_0 a quantity independent of energy, we obtain

$$\langle\tau\rangle = \tau_0(kT)^p \frac{\Gamma[(5/2)+p]}{\Gamma(5/2)} \tag{2.48}$$

where the integrals in equation (2.46) are given by the Γ functions.[4] From equations (2.48) and (2.45) we get

$$\mu = \langle\mu\rangle = \frac{e\tau_0 k^p}{m^*}\frac{\Gamma[(5/2)+p]}{\Gamma(5/2)}T^p \tag{2.49}$$

For the case of scattering by thermal lattice vibrations, $p = -1/2$. The mobility is therefore expressed by

$$\mu(T) = \frac{\pi^{3/2}\hbar^4 ec_{LL}}{3.2^{1/2}N\Delta_c^2 k^{3/2}}\frac{1}{m^{*5/2}T^{3/2}} \tag{2.50}$$

where e, $\hbar$, and k are universal constants; c_{LL}, N, Δ_c, and m^* are properties of the solid; and T is the temperature. In a given solid, the mobility depends only on temperature as

$$\mu = \mu_0 T^{-3/2} \tag{2.51}$$

In the case of scattering by optical lattice vibrations,

$$\mu \propto m^{*-3/2}[\exp(\hbar\omega_0/kT) - 1] \tag{2.52}$$

where ω_0 is the optical frequency.

The mobility in the case of impurity ion scattering is given by

$$\mu_I = \frac{8(2^{1/2}k^{3/2}\varepsilon^2 T^{3/2})}{\pi^{3/2}e^3 Z^2 N_I m^{*1/2} \ln[1 + (3\varepsilon kT/N_I^{1/3} Ze^2)^2]} \tag{2.53}$$

where ε and m^* are properties of the solid and N_I is the number of impurity ions and Z its atomic number. At high temperatures, we can neglect the logarithmic dependence of μ_I on temperature and, therefore, mobility increases with temperature as

$$\mu_I \simeq \mu_{0I} T^{3/2} \tag{2.54}$$

As the temperature decreases, the mobility due to impurity ion scattering also decreases as

$$\mu_I \simeq 1/T^{1/2} \tag{2.55}$$

For scattering by neutral centers, the relaxation time is independent of the energy. Thus,

$$\tau = m^{*2}e^2/20\varepsilon\hbar^3 N_n \tag{2.56}$$

and the corresponding mobility has the form

$$\mu_n = e^3 m^*/20\varepsilon\hbar^3 N_n \tag{2.57}$$

where N_n is the number of neutral centers.

For dislocation scattering $\tau \sim E^{-1/2}$ and, therefore,

$$\mu \sim 1/T^{1/2} \tag{2.58}$$

If all the mechanisms are assumed to be independent of each other, the total relaxation time τ is of the form

$$\tau = \left(\sum_i \tau_i^{-1} \right)^{-1} \tag{2.59}$$

The electron and hole mobilities (at 300 K) of some selected semiconductors are listed in Table 2.2.

An important parameter associated with mobility is the carrier diffusion coefficient D. In thermal equilibrium, the relationship between D_n and μ_n for electrons (D_p and μ_p for holes) is given by

$$D_n = 2\frac{[(kT/q)\mu_n]F_{1/2}[(E_c - E_f)/kT]}{F_{-1/2}[(E_c - E_f)/kT]} \tag{2.60}$$

where $F_{1/2}$ and $F_{-1/2}$ are Fermi–Dirac integrals. For nondegenerate semiconductors,

$$D_n = (kT/q)\mu_n \tag{2.61}$$

$$D_p = (kT/q)\mu_p \tag{2.62}$$

Equations (2.61) and (2.62) are known as the Einstein relations. The diffusion length of a carrier, i.e., the average distance traveled by a carrier before it recombines, is expressed by

$$L = (D\tau)^{1/2} \tag{2.63}$$

where τ is the free lifetime of the carrier.

The resistivity ρ, the reciprocal of the conductivity σ, is defined by the relation

$$\rho = 1/\sigma = 1/ne\mu \tag{2.64}$$

where n is the number of charge carriers (cm^{-3}). The current density J is related to the conductivity through the electric field E by

$$J = \sigma E \tag{2.65}$$

For a semiconductor with both types of carriers, we obtain the following expression for resistivity:

$$\rho = 1/\sigma = 1/[e(\mu_n n + \mu_p p)] \tag{2.66}$$

Semiconductor films used in photovoltaic devices are, in general, polycrystalline and have a high density of inhomogeneities incorporated in them in the form of grain boundaries, different phases, and, in some cases, different material components. The conduction processes in these films are markedly different from those in single-crystal homogeneous materials. We present in the following sections the various theories which have been proposed to explain the conductivity in such materials.

2.3.2. *Polycrystalline Materials*

Two models have been used to explain the conduction mechanism in polycrystalline films. These are (i) Petritz's barrier model[27] for semiconductors, in which the grain conductivity is much greater than the conductivity of the intergrain regions and (ii) van den Broek's theory[28] for high-resistivity PbO films. Each of these models holds only under specific conditions. A more general theory in which one model can be transformed into the other and vice versa was developed by Snejdar et al.[29] and is presented below.

The model formulates the existence of an isotype heterojunction (see Chapter 3 for a discussion on heterojunctions) with certain interface states at the interface between a grain and the intergrain region. It is assumed that the grains are cubic and of identical dimensions. There are three possible ways in an n-type semiconductor in which the electrons can penetrate the intergrain barriers: (i) by tunneling, (ii) by thermal emission over the barrier, and (iii) through the ohmic conductivity of the heterojunction. Mathematical analysis of the physical phenomena involved gives an expression for the conductivity of the form

$$\sigma = \sigma_{\mathrm{Bb}} \left[\frac{d_G d_B}{(d_G + d_B)^2} + \frac{d_B}{d_G + d_B} \right] + [(d_G + d_B) R_N]^{-1} \tag{2.67}$$

The first term represents the conductivity of intergrain domains parallel to the current flow and the second term gives the conductivity of the series-connected grains and intergrain domains. R_N is the resistance of the system and is a function of the bulk properties of the semiconductor, the interface properties, and the geometrical dimensions of the grain d_G and the intergrain domain d_B.

The resistance R_N contains a large number of physical parameters. Some of these parameters can be neglected in certain cases. Other parameters can be determined experimentally. For example, when the electron transfer through the intergrain region is dominated by thermal emission, i.e., tunneling of carriers can be neglected and barrier scattering is predominant, the electron mobility μ_B can be obtained from equation (2.67) and is given by

$$\mu_B = \{[e\langle v_{\mathrm{th},G}\rangle(1 - Q_G)]/4kT\}\bar{d}_G \exp(-\phi_B/kT) \tag{2.68}$$

where $\langle v_{\text{th},G} \rangle$ is the electron thermal velocity in the grains, Q_G is the reflection coefficient of electrons at the barrier, and ϕ_B is the barrier height. $\bar{d}_G$ is the mean value of the grain size, expressed as

$$(\bar{d}_G)^{-1} = N^{-1} \sum_{i=1}^{N} (d_{G_i})^{-1} \tag{2.69}$$

where N is the number of systems along the sample length and the d_{G_i} values are the crystallite sizes determined experimentally. The value of ϕ_B, the barrier height in equation (2.68), can be determined experimentally from the $1/T$ dependence of the measured effective Hall mobility.

2.3.3. *Heterogeneous Materials*

In this section we deal with the multicomponent granular systems which are essentially composites of metals and semiconductors/insulators. These materials have several unique properties, most notably high resistivities and low temperature coefficients of resistivity (TCR).

Granular metal films exist in three distinct structural regimes which are classified as (i) metallic regime, in which the volume fraction of the metal, x, is large, and the metal grains touch and form a metallic continuum with dielectric inclusions; (ii) dielectric regime, which is an inversion of the metallic regime with small, isolated metal particles dispersed in a dielectric continuum; and (iii) transition regime, in which the structural inversion between the metallic and the dielectric regime takes place.

In the metallic regime, phenomena depending on electron transfer are retained whereas properties depending on electron mean free path are drastically modified owing to the strong electron scattering from dielectric inclusions and grain boundaries. Typical examples are the decrease in the electrical conductivity by orders of magnitude from the corresponding crystalline value and a smaller (although positive) temperature coefficient of resistivity than in pure metals.

In the transition regime, the electrical conductivity is due to percolation along the metallic maze. With further decrease in x, the metallic maze breaks up into isolated metal particles dispersed in a dielectric continuum. Conduction then occurs by tunneling between isolated metal particles. The TCR changes sign and becomes negative at a composition and temperature where thermally assisted tunneling and percolation contribute equally to electrical conductivity.

Below a composition x_c, called the percolation threshold composition, only tunneling contributes to the electrical conductivity.

Abeles et al.[31] have treated the case of composite systems in detail. We present the final equations derived from their model.

2.3.3.1. *Metallic and Transition Regimes*

In the metallic and transition regimes, percolation conductivity and tunneling have been employed to study the conduction processes.

From Landauer's effective medium theory,[32] the conductivity in the percolation regime can be expressed by

$$\sigma(x)/\sigma(1) = (x - x_c)/(1 - x_c) \tag{2.70}$$

where $\sigma(x)/\sigma(1)$ is the normalized conductivity.

In the tunneling regime, assuming that the particle size follows a distribution function $D(d)$, the charging energy E_c necessary to place an electron on an isolated particle of size d, is of the order of e^2/Kd, where K is the effective dielectric constant of the granular metal. At the granular metal/insulator interface, the charging energy E_j needed for an electron to transfer from one electrode to the other electrode of a junction is given by $E_j = e^2/2C_e$, where C_e is an effective junction capacity. When the electrodes consist of ordinary metals, the effective capacity C_e, given by the junction capacity $C_j = \varepsilon A/4\pi L_j$, where ε is the dielectric constant of the insulator, L_j is the thickness of the insulator, and A is the area of the junction, is very large and E_j becomes negligible.

For tunneling into an isolated grain of a granular metal, the effective junction area for the tunneling electron is $\pi d^2/4$ and the effective capacity C_e becomes equal to $C_j\pi d^2/4A$. In this case, E_j is no longer negligible. The tunneling electrons, therefore, have to possess energies at least E_j above the Fermi level of the metal continuum. Similarly, for a reverse voltage under which holes are injected from isolated particles, the holes must have energies at least E_j below the Fermi level. Consequently, the effective density of localized states for tunneling exhibits a gap about the Fermi energy.

2.3.3.2. *Dielectric Regime*

Transport of electrons and holes by tunneling from one isolated metallic grain to the next is the mode of electrical conduction in the dielectric regime of a granular metal film. The charging energy E_c has the form

$$E_c = (e^2/d)F(s/d) \tag{2.71}$$

where d is the size of the grain, s is the separation between grains, and F is a function whose form depends on the shape and arrangement of the grains and on the interaction between the pair of charges. We note here that although d, s, and E_c are not constant in a granular metal, their variations are so correlated that the quantity sE_c is a constant whose magnitude depends only on the volume fraction of metal, x, in the sample and the dielectric constant of the insulator.

The conductivity behavior in the dielectric regime can be divided into two regions: (i) the low-field region, when the voltage difference between neighboring metal grains, ΔV, is much less than kT/e and (ii) the high-field region. The total low-field conductivity σ_L is the sum of the products of mobility, charge, and density of charge carriers over all possible percolation paths and is given by

$$\sigma_L = \int_0^\infty \beta(s)\exp[-2\chi s - (C/2\chi skT)]ds \tag{2.72}$$

where $C \equiv 2\chi sE_c$ is a constant and $\beta(s)$ is the density of percolation paths associated with the value s. χ is given by the expression

$$\chi = (2m\phi/\hbar^2)^{1/2} \tag{2.73}$$

where m is the electron mass and ϕ is the effective barrier height.

With $e\Delta V \gtrsim kT$, we have the high-field region, and the current density is expressed by

$$j_H = \frac{2eNl\gamma}{h}\int_0^\infty \exp(-2\chi s)g(s)D_0(s)\frac{[e\mathscr{E}W - (C/2\chi s)]}{1 - \exp[(C/2\chi s - e\mathscr{E}W)/kT]}\,ds \tag{2.74}$$

Here N is the number of metal grains per unit volume, l is the recombination length of electrons and holes, γ is a geometric factor to account for tunneling across nonplanar barriers, $\chi = [2m(\phi - E)/\hbar^2]^{1/2}$, where E is the electronic energy, ϕ is the effective barrier height, $C = 2\chi sE_c$, and $g(s)$ and $D_0(s)$ are related by the distribution function $D(s)$ of s as

$$D(s) = g(s)D_0(s) \tag{2.75}$$

$D_0(s)$ is given by the expression

$$D_0(s) = (s/s_0^2)\exp(-s/s_0) \tag{2.76}$$

where s_0 is the most probable value of s. $\mathscr{E}$ in equation (2.74) is the applied electric field such that $\Delta V = \mathscr{E}W$.

We should point out that owing to the quantization of electronic motion within every metal particle, the electronic energy levels are separated by a discrete amount $\delta = E_f/(\pi/6)d^3n$, where n is the number of conduction electrons per unit volume. No direct tunneling can occur and the electron must emit or absorb a phonon during the tunneling process in order to reach the final state. The probability of a phonon-assisted tunneling process is expected to increase with temperature, approaching a high-temperature limit when $kT \gtrsim \delta$.

2.3.4. *Amorphous Materials*

The dc conductivity σ of several amorphous semiconductors has the general form

$$\sigma = \sigma_0 \exp(-\Delta E/kT) \tag{2.77}$$

where ΔE[33] is an activation energy with a value generally between 0.5 to 1.0 eV. σ_0 has a value of approximately 1000 $\Omega^{-1}\,\text{cm}^{-1}$ for many amorphous chalcogenide semiconductors. The resulting values of the conductivity at 300 K lie in the range of 10^{-3} to 10^{-13} $\Omega^{-1}\,\text{cm}^{-1}$. If the carrier mobility $\mu(E)$ is plotted as a function of the energy E, the curve shows a sharp decrease at energies E_c and E_v, as shown in Figure 2.4a. These decreases are called the mobility edges, and for $T \neq 0$, $\mu(E)$ has a small nonzero value between E_c and E_v. The energy region between E_c and E_v is termed the mobility gap and the activation energy ΔE in equation (2.77) is related to the mobility gap in that it denotes the energy necessary to excite the carrier across the mobility gap.

2.3.4.1. *Band Model*

Several workers have proposed different energy band models for amorphous semiconductors in order to explain the electronic conduction

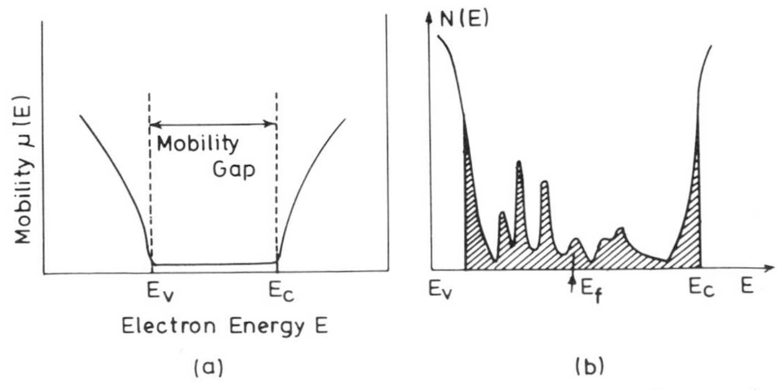

Figure 2.4. (a) Mobility as a function of electron energy in an amorphous semiconductor. (b) Schematic representation of a typical band structure in an amorphous semiconductor.

processes in these materials. All the models invoke the concept of localized states in the band tails. The formation of localized states occurs because of the spatial fluctuations in the potential caused by the configurational disorder in amorphous materials.

In the Cohen–Fritzsche–Ovshinsky (CFO) model,[34] proposed specifically for chalcogenide glasses, the authors postulate that tail states extend across the gap in a structureless distribution and the disorder is sufficiently great to cause the tails of the valence and conduction bands to overlap, leading to an appreciable density of states in the middle of the gap. As a result of band overlapping, a redistribution of electrons occurs, forming negatively charged, filled states in the conduction band tail and positively charged, empty states in the valence band tail. This leads to self-compensation and pinning of the Fermi level close to the middle of the gap, a feature required to explain the electrical properties of these materials. However, the high transparency of the amorphous chalcogenides below a well-defined absorption edge suggests a very limited degree of tailing in chalcogenides. The model may be more applicable to a-Si in which the absorption is markedly higher than in the crystalline counterpart.

In the Davis–Mott model,[35] the tails of localized states are narrow and extend a few tenths of an electron volt into the forbidden gap. Defects in the random network, such as dangling bonds and vacancies, give rise to a band of compensated levels near the middle of the gap. Moreover, the center band may be split into a donor and an acceptor band pinning the Fermi level at the center. At the transition from extended to localized states, the mobility decreases by several orders of magnitude producing a mobility edge. The interval between E_c and E_v acts as a pseudogap and is defined as the mobility gap. Cohen,[36] however, suggests a continuous rather than an abrupt drop in mobility occurring in the extended states just inside the mobility edge.

It should be noted that the density of states of a real amorphous semiconductor does not decrease monotonically into the gap but may show many well-separated peaks at well-defined energies in the gap, as shown in Figure 2.4b. Such localized gap states are produced by defect centers, the nature of which is not always clear. The position of the Fermi level in such a case is largely determined by the charge distribution in the gap states.

Emin[37] has proposed the Small–Polaron model in which he states that the presence of disorder in a noncrystalline solid tends to slow down a carrier leading to a localization of the carrier at an atomic site. If the carrier stays at the site for a time sufficiently long for atomic rearrangements to occur, atomic displacements in the immediate vicinity of the carrier may be induced, forming a small polaron.

2.3.4.2. *Electronic Properties*

Based on the Davis–Mott model, three conduction processes in amorphous semiconductors have been proposed, each process contributing in different temperature ranges. At very low temperatures, thermally assisted tunneling between states at the Fermi level is responsible for conduction. At intermediate temperatures, charge carriers are excited into the localized states of the band tails and the carriers in these localized states are transported by hopping. At high temperatures, carriers are excited across the mobility gap into the extended states.

2.3.4.2a. *Conduction in the Extended States.* In the nondegenerate case and assuming a constant density of states and constant mobility, the conductivity can be expressed as

$$\sigma = eN(E_c)kT\mu_c \exp[-(E_c - E_f)/kT] \tag{2.78}$$

where μ_c is the average mobility and $N(E_c)$ is the density of states at the conduction band edge. Mott[38] calculated the mobility to be

$$\mu_c = 0.078ea^2B/\hbar kT \tag{2.79}$$

where a is the interatomic distance and B is the bandwidth. For $a = 2$ Å and $B = 5$ eV, $\mu_c = 10$ cm^2 V^{-1} s^{-1} at room temperature. This value corresponds to a mean free path comparable to the interatomic distance.

Cohen[36] suggested a Brownian-type motion and thus from Einstein's relation

$$\mu_c = \tfrac{1}{6}(ea^2/kT)\nu = De/kT \tag{2.80}$$

where the diffusion coefficient D is written as

$$D = \tfrac{1}{6}\nu a^2 \tag{2.81}$$

and ν is the jump frequency. It may be noted that μ_c has the same temperature dependence in equations (2.79) and (2.80).

Since μ_c varies as $1/kT$, one expects the conductivity to follow the relation

$$\sigma = \sigma_0 \exp[-(E_c - E_f)/kT] \tag{2.82}$$

Assuming a linear temperature dependence of $E_c - E_f$, we have

$$E_c - E_f = E(0) - \gamma T \tag{2.83}$$

where $E(0)$ is the value of $(E_c - E_f)$ at $T = 0$ K. The conductivity is then given by

$$\sigma = \sigma_0 \exp(\gamma/k)\exp[-E(0)/kT]$$
$$= C_0 \exp[-E(0)/kT] \tag{2.84}$$

where $C_0 = eN(E_c)kT\mu_c \exp(\gamma/K)$. Typically, σ_0 lies between 10 and 10^3 $\Omega^{-1}\,\text{cm}^{-1}$ in most amorphous semiconductors and γ generally has values between 2×10^{-4} and 4×10^{-4} eV deg^{-1} for chalcogenide glasses.

Hindley[39] and Friedman[40] have derived an expression for the dc conductivity in the extended states on the basis of a random phase model:

$$\sigma = (2\pi e^2/3\hbar a)\{Za^6J^2[N(E_c)]^2\}\exp[-(E_c - E_f)/kT] \tag{2.85}$$

where Z is the coordination number and J the electronic transfer integral. Equation (2.85) can be expressed in the form

$$\sigma = \sigma_0 \exp[-(E_c - E_f)/kT] \tag{2.86}$$

The conductivity mobility is given by

$$\mu_c = (2\pi ea^2/3\hbar)Z[(J/kT)a^3JN(E_c)] \tag{2.87}$$

2.3.4.2b. *Conduction in Band Tails.* Since the wavefunctions are localized, $\sigma(E) = 0$ and conduction can only occur by thermally activated hopping. The hopping mobility is given by

$$\mu_{\text{hop}} = \mu_0 \exp[-W(E)/kT] \tag{2.88}$$

The preexponential factor μ_0 can be expressed by the relation

$$\mu_0 = \tfrac{1}{6}\nu_{\text{ph}}eR^2/kT \tag{2.89}$$

where ν_{ph} is the phonon frequency and R is the distance covered in one hop. Putting values of $\nu_{\text{ph}} = 10^{13}$ s^{-1} and $W \simeq kT$, we obtain $\mu_{\text{hop}} \sim 10^{-2}$ cm^2 V^{-1} s^{-1} at room temperature, which is about a factor of 100 less than μ_c. If we assume a power law for the density of states of the form

$$N(E) = [N(E_c)/(\Delta E)^s](E - E_A)^s \tag{2.90}$$

with $\Delta E = E_c - E_A$, then

$$\sigma_{\text{hop}} = \sigma_{0\,\text{hop}}(kT/\Delta E)^s C \exp[-(E_A - E_f + W)/kT] \tag{2.91}$$

where

$$\sigma_{0\,\mathrm{hop}} = \tfrac{1}{6}\nu_{\mathrm{ph}} e^2 R^2 N(E_c) \tag{2.92}$$

and

$$C = s! - \left(\frac{\Delta E}{kT}\right)^s \exp\left(-\frac{\Delta E}{kT}\right)\left[1 + s\left(\frac{kT}{\Delta E}\right) + s(s-1)\left(\frac{kT}{\Delta E}\right)^2 + \cdots\right] \tag{2.93}$$

W is the energy difference between one site and the next.

For a linear variation, $s = 1$. The conductivity is then given by

$$\sigma_{\mathrm{hop}} = \sigma_{0\,\mathrm{hop}}(kT/\Delta E)C_1 \exp[-(E_A - E_f + W)/kT] \tag{2.94}$$

where

$$C_1 = 1 - \exp(-\Delta E/kT)[1 - (\Delta E/kT)] \tag{2.95}$$

2.3.4.2c. *Conduction in the Localized States at the Fermi Level*. Suppose an electron is scattered by phonons from one localized state to another. If the energy difference between the states is denoted by W, the probability that an electron will jump from one state to another is given by[41]

$$p = \nu_{\mathrm{ph}} \exp(-2\alpha R - W/kT) \tag{2.96}$$

where R is the jumping distance that at high temperatures equals the interatomic spacing and α is representative of the rate of fall-off of the wavefunction at a site. From Einstein's relation [equation (2.61)] with $D = (\frac{1}{6})pR^2$, we obtain for the conductivity the relation

$$\sigma = \tfrac{1}{6}e^2 p R^2 N(E_f) \tag{2.97}$$

where $N(E_f)$ is the density of states at the Fermi level. Substituting the value of p from equation (2.96), we get

$$\sigma = \tfrac{1}{6}e^2 R^2 \nu_{\mathrm{ph}} N(E_f)\exp(-2\alpha R)\exp(-W/kT) \tag{2.98}$$

At lower temperatures, the number and energy of phonons decrease and hence the more energetic phonon-assisted hops become progressively less favorable. Thus carriers, in order to find sites which lie energetically closer than the nearest neighbors, tend to hop larger distances. This mechanism is the so-called variable range hopping. The factor $\exp[-2\alpha R - (W/kT)]$ is not a maximum for nearest neighbors. Mott[42] has obtained an expression for the most probable jump distance, namely

$$R = [9/8\pi\alpha N(E_f)kT]^{1/4} \tag{2.99}$$

The jump frequency then takes the form

$$p = \nu_{ph} \exp(-A/T^{1/4}) \tag{2.100}$$

where $A = 2.1[\alpha^3/kN(E_f)]^{1/4}$. This leads to a temperature dependence for the conductivity of the form

$$\sigma = \tfrac{1}{6}R^2 e^2 \nu_{ph} N(E_f) \exp(-A/T^{1/4}) \tag{2.101}$$

Alternatively,

$$\sigma = \sigma_0(T) \exp(-A/T^{1/4}) \tag{2.102}$$

where the preexponential factor $\sigma_0(T)$ is given by

$$\sigma_0(T) = \frac{e^2}{2(8\pi)^{1/2}} \nu_{ph} \left[\frac{N(E_f)}{\alpha kT}\right]^{1/2} \tag{2.103}$$

It should be noted that Mott's derivation of variable range hopping is based on several simplifying assumptions and does not take into account: (i) the energy dependence of the density of states at E_f, (ii) correlation effects in the tunneling process, (iii) multiphonon processes, and (iv) electron–phonon interaction. Although the $T^{-1/4}$ dependence of $\ln \sigma$ is widely observed experimentally, the values of $N(E_f)$ calculated from $\sigma_0(T)$ are unreasonably high. Several authors have clearly demonstrated the major importance of the energy distribution of the density of states in the theory of variable range hopping.

2.3.5. *Thin Films*

We shall now discuss conduction processes in thin semiconducting and insulating films.

2.3.5.1. *Semiconducting Films*

When transport occurs through thin specimens, the carriers are subjected to considerable scattering by the boundary surfaces in addition to the normal bulk scattering. The effective carrier mobility is thus reduced below the bulk value because this additional scattering gives rise to conductivity size effects. Analysis of size effects in semiconductors is modified from that in metals owing to the additional feature of surface space charge and the resultant surface potential barrier in semiconductors.

If we assume that the band edges are flat up to the surfaces and the electron density n is uniform throughout the sample and equal to the bulk value n_B, the effect of surface scattering can be incorporated in the form of an average collision time τ_s.[10] The bulk scattering is characterized by the relaxation time τ_B (usually $\sim 10^{-12}$ s). If the bulk and the surface scattering processes are additive, the average relaxation time τ_F for electrons in thin films is given by

$$1/\tau_F = 1/\tau_s + 1/\tau_B \tag{2.104}$$

τ_s can be estimated to be

$$\tau_s = t/v_z = (t/l)\tau_B = \gamma\tau_B \tag{2.105}$$

where t is the mean distance of a carrier from the surface (thickness of the film), v_z is the unilateral mean velocity, l is the mean free path (mfp), and $\gamma \equiv t/l$. The mfp is defined by

$$l \equiv \tau_B v_z = \mu_B (h/e)[(3/8\pi)n_B]^{1/3} \tag{2.106}$$

where μ_B is the bulk carrier mobility. The average electron mobility μ_F is given by

$$\mu_F = e\tau_F/m^* = \mu_B/(1 + 1/\gamma) \tag{2.107}$$

where m^* is the effective mass of the carrier. The average mobility decreases with decreasing film thickness and approaches bulk value for $\gamma \gg 1$. We emphasize that equation (2.107) is valid for: (i) thick films in the flat-band approximation only and (ii) sufficiently thin films such that t is small compared with the effective Debye length L_D. For an intrinsic semiconductor,

$$L_D = [4\pi\varepsilon kT/e^2(n_B + p_B)]^{1/2} \tag{2.108}$$

where ε is the static dielectric constant and n_B and p_B are the densities of the negative and positive carriers, respectively.

We have assumed diffuse scattering at the boundary in the above treatment. If p is the scattering coefficient, i.e., the probability that an electron is specularly reflected, only a fraction $(1-p)$ of the electrons is scattered diffusely. We therefore replace $1/\tau_s$ by $(1-p)/\tau_s$ and equation (2.107) becomes

$$\mu_F = \mu_B/[1 + (1-p)/\gamma] \tag{2.109}$$

In thick films, where $t \gg L_D$, the carrier transport is dominated by the size effect because of the surface space charge. With an accumulation layer and symmetrical potential barriers at the two surfaces, we can consider the carriers as moving in a thin film with one surface, $z = 0$, a diffuse scatterer and the other, $z = L_c$, a specular reflector. L_c is the effective carrier distance from the surface to the center of the space charge and is related to L_D by the relation

$$L_c = (|\nu_s|/F_s)L_D \tag{2.110}$$

where $\nu_s = eV_s/2kT$ is the normalized value of the surface potential V_s, and F_s is a space-charge distribution function.[22] F_s decreases rapidly with increasing ν_s. For small ν_s (flat band), $L_c \rightarrow L_D$. The electron surface mobility is given by

$$\mu_s = \mu_B/[1 + (l/L_c)(1 + \nu_s)^{1/2}] \tag{2.111}$$

For holes, the relation is the same with the sign of ν_s changed. For partially specular reflection of electrons, L_c is replaced by $L_c/(1-p)$:

$$\mu_s = \mu_B/[1 + (l/L_c)(1-p)(1 + \nu_s)^{1/2}] \tag{2.112}$$

As $\nu_s \rightarrow 0$, $\mu_s/\mu_B \rightarrow 1/\{1 + [l(1-p)]/L_c\}$.

In the case of a depletion layer at the boundary, only those electrons possessing sufficient energy to surmount the potential barrier are able to reach the surface. The expression for the surface mobility is thus complicated. However, μ_s approaches μ_B for large negative values of ν_s. For $\nu_s \rightarrow 0$, $\mu_s/\mu_B \rightarrow \{1 - [l(1-p)]/L_c\}$. For $l \ll L_c$, the size effect is absent in both accumulation and depletion layers.

The conductivity of thin films in the flat-band approximation is given by[43]

$$\sigma_F = en_B\mu_B[1 - (3/8\gamma)\sin^4\theta_0] \qquad \text{for } \gamma \gg 1 \tag{2.113}$$

and

$$\sigma_F = en_B\mu_B[\tfrac{3}{2}\cos\theta_0(1 - \cos^2\theta_0/3)] \qquad \text{for } \gamma \ll 1 \tag{2.114}$$

where θ_0 defines a boundary condition such that all carriers incident at angles less than θ_0 to the surface normal are diffusely scattered whereas all carriers incident at greater angles are specularly reflected.

In small magnetic fields, the Hall coefficient R_{HF} of a thin film ($t \ll L_D$) is given by[10]

$$R_{HF} = R_{HB}\eta(1/\gamma) \tag{2.115}$$

where R_{HB} is the bulk Hall coefficient and $\eta(1/\gamma)$ is the correction factor which has been calculated numerically by Amith.[44] η approaches unity when $\gamma \gg 1$.

When the thickness of the semiconducting film is comparable to or smaller than both the mfp and the effective deBroglie wavelength of the carriers, the transverse component of the quasi-momentum is quantized and, consequently, the electron states assume quasi-discrete energy values. Under conditions satisfactory for the occurrence of quantum size effects, all transport properties exhibit oscillatory behavior as a function of the film thickness with a period

$$\Delta t = h/(8m^* E_f) \tag{2.116}$$

where $1/m^* = (1/m_e^*) + (1/m_h^*)$ and E_f is the Fermi energy.

The electron transport properties of thick epitaxial semiconducting films with bulk or nearly bulk-like properties may be characterized by the resistivity, Hall coefficient, and Hall mobility. Electrical conduction, in general, is determined by various overlapping scattering mechanisms. By analogy with Mathiessen's rule, the mobility is given by

$$1/\mu_F = 1/\mu_L + 1/\mu_I + 1/\mu_S \tag{2.117}$$

where μ_L, μ_I, and μ_S are contributions due to lattice, impurity, and surface (including grain boundary) scattering, respectively. In general, the additive effect of various overlapping scattering mechanisms gives rise to a complicated temperature dependence.

2.3.5.2. *Insulating Films*

Carriers may be generated or modulated inside the insulator film (bulk-limited processes) or injected from the metal electrode (injection-limited processes). The various mechanisms of current transfer through a thin insulator film sandwiched between two metal electrodes are thermionic emission, Schottky emission of electrons over the metal/insulator interface barrier into the conduction band of the insulator, direct quantum mechanical tunneling of electrons from one metal electrode to another, and the tunneling of carriers through the insulator barrier gap at high applied fields (field or cold emission). These are injection processes. Bulk-limited processes include space-charge-limited current and Poole–Frenkel emission.

The thermionic emission current over a barrier ϕ is given by

$$J = AT^2 \exp(-\phi/kT) \tag{2.118}$$

where A is the Richardson constant. In a sandwich structure, we add the thermionic current in both directions. The net current density is

$$J = AT^2[\exp(-\phi_1/kT) - \exp(-\phi_2/kT)] \qquad (2.119)$$

where ϕ_1 and ϕ_2 are the maximum barrier heights above the Fermi level of the negative and positive electrodes, respectively. When a potential V exists between electrodes such that $\phi_2 = \phi_1 + eV$, then

$$J = AT^2 \exp(-\phi_1/kT)[1 - \exp(-eV/kT)] \qquad (2.120)$$

For a symmetrical barrier, applying a parabolic image-force correction for one barrier yields the Richardson–Schottky formula

$$J = AT^2 \exp\{[-\phi - (14.4eV/\varepsilon s)^{1/2}]/kT\} \qquad (2.121)$$

where ε is the dielectric constant and s is the effective thickness of the insulator barrier in angstroms. In equation (2.121), $e^2 = 14.4eV$-Å is used as a convenient unit.

Simmons[45] has shown, on the basis of parametric calculations of the $I - V$ characteristics of a tunnel structure, that at a temperature of 300 K and for a barrier thickness $s < 40$ Å, tunneling predominates. However, for $s > 40$ Å, either the thermionic or the tunnel mechanism can predominate, depending upon the barrier height and the applied voltage. Simmons[45] has obtained a generalized formula for the tunnel current density from electrode 2 to electrode 1 (at $T = 0$ K) through a barrier of width Δs and mean barrier height ϕ of the form

$$J = \frac{6.2 \times 10^{10}}{(\Delta s)^2} \{\bar{\phi} \exp(-1.025\, \Delta s \bar{\phi}^{1/2}) - (\bar{\phi} + V)\exp[-1.025\, \Delta s(\bar{\phi} + V)^{1/2}]\} \qquad (2.122)$$

where V is the positive voltage applied to electrode 2 relative to electrode 1.

For low and intermediate voltages ($0 \leqslant V \leqslant \phi$), J for a rectangular barrier of height ϕ_0 with similar electrodes can be obtained by substituting $\Delta s = s$ (where s is the insulator thickness) and $\bar{\phi} = \phi_0 - eV/2$ in equation (2.122). For $V > \phi_0/e$, $\Delta s = s\phi_0/eV$ and $\bar{\phi} = \phi_0/2$. J is then given by

$$J = 3.38 \times 10^{10}(F^2/\phi_0)\{\exp[-0.689(\phi_0^{3/2}/F)] - (1 + 2V/\phi_0)\exp[-0.689(\phi_0^{3/2}/F)(1 + 2V/\phi_0)^{1/2}]\} \qquad (2.123)$$

where $F = V/s$. When $V > (\phi_0 + E_f)/e$, the last term in equation (2.123) is negligible compared with the first and we obtain the Fowler–Nordheim equation, namely

$$J \propto F^2 \exp(-0.689\, \phi_0^{3/2}/F) \tag{2.124}$$

Applying an image-force correction to the general tunnel equation, we get for a rectangular barrier for all values of V a tunnel current density of the form

$$J = \frac{6.2 \times 10^{10}}{\Delta s^2} \{\phi_1 \exp(-1.025\, \Delta s \phi_1^{1/2}) - (\phi_1 + V)\exp[-1.025\, \Delta s(\phi_1 + V)^{1/2}]\} \tag{2.125}$$

J is in A cm^{-2}, ϕ_0 in V, and s_1, s_2, and s in angstroms.

$$\phi_1 = \phi_0 - \frac{V}{2s}(s_1 + s_2) - \frac{5.75}{\varepsilon(s_2 - s_1)} \ln\left[\frac{s_2(s - s_1)}{s_1(s - s_2)}\right]$$

and

$$\left.\begin{aligned} s_1 &= 6/\varepsilon\phi_0 \\ s_2 &= s\left(1 - \frac{46}{3\phi_0\varepsilon s + 20 - 2V\varepsilon s}\right) + \frac{6}{\varepsilon\phi_0} \end{aligned}\right\} \text{for } V < \phi_0$$

$$\left.\begin{aligned} s_1 &= 6/\varepsilon\phi_0 \\ s_2 &= \frac{\phi_0\varepsilon s - 28}{\varepsilon V} \end{aligned}\right\} \text{for } V > \phi_0$$

ε is the optical dielectric constant.

The above expressions for tunnel current have been derived for $T = 0$ K. The expression[10] for the tunnel current density $J(V, T)$ at a temperature T, based on the temperature dependence of the distribution function, is

$$\frac{J(V,T)}{J(V,0)} = \frac{\pi BkT}{\sin(\pi BkT)}$$

where $B = (P/2)\bar{\phi}^{1/2}$ or

$$\frac{J(V,T)}{J(V,0)} \approx 1 + \frac{3 \times 10^{-9}(\Delta sT)^2}{\bar{\phi}} + QT^4 + \cdots \tag{2.126}$$

P and Q are constants and Δs and $\bar{\phi}$ are expressed in angstroms and volts, respectively. We note here that the tunnel current can also vary with temperature because of the temperature dependence of the barrier height.

At low applied biases, if the injected carrier density is lower than the thermally generated free-carrier density, Ohm's law is obeyed. When the injected-carrier density is greater than the free-carrier density, the current becomes space-charge-limited. For the case of one-carrier trap-free space-charge-limited current, the Mott–Gurney[10] relation gives

$$J = 10^{-13} \mu \varepsilon V^2 / t^3 \tag{2.127}$$

where V is the applied voltage, t is the insulator thickness, μ is the drift mobility of the charge carriers, and ε is the dielectric constant. In the presence of shallow traps

$$J = 10^{-13} \theta \mu \varepsilon V^2 / t^3 \tag{2.128}$$

where θ is the ratio of the free to the trapped charges.

For an exponential trap distribution,[10]

$$J \propto V^{n+1} / t^{2n+1} \tag{2.129}$$

where n is a parameter characteristic of the trap distribution.

Depending on the number and level of traps, thermal and/or field ionization of traps can contribute significantly to the current transport in thin insulator films. The current–voltage relation in the case of field-enhanced thermal excitation of trapped electrons into the conduction band, known as the Poole–Frenkel effect, is given by

$$J \propto \exp[(57.7 \times eV/\varepsilon s)^{1/2} - E_t]/kT \tag{2.130}$$

where E_t is the depth of the trap potential well and s is the electrode spacing in angstroms. Instead of traps, if impurities are ionized, the same relation holds with E_t being the energy required to ionize an impurity under zero-field conditions.

2.3.6. *High-Field Effects*

At low electric fields, the drift velocity of carriers in a semiconductor is proportional to the electric field, and mobility is independent of the electric field, as discussed previously. However, at reasonably high fields, nonlinearities in mobility are observed and in some cases the drift velocity is found to saturate. The most frequent scattering process at high fields is the emission of optical phonons. On the average, the carriers acquire more energy than they possess at thermal equilibrium. With an increase in the field, the average energy of the carriers also increases and they acquire an effective temperature T_e which is higher than the lattice temperature T. The rate at which energy is fed from the field to the carriers is balanced by an equal rate of loss of energy from carriers to the lattice. It can be shown[9] that for Ge and Si

$$T_e/T = \tfrac{1}{2}\{1 + [1 + (3\pi/8)(\mu_0 E/C_s)^2]^{1/2}\} \tag{2.131}$$

and

$$v_d = \mu_0 E (T/T_e)^{1/2} \tag{2.132}$$

where μ_0 is the low-field mobility, E the electric field, and C_s the velocity of sound. At $\mu_0 \ll C_s$, equations (2.131) and (2.132) reduce to

$$T_e \simeq T[1 + (3\pi/32)(\mu_0 E/C_s)^2] \tag{2.133}$$

and

$$v_d \simeq \mu_0 E[1 - (3\pi/64)(\mu_0 E/C_s)^2] \tag{2.134}$$

At sufficiently high fields, the drift velocity attains a saturation limit given by

$$v_{d_s} = (8E_p/3\pi m_0)^{1/2} \sim 10^7 \text{ cm s}^{-1} \tag{2.135}$$

where E_p is the optical phonon energy.

In materials like GaAs which have a high-mobility ($\mu \sim 4000$ to 8000 $cm^2 V^{-1} s^{-1}$) valley and low-mobility ($\mu \sim 100$ $cm^2 V^{-1} s^{-1}$), high-energy satellite valleys [see Figure 2.1d], the velocity–field relationship is more complicated. The density of states in the upper valley is about 70 times that in the lower valley. At high fields, the electrons in the lower valley are field-excited to the normally unoccupied upper valley leading to a differential negative resistance. The steady-state conductivity is given by

$$\sigma = e(\mu_1 n_1 + \mu_2 n_2) = en\bar{\mu} \tag{2.136}$$

where n_1 and n_2 are the electron densities in the lower and upper valley and μ_1 and μ_2 are the corresponding mobilities. The average mobility $\bar{\mu}$ is given by

$$\bar{\mu} = (\mu_1 n_1 + \mu_2 n_2)/(n_1 + n_2) \tag{2.137}$$

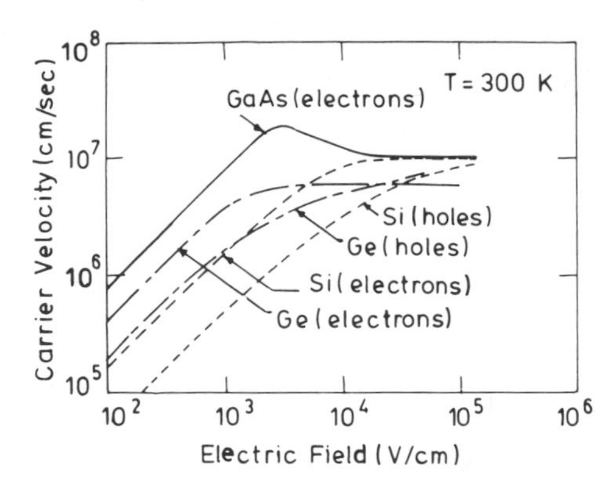

Figure 2.5. Carrier velocity as a function of electric field for Si, Ge, and GaAs (after Sze[9]).

The current density is then

$$J = ne\bar{\mu}E = nev \tag{2.138}$$

where n is the total carrier density ($n = n_1 + n_2$) and v is the average electron velocity. The condition for negative differential conductance is

$$dv/dE \equiv \mu_D < 0 \tag{2.139}$$

Assuming that n_2/n_1 can be expressed by the relation

$$n_2/n_1 = (E/E_0)^k \equiv F^k \tag{2.140}$$

where k is a constant and E_0 is the field at which $n_1 = n_2$, and defining a constant $B = \mu_2/\mu_1$, we obtain for a given field the following expression for the average velocity

$$v = \mu_1 E(1 + BF^k)/(1 + F^k) \tag{2.141}$$

Figure 2.5 shows the measured carrier velocity as a function of electric field for Si, Ge, and GaAs.

2.4. *Optical Interactions*

We shall now discuss the various optical interactions which lead to carrier generation and associated photogenerated processes.

2.4.1. *Reflectance and Transmittance at an Interface*

The reflectance R and transmittance T at normal incidence at an interface between two media, with refractive indices n_0 and n_1 (which are complex in the most general case), are given by[46,47]

$$R = \left(\frac{n_0 - n_1}{n_0 + n_1}\right)\left(\frac{n_0 - n_1}{n_0 + n_1}\right)^* \tag{2.142}$$

$$T = \left(\frac{2n_0 n_1}{n_0 + n_1}\right)\left(\frac{2n_0 n_1}{n_0 + n_1}\right)^* \tag{2.143}$$

where the asterisk denotes a complex conjugate. $[(n_0 - n_1)/(n_0 + n_1)]$ and $[2n_0 n_1/(n_0 + n_1)]$ are called reflectivity r and transmittivity t, respectively, or Fresnel coefficients.

At angles of incidence other than normal, r and t depend on the polarization of the incident wave. For a wave with the electric vector in the plane of incidence (TM or p polarized wave)

$$r_p = \frac{n_0 \cos\theta_1 - n_1 \cos\theta_0}{n_0 \cos\theta_1 + n_1 \cos\theta_0} \tag{2.144}$$

$$t_p = \frac{2n_0 \cos\theta_0}{n_0 \cos\theta_1 + n_1 \cos\theta_0} \tag{2.145}$$

where θ_0 and θ_1 are the angles of incidence and refraction, respectively. Obviously, if n_1 is complex, θ_1 will be complex and, therefore, will not represent the angle of refraction, except for $\theta_1 = \theta = 0$. Only for this case will the Fresnel coefficients be defined.

For a wave with the electric vector normal to the plane of incidence (TE or s polarized wave)

$$r_s = \frac{n_0 \cos\theta_0 - n_1 \cos\theta_1}{n_0 \cos\theta_0 + n_1 \cos\theta_1} \tag{2.146}$$

$$t_s = \frac{2n_0 \cos\theta_0}{n_0 \cos\theta_0 + n_1 \cos\theta_1} \tag{2.147}$$

2.4.1.1. *Thin Films*

For a thin film on a substrate we have to consider two interfaces: the incident medium (refractive index n_i)/film (refractive index n_f) interface 1

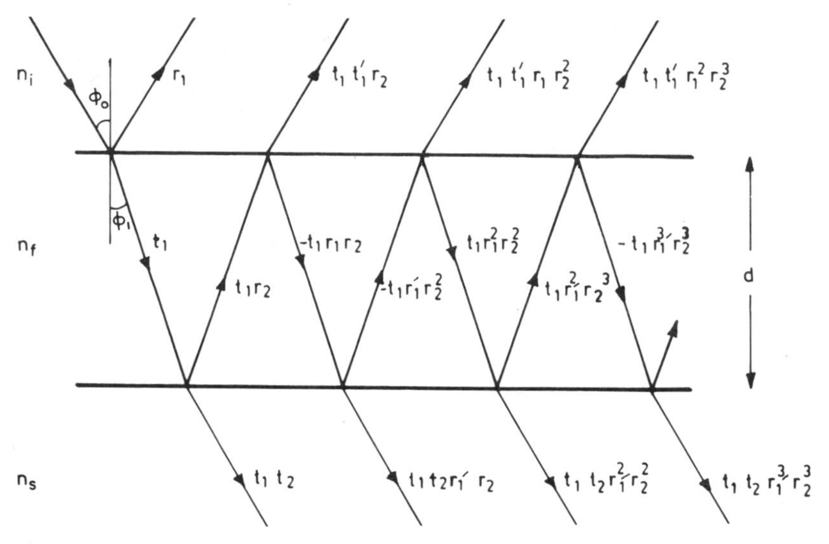

Figure 2.6. Multiple reflections in a thin film.

and film/substrate (refractive index n_s) interface 2.[47,48] Taking into account the multiple reflections at the two interfaces (Figure 2.6),

$$r = \frac{r_1 + r_2 \exp(-2i\delta_1)}{1 + r_1 r_2 \exp(-2i\delta_1)} \tag{2.148}$$

$$t = \frac{t_1 t_2 \exp(-i\delta_1)}{1 + r_1 r_2 \exp(-2i\delta_1)} \tag{2.149}$$

with $\delta_1 = (2\pi/\lambda) n_f d \cos\theta_1$, d being the film thickness.

The reflectance and transmittance are thus given by

$$R = \frac{r_1^2 + 2r_1 r_2 \cos 2\delta_1 + r_2^2}{1 + 2r_1 r_2 \cos 2\delta_1 + r_1^2 r_2^2} \tag{2.150}$$

$$T = \frac{n_s}{n_i}\left(\frac{t_1^2 t_2^2}{1 + 2r_1 r_2 \cos 2\delta_1 + r_1^2 r_2^2}\right) \tag{2.151}$$

Equations (2.150) and (2.151) can be solved in terms of n_i, n_f, and n_s (which can be complex) to yield

$$\begin{aligned} R = {} & [(n_i^2 + n_f^2)(n_f^2 + n_s^2) - 4n_i n_f^2 n_s + (n_i^2 - n_f^2)(n_f^2 - n_s^2)\cos 2\delta_1] \\ & \times [(n_i^2 + n_f^2)(n_f^2 + n_s^2) + 4n_i n_f^2 n_s + (n_i^2 - n_f^2)(n_f^2 - n_s^2)\cos 2\delta_1]^{-1} \end{aligned} \tag{2.152}$$

$$T = \frac{8 n_i n_f^2 n_s}{(n_i^2 + n_f^2)(n_f^2 + n_s^2) + 4n_i n_f^2 n_s + (n_i^2 - n_f^2)(n_f^2 - n_s^2)\cos 2\delta_1} \tag{2.153}$$

If the Fresnel coefficients are small enough so that their product can be neglected or if the absorption in the film is high and multiple reflections can be neglected, then equation (2.149) reduces to

$$t = t_1 t_2 \exp(i\delta_1) \tag{2.154}$$

Thus, for a film with a complex refractive index ($n_f \equiv n_f - ik_f$), T can be expressed as

$$T = (1 - R_1)(1 - R_2)\exp(-\alpha d) \tag{2.155}$$

where $\alpha = 4\pi k/\lambda$ is the absorption coefficient. Equation (2.155) can be further simplified to give

$$\alpha = (1/d)\ln[(1 - R)^2/T] \tag{2.156}$$

from whence α can be calculated if R and T are known.[10]

In an alternative approach, one can use a matrix formulation to calculate R and T. The optical admittance Y (defined as $H/k \times E$) for the assembly is given by $Y = C/B$ where

$$\begin{pmatrix} B \\ C \end{pmatrix} = \begin{pmatrix} \cos\delta_1 & i\sin\delta_1/n_f \\ in_f\sin\delta_1 & \cos\delta_1 \end{pmatrix} \begin{pmatrix} 1 \\ n_s \end{pmatrix} \tag{2.157}$$

For $\theta \neq 0$, $n_{f_p} = n_f \cos\theta$ and $n_{f_s} = n_f/\cos\theta$ should be used instead of n_f in the calculations. Thus, the assembly behaves like an interface between the incident medium and a medium with an effective refractive index Y, and R and T are given by

$$R = \left(\frac{n_i - Y}{n_i + Y}\right)\left(\frac{n_i - Y}{n_i + Y}\right)^* \tag{2.158}$$

$$T = \frac{4n_i n_s}{(n_i + Y)(n_i + Y)^*} \tag{2.159}$$

This approach is particularly useful in describing a stack of multilayer coatings, since, mathematically, for a stack with l layers[49]

$$\begin{pmatrix} B \\ C \end{pmatrix} = \left[\prod_{j=1}^{l} \begin{pmatrix} \cos\delta_j & i\sin\delta_j/n_j \\ in_j\sin\delta_j & \cos\delta_j \end{pmatrix}\right] \begin{pmatrix} 1 \\ n_s \end{pmatrix} \tag{2.160}$$

R and T can easily be calculated thereafter.

2.4.1.2. *Grain Boundaries*

Polycrystalline thin films have a large concentration of dislocations, vacancies, and other defects along grain boundaries. Hunderi[55] has treated grain boundaries in metal films as a disordered array of polarizable scatterers embedded in a homogeneous medium. Further, the grain boundaries have been physically represented as flat ellipsoidal disks. For light incident normal to the sample surface, the reflectivity r, obtained by summing over all the reflected waves and taking into account multiple reflections between the scatterers and the sample surface but neglecting reflections between the ellipsoids, is given by

$$r = r_0 \left| \frac{1 - \frac{1}{4} v_f \alpha_f [(1+n)/(1-n)]}{1 - \frac{1}{4} v_f \alpha_f [(1-n)/(1+n)]} \right|^2 \tag{2.161}$$

where r_0 is the reflectivity of the sample in the absence of scatterers, n is the refractive index, v_f is the fractional volume of the sample occupied by the ellipsoids, and α_f is the polarizability of the ellipsoids.

2.4.1.3. *Inhomogeneous Systems*

Several theories have been developed to explain the optical behavior of inhomogeneous or composite systems which consist of two or more distinct phases. According to the Mie scattering theory,[56,57] based on the physical model of N identical particles per unit volume placed far apart so that scattering from different particles are independent of each other and no multiple scattering occurs, the extinction coefficient k (equal to $\alpha\lambda/4\pi$) is given by

$$k = 18\pi N V m_0^3 \frac{\varepsilon_2}{\lambda[(\varepsilon_1 + 2m_0^2)^2 + \varepsilon_2^2]} \tag{2.162}$$

where NV gives the total volume of the particles per unit volume of the sample, ε_1 and ε_2 are the real and imaginary parts of the dielectric constant, respectively, of the material of the particles, and λ is the vacuum wavelength of light.

In the Maxwell–Garnett (MG) theory,[58] the composite is treated as identical spherical particles of permeability $\varepsilon_p(\omega)$ embedded in a surrounding medium of permeability $\varepsilon_m(\omega)$. The effective dielectric permeability $\bar{\varepsilon}^{\text{MG}}$ can then be expressed as

$$\bar{\varepsilon}^{\text{MG}} = \varepsilon_m \frac{1 + \frac{2}{3} f \alpha'}{1 - \frac{1}{3} f \alpha'} \tag{2.163}$$

where f denotes the volume fraction occupied by the spheres and α' for spheres can be written as

$$\alpha' = \frac{\varepsilon_p - \varepsilon_m}{\varepsilon_m + \frac{1}{3}(\varepsilon_p - \varepsilon_m)} \tag{2.164}$$

For randomly oriented ellipsoids, Polder and van Santen (PVS) have given the expression[59]

$$\alpha' = \frac{1}{3} \sum_{i=1}^{3} \frac{\varepsilon_p - \varepsilon_m}{\varepsilon_m + L_i(\varepsilon_p - \varepsilon_m)} \tag{2.165}$$

where L_i is the triplet depolarization factor.

Bruggeman[60] has developed a physical model in which both the matrix and inclusions are considered to be small spheres (or ellipsoids in a more general case) occupying the whole volume of the material. Under these conditions, the effective dielectric permeability $\bar{\varepsilon}^{\mathrm{BR}}$ can be expressed by[58]

$$\bar{\varepsilon}^{\mathrm{BR}} = \varepsilon_m \frac{1 - f + \frac{1}{3}f\alpha}{1 - f - \frac{2}{3}f\alpha} \tag{2.166}$$

2.4.2. *Absorption Phenomena in Semiconductors*

We now turn to a consideration of the electron–photon interaction in a semiconductor in terms of the band diagram. It should be remembered that only those absorption processes contribute to the photovoltaic effect in which electron–hole generation occurs. However, we shall note all possible absorption mechanisms since these constitute phenomena by which photons are indirectly lost for the purpose of generation, and such loss mechanisms are important in analyzing the performance of a photovoltaic device. We shall present only the final equations for the absorption coefficients for different absorption processes. For a more detailed description of the processes, the reader is referred to Moss et al.[7]

For a semiconductor in which the minimum of the conduction band and the maximum of the valence band occur at the same value of k, absorption begins at $h\nu = E_g$ and the electron is transferred vertically between the two bands without a change in momentum (Figure 2.1d). Nonvertical transitions are normally forbidden in this case. The optical absorption coefficient for a direct transition is given by

$$\alpha = A(h\nu - E_g)^{1/2} \tag{2.167}$$

where for allowed transitions

$$A = 3.38 \times 10^7 n^{-1}(m_e/m_0)^{1/2}(E_g/h\nu) \tag{2.168}$$

where m_0 is the free-electron mass, ν is the frequency of the radiation, and n is the refractive index. For forbidden transitions in the case of simple parabolic bands

$$\alpha = A'(h\nu - E_g)^{3/2} \tag{2.169}$$

where A' is a slowly varying function of energy.

In semiconductors where the conduction band minimum and the valence band maximum occur at different k values, optical transitions from the latter to the former require the participation of phonons in order to conserve momentum because of the change in the electron wavevector (Figure 2.1c). Phonons are either emitted or absorbed. The absorption coefficient for an allowed indirect transition is given by

$$\alpha = \frac{C(h\nu + E_p - E_g)^2}{\exp(E_p/kT) - 1} + \frac{C(h\nu - E_p - E_g)^2}{1 - \exp(-E_p/kT)} \tag{2.170}$$

where the first term represents the contribution of phonon absorption and must be taken to be zero if $h\nu < E_g - E_p$, while the second term represents the contribution of phonon emission and is to be taken to be zero for $h\nu < E_g + E_p$. E_p is the phonon energy and C is a constant. Indirect transitions occur with a lower probability and give rise to an absorption edge that is less steep than for direct transitions.

For forbidden indirect transitions

$$\alpha = C'(h\nu \pm E_p - E_g)^3 \tag{2.171}$$

where C' is a slowly varying function of temperature and energy.

In the presence of excitons, at $h\nu = E_g$, the absorption coefficient for allowed direct transitions takes the constant value

$$\alpha_0(E_g) = 2\pi A R^{1/2} \tag{2.172}$$

where R is the Rydberg constant of the exciton. For $h\nu \gg E_g$,

$$\alpha_{ex}(h\nu) = \pi A R^{1/2} \exp(\pi\gamma)/\sinh \pi\gamma \tag{2.173}$$

where $\gamma = [R/(h\nu - E_g)]^{1/2}$. For forbidden direct transitions, at $h\nu = E_g$,

$$\alpha_0(E_g) = 2\pi A' R^{3/2} \tag{2.174}$$

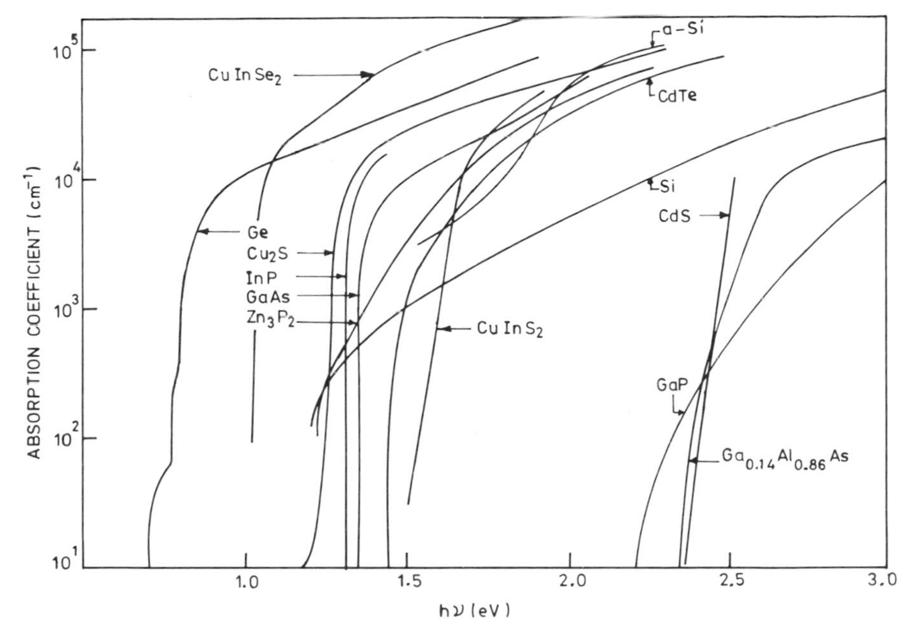

Figure 2.7. Absorption spectra of various semiconductors.

and for $h\nu \gg E_g$,

$$\alpha_{ex}(h\nu) = \pi A' R^{3/2}(1 + 1/\gamma^2)\exp(\pi\gamma)/\sinh \pi\gamma \tag{2.175}$$

For indirect transitions, no exciton line spectra are obtained. Instead, continuous absorption begins at $h\nu = E_g - E_p - R$ and the shape of the absorption curve for the lowest exciton band ($l = 1$) has a dependence of the form

$$\alpha \propto (h\nu - E_g \pm E_p + R)^{1/2} \tag{2.176}$$

For larger photon energies, an approximate expression for the absorption at $h\nu \simeq E_g \pm E_p$ can be written as

$$\alpha \propto (h\nu - E_g \pm E_p)^{3/2} \tag{2.177}$$

Figure 2.7 shows the absorption spectra of several useful semiconductors.

Application of an electric field F shifts the onset of absorption to energies lower than E_g (Franz–Keldysh effect). For direct allowed transitions and $h\nu < E_g$,

$$\alpha = \frac{A\hbar eF}{8(2m_r)^{1/2}(E_g - h\nu)} \exp\left[\frac{-4(2m_r)^{1/2}(E_g - h\nu)^{3/2}}{3\hbar eF}\right] \tag{2.178}$$

where m_r is the reduced mass given by $(1/m_r) = (1/m_e) + (1/m_h)$. An identical expression is also applicable for direct forbidden transitions. For $h\nu > E_g$ and for direct, allowed transitions

$$\alpha = A(h\nu - E_g)^{1/2}[1 + G(h\nu, F)] \tag{2.179}$$

where $G(h\nu, F)$ reflects the properties of the Airy integrals and is thus an oscillatory function of photon energy and electric field. For a one-phonon-assisted, indirect absorption process, the general behavior is similar to that for direct transitions.

In heavily doped semiconductors, the absorption edge lies at much shorter wavelengths as compared to the intrinsic case. This effect, known as the Moss–Burnstein shift, is depicted in Figure 2.8 for SnO_x : Sb films with different doping levels.[50] For direct transitions in an n-type semiconductor

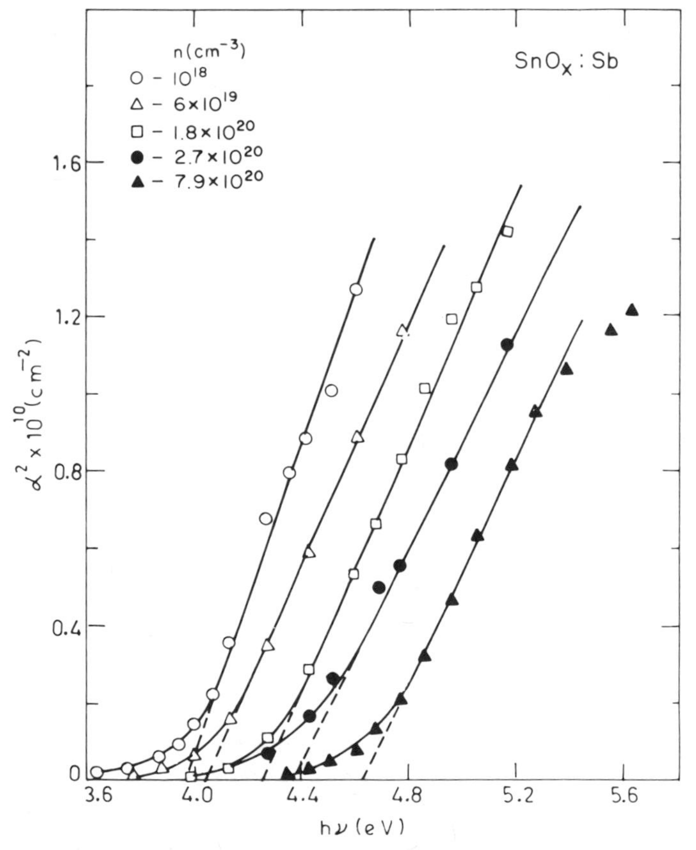

Figure 2.8. Plots of the square of the absorption coefficient vs. photon frequency for SnO_x : Sb films at different doping levels (after Shanthi et al.[50]).

with parabolic energy bands unaltered by the addition of donor impurities, the photon energy at which the absorption coefficient is reduced in the ratio $\alpha(h\nu)/\alpha_0(h\nu)$ is given by

$$h\nu = E_g + \left\{E_f - kT \ln\left[\frac{\alpha_0(h\nu)}{\alpha(h\nu)} - 1\right]\right\}(1 + m_e/m_h) \tag{2.180}$$

If the carrier concentration n is known then m_e can be calculated from the expression

$$n - 4\pi(2m_e kT/h^2)^{3/2} F_{1/2}(E_f/kT) \tag{2.181}$$

and thus m_h can be determined from equation (2.180).

For p-type semiconductors, the absorption thresholds occur in the opposite order. For indirect transitions, too, an increase in carrier concentration produces a shift of the absorption edge to higher energy but no sharp absorption thresholds are obtained. No analytical expressions are available for the shift in absorption edge in this case.

At low impurity concentrations, shallow impurities form discrete localized energy levels in the forbidden gap. With an increase in the impurity concentration an impurity band is formed in which electrons and holes are free to move. With a further increase in impurity concentration, the impurity band broadens, overlaps the main band edge, and forms a density-of-state tail extending into the forbidden gap. The concentration at which the overlap with the main band takes place is given by

$$N_{\min} \simeq 3 \times 10^{23}(m^*/m_0\varepsilon)^3 \tag{2.182}$$

where m^* is the effective mass for carriers in the main band and ε is the dielectric constant. Kane(51) formulated the impurity tail to be Gaussian whereas Halperin and Lax(52) derived the density of states to be exponential. Franz(53) calculated that for an exponential edge, defined by $\alpha = \alpha_0 \exp K(h\nu - E_0)$, a displacement occurs on the application of a field such that

$$\alpha = \alpha_0 \exp K(h\nu - E_0 + K^2 e^2 \hbar^2 F^2/12 m_r) \tag{2.183}$$

We should note that exponential absorption edges can also result from: (i) internal electric fields within the crystal, (ii) deformation of the lattice owing to strain caused by the presence of imperfections, and (iii) inelastic scattering of charge carriers by phonons. Owing to phonon scattering, several materials exhibit an absorption edge for which

$$d(\ln\alpha)/d(h\nu) = -1/kT \tag{2.184}$$

This is often referred to as Urbach's rule.(54)

The absorption coefficient for transitions between localized acceptors and the conduction band (for direct transitions) is given by

$$\alpha = \frac{N_I A_I (h\nu - E_g + E_A)^{1/2}}{[1 + 2m_e a^{*2}(h\nu - E_g + E_A)/\hbar^2]^4} \qquad (2.185)$$

where A_I is a constant, N_I is the concentration of ionized acceptors, a^* is a modified Bohr radius, and E_A is the ionization energy for the impurity ground state. Equation (2.185) takes into account the fourfold degeneracy of the valence band at $k = 0$. For transitions between a single valence band and a discrete donor level, equation (2.185) holds when E_A is replaced by the donor ionization energy and m_e is replaced by m_h appropriate to the valence band being considered.

2.4.3. *Carrier Generation and Recombination*

In an ideally pure semiconductor, interaction with a sufficiently energetic photon ($h\nu \geqslant E_g$) will produce a transition giving an electron in the conduction band where it is free to move. Simultaneously, a mobile hole will be produced in the valence band. Under some conditions, these two carriers will move in the same direction; under other conditions they will move in opposite directions. In general, either carrier may have the higher mobility. Moreover, in complex situations the numbers of each carrier generated and their lifetimes are not equal.

In most cases, photogeneration of carriers will mirror the absorption spectrum with three imporant exceptions: (i) photoeffects arise from active processes; inactive processes such as scattering and free-carrier absorption can be ignored; (ii) photoeffects depend on the distribution of the generated carriers; the generation function is exponential. In limiting cases, the distribution is essentially uniform at very low absorption levels or effectively a surface generation case at very high levels, with consequent carrier flow in the direction of the irradiated surface and/or surface recombination effects; and (iii) in general, the quantum efficiency, i.e., the number of electron–hole pairs created per absorbed photon, is unity or less than unity. However, for very shortwave radiation, multiple carrier generation becomes a possibility. Particular examples are Ge[61] and InSb.[62]

In impure or doped semiconductors, excitation of the appropriate impurity level produces an extrinsic photoresponse with only one type of carrier being generated. In n-type materials, donor levels may be excited to produce free electrons, while in p-type materials, free holes may be generated by ionizing acceptor levels. Amphoteric semiconductors with appropriate doping will show either effect. Higher-energy gap materials usually produce only one of the effects. Obviously, for extrinsic photoeffects to occur, the minimum photon energy required is less than the energy gap. The magnitude of the absorption is directly proportional to the doping concentration and is, in general, quite low. As a result, variations in the concentration of photocarriers are small and can be considered to be approximately uniform in samples of moderate thickness. These processes are useful in the photoconductivity mode and are not particularly appropriate, except indirectly, to photovoltaic operation.

Analogous to multiple carrier generation, a photocarrier, once generated, may attain sufficient energy through acceleration in a high electric field and excite other carriers by collision processes. The multiplication factor M increases with the field and at a critical breakdown field E_c, M becomes infinite:

$$M^{-1} = 1 - (E/E_c)^m \tag{2.186}$$

where m is usually about 2, but can range from 1.4 to 4.

2.4.3.1. *Transport of Photocarriers*

During their lifetime, the photocarriers will move under the influence of internal or external fields and concentration gradients. We have the following effects: (i) in a homogeneous material, for no applied field, photodiffusion in the direction of irradiation and photovoltage at point contacts can occur; (ii) photoconductivity will be observed in homogeneous materials under an applied electric field; (iii) photoelectromagnetic (PEM) effects arise under an applied magnetic field in homogeneous materials; and (iv) photovoltaic phenomena in internal barrier junctions in homogeneous as well as inhomogeneous/heterogeneous systems.

The internal photoeffects can be described by three types of equations described below.

2.4.3.1a. *Maxwell Equations*

$$\nabla \times \mathbf{E} = -(\partial \mathbf{B}/\partial t) \tag{2.187a}$$

$$\nabla \times \mathbf{E} = (\partial \mathbf{D}/\partial t) + \mathbf{J}_{\text{cond}} = \mathbf{J}_{\text{tot}} \tag{2.187b}$$

$$\nabla \cdot \mathbf{D} = \rho(x, y, z) \tag{2.187c}$$

$$\nabla \cdot \mathbf{B} = 0 \tag{2.187d}$$

$$\mathbf{B} = \mu_0 \mathbf{H} \tag{2.187e}$$

$$\mathbf{D} = \varepsilon_s \mathbf{E} \tag{2.187f}$$

where **E** and **D** are the electric field and displacement vector, respectively, **H** and **B** are the magnetic field and induction vector, respectively, ε_s and μ_0 are the permittivity and permeability, respectively, $\rho(x, y, z)$ is the total charge density, $\mathbf{J}_{\text{cond}}$ is the conduction current density and $\mathbf{J}_{\text{tot}}$ is the total current density.

2.4.3.1b. *Current Density Equations*

$$\mathbf{J}_n = e\mu_n n\mathbf{E} + eD_n \nabla n \tag{2.188a}$$

$$\mathbf{J}_p = e\mu_p p\mathbf{E} - eD_p \nabla p \tag{2.188b}$$

where $\mathbf{J}_n$ and $\mathbf{J}_p$ are the electron and hole current density, respectively, D_n and D_p are the carrier diffusion constants, μ_n and μ_p are the respective electron and hole mobilities, and n and p are the electron and hole densities. The first terms on the right-hand side give the drift component from the field and the second terms give the diffusion component from the carrier concentration gradient.

2.4.3.1c. *Continuity Equations*

$$\partial n/\partial t = G_n - U_n + (1/e)\nabla \cdot \mathbf{J}_n \tag{2.189a}$$

$$\partial p/\partial t = G_p - U_p - (1/e)\nabla \cdot \mathbf{J}_p \tag{2.189b}$$

where G_n and G_p, in $\text{cm}^{-3}\,\text{s}^{-1}$, are, respectively, the electron and hole generation rate owing to external influences such as optical excitation with photons or impact ionization under large electric fields, U_n is the electron recombination rate in a p-type semiconductor, and U_p is the hole recombination rate in an n-type semiconductor.

The generation rate G for optical excitation is given by

$$G(x) = \Phi_0 \alpha \exp(-\alpha x) \tag{2.190}$$

where Φ_0 is the number of photons ($\text{cm}^{-2}\,\text{s}^{-1}$) and α is the absorption coefficient, which is a function of wavelength.

The recombination rate is determined by several factors. Recombination processes are means by which a physical system is restored to thermal equilibrium ($pn = n_i^2$) whenever the equilibrium condition is disturbed, i.e., when $pn \neq n_i^2$. Recombination processes[64–66] in the bulk of the semiconductor can be of several types: (i) band-to-band recombination, (ii) single-level recombination, and (iii) multiple-level recombination. These three types of recombination processes are depicted in Figure 2.9. The band-to-band recombination, in which an electron–hole pair recom-

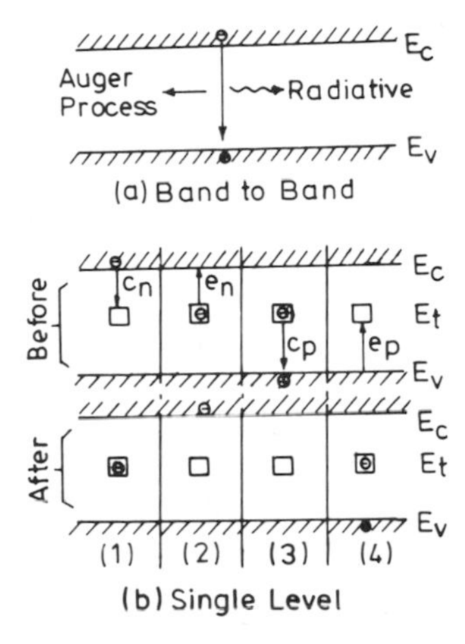

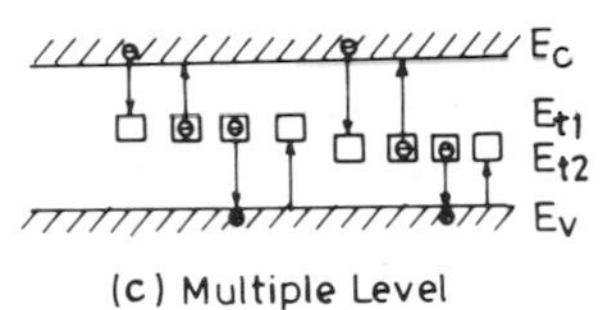

Figure 2.9. Schematic representation of different types of recombination processes: (a) band-to-band, (b) single-level, and (c) multiple-level. In (b), 1 refers to electron capture, 2 to electron emission, 3 to hole capture, and 4 to hole emission (after Sah et al.[64]).

bines, is accomplished either by the emission of a photon (radiative process) or by transfer of the energy to another free electron or hole (Auger or nonradiative process). The single-level recombination consists of (1) electron capture, (2) electron emission, (3) hole capture, and (4) hole emission. The recombination rate, $U(\mathrm{cm}^{-3}\,\mathrm{s}^{-1})$, for a single-level process is expressed by

$$U = \frac{\sigma_p \sigma_n v_{\text{th}}(pn - n_i^2)N_t}{\sigma_n\{n + n_i \exp[(E_t - E_i)/kT]\} + \sigma_p\{p + n_i \exp[(-E_t - E_i)/kT]\}} \tag{2.191}$$

where σ_p and σ_n are the hole and electron capture cross-sections, respectively, v_{th} is the carrier thermal velocity equal to $(3kT/m^*)^{1/2}$, N_t is the trap density and E_t the corresponding trap energy level, and E_i is the intrinsic Fermi level. It is obvious from equation (2.191) that U approaches a maximum at $E_t \approx E_i$.

Under low-injection conditions, i.e., $\Delta n = \Delta p < n$ or p the majority carriers,

$$U = (p_n - p_{n0})/\tau_p \tag{2.192}$$

where p is the equilibrium minority carrier concentration, $p_n = \Delta p + p_{n0}$, and τ_p is the minority carrier lifetime given by the expression

$$\tau_p = 1/\sigma_p v_{\text{th}} N_t \tag{2.193}$$

for an n-type semiconductor. For a p-type semiconductor,

$$\tau_n = 1/\sigma_n v_{\text{th}} N_t \tag{2.194}$$

Analogous to the bulk recombination, excess carriers created at the surface of a semiconductor, by optical generation or electrical injection, recombine at the surface to restore thermal equilibrium. Under the assumptions that single-charge centers with a single discrete energy level in the forbidden gap are responsible for the surface states, that the homogeneous semiconductor is under uniform and steady excitation, and that bulk trapping is negligible and the excess electron and hole densities in the bulk are equal, we have for small disturbances

$$s_{rp} = \sigma_p v_{\text{th}} N_{\text{st}} \tag{2.195}$$

where s_{rp} is the hole surface recombination velocity, σ_p is the hole capture cross-section, and N_{st} is the number of surface trapping centers per unit area. Recombination via multicharge centers and via single-charge centers with excited levels have been treated by several authors and a detailed discussion of these effects is given in Many et al.[22]

2.4.3.2. *Photoconductivity*

A change in conductivity, $\Delta\sigma$, on illumination, termed photoconductivity, results when absorption of light increases the values of the dark free-carrier densities n and p, and/or the dark mobilities μ_n and μ_p. Thus

$$\Delta\sigma = e(\Delta n \mu_n + \Delta p \mu_p) \tag{2.196}$$

$$\Delta\sigma = e(n \Delta\mu_{\text{b}n}^* + p \Delta\mu_{\text{b}p}^*) \tag{2.197}$$

The change in mobility on illumination can occur if the material is inhomogeneous in which case n and p are not uniform throughout, for example, a material with high-conductivity regions separated by narrow low-conductivity regions in the dark. The low-conductivity regions, which act as barriers to the flow of current between high-conductivity regions, limit the conductivity. Light absorbed in the low-conductivity regions reduces the resistance of these barriers and the flow of current through the material is much greater than in the dark. This barrier effect can be described in terms of an effective mobility μ_{b}^*, i.e., the mobility if the material is considered a homogeneous material with the free-carrier density equal to that found in the high-conductivity regions.

Two conditions are generally assumed while discussing photoconductivity: (i) only one of the carriers dominates the conductivity and the contribution of the other can be effectively neglected and (ii) during the photoconductivity process, the material stays neutral, i.e., $\Delta n = \Delta p$. Δn and Δp can be expressed in terms of the free lifetimes of the carriers τ_n and τ_p:

$$\Delta n = f\tau_n \quad \text{and} \quad \Delta p = f\tau_p \tag{2.198}$$

where f is the number of electron–hole pairs created per second per unit volume of the photoconductor. The photosensitivity of a photoconductor can be defined in terms of the number of charge carriers that pass between the electrodes per second for each photon absorbed per second:

$$\Delta I / e = GF \tag{2.199}$$

where ΔI is the photocurrent, F is the total number of electron–hole pairs created per second, and G is the photoconducting gain (ratio of light to dark resistance). The photosensitivity of a material, i.e., the change in conductivity from optical excitation divided by the excitation intensity, is often increased by incorporating centers which capture minority carriers rapidly, but have a much smaller probability of capturing majority carriers to bring about recombination. This is termed extrinsic photoconductivity.

At low excitations, $n \gg \Delta n$, the photocurrent varies linearly with excitation levels, $n \ll \Delta n$,

$$\Delta n = G_n / 2C_n n \quad \text{(monomolecular recombination)} \tag{2.200}$$

where G_n is the generation rate and C_n is the capture coefficient. At high excitation levels, $n \ll \Delta n$,

$$\Delta n = (G_n / C_n)^{1/2} \quad \text{(bimolecular recombination)} \tag{2.201}$$

and the photocurrent is proportional to the square root of the light intensity.

An excellent description of photoconductivity in solids is due to Bube[11] and the reader is referred thereto for further details.

2.5. *Metal Films*

In this section, our primary purpose is to familiarize the reader with the conduction processes and transport parameters of interest for solar cell applications of metal films. Electrical conduction in metal films depends on whether the film is granular or island-like, porous or network type, or physically continuous.

2.5.1. *Granular Films*

Electrical conductivity of a granular film is many orders of magnitude smaller than that of the corresponding bulk and is characterized by a negative temperature coefficient of resistivity (TCR). At large (> 100 Å) separations d between metal particles, conduction is dominated by thermionic emission and the conductivity is given by

$$\sigma = (AeT/k)\exp - [(\phi - Be^2/d)/kT] \tag{2.202}$$

where A is a constant characteristic of the film, ϕ is the work function of the metal particles, and the term Be^2/d (B is a constant) is the image force reduction of the work function. At high electric fields E applied between metal particles, the effective work function is further lowered to

$$\phi_{\text{eff}} = \phi - Be^2/d - e^{3/2}(E)^{1/2} \tag{2.203}$$

This, of course, yields the Schottky emission with the characteristic $\exp(-1/T)$ and $\exp(E^{1/2})$ dependence.

At small d values ($\leqslant 50$ Å) and low temperatures, thermally activated tunneling of electrons between the particles determines the conductivity, which is given by

$$\sigma \propto (d\phi^{1/2})\exp[(-e^2/\varepsilon rkT) - (4\pi d/h)(2m\phi)^{1/2}] \tag{2.204}$$

where r is the linear dimension of the particle, ε the dielectric constant, and m the effective mass. Note that because of the strong dependence on d, any changes therein from the straining of the film would yield a large change in the conductivity of these films.

Tunneling may also take place via substrate and associated traps. For a hopping process, the current I versus voltage V behavior is given by

$$I = AV + BV^n \tag{2.205}$$

where A and B are constants and the value of n depends on the energy distribution of traps.

2.5.2. *Network Films*

The conductivity of such films is controlled by the metal bridges and percolation across the voids. The temperature dependence of the resistivity of such films follows an empirical relation of the type

$$\rho = \rho_0[1 + \alpha(T - 273)] + C\exp(E_A/kT) \tag{2.206}$$

where $\alpha = \alpha_m - \alpha_s$ (α_m and α_s are the thermal expansion coefficients of the metal and substrate, respectively) and E_A is an activation energy term. C is a constant.

2.5.3. *Continuous Films*

The conductivity of physically continuous films is controlled by various scattering processes and the resistivities from the different processes are additive (Matthiessen's rule). Thus the resistivity of a film (ρ_F) is given by

$$\rho_F = \rho_B + \rho_s + \rho_{GB} + \rho_D + \rho_I \tag{2.207}$$

The subscripts refer to electron–phonon (bulk) scattering (B) and scattering by the free surfaces (s) of the film, grain boundaries (GB) or internal surfaces, structural defects (D), and impurities (I) if present.

In the case of free surface scattering, a simplified expression for the relative contribution is (see Chopra[10])

$$\rho_F/\rho_B = \sigma_B/\sigma_F = 1 + (3/8\gamma)(1-p) \qquad \text{for } \gamma \geqslant 1 \tag{2.208a}$$

$$\rho_F/\rho_B \sim \tfrac{4}{3}[1/\gamma(1+2p)][1/\ln(1/\gamma)] \qquad \text{for } \gamma \ll 1,\ p < 1 \tag{2.208b}$$

where $\gamma \equiv t/l$ is the film thickness normalized with respect to the mfp (l) and p is the fraction of electrons scattered specularly from the surface. Figure 2.10a shows the variation of ρ_F with γ for different values of p.

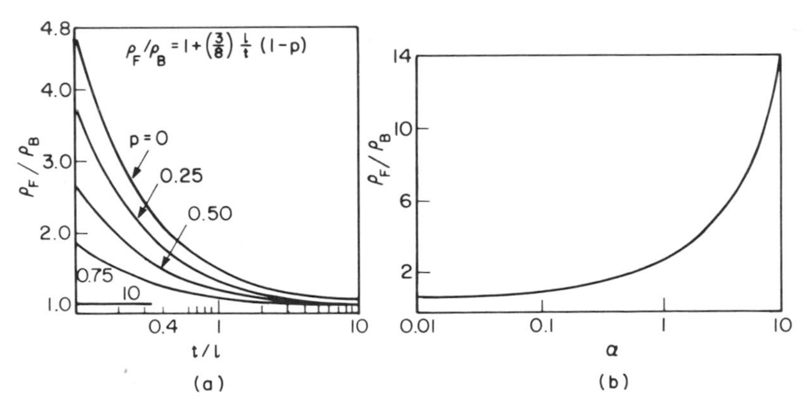

Figure 2.10. (a) Theoretical size dependence of thin-film conductivity. (b) The effect of grain boundary scattering on the conductivity (after Mayadas and Shatzkes[67]). See text for discussion.

The grain boundary contribution depends on the average grain diameter d as well as on the reflection coefficient R at the grain boundary.[67] The relative contribution ρ_F/ρ_B as a function of $\alpha = (1/d)[R/(1-R)]$ is shown in Figure 2.10b. Tellier and Tosser[68] have recently propounded a unified theory which takes into account both the surface and grain boundary scattering. However, this theory does not allow for scattering by defects and impurities at the grain boundaries, which processes are more significant than geometrical scattering.

The ρ_D term arises primarily from scattering by the vacancies and vacancy clusters. Note, however, that scattering from both point and line defects can, and indeed does, have a significant effect on other electron transport properties. This contribution is estimated variously as 1 to 2 $\mu\Omega$ cm per 1 at.% vacancy concentration. Thin films are notorious for a high concentration of frozen-in structural defects of a wide variety and thus ρ_D can be a very significant contribution.

The mean-free-path limited resistivity allows us to calculate l and p from the thickness dependence of ρ_F. Unfortunately, the microstructure of thin films is also a very sensitive function of film thickness and deposition conditions. Consequently, we observe a thickness dependence of ρ_F even for $t \gg l$, with dominant contributions from structural defects and active grain boundaries. Therefore, the use of the size effect theories for calculating ρ_F must be done with prudence.

Finally, the impurity contributions are linearly (with respect to composition) additive to the overall resistivity. If a disordered AB alloy is formed, the resistivity follows the Mott–Jones relationship

$$\rho = Kx(1-x) \tag{2.209}$$

where x is the at.% of A atoms and K is a constant for the system. When ordered, the resistivity should exhibit a sharp minimum.

The Hall coefficient of continuous metal films is nearly the same as that of the corresponding bulk. It is generally reduced as the film structure becomes micropolycrystalline or amorphous, tending to a free-electron gas value. These changes are due to changes in the Fermi surface. The density of conduction electrons, however, is expected to remain the same as in the bulk.

3

Photovoltaic Behavior of Junctions

3.1. Introduction

As noted in Chapter 2, a generalized photovoltaic device is composed of three functional elements, namely, an absorber, a junction region or converter, and a collector. A description of the basic physical processes that may occur in the absorber/generator was given in Chapter 2. In this chapter, we focus our attention on the converter/junction region. Our main interest is to gain a clear perception of the physics underlying solar cell operation. Throughout the text, therefore, the emphasis will be on a qualitative discussion of the physical effects, rather than on quantitative derivations. Accordingly, except in specific cases, we shall present only the final equations and refer the reader to the appropriate literature for the rigorous theory.

It is convenient to view the converter and collector in combination for the purpose of understanding their roles in solar cells. The primary role of the converter is to collect the minority carriers generated in the absorber and convert them into majority carriers at the maximum voltage limited by the excitation energy of the minority carriers. The collector distributes the majority carriers into the external load. The converter is usually a junction, caused by an inhomogeneity in the system. There are three basic types of junctions: (i) a p/n homojunction, (ii) a heterojunction, and (iii) a metal/semiconductor or conductor/insulator/semiconductor (MS or CIS) junction.

We shall first discuss the basic static and dynamic characteristics of these junctions in dark (Section 3.2). Next, in Section 3.3, we shall examine the effect of illumination on the behavior of the junctions and their operation as photovoltaic devices.

3.2. Junctions in Dark

We shall now formulate the energy band diagrams and the current transport equations describing the various junctions. A homojunction provides the most convenient starting point, not only because it is structurally and electronically the simplest junction and, hence, well understood and theoretically well developed,[1] but also because most of the concepts and results associated with a homojunction are equally applicable to heterojunctions and CIS devices.

3.2.1. Homojunctions

A p/n homojunction is essentially one semiconductor with two regions of different conductivity type, n- and p-type. The junction is formed at the region where the conductivity changes from one type to another. Depending on whether the impurity concentration in the semiconductor changes abruptly or gradually from acceptor impurities N_A to donor impurities N_D, we term the junction abrupt or graded, respectively.

3.2.1.1. Energy Band Diagram and Static Characteristics

In thermal equilibrium, the impurity distribution for an abrupt p/n homojunction is shown in Figure 3.1a. If $N_A \gg N_D$, a one-sided abrupt or p^+/n junction is formed. Conversely, when $N_D \gg N_A$, an n^+/p junction is formed.

From the current density equation (2.188), at thermal equilibrium, i.e., with no applied voltage and no current flow, we obtain

$$J_n = 0 = q\mu_n[n\mathscr{E} + (kT/q)(\partial n/\partial x)] = \mu_n n(\partial E_F/\partial x) \tag{3.1}$$

where $E_F - E_i = kT\ln(n_{n0}/n_i)$, as given in equation (2.27). Thus, we have from equation (3.1),

$$\partial E_F/\partial x = 0 \tag{3.1a}$$

Similarly, from equations (2.188) and (2.30),

$$J_p = 0 = q\mu_p[p\mathscr{E} - (kT/q)(\partial p/\partial x)] = \mu_p p(\partial E_F/\partial x) \tag{3.2}$$

It is obvious that the condition for zero net electron and hole currents imposes the requirement that the Fermi level must be the same throughout the sample.

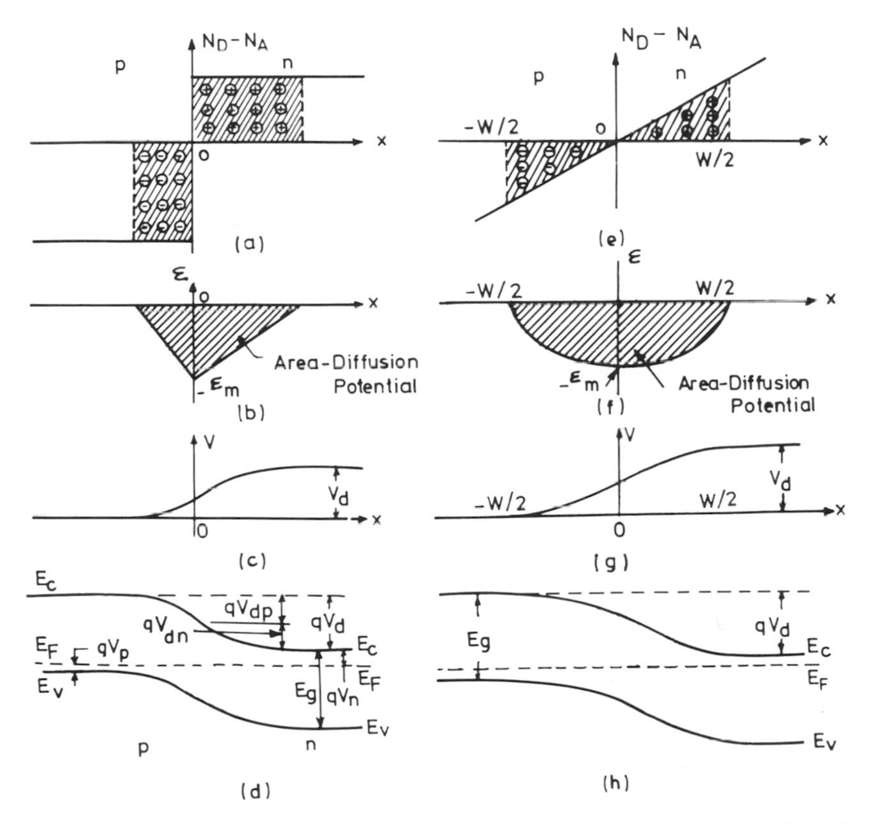

Figure 3.1. (a) Impurity distribution, (b) field distribution, (c) potential variation with distance, (d) energy band diagram for an abrupt p/n junction in thermal equilibrium; (e), (f), (g), and (h) are the corresponding plots for a linearly graded junction (after Shockley[4]).

The diffusion potential V_d, also termed the built-in potential, depicted in Figure 3.1c, can be expressed by the relation

$$\begin{aligned} qV_d &= E_g - (qV_n + qV_p) \\ &= kT \ln\left(\frac{N_c N_v}{n_i^2}\right)^{1/2} - \left[kT \ln\left(\frac{N_c}{n_{n0}}\right) + kT \ln\left(\frac{N_v}{p_{p0}}\right)\right] \\ &= kT \ln\left(\frac{n_{n0} p_{p0}}{n_i^2}\right) \simeq kT \ln\left(\frac{N_A N_D}{n_i^2}\right) \end{aligned} \tag{3.3}$$

where N_c and N_v are defined by equations (2.7) and (2.10). At equilibrium, $n_{n0} p_{n0} = n_{p0} p_{p0} = n_i^2$ (where the subscripts n and p refer to the type of semiconductor and 0 refers to the thermal equilibrium value). Therefore,

$$V_d = (kT/q)\ln(p_{p0}/p_{n0}) = (kT/q)\ln(n_{n0}/n_{p0}) \tag{3.4}$$

Thus, the hole and electron densities, on either side of the junction, are related by

$$p_{n0} = p_{p0} \exp[-q(V_d/kT)] \tag{3.5a}$$

and

$$n_{p0} = n_{n0} \exp[-q(V_d/kT)] \tag{3.5b}$$

From the condition for charge neutrality,

$$N_D x_n = N_A x_p \tag{3.6}$$

Solution of Poisson's equation yields,[1] for abrupt approximation, the following relations for the electric field:

$$\mathscr{E}(x) = -[qN_A(x + x_p)]/\varepsilon_s \qquad \text{for } -x_p \leq x < 0 \tag{3.7a}$$

and

$$\begin{aligned}\mathscr{E}(x) &= -\mathscr{E}_m + (qN_D x)/\varepsilon_s \\ &= q(N_D/\varepsilon_s)(x - x_n) \qquad \text{for } 0 < x \leq x_n\end{aligned} \tag{3.7b}$$

where ε_s is the permittivity of the semiconductor and $\mathscr{E}_m$ is the maximum field which exists at $x = 0$ and is given by

$$|\mathscr{E}_m| = (qN_D x_n)/\varepsilon_s = (qN_A x_p)/\varepsilon_s \tag{3.8}$$

The built-in potential V_d and the potential distribution $V(x)$ are expressed by

$$V(x) = \mathscr{E}_m(x - x^2/2W) \tag{3.9}$$

and

$$V_d = \tfrac{1}{2}\mathscr{E}_m W \equiv \tfrac{1}{2}\mathscr{E}_m(x_n + x_p) \tag{3.10}$$

where W is the total depletion width. From equations (3.8) and (3.10), we obtain, for a two-sided abrupt junction,

$$W = [V_d(2\varepsilon_s/q)(N_A + N_D)/(N_A N_D)]^{1/2} \tag{3.11a}$$

For a one-sided abrupt junction,

$$W = (2\varepsilon_s V_d / qN_B)^{1/2} \tag{3.11b}$$

where $N_B = N_D$ or N_A, depending on whether $N_A \gg N_D$ or $N_A \ll N_D$.

When the junction is reverse biased to a voltage V (positive voltage V on n-region with respect to the p-region), the total electrostatic potential variation across the junction is $V_d + V$. In the converse case of forward bias, $V_d - V$ is the potential variation across the junction. The depletion layer width as a function of the applied voltage V is obtained from

$$W = [(V_d \pm V)(2\varepsilon_s/q)(N_A + N_D)/N_A N_D]^{1/2} \tag{3.12a}$$

for a two-sided abrupt junction, and

$$W = [2\varepsilon_s (V_d \pm V)/qN_B]^{1/2} \tag{3.12b}$$

for a one-sided abrupt junction. The positive and negative signs are for the reverse and forward bias conditions, respectively. For a one-sided abrupt n^+/p Si junction, $V_d \sim 1.0$ V and $W \sim 0.1$ μm (at zero bias) for $N_B = 10^{17}$ cm^{-3}.

The depletion-layer capacitance per unit area, defined by $C \equiv (dQ_c/dV)$, where dQ_c is the incremental increase in charge per unit area for an incremental change of the applied voltage dV, is given by (for a one-sided abrupt junction)

$$C \equiv \frac{dQ_c}{dV} = \frac{d(qN_B W)}{d(qN_B W^2/2\varepsilon_s)} = \frac{\varepsilon_s}{W} = \left[\frac{q\varepsilon_s N_B}{2(V_d \pm V)}\right]^{1/2} \tag{3.13}$$

or

$$1/C^2 = 2(V_d \pm V)/q\varepsilon_s N_B \tag{3.13a}$$

or

$$d(1/C^2)/dV = 2/q\varepsilon_s N_B \tag{3.13b}$$

We should point out that more accurate considerations lead to the expression $V_d \pm V - (2kT/q)$ instead of $V_d \pm V$ in equation (3.13). We can replace the permittivity ε_s in the above equations by the term $\varepsilon\varepsilon_0$, where ε is the dielectric constant of the semiconductor and ε_0 is the permittivity of free space.

For a two-sided abrupt junction, the capacitance per unit area is expressed as[2]

$$C = \left[\frac{qN_D N_A \varepsilon_n \varepsilon_p \varepsilon_0}{2(\varepsilon_n N_D + \varepsilon_p N_A)(V_d \pm V)}\right]^{1/2} \tag{3.14a}$$

The relative voltage supported in each region is given by

$$(V_{dn} \pm V_n)/(V_{dp} \pm V_p) = N_A \varepsilon_p / N_D \varepsilon_n \tag{3.14b}$$

where V is the applied voltage and V_n and V_p are the voltage drop in the n- and p-region, respectively, with $V = V_n + V_p$. The width of each of the space-charge regions, as a function of the applied voltage V, can be expressed by

$$x_n = \left[\frac{2N_A \varepsilon_n \varepsilon_p \varepsilon_0 (V_d \pm V)}{qN_D(\varepsilon_n N_D + \varepsilon_p N_A)}\right]^{1/2} \tag{3.14c}$$

and

$$x_p = \left[\frac{2N_D \varepsilon_n \varepsilon_p \varepsilon_0 (V_d \pm V)}{qN_A(\varepsilon_n N_D + \varepsilon_p N_A)}\right]^{1/2} \tag{3.14d}$$

At forward bias, in addition to the depletion capacitance discussed above, there is a diffusion capacitance, which at low frequencies can be written as[1]

$$C_d = \frac{q}{kT}\left[\frac{qL_p p_{n0} + qL_n n_{p0}}{2}\right] \exp\left(\frac{qV}{kT}\right) \tag{3.15}$$

where V is the forward bias and $L_{n,p}$ is the minority carrier diffusion length.

The energy band diagram of an abrupt p/n homojunction at thermal equilibrium is shown in Figure 3.1d.

The impurity distribution for a linearly graded junction is depicted in Figure 3.1e. The field distribution, obtained from solution of Poisson's equation, is

$$\mathscr{E}_x = -(qa/2\varepsilon\varepsilon_0)[(W/2)^2 - x^2] \tag{3.16}$$

with the maximum field at $x = 0$ given by

$$|\mathscr{E}_m| = qaW^2/8\varepsilon\varepsilon_0 \tag{3.17}$$

where a is the impurity gradient in cm^{-4}.

The built-in potential V_d, shown in Figure 3.1g, can be expressed by

$$V_d = qaW^3/12\varepsilon\varepsilon_0 \tag{3.18}$$

or

$$W = (12\varepsilon\varepsilon_0 V_d/qa)^{1/3} \tag{3.19}$$

The depletion-layer capacitance relation in this case is

$$C \equiv dQ_c/dV = \frac{d(qaW^2/8)}{d(qaW^3/12\varepsilon\varepsilon_0)} = \varepsilon\varepsilon_0/W = \left[\frac{qa\varepsilon^2\varepsilon_0^2}{12(V_d \pm V)}\right]^{1/3} \tag{3.20}$$

The energy band diagram of a linearly graded p/n homojunction at thermal equilibrium is shown in Figure 3.1h.

3.2.1.2. *Current Transport*

3.2.1.2a. *Ideal Diode Law.* We shall now derive the current–voltage relationship for an ideal p/n junction. The following assumptions form the basis of the derivation: (i) abrupt depletion-layer approximation, (ii) Boltzmann relations are valid throughout the depletion layer, (iii) low-injection conditions, i.e., the injected minority carrier densities are small compared with the majority carrier densities, and (iv) no generation currents exist in the depletion layer and the electron and hole currents are constant through the depletion layer.

At thermal equilibrium, the Boltzmann relation is

$$n = n_i \exp\left(\frac{E_F - E_i}{kT}\right) \equiv n_i \exp\left[\frac{q(\psi - \phi)}{kT}\right] \tag{3.21a}$$

$$p = n_i \exp\left(\frac{E_i - E_F}{kT}\right) \equiv n_i \exp\left[\frac{q(\phi - \psi)}{kT}\right] \tag{3.21b}$$

where ψ and ϕ are the potentials corresponding to the intrinsic level and the Fermi level, respectively. In other words, $\psi \equiv -E_i/q$ and $\phi \equiv -E_F/q$. When a voltage is applied, thermal equilibrium is disturbed and minority carrier densities on both sides of the junction are changed. Thus $pn \neq n_i^2$. Denoting the quasi-Fermi levels for electrons and holes as ϕ_n and ϕ_p, respectively, we define them by

$$n \equiv n_i \exp[q(\psi - \phi_n)/kT] \tag{3.22a}$$

and

$$p \equiv n_i \exp[q(\phi_p - \psi)/kT] \tag{3.22b}$$

Alternative expressions for ϕ_n and ϕ_p are

$$\phi_n \equiv \psi - (kT/q)\ln(n/n_i) \tag{3.23a}$$

and

$$\phi_p \equiv \psi + (kT/q)\ln(p/n_i) \tag{3.23b}$$

Thus,

$$pn = n_i^2 \exp[q(\phi_p - \phi_n)/kT] \tag{3.24}$$

For forward bias, $(\phi_p - \phi_n) > 0$ and $pn > n_i^2$. At reverse bias, $(\phi_p - \phi_n) < 0$ and $pn < n_i^2$.

From equation (3.22a) and the fact that $\mathscr{E} \equiv -\nabla\psi$, we obtain

$$\begin{aligned} J_n &= q\mu_n[n\mathscr{E} + (kT/q)\nabla n] \\ &= q\mu_n n(-\nabla\psi) + q\mu_n(kT/q)[(qn/kT)(\nabla\psi - \nabla\phi_n)] \\ &= -q\mu_n n\nabla\phi_n \end{aligned} \tag{3.25a}$$

Similarly,

$$J_p = -q\mu_p p\nabla\phi_p \tag{3.25b}$$

In Figure 3.2 we have shown the idealized energy band diagram and the potential distributions in a p/n junction, under forward and reverse biased conditions.

The electron density n varies by several orders of magnitude from the n side to the p side, whereas the electron current J_n is almost constant. Therefore, ϕ_n must also be constant over the depletion layer. The electrostatic potential difference across the junction is written as

$$V = \phi_p - \phi_n \tag{3.26}$$

Combining equations (3.24) and (3.26), we obtain an expression for the electron density at the boundary of the depletion-layer region on the p-side ($x = -x_p$) as

$$n_p = (n_i^2/p_p)\exp(qV/kT) = n_{p0}\exp(qV/kT) \tag{3.27a}$$

Similarly, for the hole density at $x = x_n$ on the n-side, we can write

$$p_n = p_{n0}\exp(qV/kT) \tag{3.27b}$$

Equations (3.27a) and (3.27b) define the boundary conditions for the ideal current–voltage equation. Utilizing the continuity equations (2.189), we obtain for the steady state:

$$-\frac{n_n - n_{n0}}{\tau_n} + \mu_n\mathscr{E}\frac{\partial n_n}{\partial x} + \mu_n n_n\frac{\partial\mathscr{E}}{\partial x} + D_n\frac{\partial^2 n_n}{\partial x^2} = 0 \tag{3.28}$$

$$-\frac{p_n - p_{n0}}{\tau_p} - \mu_p \mathscr{E} \frac{\partial p_n}{\partial x} - \mu_p p_n \frac{\partial \mathscr{E}}{\partial x} + D_p \frac{\partial^2 p_n}{\partial x^2} = 0 \tag{3.29}$$

Imposing the condition that $(p_n - p_{n0})/\tau_p = (n_n - n_{n0})/\tau_n$, the above equations yield the following solution at $x = x_n$:

$$J_p = -qD_p \frac{\partial p_n}{\partial x}\bigg|_{x_n} = \frac{qD_p p_{n0}}{L_p}[\exp(qV/kT) - 1] \tag{3.30a}$$

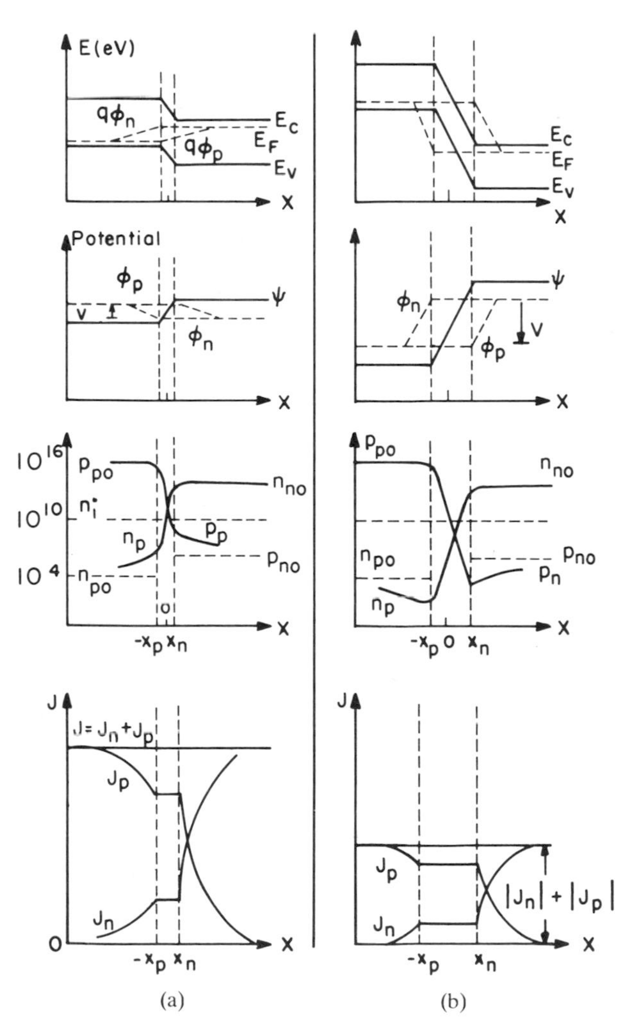

Figure 3.2. Energy band diagram, intrinsic Fermi level Ψ, quasi-Fermi level (ϕ_n for electrons and ϕ_p for holes), carrier distributions, and current densities for (a) forward- and (b) reverse-bias condition (after Shockley[4]).

where D_p is the diffusion coefficient for holes and is given by $L_p \equiv (D_p \tau_p)^{1/2}$. One can obtain a similar relation for J_n for the p-side ($x = -x_p$),

$$J_n = qD_n \frac{\partial n_p}{\partial x}\bigg|_{-x_p} = \frac{qD_n n_{p0}}{L_n}[\exp(qV/kT) - 1] \tag{3.30b}$$

The carrier densities and the current densities for the forward and reverse biased p/n junction are shown in Figure 3.2.

The total current is expressed by the celebrated Shockley equation[3,4]

$$J = J_p + J_n = J_s[\exp(qV/kT) - 1] \tag{3.31a}$$

where

$$J_s \equiv (qD_p p_{n0}/L_p) + (qD_n n_{p0}/L_n) \tag{3.31b}$$

The ideal current–voltage (J–V) relation is shown graphically in Figures 3.3a and b in linear and semilog plots, respectively. For a forward bias greater than $3kT/q$, the rate of increase in the current is constant; at 300 K, there is a 59.5 mV ($= 2.3kT/q$) change in voltage for every decade change of current. In the reverse-biased situation, the current density saturates at $-J_s$.

For a one-sided p^+/n abrupt junction (donor concentration N_D), $p_{n0} \gg n_{p0}$ and the second term in equation (3.31b) can be neglected. Assuming $D_p/\tau_p \propto T^\gamma$, where γ is a constant, we obtain the temperature dependence of J_s as[1]

$$J_s \sim [T^3 \exp(-E_g/kT)]T^{\gamma/2} = T^{(3+\gamma/2)}\exp(-E_g/kT) \tag{3.32}$$

3.2.1.2b. *Generation–Recombination.* We will now examine some departures from the ideal diode law. Most practical diodes have other current transport mechanisms operating in certain voltage regions in addition to the ideal current. These nonideal currents can arise from: (i) surface effects owing to ionic charges on or outside the semiconductor surface causing surface depletion layers, (ii) generation and recombination of carriers in the depletion region, (iii) tunneling of carriers between states in the bandgap, and (iv) high-injection conditions which may occur even at relatively small forward bias. Surface effects are discussed separately in Section 3.3.4. We deal first with the generation–recombination currents.

Under the conditions $p < n_i$ and $n < n_i$, the generation rate of electron–hole pairs under a reverse-biased condition is given by equation (2.192) and can be written as

$$U \equiv -n_i/\tau_e \tag{3.33}$$

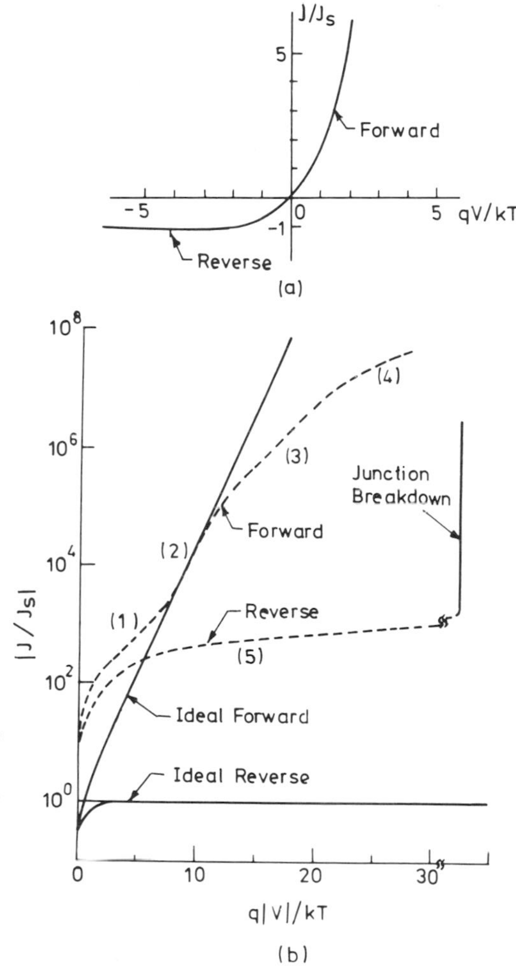

Figure 3.3. (a) Ideal current–voltage characteristics and (b) semilog plot of the current–voltage characteristics of a real diode exhibiting: (1) generation–recombination region, (2) diffusion current region, (3) high-injection region, (4) series resistance effect, and (5) reverse leakage current owing to generation–recombination and surface effect (after Sze[1]).

where τ_e is the effective lifetime. The generation current in the depletion region is thus given by

$$J_{\text{gen}} = \int_0^W q\,|U|\,dx \approx q\,|U|\,W = \frac{qn_iW}{\tau_e} \tag{3.34}$$

where W is the depletion width. At a given temperature, W is dependent on the applied reverse bias. Thus, for abrupt junctions,

$$J_{\text{gen}} \sim (V_d + V)^{1/2} \tag{3.35a}$$

and for linearly graded junctions,

$$J_{\text{gen}} \sim (V_d + V)^{1/3} \tag{3.35b}$$

The total reverse current, for $p_{n0} \gg n_{p0}$ and $|V| > 3kT/q$, is then the sum of the diffusion current in the neutral region and the generation current in the depletion region:

$$J_R = q(D_p/\tau_p)^{1/2}(n_i^2/N_D) + (qn_iW/\tau_e) \tag{3.36}$$

At forward bias, capture processes in the depletion region are dominant and the recombination current has the form[1]

$$J_{\text{rec}} = \int_0^W qU dx \approx \tfrac{1}{2} qW\sigma v_{\text{th}} N_t n_i \exp\left(\frac{qV}{2kT}\right) \sim n_i N_t \tag{3.37}$$

where $\sigma = \sigma_n = \sigma_p$ and v_{th} and N_t have the same meaning as in equation (2.191). In the above derivation of J_{rec}, it is assumed that the energy level of the capture center E_t lies at E_i. The total forward current is given by the sum of equations (3.31a) and (3.37), for $p_{n0} \gg n_{p0}$ and $V > kT/q$, as

$$J_F = \left(\frac{qD_p}{\tau_p}\right)^{1/2} \frac{n_i^2}{N_D} \exp\left(\frac{qV}{kT}\right) + \frac{q}{2} W\sigma v_{\text{th}} N_t n_i \exp\left(\frac{qV}{2kT}\right) \tag{3.38}$$

For practical diodes, the experimental results can be well represented by the empirical relation

$$J_F \sim \exp(qV/nkT) \tag{3.39}$$

where the factor n, termed the diode factor, is equal to 1 when the diffusion current dominates and equal to 2 when the recombination current dominates. When both the mechanisms are comparable in magnitude, n lies between 1 and 2.

3.2.1.2c. *Tunneling.* If energy states are introduced within the bandgap, then tunneling currents could dominate the transport processes. The tunneling currents have the form[5]

$$J_{\text{tun}} = K_1 N_t \exp(BV) \tag{3.40}$$

where K_1 is a constant containing the effective mass, the built-in potential, the doping level, the dielectric constant, and Planck's constant. N_t is the density of states available for an electron or hole to tunnel into and B is another constant given by

$$B = \frac{4}{3\hbar}(m^* \varepsilon / N_{D,A})^{1/2} \tag{3.41}$$

Since the tunneling current also varies exponentially with voltage, the only method for differentiating tunneling currents from thermal currents such as injection or generation–recombination currents is from the temperature dependence. Tunneling currents are very insensitive to temperature.

3.2.1.2d. *Nonideal Diodes.* When more than onc dark current component is present, the $I-V$ characteristics are given by

$$I(V) = \sum_i I_{s_i}[\exp(A_i V) - 1] \tag{3.42}$$

where the sum over i indicates the various possible contributions to the diode current and $A_i = q/n_i kT$. I_s is the reverse saturation current.

Yet another nonideal behavior that must be taken into consideration while dealing with practical diodes is the effect of series and shunt resistances. The series resistance R_s and the shunt resistance R_{sh} modify equation (3.42) to give the most generalized $I-V$ characteristics as

$$I - (V - IR_s)/R_{sh} = \sum_i I_{s_i}\{\exp[A_i(V - IR_s)] - 1\} \tag{3.43}$$

In Figure 3.3b, we have also shown the $I-V$ characteristics of a practical diode exhibiting different transport mechanisms in different voltage regions.

Before concluding this section, we should add a word on the high-injection condition. At high current densities under forward bias, when the minority carrier density is comparable with the majority concentration, both drift and diffusion currents must be considered. J_p and J_n are then given by equations (3.25a) and (3.25b).

3.2.2. *Heterojunctions*

A junction formed between two semiconductors having different energy bandgaps is termed a heterojunction. If the conductivity type is the same in the two semiconductors, the heterojunction is called an isotype heterojunction. An anisotype heterojunction is one in which the conductivity type is different in the two semiconductors. Some of the requirements for forming a good quality heterojunction are: (i) the lattice constant of the two materials should be nearly equal, (ii) the electron affinities should be compatible, and (iii) the thermal expansion coefficients should be close. Mismatch of lattice constants and thermal expansion coefficients leads to

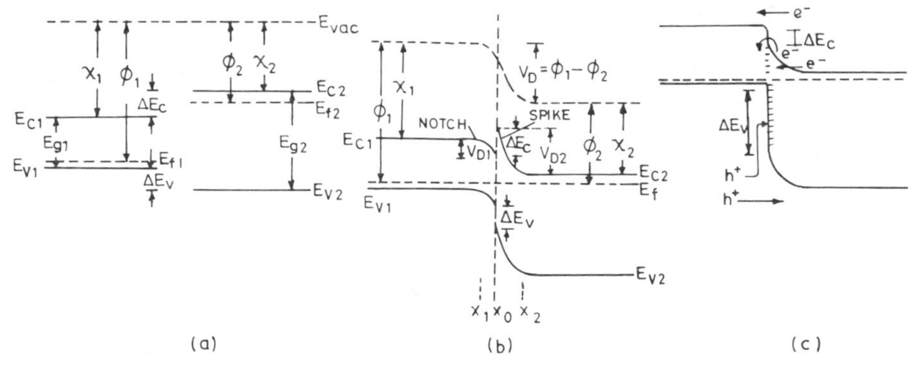

Figure 3.4. Equilibrium energy band diagram (a) before and (b) after the formation of an abrupt heterojunction. (c) Various current transport mechanisms in a heterojunction.

interfacial dislocations at the heterojunction interface, giving rise to interface states which act as trapping centers. Differences in electron affinity between the two materials can result in energy discontinuities in the form of a notch or a spike, in one or both of the energy bands in an abrupt heterojunction.

3.2.2.1. *Energy Band Diagram and Static Characteristics*

Consider two isolated semiconductors 1 and 2, as shown in Figure 3.4a, with bandgaps E_{g1} and E_{g2}, permittivities ε_{s1} and ε_{s2}, work functions ϕ_1 and ϕ_2, and electron affinities χ_1 and χ_2. The work function ϕ and the electron affinity χ are defined, respectively, as the energy required to remove an electron from the Fermi level E_f and from the conduction band edge E_c to the vacuum level. The difference in the energy of the conduction band edges in the two semiconductors, or the electron affinity difference, is denoted by ΔE_c and the hole affinity difference by ΔE_v.

The energy band diagram of a heterojunction formed between these semiconductors is shown in Figure 3.4b. At thermal equilibrium, the Fermi level coincides on both sides of the junction and the vacuum level is everywhere parallel to the band edges and is continuous. If E_g and χ are independent of doping, i.e., for nondegenerate semiconductors, ΔE_c and ΔE_v are invariant with doping. The total built-in potential V_d is given by

$$\begin{aligned} V_d &= \phi_1 - \phi_2 = V_{d1} + V_{d2} \\ &= E_{g2} - (E_f - E_{v2}) + \chi_2 - \chi_1 - (E_{c1} - E_f) \end{aligned} \tag{3.44a}$$

or

$$\begin{aligned} V_d &= E_{g2} - (E_f - E_{v2}) - (E_{c1} - E_f) + \Delta E_c \\ &= V_{d2} + \Delta E_c \end{aligned} \tag{3.44b}$$

The depletion widths and capacitance can be obtained by solving Poisson's equation, as in the case of the homojunction, and results are similar to equations (3.14c), (3.14d), and (3.14a):

$$C = \left[\frac{qN_{D1}N_{A2}\varepsilon_1\varepsilon_2\varepsilon_0}{2(\varepsilon_1 N_{D1} + \varepsilon_2 N_{A2})(V_D \pm V)}\right]^{1/2} \tag{3.45}$$

$$x_1 = \left[\frac{2N_{A2}\varepsilon_1\varepsilon_2\varepsilon_0(V_d - V)}{qN_{D1}(\varepsilon_1 N_{D1} + \varepsilon_2 N_{A2})}\right]^{1/2} \tag{3.46}$$

and

$$x_2 = \left[\frac{2N_{D1}\varepsilon_1\varepsilon_2\varepsilon_0(V_d - V)}{qN_{A2}(\varepsilon_1 N_{D1} + \varepsilon_2 N_{A2})}\right]^{1/2} \tag{3.47}$$

The relative voltage supported in each of the semiconductors is

$$\frac{V_{d1} \pm V_1}{V_{d2} \pm V_2} = \frac{N_{A2}\varepsilon_2}{N_{D1}\varepsilon_1} \tag{3.48}$$

where $V = V_1 + V_2$.

We have treated above only one specific example of a heterojunction. Depending on the relative values of E_g, ϕ, and χ in the two semiconductors, a large variety of heterojunction energy band profiles are possible. These have been described by Sharma and Purohit[6] and Milnes and Feucht.[7]

3.2.2.2. *Current Transport*

In the preceding discussion, we assumed that no interface states exist. However, in real devices, there may be a large density of interface states if there is considerable lattice mismatch. The interface states density varies[8] from on the order of 10^{10} cm^{-2} for an almost perfect lattice match to on the order of 10^{14} cm^{-2} for several percent mismatch.

Several transport mechanisms may be operative at the interface. These are shown schematically in Figure 3.4c and include (i) the ideal diffusion or emission currents for electrons, (ii) recombination–generation currents, (iii) recombination through interface states at the junction, (iv) tunneling from band states to localized defect states in the gap, across the interface, and (v) band-to-band tunneling.

Severl authors[9–29] have proposed different models based on experimental data as well as theoretical considerations. We shall discuss three main models for abrupt anisotype heterojunctions, namely, (i) Anderson's model, (ii) the tunneling model, and (iii) the interface recombination model.

3.2.2.2a. *Anderson's Model.* Anderson[14,15] in his model neglected the effects of dipoles and interface states. Further, he assumed that owing to discontinuities in the band edges at the interface, the diffusion current consists only of electrons or holes. In the p/n heterojunction shown in Figure 3.4b, the predominant current carriers are electrons because of the lower barrier for electrons as compared to that for holes. The predicted current–voltage relation, in the absence of generation–recombination currents, is given by

$$J = A \exp(-qV_{d2}/kT)[\exp(qV_2/kT) - \exp(-qV_1/kT)] \tag{3.49}$$

where $A = aqXN_{a2}(D_{n1}/\tau_{n1})^{1/2}$, with X being the transmission coefficient for electrons across the interface, a the junction area, and D_{n1} and τ_{n1} the diffusion coefficient and lifetime of minority carriers (electrons), respectively, in the p-type material. The treatment for the n/p heterojunction[15] is completely analogous. An important feature of equation (3.49) is that the first term in the square bracket is important for forward bias, whereas the second term is important for reverse bias.

Perlman and Feucht[17] have included the effects owing to a spike in the conduction band edges in the emission model. For the case of an abrupt p/n heterojunction, where the charge transport is mainly due to electrons, the current–voltage relation, neglecting generation–recombination within the space-charge region, is expressed as

$$I = \frac{I_s[\exp(qV/kT) - 1]}{(I + I_s/I_d)} \tag{3.50}$$

where $I_s = aqN_{D1}(D_{n1}/\tau_{n1})^{1/2}$, identical to the case for a homojunction (equation 3.31b), and a is the junction area.

$$I_d = \tfrac{1}{2}aqX_m N_{D2}\bar{v}_{x_{e2}} \exp[-(q/kT)(V_F + V)] \tag{3.51}$$

is the emission limited current. $\bar{v}_{x_{e2}} = (2kT/\pi m_n^*)^{1/2}$, where m_n^* is the effective mass of electrons in the n-type material, $\bar{v}_{x_{e2}}$ is the x component of the average speed of electrons in the n-type material, V_F is the forward barrier, and X_m is the transmission coefficient. The ratio I_s/I_d determines the mode of operation. If $I_s \ll I_d$, the total current is as predicted by the homojunction model, whereas for $I_s \gg I_d$, the current is analogous to the Schottky diode current (see Section 3.2.3.2).

For an abrupt isotype heterojunction, Anderson's emission model,[15] neglecting interface states, gives

$$I = B \exp(-qV_{d2}/kT)[\exp(qV_2/kT) - 1] \tag{3.52}$$

where $B = aqXN_{D2}(kT/2\pi m_n^*)^{1/2}$, with X as the transmission coefficient for electrons across the interface, a as the junction area, and m_n^* as the effective mass of electrons in the wide bandgap semiconductor.

Kumar[25] has analyzed the isotype heterojunction on the basis of diffusion theory and obtained an analytical expression for the I–V relationship of the form

$$I = A\left\{\frac{N_{D1}}{N_{D2}}\exp\left(\frac{qV_{d1}}{kT}\right)\left[1-\exp\left(\frac{-qV_1}{kT}\right)\right] + \exp\left(\frac{-qV_{d2}}{kT}\right)\left[\exp\left(\frac{qV_2}{kT}\right)-1\right]\right\} \tag{3.53}$$

where A is given by $(\mu kTN_{D2}/L_{nn})$, with μ the electron mobility in semiconductor 2 and L_{nn} a parameter weakly dependent on V and T.

3.2.2.2b. *Tunneling Model.* Rediker et al.[20] have proposed a model for an abrupt anisotype heterojunction where electrons have to surmount or tunnel through the potential barrier in the n-type wide bandgap material in order to flow from n-type to p-type or vice versa. The properties of the n-type semiconductor largely determine whether thermal emission over the barrier or tunneling through the barrier dominates electron flow. If tunneling through the barrier greatly exceeds thermal emission over the barrier, then

$$I = I_s(T)\exp(V/V_0) \tag{3.54}$$

where V_0 is a constant and $I_s(T)$ is a weakly increasing function of temperature. Newman[21] observed that $I_s(T)$ is empirically proportional to $\exp(T/T_0)$. Thus the expression for I can be rewritten as

$$I = I_{s0}\exp(T/T_0)\exp(V/V_0) \tag{3.55}$$

where I_{s0}, T_0, and V_0 are constants.

3.2.2.2c. *Interface Recombination.* Rothwarf[30] has developed an interface recombination model, particularly for the Cu_2S/CdS heterojunction. The current density for the interface recombination case, when the smaller bandgap p-type material (Cu_2S) is degenerate, has the form

$$J = qN_{c2}S_I\exp\left[\frac{-(qV_{d2}+\delta_2)}{kT}\right]\left\{\exp\left[\frac{q(V-JR_sA_\perp)}{kT}\right]-1\right\} \tag{3.56}$$

where N_{c2} is the effective density of states in the n-type wide bandgap semiconductor (CdS), S_I is the effective interface recombination velocity, V_{d2} is the diffusion voltage in CdS, δ_2 is equal to $E_{c2} - E_f$, R_s is the series resistance, and $A_\perp$ is the normal area of the diode. S_I is related to the density and energy distribution of interface states, which, in turn, depend upon the lattice mismatch between the p-type and n-type semiconductors. The explicit form of S_I is given approximately by the relation

$$S_I = v_{\text{th}} \sigma N_I^* \tag{3.57}$$

where v_{th} is the thermal velocity of electrons, σ is the capture cross-section of interface states, and N_I^* is the density of empty interface states. S_I can also be written as[8]

$$S_I = v_{\text{th}} \sigma \Delta a / a^3 \tag{3.58}$$

where Δa is the difference in lattice constant between the two materials and a is the average lattice constant in the plane of the junction.

3.2.3. *MS and CIS Junctions*

The metal/semiconductor (MS) or Schottky barrier is conceptually simpler than either the homojunction or the heterojunction. Unfortunately, in practice, the mechanisms governing the behavior of this device are not as well understood as those of the homojunction. The primary consideration is the condition of the surface of the semiconductor, in particular the presence or absence of an oxide or other insulating layer between the metal and the semiconductor and defects. The metal can be replaced by transparent, wide bandgap, highly conducting (degenerate), oxide semiconductors as in ITO/Si devices. The properties of the device are largely controlled by the properties of this layer.

3.2.3.1. *Energy Band Diagram and Static Characteristics*

The detailed energy band diagram[1] of an MIS junction, with an interfacial layer of the order of atomic distance, is shown in Figure 3.5a. The various terms are: ϕ_M is the work function of the metal, ϕ_{BN} is the metal/semiconductor barrier height, ϕ_{BO} is the asymptotic value of ϕ_{BN} at zero electric field, ϕ_0 is the energy level at the semiconductor surface, $\Delta\phi$ is the image-force barrier lowering, Δ is the potential across the interfacial layer, χ is the electron affinity of the semiconductor, V_d is the built-in potential, ε_s is the permittivity of the semiconductor, ε_i is the permittivity of the interfacial layer, δ is the thickness of the interfacial layer, Q_{sc} is the space-charge density in the semiconductor, Q_{ss} is the surface-state density on the semiconductor, and Q_{M} is the surface-charge density on the metal.

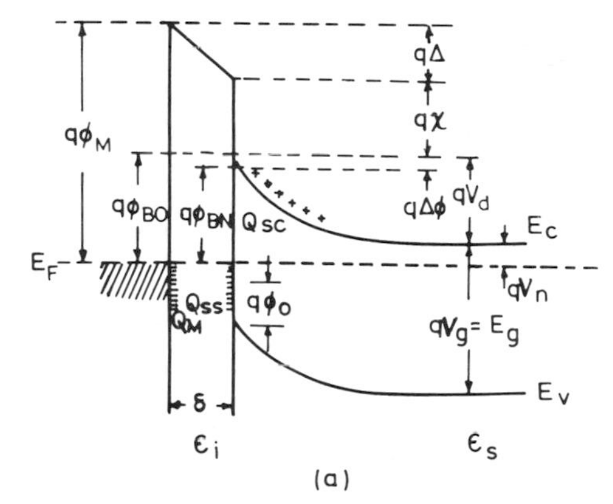

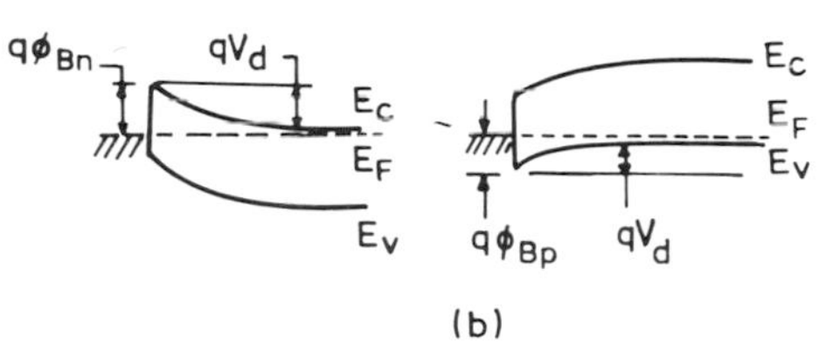

Figure 3.5. (a) Energy band diagram of an MIS junction. (b) Energy band diagrams of metal-semiconductor (MS) junctions for n-type and p-type semiconductors.

In the limit $\delta = 0$, we have the Schottky barrier energy band profile, as shown in Figure 3.5b. For an abrupt junction, i.e., $\rho \simeq qN_D$ for $x < W$ and $\rho \approx 0$, $dV/dx \approx 0$ for $x > W$, where W is the depletion width, we have the following relationships[1]:

$$W = \left[\frac{2\varepsilon_s}{qN_D}\left(V_d - V - \frac{kT}{q}\right)\right]^{1/2} \tag{3.59}$$

$$|\mathscr{E}(x)| = \frac{qN_D}{s}(W - x) = \mathscr{E}_m - \frac{qN_D}{\varepsilon_s}x \tag{3.60}$$

and

$$V(x) = \frac{qN_D}{\varepsilon_s}\left(Wx - \tfrac{1}{2}x^2\right) - \phi_{\mathrm{BN}} \tag{3.61}$$

$\mathscr{E}_m$ is given by

$$\mathscr{E}_m = \mathscr{E}(x = 0) = \frac{2(V_d - V - kT/q)}{W} \tag{3.62}$$

The depletion-layer capacitance can be expressed by

$$C \equiv \left[\frac{q\varepsilon_s N_D}{2(V_d - V - kT/q)}\right]^{1/2} = \frac{\varepsilon_s}{W} \tag{3.63}$$

3.2.3.2. *Current Transport*

The current transport in Schottky barriers is mainly due to majority carriers. Bethe[31] proposed a simple isothermal thermionic emission model, whereas Schottky[32] propounded the isothermal diffusion theory. Crowell and Sze[33] synthesized the two models into a single more generalized theory and derived the following current–voltage relationship:

$$J = \frac{qN_c v_R}{1 + v_R / v_D} \exp\left[-\frac{q\phi_{BN}}{kT}\right]\left[\exp\left(\frac{qV}{kT}\right) - 1\right] \tag{3.64}$$

where v_D is an effective diffusion velocity associated with the transport of electrons from the edge of the depletion layer at W to the potential energy maximum, N_c is the effective density of states in the conduction band, and v_R is the thermionic recombination velocity near the metal/semiconductor interface. For a Maxwellian electron distribution for $x \gg x_m$ and if no electrons return from the metal, other than those associated with the current density qn_0v_R (n_0 is the quasi-equilibrium electron density at x_m),

$$v_R = A^*T^2/qN_c \tag{3.65}$$

where A^* is the effective Richardson constant given by

$$A^* = 4\pi qm^*k^2/h^3 \tag{3.66}$$

If $v_D \gg v_R$, then from equation (3.64), v_R dominates the pre-exponential term and the emission process governs current transport. Thus

$$J = A^*T^2 \exp(-q\phi_{BN}/kT)[\exp(qV/kT) - 1] \tag{3.67}$$

For $v_D \ll v_R$, the diffusion process applies. If image-force effects are neglected and electron mobility is assumed to be independent of the electric field, $v_D = \mu\mathscr{E}$, where $\mathscr{E}$ is the electric field in the semiconductor near the boundary. We thus are led to the Schottky diffusion result:

$$J \simeq qN_c\mu\mathscr{E} \exp\left(-\frac{q\phi_{BN}}{kT}\right)\left[\exp\left(\frac{qV}{kT}\right) - 1\right] \tag{3.68}$$

For MIS or CIS devices, at least two terms are needed for the current–voltage relationship. These are a combination of terms, one with diode factor $n = 1$ and the other with diode factor $n > 1$. The reverse saturation current density for the $n = 1$ term is

$$J_s(n = 1) = C_1 q \frac{D_p p_{n0}}{L_p} \tag{3.69}$$

where $1 \geqslant C_1 \geqslant 0$. For the $n > 1$ term,

$$J_s(n > 1) = C_2 A T^2 \exp\left(-\frac{\phi_{BN}}{kT}\right) \tag{3.70}$$

with $1 \geqslant C_2 \geqslant 0$.

We should point out that the above equations are empirical. Clearly understood equations for CIS devices have not yet been fully developed.

3.3. *Effect of Illumination on Junction Behavior*

We now consider the effect of illumination on the behavior of junctions described in the preceding sections and the operation of these devices as photovoltaic devices and solar cells. The most direct effect of illumination is observed in the current–voltage ($I-V$) characteristics of the junction. When a p/n junction (homo- or hetero-), CIS, or MS junction is exposed to light, a current can flow at zero external applied voltage. The equivalent circuit of the solar cell is shown in Figure 3.6a, where the light-generated current I_L is represented by a constant current source. The $I-V$ relationship of the illuminated junction is simply equation (3.43) shifted by I_L. This is called the superposition principle. Thus,

$$I - \frac{V - IR_s}{R_{sh}} = \sum_i I_{s_i}\{\exp[A_i(V - IR_s)] - 1\} - I_L \tag{3.71}$$

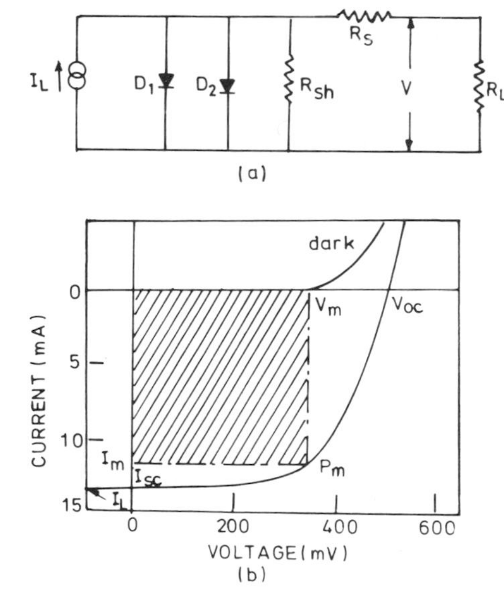

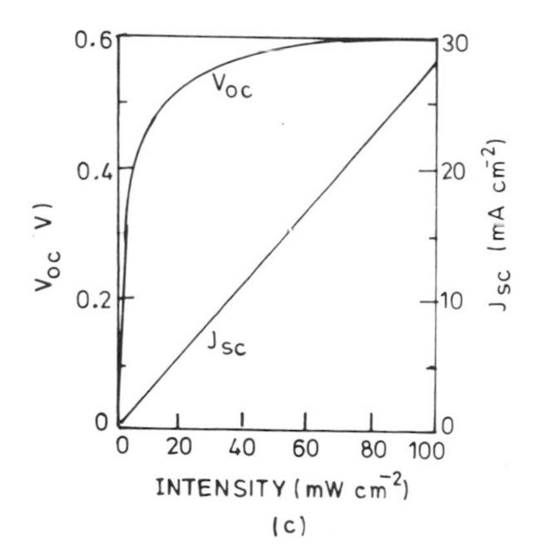

Figure 3.6. (a) Equivalent circuit of a solar cell including series and shunt resistances. (b) Typical light and dark $I-V$ characteristics of a solar cell. (c) V_{oc} and J_{sc} as functions of light intensity.

In Figure 3.6b, a typical $I-V$ curve in both light and dark is shown. In the above equation, I_{s_i}, A_i, R_s, and R_{sh} usually are dependent on light intensity and wavelength.[34–38]

3.3.1. *Photovoltaic Parameters*

In order to use the $I-V$ characteristics to calculate the cell power output and conversion efficiency, we define some quantities called the photovoltaic parameters.

3.3.1.1. *Short-Circuit Current*

The short-circuit current I_{sc} is the current that flows through the junction under illumination at zero applied bias and in the ideal case (when R_s and R_{sh} resistance effects are not present) is equal to the light-generated current I_L and proportional to the incident number of photons, i.e., the illumination intensity.

3.3.1.2. *Open-Circuit Voltage*

The open-circuit voltage V_{oc} defined at zero current through the device is given by, from equation (3.71), assuming only one current mechanism,

$$V_{oc} = (1/A)\ln[(I_{sc}/I_s) + 1] \tag{3.72}$$

where $A = q/nkT$. The variation of I_{sc} and V_{oc} with illumination intensity is shown in Figure 3.6c.

It might be assumed from equation (3.72) that high values of n would be desirable in obtaining high open-circuit voltages, but this is not actually the case, since high n is usually associated with high values of I_s. V_{oc} for p/n junctions is always higher for low values of n (close to 1).

3.3.1.3. *Power Output*

The output power is given by

$$P = IV = I_s V[\exp(AV) - 1] - I_{sc}V \tag{3.73}$$

The condition for maximum power can be obtained by setting $\partial P/\partial V = 0$; we then obtain

$$I_m = (I_{sc} + I_s)\frac{qV_m/nkT}{1 + qV_m/nkT} \tag{3.74}$$

as the current output at maximum power and

$$\exp\left(\frac{qV_m}{nkT}\right)\left[1+\left(\frac{qV_m}{nkT}\right)\right]=\left(\frac{I_{sc}}{I_s}\right)+1=\exp\left(\frac{qV_{oc}}{nkT}\right) \tag{3.75}$$

allows the voltage at maximum power point, V_m, to be calculated. The maximum output power is given by $P_m = V_m I_m$.

3.3.1.4. *Fill Factor*

The fill factor FF, which is defined as $V_m I_m / V_{oc} I_{sc}$, measures the squareness of the I–V curve, and is expressed by the relation

$$\mathrm{FF} = V_m \left\{\frac{1-(I_s/I_{sc})[\exp(qV_m/nkT)-1]}{(nkT/q)\ln[(I_{sc}/I_s)+1]}\right\} \tag{3.76}$$

$$= \frac{V_m}{V_{oc}}\left\{1-\frac{\exp(qV_m/nkT)-1}{\exp(qV_{oc}/nkT)-1}\right\} \tag{3.77}$$

The relationships expressed by equations (3.72)–(3.77) apply only when there are no series or shunt resistance effects and where the current can be expressed by a single exponential. The two ratios V_m/V_{oc} and I_m/I_{sc} and the FF all improve with increasing values of V_{oc} and with decreasing values of n and T. Higher bandgap materials yield higher ratios and fill factors because of their higher open-circuit voltages. The variation of V_m/V_{oc}, I_m/I_{sc}, and FF as a function of the normalized V_{oc} are shown in Figure 3.7.

When series and shunt resistance effects cannot be neglected, the two ratios V_m/V_{oc} and I_m/I_{sc} and the FF are all reduced below the values shown in Figure 3.7. The lowering in fill factor from resistive effects is discussed in greater detail in Section 3.3.4.3f.

3.3.1.5. *Efficiency*

The efficiency η of a solar cell in converting light into useful power is given by

$$\eta = V_m I_m / P_{in} \tag{3.78}$$

The input power is

$$P_{in} = A_t \int_0^\infty F(\lambda)(hc/\lambda)\,d\lambda \tag{3.79}$$

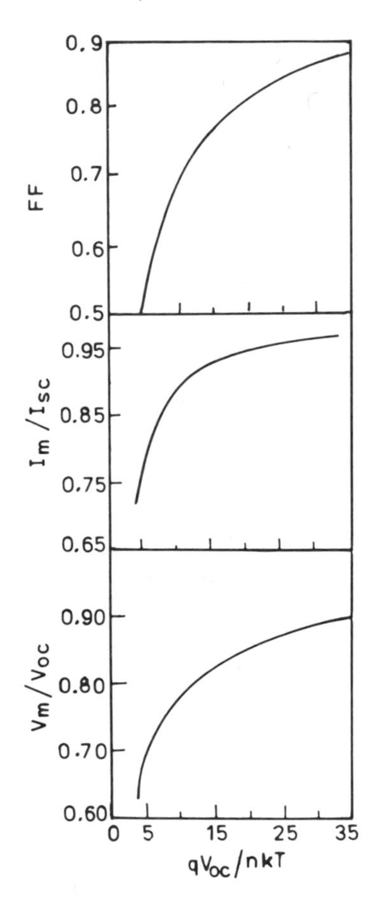

Figure 3.7. Variation of V_m/V_{oc}, I_m/I_{sc}, and FF as functions of V_{oc}/nkT (after Hovel[5]).

where A_t is the total device area, $F(\lambda)$ is the number of photons per square centimeter per second per unit bandwidth incident on the device at wavelength λ, and hc/λ is the energy associated with each photon. The various air mass solar spectra are shown in Figure 1.3 and their spectral distribution is given in Appendix A. The maximum output power P_m is given by

$$P_m = V_m I_m = V_{oc} I_{sc} \mathrm{FF} \tag{3.80}$$

Therefore, the conversion efficiency of a solar cell is expressed by

$$\eta = (V_{oc} I_{sc} \mathrm{FF}/P_{in}) \times 100\% \tag{3.81}$$

Ignoring series and shunt resistance losses and assuming a single exponential dark forward $I-V$ characteristic, we can write an analytical expression for the efficiency as

$$\eta = \frac{\mathrm{FF}(nkT/q)\ln[(I_{sc}/I_s)+1]q\int_0^\infty F(\lambda)\mathrm{SR}(\lambda)_{ext}d\lambda}{A_t\int_0^\infty F(\lambda)(hc/\lambda)d\lambda} \tag{3.82}$$

where $\mathrm{SR}(\lambda)$ is the spectral response defined in the next section. An alternative expression for η is

$$\eta = Q(I_m/I_{sc})[V_m q n_{ph}(E_g)/N_{ph}E_{av}](1-R) \tag{3.83}$$

where N_{ph} is the total number of photons per square centimeter per second in the source spectrum, E_{av} is their average energy, $n_{ph}(E_g)$ is the number of photons per centimeter square per second with energy greater than the bandgap, R is the average reflectivity, and Q is the average collection efficiency, i.e., the ratio of the number of carriers actually collected to $n_{ph}(E_g)$, the number capable of being collected. The spectral response and collection efficiency are related by

$$qQn_{ph}(E_g)(1-R) = q\int_0^\infty F(\lambda)\mathrm{SR}(\lambda)_{ext}d\lambda = J_{sc} \tag{3.84}$$

where J_{sc} is the short-circuit current density. If monochromatic light is incident, SR and Q are identical. In the above discussion, one assumption is implicit: the quantum yield, defined as the number of electron–hole pairs created per photon absorbed, is unity.

3.3.1.6. *Spectral Response*

The photocurrent collected at each wavelength relative to the number of photons incident on the surface at that wavelength determines the spectral response of the device. The internal spectral response $\mathrm{SR}(\lambda)$ is the number of electron–hole pairs collected under short-circuit conditions relative to the number of photons entering the material and is given by

$$\mathrm{SR}(\lambda) = \frac{J_p(\lambda)}{qF(\lambda)[1-R(\lambda)]} + \frac{J_n(\lambda)}{qF(\lambda)[1-R(\lambda)]} + \frac{J_{dr}(\lambda)}{qF(\lambda)[1-R(\lambda)]} \tag{3.85}$$

while the external response $\mathrm{SR}(\lambda)_{ext}$ is the internal response modified by reflection losses from the surface of the device

$$\mathrm{SR}(\lambda)_{ext} = \mathrm{SR}(\lambda)[1-R(\lambda)] \tag{3.86}$$

$J_p(\lambda)$, $J_n(\lambda)$, and $J_{dr}(\lambda)$ are the hole diffusion, electron diffusion, and drift contributions, respectively, to the total photocurrent density J_L. Quantitative expressions for these terms are presented later in Section 3.3.4.

3.3.2. *Superposition Principle*

The foregoing discussion on the light $I-V$ characteristics is based on the validity of the superposition principle, which states that the current flowing in an illuminated device subject to a bias V is given by the superposition of the short-circuit photocurrent and the current that would flow at bias V in the dark. However, this principle is not valid in all cases. Typical examples where the superposition principle fails are Cu_2S/CdS heterojunction solar cells and amorphous Si solar cells. Several authors[39–41] have discussed and analyzed the application of the superposition principle to solar cells. Lindholm et al.[39] have opined that if considerable photocurrent and considerable dark thermal recombination current both occur within the junction space-charge region, then the $I-V$ characteristic shifting approximation is invalid. The shifting approximation is found to be invalid also if low-injection concentrations of holes and electrons are not maintained throughout the quasi-neutral regions and if sizable series resistance is present. However, Tarr and Pulfrey[40] have concluded that the superposition can apply at one sun illumination even for devices in which the dark current is dominated by recombination in the depletion region and for which substantial photogeneration also occurs in that region, as in GaAs solar cells.

For a-Si cells, the Cu_2S/CdS heterojunction cell, and several other heterojunction and Schottky cells, the failure of the superposition is manifested through the photocurrent I_L being voltage-dependent. The detailed mechanisms vary from device to device. In some cases, the dark current is modified by the presence of light. Rothwarf[41] has proposed a generalized form of the superposition principle which adequately describes most solar cells

$$I(V) = I_D'(V) - I_L(V) \tag{3.87}$$

where $I_D'(V)$ is the current for the diode that would exist in the dark if the same conditions present in the light could be maintained and $I_L(V)$ is the voltage-dependent light-generated current. Explicit expressions for $I_L(V)$ are presented in Section 3.3.4.2.

3.3.3. *Idealized Model for Efficiency Limit Predictions*

Early workers in the field carried out efficiency limit calculations for homojunctions of different materials, under the idealized conditions of unity collection efficiency, no reflection loss, and absence of series and shunt resistance effects. Although these conditions are never fulfilled in real devices, nonetheless, the calculations are instructive in that they relate the photovoltaic parameters to the intrinsic material properties and thus predict which material systems are potentially suitable for achieving high conversion efficiencies. The works of Prince,[42] Loferski,[43] Wysocki and Rappaport,[44] Wolf,[45] and Shockley and Queisser[46] are classics in this field. More recently, deVos[47] and deVos and Pauwels[48] developed a similar treatment for single heterojunction solar cells. This generalized model treats the homojunction as a special case of the heterojunction. We shall now present deVos's theory in detail to gain some insight into the materials parameters that affect conversion efficiency in a solar cell.

3.3.3.1. *Assumptions*

The maximum attainable efficiency η is calculated as a function of the two semiconductor bandgaps E_{g1} and E_{g2} (Figure 3.8a). The efficiency is split into three factors: (a) the spectrum factor η_1, (b) the voltage factor η_2, and (c) the curve factor η_3. The calculations have been carried out under the following assumptions:

1. The temperature T is equal to 293 K.
2. The energy spectrum of the incident illumination is the curve proposed by Moon,[49] which is approximated by the simple analytic equation

$$P(\lambda) = 0 \qquad \text{for } \lambda \ll \lambda_0 \tag{3.88a}$$

$$P(\lambda) = P_0\left[\exp\left(-\frac{\lambda}{\lambda_1}\right) - Q_0 \exp\left(-\frac{\lambda}{\lambda_2}\right)\right] \qquad \text{for } \lambda \gg \lambda_0 \tag{3.88b}$$

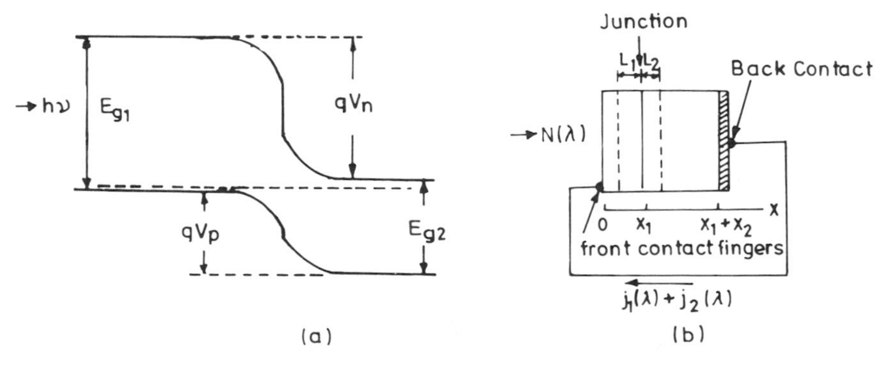

Figure 3.8. (a) Energy band model of a p/n heterojunction and (b) solar cell geometry.

with

$$\lambda_0 = \frac{\lambda_1 \lambda_2}{\lambda_1 - \lambda_2} \ln Q_0 \tag{3.88c}$$

The real curve can be matched by the analytic curve for the following values of the parameters:

$$P_0 = 3.82 \times 10^9 \text{ W/m}^3$$

$$Q_0 = 120$$

$$\lambda_1 = 5.90 \times 10^{-7} \text{ m}$$

$$\lambda_2 = 6.30 \times 10^{-4} \text{ m}$$

This gives an incident power of 1136 W/m^2.

3. The illumination is incident perpendicularly on semiconductor 1, with semiconductor 2 at the rear.

4. Each semiconductor has an absorption coefficient α_i $(i = 1, 2)$ obeying the relations

$$\alpha_i(\lambda) = 0 \qquad \text{for } \lambda > hc/E_{gi} \tag{3.89}$$

$$\alpha_i(\lambda) = \alpha_{0i} \qquad \text{for } \lambda < hc/E_{gi} \tag{3.90}$$

where E_{gi} is the bandgap of the semiconductor. If the length of the light path through semiconductor i is l_i, the fraction of energy absorbed $[1 - \exp(-\alpha_i l_i)]$ is thus zero for $\lambda > hc/E_{gi}$ and a constant k_i for $\lambda < hc/E_{gi}$. To obtain the maximum attainable efficiency, the optimum values of k_i will be used.

5. The dark current–voltage characteristics follow the ideal Shockley diode equation [equation (3.31a)]. The saturation current density is given by

$$J_s = q \sum_{i=1}^{2} \left(\frac{D_i}{\tau_i}\right)^{1/2} \frac{N_{ci} N_{vi}}{N_i} \exp\left(\frac{-E_{gi}}{kT}\right) \tag{3.91}$$

which is the same as equation (3.31b).

6. The saturation current density J_s can be rewritten as

$$J_s = \frac{q}{hc} \sum_{i=1}^{2} a_i \exp\left(\frac{-E_{gi}}{kT}\right) \tag{3.92}$$

where a_i denotes

$$a_i = \frac{h_c}{N_i} \left(\frac{D_i}{\tau_i}\right)^{1/2} N_{ci} N_{vi} = hc \left(\frac{kT}{q}\right)^{1/2} \left(\frac{\mu_i}{\tau_i}\right)^{1/2} \frac{N_{ci} N_{vi}}{N_i} \tag{3.93}$$

For different semiconductor materials,

$$10^{-2} \leqslant \mu_1 \leqslant 10 \text{ m}^2\text{ V}^{-1}\text{ s}^{-1}$$

$$10^{-10} \leqslant \tau_i \leqslant 10^{-5} \text{ s}$$

$$10^{47} \leqslant N_{ci}N_{vi} \leqslant 10^{51} \text{ m}^{-6}$$

$$10^{21} \leqslant N_i \leqslant 10^{25} \text{ m}^{-3}$$

$$10^{-2} \leqslant a_i \leqslant 10^{10} \text{ W/m}$$

7. The Fermi level lies at the band edges in the respective semiconductors such that the barrier heights V_p and V_n of the junction (Figure 3.8a) are equal to the respective energy gaps of the semiconductors.

8. Reflection losses and series and shunt resistance losses are neglected. There is assumed to be no recombination, so collection efficiency is unity.

3.3.3.2. *Spectrum Factor* η_1

Let j_{L1} and J_{L2} denote the number of light-generated pairs per unit area and unit time in semiconductor 1 and semiconductor 2, respectively. Then

$$\eta_1 = \frac{J_{L1}E_{g1} + j_{L2}E_{g2}}{\int_0^\infty P(\lambda)d\lambda} \tag{3.94}$$

(a) For the case $E_{g1} \gg E_{g2}$, we have

$$j_{L1} = k_1 \int_0^{hc/E_{g1}} P(\lambda)\frac{\lambda}{hc}\, d\lambda \tag{3.95}$$

and

$$j_{L2} = k_2(1-k_1)\int_0^{hc/E_{g1}} P(\lambda)\frac{\lambda}{hc}\, d\lambda + k_2\int_{hc/E_{g1}}^{hc/E_{g2}} P(\lambda)\frac{\lambda}{hc}\, d\lambda \tag{3.96}$$

η_1 will be maximal in the case $k_1 = k_2 = 1$ and will be given by

$$\eta_1 = \frac{E_{g1}\int_0^{hc/E_{g1}} P(\lambda)\lambda d\lambda + E_{g2}\int_{hc/E_{g1}}^{hc/E_{g2}} P(\lambda)\lambda d\lambda}{hc\int_0^\infty P(\lambda)d\lambda} \tag{3.97}$$

It is thus clear from equation (3.97) that the spectrum factor η_1 is optimal if all photons with an energy E larger than E_{g1} are absorbed by semiconductor 1 and all photons with energy between E_{g2} and E_{g1} are absorbed by semiconductor 2.

(b) For $E_{g1} = E_{g2} = E_g$, we obtain the spectrum factor η_1 for a homojunction as

$$\eta_1(E_g) = \frac{(E_g/hc)\int_0^{hc/E_g} P(\lambda)\lambda d\lambda}{\int_0^{\infty} P(\lambda)d\lambda} \tag{3.98}$$

(c) When $E_{g1} \leqslant E_{g2}$, we have

$$j_{L1} = k_1 \int_0^{hc/E_{g1}} P(\lambda)\frac{\lambda}{hc}\, d\lambda \tag{3.99}$$

$$j_{L2} = k_2(1-k_1)\int_0^{hc/E_{g2}} P(\lambda)\frac{\lambda}{hc}\, d\lambda \tag{3.100}$$

(i) If

$$E_{g1}\int_0^{hc/E_{g1}} P(\lambda)\lambda d\lambda - E_{g2}\int_0^{hc/E_{g2}} P(\lambda)\lambda d\lambda < 0$$

then maximum energy is used for $k_1 = 0$ and $k_2 = 1$, and the spectrum factor can be written as

$$\eta_1 = \frac{E_{g2}}{hc}\frac{\int_0^{hc/E_{g2}} P(\lambda)\lambda d\lambda}{\int_0^{\infty} P(\lambda)d\lambda} \tag{3.101}$$

(ii) For

$$E_{g1}\int_0^{hc/E_{g1}} P(\lambda)\lambda d\lambda - E_{g2}\int_0^{hc/E_{g2}} P(\lambda)\lambda d\lambda > 0$$

maximum energy is used for $k_1 = 1$ and k_2 arbitrary. η_1 is written as

$$\eta_1 = \frac{E_{g1}}{hc}\frac{\int_0^{hc/E_{g1}} P(\lambda)\lambda d\lambda}{\int_0^{\infty} P(\lambda)d\lambda} \tag{3.102}$$

Thus the spectrum factor is optimal if all photons are absorbed in the most appropriate semiconductor.

In Figure 3.9a, the (E_{g1}, E_{g2}) plane is divided into three regions, each with an optimal assumption for k_1 and k_2. Figure 3.9b shows $\eta_1\,(E_{g1}, E_{g2})$ of a heterojunction solar cell. Lines of constant spectrum factor η_1 are marked by the value of η_1 in percent. Figure 3.9c, the spectrum factor of a homojunction, shows that the maximum attainable η_1 of a single homojunction solar cell is 44.4% for a bandgap $E_g \sim 0.92$ eV. Wolf[45] calculated a maximum of 46% at $E_g = 0.9$ eV. From Figure 3.9b, we see that $\eta_1 > 44.4\%$ cannot be attained for $E_{g1} \leqslant E_{g2}$. For $E_{g1} > E_{g2}$, higher spectrum factors η_1 are possible. The highest η_1 is 62.2% for a combination of $E_{g1} - 1.48$ eV and $E_{g2} = 0.65$ eV. Figures 3.9d and e show the parts of the incident solar energy utilizable in a heterojunction solar cell and a homojunction solar cell, respectively. Obviously, the two-bandgap system is better matched to the solar spectrum than the one-bandgap system.

3.3.3.3. *Voltage Factor* η_2

The light-generated current density J_L is given by

$$J_L = qj_{L1} + qj_{L2} \tag{3.103}$$

The two current densities need energies E_{g1} and E_{g2}, respectively, to be generated. The voltage factor η_2 is thus defined as

$$\eta_2 = \frac{(j_{L1} + j_{L2})qV_{oc}}{j_{L1}E_{g1} + j_{L2}E_{g2}} \tag{3.104}$$

where V_{oc} is the open-circuit voltage described by equation (3.72)

(a) For $E_{g1} \geqslant E_{g2}$, the current densities are given by equations (3.95) and (3.96), so we get

$$\eta_2 = kT \frac{[k_1 + (1-k_1)k_2] \int_0^{hc/E_{g1}} P(\lambda)\lambda d\lambda + k_2 \int_{hc/E_{g1}}^{hc/E_{g2}} P(\lambda)\lambda d\lambda}{[E_{g1}k_1 + E_{g2}(1-k_1)k_2] \int_0^{hc/E_{g1}} P(\lambda)\lambda d\lambda + \left[E_{g2}k_2 \int_{hc/E_{g1}}^{hc/E_{g2}} P(\lambda)\lambda d\lambda\right]}$$
$$\times \ln\left[1 + \frac{[k_1 + (1-k_1)k_2] \int_0^{hc/E_{g1}} P(\lambda)\lambda d\lambda + k_2 \int_{hc/E_{g1}}^{hc/E_{g2}} P(\lambda)\lambda d\lambda}{\sum_1^2 a_i \exp(-E_{gi}/kT)}\right] \tag{3.105}$$

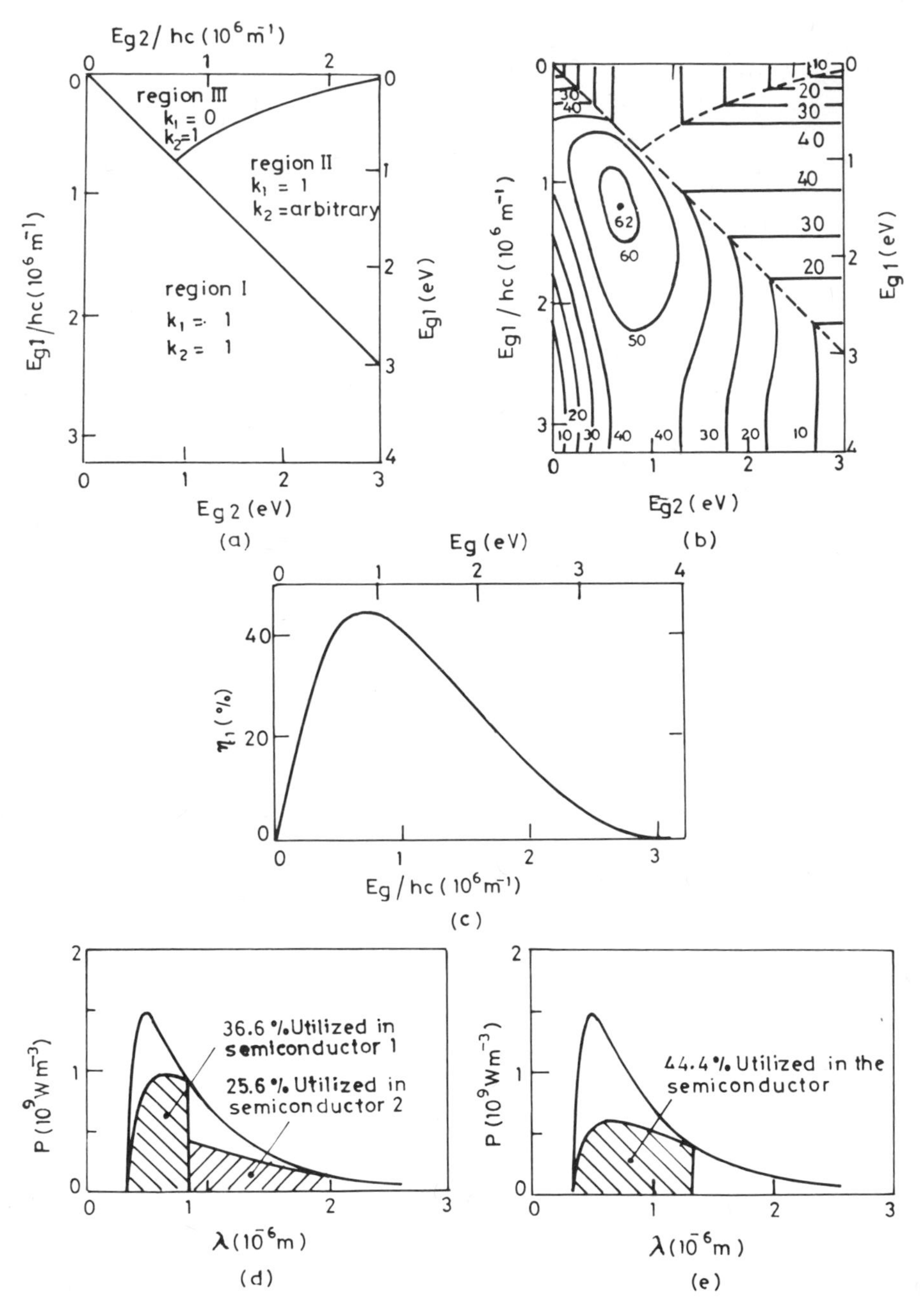

Figure 3.9. (a) Choice of k_1 and k_2 needed for optimal spectrum factor η; (b) optimal $\eta_1(E_{g1}, E_{g2})$ of a heterojunction solar cell; (c) $\eta_1(E_g)$ of a homojunction solar cell. Fraction of the incident solar energy utilized in (d) a heterojunction solar cell and (e) a homojunction solar cell (after deVos[47]).

The first factor of this expression has a maximum value of kT/E_{g2} for $k_1 = 0$ and k_2 arbitrary; the logarithmic factor is a maximum for $k_2 = 1$ and k_1 arbitrary; therefore η_2 is maximal for $k_1 = 0$ and $k_2 = 1$ and can be written as

$$\eta_2 = \frac{kT}{E_{g2}} \ln \left[1 + \frac{\int_0^{hc/E_{g2}} P(\lambda)\lambda d\lambda}{\sum_1^2 a_i \exp(-E_{gi}/kT)} \right] \tag{3.106}$$

For $k_2 = 1$, V_{oc} is independent of k_1. Therefore, η_2 is optimal for $k_1 = 0$, because V_{oc} is then governed by the smaller of E_{g1}/q and E_{g2}/q. The voltage factor in the conditions for optimal spectrum factor ($k_1 = k_2 = 1$) is given by

$$\eta_2 = \frac{kT \int_0^{hc/E_{g2}} P(\lambda)\lambda d\lambda}{E_{g1} \int_0^{hc/E_{g1}} P(\lambda)\lambda d\lambda + E_{g2} \int_0^{hc/E_{g2}} P(\lambda)\lambda d\lambda} \times \ln \left[1 + \frac{\int_0^{hc/E_{g2}} P(\lambda)\lambda d\lambda}{\sum_1^2 a_i \exp(-E_{gi}/kT)} \right] \tag{3.107}$$

(b) For the case $E_{g1} \leq E_{g2}$, the light-generated current densities are given by equations (3.99) and (3.100). Thus,

$$\eta_2 = kT \frac{[k_1 + (1-k_1)k_2] \int_0^{hc/E_{g2}} P(\lambda)\lambda d\lambda + k_1 \int_{hc/E_{g2}}^{hc/E_{g1}} P(\lambda)\lambda d\lambda}{[E_{g1}k_1 + E_{g2}(1-k_1)k_2] \int_0^{hc/E_{g2}} P(\lambda)\lambda d\lambda + E_{g1}k_1 \int_{hc/E_{g2}}^{hc/E_{g1}} P(\lambda)\lambda d\lambda} \times \ln \left[1 + \frac{[k_1 + (1-k_1)k_2] \int_0^{hc/E_{g2}} P(\lambda)\lambda d\lambda + k_1 \int_{hc/E_{g2}}^{hc/E_{g1}} P(\lambda)\lambda d\lambda}{\sum_1^2 a_i \exp(-E_{gi}/kT)} \right] \tag{3.108}$$

Both the first factor and the logarithmic term are maximal for $k_1 = 1$ and k_2 arbitrary. The voltage factor is then maximal and equal to

$$\eta_2 = \frac{kT}{E_{g1}} \ln \left[1 + \frac{\int_0^{hc/E_{g1}} P(\lambda)\lambda d\lambda}{\sum_1^2 a_i \exp(-E_{gi}/kT)} \right] \tag{3.109}$$

For the case $k_1 = 0$ and $k_2 = 1$,

$$\eta_2 = \frac{kT}{E_{g2}} \ln \left[1 + \frac{\int_0^{hc/E_{g2}} P(\lambda)\lambda d\lambda}{\sum_1^2 a_i \exp(-E_{gi}/kT)} \right] \quad (3.110)$$

Figure 3.10a shows the (E_{g1}, E_{g2}) plane, indicating optimal values of k_1 and k_2 to obtain a maximal η_2. Optimal $\eta_2(E_{g1}, E_{g2})$ are plotted in Figure 3.10b for $a_1 = a_2 = 10^4$ W/m. Figure 3.10c shows $\eta_2(E_{g1}, E_{g2})$ for the same a_1 and a_2 values, but for k values chosen to maximize the spectrum factor. From Figure 3.10b, it is clear that the largest attainable voltage factor η_2 of a heterojunction solar cell is somewhat larger but very close to the voltage factor of the homojunction with bandgap equal to the smaller of the bandgaps E_{g1} and E_{g2}. The curve showing η_2 as a function of the homojunction gap E_g is shown in Figure 3.10d and this curve agrees well with that obtained by Wolf.[45]

3.3.3.4. *Curve Factor* η_3

The curve factor is defined by

$$\eta_3 = -\frac{J(V_m) \cdot V_m}{J_L \cdot V_{oc}} \quad (3.111)$$

The voltage V_m is determined by the condition that the product $J(V) \cdot V$ is maximum for $V = V_m$. This gives the following transcendental equation

$$(1 + qV_m/kT)\exp(qV_m/kT) - 1 = J_L/J_S \quad (3.112)$$

Putting

$$v = qV_m/kT \quad (3.113)$$

and taking into account equations (3.43) and (3.72), we get

$$\eta_3 = \frac{-v(\exp v - 1 - J_L/J_S)}{(J_L/J_S)\ln(1 + J_L/J_S)} \quad (3.114)$$

with

$$(1 + v)\exp v - 1 = J_L/J_S \quad (3.115)$$

From equations (3.114) and (3.115), it follows that

$$\eta_3 = \frac{v^2 \exp v}{[(1 + v)\exp v - 1][v + \ln(1 + v)]} \quad (3.116)$$

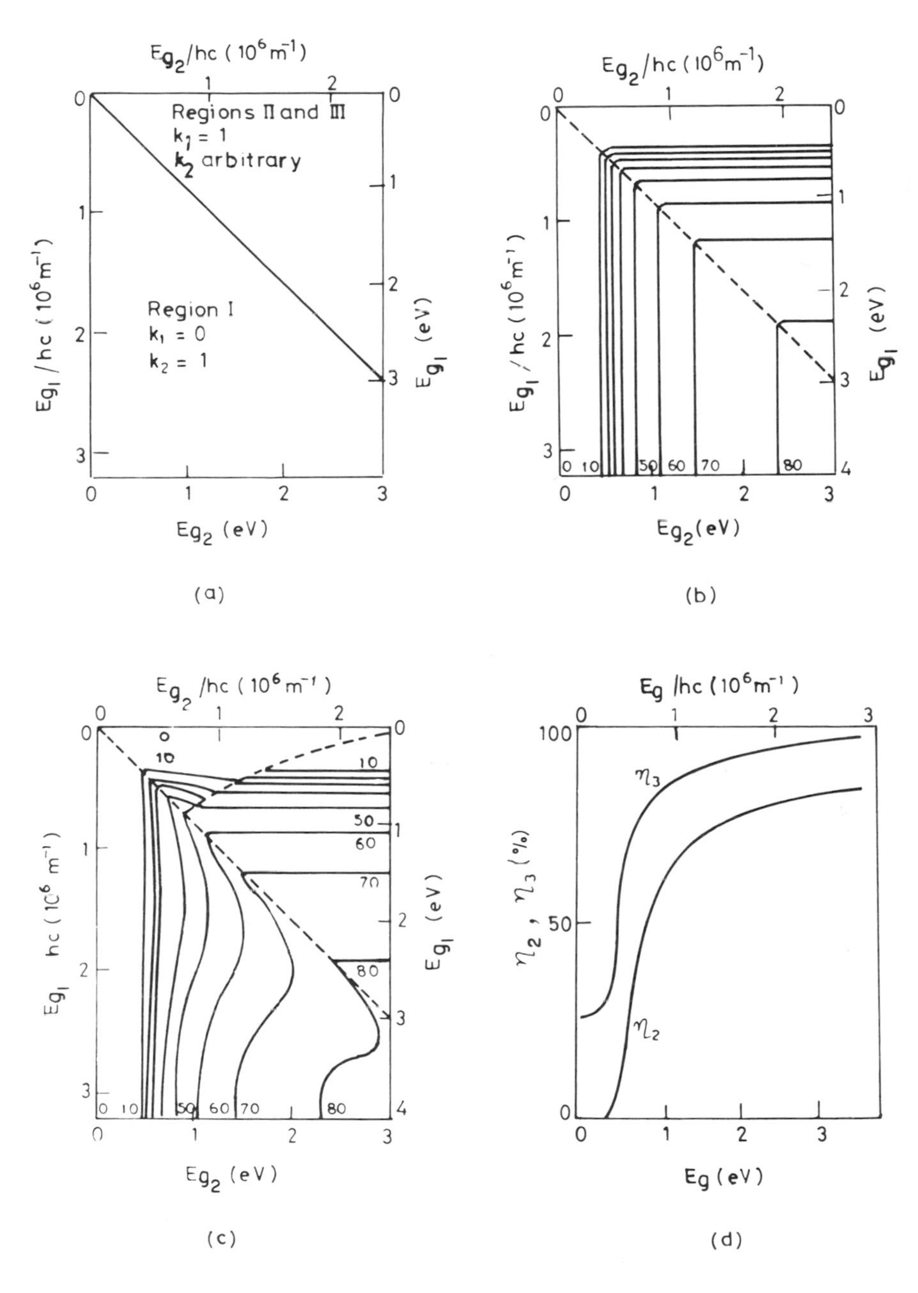

Figure 3.10. (a) Choice of k_1 and k_2 for optimal voltage factor η_2; (b) optimal $\eta_2(E_{g1}, E_{g2})$ of a heterojunction solar cell; (c) nonoptimal $\eta_2(E_{g1}, E_{g2})$ of a heterojunction solar cell; and (d) voltage factor $\eta_2(E_g)$ and the curve factor $\eta_3(E_g)$ of a homojunction solar cell (after de Vos[47]).

The curve factor η_3 will be optimal when v is maximal. This condition occurs when J_L/J_S is maximal, since the left-hand side of equation (3.115) increases monotonically with v in the range $0 \leqslant v < +\infty$.

(a) For $E_{g1} \geqslant E_{g2}$, from equations (3.95), (3.96), and (3.103), we get

$$\frac{J_L}{J_S} = \frac{[k_1 + (1-k_1)k_2] \int_0^{hc/E_{g1}} P(\lambda)\lambda d\lambda + k_2 \int_{hc/E_{g1}}^{hc/E_{g2}} P(\lambda)\lambda d\lambda}{\sum_1^2 a_i \exp(-E_{gi}/kT)} \tag{3.117}$$

This expression will be maximal for $k_2 = 1$ and k_1 arbitrary and will then be given by

$$\frac{J_L}{J_S} = \frac{\int_0^{hc/E_{g2}} P(\lambda)\lambda d\lambda}{\sum_1^2 a_i \exp(-E_{gi}/kT)} \tag{3.118}$$

It may be noted here that the conditions $k_2 = 1$ and k_1 arbitrary are consistent with the conditions for maximal product $\eta_1\eta_2$.

(b) When $E_{g1} = E_{g2} = E_g$,

$$\frac{J_L}{J_S} = \frac{\int_0^{hc/E_g} P(\lambda)\lambda d\lambda}{(a_1 + a_2)\exp(-E_g/kT)} \tag{3.119}$$

which is the case for a homojunction solar cell.

(c) For $E_{g1} \leqslant E_{g2}$, we get from equations (3.99), (3.100), and (3.103)

$$\frac{J_L}{J_S} = \frac{[k_1 + (1-k_1)k_2] \int_0^{hc/E_{g2}} P(\lambda)\lambda d\lambda + k_1 \int_{hc/E_{g2}}^{hc/E_{g1}} P(\lambda)\lambda d\lambda}{\sum_1^2 a_i \exp(-E_{gi}/kT)} \tag{3.120}$$

The optimal conditions are $k_1 = 1$ and k_2 arbitrary. For $k_1 = 1$, we get

$$\frac{J_L}{J_S} = \frac{\int_0^{hc/E_{g1}} P(\lambda)\lambda d\lambda}{\sum_1^2 a_i \exp(-E_{gi}/kT)} \tag{3.121}$$

For $k_1 = 0$ and $k_2 = 1$,

$$\frac{J_L}{J_S} = \frac{\int_0^{hc/E_{g2}} P(\lambda)\lambda d\lambda}{\sum_1^2 a_i \exp(-E_{gi}/kT)} \tag{3.122}$$

Figure 3.11a indicates k_1 and k_2 values for optimal curve factor η_3 while Figure 3.11b shows the optimal $\eta_3(E_{g1}, E_{g2})$ for $a_1 = a_2 = 10^4$ W/m. For the same values of a_1 and a_2 but with k_1 and k_2 chosen to optimize the spectrum factor, η_3 is plotted in Figure 3.11c. Figure 3.11b shows that the largest attainable η_3 of a heterojunction solar cell is approximately the curve factor of the homojunction with the bandgap E_g equal to either E_{g1} or E_{g2}, whichever is smaller.

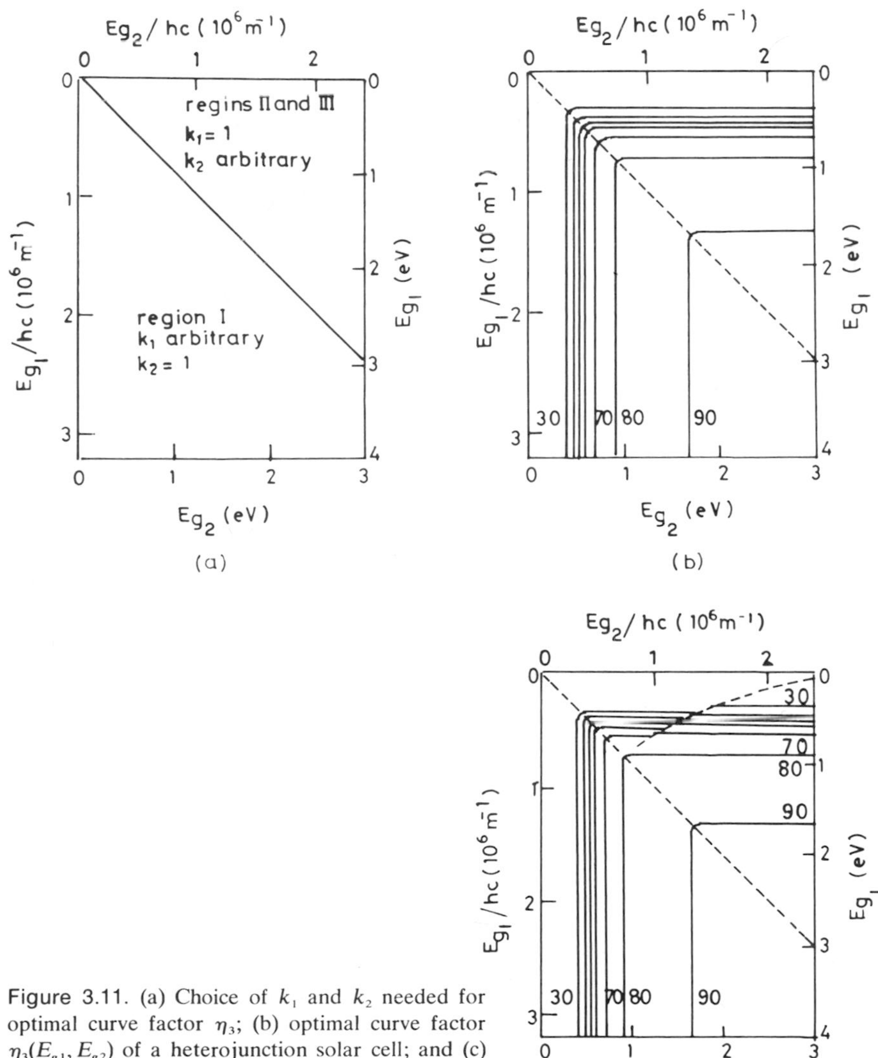

Figure 3.11. (a) Choice of k_1 and k_2 needed for optimal curve factor η_3; (b) optimal curve factor $\eta_3(E_{g1}, E_{g2})$ of a heterojunction solar cell; and (c) nonoptimal curve factor $\eta_3(E_{g1}, E_{g2})$ of a heterojunction solar cell (after deVos[47]).

3.3.3.5. *Efficiency* η

The product of spectrum factor, voltage factor, and curve factor gives the maximum attainable efficiency η of a solar cell.

$$\eta = \eta_1 \eta_2 \eta_3 \tag{3.123}$$

From equations (3.94), (3.104), and (3.111), we find that

$$\eta = -\frac{J(V_m) \cdot V_m}{\int_0^\infty P(\lambda) d\lambda} \tag{3.124}$$

with V_m obeying equation (3.112). Thus, we get

$$\eta = \frac{kT}{q} \cdot \frac{J_S v^2 \exp v}{\int_0^\infty P(\lambda) d\lambda} \tag{3.125}$$

It is clear that η will be maximum for maximal v and thus for maximal J_L/J_S, which is the condition for optimal η_3. Thus the required absorption conditions are as shown in Figure 3.11a. Figure 3.12a shows the function $\eta(E_{g1}, E_{g2})$ under the optimum conditions described by Figure 3.11a, for the case $a_1 = a_2 = 10^4$ W/m. This shows that the maximum attainable efficiency of a heterojunction cell does not differ much from that of a homojunction cell with a bandgap E_g equal to the smaller of the bandgaps E_{g1} and E_{g2}. The maximum attainable efficiency $\eta(E_g)$ of a homojunction cell is plotted in Figure 3.12b. It shows a maximum $\eta_{\max}$ of 22.9% occurring at a bandgap of $E_{g_{\max}} = 1.32$ eV.

In the general case, with given a_1 and a_2 values, the most efficient homojunction has a bandgap

$$E_g = E_{g_{\max}}(a_1 + a_2) \tag{3.126}$$

attaining an efficiency

$$\eta = \eta_{\max}(a_1 + a_2) \tag{3.127}$$

The most efficient heterojunction cell has one bandgap

$$E_g = E_{g_{\max}}[\min(a_1, a_2)] \tag{3.128}$$

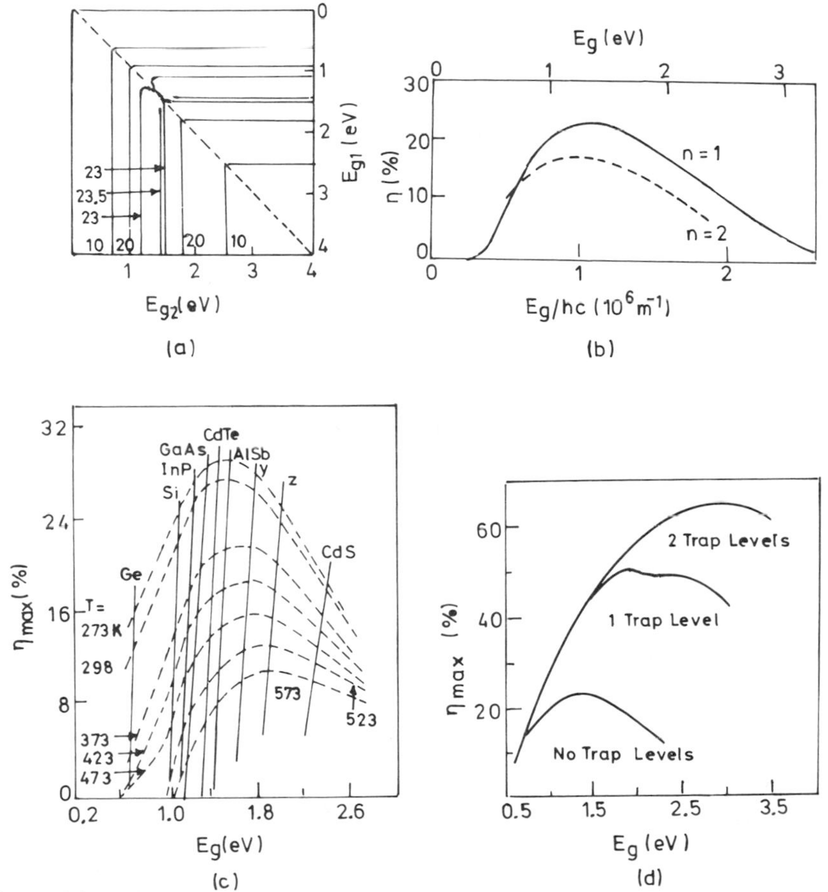

Figure 3.12. (a) Maximum attainable efficiency $\eta(E_{g1}, E_{g2})$ of a heterojunction solar cell; (b) maximum attainable efficiency $\eta(E_g)$ of a homojunction solar cell, with the curve for $n = 2$ (from Loferski[43]) also shown for comparison; (c) maximum attainable efficiency $\eta(E_g)$ of a homojunction solar cell for different temperatures (from Wysocki and Rappaport[44]); and (d) maximum attainable efficiency as a function of number of traps (from Wolf[45]).

and another bandgap

$$E_g > E_{g_{max}}[\min(a_1, a_2)] + kT\,|\ln(a_1/a_2)| \tag{3.129}$$

and attains an efficiency

$$\eta = \eta_{max}[\min(a_1, a_2)] \tag{3.130}$$

This results in an advantage for the heterojunction solar cell because

$$\eta_{max}[\min(a_1, a_2)] > \eta_{max}(a_1 + a_2) \tag{3.131}$$

In a homojunction cell, the largest attainable efficiency is determined by the sum a_1 and a_2 and thus approximately by $\max(a_1, a_2)$. In a heterojunction the condition is $\min(a_1, a_2)$. For a heterojunction solar cell, only one of the parameters a_1 and a_2 has to be small to obtain large efficiencies. For a homojunction cell, both a_1 and a_2 have to be small.

In Figure 3.12b we have also shown the $\eta(E_g)$ vs. E_g curve calculated for a diode factor $n = 2$.[43] Wysocki and Rappaport[44] showed the effect of temperature on $\eta(E_g)$ and their results are given in Figure 3.12c. The limit efficiency obtainable without, with one, and with two trap levels in the forbidden gap[45] is shown in Figure 3.12d.

3.3.3.6. *Collection Efficiency*

In the preceding discussion, the collection efficiency Q is assumed to be unity. However, the collection efficiency will depend, among other material parameters, on the energy gaps E_{g1} and E_{g2}. deVos and Pauwels[48] have calculated the collection efficiency of heterojunction solar cells as a function of the energy gaps.

Consider the solar cell geometry shown earlier in Figure 3.8b. Parameters of the semiconductor on the illuminated side of the junction are denoted by the subscript 1 and those of the rear side semiconductor by the subscript 2. As in the preceding analysis, the ideal diode equation is assumed to hold and the energy barriers at the junction are such as not to affect the collection of minority carriers. Also, the absorption is assumed to follow an exponential law.

Photons incident on the front surface ($x = 0$) with energies greater than E_{g1} at a rate N_1 per second per square meter will generate minority carriers in the front semiconductor at a rate G_1 (in $s^{-1}\,m^{-2}$) and the ratio of the two rates gives the generation efficiency of the front semiconductor:

$$k_1 = G_1/N_1 \tag{3.132}$$

Similarly,

$$k_2 = G_2/N_2 \tag{3.133}$$

where k_2 is the generation efficiency of the rear semiconductor, G_2 is the minority carrier generation rate in the second semiconductor, and N_2 is the rate of photons, with energies greater than E_{g2}, incident on the rear semiconductor ($x = x_1$). Thus, the total generation efficiency k of the solar cell is

$$k = \frac{G_1 + G_2}{N} = \frac{k_1 N_1 + k_2 N_2}{N} = \frac{N_1}{N}\left(k_1 + k_2 \frac{N_2}{N_1}\right) \tag{3.134}$$

Here N is the rate of photons incident on the front surface with energies larger than the smaller bandgap (E_{g1}, E_{g2}). We now define a quantity Q_1, the collection efficiency of the front semiconductor, as that fraction of the minority carriers G_1 which is collected at the junction to give a contribution J_1 to the short-circuit carrier current (in $s^{-1}\,m^{-2}$) of the solar cell. Then

$$Q_1 = J_1/G_1 \tag{3.135}$$

Similarly, for the rear semiconductor,

$$Q_2 = J_2/G_2 \tag{3.136}$$

The collection efficiency Q of the solar cell is given by

$$Q = \frac{J_1 + J_2}{G_1 + G_2} = \frac{Q_1 G_1 + Q_2 G_2}{G_1 + G_2} = \frac{Q_1 k_1 N_1 + Q_2 k_2 N_2}{k_1 N_1 + k_2 N_2} = \frac{Q_1 k_1 + Q_2 k_2 (N_2/N_1)}{k_1 + k_2 (N_2/N_1)} \tag{3.137}$$

The current efficiency σ is defined as the ratio of the short-circuit carrier current to the flux producing the short-circuit carrier current:

$$\sigma = (J_1 + J_2)/N = kQ \tag{3.138}$$

If the absorption characteristics of the semiconductors are ideal, i.e., photons with energies $E < E_g$ are not absorbed and the absorption coefficient α_i $(i = 1, 2)$ for photons with $E > E_g$ is independent of the wavelength, and the generation and collection efficiencies under this condition do not depend on the spectral composition of the incident light and the bandgaps of the two semiconductors but only on the geometry and material parameters of both semiconductors, the dependence of the generation and the collection efficiency of the heterojunction solar cell on the composition of the incident light and on the bandgaps of both the semiconductors comes through the terms N_1/N and N_2/N_1:

$$k_1 = 1 - \exp(-\alpha_1 x_1) \tag{3.139}$$

and

$$k_2 = 1 - \exp(-\alpha_2 x_2) \tag{3.140}$$

Q_1 can be obtained from the solution of the diffusion equation

$$\frac{d^2 m_1}{dx^2} - \frac{m_1}{L_1^2} + \frac{\alpha_1 N_1}{D_1} \exp(-\alpha_1 x) = 0 \tag{3.141}$$

where m is the excess minority carrier concentration, D is the diffusion coefficient, $L = (D\tau)^{1/2}$ is the diffusion length, and τ is the minority carrier lifetime. The boundary conditions for equation (3.141) are determined by recombination at the front surface,

$$D_1 \frac{dm_1}{dx}\bigg|_{x=0} = S_1 m_1(0) \tag{3.142}$$

and the short-circuit condition at the junction,

$$m_1(x_1) = 0 \tag{3.143}$$

S is the surface recombination velocity.

Q_2 can be found by solving a similar diffusion equation

$$\frac{d^2 m_2}{dx^2} - \frac{m_2}{L_2^2} + \frac{\alpha_2 N_2}{D_2} \exp[-\alpha_2(x - x_1)] = 0 \tag{3.144}$$

under boundary conditions determined by recombination at the rear surface,

$$-D_2 \frac{dm_2}{dx}\bigg|_{x=x_1+x_2} = S_2 m_2(x_1 + x_2) \tag{3.145}$$

and short-circuit condition at the junction,

$$m_2(x_1) = 0 \tag{3.146}$$

Q_1 and Q_2 can then be expressed as

$$Q_1 = -D_1(dm_1/dx)\big|_{x=x_1}/N_1[1 - \exp(-\alpha_1 x_1)] \tag{3.147}$$

and

$$Q_2 = D_2(dm_2/dx)\big|_{x=x_1}/N_2[1 - \exp(-\alpha_2 x_2)] \tag{3.148}$$

Introducing the following dimensionless parameters

$$b_i = \alpha_i L_i \qquad (i = 1, 2) \tag{3.149}$$

$$c_i = S_i L_i / D_i \qquad (i = 1, 2)$$

and geometric parameters

$$y_i = x_i / L_i \qquad (i = 1, 2) \tag{3.150}$$

we obtain

$$Q_1 = \left(\frac{b_1}{b_1^2 - 1}\right)$$
$$\times \frac{(b_1 + c_1)\{1 - [\exp(-b_1 y_1)]\cosh y_1\} - (1 + b_1 c_1)[\exp(-b_1 y_1)]\sinh y_1}{[1 - \exp(-b_1 y_1)](\cosh y_1 + c_1 \sinh y_1)} \quad (3.151)$$

and

$$Q_2 = \left(\frac{b_2}{b_2^2 - 1}\right) \frac{(b_2 c_2 - 1)\sinh y_2 + (b_2 - c_2)[\cosh y_2 - \exp(-b_2 y_2)]}{[1 - \exp(-b_2 y_2)](\cosh y_2 + c_2 \sinh y_2)} \quad (3.152)$$

Figures 3.13a and b represent the expressions for Q_1 and Q_2 for the case $c_1 = 0$ (zero recombination at the front surface) and $c_2 = \infty$ (ohmic contact at the rear surface).

The generation efficiencies can be expressed as

$$k_1 = 1 - \exp(-b_1 y_1) \quad (3.153)$$

and

$$k_2 = 1 - \exp(-b_2 y_2) \quad (3.154)$$

We define a quantity u as

$$u = \frac{\int_0^{hc/E_{g2}} N(\lambda)d\lambda}{\int_0^{hc/E_{g1}} N(\lambda)d\lambda} \quad (3.155)$$

Now

$$N(\lambda) = P(\lambda)\lambda/hc \quad (3.156)$$

where $P(\lambda)$ has already been defined. Using equation (3.156), we can calculate $u(E_{g1}, E_{g2})$, and lines of constant u in the (E_{g1}, E_{g2}) plane are shown in Figure 3.13c. N_1 is given by

$$N_1 = \int_0^{hc/E_{g1}} N(\lambda)d\lambda \quad (3.157)$$

(i) If $E_{g2} \geq E_{g1}$ $(u \leq 1)$, then

$$N = \int_0^{hc/E_{g1}} N(\lambda)d\lambda \quad (3.158)$$

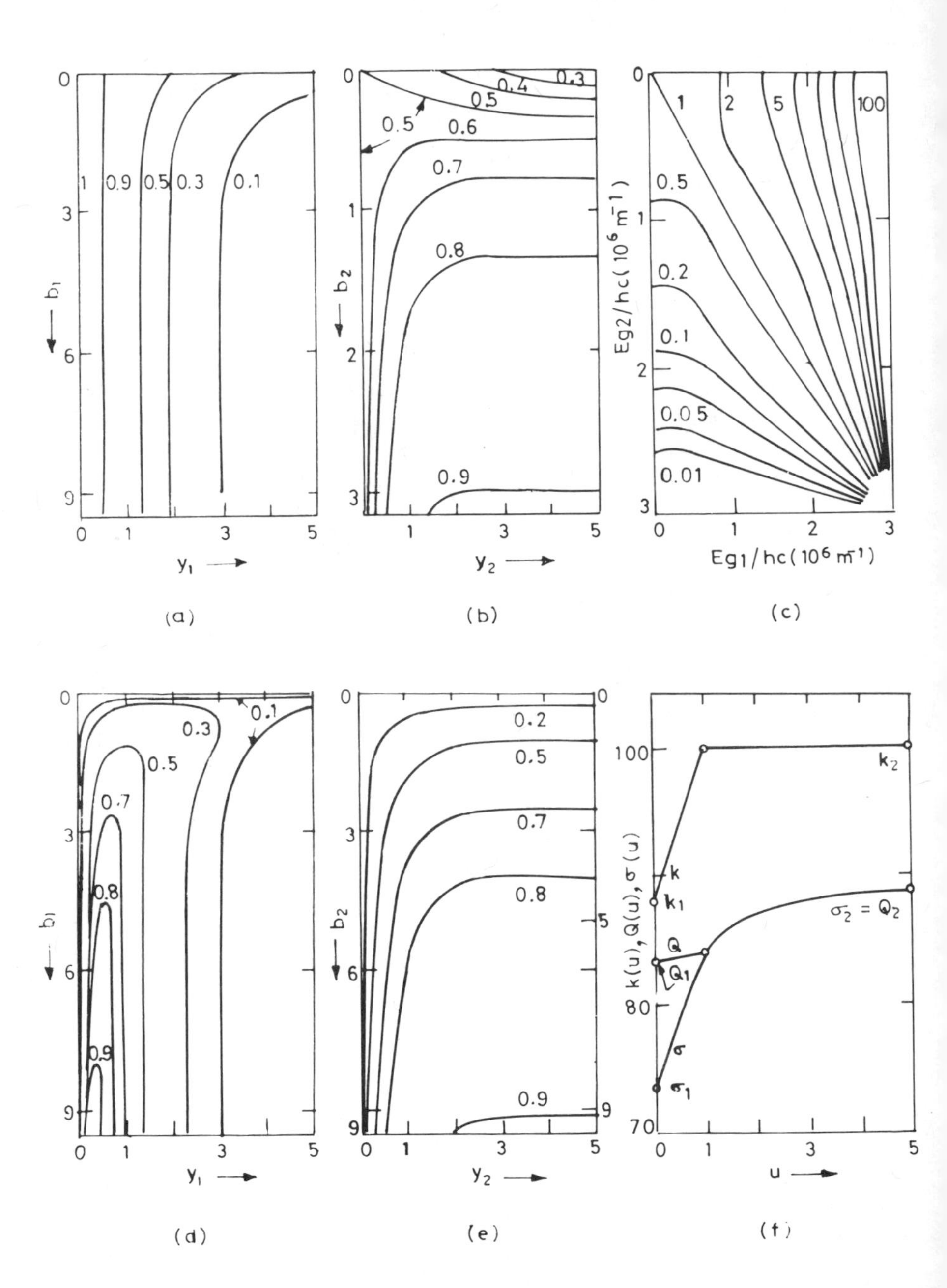

Figure 3.13. Representation of the function (a) $Q_1(b_1, y_1)$; (b) $Q_2(b_2, y_2)$; (c) $u(E_{g1}, E_{g2})$; (d) $\sigma_1(b_1, y_1)$; and (e) $\sigma_2(b_2, y_2)$. (f) $k(u)$, $Q(u)$, and $\sigma(u)$ calculated for $b_2 = 9$, $y_1 = 0.7$, $y_2 = 3$, $c_1 = 0$, and $c_2 = \infty$ (after deVos and Pauwels[48]).

and

$$N_2 = \exp(-b_1 y_1) \int_0^{hc/E_{g2}} N(\lambda) d\lambda \tag{3.159}$$

Thus,

$$N_1/N = 1 \tag{3.160}$$

and

$$N_2/N_1 = u \exp(-b_1 y_1) = u(1 - k_1) \tag{3.161}$$

(ii) If $E_{g2} \leqslant E_{g1}$ ($u \geqslant 1$), then

$$N = \int_0^{hc/E_{g2}} N(\lambda) d\lambda \tag{3.162}$$

and

$$N_2 = \exp(-b_1 y_1) \int_0^{hc/E_{g1}} N(\lambda) d\lambda + \int_{hc/E_{g1}}^{hc/E_{g2}} N(\lambda) d\lambda \tag{3.163}$$

We have, therefore,

$$N_1/N = 1/u \tag{3.164}$$

and

$$N_2/N_1 = \exp(-b_1 y_1) + u - 1 = u - k_1 \tag{3.165}$$

From equations (3.134), (3.160), (3.161), (3.164), and (3.165), it follows that

$$k = k_1 + u(1 - k_1)k_2 \qquad u \leqslant 1 \tag{3.166}$$

$$= k_2 + u^{-1}(1 - k_2)k_1 \qquad u \geqslant 1 \tag{3.167}$$

$k(u)$ increases monotonically from k_1 at $u = 0$ to $k_1 + k_2 - k_1 k_2$ at $u = 1$ (i.e., $E_{g1} = E_{g2}$) and then increases monotonically to k_2 at $u = +\infty$. This implies that all heterojunction solar cells ($u \neq 1$) show a smaller generation efficiency k than a homojunction solar cell ($u = 1$) with the same k_1 and k_2.

From equations (3.137), (3.160), (3.161), (3.164), and (3.165), we obtain

$$Q = \frac{Q_1 k_1 + Q_2 k_2 (1 - k_1) u}{k_1 + k_2 (1 - k_1) u} \qquad u \leqslant 1 \tag{3.168}$$

$$= \frac{Q_1 k_1 + Q_2 k_2 (u - k_1)}{k_1 + k_2 (u - k_1)} \qquad u \geqslant 1 \tag{3.169}$$

$Q|u|$ varies monotonically from Q_1 at $u=0$ to Q_2 at $u=+\infty$. It increases monotonically if $Q_1 < Q_2$. The function decreases monotonically if $Q_1 > Q_2$. It is clear that if $Q_1 < Q_2$ a heterojunction cell with $u > 1$ (i.e., with its larger bandgap at the illuminated side) will have a larger collection efficiency Q than a homojunction cell with the same Q_1 and Q_2. Similarly, if $Q_1 > Q_2$ a heterojunction cell with $u < 1$ (smaller bandgap at the illuminated side) will have a larger collection efficiency than a homojunction cell with the same Q_1 and Q_2. In effect, for optimum collection efficiency of the solar cell, the semiconductor with the larger collection efficiency should have the smaller bandgap. The physical reason is that in a heterojunction solar cell it is advantageous to absorb more photons where the collection is better.

From equation (3.138) we have

$$\sigma = kQ = \sigma_1 + u(1-k_1)\sigma_2 \qquad u \leqslant 1 \tag{3.170}$$

$$= \sigma_2 + u^{-1}(\sigma_1 - k_1\sigma_2) \qquad u \geqslant 1 \tag{3.171}$$

where σ_1 and σ_2 are the current efficiencies of the front and rear semiconductor, respectively,

$$\sigma_1 = k_1 Q_1 \tag{3.172}$$

$$\sigma_2 = k_2 Q_2 \tag{3.173}$$

Figures 3.13d and e represent the functions $\sigma_1(b_1, y_1)$ and $\sigma_2(b_2, y_2)$ for the case $c_1 = 0$ and $c_2 = \infty$. $\sigma(u)$ increases monotonically from σ_1 at $u = 0$ to $\sigma_1 + (1-k_1)\sigma_2$ at $u = 1$. For $u \to 1$ to $+\infty$, $\sigma(u)$ increases (if $\sigma_2 > Q_1$) or decreases (if $\sigma_2 < Q_1$) monotonically from $\sigma_1 + (1-k_1)\sigma_2$ to σ_2. The implication is that a heterojunction solar cell with the smaller bandgap at the illuminated side ($u < 1$) always has a smaller current efficiency σ than a homojunction solar cell ($u = 1$) with the same k_1, k_2, Q_1, and Q_2. On the other hand, a heterojunction solar cell with the larger bandgap on the illuminated side of the junction ($u > 1$) can show a larger as well as a smaller current efficiency than a corresponding homojunction ($u = 1$). The larger current efficiency for a heterojunction as compared to a homojunction results from the fact that under appropriate conditions ($\sigma_2 > Q_1$) more of the carriers are generated immediately behind the junction, where they are better collected. In this case, the first semiconductor acts as a window for some of the incident photons that are absorbed in the second semiconductor close to the junction, resulting in high collection efficiency. Figure 3.13f gives the curves for $k(u)$, $Q(u)$, and $\sigma(u)$ for specified parametric values.

In deriving the expression for collection efficiency we have assumed that the ideal diode law is obeyed. Implicit in that assumption is the condition that bulk-material properties on either side of the heterojunction determine the reverse saturation current. Also there should be no discontinuity in the conduction or valence band edges which can reduce the collection efficiency. In order to achieve the above conditions in practice, it is necessary that the two materials have similar crystal structures and lattice spacing. If there is a lattice mismatch between the two materials comprising the heterojunction, misfit dislocation present on both sides of the junction will give rise to dangling bonds that will produce interface states leading to nonideal diode behavior.

As we have noted before, while discussing the dark $I-V$ characteristics of heterojunctions, differences in electron affinity between the two semiconductors in a heterojunction result in discontinuities in the energy bands. Thus, for high carrier collection, the semiconductors must have closely matched electron affinities.

3.3.4. *Design and Optimization of Solar Cells*

Although idealized models, like the one described above, are useful in predicting the limit to the conversion efficiency of a particular material or a combination of materials, they do not aid in the design and optimization of actual devices. To realize the optimum performance of a device, we must take into account the actual transport mechanisms operative at the junction, the presence of surface recombination, the effect of drift fields, and the contribution to the current from regions of different properties. Such analysis leads to the formulation of more realistic models that are useful for establishing design criteria for solar cells.

3.3.4.1. *Homojunctions*

The generation rate of electron–hole pairs as a function of distance x from the surface is

$$G(\lambda) = \alpha(\lambda)\phi(\lambda)[1 - R(\lambda)]\exp[-\alpha(\lambda)x] \tag{3.174}$$

where λ is the wavelength of the incident light, $\phi(\lambda)$ is the number of incident photons per square centimeter per second per unit bandwidth, and R is the number of photons reflected from the surface. The photocurrent produced by these carriers and the spectral response can be determined for low-injection levels by using the minority carrier continuity equations [equations (2.189)] and by applying appropriate boundary conditions. Homojunctions can be represented by one of several physical models, depending on how these are fabricated. We shall discuss three cases in the following sections.

3.3.4.1a. *Uniform Doping.* When the two sides of the junction are uniformly doped, no electric fields exist outside the depletion region. Considering a uniformly doped n/p junction, the hole photocurrent density per unit bandwidth collected at the junction edge ($x = x_j$) from the top layer for a given wavelength is[5,50]

$$J_p = \left[\frac{q\phi(1-R)\alpha L_p}{\alpha^2 L_p^2 - 1}\right] \times \left[\frac{[(S_p L_p/D_p) + \alpha L_p] - \exp(-\alpha x_j)[(S_p L_p/D_p)\cosh(x_j/L_p) + \sinh(x_j/L_p)]}{(S_p L_p/D_p)\sinh(x_j/L_p) + \cosh(x_j/L_p)} - \alpha L_p \exp(-\alpha x_j)\right] \tag{3.175}$$

It should be noted that the effect of surface recombination has been included in the above expression through the surface recombination velocity term S_p.

The electron photocurrent density at the junction edge from the base layer of the cell is

$$J_n = \frac{q\phi(1-R)\alpha L_n}{\alpha^2 L_n^2 - 1} \exp[-\alpha(x_j + W)] \times \left\{\alpha L_n - \frac{(S_n L_n/D_n)[\cosh(H'/L_n) - \exp(-\alpha H')]}{(S_n L_n/D_n)\sinh(H'/L_n) + \cosh(H'/L_n)} + \frac{\sinh(H'/L_n) + \alpha L_n \exp(-\alpha H')}{(S_n L_n/D_n)\sinh(H'/L_n) + \cosh(H'/L_n)}\right\} \tag{3.176}$$

where W is the depletion width, $H' = H - (x_j + W)$, H is the total cell thickness, and S_n is the surface recombination velocity at the back of the cell. For an ohmic contact at the back, S_n can be taken as infinite.

The photocurrent collected from the depletion region is given by

$$J_{\rm dr} = q\phi(1-R)\exp(-\alpha x_j)[1 - \exp(-\alpha W)] \tag{3.177}$$

It is assumed that the electric field in the depletion region is high enough so that the photogenerated carriers are accelerated out of the region before recombination occurs.

The total short-circuit photocurrent density at a wavelength λ is then

$$J = J_n + J_p + J_{\rm dr}$$

We should point out that the treatment for a p/n junction is analogous and the photocurrent expression can be obtained by interchanging L_n, D_n, and S_n with L_p, D_p, and S_p, respectively.

The reverse saturation current density, taking into account surface recombination at the top and at the back surfaces, can be expressed by the relation[5]

$$J_s = q\frac{D_p}{L_p}\frac{n_i^2}{N_d}\left[\frac{(S_pL_p/D_p)\cosh(x_j/L_p)+\sinh(x_j/L_p)}{(S_pL_p/D_p)\sinh(x_j/L_p)+\cosh(x_j/L_p)}\right]$$
$$+q\frac{D_n}{L_n}\frac{n_i^2}{N_a}\left[\frac{(S_nL_n/D_n)\cosh(H'/L_n)+\sinh(H'/L_n)}{(S_nL_n/D_n)\sinh(H'/L_n)+\cosh(H'/L_n)}\right] \quad (3.178)$$

3.3.4.1b. *Constant Electric Fields.* Electric fields may be present in one or both regions of a cell owing to concentration gradients of donors or acceptors. The electric field causes the energy band edges to be sloped and the field at any point is given by the slope

$$E = \frac{1}{q}\frac{dE_c}{dx} = \frac{1}{q}\frac{dE_v}{dx} = \frac{D}{\mu}\frac{1}{N}\frac{dN}{dx}$$

where N is the ionized impurity concentration. Wolf[51] and Ellis and Moss[52] have derived expressions for the photocurrent, assuming the electric field, mobility, and lifetime to be constant across the diffused region. The photocurrent from the n-type top-diffused layer, at any wavelength, is given by[51]

$$J_p = \frac{q\phi(1-R)\alpha L_{pp}}{(\alpha+E_{pp})^2L_{pp}^2-1}$$
$$\times\left\{\frac{(\alpha+E_{pp})L_{pp}\exp(E_{pp}x_j)-\exp(x_j/L_{pp})\exp(-\alpha x_j)}{(S_pL_{pp}/D_p+E_{pp}L_{pp})\sinh(x_j/L_{pp})+\cosh(x_j/L_{pp})}\right.$$
$$+\frac{(S_pL_{pp}/D_p+E_{pp}L_{pp})[\exp(E_{pp}x_j)-\exp(x_j/L_{pp})\exp(-\alpha x_j)]}{(S_pL_{pp}/D_p+E_{pp}L_{pp})\sinh(x_j/L_{pp})+\cosh(x_j/L_{pp})}$$
$$\left.-\exp(-\alpha x_j)[(\alpha+E_{pp})L_{pp}-1]\right\} \quad (3.179)$$

where $E_{pp} = qE/2kT$ and L_{pp} is an effective diffusion length given by $L_{pp}^{-1} = [E_{pp}^2 + (1/L_p^2)]^{1/2}$. The above equation is valid under the condition that the carriers accelerated by the field do not reach their saturation drift velocity.

The photocurrent from the p-type base, at a given wavelength, assuming the field, lifetime, and mobility to be constant in the base, is[5]

$$J_n = \frac{q\alpha\phi(1-R)L_{nn}\exp(-E_{nn}x_j)\exp(-\alpha W)}{(\alpha - E_{nn})^2L_{nn}^2 - 1}$$
$$\times\{[(\alpha - E_{nn})L_{nn} - 1]\exp[-(\alpha - E_{nn})x_j] + \exp[-(\alpha - E_{nn})(H-W)]\}$$
$$\times\left\{\frac{\exp(H'/L_{nn})\exp(\alpha - E_{nn})H' - (\alpha - E_{nn})L_{nn}}{(E_{nn} + S_n/D_n)L_{nn}\sinh(H'/L_{nn}) + \cosh(H'/L_{nn})}\right.$$
$$\left. - \frac{(E_{nn} + S_n/D_n)L_{nn}[\exp(-H'/L_{nn})\exp(\alpha - E_{nn})H' - 1]}{(E_{nn} + S_n/D_n)L_{nn}\sinh(H'/L_{nn}) + \cosh(H'/L_{nn})}\right\} \quad (3.180)$$

E_{nn} and L_{nn} have expressions analogous to E_{pp} and L_{pp}, respectively.

Analytical solutions cannot be obtained when the electric field, mobility, and lifetime are not constant, in either the top layer or the base layer. Numerical computations[53–57] have to be employed to obtain photocurrent densities.

The reverse saturation current density in the case of uniform electric fields takes the form[52]

$$J_s = q\frac{D_p}{L_{pp}}\frac{n_i^2}{N_d}$$
$$\times\left[\frac{\sinh(x_j/L_{pp}) + [(S_pL_{pp}/D_p) + E_{pp}L_{pp}]\cosh(x_j/L_{pp})}{[(S_pL_{pp}/D_p) + E_{pp}L_{pp}]\sinh(x_j/L_{pp}) + \cosh(x_j/L_{pp})} - E_{pp}L_{pp}\right]$$
$$+ q\frac{n_i^2}{N_a}\left[\frac{D_n}{L_{nn}}\coth\frac{H-(x_j+W)}{L_{nn}} - E_{nn}D_n\right] \quad (3.181)$$

An infinite surface recombination velocity (metallic ohmic contact) is assumed at the back of the cell.

3.3.4.1c. *Back Surface Field.* The incorporation of a back surface field[58] at the back of the base region helps to improve the short-circuit current density as well as the open-circuit voltage of the cell due to reduced recombination at the back surface. The energy band diagram of a back surface field (BSF) solar cell is shown in Figure 3.14. Region 1 is the

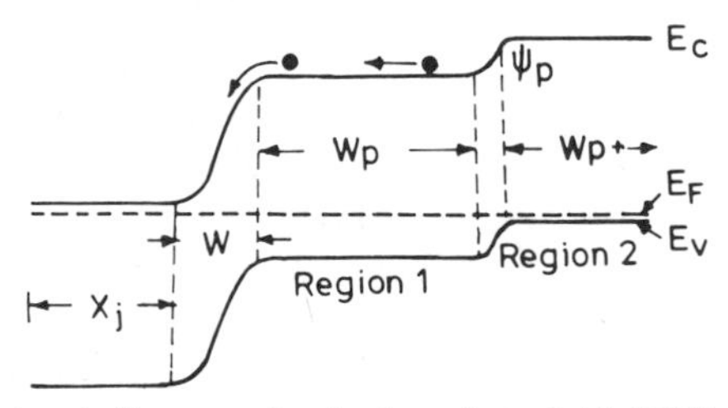

Figure 3.14. Energy band diagram of a back surface field (BSF) device (after Hovel[5]).

normal high-resistivity ($1-10\ \Omega\,\text{cm}$) portion while region 2 is the heavily doped layer adjacent to the back contact. The potential energy barrier ψ_p between the two base regions tends to confine the minority carriers in the more lightly doped region, away from the ohmic contact.

As a first-order approximation, the BSF cell can be described as a normal cell of width $x_j + W + W_p$ exhibiting a very small recombination velocity at the back. We can then use equation (3.122) or (3.126) with $S_n = 0$, provided $W_p \gg W_p^+$.

The reverse saturation current density component from the base of a BSF cell is given by[59]

$$J_s\,(\text{base}) = q\frac{D_n}{L_n}\frac{n_i^2}{N_a}\left[\frac{(SL_n/D_n)\cosh(W_p/L_n)+\sinh(W_p/L_n)}{\cosh(W_p/L_n)+(SL_n/D_n)\sinh(W_p/L_n)}\right] \qquad (3.182)$$

where

$$S = \frac{N_a}{N_a^+}\frac{D_n^+}{L_n^+}\left[\frac{(S_nL_n^+/D_n^+)\cosh(W_p^+/L_n^+)+\sinh(W_p^+/L_n^+)}{\cosh(W_p^+/L_n^+)+(S_nL_n^+/D_n^+)\sinh(W_p^+/L_n^+)}\right] \qquad (3.183)$$

and N_a, D_n, L_n and N_a^+, D_n^+, L_n^+ refer to the properties of region 1 and region 2, respectively. Equation (3.182) is valid for the condition that no drift fields exist in the cell.

3.3.4.2. *Heterojunctions*

The heterojunction solar cell has been the subject of extensive research interest and several workers[47,48,60–70] have treated the photovoltaic properties of heterojunctions in considerable detail. Fonash[60] has derived a general formulation for the current–voltage characteristic of a p/n heterojunction solar cell of the form window/absorber. The total voltage V developed across the cell is divided into two parts: (i) a part V_1 developed in the window material and (ii) a part V_2 developed in the absorber material, i.e., $V = V_1 + V_2$. A schematic diagram of a window/absorber p/n heterojunction solar cell configuration is shown in Figure 3.15. The total current I produced in the cell under illumination may be written as

$$I = A\left(eD_n\frac{dn}{dx}\bigg|_{x=w_2} + J_{\text{maj}}\big|_{x=w_2}\right) \qquad (3.184)$$

where A is the cross-sectional area of the device. The first term in the bracket is the contribution to the minority-carrier current at $x = w_2$ from electron diffusion. The majority carrier current AJ_{maj} at $x = w_2$ consists of:

(i) $AJ_L(D2)$, a photogenerated hole current developed in the depletion region $D2(w_2 \geqslant x \geqslant 0)$, (ii) an opposing current $A(J_1 + J_2)$ that is due to the interface recombination paths 1 and 2, (iii) $AJ_R(D2)$, arising from recombination in region $D2$, (iv) $AJ_R(D1)$, arising from recombination in the depletion region $D1$ $(-w_1 \leqslant x \leqslant 0)$, and (v) an opposing current $[-A_eD_p(dp/dx)$ at $x = -w_1]$ that is due to hole diffusion. However, owing to the large bandgap of the window (material 1), J_R $(D1)$ and $-eD_p dp/dx$ at $x = -w_1$ are negligible compared to the other terms. Thus

$$J_{\text{maj}}\big|_{x=w_2} = J_L(D2) - J_1 - J_2 - J_R(D2) \tag{3.185}$$

Evaluating the minority current term of equation (3.130) by solving the continuity equations for minority carriers in the p-type absorber under appropriate boundary conditions, we obtain the relation[60]

$$\begin{aligned} I = A\Bigg\{ & F(D1)F(I)F(D2)J_L - F(D1)F(I)F(D2)J_{D1F}\left[\exp\left(\frac{V}{kT}\right) - 1\right] \\ & - F(D1)eS_1 n_{n0}\exp\left(\frac{-V_{d1}}{kT}\right)\left[\exp\left(\frac{V}{kT}\right) - 1\right] \\ & - F(D1)F(I)eS_2 n_{p0}\exp\left(\frac{V_{d2}}{kT}\right)\left[\exp\left(\frac{V}{kT}\right) - 1\right] - J_R(D2)\Bigg\} \end{aligned} \tag{3.186}$$

$F(D1)$ and $F(D2)$ are the collection factors in the depletion region 1 and 2, respectively, and $F(I)$ characterizes transport across the metallurgical

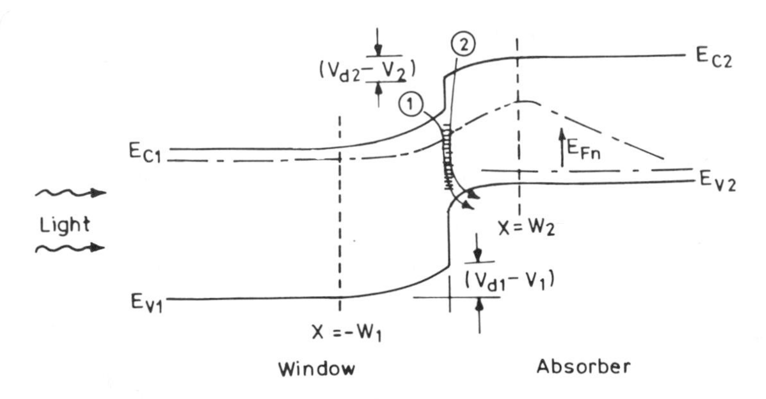

Figure 3.15. Energy band diagram of a window/absorber p/n heterojunction solar cell under illumination (after Fonash[60]).

junction. $J_L \equiv J_L(D2) + J_L(B)$, where $J_L(B)$ is the photocurrent of electrons diffusing out of the base of the absorber. $J_{D1F}[\exp(V/kT) - 1]$ is the bucking current component arising from electron diffusion in the base of the absorber. S_1 and S_2 are interface recombination velocities associated with paths 1 and 2, respectively. $F(D1)$, $F(D2)$, and $F(I)$ are defined by the relations

$$F(D1) \equiv \left[e\mu_{n1} n_{n0} \exp\left(\frac{-V_{d1}}{kT}\right) \exp\left(\frac{-V_2}{kT}\right) \xi_{M1} \right] \times \left[e\mu_{n1} n_{n0} \exp\left(\frac{-V_{d1}}{kT}\right) \exp\left(\frac{-V_2}{kT}\right) \xi_{M1} + eS_1 n_{n0} \exp\left(\frac{-V_{d1}}{kT}\right) + F(I) eS_2 n_{p0} \exp\left(\frac{V_{d2}}{kT}\right) + F(I)F(D2)J_{D1F} \right]^{-1} \tag{3.187}$$

$$F(D2) \equiv \frac{e\mu_{n2} n_{p0}(2kT/\pi^{1/2} L_2)}{e\mu_{n2} n_{p0}(2kT/\pi^{1/2} L_2) + J_{D1F}} \leqslant 1 \tag{3.188}$$

$$= \left(\frac{2}{\pi^{1/2}} \frac{L_n}{L_2}\right) \left[\left(\frac{2}{\pi^{1/2}} \frac{L_n}{L_2}\right) + 1 \right]^{-1} \tag{3.188a}$$

if $J_{D1F} = en_{p0} D_n / L_n$, and

$$F(I) \equiv J_{\text{TH}} \exp\left(\frac{-V_2}{kT}\right) \left[J_{\text{TH}} \exp\left(\frac{-V_2}{kT}\right) + eS_2 n_{p0} \exp\left(\frac{V_{d2}}{kT}\right) + F(D2) J_{D1F} \right]^{-1} \leqq 1 \tag{3.189}$$

ξ_{M1} is the maximum electrostatic field in region $D1$ when a voltage V_1 is being developed in the window material. J_{TH} is discussed extensively in terms of material properties and the actual mechanism of electron transfer in Fonash[71] and Fonash and Ashok.[72].

The collection factors provide convenient criteria for the design of heterojunctions, since these illustrate the interplay of materials and junction properties and aid in determining the optimum voltage distribution between the window and absorber.

The equations presented in Sections 3.3.4.1 and 3.3.4.2 relate the magnitude of the photogenerated current density to the material properties, the junction characteristics, and the physical configuration of the device. Thus these serve as useful aids in the optimization of a particular device structure for maximum conversion efficiency. In the light of the above models, let us now try to get a feel of the critical material properties which influence the photovoltaic performance of solar cells.

3.3.4.3. *Cell Parameters and Relation to Material Properties*

3.3.4.3a. *J_{sc} and V_{oc}*. The short-circuit current density J_{sc} is related to the light-generated current density J_L through the shunt resistance R_{sh} and series resistance R_s and, for high-quality cells with high R_{sh} and low R_s, the values of J_{sc} and J_L are identical. J_L, at any wavelength, is determined primarily by the optical excitation rate $G(\lambda)$, and hence by the photon flux $\phi(\lambda)$, the absorption coefficient $\alpha(\lambda)$, the thickness of the absorber t, and the reflectivity $R(\lambda)$ of the active layer. It is also determined by the diffusion length L, and hence by the carrier lifetime τ, and mobility μ. The open-circuit voltage, in turn, depends on J_L and can be approximated as the difference of the two majority carrier quasi-Fermi levels in the bulk of the p- and n-type regions.

3.3.4.3b. *α, t, and L*. We have discussed the factors that determine the absorption coefficient in Chapter 2. The absorption coefficient, the optimum thickness of the absorber layer, and the diffusion length of minority carriers are closely connected. In the intrinsic range, $\alpha(\lambda)$ and its slope near the band edge are 10–100 times larger in direct-bandgap than in indirect-bandgap materials. Therefore, the thickness of the absorber layer needed to absorb a substantial fraction ($\sim 90\%$) of the incident light varies from about 0.3 μm in Cu_2S or GaAs to more than 50 μm in Si. In Figure 3.16 we show the dependence of J_L on thickness for GaAs, Si, and Cu_2S/CdS solar cells, for the case of zero losses.[73] Obviously, higher current densities are obtained for larger thicknesses, as a result of more complete absorption. However, the photogenerated charge carriers have to

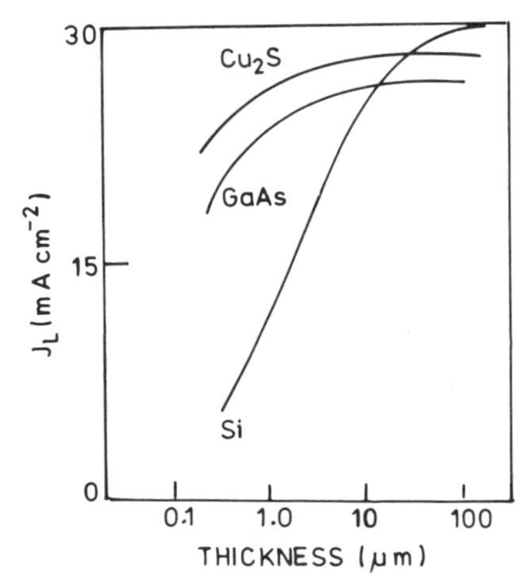

Figure 3.16. Dependence of photocurrent J_L on the thickness of the absorber layer in some typical solar cell materials for AM1 condition with no losses (after Woodall and Hovel[73]).

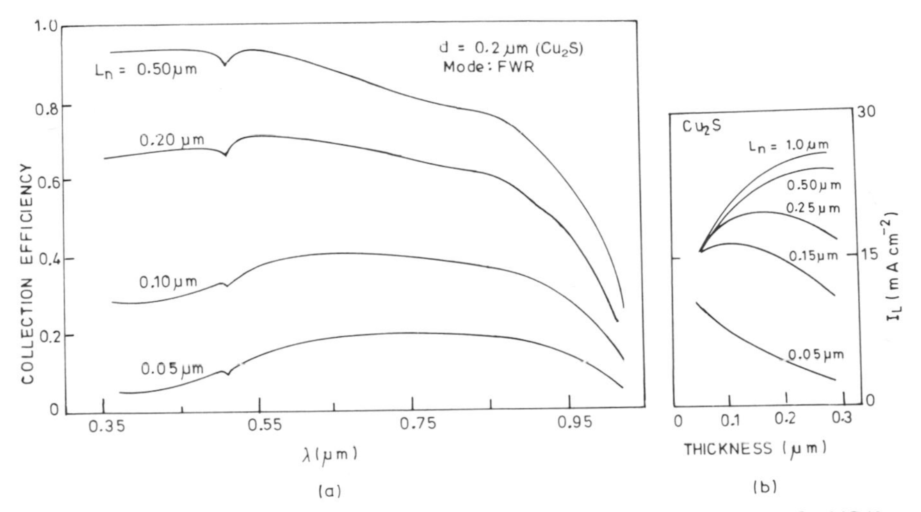

Figure 3.17. (a) Calculated spectral response in front-wall reflection mode for Cu_2S/CdS thin film solar cells and (b) calculated AM1 photocurrent densities for Cu_2S/CdS solar cells as a function of thickness of Cu_2S with diffusion length L_n (in Cu_2S) as a parameter (after Rothwarf[74]).

be collected and transported across the junction, and the collection efficiency is determined critically by the diffusion length of the carriers. For sufficient carrier collection, a minority carrier diffusion length exceeding twice the thickness of the absorber is necessary. Several workers[74–80] have carried out a parametric study of the photovoltaic parameters of solar cells in order to establish design criteria. Here, and in subsequent sections, we give a few illustrative examples of the results of such calculations. Figure 3.17a shows the calculated spectral response of Cu_2S/CdS solar cells[74] for different electron diffusion lengths in Cu_2S. The total AM1 current densities as a function of thickness, with L_n as a parameter, are shown in Figure 3.17b. In both cases, interference effects are excluded.

The expression relating the diffusion length to the mobility and lifetime of carriers is given in Chapter 2. The mobility and lifetime are determined by several factors, as previously discussed (Section 2.3), and the behavior varies from material to material. In Si, τ of about 1 μs in 0.01-Ω cm and up to 25 μs in 10-Ω cm material are typical, leading to a diffusion length of several hundred micrometers.[81] Thus, for Si solar cells, good collection efficiency is achieved even for the large active layer thicknesses (~200 μm) required for complete light absorption. In GaAs and Cu_2S, where the major photon absorption occurs in a thickness of a few tenths of a micrometer, shorter diffusion lengths (~0.2–0.7 μm for Cu_2S and ~1–5 μm for GaAs) and lifetimes (~10^{-9} s for Cu_2S and ~10^{-8} s for GaAs) are tolerable.[81]

3.3.4.3c. *Surface Recombination.* A high surface recombination velocity (2.4.3.1) can cause more than 50% of the carriers photogenerated close to the surface to be captured, thus effecting a reduction in J_{sc} and to some extent in V_{oc}. The effect of S_n on the photogenerated current for a Cu_2S/CdS solar cell[75] is shown in Figure 3.18a, for different values of L_n. Surface recombination can be reduced by proper doping of the region near a surface to produce a drift field counteracting minority carrier diffusion to the surface,[51] or by creating a surface layer of a wide bandgap material with small interface recombination (heteroface solar cell).[82] The effect of a surface layer on J_L for a GaAs solar cell is shown in Figure 3.18b.

3.3.4.3d. *Drift Field.* A drift field incorporated in the absorber layer assists in minority carrier diffusion toward the junction, and hence in an enhancement in J_{sc}. However, drift fields reduce V_{oc} by the voltage drop across the layer in which they act.[83] Thus very large drift fields are counterproductive and must be avoided.

3.3.4.3e. *Grain Size.* Grain boundaries in the absorber layer act like internal surfaces with high recombination velocities, causing a reduction in both J_{sc} and V_{oc}. The influence of grain boundaries on the photovoltaic parameters has been analyzed by several workers for homojunction,[84,85] Schottky barrier,[84,86] and heterojunction[74,87–90] solar cells. Card and Yang[84] have calculated the dependence of the minority carrier lifetime τ on the doping concentration N_d, grain size d, and interface state density N_{is} at the grain boundaries in n-type polycrystalline semiconductors. The recombination velocity at grain boundaries is enhanced by the diffusion

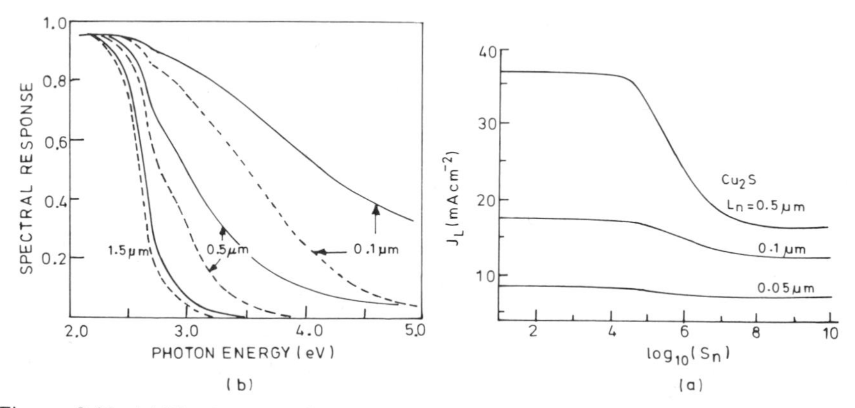

Figure 3.18. (a) Photogenerated current of a Cu_2S/CdS thin film solar cell as a function of outer-surface recombination velocity, for different values of L_n (after Hill[75]); (b) spectral response curves for a $Ga_{1-x}Al_xAs$/GaAs solar cell with (———) and without (– – –) the $Ga_{1-x}Al_xAs$ contribution for different junction depths (after Hovel and Woodall[79]).

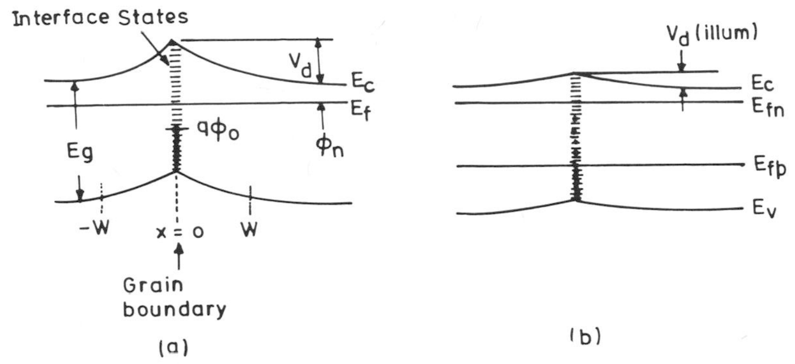

Figure 3.19. Energy band diagram at a grain boundary (a) in dark and (b) under illumination (after Card and Yang[84]).

potential V_d adjacent to the boundaries and is in the range from about 10^2 to 10^6 cm s^{-1}, depending on N_{is} and N_d. The energy band diagram in the dark in a region where an n-type semiconductor surrounds a grain boundary is shown in Figure 3.19a. V_d for any given interface state density, assuming N_{is} (states cm^{-2} eV^{-3}) to be constant with energy over the energy gap, is obtained from the relation

$$qN_{is}(E_f/q - \phi_0) = (8q\varepsilon_s N_d V_d)^{1/2} \tag{3.190}$$

where $q\phi_0$ defines a neutral level such that for $E_f = q\phi_0$, the net charge in these states is zero. The magnitude of N_{is} depends on the degree of orientational mismatch at the grain boundary, and for high-angle grain boundaries, the density of interface states and the resulting diffusion potential are large. Illumination alters the population of the interface states considerably and, consequently, V_d decreases to that value which maximizes recombination. The energy band diagram at a grain boundary under illumination is shown in Figure 3.19b. The decrease in V_d on illumination causes τ to decrease with increasing N_d. Card and Yang[84] have derived the following expressions for the recombination current J_r, the recombination velocity at the grain boundary $S(W)$, and the minority carrier lifetime τ_p:

$$J_r \simeq qN_{is}\sigma v \frac{p(0)}{2}(E_{fn} - E_{fp}) \qquad \text{for } n(0) \simeq p(0) \tag{3.191}$$

$$S(W) \simeq \tfrac{1}{2}S(0)\exp\left(\frac{qV_d}{kT}\right) \simeq \tfrac{1}{4}N_{is}\sigma v(E_{fn} - E_{fp})\exp\left(\frac{qV_d}{kT}\right) \tag{3.192}$$

and

$$\tau_p = \frac{1}{\sigma v N_r} = \frac{2d\exp(-qV_d/kT)}{3\sigma v N_{is}(E_{fn} - E_{fp})} \tag{3.193}$$

where $p(0)$ and $n(0)$ are hole and electron densities at $x = 0$, $\sigma = \sigma_n = \sigma_p$ is the capture cross-section of the interface state, v is the thermal velocity of carriers in either the conduction or valence bands, $S(0)$ is the recombination velocity at $x = 0$, d is the dimension of a cubic grain, and N_r is the effective volume concentration of recombination centers. The calculated J_{sc} and V_{oc} for Schottky barrier and p^+/n polycrystalline silicon solar cells as a function of grain size are shown in Figure 3.20a.

In polycrystalline silicon, for low-angle grain boundaries with $N_{is} \simeq 10^{11}$ cm^2 eV^{-1}, τ decreases from 10^{-6} to 10^{-10} s as the grain size reduces from 1000 to 0.1 μm ($N_d = 10^{16}$ cm^{-3}). For a given grain size, τ decreases with

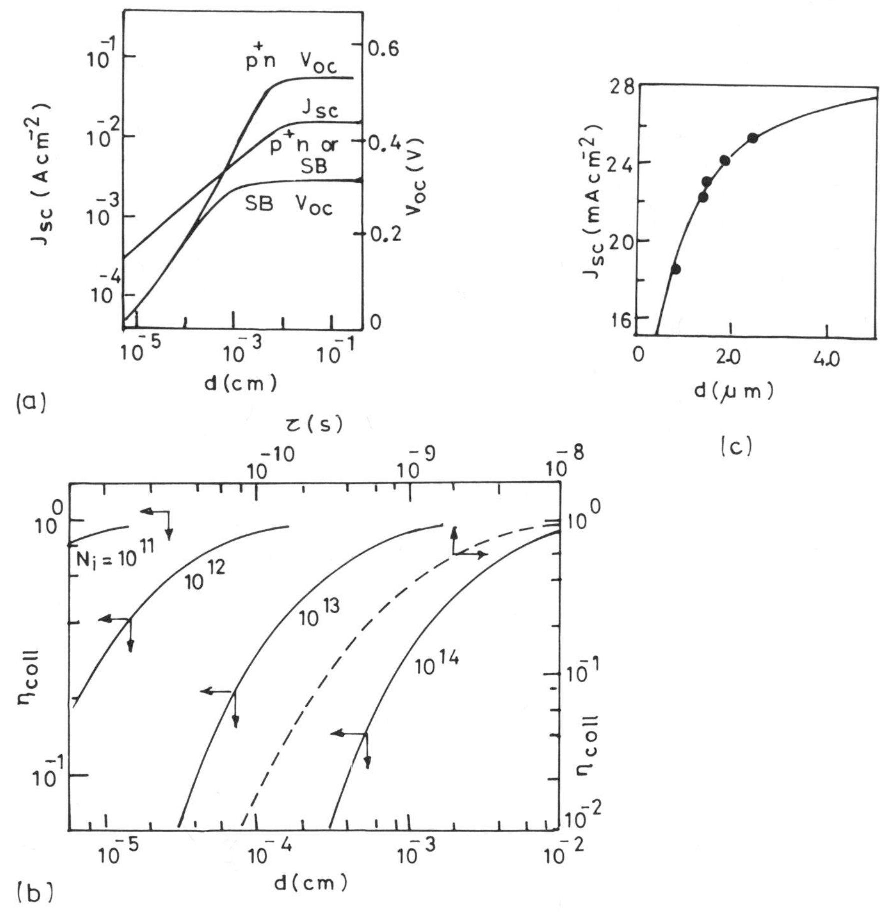

Figure 3.20. (a) Dependence of the J_{sc} and V_{oc} of polycrystalline Si Schottky barrier and p^+/n solar cells on grain size d (after Card and Yang[84]); (b) the collection efficiency of a $CuInSe_2$/CdS heterojunction as a function of minority carrier lifetime τ and grain diameter d; (c) dependence of J_{sc} on d for a CdS/$CuInSe_2$ thin film device (after Kazmerski et al.[90]).

increasing N_d. The V_{oc} of p/n homojunction cells decreases for $\tau \leqslant 10^{-7}$ s, whereas for Schottky barrier cells, the decrease occurs at $\tau \leqslant 10^{-8}$ s.

Fraas[88] has presented a qualitative description of grain boundary effects in polycrystalline heterojunction solar cells and has treated the grain boundary as a surface in accumulated, depleted, or inverted condition. Rothwarf[74,87] and Kazmerski[89] have examined the effects of grain boundaries in compound semiconductor thin film photovoltaic devices and derived interrelationships between minority carrier recombination velocity, grain size, V_{oc}, and J_{sc}. In a recent publication, Kazmerski et al.[90] have evaluated the dependence of the minority carrier lifetime on grain diameter and used these data to determine the dependence of the collection efficiency on grain size for various grain boundary conditions (low-, medium-, and high-angle grain boundaries) in heterojunction solar cells. In a cylindrical grain model,[87] all carriers generated closer to the grain boundary than the junction are lost and the collection efficiency is given by

$$\eta_{\text{coll}} = 1 + \frac{4}{\alpha d}\left(\frac{2}{\alpha d} - 1\right) + \frac{\exp(-\alpha t)}{1 - \exp(-\alpha t)}\frac{4t}{d}\left(1 - \frac{t}{d} - \frac{2}{\alpha d}\right) \tag{3.194}$$

where α is the absorption coefficient, d is the grain diameter, and t is the film thickness. It is assumed that the cell is illuminated through the higher bandgap semiconductor with $\frac{1}{2}d > t$. Equation (3.194) can be rewritten in terms of τ as

$$\eta_{\text{coll}} = \frac{\{\exp[-t/(D_1\tau_1)^{1/2}] - \exp(-\alpha t)\}[\alpha - 1/(D_2\tau_2)^{1/2}]}{\{\exp[-t/(D_2\tau_2)^{1/2}] - \exp(-\alpha t)\}[\alpha - 1/(D_1\tau_1)^{1/2}]} \tag{3.195}$$

where the subscript 1 refers to the polycrystalline case and the subscript 2 to the single crystal case. The dependence of η_{coll} on τ and d is shown in Figure 3.20b, for various interface state densities, in the case of $CdS/CuInSe_2$ thin film solar cells. J_{sc} of a heterojunction solar cell can be expressed as[30]

$$J_{sc} = \frac{\mu F(\phi')}{S_I + \mu F(\phi')}\int_\lambda \phi_0(\lambda)T_g R'(\lambda)A'(\lambda)\eta_{\text{coll}}(\lambda,\tau)\,d\lambda \tag{3.196}$$

where F is the electric field at the junction, ϕ' is the photon flux at the junction, ϕ_0 is the incident photon flux, T_g is the optical transmission of the top electrode, $R'(\lambda)$ and $A'(\lambda)$ are functions of the reflection and absorption parameters, and S_I is the interface recombination velocity. Figure 3.20c shows J_{sc} as a function of grain diameter for $CdS/CuInSe_2$ devices.

3.3.4.3f. *Resistive Effects.* The fill factor of a solar cell can be degraded by high series resistance and low shunt resistance.[91–95] An

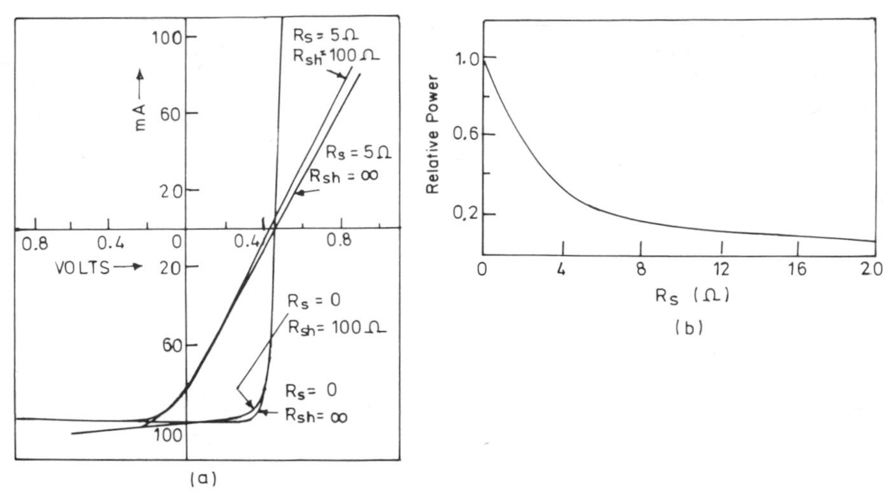

Figure 3.21. (a) Effect of series resistance R_s and shunt resistance R_{sh} on the $I-V$ characteristics of a solar cell. (b) Relative maximum available power from a solar cell as a function of R_s (after Prince[42]).

analytical expression for the fill factor has been derived by Rothwarf[91]:

$$\mathrm{FF} = \mathrm{FF}_0 - CI_{sc}\frac{R_s}{V_{oc}} - \frac{V_m}{V_{oc}}\frac{V_m}{I_{sc}R_{sh}} - \frac{V_m}{V_{oc}}S_I\frac{[1 - F_2(V_m)/F_2(0)]}{S_I + \mu_2 F_2(V_m)} \tag{3.197}$$

FF_0 is the lossless fill factor for an ideal diode as a function of V_{oc}/kT, the second term on the right-hand side is the linearized series resistance loss with R_s the series resistance and C a numerical parameter weakly dependent on V_{oc}/kT. The third term is due to the shunt resistance R_{sh}, which causes a reduction in I_m by V_m/R_{sh}. The fourth term results from the voltage dependence of I_L and is not a typical feature of all cells but has been observed in Cu_2S/CdS and CdTe solar cells. F_2 is the field at the junction.

The series resistance of the cell can arise from the following causes: (i) grid bus bar metal resistance, (ii) grid finger metal resistance, (iii) front surface contact resistance between the metal grid and the semiconductor, (iv) sheet resistance of the semiconductor layer at the surface, (v) base layer bulk resistance, and (vi) back surface contact resistance between the metal and the semiconductor. The effect of series and shunt resistances on the $I-V$ characteristics of solar cells is shown in Figure 3.21a. The maximum power as a function of series resistance[95] is shown in Figure 3.21b. The effect of an adverse series resistance is more prominent at high levels of illumination, whereas the degradation caused by a poor shunt is more pronounced at low illumination levels.[96]

The series resistance can be reduced by proper design of the grid pattern and suitable choice of metals to provide ohmic contacts. Grid design is discussed in greater detail in Appendix C.

3.3.4.3g. *Mode of Operation.* The photocurrent generated in a solar cell also depends on whether the cell is operated in the front-wall (light incident through the absorber) or back-wall (cell illuminated through the window) mode,[74] particularly in heterojunction solar cells. Moreover, by using high-reflection coatings (back electrodes) in either homo- or heterojunctions, the thickness of absorber material required to absorb the photons can be reduced[97,98] and the collection efficiency can thus be increased without increasing the minority carrier diffusion length. By appropriate shaping of the back surface and the use of reflecting electrodes, the light may be trapped by repeated internal reflection.[99] We can thus envisage four modes of operation, namely, (i) front-wall (FW), (ii) back-wall (BW), (iii) front-wall reflected (FWR), and (iv) back-wall reflected (BWR). The calculated spectral response[74] for the four modes of operation (including reflection losses) of a Cu_2S/CdS solar cell are shown in Figure 3.22.

3.3.4.3h. *Junction Roughness.* In thin film solar cells, particularly Cu_2S/CdS, the surface of the cell is very rough and the surface topography is usually made up of pyramids.[100] Consequently, the junction interface is also highly convoluted and nonplanar. The actual area of the junction is,

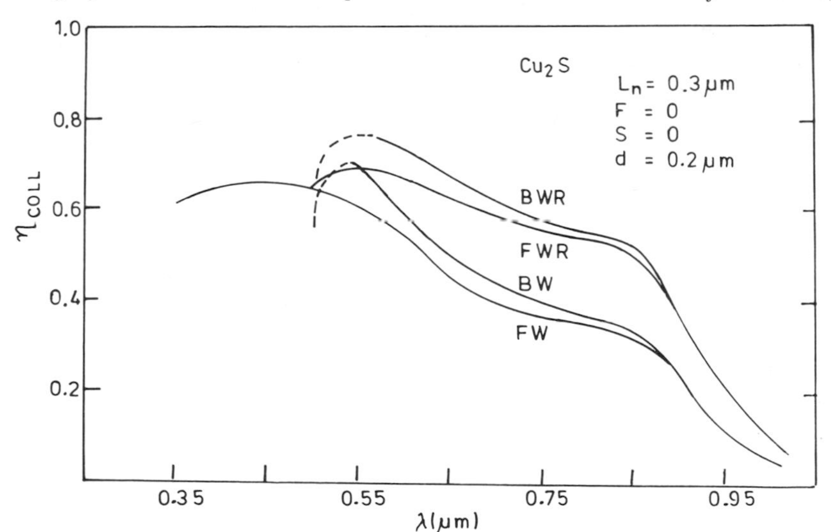

Figure 3.22. Calculated spectral response for four modes of operation of Cu_2S/CdS thin film solar cells (after Rothwarf[74]). *F* and *S* refer to the drift field and surface recombination velocity, respectively, in Cu_2S.

therefore, much larger than the corresponding geometrical flat-plane area. Since such a surface is nonreflecting, J_L is increased. However, the reverse saturation current J_s also increases with the junction surface, lowering V_{oc} and FF. Several workers have studied the influence of the rough-shaped junction profile on the performance of these cells.[74,101–102] The lowering in the V_{oc}, ΔV_{oc}, is given by[74]

$$\Delta V_{oc} = (kT/q)\ln(A_j/A_\perp) \tag{3.198}$$

where A_j is the actual junction area and $A_\perp$ is the perpendicular geometrical area. The calculated effect of grain boundary growth induced rough junction profile on V_{oc} for Cu_2S/CdS solar cells is shown in Figure 3.23. The junction profile model is shown schematically in the inset.

3.3.4.3i. *Heterojunction Interface.* In heterojunctions, a dislocation field at the junction interface, arising from lattice mismatch between the two materials, can counteract a space charge and/or act as a recombination surface. The charge state, capture cross-section, and density of these interface states strongly influence the photocurrent through the junction and determine the magnitude of the field and/or potential jumps. The interface recombination velocity S_I is given approximately by equation (3.58).

Of even greater importance than matching of lattice constants is the

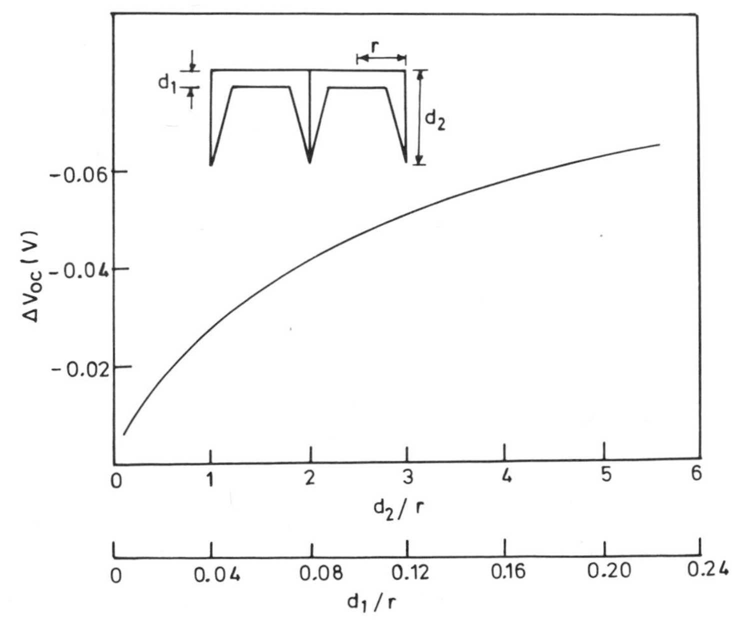

Figure 3.23. Effect of grain boundary growth of Cu_2S on the V_{oc} of a Cu_2S/CdS thin film cell (after Rothwarf[74]).

Table 3.1. Optimum Heterojunction Structures (from Vanhoutte and Pauwels[62])[a]

Heterojunction structure	Type	Heterojunction structure	Type
n-Si(C)/p-CdTe(a)	(i), (ii)	n-InP(C)/p-CdTe(a)	(iii)
n-Si(C)/p-GaAs(a)	(i), (ii)	n^+-In_2O_3(C)/p-$CuInSe_2$(a)	(i), (ii)
n-GaAs(C)/p-InP(a)	(i), (ii)	n-GaAs(C)/p-CdTe(a)	(i), (ii)
n-GaAs(C)/p-Si(a)	(iii)	n^+-In_2O_3(C)/p-InP(a)	(iii)
n-ZnSe(C)/p-InP(a)	(i), (ii)	n-ZnO(C)/p-$CuInSe_2$(a)	(i), (ii)
n-ZnSe(C)/p-CdTe(a)	(i), (ii)	n-ZnO(C)/p-Cu_2S(a)	(i), (ii)
n-ZnSe(C)/p-GaAs(a)	(iii)	n-ZnO(C)/p-InP(a)	(iii)
n-ZnSe(C)/p-Si(a)	(iii)	n^+-SnO_2(C)/p-$CuInSe_2$(a)	(i), (ii)
n^+-ITO(C)/p-InP(a)	(i), (ii)	n^+-SnO_2(C)/p-Cu_2S(a)	(iii)
n^+-ITO(C)/p-CdTe(a)	(i), (ii)	n^+-SnO_2(C)/p-InP(a)	(iii)
n^+-ITO(C)/p-GaAs(a)	(iii)	n-$CuInSe_2$(C)/p-Cu_2S	(iii)
n^+-ITO(C)/p-Si(a)	(iii)	n-CdS(C)/p-$CuInSe_2$(a)	(iii)
n-CdTe(C)/p-Cu_2S(a)	(i), (ii)	n-CdS(C)/p-Cu_2S(a)	(iii)
n-CdTe(C)/p-InP(a)	(i), (ii)	n-CdSe(C)/p-$CuInSe_2$(a)	(iii)
n-CdTe(C)/p-GaAs(a)	(iii)	p^+-AlSb(C)/n-Si(a)	(iv), (v)
n-GaP(C)/p-Cu_2S(a)	(i), (ii)	p-GaAs(C)/n-InP(a)	(vi)
n-GaP(C)/p-InP(a)	(i), (ii)	p-GaAs(C)/n-$CuInSe_2$(a)	(vi)
n-GaP(C)/p-CdTe(a)	(iii)	p-GaAs(C)/n-CdTe(a)	(vi)
n-GaP(C)/p-GaAs(a)	(iii)	p-InP(C)/n-GaAs(a)	(iv), (v)
n-InP(C)/p-$CuInSe_2$(a)	(i), (ii)	p-InP(C)/n-$CuInSe_2$(a)	(vi)
n-InP(C)/p-Cu_2S(a)	(i), (ii)	p-InP(C)/n-CdTe(a)	(vi)
p-CdTe(C)/n-GaAs(a)	(iv), (v)	p-Cu_2S(C)/n-InP(a)	(iv), (v)
p-CdTe(C)/n-InP(a)	(iv), (v)	p-Cu_2S(C)/n-$CuInSe_2$(a)	(iv), (v)
p-CdTe(C)/n-InP(a)	(iv), (v)	p-Cu_2S(C)/n-$CuInSe_2$(a)	(iv), (v)
p-CdTe(C)/n-$CuInSe_2$(a)	(iv), (v)	p-Cu_2S(C)/n-CdTe(a)	(iv), (v)
p-Cu_2S(C)/n-GaAs(a)	(iv), (v)	p-GaP(C)/n-CdSe(a)	(iv), (v)

[a] (C) denotes a collector, (a) an absorber.

matching of electron affinities.[8,62] The barrier height is reduced by the degree of band mismatch according to[8]

$$\phi = E_{g1} - \delta_1 - qV_{D1} - (\chi_2 - \chi_1) \tag{3.199}$$

where χ_1 and χ_2 are the respective electron affinities, δ_1 is the separation between the valence band edge and the Fermi level in the p-type absorber material, and V_{D1} is the diffusion voltage in that material. A lower ϕ leads to a higher J_s and consequently lower V_{oc}. Several authors have proposed optimum heterojunction structures with well matched absorber and window components. Vanhoutte and Pauwels[62] pointed out that the band mismatch at the interface is the more critical factor and that often an insulating layer of a few angstroms is present at the interface which grades one crystal structure into the other such that large lattice mismatch does not necessarily mean high interface recombination velocity. Based on the above considerations, Vanhoutte and Pauwels[62] indicated near optimum heterojunction solar cell structures, which are listed in Table 3.1. The heterojunctions have been classified according to the discontinuities at the interface in the conduction or valence band, and the energy band diagrams are shown in Figure 3.24. Only thin film heterojunctions, actual or potential, have been considered. Two types of optimum structures have been proposed.[63] The first type behaves as a Schottky diode solar cell, with all photons being absorbed in the weakly doped semiconductor, and its efficiency is maximum if this semiconductor is inverted at the interface. In the second type, the photons should be absorbed in the heavily doped semiconductor, and its efficiency is maximum if the energy barrier at the interface is zero.

3.3.4.3j. *Incident Spectrum.* The total short-circuit current density depends sensitively on the intensity and spectral distribution of the incident radiation. Different spectra give rise to different carrier generation profiles and, hence, different photocurrent magnitudes. Moreover, in certain cases, the spectral content may modulate the field at the junction and, consequently, the collection efficiency.[30] The solar spectrum at any particular location depends on the local atmospheric conditions such as total atmospheric mass, water vapor, ozone, dust, and carbon dioxide content. Different solar spectra are given in Appendix A.

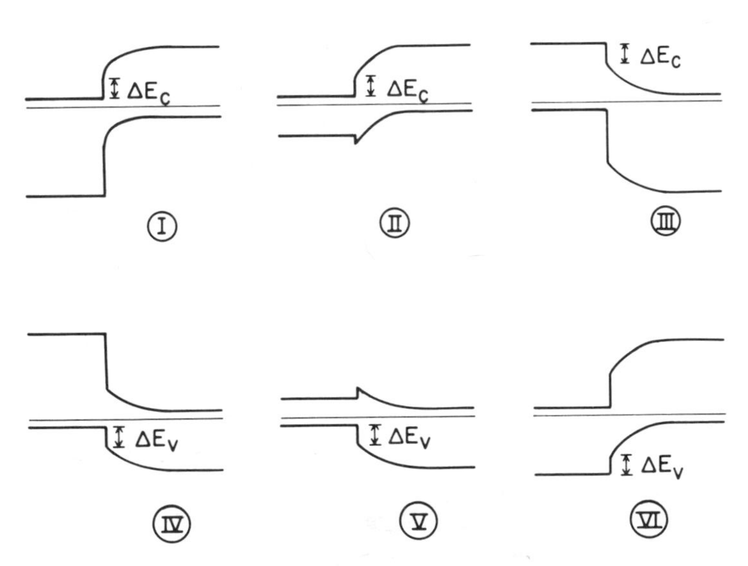

Figure 3.24. Several types of nearly optimum heterojunction structures, with one layer acting as the absorber and the other acting as the collector: (i) n^+/p (absorber), $\Delta E_v > 0$; (ii) n^+/p (absorber), $\Delta E_v < 0$ (MS type); (iii) p^+ (absorber)/n, $\Delta E_c > 0$; (iv) p^+/n (absorber), $\Delta E_c > 0$; (v) p^+/n (absorber), $\Delta E_c < 0$ (MS type); and (vi) n^+ (absorber)/p, $\Delta E_v > 0$ (after Vanhaotte and Pauwels[62]).

3.3.4.4. *Loss Mechanisms*

From the preceding discussion, it is clear that the design and optimization of a photovoltaic device of a particular material requires a delicate balance between several conflicting requirements. Further, since the basic material properties differ from material to material, no single design is applicable to all photovoltaic systems and each device design has to be optimized in consonance with the material properties. Despite optimization of a device, the efficiencies achieved for practical devices fall short of the theoretically predicted performance. This mismatch between experimental results and theoretical predictions arises because of the various losses that occur in a solar cell. These losses can be divided into two groups: basic loss mechanisms, which are fixed and cannot be reduced for a given solar cell material, and design losses against which certain improvements can be expected. The various loss mechanisms can be classified as: (i) photon losses, (ii) carrier losses, (iii) voltage losses, and (iv) power losses. In Table 3.2, we have enumerated these losses and indicated their causes. Figure 3.25 gives bar charts of AM1 energy losses in typical Si and Cu_2S/CdS solar cells.[2]

3.3.4.5. *Design Criteria*

Based on considerations enumerated in Sections 3.3.4.3 and 3.3.4.4, we can provide certain general rules for design requirements in different types of cells.

3.3.4.5a. *Homojunctions.* The rear contact should be ohmic with sufficient current-carrying capacity and good reflection of long wavelengths. The base region should possess long minority carrier diffusion length, appropriate thickness for sufficient photon absorption, and proper back surface field to reduce surface recombination. Heavy doping is favorable for the top layer; however, the lifetime should not be severely affected. Moreover, low surface recombination velocity and sufficient thickness to prevent undue series resistance are desirable. The top contact grid should have sufficient number of grid lines to reduce R_s, yet maintain high optical transmission and good ohmic contact. An antireflection coating (see Appendix B) or a textured surface is imperative to reduce reflection losses. Finally, the encapsulant material should be transparent in the active range of cell and solar spectrum and should be impermeable to water vapor and oxygen.

3.3.4.5b. *Heterojunctions.* The substrate (in cases of thin fim devices) should have a thermal expansion coefficient compatible with the base

Table 3.2. Energy Loss Mechanisms in Solar Cells (after Rothwarf and Boer[2])

Region	Mechanisms	Form of energy lost
Photon losses		
Protective layers Antireflection layers	Reflection Absorption	Photon radiation Heat generation
Grid	Shading	Photon radiation and heat generation
Active layers	Reflection at front and back surface	Photon radiation
	Absorption at back contact	Heat generation
	Nongeneration free carrier absorption and/or transitions among defect levels	Heat generation
	Incomplete utilization of photon energy $E = h\nu - E_g$ is wasted	Heat generation
Carrier Losses		
Front and rear surfaces	Surface recombination	Heat generation
Active layers	Bulk recombination	Heat generation
Junction	Recombination of light generated carriers	Heat generation
	Recombination of forward current carriers	Heat generation
Power Losses		
Active layers	Resistive ohmic losses	Heat generation
Contacts	Resistive ohmic losses	Heat generation

semiconductor material, good thermal conductivity, and, for back-wall cells, should be transparent in the solar spectrum. The rear contact should have good reflection for long-wavelength photons and good lattice parameters to facilitate proper nucleation of base material. In case of back-wall cells, the rear contact should be transparent and highly conducting. The base material should have a long minority carrier diffusion length if it absorbs the incident photons significantly and sufficient thickness to

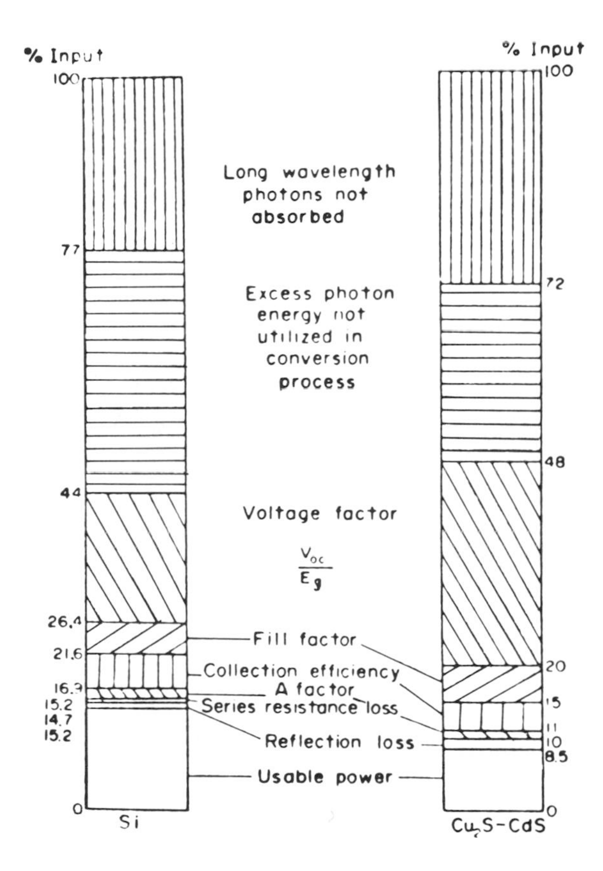

Figure 3.25. Bar chart of AM1 losses for Si and Cu_2S/CdS solar cells (after Rothwarf and Boer[2]).

prevent shorts through grain boundaries. The layer should be properly doped to allow development of the desired space-charge region and a good electron affinity match with the top absorber layer. The absorber layer should have low surface recombination, long minority carrier diffusion length, and sufficient thickness to absorb completely near bandgap photons after reflection from the rear contact. Moreover, the sheet resistance should be low to avoid series resistance and/or gridding problem and the lattice parameters should be closely matched with those of the base layer. The requirements for the grid contact, antireflection coating, and encapsulant are the same as for homojunction cells.

Thus far we have discussed homojunction and heterojunction photovoltaic devices. We now proceed to a description of CIS solar cells.

3.3.4.6. *CIS Solar Cells*

In this section, we examine the photovoltaic operation of CIS devices. An excellent review[103] of CIS solar cells has appeared recently in the literature, and the following discussion is extracted from this review.

The CIS photovoltaic device incorporates an ultrathin ($\sim$10–30 Å) interfacial layer in a metal/semiconductor junction (Schottky diode) or oxide-semiconductor/base-semiconductor heterojunction diode. In MIS solar cells, the top layer is either a thin conducting metal film or a metallic compound, e.g., $(SN)_x$. In SIS solar cells, the top layer is a transparent, conducting semiconductor (e.g., SnO_x : Sb), which is degenerate and has a wide bandgap ($>$3.2 eV) to ensure negligible photogeneration of carriers in the top layer. In an optimized device, the top contact is determined by its work function and the conductivity type of the base-semiconductor. The interfacial layer can be a naturally grown or a deposited dielectric.

The simple equilibrium energy band diagrams for *p*-type MIS and SIS solar cells are shown in Figure 3.26a and b. E_{gos}, E_{gi}, and E_{gs} are the bandgaps of the oxide-semiconductor, insulator, and base-semiconductor, respectively. The energy difference between the conduction band edges of the base-semiconductor and the insulator is denoted by ϕ_{si}. The metal to insulator and oxide-semiconductor to insulator barrier height, ϕ_{mi} and ϕ_{osi},

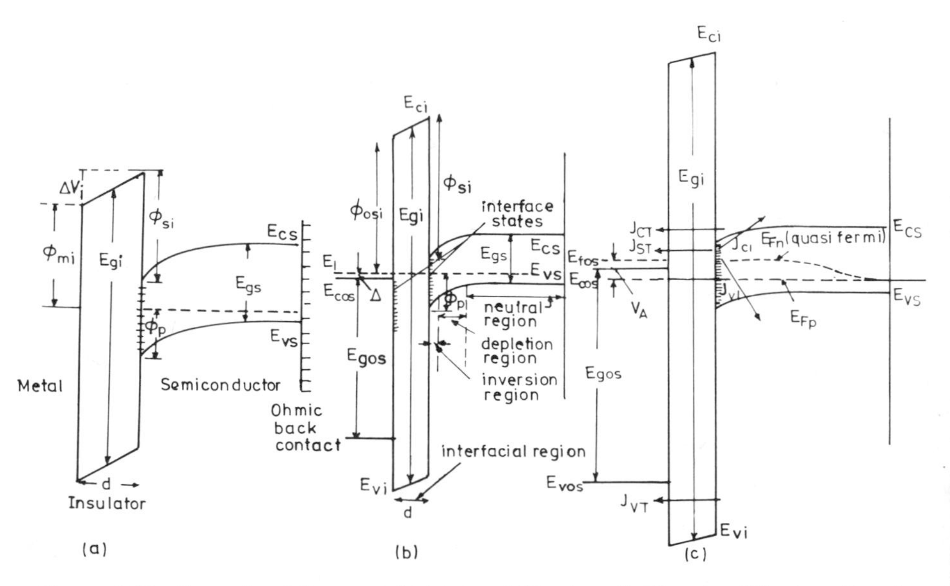

Figure 3.26. Energy band diagrams of (a) *p*-type MIS solar cell, (b) *p*-type SIS solar cell, and (c) *p*-type SIS solar cell in dark and biased positively with respect to the top layer (after Singh et al.[103]).

respectively, are related to the work function of the top layer. The degree of degeneracy determines whether the Fermi level in the oxide-semiconductor is above or below the conduction band edge. Δ denotes the difference between the conduction band edge and the Fermi level in the oxide-semiconductor. The barrier height ϕ_p governs the open-circuit voltage of the device.

As shown in the Figure 3.26, the surface of the p-type base-semiconductor can be inverted at zero bias if the work function of the top contact is less than or equal to the electron affinity of the base-semiconductor. For n-type base-semiconductors, inversion at the surface may be ensured by keeping the top layer work function higher than or equal to the sum of the electron affinity and bandgap of the base-semiconductor. With an inverted semiconductor surface, the dominant dark current in CIS diodes, near zero bias, will be due to minority carriers. Thus, if the interfacial layer is thin and the top layer is such as to produce inversion at the base-semiconductor surface, the electronic properties of the CIS device will be identical to an abrupt one-sided p/n junction. This then is the advantage of a CIS diode: without conventional doping by diffusion or ion-implantation, the conductivity of the base-semiconductor near the surface is altered by suitable choice of the top layer and the base-semiconductor.

The transport through the thin interfacial layer is expected to be predominantly by tunneling. The device output characteristics are not controlled by the interface between the top contact and the interfacial layer, which is, therefore, neglected in device modeling. The energy band diagram of a p-type SIS solar cell in dark, biased positively with respect to the top layer, is shown in Figure 3.26c. The current flow from the conduction and valence bands of the base-semiconductor to the top contact are represented by $J_{c\mathrm{T}}$ and $J_{v\mathrm{T}}$, respectively. $J_{s\mathrm{T}}$ denotes the net current flow to the top contact from localized surface states at the semiconductor/insulator interface and $J_{c\mathrm{I}}$ and $J_{v\mathrm{I}}$ are the effective current flows due to the interchange of charges between the interface states and the conduction and valence bands, respectively, of the base-semiconductor from recombination–generation processes. Under conditions of inversion, the photovoltaic conversion efficiencies of the MIS and SIS diodes are identical and the results for both can be obtained by the following substitution:

$$\phi_{\mathrm{mi}} = \phi_{\mathrm{osi}} \pm \Delta \qquad (3.200)$$

A detailed description of the optimization of CIS solar cells is given elsewhere.[104,105] The performance of CIS solar cells is determined mainly by the parameters discussed below.

3.3.4.6a. ***Interfacial Layer Thickness.*** As noted above, the appropriate interfacial layer thickness, particularly for Si CIS devices, is from about 10 to 30 Å. Beyond the upper limit, the Si CIS diode would revert to the equilibrium mode operation(106) owing to tunneling current, and beyond the lower limit, Schottky diode or heterojunction diode behavior would set in.(107) The dark $I-V$ characteristics computed for an MIS diode(106) with interfacial layer thickness as a parameter are shown in Figure 3.27a. With decrease in the insulator thickness, the characteristics tend to converge to an asymptotic curve, in both forward and reverse bias, taking the form of the ideal Shockley diode [equation (3.31)]. This convergence of the reverse

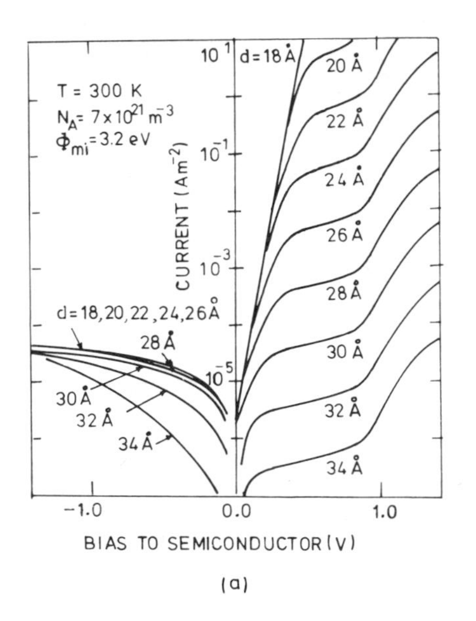

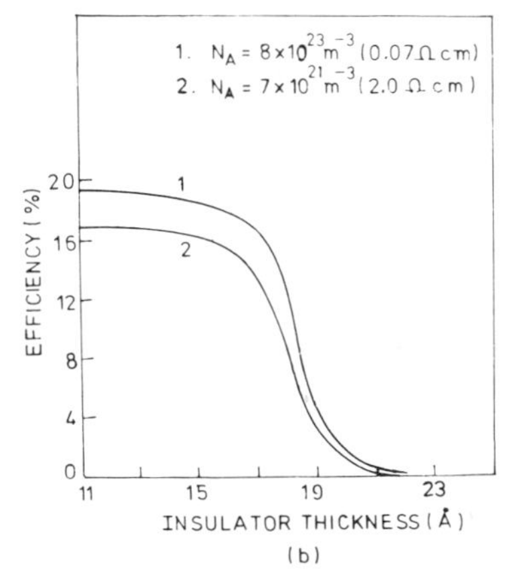

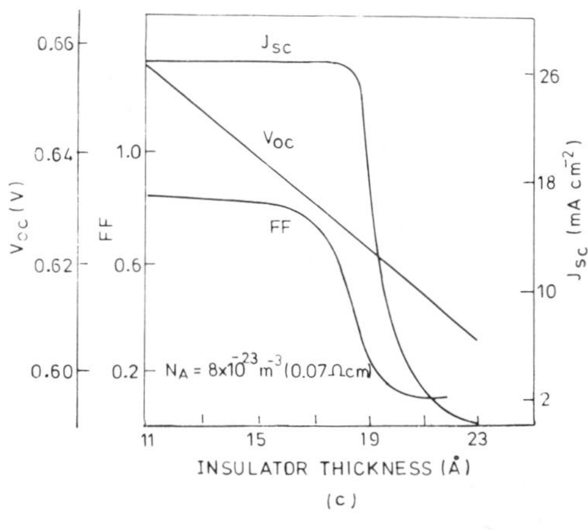

Figure 3.27. (a) Calculated dark $I-V$ characteristics of an MIS device; (b) efficiency of an MIS solar cell as a function of insulator layer thickness; and (c) J_{sc}, FF, and V_{oc} as functions of insulator thickness (after Singh et al.(103)).

characteristics on the Shockley diode curve at about 28 Å indicates the transition from the equilibrium to the nonequilibrium mode of operation. In forward bias, the departure from the ideal diode locus occurs at higher biases with decreasing interfacial layer thickness and indicates a changeover from semiconductor-limited to tunneling-limited operation. The normal AM1 short-circuit current density of Si p/n junction solar cells is about 40 mA cm^{-2}. Thus, from Figure 3.27a, we would not expect a significant conversion efficiency for insulator thicknesses much above 20 Å, since a large degree of photocurrent suppression would occur because of the $I-V$ characteristic, similar to that observed in p/n junction solar cells with a large series resistance. Indeed, we may consider tunnel-limited regions as equivalent to series resistance. Similar calculations for an SIS diode[108] show that for a solar cell with J_{sc} equal to about 40 mA cm^{-2}, interfacial layers of less than 16 Å are required to avoid tunnel-limited operation.

The conversion efficiency of MIS solar cells as a function of insulator thickness for two typical values of substrate resistivity is shown in Figure 3.27b. For an interfacial layer thickness greater than about 20 Å, the efficiency is very low but rises rapidly with decreasing thickness and then saturates at about 15 Å.[109] In the region from 15 to 20 Å, the decrease in efficiency with thickness of the interfacial layer can be attributed to an increase in the effective series resistance arising from: (i) the exponential change in tunnel current with thickness and (ii) the change in the $I-V$ characteristics owing to the transition from the semiconductor-current-limited to the tunnel-current-limited mode. This conclusion is supported by the calculations for J_{sc} and FF, illustrated graphically in Figure 3.27c, both of which are markedly similar in form to the efficiency curves. The saturation of the efficiency below about 15 Å is explained by the fact that at that point tunneling occurs at a sufficiently high rate such that the effective resistance via this process becomes insignificant and all possible current from the semiconductor is extracted.

The increase in the efficiency with an increase in doping, as shown in Figure 3.27b, is attributed to a decrease in the reverse saturation current density leading to a concomitant increase in V_{oc}. The slow and apparently linear variation of the V_{oc} with insulator thickness can be explained by considering the expression for the barrier height ϕ_p that controls V_{oc}:

$$\phi_p = E_{gs} + \phi_{si} - \phi_{mi} - F_s d + \frac{q(Q_{ss} + Q_i)d}{\varepsilon_i} \tag{3.201}$$

Q_{ss} and Q_i denote the charge of surface states and the interfacial insulator, respectively. F_s is the field at the interface, d is the interfacial layer thickness, and ε_i is its dielectric constant.

3.3.4.6b. *Work Function of the Top Layer.* From equation (3.201), it is clear that to obtain the maximum barrier height for a *p*-type base-semiconductor, the top layer work function should be less than the electron affinity of the base-semiconductor. Conversely, for an *n*-type base-semiconductor, the top layer work function should be greater than the electron affinity of the base-semiconductor. It should be pointed out that the work function depends on the surface properties of the material, as well as on the method of formation. However, it is the bulk values that are usually used in calculations. For *p*-type MIS solar cells, Al, Cr, Ta, Ti, Mg, and Be are possible metals, whereas Ag, Au, Pt, and Pd are appropriate for *n*-type MIS solar cells. The work function of metals has been compiled by Michaelson[110] (see Table 6.4). In SIS solar cells, ZnO is appropriate for *p*-type base-semiconductors and SnO_x is suitable for *n*-type base-semiconductors. Indium tin oxide (ITO) has been found suitable for fabricating SIS solar cells both on *n*- and *p*-type base-semiconductors, depending on the method of deposition. Several other oxide semiconductors are known, e.g., CdO, Bi_2O_3, and Tl_2O_3, but either their bandgap is not suitable or the work function is inappropriate.

The dependence of the calculated η, J_{sc}, V_{oc}, and FF of MIS solar cells on the metal work function is shown in Figure 3.28a for a *p*-type base-semiconductor. For ϕ_{mi} less than 2.7 eV, the efficiency saturates. Above ϕ_{mi} equal to 3.4 eV, the efficiency decreases rapidly owing to the reduced barrier height, and at $\phi_{mi} \geqslant 3.9$ eV, these become majority carrier diodes, unsuitable for photovoltaic purposes. The slight increase in efficiency for ϕ_{mi} less than 3.2 eV is due to a slight increase in V_{oc}. For $\phi_{mi} > 3.2$ eV, the barrier height ϕ_p is less than that required for optimum operation and V_{oc} decreases. The fill factor decreases with an increase in metal work function owing to an increase in the dark current of the device as well as in the diode factor. For $\phi_{mi} > 3.4$ eV, FF decreases rapidly because of the substantial decrease in V_{oc}. J_{sc} does not increase for $\phi_{mi} \leqslant 3.2$ eV, since essentially all the semiconductor current is being extracted. Up to $\phi_{mi} \approx 3.6$ eV, the majority carrier current near zero bias is small and, hence, J_{sc} is constant in this range. For $\phi_{mi} > 3.6$ eV, J_{sc} decreases because of increased recombination at interface states and decreased minority carrier collection efficiency owing to the reduced barrier height.

Apart from the fact that the efficiency of a *p*-type MIS solar cell does not increase substantially with $\phi_{mi} < 3.2$ eV, the main advantage of a low-work-function metal is that the device performance is less sensitive to defect density and insulator charge, since lower values of ϕ_{mi} favor the formation of a minority carrier diode, even at high doping levels. The surface states are heavily occupied by the large number of minority carriers at the surface, thus effectively reducing recombination at the surface states.

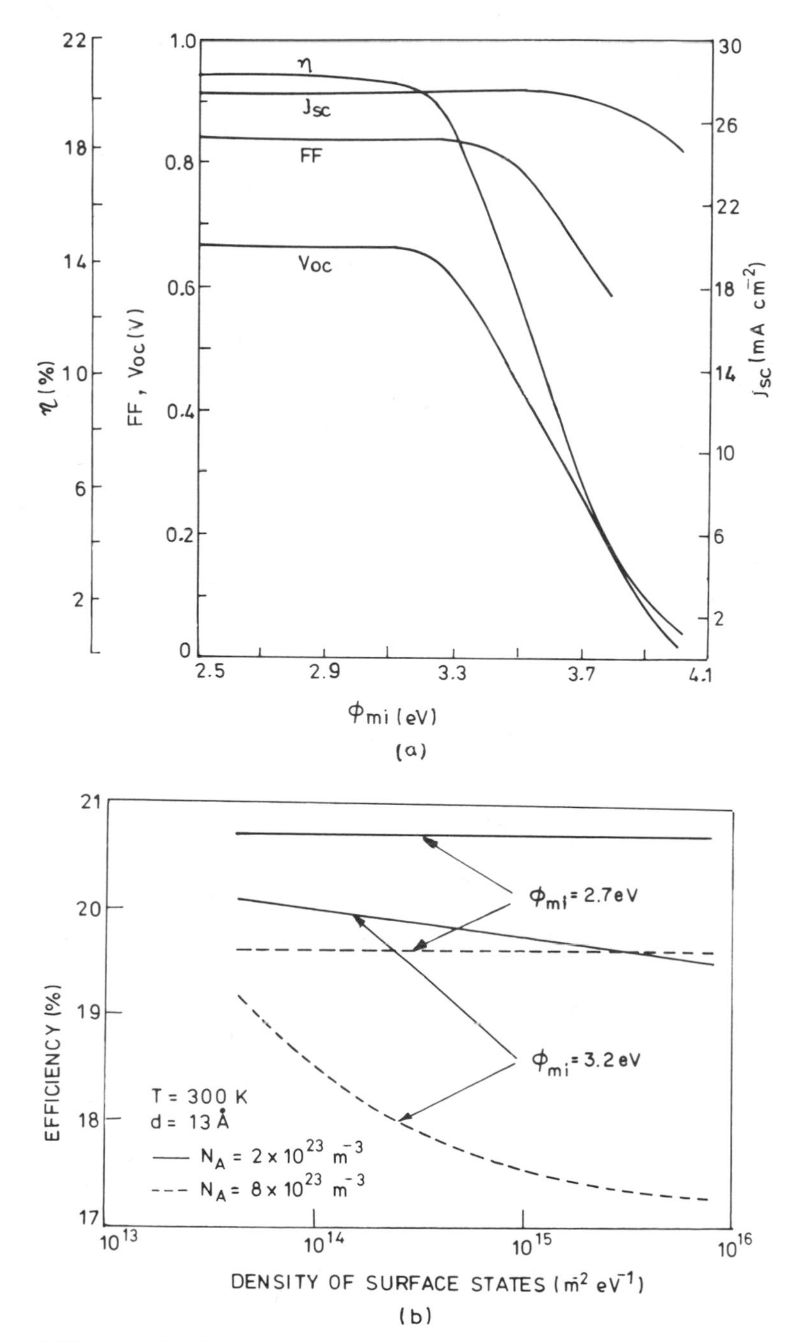

Figure 3.28. (a) η, J_{sc}, V_{oc}, and FF as functions of metal work function ϕ_{mi} for an MIS solar cell. (b) Effect of surface state density on the conversion efficiency (after Singh et al.[103]).

The effect of surface states on the efficiency of the device is shown in Figure 3.28b with ϕ_{mi} and substrate doping density as parameters. It is clear that, for low ϕ_{mi}, the surface state density has a negligible effect on conversion efficiency. Thus, with a low-work-function metal, minority carrier diodes can be formed even with lower resistivity substrates, resulting in MIS solar cells with efficiencies equal to those predicted for ideal p/n junctions.

3.3.4.6c. *Base-Semiconductor Doping.* The doping density affects the various transport parameters. The calculated dependence of η, J_{sc}, V_{oc}, and FF of MIS solar cells on the base-semiconductor doping density is illustrated in Figure 3.29. V_{oc} increases with an increase in doping level because of a decrease in J_s. Beyond a certain doping density ($\sim 4 \times 10^{17}$ cm^{-3}), V_{oc} decreases, since the top layer Fermi level is no longer pinned to the minority carrier quasi-Fermi level.

J_{sc} initially remains constant and then decreases slowly with doping level, owing to the degradation in the diffusion length of minority carriers. At high doping levels, the device becomes unpinned and J_{sc} starts

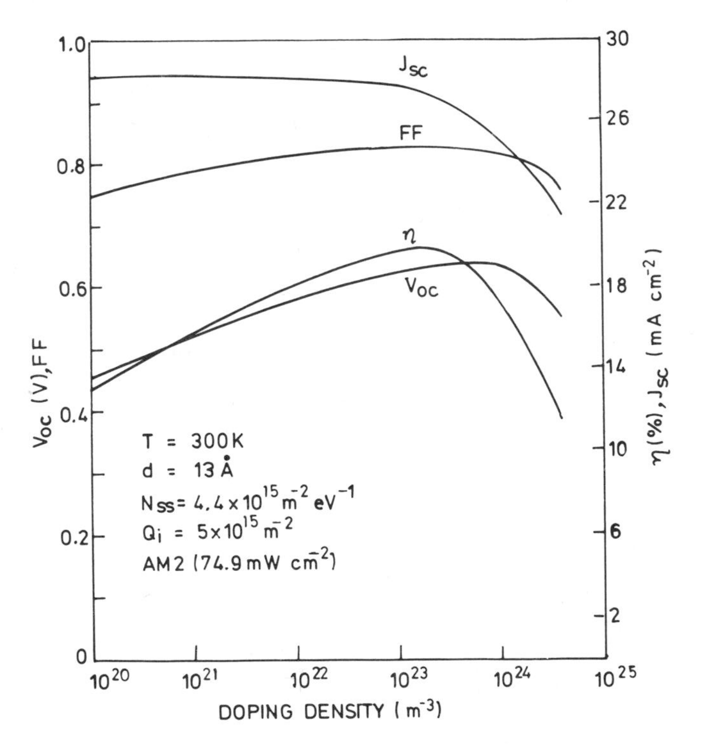

Figure 3.29. Effect of base layer doping density on J_{sc}, V_{oc}, FF, and η of an MIS solar cell (after Singh et al.[103]).

decreasing substantially. The fill factor initially rises and then decreases with doping level, following the behavior of V_{oc}. The increase in the diode factor to values higher than unity also affects FF adversely. However, in the present case, the V_{oc} effect is predominant.

The net effect of the behavior of V_{oc}, J_{sc}, and FF with doping level is a peak efficiency at about 2×10^{17} cm^{-3}. Beyond this doping density, η decreases because of the degradation in V_{oc}, J_{sc}, and FF. We should mention that the value of the doping level at which peak efficiency occurs depends on the top layer work function.

3.3.4.6d. *Surface States and Insulator Charge.* A positive surface charge and donor-type surface states have a beneficial effect on the performance of p-type MIS solar cells. Conversely, a negative surface charge and acceptor-type surface states will enhance the performance of n-type MIS solar cells. As seen from equation (3.201) above, both surface states and insulator charge can affect V_{oc}. J_{sc} may also be affected by a large density of surface states.

Figure 3.30a, which shows the effect of surface states and surface charge on the efficiency, indicates that for insulator thicknesses less than 14 Å, maximum conversion efficiency is obtained for minimum surface states and surface charge. This assumes that the surface state density N_{ss} and the surface charge are interrelated. Otherwise, the photovoltaic efficiency of

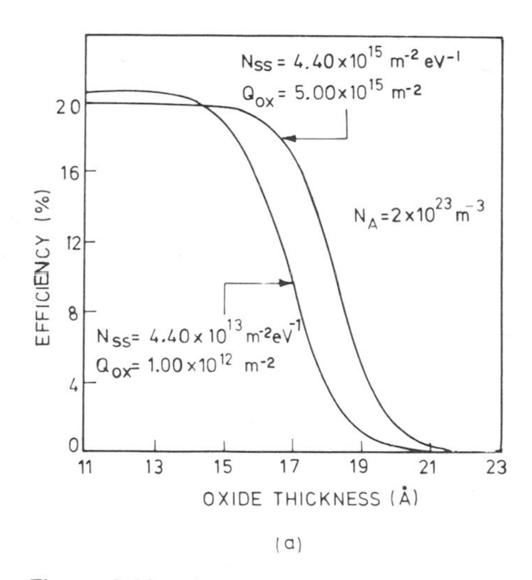

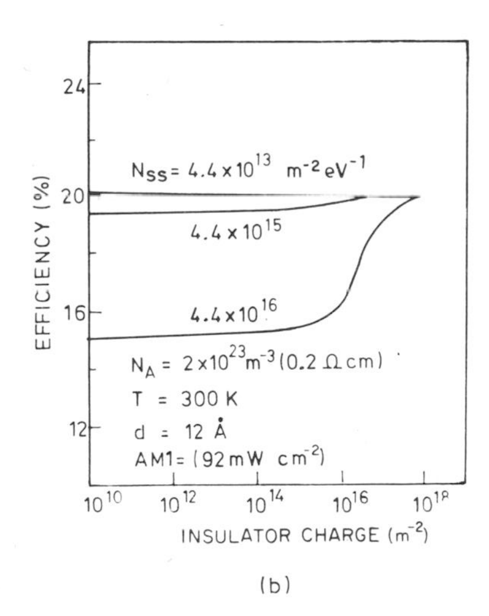

Figure 3.30. (a) Efficiency of an MIS device as a function of insulator thickness for different values of surface state density and surface charge. (b) Efficiency of an MIS device as a function of insulator charge, with the surface states density as a variable parameter (after Singh et al.[103]).

p-type MIS solar cells increases with a higher positive value of insulator charge. It is only when both the surface state density and the surface charge have negligible contributions that minimum reduction in V_{oc} is possible. For insulator thicknesses less than 14 Å, J_{sc} is not significantly affected by the surface states and changes in η arise from changes in V_{oc}. For insulator thicknesses greater than 14 Å, a large surface state density results in a higher conversion efficiency, since the tunneling component of the current is supplemented by a recombination–generation current. It should be noted, however, that this higher efficiency is still less than the maximum obtainable efficiencies for MIS solar cells.

The dependence of η on insulator charge for different values of surface state density is shown in Figure 3.30b. The lowest efficiency occurs for a combination of low surface charge and high surface state density. In the case of low surface state density, η is not significantly affected by the higher value of insulator charge. The saturation is attributed to the fact that maximum barrier heights are obtained at intermediate values of surface state density and surface charge, and V_{oc} is not increased by any increase in surface charge. We should note that the dynamic effects of surface states are negligible for inversion layer devices.[111] However, in the absence of an inversion layer, the surface states control the device properties if their density exceeds 10^{16} $m^{-2} eV^{-1}$.

3.3.4.6e. *Crystallographic Orientation.* The density of surface states depends on crystallographic orientation ($N_{ss}\langle 111\rangle > N_{ss}\langle 110\rangle > N_{ss}\langle 100\rangle$) as do the tunneling probability and tunnel currents. However, as noted above, for optimum insulator thickness, the device current is semiconductor-limited and, therefore, the effect of surface states as a function of crystal orientation will predominate over the effect of tunnel currents. Also, for *p*-type Si, the minority carrier lifetime is a maximum for the ⟨100⟩ orientation and, therefore, the ⟨100⟩ orientation will yield the highest conversion efficiency for *p*-type Si-based CIS solar cells. However, in *n*-type Si, the minority carrier lifetime is maximum for the ⟨111⟩ orientation,[112] but in most cases, the advantage of higher carrier lifetime more than compensates for the disadvantage resulting from higher surface state density. Thus, the ⟨111⟩ orientation is preferable for *n*-type Si-based solar cells.

3.3.4.6f. *Bandgap.* Higher efficiencies can be obtained with semiconductors other than Si. InP- and GaAs-based CIS solar cells can exhibit up to 24 and 25% efficiencies, respectively, at AM1,[105] with a suitable top layer work function and insulator thickness. It may be noted that the range of useful insulator thickness increases with an increase in the bandgap of the base-semiconductor. This is due to the fact that the photocurrent remains unaltered by the presence of the interfacial layer provided that the rate at which minority carriers arrive at the interface is lower than the rate that can be supported by the tunneling process. Since electron–hole pairs in the base-semiconductor can be created only by photons with energy $h\nu > E_g$, the rate at which minority carriers arrive at the interface will be lower in the high-bandgap semiconductor.

3.3.4.6g. *Temperature.* Temperature effects are governed mainly by the transport properties of the semiconductor in CIS solar cells with ultrathin interfacial layers. Further, the tunneling barrier depends on the temperature. As noted earlier, at low forward bias, the current flow in dark is semiconductor-limited and the temperature dependence is similar to a p/n junction device. In the low-bias region, the diode current is limited by recombination processes in the depletion region and n is approximately equal to 2. In the intermediate-bias regions, current flow is predominantly due to minority carrier diffusion and n approaches 1. At still higher forward bias, the device performance changes from a semiconductor-limited to a tunnel-limited mode. The dark I-V characteristics in reverse bias follow a behavior similar to a p/n junction. At lower temperatures, recombination–generation is predominant and at higher temperatures current flow is due to diffusion.

The calculated temperature dependences of η, V_{oc}, J_{sc}, and FF for a p-type SIS solar cell[108] are shown in Figure 3.31. The linear decrease of V_{oc} with increasing temperature can be explained on the basis of an increase in the dark current with temperature. J_{sc} remains practically constant with temperature since the diffusion length and absorption coefficient are assumed to be constant in the range of temperature considered. However, if the temperature dependence of the absorption coefficient of Si[113] is taken into account, J_{sc} will increase slightly with temperature. The fill factor decreases with temperature in a fashion similar to that in p/n junction solar cells. The efficiency decreases linearly with temperature owing to the decrease of V_{oc} and FF, the improvement in J_{sc} being overshadowed. Thus, the temperature dependence of CIS solar cells is similar to that of p/n junction solar cells.

3.3.4.6h. *Intensity.* The calculated efficiencies of p-type Si-based CIS solar cells as a function of illumination intensity is shown in Figure 3.32, for various interfacial layer thicknesses. Losses resulting from reflection from the top layer have been taken into account.[105] The increase in the optimum intensity with decrease in the interfacial layer thickness can be explained by the fact that with increase in intensity, the device characteristics shift from a semiconductor-limited to a tunnel-limited mode. For thicker oxides and at high levels of illumination, a large degree

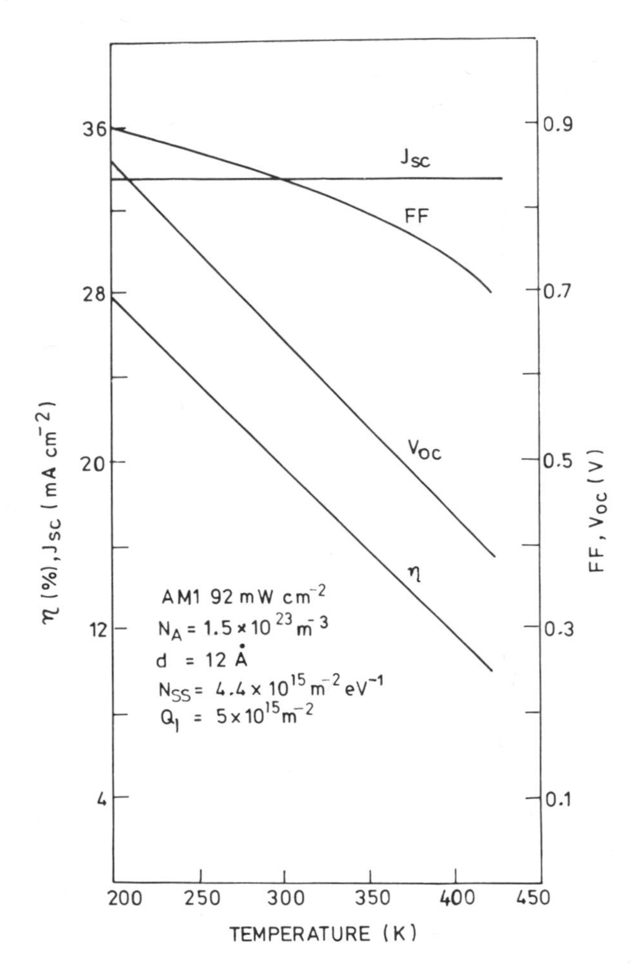

Figure 3.31. Temperature dependence of η, V_{oc}, J_{sc}, and FF of p-type SIS solar cells (after Singh et al.[103]).

of photocurrent suppression is observed, as in a p/n junction solar cell with large series resistance. At low intensity levels, the resistance offered by the interfacial layer, which is nonlinear in nature, is like an ordinary series resistance. It should be noted that it is possible to control the oxide layer thickness[103] to ± 1 Å, and if the top layer and back contact do not contribute appreciably to the series resistance, CIS solar cells can be profitably operated at high intensities.

3.3.4.6i. *Spectral Response.* Green[114] has developed a quantitative treatment for the spectral response of MIS cells. A fall-off in the short-wavelength response of the cell can occur with certain combinations of insulator thickness, surface state density, and barrier height giving rise to a dead layer extending up to about 750 Å. However, if the minority carrier concentration at the base-semiconductor surface is large and the surface strongly inverted, the above effects become negligible and all the carriers generated near the surface can be collected. In such a case, the presence of the thin interfacial layer not only enhances the open-circuit voltage but also improves the short-wavelength response.

3.3.4.7. *Inversion Layer Grating MIS Cells*

An alternate approach to the conventional MIS structure is the inversion layer (IL), grating MIS cell.[116,117] The basic structure of an IL cell consists of a sintered aluminum back contact on a p-Si substrate, a photolithographically defined Al MIS front contact in the form of a grating, and a thick oxide/antireflection coating. The oxide, containing a high density of positive charge, inverts the p-Si surface thereby inducing an ideal

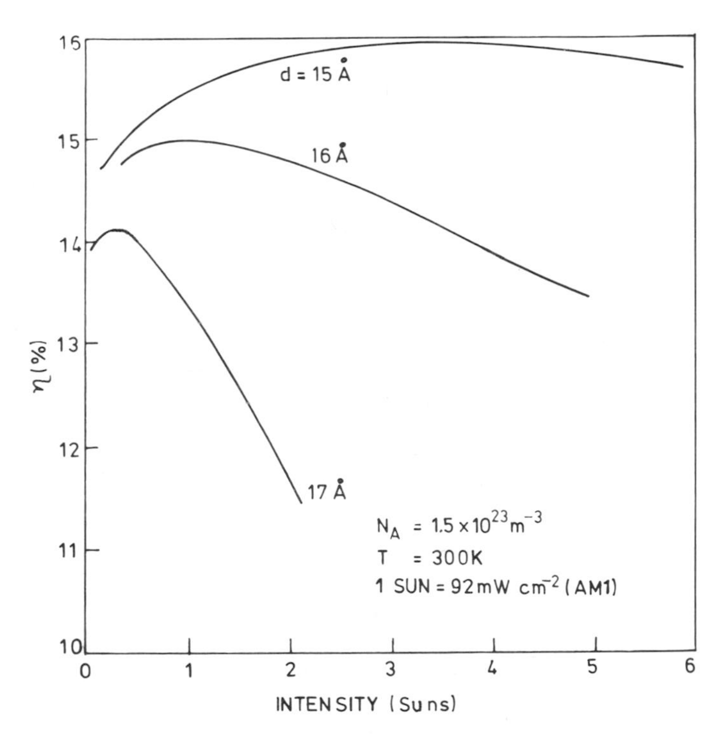

Figure 3.32. Illumination intensity dependence of the efficiency of a *p*-type SIS solar cell (after Singh et al.[103]).

Table 3.3. Photovoltaic Characteristics of Various Bulk Solar Cells (from Singh et al.[103] and Bucher[115])[a]

Active material	Device configuration	V_{oc} (V)	J_{sc} (mA cm^{-2})	FF	η (%)	Remarks
Homojunctions						
Si (S)	*p*-Si/*n*-Si	0.59	46		15.5	Ta_2O_5 AR; "Comsat" nonreflecting (CNR) cell
Si (S)	*p*-Si/*n*-Si				⩾20	Ta_2O_5AR; CNR cell; under terrestrial conditions
Si (S)	*p*-Si/*n*-Si				~18	Ta_2O_5 AR; "Violet" cell
Si (S)	*p*-Si/*n*-Si ($p^+/p/n^+$ diode)	0.61	4.29		15.2	Ta_2O_5 AR; "Helios" cell
Si (S)	*n*-Si/*p*-Si (n^+/graded $n/p/p^+$ epi.)	0.636	24.5	0.79	12.6	Al_2O_3 AR
Si (S)	*p*-Si/*n*-Si	0.53	29.7		10.7	No AR; inversion layer cell
Si (S)	*n*-Si/*p*-Si ($n^+/n/p^+$)	0.60	37.1	0.775	17.2	TiO_2, SiO_2, ZrO_2 AR
Si (R)	*p*-Si/*n*-Si	0.572	27.4	0.754	11.8	EFG ribbon
Si (W)	*p*-Si/*n*-Si	0.55	35.0	0.753	10.7	Web process
Si (P)	*p*-Si/*n*-Si	0.551	37.8	0.755	11.6	TiO_x AR; grain size ~10^3 μm
Si (P)	*p*-Si/*n*-Si	0.57	25.1		10	Grain size ~2–3 mm
GaAs (S)	*p*-GaAs/*n*-GaAs	0.97	25.6	0.81	20.5	Anodized n^+ layer + SiO + MgF_2 AR
GaAs (S)	*n*-GaAs/*p*-GaAs				13–14	With AR
GaAs (S)	*n*-GaAs/*p*-GaAs/*p*-$Ga_{1-x}Al_xAs$	0.976	27.8	0.76	21.9	TiO_2 AR
GaAs (S)	*n*-GaAs/*p*-GaAs/*p*-$Ga_{1-x}Al_xAs$	1.015	33.1	0.745	18.5	TiO_2 AR
Heterojunctions						
Si (S)	*p*-Si/*n*-CdS	0.49	24	0.61	7.2	No AR
Si (S)	*n*-Si/*p*-Cu_2S	0.42	20	0.6	4.2	No AR; evap. Cu_2S
GaAs (S)	*p*-GaAs/*n*-AlAs	0.78	28	0.82	18.5	Anodized AlAs Oxide AR
GaAs (S)	*n*-GaAs/*p*-GaAs/*p*-AlAs	1.00	23	0.80	13.5	SiO_x AR
GaAs (S)	*p*-GaAs/*n*-$Ga_{0.5}Al_{0.5}As$	0.88	16.9	0.77	13.5	
GaAs (S)	*p*-GaAs/*n*-$Ga_{1-x}Al_xAs$	0.88	27.7	0.76	13.6	Si_3N_4 AR
GaAs (S)	*n*-GaAs/*p*-GaP	0.5			8	
GaAs (P)	*n*-GaAs/*p*-$Cu_{1.8}Se$	0.54	15	0.52	4.26	Krylon AR
InP (S)	*p*-InP/*n*-CdS	0.79	18.7	0.75	15.0	SiO_x AR
InP (S)	*p*-InP/*n*-CdS	0.807	18.6	0.74	14.4	SiO_x AR
CdTe (S)	*p*-CdTe/*n*-CdTe/*n*-CdS	0.67	20.4	0.60	10.5	No AR
CdTe (S)	*p*-CdTe/*n*-CdS	0.63	16.1	0.658	7.9	Clycerol AR; evap. CdS

$Cu_{2-x}S$ (S)	n-CdS/p-$Cu_{2-x}S$				8	Cu_2S by solid-state reaction
Cu_2S (P)	n-$Cd_xZn_{1-x}S$/p-Cu_2S	0.64	12.7	0.623	6.29	No AR
CdTe (S)	n-CdTe/p-$Cu_{2-x}Te$	0.75			$\leqslant 7.5$	No AR
$CuInSe_2$ (S)	p-$CuInSe_2$/n-CdS	0.49	38	0.60	12	SiO AR
CIS Junctions						
Si (S)	SiO/Al (g)/SiO_2/p-Si or SiO/Mg(g)/SiO_x/p-Si	0.621	36.5	0.81	18.3	SiO AR
Si (S)	SiO/Cr(t)/SiO_x/p-Si	0.570	30.9	0.68	12.0	SiO AR
Si (S)	TiO_x/Ti(t)/SiO_x/p-Si	0.550	33.0	0.65	11.7	TiO_x AR
Si (S)	Ta_2O_5/Al(t)/SiO_x/p-Si	0.598	30.2	0.73	13.2	Ta_2O_5 AR
Si (S)	SiO_x/Al(g)/SiO_x/p-Si	0.485	21.3	0.72	9.8	SiO_x AR
GaAs (S)	Ta_2O_5/Ag(t)/Sb_2O_3/n-GaAs	0.795	25.6	0.81	15.3	Ta_2O_5 AR
GaAs (S)	Ta_2O_5/Ag(t)/SiO_2/n-GaAs	0.740	25.6	0.81	15.3	Ta_2O_5 AR
InP (S)	Au(t)/P_xO_y/n-InP	0.460	17.2	0.76	6.0	No AR
CdTe (S)	Au(t)/Cadmium stearate/n-CdTe	0.305	2.0	0.5	0.3	No AR
Si (S)	ITO/SiO_x/p-Si	0.526	28.7	0.79	11.9	Ion beam deposited ITO; no AR
Si (S)	ITO/SiO_x/n-Si	0.500	32.0	0.65	10.4	Sprayed ITO; no AR
Si (S)	ITO/SiO_x/n-Si	0.605	32.1	0.67	13.4	Sprayed ITO; no AR
Si (S)	SnO_2/SiO_x/n-Si	0.615	29.1	0.685	12.1	EB evap. SnO_2; no AR
InP (S)	MgF_2/ITO/P_2O_5/p-InP	0.760	21.6	0.65	14.4	Ion beam deposited ITO; MgF_2 AR
GaAs (S)	ITO/Ga_2O_3/p-GaAs	0.520	13.0	0.55	5.0	Ion beam deposited ITO; no AR
CdTe (S)	ITO/TeO_2/p-CdTe	0.820	14.5	0.55	8.0	rf sputtered ITO; no AR
CdTe (S)	ZnO/TeO_2/p-CdTe				8.8	Sprayed ZnO; no AR
$CuInSe_2$ (S)	ITO/CuO/p-$CuInSe_2$	0.500	30.2	0.55	8.3	EB evap. ITO; no AR
Si (P)	SiO/Al(g)/SiO_x/p-Si	0.540	32.7	0.755	13.3	SiO AR
Si (P)	SiO/Cr(t)/SiO_x/p-Si	0.540	26.4	0.67	9.5	SiO AR
Si (P)	ITO/SiO_x/n-Si	0.500	29	0.71	10.3	Sprayed ITO
Si (P)	SnO_2/SiO_x/n-Si	0.520	29.4	0.65	9.9	Sprayed SnO_2
Si (P)	ITO/SiO_x/p-Si	0.542	24.9	0.71	9.6	Ion beam deposited ITO
Si (P)	ITO/SiO_x/p-Si	0.522	28.1	0.79	11.5	Ion beam deposited ITO
Si (P)	ITO/SiO_x/p-Si	0.522	26.0	0.70	9.5	Ion beam deposited ITO
Si (P)	ITO/SiO_x/p-Si	0.498	24.7	0.77	9.5	Ion beam deposited ITO
Si (P)	ITO/SiO_x/p-Si/ceramic	0.483	24.7	0.75	8.9	Ion beam deposited ITO
GaAs (P)	Sb_2O_3/Ag(t)/SiO_x/GaAs	0.700	25.5	0.785	14.0	Sb_2O_3 AR

[a] (S) denotes single crystal; (P), polycrystalline; (R), ribbon; (W), web process. (g) refers to grating-type metal barrier and (t) refers to transparent-type metal barrier.

n^+/p junction. The minority carriers (electrons) generated in the bulk of the p-Si are collected vertically and then flow along the inversion layer to the MIS contact and removed from the cell via the metal grid. Obviously, for high minority carrier collection efficiency, the MIS contacts have to be very closely spaced, typically of the order of the diffusion length of the minority carriers (50–120 μm). Godfrey and Green[117] have fabricated IL grating cells of greater than 17.5% AM1 efficiency, calculated for the active area.

The photovoltaic characteristics of various bulk semiconductor solar cells are given in Table 3.3.

4

Photovoltaic Measurements, Junction Analysis, and Material Characterization

4.1. *Introduction*

Successful production of an efficient solar cell requires the coupling of fabrication techniques with a basic understanding of the device. It is thus of fundamental interest to be able to correlate the cell performance with the basic structural, electronic, and optical properties of the semiconductor in order to evaluate precisely their respective contributions to the junction behavior and identify avenues of further improvement in conversion efficiency. Therefore, a detailed characterization of the material parameters of the various components of the solar cell is imperative. The junction must be evaluated in terms of current–voltage ($I-V$) and capacitance–voltage ($C-V$) characteristics and spectral response to obtain essential functional parameters such as reverse saturation current density J_s, diode factor n, ionized impurity concentration N_D, diffusion voltage V_D, barrier height ϕ_B, depletion width W, and junction field F. Based on these data, an energy band diagram for the junction can be formulated and a physical model proposed to describe the basic mechanisms governing cell performance. Coupled with carrier and photon loss measurements and analyses, device modeling can be a very effective aid in the optimization of a photovoltaic device. This chapter deals with the basic experimental techniques of evaluating solar cell performance and related diagnostic measurements, with special emphasis on techniques developed specifically for thin film solar cells.

In the subsequent sections, each dealing with a particular measurement technique, we first describe the experimental setup and the method of data acquisition. This is followed by a discussion of the analysis and interpretation of data for obtaining material- and junction-related parameters of interest.

4.2. *Photovoltaic Measurements*

The primary evaluation of a solar cell is done by determining the load characteristics of the device under illumination and obtaining therefrom the various photovoltaic performance characteristics, namely, η, V_{oc}, I_{sc}, I_L, FF, V_m, I_m, P_m, R_s, and R_{sh}. Spectral response measurements and optical scanning are necessary to determine the wavelength and spatial dependence of the photogenerated current.

4.2.1. *I–V Characteristics*

In view of the strong dependence of solar cell performance on the intensity and spectral distribution of the incident light and the temperature, there has long been a need for a set of standard measurement procedures, which would provide a common basis for comparing solar cells. As a result of several workshops, terrestrial photovoltaic measurement procedures have been evolved and are compiled in a report by NASA.[1] The recommendations of the report are recounted below.

The only accepted method for measuring the incident intensity is the reference cell method. The hermetically sealed reference cell must be made from the same material and possess essentially the same spectral response as the cells or array of cells being tested. The reference cell is calibrated at the photovoltaic testing facility at NASA Lewis Research Center, Cleveland, Ohio, using a normal-incidence pyrheliometer (NIP), in units of short-circuit current output per unit of radiant energy input ($\mathrm{A\,W^{-1}\,m^{-2}}$). The standard test conditions (STC) are: cell temperature equal to $28 \pm 2\,°\mathrm{C}$ and irradiance equal to $1000\ \mathrm{W\,m^{-2}}$ as measured by the reference cell. A digital voltmeter, potentiometric recorder, or other measuring instruments capable of measuring with an error less than $\pm 0.5\%$ over the range 0 to 100 mV are employed to measure the reference cell short-circuit current with a 0.1% precision resistor at a voltage less than 20 mV across the cell. In case preamplifiers are used to match an automatic data recording system, the $\pm 0.5\%$ error requirement must be fulfilled by the system. The reference cell holder must be fitted with a suitable thermocouple or thermistor to facilitate temperature setting at standard conditions. The thermocouple and associated measuring equipment must be capable of 1 °C accuracy. The reference cell temperature must always be maintained at 28 ± 2 °C, since the calibration is done at that temperature.

Solar cell measurements can be carried out either in natural sunlight or indoors using solar simulators. The recommended light sources for solar simulators used in terrestrial photovoltaic measurements are: (i) a dichroic filtered ELH-type tungsten lamp, (ii) a short-arc steady-state xenon lamp, and (iii) a long-arc pulsed xenon lamp. The spectral distributions of these three sources match reasonably that of terrestrial sunlight and, therefore, the consequences of not exactly matching the spectral responses of test and reference cell are not severe ($<1\%$). The spectra of these commercially available light sources are shown in Figure 4.1.

The sunlight simulator must be capable of at least 1000 W m^{-2}, as measured with a reference solar cell. Moreover, the nonuniformity of total irradiance, defined as

$$\frac{\text{maximum irradiance} - \text{minimum irradiance}}{2 \times \text{average irradiance}} \times 100$$

where the irradiances are in the plane of the test cell, should be less than 2%. The temporal stability of irradiance, defined in a similar manner to the nonuniformity of total irradiance, should also be within 2% over the period of time in which cell measurements are made. Finally, the solar beam subtense angle at any point on the test cell must be less than 30°.

The steady-state xenon arc lamp can uniformly illuminate areas up to 30×30 cm^2. The pulsed xenon arc lamps can uniformly illuminate areas up to 5.5 m in diameter. The uniformity achievable with these sources is about $\pm 2\%$ across the test plane. With appropriate electronic controls, the intensity stability of these sources is good.

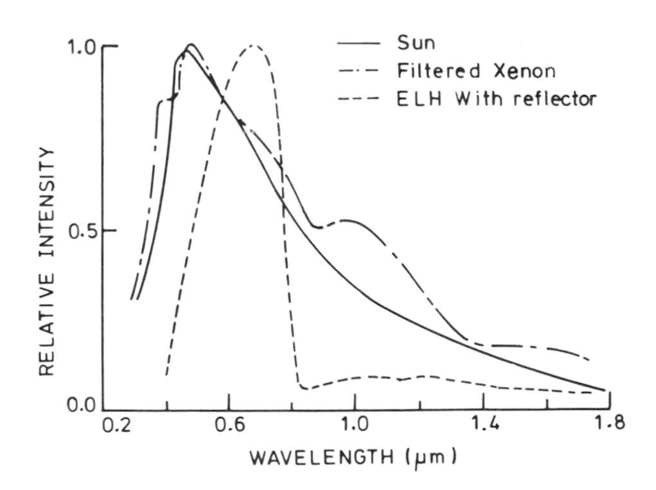

Figure 4.1. Spectra of different light sources.

While making measurements in natural sunlight, the following procedure should be adopted: the reference cell and the test cell or array should be aligned perpendicular to the sun, with the reference cell coplanar with the test cells. The entire cell or array and the reference cell should be fully and uniformly illuminated. The surrounding area must be free of highly reflective surfaces which could significantly increase the solar radiation falling onto the cell. The current–voltage ($I-V$) characteristic of the cell (or array) should be recorded simultaneously with the reference cell output. It should be ensured that the solar intensity, as measured by the reference cell, remains constant within 0.5% during the period of measurement and is at least 800 W m^{-2}.

The test cell should be mounted on a fixture that has vacuum holddown, a temperature-controlled block, four terminal contacts (current + and −; voltage + and −), and a thermocouple.

In general, in outdoor measurements, the solar irradiance is not 1000 Wm^{-2} and the cell/array temperature, unless controlled, is not 28 ± 2 °C. The measured $I-V$ curve can be translated to that corresponding to the curve at STC using the following equations:

$$\Delta I = I_{sc1}[(J_2/J_1) - 1] + \alpha(T_2 - T_1)A \quad (4.1)$$

$$I_2 = I_1 + \Delta I \quad (4.2)$$

$$V_2 = V_1 + \beta(T_2 - T_1) - \Delta I R_s - K(T_2 - T_1)I_2 \quad (4.3)$$

where I_2, V_2, J_2, and T_2 are current, voltage, irradiance, and temperature at STC, I_1, V_1, J_1, and T_1 are the measured values, α and β are the current and voltage temperature coefficients, R_s is the series resistance, K is a curve correction factor, and A is the area. R_s and K have to be experimentally determined. It should be noted that where the temperature is not controlled, the cell/array should be shadowed. The shadow should be removed just prior to the measurement and the data recorded quickly while the cell/array is still near the ambient temperature.

In indoor measurements using the steady-state method, the light source should first be turned on and stabilized. The source intensity should be adjusted to 1000 W m^{-2} (as determined by the reference solar cell held at 28 ± 2 °C) by varying the source-to-cell distance. Feeding less or more than the specified power to the source may change the spectrum and, therefore, should not be resorted to as a means of intensity adjustment. The reference cell should be replaced by the test fixture and the cell temperature set to 28 ± 2 °C using a dummy solar cell. After all the adjustments are complete, the test cell should be placed in the fixture and the $I-V$ characteristics recorded.

In the pulsed method, the reference cell and the test cell/array should be mounted coplanar and perpendicular to the pulsed beam. The temperature of the test cell/array should be measured and entered into the pulsed simulator data system. For other adjustments the procedures recommended by the manufacturer should be followed.

While making measurements with concentrator systems certain additional precautions are necessary: the cell performance and system performance should be measured separately; the cell area designed to be illuminated by the concentrator should be used while calculating the efficiency of a concentrator cell; the temperature of the cell should be maintained at 28 ± 2 °C; the nonuniformity of irradiance in the test plane should be within $\pm 20\%$, and the angle of incidence of the concentrated beam on the cell should be within a cone of full angle 60°.

The general types of equipment used to obtain a solar cell $I - V$ curve are: (i) Fixed-load resistors — these allow only discrete points along the $I–V$ curve to be obtained. The voltage drop across the cell when the short-circuit current is being recorded should not exceed 20 mV and the open-circuit voltage should be measured with a voltmeter having an internal resistance of at least 20 kΩ/V. (ii) Variable power supply — with this equipment, a continuous $I–V$ curve can be obtained. A sinusoidal signal or a ramp is imposed on the device and the voltage across the cell and the current through the cell are fed to the X- and Y-axes, respectively, of an $X–Y$ recorder or oscilloscope, after converting the current to a voltage signal using a current-to-voltage converter. This approach permits biasing of the cell in the forward and reverse directions, enabling additional diagnostic information to be obtained. (iii) Electronic load — this is basically an automatic power supply specifically tailored for solar cell use. (iv) Microprocessor-based data acquisition system — this also employs a power supply to obtain the data. However, the power supply is regulated by the programmed microprocessor which also stores the data and reduces the data to give relevant information. This system has the distinct advantage of speed in data reduction and is particularly suitable, indeed essential, when a large number of cells have to be routinely measured as, for example, in a production facility. A block diagram of such a system is shown in Figure 4.2.

We should caution that great care must be exercised in making electrical connections, because any wire resistance developed between the test device and the instrumentation, in any of the above modes of measurement, could cause losses in power output and result in an inaccurate $I–V$ curve.

The shape of the $I–V$ curves can reveal the various failure modes,[2] as shown in Figure 4.3, which shows several curves representing idealized cases, each curve corresponding to a single failure mode. We should point out that, in practice, several of these failure modes can occur simultane-

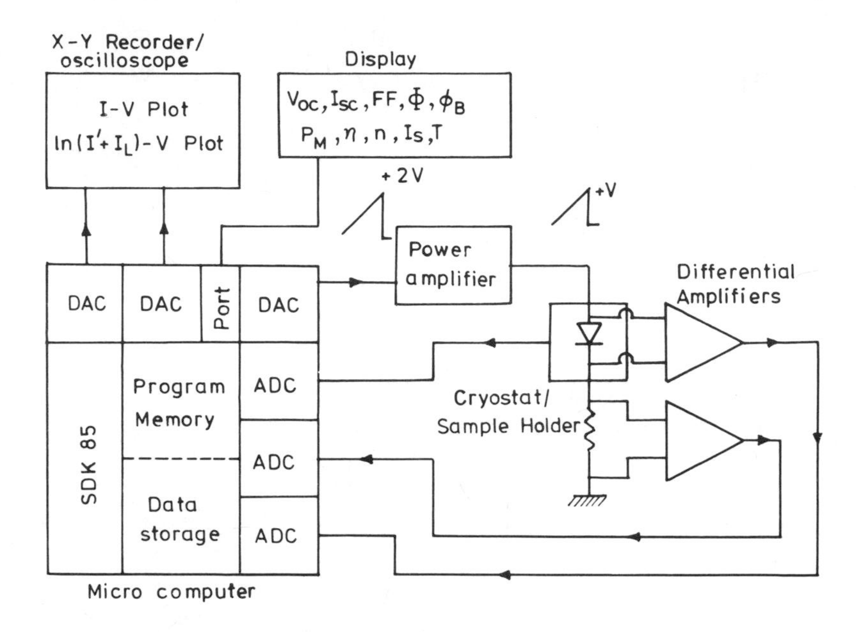

Figure 4.2. Block diagram of a microprocessor based $I-V$ data acquisition and analysis system.

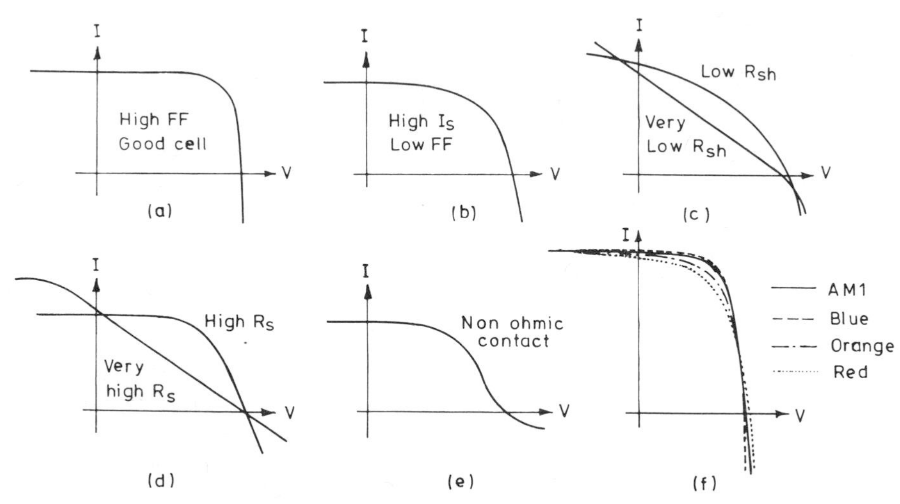

Figure 4.3. Typical $I-V$ characteristics for: (a) good cell, (b) poor junction with high I_s, (c) cell with low shunt resistance (R_{sh}), (d) cell with high series resistance (R_s), and (e) cell with nonohmic contact. (f) Effect of the spectral content of the incident light on the $I-V$ curves of a Cu_2S/CdS cell.

ously. As seen from Figure 4.3, different failure mechanisms can give $I-V$ curves with similar shapes in the power (fourth) quadrant. Therefore, to distinguish the failure modes, the ability to bias the cell in both the forward and reverse direction is essential. Figure 4.3a represents the square shape of a good cell with high shunt resistance R_{sh}, low series resistance R_s, and good fill factor FF. The curve of a junction with excessive diode saturation current I_s and, hence, poor FF, is shown in Figure 4.3b. The V_{oc} is not usually affected except in extreme cases. A low R_{sh} (Figure 4.3c) and high R_s (Figure 4.3d) both reduce the FF, with the V_{oc} also being affected in the former case and the I_{sc} being affected in the latter case under extreme conditions. Biasing the cell in the reverse direction indicates the true light-generated current in such extreme cases. We should note here that the effect of a high R_s becomes more pronounced at higher illumination intensities while the effect of a poor shunt is predominant at lower illumination levels. The intensity dependence of the $I-V$ curve, therefore, provides a means of distinguishing and separating the R_s and R_{sh} effects. Figure 4.3e shows the effect on the $I-V$ curve of a nonohmic contact, which manifests itself in the form of a curvature near the V_{oc} or the I_{sc} point. In Figure 4.3f, the result of inadequate matching of spectral distribution to the Cu_2S/CdS thin film solar cell is shown. The use of a red-wavelength-rich spectrum can severely distort the $I-V$ characteristics.[4] This underscores the importance of properly matching the spectral distribution of the source to the solar cell.

Once a reliable $I-V$ curve is obtained, the V_{oc} and the I_{sc} can be read directly from the curve. The voltage V_m and the current I_m at the maximum power point can be determined by computing the IV product at various points along the curve and selecting the point where the product is a maximum. Thus the maximum power P_m and the FF can be obtained from the relations $P_m = V_m I_m$ and $\mathrm{FF} = V_m I_m / V_{oc} I_{sc}$. The efficiency can then be calculated in a straightforward manner from the relations $\eta = (V_{oc} I_{sc} \mathrm{FF}/P_{in}) \times 100$ or $\eta = (P_m/P_{in}) \times 100$, where P_{in} is the illumination intensity incident on the cell as measured by the reference cell. Alternatively, the $I-V$ curves can be plotted on a graph having isoefficiency contours and/or isopower contours and the efficiency/maximum power read directly from the graph. Obviously, this method requires that the cell area, the cell temperature, and the illumination level be maintained at the same values from sample to sample and that these should correspond to values for which the isoefficiency/isopower contours have been drawn. This method is suitable when the cell design, particularly the area, has been standardized and a large number of cells have to be routinely characterized as, for example, in a production line.

The R_{sh} value of the device can be obtained readily from the inverse of

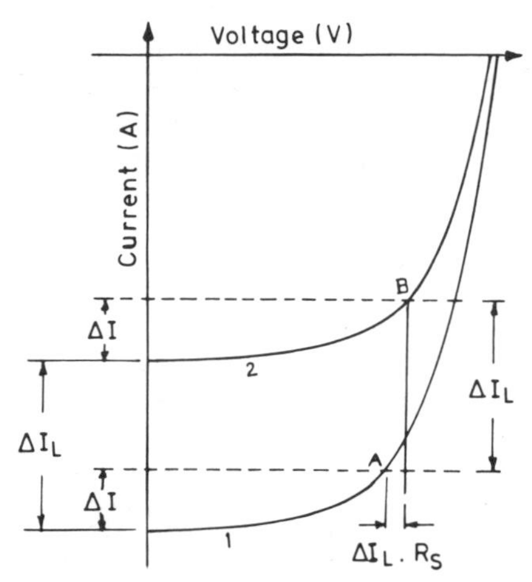

Figure 4.4. Method for determining the series resistance of a cell from $I-V$ curves at two different intensities (after Wolf and Rauschenbach[3]).

the slope of the $I-V$ curve in the third quadrant. To determine R_s, a convenient technique is to use the far forward characteristics (in the first quadrant), where the $I-V$ curve becomes linear, and measure the slope, the inverse of which yields R_s. A more rigorous technique, described by Wolf and Rauschenbach,[3] is illustrated in Figure 4.4. The $I-V$ curves are recorded at two different intensities. It is not necessary to know the magnitudes of the intensities. We choose an arbitrary interval ΔI from the I_{sc} point and obtain the point A on Curve 1, near the knee of the characteristic. Similarly, taking the same ΔI value, we obtain the point B on Curve 2. The points A and B are displaced relative to each other on the X-axis by an amount $\Delta V = \Delta I_L R_s$, whence the value of R_s can be easily calculated. However, by virtue of being based on the superposition principle, this technique is applicable only to those devices in which the superposition principle holds.

In the microprocessor-controlled $I-V$ data system, with appropriate programs, the various photovoltaic parameters are displayed directly on the digital readout.

Before concluding this section we wish to note that some workers tend to quote conversion efficiency values for a monochromatic wavelength or for a very narrow wavelength region where the efficiency of the cell is highest. The wavelength or the wavelength range is not explicitly mentioned in the headlines but is hidden in the text. This meaningless practice, obviously prompted by the desire to quote attractive numbers, must be strongly discouraged, since the utility of a particular device has to be judged by its performance over the entire solar spectrum and not by its performance at a particular portion of the spectrum. A similar, though less serious, practice is the reporting of efficiency values for active areas where the area of the cell shadowed by the grid lines and bus bars is subtracted from the total area for calculating efficiencies.

4.2.2. *Spectral Response*

The measurement of the wavelength dependence of the photocurrent is valuable not only as a characterization technique but also as an effective diagnostic and analytical tool. Any change in the device configuration, in the material properties, or in the electronic structure of the junction designed to produce a change in the photocurrent inevitably shows up in the spectral response curve and a careful analysis of the data can provide considerable qualitative and quantitative information on the nature of the change. Thus, spectral response measurements can be utilized to direct cell performance improvement strategy.

A block-diagram of a spectral response measurement setup is shown in Figure 4.5. The solar cell is illuminated by monochromatic light from the monochromator and the cell output is amplified and phase-sensitively detected by a lock-in amplifier. The mechanical chopper in front of the light source modulates the light beam and the reference cell provides the reference signal to the lock-in amplifier to facilitate phase sensitive detection. Such an ac measurement is desirable for two reasons: (i) the effect of stray light is eliminated and (ii) bias light effects can be investigated. In certain devices, notably the Cu_2S/CdS cell,[5-7] the spectral response exhibits a hysteresis, being higher while traversing from shorter wavelengths to longer wavelengths and markedly reduced when traversing

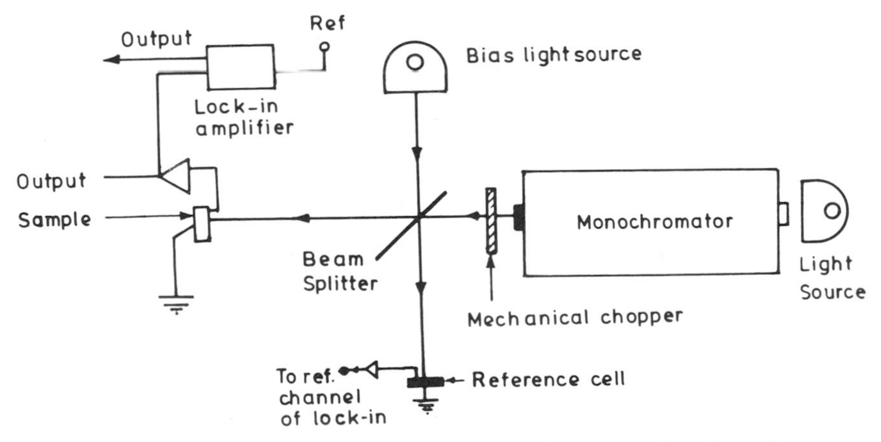

Figure 4.5. Setup for measuring the spectral response of a solar cell.

from longer wavelengths to shorter wavelengths. On application of a white light bias, the hysteresis disappears and the response is enhanced. This effect is attributed to the presence of deep traps in the depletion region of CdS. The traps, which are emptied or filled by different wavelengths, modulate the field at the junction affecting the photocurrent. In such cases, the addition of a white light bias is necessary to saturate the traps and render them ineffective and also to provide the actual illumination conditions in which the solar cell is designed to perform; then the measured spectral response is the true response of the cell.

By feeding the output of the lock-in amplifier to a strip chart recorder and coupling the wavelength drive of the monochromator to the recorder drive, a continuous spectral response curve can be obtained. In the event a monochromator is not available or large areas need to be illuminated, discrete wavelength interference filters may be used to obtain the spectral response. To normalize the measured spectral response data, the photon flux or the power density incident on the cell at each wavelength has to be determined. The photocurrent at a particular wavelength has to be divided by the photon flux or the power density at that wavelength. The normalized spectral response, in terms of photocurrent per unit photon or photocurrent per unit power thus obtained, is then ready for further analysis.

Using the normalized spectral response, we can obtain the total light-generated current for different illumination spectra (AM0, AM1 etc.) by convoluting the spectral response with the desired AMn spectrum.[8] Moreover, the photoresponse of a Schottky-type cell can yield an accurate and direct estimation of the barrier height.[9–11] The photocurrent R per absorbed photon is given by

$$R \sim \frac{T^2}{(E_s - h\nu)^{1/2}} \left[\frac{x^2}{2} + \frac{\pi^2}{6} - \left(e^{-x} - \frac{e^{-2x}}{4} + \frac{e^{-3x}}{9} - \cdots \right) \right] \quad \text{for } x \geqslant 0 \qquad (4.4)$$

where $x \equiv h(\nu - \nu_0)/kT$, $h\nu_0$ is the barrier height $q\phi_B$, and E_s is the sum of $h\nu_0$ and the Fermi energy measured from the bottom of the metal conduction band. For $E_s \gg h\nu$ and $x > 3$, equation (4.4) reduces to

$$R \sim (h\nu - h\nu_0)^2 \qquad (4.5a)$$

or

$$(R)^{1/2} \sim h(\nu - \nu_0) \qquad (4.5b)$$

A plot of the square root of the photoresponse as a function of photon

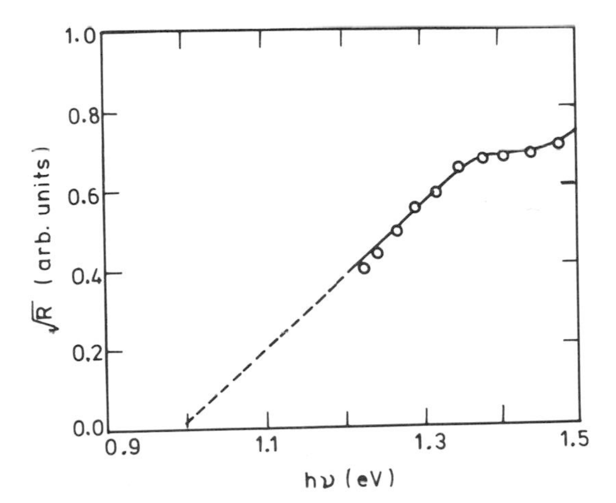

Figure 4.6. Square root of photoresponse as a function of photon energy for a Cu_2S/CdS solar cell.

energy is a straight line whose extrapolated intercept on the energy axis yields the barrier height, as shown in Figure 4.6 for a Cu_2S/CdS solar cell.

With certain simplifying assumptions, it is possible to estimate the minority carrier diffusion lengths from spectral response curves.[12–18] The assumptions involved are: (i) an absorber layer thickness greater than or equal to the diffusion length, (ii) a well-defined planar slab (or junction) of the absorber, (iii) carrier generation only in the absorber (and not in the window in case of a heterojunction), and (iv) a single light pass through the absorber. Taking the ratio of the responses at two adjacent wavelengths, we obtain the expression[13]

$$\frac{R_1(1-R_2')\exp(2\alpha_1 l)}{R_2(1-R_1')\exp(2\alpha_2 l)} = \exp[(\alpha_1-\alpha_2)L_D]\,\frac{\exp(\alpha_1 L_D)-1}{\exp(\alpha_2 L_D)-1} \tag{4.6}$$

where R_1 and R_2 denote the response at wavelengths λ_1 and λ_2, respectively, R_1' and R_2' are the reflectance and α_1 and α_2 are the absorption coefficients at λ_1 and λ_2, l is the absorber layer thickness, and L_D is the minority carrier diffusion length. Thus, if l, α, R', and R values at different wavelengths are known, L_D can be determined by a simple program.[13,16]

The diffusion length can also be obtained by plotting $R/\lambda \exp(-\alpha d)$ vs. $R/\alpha\lambda \exp(-\alpha d)$. The gradient of this linear curve gives the diffusion length in the base semiconductor. Here d is the thickness of the top layer.[14,19]

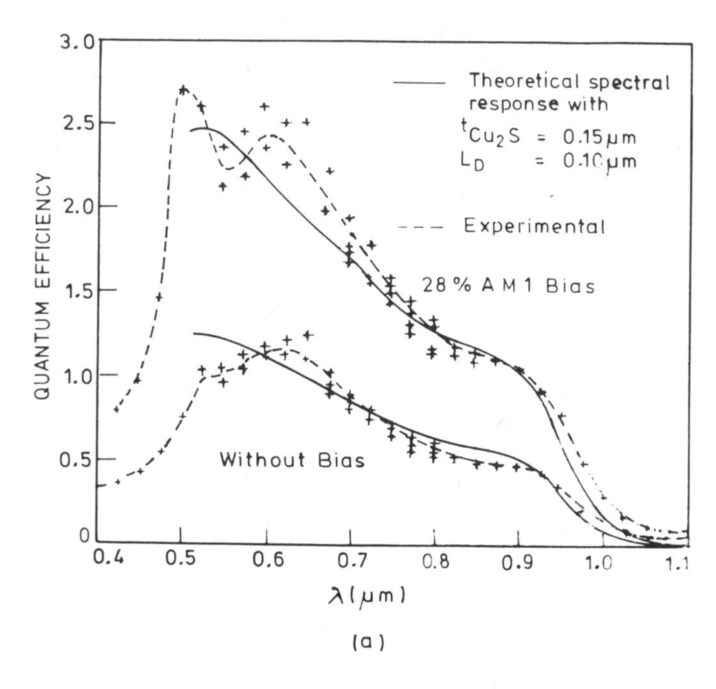

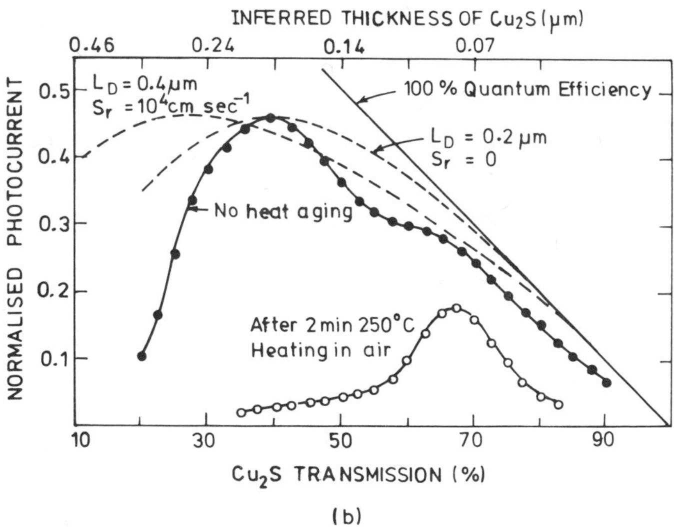

Figure 4.7. (a) Spectral response of a Cu_2S/CdS thin film cell (after Anderson and Jonath[17]). (b) Photocurrent (calculated and measured) as a function of Cu_2S transmission at a fixed wavelength for a Cu_2S/CdS solar cell (after Szedon et al.[18]).

Indirect estimates of L_D can be obtained by computer fitting of the experimental data.[12,17,18] Theoretical spectral response curves are derived for different values of L_D, S_r (surface recombination velocity), and thickness, and matched with the experimental data to determine L_D and S_r. Figure 4.7a illustrates this technique for a Cu_2S/CdS cell. The measured spectral responses, with and without bias, are compared with theoretical curves obtained with a Cu_2S thickness of 0.15 μm and electron diffusion length in Cu_2S of 0.10 μm. Reasonable agreement is obtained for $0.51 < \lambda < 1.0$ μm.

Another variation of this technique is illustrated in Figure 4.7b, where the measured normalized photocurrent at a fixed wavelength is plotted as a function of the Cu_2S thickness, inferred from measurements of Cu_2S transmission.[18] The photocurrent-vs.-transmission curve was obtained by scanning a laser spot across a Cu_2S/CdS cell that had a tapered Cu_2S thickness. Also shown in the figure are the theoretically calculated response-vs.-transmission curves obtained by solving, with appropriate boundary conditions, the equation

$$D_n \frac{d^2(\Delta n)}{dx^2} - \frac{\Delta n}{\tau} = -\alpha I_0 \exp(-\alpha x) \tag{4.7}$$

where D_n is the diffusion constant for electrons, τ is the electron lifetime, Δn is the excess electron concentration, and I_0 is the photon flux into the Cu_2S. The film thickness t can be converted to transmission T through the Cu_2S layer by the expression

$$T = T_0 \exp(-\alpha t) \tag{4.8}$$

For no electronic losses, i.e., infinite diffusion length and $S_r = 0$, a straight line is predicted, shown by the solid line in Figure 4.7b. For $S_r = 0$, the peak response occurs at progressively larger transmission values with decrease in L_D, for lower transmission (higher Cu_2S thickness) values. At high values of transmission (small film thicknesses), the rate of change of photocurrent with transmission is not sensitively influenced by the value of L_D. For a fixed value of L_D, increasing S_r suppresses the peak amplitude of the response, changes the rate of photocurrent decrease with transmission (in the transmission range above 70%), and shifts the transmission value for peak response. From the figure, the peak response of the measured curve corresponds to that of the theoretical curve with $S_r = 0$ and $L_D = 0.2$ μm.

We should mention here that estimates of L_D obtained from spectral response data must be viewed with caution since the assumptions involved are generally not met with in actual devices.

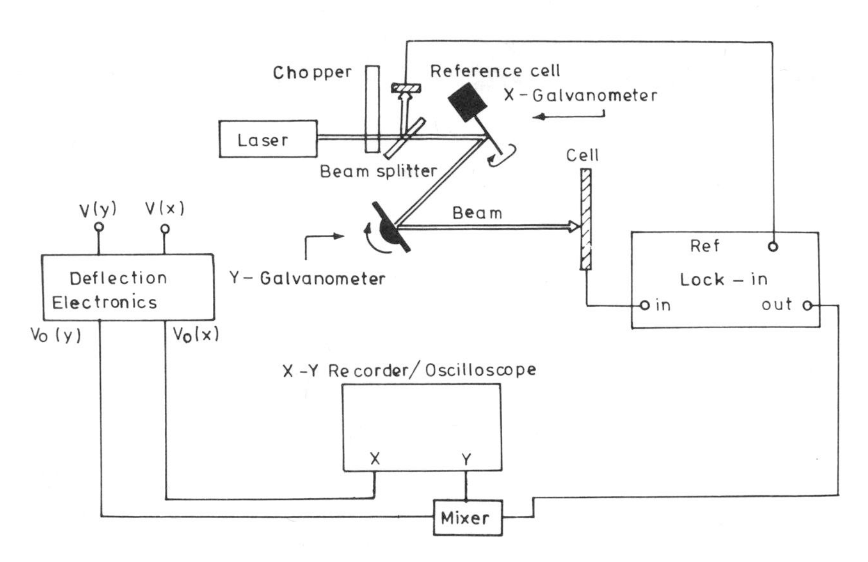

Figure 4.8. Schematic diagram of an optical scanning experimental setup.

4.2.3. *Optical Scanning*

In recent years, optical spot scanning of solar cells[20–25] has emerged as a versatile diagnostic technique for monitoring cell uniformity and detecting anomalously good or bad regions within a single cell. A schematic diagram of the experimental apparatus is shown in Figure 4.8. A finely focused light beam from a white light or a monochromatic source is moved back and forth over the cell in a raster fashion, using a two-mirror optical scanning system, resulting in coverage of the entire area of the cell. The two mirrors, attached to two galvanometers, oscillate synchronously about two axes which are orthogonal relative to one other. The deflection circuitry controls the synchronous motion of the two mirrors. The photocurrent from the cell is amplified, converted to a voltage signal, and fed to the Y-axis input of an X–Y recorder, whose pen follows the raster pattern generated by the light spot. A Y-modulated photocurrent image of the cell is thus obtained, an example of which is shown in Figure 4.9 for a Cu_2S/CdS solar cell. A scratch on the cell surface is revealed by the dip in the photocurrent response in the scratched region.

Alternatively, the same flying spot arrangement can be utilized to obtain a brightness-modulated photoresponse image of the cell, also called light beam induced current (LBIC) pictures. In this technique, the photocurrent signal from the cell is fed to an oscilloscope to modulate the brightness of the beam. To increase spatial resolution, a laser beam with a beam spot diameter of about 1 μm can be used to scan the cell. We should note that it is possible to obtain a reflection image of the cell using the same arrangement.[19,25] The electrical analog of LBIC is the electron beam induced current (EBIC) micrograph obtained by scanning the electron beam of a scanning electron microscope over the cell surface and measuring the cell response. The EBIC can be used in a similar fashion to yield either Y-modulated or brightness-modulated images of the cell. Since the electron beam diameter is about 50 Å, very high-resolution images are obtained.

LBIC and EBIC measurements have been applied to study a variety of effects in solar cells and diagnose cell performance. A few noteworthy examples are: (i) study of cell uniformity over large areas,[22,23] (ii) the effect of grain boundaries in polycrystalline solar cells,[14,19] (iii) identification of geometrically different layers of a solar cell device and delineation of the active area of the cell,[23] (iv) identification of wavelength-sensitive inhomogeneities in cell structure,[22] (v) correlation of inhomogeneity in cell performance with microstructural features[24,25] by comparison of LBIC with photomicrographs/scanning electron microscope micrographs, (vi) inhomogeneities in electric field distributions, e.g., in edge regions,[21] and (vii) study of bias effects in photodetectors.[20]

One of the most powerful applications of LBIC and EBIC measurements is the direct determination of minority carrier diffusion length. This is discussed in detail later in Section 4.4.4.

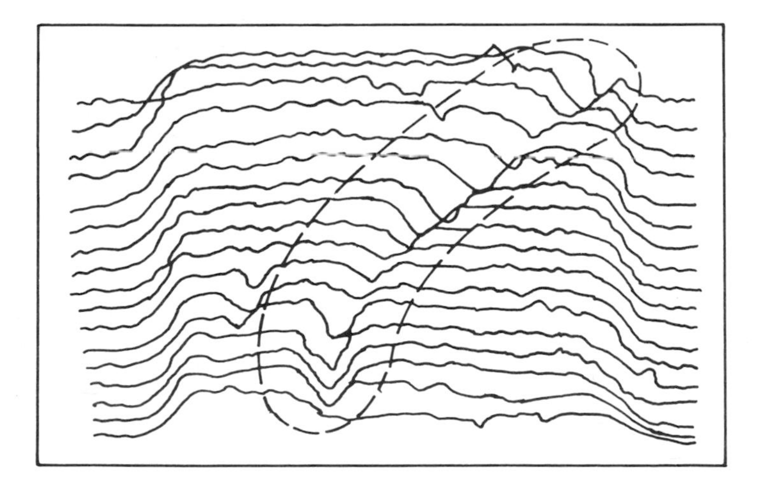

Figure 4.9. Y-modulated photocurrent image of a Cu_2S/CdS solar cell (obtained by optical scanning) revealing a poor response region.

4.3. Junction Analysis

In this section we shall deal with those measurements which yield information about the junction and the quality of the diode. These include primarily the temperature and illumination intensity dependence of the $I-V$ characteristics and the $C-V$ characteristics. We have also discussed the deep-level transient spectroscopy (DLTS) technique which involves measuring the transient capacitance of a reverse-biased junction and provides valuable information on traps.

4.3.1. I–V Analysis

An analysis of the temperature and illumination intensity dependence of the $I-V$ characteristics of a solar cell can yield considerable information on the junction quality and the transport mechanism occurring at the junction.[26–36] However, the measured $I-V$ curves have to be corrected for series and shunt resistive effects before further quantitative analysis is possible. Recalling equation (3.71), we can rewrite it in the form

$$I' = \sum_i I_{si}\{\exp[q(V')/n_i kT] - 1\} - I_L \tag{4.9}$$

where $I' = I - (V'/R_{sh})$ and $V' = V - IR_s$. Equation (4.9) is then the same as equation (3.71) with the resistive effects corrected out. The diode parameters I_{si} and n_i can then be determined experimentally from the intercept on the y-axis and the slope, respectively, of the $\ln(I' + I_L)$ vs. V' plot. For a solar cell exhibiting double-diode behavior, the $\ln(I' + I_L)$ vs. V' plot will yield regions with two slopes, and consequently two n values, corresponding to the two different diodes, and also two I_s values, which can be obtained by the extrapolated intercepts on the y-axis, as shown in Figure 4.10 for a Cu_2S/CdS solar cell[10,36] at different temperatures. By determining J_s, as described above, at different temperatures, we can obtain the temperature variation of J_s, which can then be fitted to one of equations (3.32), (3.36), (3.49), (3.50), (3.52), (3.53), (3.55), (3.56), (3.67), or (3.68) to infer the dominant transport mechanism and deduce values of E_g, ϕ_B, V_D, A^*, and S_I. The $\log I_s$ vs. $1/kT$ plots for $Cu_2S/Zn_xCd_{1-x}S$ solar cells are shown in Figure 4.11, for different x values.

For a diode exhibiting ideal Shockley behavior, values of the minority carrier diffusion length can be estimated[35] from J_s through equation (3.31b). Rohatgi et al.[34] have studied the effects of impurities on solar cells by analyzing the $I-V$ curves.

Another method which can be used to derive junction parameters is the V_{oc}–Intensity analysis.[33] If the diode parameters are not intensity-

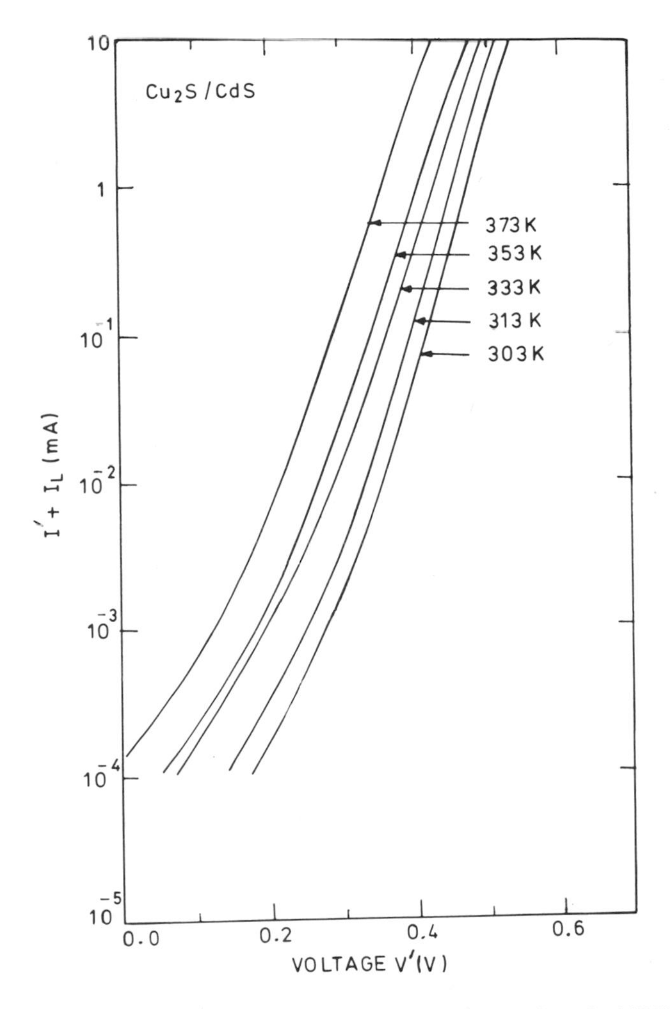

Figure 4.10. Plots of $\log(I' + I_L)$ vs. V' at different temperatures for a Cu_2S/CdS solar cell.

dependent, a linear V_{oc} vs. ln(Intensity) relationship will hold as a consequence of equation (3.72), since J_{sc} is linear with light intensity. This is shown in Figure 4.12 for a Cu_2S/CdS thin film solar cell. The values of n can be estimated from the slope of the plot. Multiple diodes show up as multiple slopes, as seen in the figure. Alternatively, one can plot V_{oc} vs. $\log I_{sc}$ to obtain the same information as well as the value of I_s from the intercept on the x-axis.

I–V analysis is most conveniently done by feeding the automatically recorded I–V data into a computer and using an appropriate program to

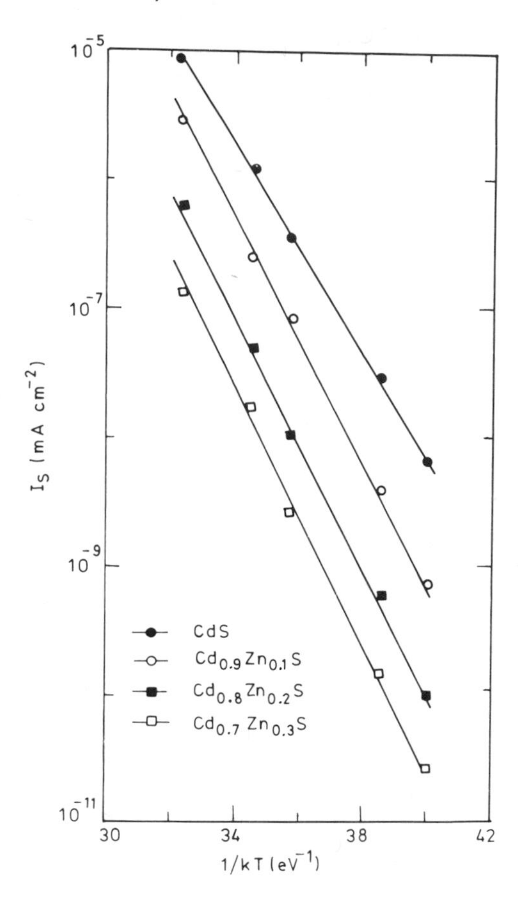

Figure 4.11. Plots of $\log I_s$ vs. $1/kT$ of $Cu_2S/Zn_xCd_{1-x}S$ solar cells for different x values.

obtain the parameters of interest. It is convenient to use an automatic logarithmic $I-V$ plotter which records the $I-V$ characteristics of a solar cell and the $\log I$ vs. V or $\log(I + I_L)$ vs. V characteristics.[32] The instrument employs a logarithmic amplifier to amplify the voltage signal, corresponding to the current through the cell, logarithmically.

It is useful for the purpose of device modeling to be able to determine quantitatively the various losses. One of the loss mechanisms, as noted earlier, the resistive losses, degrade the fill factor. The measured values of R_s and R_{sh} can be substituted in equation (3.197) to enable a quantitative evaluation of their contribution to the degradation of the fill factor.

4.3.2. *Capacitance Measurements*

The voltage dependence of the junction capacitance is most commonly employed to study junction behavior and determine junction-related

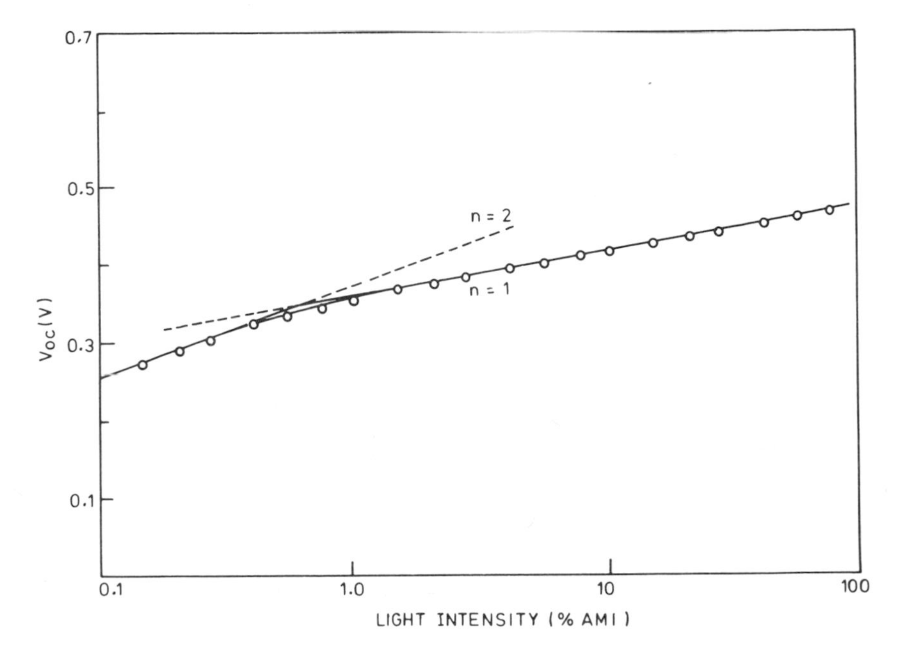

Figure 4.12. V_{oc} as a function of illumination intensity for a Cu_2S/CdS solar cell.

parameters.[36–46] $C-V$ measurements can be made point by point at different externally imposed bias values using a capacitance bridge. However, it is more convenient to obtain a continuous $C-V$ plot using an arrangement of the type shown in Figure 4.13. The operational principle of the automatic $C-V$ plotter is a follows: a sinusoidal signal at 100 kHz (or at any desired frequency) of about 20-mV peak-to-peak amplitude from the oscillator is applied along with the ramp voltage to the impedance (solar cell) to be measured through a mixer. The current-to-voltage converter

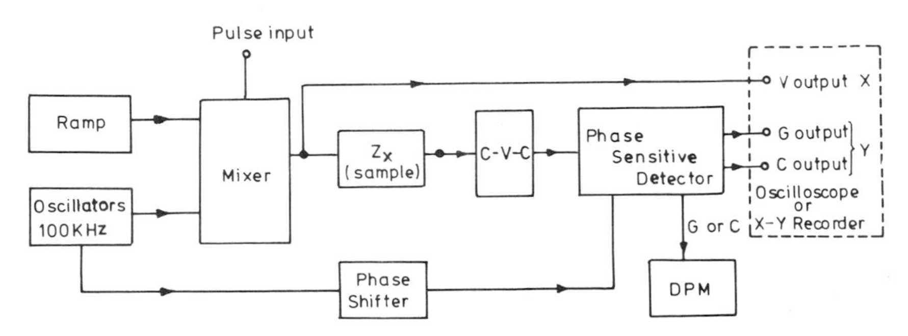

Figure 4.13. Block diagram of an automatic $C-V$ plotter.

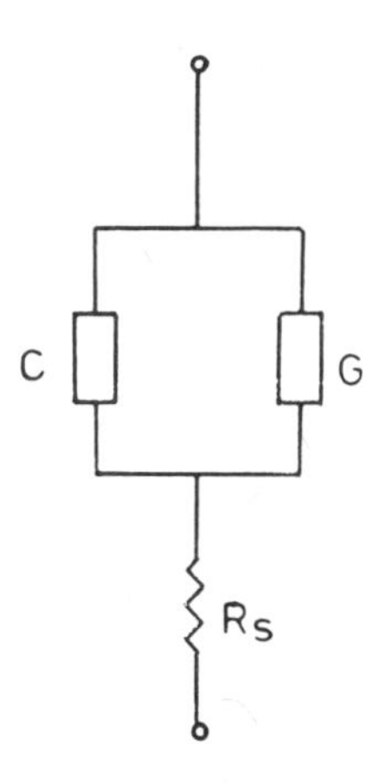

Figure 4.14. Equivalent impedance circuit of a solar cell.

(C–V–C) converts and amplifies the current passing through the impedance to a proportional voltage. This voltage, consisting of two components, an in-phase component proportional to the resistive part and a 90° out-of-phase component proportional to the capacitative part, is fed to two phase-sensitive detectors (PSD) to resolve the individual components. The output of the two PSDs can be fed either to an X–Y recorder or to an oscilloscope to obtain a C–V or a G–V (conductance–voltage) plot. The digital panel meter (DPM) reads either the capacitance or the conductance through a selector switch. Such automatic C–V/G–V plotters are commercially available. However, their measuring range is limited to 2 to 2000 pF. In the authors' laboratory, an automatic C–V plotter has been developed which is capable of measuring capacitance values from 2 pF to 200 nF.[47]

While making C–V measurements, it is important that we keep the measuring frequency in the range where the oscillating charge in the junction region is able to respond. In certain cells containing deep trap levels, for example Cu_2S/CdS, the measuring frequency should be selected such that these traps do not respond. The solar cell can be reverse biased or forward biased. However, in the forward-biased mode the applied voltage should not exceed approximately 200 mV, beyond which the conductance of the cell is high.

Before the C–V data can be processed further, the measured capacitance values need to be corrected for conductance effects. Considering a parallel equivalent circuit of the type shown in Figure 4.14, we can relate the measured capacitance C_M and conductance G_M values to the actual capacitance C and conductance G values by the expressions[46]

$$C_M = \frac{C}{(1 + GR_s)^2 + (\omega R_s C)^2} \tag{4.10}$$

$$G_M = \frac{G + R_s(G^2 + \omega^2 C^2)}{(1 + GR_s)^2 + (\omega R_s C)^2} \tag{4.11}$$

It should be noted that for $(\omega R_s C) \gg (1 + GR_s)$, the relation $C_M \approx [1/(\omega R_s)^2](1/C)$, i.e., $C_M \sim 1/C$, holds. Since $C \sim (V)^{-1/2}$, R_s, the series resistance of the cell, can be obtained from the slope of a C_M^{-2} vs. V^{-1} curve.

From the above relations, C and G can be obtained using a computer program. One can then plot C^{-2} vs. V curves to derive values of the V_D, N_D, W, and F using equations (3.12), (3.13), and (3.14). However, these equations are derived for the simple case of a uniform space charge. Various other space-charge profiles are possible[44,38] and these are shown in Figures 4.15a–d along with the corresponding expected C^{-2} vs. V plots. Figure 4.15a corresponds to a uniform space-charge profile. For the

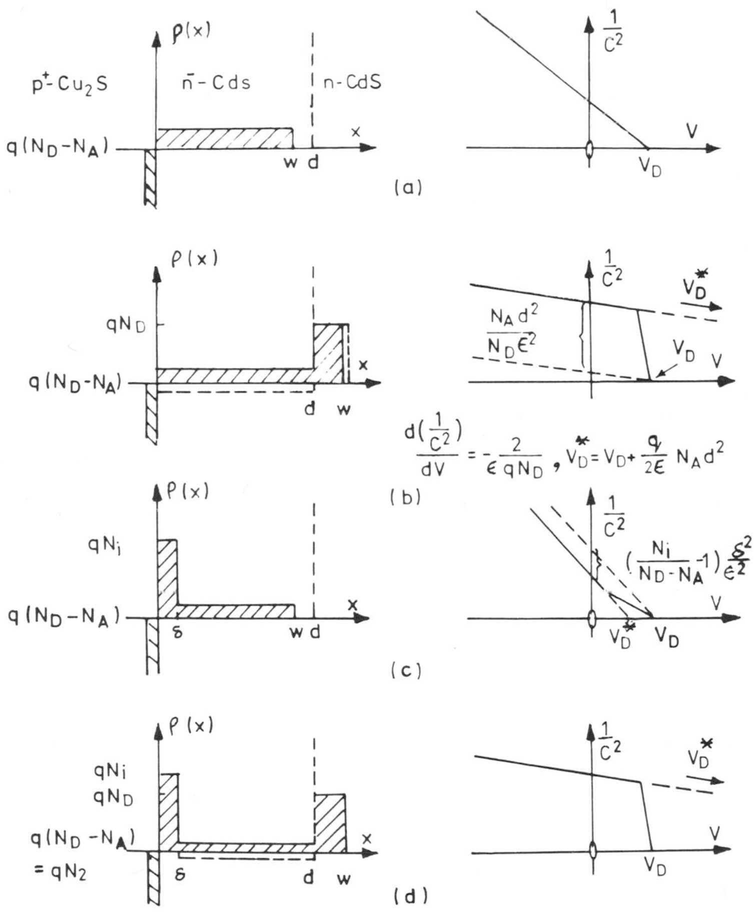

Figure 4.15. Various possible space-charge profiles and corresponding C^{-2} vs. V characteristics.

space-charge profile of Figure 4.15b, usually observed in heat-treated Cu_2S/CdS solar cells, we have the following relationships:

$$\frac{d(1/C^2)}{dV} = \frac{-2}{\varepsilon q N_D} \tag{4.12}$$

and

$$V_D^* = V_D + \left(\frac{q}{2\varepsilon}\right) N_A d^2 \tag{4.13}$$

where V_D^* is the extrapolated intercept on the voltage axis and V_D, the intercept on the voltage axis, is the diffusion voltage. ε is the permittivity of the semiconductor. For the most general case of Figure 4.15d, Pfisterer[38] has derived the following expressions:

$$W^2 = \frac{2\varepsilon(V_D - V)}{qN_D} + d^2 - \frac{N_i\delta^2 - (N_D - N_A)(d^2 - \delta^2)}{N_D} \tag{4.14}$$

$$\frac{1}{C^2} = \frac{W^2}{\varepsilon^2} = \frac{2(V_D - V)}{\varepsilon q N_D} + \frac{d^2}{\varepsilon^2} - \frac{N_i\delta^2 - (N_D - N_A)(d^2 - \delta^2)}{\varepsilon^2 N_D} \tag{4.15}$$

and

$$V_D^* = V_D - \frac{q}{2\varepsilon}[N_i\delta^2 + (N_D - N_A)(d^2 - \delta^2) - N_D d^2] \tag{4.16}$$

Thus, depending on the form of the C^{-2} vs. V curve, it is possible to deduce V_D, N_D, N_A, and W. By plotting N_D vs. X (distance from the junction), we can obtain the impurity profile.[10,45] From the measured value of N_D, the Fermi level can be determined from equation (2.21). The field F at the junction is given by[42]

$$F(V) = \frac{2(V_D - V)}{W(V)} = \frac{qN_D W(V)}{\varepsilon} \tag{4.17}$$

Photocapacitance measurements and spectral dependence of capacitance can provide very useful information on deep-levels-related effects on junction behavior. Rothwarf et al.[42] have investigated the photocapacitance of Cu_2S/CdS junctions and determined the effects of the illumination spectral content on the junction field. By combining the photocapacitance and spectral response (collection efficiency) data, the above workers have deduced the interface recombination velocity.

While interpreting C–V data, we must take into account the influence of the geometrical structure of the junction on the measured capacitance.

The area factor, defined as the ratio of the real junction area to the geometrical area of the device, can be, in some cases, as high as 5 to 6, as for example in Cu_2S/CdS thin film solar cells processed by the wet technique.[38] This area factor decreases with increase in reverse bias voltage owing to smoothing of the space-charge region. Prolonged heat treatment[38,43] also produces the same effect because of increasing thickness of the compensated CdS layer. The result of a large area factor is a steeper slope of the C^{-2} vs. V curve and a stronger shift to larger C^{-2} values with heat treatment, as compared to a planar junction. Consequently, the capacitance measurements simulate a lower doping concentration and a higher copper diffusion coefficient than is actually present. In this connection, we can note that Pfisterer[38] and Hall and Singh[43] have utilized C–V measurements on heat-treated Cu_2S/CdS solar cells to estimate the diffusion coefficient of Cu into CdS. Deviation from rectangular space-charge profiles can also alter the C^{-2} vs. V curve.[38] As pointed out by Pfisterer,[38] the determination of doping concentrations from C^{-2} vs. V plots is not possible in either of the two cases of nonplanar junction and/or nonrectangular space-charge profile.

Neugroschel et al.[40] have discussed a technique to determine the minority carrier diffusion length and lifetime from forward-biased capacitance measurements of p/n junctions. The forward-biased capacitance, C_{QNB}, of the quasi-neutral base region of a p^+/n junction (arguments are analogous for n^+/p junctions) is given by the relation

$$C_{QNB} = \frac{q}{kT}\frac{Aqn_i^2}{2N_D} L_p \left[\exp\left(\frac{qV}{kT}\right) - 1\right] \tag{4.18}$$

where N_D is the base doping concentration (as determined from the reverse-biased capacitance vs. voltage characteristics), L_p is the minority carrier diffusion length in the n-type base, and A is the area of the device. Thus, if C_{QNB} can be experimentally determined, equation (4.18) would yield L_p, and from the relation $L_p = (D_p\tau_p)^{1/2}$ the lifetime could be deduced, provided D_p is known. However, the capacitance C measured at the terminals contains other components in addition to C_{QNB}:

$$C = C_I + C_s + C_{sc} + C_{QNE} + C_{QNB} \tag{4.19}$$

where the subscript s refers to the surface region, sc to the junction space-charge region, QNE to the quasi-neutral emitter region, and I to the ionized impurities in the junction space-charge region. C_s can be neglected for devices with low concentrations of surface states and of charge in the oxide over the surface. C_{QNE} and C_{QNB} are both proportional to $[\exp(qV/kT) - 1]$.

To extract C_{QNB}, $C-V$ measurements are made at relatively high as well as at low frequencies. At low frequencies, the mobile carriers in all regions of the device respond to the signal but at higher frequency, the minority carriers in the base region cannot follow the signal and C_{QNB} contributes negligibly to the total capacitance. In contrast, the mobile carriers associated with C_I and C_{sc} respond to the signal and contribute to the measured capacitance. Thus, subtraction of the high-frequency capacitance from the low-frequency capacitance yields a capacitance associated with the quasi-neutral regions. Under the condition that the lifetime in the emitter region, τ_e, is much shorter than τ_p and the signal frequency f is such that $\tau_e \ll (1/f) \ll \tau_p$, the capacitance obtained after subtraction is equal to C_{QNB}. Thus, L_p can be determined from equation (4.18). A limitation of this method is that the neutral base capacitance must contribute significantly to the total capacitance being measured and, hence, this technique may not be applicable to devices having very small diffusion lengths.

4.3.3. *DLTS Technique*

The DLTS technique is capable of providing information about the energy, density, and capture cross-section of recombination centers arising from the presence of deep levels in the bandgap of a material[48–55] and is able to distinguish between minority and majority carrier traps. Moreover, both radiative and nonradiative centers, spread over a wide range within the bandgap, can be detected by this technique. The method has high sensitivity and resolution and deep-level impurity concentrations as small as 10^{-4} to 10^{-5} of the shallow impurity doping can be detected with an energy resolution of about 0.03 eV. It is also possible to determine the spatial distribution of the deep level impurities. Finally, in contrast to other thermally stimulated measurements, the detected signal here is independent of the heating or cooling rate. The observed peaks always occur at the same temperature and are dependent only on the measuring frequency.

The block diagram of a DLTS system is shown in Figure 4.16a. The lock-in amplifier periodically triggers a $C-V$ plotter in a pulsed bias mode. If deep-level impurities exist within the depletion region of the sample, which can be either a p/n junction or a Schottky barrier, an exponential capacitance transient is detected by the lock-in. The magnitude of the transient is proportional to the number of deep impurities present and the time constant of the transient is governed by the thermal emission probability. The schematic diagram of the lock-in reference signal and the input signal to the lock-in amplifier is shown in Figure 4.16b. The measurement involves varying the temperature of the sample from liquid N_2 (77 K) to room temperature (or higher) at a fixed lock-in frequency. At a particular temperature, where the emission rate from the traps bears the

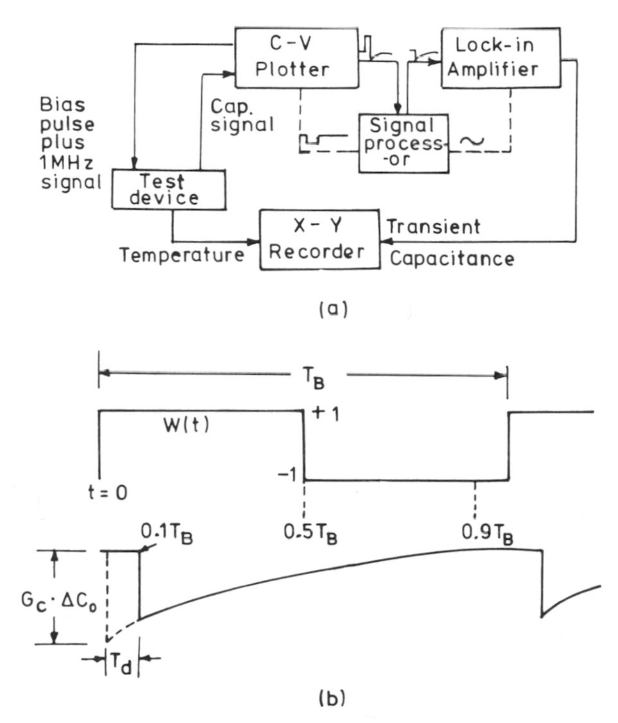

Figure 4.16. (a) Block diagram of a DLTS system. (b) Schematic representation of the weighting function and the input signal to the lock-in amplifier (after Rohatgi[49]).

proper relationship to the repetition frequency, a peak, whose magnitude is proportional to the density of traps, is observed. Thus, the temperature uniquely identifies the trap responsible for the transient. The DLTS spectrum, which is a plot of the lock-in output vs. temperature, is recorded on an X–Y recorder. Typical DLTS spectra[50] for polycrystalline Cu_2S/CdS heterojunctions are shown in Figure 4.17a. The peaks corresponding to different traps are labeled B, D, and F.

The activation energy of the trap can be determined by selecting a different value of the time base for each thermal scan, which moves the peak to a different temperature.[49] The emission rate e at the peak temperature, for each peak, is given by

$$e = \left[\frac{N_c \sigma V_{th}}{g}\right] \exp\left(\frac{-\Delta E}{kT}\right) \tag{4.20}$$

where g is the degeneracy of the level, N_c (or N_v) is the effective density of states in the band with which the trapped carriers interact, σ is the capture

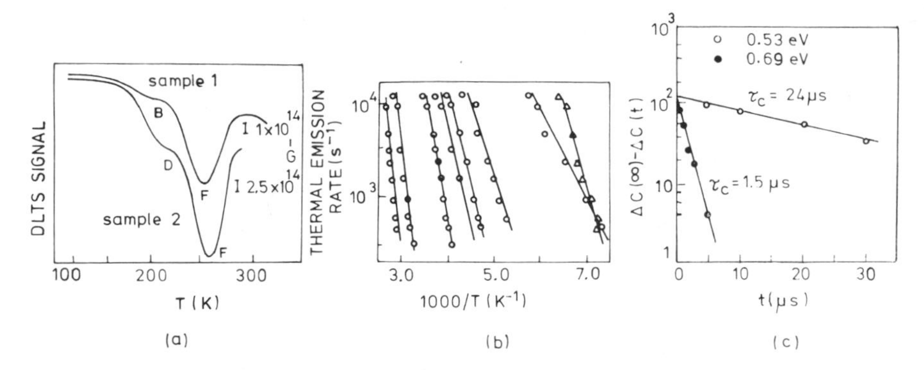

Figure 4.17. (a) DLTS spectra of two polycrystalline Cu/CdS junctions. (b) Arrhenius plots of the emission rate constant for different traps observed in polycrystalline CdS. (c) Dependence of filled-trap concentration on pulse duration for two traps in polycrystalline CdS (after Besomi and Wessels[50]).

cross-section, V_{th} is the thermal velocity, and ΔE represents the energy of the trap, provided σ is independent of temperature. Since the product $N_c V_{th}$ is proportional to T^2, the activation energy of the trap can be determined from the plot of $\log e/T^2$ vs. $1/T$, as illustrated in Figure 4.17b for several traps observed in polycrystalline CdS.[50] Experimentally, $e = 1/\tau$ is obtained from the relation[49]

$$L = \frac{G_c G_L \tau}{T_B} \Delta C_0 \exp\left(\frac{-T_d}{\tau}\right)\left[1 - \exp\left(\frac{-Te}{2\tau}\right)\right]^2 \tag{4.21}$$

where L is the output of the lock-in amplifier, G_c is the gain of the capacitance meter, G_L is the gain of the amplifier and T_d, T_B, and ΔC_0 are as shown in Figure 4.16b. The time constant τ_p of the transient for which the lock-in output is maximum is given by[49]

$$\left(1 + \frac{T_d}{\tau_p}\right) - \left[\exp\left(\frac{-Te}{2\tau_p}\right)\right]\left[\frac{Te}{\tau_p} + \left(1 + \frac{T_d}{\tau_p}\right)\right] = 0 \tag{4.22}$$

It is obvious from equations (4.21) and (4.22) that e or τ of the transient at the peak temperature is fixed by the values of T_B and T_d.

If the pulse is large enough to fill all the traps, the trap and doping density are spatially uniform, and the trap density $N_T \ll N_A - N_D$, then

$$N_T = 2(\Delta C_0/C)(N_A - N_D) \tag{4.23}$$

where ΔC_0 is the capacitance change at the end of the bias pulse, C is the reverse-biased capacitance, and $N_A - N_D$ is the net doping. ΔC_0 can be obtained from the measured peak height,[49] L_{max}, by substituting the values of τ_p, obtained from equation (4.22), in equation (4.21).

The capture rate of the trap, c, is expressed by[49]

$$N(t) = N_T [1 - \exp(-ct)] \tag{4.24}$$

where $N(t)$ is the density of traps filled by a bias pulse of width t and N_T is the total trap density. $N(t)$ values are obtained from peak heights measured at different bias pulse widths and N_T corresponds to the pulse width which saturates all the traps. A plot of $\ln\{[N_T - N(t)]/N_T\}$ vs. t (Figure 4.17c) yields the value of c from which the capture cross-section σ can be determined through the relation $c = \sigma N V_{th}$. In Figure 4.17c, N_T and $N(t)$ are measured in terms of the change in capacitance $\Delta C(\infty)$ and $\Delta C(t)$. N is the injected carrier density and is equal to the doping density for the majority carrier pulse. It is, however, not easy to determine N for the minority carrier injecting pulse.

A majority carrier pulse injects majority carriers only, which reduce the reverse bias as well as the depletion region width. However, photoinjection or forward biasing of the device introduces both majority and minority carriers, whose relative ratio can be controlled by the magnitude of the injection current. For high-level injection conditions, this ratio approaches unity and the trapped minority carrier population approaches $\sigma_{min}/(\sigma_{min} + \sigma_{maj})$. It should be pointed out that minority carrier capture measurements are usually less reliable and more complicated.[49]

4.4. *Material Characterization*

In this section we deal with the various experimental techniques that are currently available for structural, compositional, electronic, optical, and optoelectronic characterization of solar cell materials. Since most of the techniques are quite standard and have been extensively used for several years, we shall not describe the experimental setup or apparatus. Instead, we shall merely emphasize the information that can be derived from these measurements.

4.4.1. *Structural Characterization*

The optical and electronic properties of thin films are very sensitively influenced by the crystallographic and microstructural characteristics of the film. Similarly, the electronic behavior of the photovoltaic junction is also affected strongly by the structural features of the interface at the junction.

Several technqiues have been developed which provide images of the morphological, crystallographic, and defect structure of solar cell components and interfaces between different layers with very fine resolution, down to atomic dimensions, although each increase in resolution is achieved at the cost of a reduction in the total area or volume of material examined. A combination of large-area–low-resolution and small-area–high-resolution studies can, in most cases, provide the necessary information.

4.4.1.1. *Crystallographic Characterization*

X-ray diffraction[56–60] is the most precise system for studying the crystal structure of solids, generally requiring no elaborate sample preparation and is essentially nondestructive. Thin surface films, up to about 1000 Å thick, can be investigated using electron diffraction.[61–63] However, sample preparation can be quite tricky since the film has to be separated from the substrate and collected on an electron microscope specimen grid. Thicker films can be characterized by reflection high-energy electron diffraction (RHEED). Analysis of the diffraction patterns obtained by these techniques and comparison with standard ASTM data can reveal the existence of different crystallographic phases in the film, their relative abundance, the lattice parameters, and any preferred orientations. From the width of the diffraction line, it is possible to estimate the average grain size in the film.[56] A relatively recent crystal-structure-sensitive technique, with a spatial resolution of about 1 μm, is the use of electron channeling patterns[64] in the scanning electron microscope (SEM). A display reflecting the crystal structure can be generated by rocking the incident electron beam around a fixed point on the sample surface. In the scanning-transmission electron microscope (STEM), diffraction information from regions as small as 50 to 100 Å can be obtained.

4.4.1.2. *Microstructural Characterization*

The simple reflection and transmission imaging modes of an optical microscope can be extended using various interference and phase contrast techniques to conveniently study the morphology and structure of a photovoltaic device.[65,66] As previously described (Section 4.2.3), optical spot scanning can be utilized to study directly the actual energy conversion operation of the cell. All the optical techniques are, however, limited in spatial resolution to a few tenths of a micrometer.

A broad range of electron beam techniques that allow extension of the resolution are available. The most versatile among these is scanning electron microscopy. The convenience of conventional scanning electron microscopy is partly due to the long working distance between the final lens and the sample surface and partly due to the ability to directly study almost any free surface. Further, the SEM offers several modes of operation.[61] The most widely employed is secondary electron imaging, which gives images of better than 100 Å resolution, almost unlimited depth of field, and good contrast between most cell components.[67–70] Relatively simple modifications give backscattered electron or cathodoluminescence images, which allow the investigator to study composition variations and/or enhance the contrast between different phases. EBIC imaging[68] and voltage contrast imaging[71] give direct insight into the electronic functioning of the device. A comparison of the various modes of operation of an SEM is given in Table 4.1.

Higher resolutions than are possible in the SEM can be achieved in the transmission electron microscope (TEM). However, the use of the TEM requires preparation of samples less than about 1000 Å thick. Sample preparation in a multilayer device is not a simple proposition, but once it is achieved, it is possible to see structural details at atomic resolutions. In recent years, ingenius sample preparation techniques have been developed that allow one to study the morphology and defect structure at the interface between two layers and/or at the active junction. We describe these sample preparation techniques below, giving examples from studies of Cu_2S/CdS thin film heterojunction solar cells to illustrate the techniques.

The Cu_2S/CdS interface morphology can be examined by etching away the Cu_2S in a KCN solution leaving behind the converted CdS surface, which can be examined in the secondary imaging mode as shown in Figure 4.18a. Observations of the Cu_2S film structure can be made by separating the Cu_2S from the host CdS film,[72] accomplished by slowly etching away

Table 4.1. Different Modes of the SEM (after Grundy and Jones[61])

Mode	Type of signal collected	Contrast information	Spatial resolution
Reflective	Reflected electrons	Compositional Crystallographic	100 nm
Emissive	Secondary emitted electrons	Topographic Voltage Magnetic and electric fields	10 nm 100 nm 1 μm
Luminescent	Photons	Compositional	100 nm
Conductive	Specimen currents	Induced conductivity	100 nm
Absorptive	Absorbed specimen currents	Topographic	1 μm
X-ray	X-ray photons	Compositional	1 μm
Transmissive	Transmitted electrons	Crystallographic	1–10 nm

Figure 4.18. Scanning electron micrographs of (a) Cu_2S/CdS interface revealed after etching away Cu_2S in KCN, (b) free-standing Cu_2S separated from a Cu_2S/CdS cell, and (c) cross-section of a Cu_2S film. Scale bars: (a) 1 μm, (b) and (c) 6 μm.

the CdS in an etching solution, for example, one part concentrated HCl to two parts H_2O. The HCl first undercuts the CdS layer at the substrate interface and subsequently dissolves the CdS until the Cu_2S floats free. The free-standing Cu_2S structure can be collected on a specimen grid and rinsed, dried, and examined in the transmission and scanning electron microscopes. Cu_2S samples prepared by this technique have been examined for spectral transmission also. An SEM micrograph of a free-standing Cu_2S sample is shown in Figure 4.18b.

In another variation,[72] the Cu_2S/CdS film cell is cemented to a glass slide using a suitable epoxy resin, with the Cu_2S surface down. The substrate, in some cases with part of the CdS film attached, is carefully peeled away. The remaining CdS film is then dissolved in HCl, leaving the Cu_2S layer cemented to the glass slide. This Cu_2S film can then be examined in the SEM.

Fractographs of the Cu_2S/CdS junction have been obtained by potting the devices with epoxy/polyester resin and cross-sectioning. The potted cross-sectioned samples are polished and ground to get below the region damaged by cross-sectioning. The CdS at the exposed surface is then etched in 1 : 2 HCl : H_2O solution to remove several micrometers of CdS from the top of the polished surface in order to reveal clearly the Cu_2S layer and the junction profile in the SEM. A secondary emission micrograph of such a sample is shown in Figure 4.18c.

Since the Cu_2S layer thickness in the usual thin film Cu_2S/CdS devices is quite low ($\sim$2000 Å), some workers have angle-lapped Cu_2S/CdS junctions at very small angles to increase the effective thickness as viewed in the microscope. Angle-lapped Cu_2S/CdS specimens have been examined in the SEM in the voltage contrast[71] and cathodoluminescence[70] modes to infer the structure at the junction. In this connection, we should mention that Titchmarsh et al.[73] have developed chemical jet thinning techniques to prepare TEM specimens of cross-sections of several multilayer heterostructures and obtained micrographs of defects present at the various interfaces.

Finally, we may note that a combination of TEM and SEM, the scanning transmission electron microscope (STEM), has the major advantages that it can be operated in either the transmission or in the scanning mode, provides better resolution for surface studies as compared to the SEM, and possesses the capability of performing small-area (50–100 Å) diffraction and chemical analysis studies.

4.4.2. *Compositional Analysis*

The general trend in compositional analysis has been toward physical rather than chemical techniques. Of particular interest are those which can be applied to completed devices.

Detection of characteristic X-ray lines forms the basis of several techniques. In X-ray fluorescence,[74] the characteristic lines are excited by short-wavelength X-ray. However, with this technique lateral resolution is poor. Submicrometer lateral and depth resolution is provided by using electron excitation (EPMA-electron probe microanalyzer) using an SEM or STEM.[75] The energy dispersion analysis of the emitted X-rays (EDAX) is done by solid state detectors that have larger acceptance angles and much higher counting efficiencies than crystal monochromators. The spatial resolution of EDAX with bulk samples is limited to about 1 μm because of the effects of elastically scattered electrons and secondary fluorescence. However, in the STEM, for thin samples, the spatial resolution of EDAX can be less than or equal to 50 Å. In conjunction with ion milling, a depth profile of elemental composition can be obtained by EPMA. The depth resolution is, for the same reasons as above, of the order of 1 μm.

Improved lateral and depth resolution is possible using low-energy (1–5 keV) electrons to stimulate emission of Auger electrons from the first few monolayers of the surface. This technique, called Auger electron spectroscopy (AES), is available in a scanning mode (scanning Auger microprobe-SAM) and yields a lateral resolution about equal to the beam size and a depth resolution of from 10 to 30 Å. In the scanning mode, a point-to-point chemical mapping of elemental concentration is possible. A particular advantage of the AES is the high sensitivity to lighter elements. By employing stationary or scanning ion beams to progressively remove material and then repeating the surface analysis, we can obtain composition gradient information.[76] Care must be exercised to eliminate errors and artifacts arising from selective sputtering, channeling, and other effects. A typical compositional depth profile of a Cu_2S/CdS solar cell, as determined by AES, is shown in Figure 4.19.

The characteristic X-ray from a Mg or Al anode is used to stimulate electron emission from the valence band of materials in a technique known as electron spectroscopy for chemical analysis (ESCA). This technique[77] is not, in general, spatially resolving, but gives a depth resolution of from 10–30 Å and can be used to determine the composition gradient in a manner similar to AES. The major utility of ESCA lies in the fact that the electron energies yield information on the chemical state of the detected element and, hence, ESCA is particularly useful in the study of alloys and compounds.

Secondary ion mass spectroscopy (SIMS) analyzes the ions emitted from a material during ion milling and is thus inherently a depth-profiling technique.[78] SIMS can detect elements in trace concentrations (parts per million) and also distinguish among different isotopes.

All three techniques — AES, ESCA, and SIMS — usually involve ultrahigh-vacuum conditions and are destructive techniques in the depth-

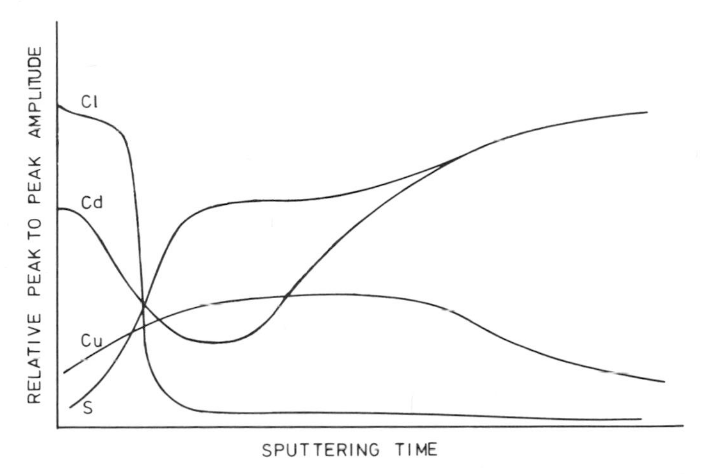

Figure 4.19. Auger depth profile of a Cu_2S/CdS thin film solar cell immediately after solid-state reaction.

profiling mode. Rutherford backscattering (RBS)[79] is a nondestructive depth-profiling technique, but it requires a particle accelerator to provide the probing beam.

In addition to these techniques, the standard analytical methods that can be applied to thin films are atomic absorption, neutron activation analysis, and spark spectrography. These are usually destructive, however, and give only average analyses of the film. Castel and Vedel[80] have developed an electrochemical analysis of semiconductor materials which yields the oxide layer thickness, stoichiometry, and equivalent thickness of the semiconductor film. As applied to Cu_2S films, the analysis is based on the voltametric endpoint detection of the following reactions:

$$Cu_2O + H_2O + 2e \rightarrow 2Cu + 2OH^- \tag{4.25}$$

$$Cu_xS + (2-x)e \rightarrow (x/2)Cu_2S + [(2-x)/2]HS^- + [(2-x)/2]H^+ \tag{4.26}$$

and

$$(x/2)Cu_2S + xe \rightarrow x\,Cu + (x/2)HS^- \tag{4.27}$$

If Q_1 and Q_2 are the charges required for completion of reactions (4.26) and (4.27), respectively, the stoichiometry factor is computed from the relation

$$x = 2Q_2/(Q_1 + Q_2) \tag{4.28}$$

The equivalent thickness of Cu_xS, δ, is obtained from

$$\delta = Q_2 W/2Fa\rho \tag{4.29}$$

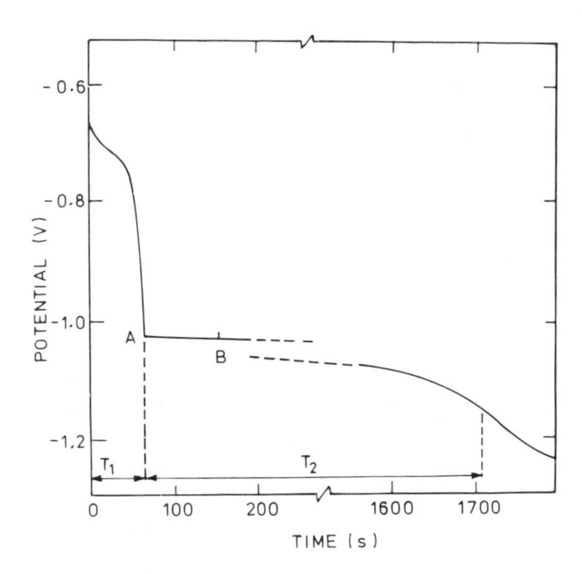

Figure 4.20. Potential vs. time plot for the electrochemical analysis of Cu_xS (after Castel and Vedel[80]).

where W is the molecular weight of Cu_2S (159.12), F is equal to one Faraday (96,486 coulombs), ρ is the density of Cu_2S (5.5 g cm^{-3}), and a is the area of the specimen. The endpoints are experimentally determined by cathodically reducing Cu_2S at constant current in a 0.1 M sodium acetate aqueous solution and recording the potential (with reference to the Ag/AgCl/KCl M electrode) variation vs. time, as shown in Figure 4.20. Typical reduction potentials are: CuO(+CuS): −0.16 V, Cu_2O: −0.45 V, Cu_xS: −0.70 V, and Cu_2S: −1.00 V. Each step in the potential-vs.-time plot denotes an endpoint. Equation (4.28) can then be written in terms of T_1 and T_2, where T_1 and T_2 are denoted in Figure 4.20,

$$x = 2T_1/(T_1 + T_2) \tag{4.30}$$

If oxides are present, the Cu_xS layer composition can be expressed by $y(CuO + CuS)$, $z\,Cu_2O$, and $(1-y-z)Cu_2S$ for which the reduction reactions are:

$$y(CuO + CuS),\ z\,Cu_2O,\ (1-y-z)Cu_2S + y\,H_2O + 2ye^- \rightarrow z\,Cu_2O,\ (1-z)Cu_2S + 2y\,OH^- \tag{4.31}$$

$$z\,Cu_2O,\ (1-z)Cu_2S + 2ze^- + z\,H_2O \rightarrow 2z\,Cu + (1-z)Cu_2S + 2z\,OH^- \tag{4.32}$$

and

$$(1-z)Cu_2S + 2(1-z)e^- + (1-z)H_2O \rightarrow 2(1-z)Cu + (1-z)OH^- + (1-z)HS^- \quad (4.33)$$

with z and y derived from the relations:

$$z = t_0'/(t_1 + t_0') \quad (4.34)$$

and

$$y = t_0/(t_1 + t_0') \quad (4.35)$$

where t_0, t_0', and t_1 are the time values for reactions (4.31), (4.32), and (4.33), respectively. The electrochemical analysis technique has recently been applied to Cu_xSe films.[81]

In Table 4.2, we have compared the capabilities and limitations of various compositional analysis techniques.

4.4.3. *Optical Characterization*

A large variety of experimental techniques has been employed to determine the optical constants (n and k) of thin films. A critical discussion of these techniques is given by Chopra.[82] The most commonly employed method involves separate determination of n and k from reflectance and transmittance measurements on the same film. Under the condition that the film thickness is such that the effects of multiple reflections are suppressed, the transmittance T of a film of index $n_1 - ik_1$ and thickness t is given by[82]

$$T = \frac{16n_0(n_1^2 + k_1^2)\exp(-4\pi k_1 t/\lambda)}{[(n_1+1)^2 + k_1^2][(n_0+n_1)^2 + k_1^2]} \qquad (n_1 > n_0) \quad (4.36)$$

where n_0 is the index of the substrate and the ambient is assumed to be air. A plot of $\log T$ vs. t would then yield the value of k_1 from the intercept as well as the slope. If interference and multiple reflections are neglected, T and R are related by $T = (1-R)\exp(-4\pi k_1 t/\lambda)$. When reflection at the film/substrate interface is taken into account, $T = (1-R)^2\exp(-4\pi k_1 t/\lambda)$ for $n_0 < n_1$, $k_0 = 0$. Since the absorption coefficient α is equal to $4\pi k/\lambda$, measurement of R and T data on the same film offers the most convenient method for determination of α.

The value of α can be determined at different wavelengths using a spectrophotometer. The spectral variation of α can be fitted to either equation (2.167) or (2.170) to determine the nature (direct or indirect) of the optical transition and the value of the optical gap E_g.

Table 4.2. Comparison of Various Techniques for Composition Determination

Method	Probe	Detected species	Probe diameter	Depth resolution	Detection limit	Quantitative analysis	Chemical state	Imaging capability	Depth profiling	Specimen destruction
AES	Electron beam	Auger electrons	0.2 μm	5–40 Å	0.1 at.%	Yes	Good	Yes	Yes	No
ESCA	X-rays	Core and valence electrons	5 mm	10–100 Å	0.5 at.%	Yes	Difficult	Good	Yes	No
SIMS	Ions	Ions	10 μm	< 10 Å	1 ppm	Very difficult	Yes	Yes	Yes	Yes
X-ray fluorescence	X-rays or γ-rays	X-rays	~ 1 cm	2–3 μm	0.01 at.%	Yes	No	No	No	No
EPMA	Electron beam	Characteristic X-rays	⩽ 100 Å	~ 1–2 μm	0.1 at.%	Yes	Yes	No	Yes	No
RBS	α-particles	α-particles	1 mm	100 Å	10^{-3} at.%	Yes	No	No	Yes	No

Determination of plasma resonance frequency from reflectance measurements of highly degenerate semiconductor films can provide values of either the carrier concentration n or the effective mass m^*, provided one of the quantities is known.[83] Measurements of the position of the extrema of the R and T of degenerate semiconductors, in particular Cu_2S, also form the basis for the determination of the stoichiometry of the films, as suggested by Rajkanan,[84] who has derived a family of $\lambda_{\min R}$- and $\lambda_{\max T}$-vs. t-curves for Cu_2S e%films of different stoichiometry. The positions of the principal reflectance minima and the principal transmittance maxima uniquely determine the film stoichiometry. Figure 4.21 shows the plot of $\lambda_{\min R}$ and $\lambda_{\max T}$ vs. t for Cu_xS films, with x as a parameter.

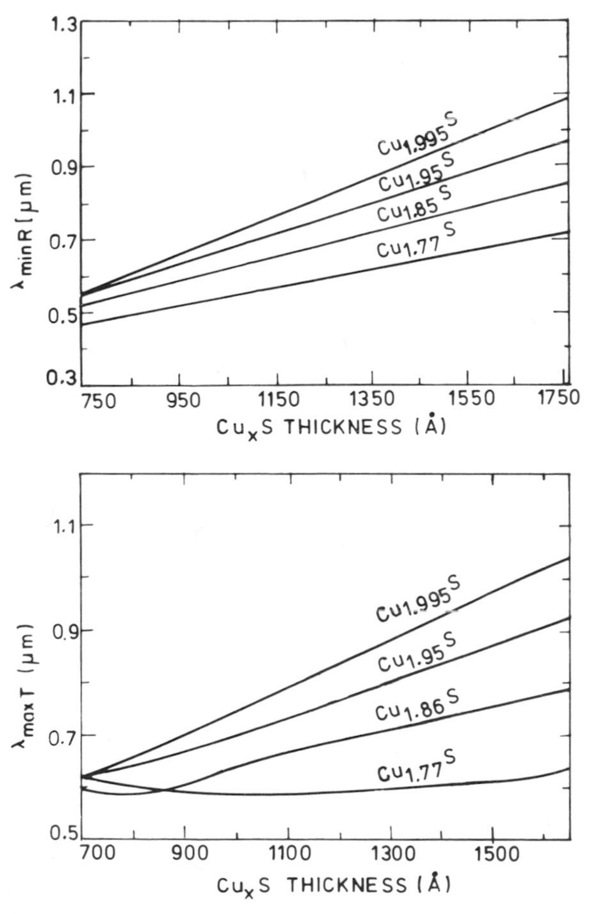

Figure 4.21. Positions of the principal reflectance minimum ($\lambda_{\min R}$) and transmittance maximum ($\lambda_{\max T}$) as a function of thickness of Cu_xS films with x as a parameter (after Rajkanan[84]).

The above optical techniques are valid for specularly reflecting and transmitting samples. However, it is important, for purposes of photon loss analysis, to be able to determine reflection and absorption losses in different layers of a solar cell as well as of the whole device. This is usually achieved[85] by measuring the total reflectance and total transmittance, using an integrating sphere attachment with the spectrophotometer.

4.4.4. *Electrical and Optoelectronic Characterization*

Electronic characterization of a semiconductor involves determination of the resistivity ρ, the mobility of the carriers (μ_n or μ_p), the carrier concentration (n or p), the Hall coefficient (R_H), the minority carrier diffusion length (L_p or L_n), the minority carrier lifetimes (τ_n or τ_p), and the temperature dependence of these parameters.

The techniques for measuring ρ, μ, and n are quite well known and have been documented in the literature.[86,87] We shall, therefore, not discuss them here, except to note that the measurement methods commonly employed include: (i) two-point method for resistivity, (ii) linear four-point probe method for resistivity, (iii) spreading resistance method for resistivity, (iv) noncollinear four-point probe method for resistivity, (v) Vander Pauw technique for resistivity, Hall coefficient, and Hall mobility, (vi) Hall effect measurements, and (vii) Haynes–Shockley method for mobility.

The temperature variation of the mobility can be fitted to one from among equations (2.49) to (2.58) to reveal the carrier transport mechanism. The temperature dependence of the carrier concentration yields values for the Fermi level and impurity levels through equations (2.8) and (2.9). Measurements under illumination give photoconductivity and photo-Hall coefficient values from which the mobility and carrier concentration in light can be determined.

Photoconductivity measurements involving response time measurements or thermally stimulated conductivity can be very useful probes for deducing the transport parameters of semiconductor films.[88,89] Information about traps is obtained from the intensity dependence of photocurrent and decay time. Thermal and optical quenching of photoconductivity yield ionization energies for impurities. Spectral response also yields information about trap levels.

One of the most important applications of photoconductivity measurements is the determination of the minority carrier lifetime by the photoconductive decay method. Excess carriers are generated by illuminating the sample with light of energy greater than the bandgap. The conductivity of the sample is directly proportional to the number of carriers, and the change in conductivity $\Delta\sigma$ owing to optical excitation is proportional to the number of excess carriers, namely,

$$\Delta\sigma = C \exp(-t/\tau) \tag{4.37}$$

where τ is the lifetime and C a proportionality constant. τ can be determined by turning off the light and recording the conductivity decay. Spatial variations in lifetime can be measured by using a laser beam and scanning the sample surface.[90]

Surface photovoltage measurements also yield the value of minority carrier diffusion length.[91–93] Under illumination, a surface photovoltage appears at the surface of a semiconductor without the presence of a p/n junction, which can be detected by capacitative coupling techniques. Surface recombination does not influence the measured lifetime value. The light intensity ϕ, expressed in terms of the surface potential, is given by

$$\phi = F(V_s)K(1 + 1/\alpha L) \tag{4.38}$$

where $F(V_s)$ is a function of the surface potential V_s, K is a constant, α is the absorption coefficient, and L is the diffusion length. If α is changed by varying the light wavelength, then the light intensity has to be adjusted to give the same value of V_s at each wavelength, provided $F(V_s)$ remains constant. A plot of ϕ vs. $1/\alpha$ extrapolated to $\phi = 0$ yields the effective value of L, from whence τ can be calculated.

A plot of V_{oc} vs. t for a device in the decay mode has up to three distinct regions corresponding to high-level injection, intermediate-level injection, and low-level injection conditions. The expressions characterizing the V_{oc} decay excitation in these three regions are[94,95]:

$$\tau = \frac{2kT}{q}\left|\frac{1}{dV_{oc}/dt}\right| \quad \text{for high injection} \tag{4.39}$$

$$\tau = \frac{kT}{q}\left|\frac{1}{dV_{oc}/dt}\right| \quad \text{for intermediate injection} \tag{4.40}$$

and

$$V_{oc} = \frac{kT}{q}\left\{\exp\left[\frac{qV(0)}{kT}\right] - 1\right\} \exp\left(\frac{-t}{\tau}\right) \quad \text{for low injection} \tag{4.41}$$

where τ is the minority carrier lifetime and t is the time. $V(0)$ is the V_{oc} at the termination of optical excitation. This analysis is based on the assumption that the contribution to V_{oc} from the heavily doped emitter is negligible. V_{oc} decay can also be observed by abruptly terminating the forward current through the solar cell.[94]

All three techniques noted above for diffusion length and lifetime measurements can be used when the diffusion lengths are greater than or equal to 10 μm and the lifetimes greater than or equal to 0.1 μs. However, in polycrystalline thin films, the diffusion length of minority carriers is generally very low, usually less than or equal to 1 μm. It is, therefore, quite difficult to determine the diffusion length by the conventional techniques discussed above. The use of the SEM with an electron beam diameter from 50 to 100 Å provides a considerable increase in resolution. The diffusion length can now be determined by utilizing the beta-voltaic effect where the minority carriers are generated by electrons. A schematic diagram of the experimental setup for this method, known as the EBIC technique, is shown in Figure 4.22a. The solar cell is fractured to reveal the junction and mounted in the SEM with the junction plane parallel to the electron beam direction. The electron beam is scanned across the junction. If we assume

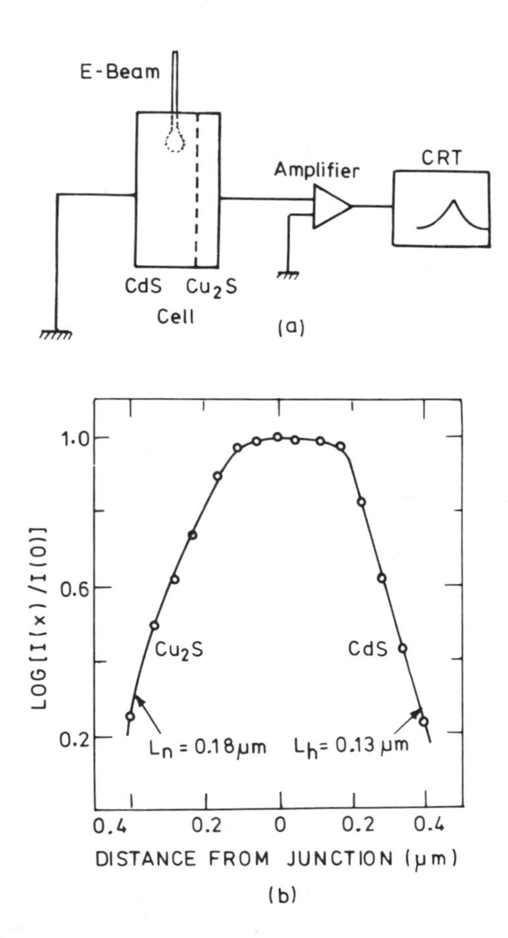

Figure 4.22. (a) Schematic diagram of an experimental setup for measuring minority carrier diffusion length by EBIC technique. (b) Plot of $\ln[I(x)/I(0)]$ vs. x for a Cu_2S/CdS thin film cell.

a semi-infinite junction collector, the EBIC I is related to the distance x from the junction by[96]

$$I = I_0 \exp(-x/L_D) \tag{4.42}$$

where I_0 is the value of I at the junction plane ($x = 0$) and L_D is the minority carrier diffusion length on the side of the junction where the beam is located. The diffusion length is obtained from the inverse of the slope of a $\ln I$-vs.-x plot, shown in Figure 4.22b. Oakes et al.[96] have pointed out that diffusion lengths can be accurately measured by the EBIC technique if $L_D \gg 0.25 R_G$, where R_G is the range of the primary electrons in the material. For $L_D \ll 0.25 R_G$, the EBIC is independent of the value of L_D but depends strongly on the electron beam energy. This implies that lower electron beam energies (~ 5 to 10 keV) are desirable for accurate L_D measurements, since the effective generation depth d (for Cu_xS) is given by

$$d = A E_0^{1.62} = 0.5 R_G \tag{4.43}$$

where E_0 is the electron beam energy in kilovolts and d is in micrometers.

In the optical analog of the EBIC technique, the LBIC technique, the electron beam is replaced by a laser beam and the diffusion length is obtained from a plot of $\ln I_{sc}$ vs. x, in a fashion similar to that described above.[94] However, resolution in the LBIC technique is limited by the beam spot size of the laser ($\gtrsim 1$ μm). Further, surface recombination has a greater influence on LBIC measurements. If surface recombination is taken into account, an effective diffusion length L_{eff} can be calculated from LBIC measurements from the following[94]:

$$\left|\frac{L_{eff}}{L_D}\right|^2 = \frac{1-\exp(-Z)}{1+R\coth(W/2)}\left\{1+\frac{RW[W\coth(W/2)-Z\coth(Z/2)]}{W^2-Z^2}\right\} \tag{4.44}$$

where $W = t/L_D$, $Z = t\alpha$, $R = S\tau/L_D$, with t being the thickness of the solar cell from illuminated surface to back surface, α the absorption coefficient at the laser wavelength, and S the surface recombination velocity.

EBIC measurements can also be made in an alternate lower-accuracy configuration to determine L_D.[96,97] In this method, the sample, say a Cu_xS/CdS cell, is positioned so that the electron beam strikes the surface of the Cu_xS and the generation volume is moved across the junction by varying E_0 to produce different effective generation depths d. The measured EBIC current I is normalized to $I_B E_0$ (where I_B is the beam current) to eliminate effects of different beam currents and the increased minority carrier generation rate at higher beam voltages. A plot of $\ln(I/I_B E_0)$ vs.

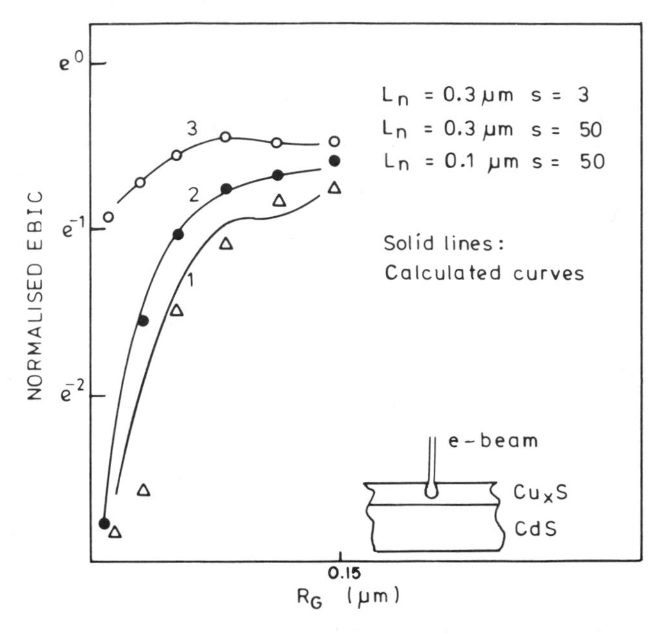

Figure 4.23. Plots of normalized EBIC as a function of average generation depth in Cu_xS for Cu_xS/CdS cells treated under different conditions. Also shown are the theoretical curves for different L_n and s (normalized surface recombination) values: (Curve 1) after dipping; (Curve 2) after heat treatment in air; (Curve 3) after H_2 glow discharge treatment and storage (after Schock[97]).

effective generation depth yields a value of L_D as long as E_0 is low enough for generation to be primarily confined to Cu_xS. Schock[97] has plotted normalized EBIC currents as a function of the primary electron range in Cu_xS and compared the results with theoretical curves to obtain values of L_D and S, as shown in Figure 4.23.

5

Thin Film Deposition Techniques

5.1. *Introduction*

As already pointed out in Chapter 1, a deposition technique and its associated process parameters have a characteristic effect on the nucleation- and growth-dominated microstructure of a thin film and thereby on its physical properties. Two-dimensional materials of thicknesses ranging from angstroms to hundreds of micrometers can be prepared by a host of so-called thin film as well as thick film techniques. The latter methods involve the preparation of thin materials from a paste or liquid form of the bulk material. The two sets of techniques yield thin film materials of widely different microstructures and properties.

Besides performance criteria, large-area thin film solar cells have to be economically viable for terrestrial applications. This makes it necessary to exploit both thin and thick film techniques that satisfy the criteria of simplicity, cost effectiveness, large-area uniform and controlled deposition, and that yield well-defined structural, metallurgical, and electro-optical properties. The deposition techniques have been reviewed extensively in the literature.[1–6] We present in this chapter a brief review of the techniques most relevant to films for solar cell applications. Some of the emerging and promising techniques that have not yet been reviewed in the literature are emphasized.

A thin film deposition process involves three steps: (i) creation of atomic/molecular/ionic species, (ii) transport of these species through a medium, and (iii) condensation of the species on a substrate. Depending on whether the vapor species has been created by a physical process (such as thermal evaporation and sputtering), by a chemical, electroless, or electrochemical process, we can broadly classify the deposition techniques under the following headings: (1) physical vapor deposition (PVD), (2) chemical vapor deposition (CVD), (3) electroless or solution growth, and (4) electrochemical deposition (ECD). By combining PVD with CVD, hybrid techniques such as reactive evaporation/sputtering and plasma deposition have been established. The principles and characteristics of some of the relevant techniques are described in the following sections.

5.2. Physical Vapor Deposition

5.2.1. Vacuum Evaporation

5.2.1.1. Kinetics

The evaporation of a material requires that it be heated to a sufficiently high temperature to produce the desired vapor pressure. The rate of free evaporation of vapor atoms from a clean surface of unit area in vacuum is given by the Langmuir–Dushman kinetic theory equation:

$$N_e = 3.513 \times 10^{22} P_e/(\mathrm{MT})^{1/2} \qquad [\text{molecules cm}^{-2}\,\text{s}^{-1}] \tag{5.1}$$

where P_e is the equilibrium vapor pressure (in torr) of the evaporant under saturated-vapor conditions at a temperature T, and M is the molecular weight of the vapor species. The vapor atoms traverse the medium and are made to condense on a substrate surface to form a thin film. The rate of condensation/deposition of the vapor atoms depends on the vapor-source–substrate geometry and the condensation coefficient on the surface under given physical conditions.

The vapor atoms are scattered by collisions with residual gas atoms in the vacuum system. The scattering probability is $\exp(-d/\lambda)$, where d is the source–substrate distance and λ is the mean free path of the gas atoms. In addition, the gas molecules impinge on the substrate surface at a rate given by equation (5.1), where, of course, the parameters P_e, T, and M refer to the gas molecules at temperature T. It is clear from the kinetic data for air given in Table 5.1 that deposition of films at normal rates of from 1 to 10 Å s^{-1} must be carried out in a vacuum better than 10^{-5} torr if significant gas contamination of the film is to be avoided. Fortunately, however, the sticking coefficient of gas atoms at elevated temperatures is a small fraction thus making pressures of 10^{-6} torr good enough for deposition of clean films, except those readily oxidizable, in which case relatively much better vacuum conditions are required.

5.2.1.2. Vapor Species

With the exception of S, Se, Te, Bi, Sb, P, and As, which vaporize in the form of polyatomic clusters depending on vaporization temperature, all

Table 5.1. Some Facts About the Residual Air at 25 °C in a Typical Vacuum Used for Film Deposition (after Chopra[2])

Pressure (torr)	Mean free path (cm) (between collisions)	Collisions/s (between molecules)	Molecules/cm²/s (striking surface)	Monolayers/s[a]
10^{-2}	0.5	9×10^{4}	3.8×10^{18}	4,400
10^{-4}	51	900	3.8×10^{16}	44
10^{-5}	510	90	3.8×10^{15}	4.4
10^{-7}	5.1×10^{4}	0.9	3.8×10^{13}	4.4×10^{-2}
10^{-9}	5.1×10^{6}	9×10^{-3}	3.8×10^{11}	4.4×10^{-4}

[a] Assuming the condensation coefficient is unity.

other elements vaporize from a solid (sublimation) or liquid phase in the form of neutral atoms. A very small fraction thereof may be charged owing to the thermal ionization of the atoms and impurities.

Vaporization of alloys and compounds is usually accompanied by dissociation or association, or both processes. If the volatilities of the various constituents are significantly different from one another, thermal decomposition takes place. If the constituents are equally volatile, "congruent" evaporation occurs. If not, evaporation is incongruent and hence the compositions of the vapor and the condensate differ from the composition of the source. This difference is further aggravated if the condensation coefficients of the constituent vapor atoms differ from one other. Generally, the tendency to dissociate is greater with higher evaporation temperatures and lower pressures. For example, binary oxides evaporating at high temperatures do so by dissociation. Only very few compounds such as MgF_2, B_2O_3, CaF_2, SiO, GeO, and SnO evaporate without dissociation.

The constituents of alloys evaporate independently of each other, mostly as single atoms even if one of the elements is known to form molecules. The application of Raoult's law to the evaporation of a liquid alloy yields the ratio of the number of evaporated atoms of A and B components as

$$\frac{N_A}{N_B} = \frac{C_A}{C_B} \times \frac{p_A}{p_B} \times \left(\frac{M_B}{M_A}\right)^{1/2} \tag{5.2}$$

where C_A and C_B are the atomic fractions of the components. Deviations of real alloy solutions from ideal behavior is taken care of by an activity coefficient in this equation.

5.2.1.3. *Reactive Evaporation*

By allowing a chemical reaction between vapor species of different elements either during their transport from source to substrate, or on the substrate surface itself, it is possible to condense films of a great variety of alloys and compounds. This reactive evaporation technique is clearly expected to be sensitively dependent on the kinetic and thermodynamic parameters of the vapor systems concerned.

The effectiveness of the reactive deposition process is considerably enhanced by making the vapor species more reactive. One way to achieve this is to ionize the vapor species by bombardment with energetic electrons from a thermally emitting source. In another version, the ions of one of the reacting species are created by a glow discharge through which the other species traverses. This technique, called "activated reactive evaporation (ARE)," has been used very sucessfully to get excellent optical and electrical quality ITO and Cu_2S films. Another method of making reactions effective is to provide the atomic forms of stable diatomic molecules, e.g., those of O_2 (for oxidation), H_2 (for hydrides), and N_2 (for nitrides). For example, alloy films of hydrogenated amorphous Si have been obtained by evaporating Si in the presence of atomic H obtained by the pyrolysis of H_2 on a hot W filament.[7]

5.2.1.4. *Evaporation Sources*

The temperature of a material for evaporation may be raised by direct or indirect heating. The simplest and most common method is to support the material in a filament or a boat which is heated electrically. Many shapes and sizes of filaments and boats[2,3] of a number of materials are commercially available to suit a range of evaporation materials and applications. The geometrical distribution of vapors from several standard sources is well known and is described in textbooks.

In the case of a uniformly emitting point source (called a Knudsen effusion cell) onto a plane receiver, the rate of deposition follows a $\cos\theta/r^2$ variation (Knudsen cosine law). Here, r is the radial distance of the receiver from the source and θ is the angle between the radial vector and the normal to the receiver direction. For evaporation from a small area onto a parallel plane receiver, the deposition rate is proportional to $\cos^2\theta/r^2$.

Effusion sources of the Knudsen and chimney type are popular for use with high-vapor-pressure materials, particularly semiconductors. Figure 5.1 shows three such sources. A narrow effusion aperture allows homogenization and equilibration of the vapor species of a compound in the source geometry. The source used in our laboratory for evaporation of CdS and similar materials consists of a quartz bottle with a narrow neck. The effusion aperture is defined by a conical Mo shield at the top of the bottle. The bottle is packed with the sintered CdS powder and heated by a cylindrical Mo heater surrounded by a heat shield consisting of fiber frax

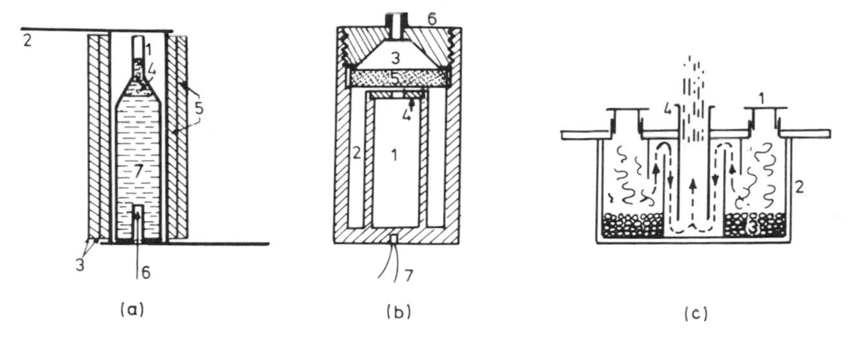

Figure 5.1. Schematic diagrams of various sources used for evaporation of: (a) CdS: 1 — quartz crucible, 2 — Mo heater, 3 — Ta heat shields, 4 — quartz wool, 5 — Al_2O_3 fiber frax thermal insulation, 6 — thermocouple, and 7 — charge; (b) $Zn_xCd_{1-x}S$ alloy: 1 — CdS cavity, 2 — ZnS annulus, 3 — mixing chamber, 4 — adjustable CdS orifice, 5 — filler, 6 — cap and nozzle assembly, and 7 — thermocouple; (c) SiO: 1 — baffle, 2 — Mo boat, 3 — charge, and 4 — chimney.

and Ta. A variant of this source used by the Delaware group for controlled evaporation of (Cd Zn)S consists of two in-line concentric graphite sources surrounded by a cylindrical Ta resistance heater. By adjusting the physical parameters of the two concentric sources, the relative evaporation rates of CdS and ZnS are controlled with a single heater.

Multiple evaporation sources are essential for multilayer technology as well as for obtaining controlled composition alloys/semiconductors. Besides controlling the vapor pressures of the constituents, the substrate temperature provides an important control over the relative sticking coefficients and thermodynamic stability of the different vapor species. Thus, a "multiple" source and temperature technique is ideally suited for preparing films of well-defined composition multicomponent semiconductors and has been used extensively for IV–VI, II–VI, and III–V compound films.

Table 5.2 lists some useful properties, evaporation parameters, and recommended vapor sources for materials of interest for solar cell applications.

5.2.1.5. *Vacuum Setups*

Vacuum evaporation requires a system with a known vacuum and its residual gas analysis. A diffusion pump backed by a rotary pump system continues to be the 10^{-6}–10^{-8} torr workhorse in thin film technology largely because of its modest price, simplicity, and high speed. By using special diffusion pump oil (e.g., polyphenyl ether), a cryogenic baffle, and an all-metal system, ultrahigh vacuum (UHV) in the range 10^{-8}–10^{-10} torr are easily obtained. The second most common UHV system is based on

Table 5.2. Properties, Evaporation Parameters, and Recommended Vapor Sources for Materials Used in Thin Film Solar Cells

Material	Density ($g\,cm^{-3}$)	Boiling point (°C)	Melting point (°C)	Temperature (°C) for vapor pressure of 10^{-4} torr	Temperature (°C) for vapor pressure of 10^{-2} torr	Boat material	Remarks[a]
Ag	10.5	2212	960.8	832	1027	Mo, Ta	
Al	2.7	2467	660.2	972	1217	graphite	
Au	19.3	2966	1063	1132	1397	W, Mo	
Cr	7.2	2482	1890	1157	1397	W	
Cu	8.9	2595	1083	1027	1257	Mo, Ta	
Fe	7.9	3000	1535	1227	1477	W	
Ge	5.3	2830	937.4	1137	1397	Mo, Ta, W	
Mo	10.2	5560	2610	2027	2527	Self-evap. fine wire	
Ni	8.9	2732	1453	1262	1527	Al_2O_3	
Pb	11.3	1744	327.5	547	715	W, Mo	
Pd	12.0	2927	1552	1197	1467	Al_2O_3	
Pt	21.5	3827	1769	1747	2097	W	
Si	2.2	2355	1410	1337	1639	BeO, C, e-beam	$T_s = 550\,°C$ $r_d = 20\,\mu m\,min^{-1}$
Ta	16.6	5425	2996	2587	3057	Self-evap. fine wire e-beam	
Ti	4.5	3260	1675	1442	1737	W, graphite	
W	19.3	5927	3410	2757	3227	self-evap. fine wire, e-beam	
CdS	4.8	sublimes	1750 at 100 atm. pressure	920	—	graphite quartz	$T_s \approx 200$–$250\,°C$ $T_{s0} = 900$–$1050\,°C$, $r_d = 1$–$2\,\mu m\,min^{-1}$, Molecular species — Cd, S
CdSe	5.8	—	> 1350	—	—	Two- source (Mo, Ta), (Ta, Mo)	$T_s > 500\,°C$
CdTe	6.2	—	1040	—	—	Two- source (Ta, Mo), Mo	$T_s \geq 250\,°C$
Cu_2O	6.0	—	1235	600		e-beam	
Cu_2S	—	—	1100	—	—	—	
Cu_2Se	6.75	—	1113			Two- source Mo	$T_s = 160\,°C$

Table 5.2 (continued)

Material	Density ($g\,cm^{-3}$)	Boiling point (°C)	Melting point (°C)	Temperature (°C) for vapor pressure of 10^{-4} torr	Temperature (°C) for vapor pressure of 10^{-2} torr	Boat material	Remarks[a]
GaAs	5.3	—	1238			Two-source, flash evap.	
In_2O_3	7.2	—	—	200	—	BeO	In presence of O_2 $T_s = 420\,°C$, $r_d = 0.8$ Å s^{-1}
InP	4.8	—	1070	730	—	Ta, W	Two-source: $T_s \simeq 225\,°C$
MgF_2	3.2	2239	1266	1540	—	W, Mo, C	
Nb_2O_5	7.5	—	1460	—	—	—	
Si_3N_4	3.44	—	1900	—	—	—	
SiO	2.1	1880	1702	600	—	W, Ta	
SiO_2	2.7	2230	1610	850	—	Decomposes	
SnO_x	0.45	—	1080 (decomposes)	—	—	W	
Ta_2O_5	8.7	—	1800	1920	—	Ta, W	
TiO_2	4.3	2500-3000	~1830	1000		Ta, W, filament	
ZnO	5.6	—	1975	—	—	W, Mo	
Zn_3P_2	4.55	1100	>420	—	—	—	$T_s \geqslant 220\,°C$ $T_{s0} \simeq 750\,°C$
ZnS	4.1	Sublimes	1850	300		Ta, Mo, C	
$CuInS_2$						Mo, W, flash. evap.	
$CuInSe_2$						Mo, W	Three-source; for two-source: $T_s = 225\,°C$ $T_{s0} \sim 1150\,°C$ ($CuInSe_2$) ~200–450 °C (Se) $r_d \sim 5$ Å s^{-1}
$Zn_xCd_{1-x}S$						Graphite quartz	Dual chamber single source: $T_s = 200–250\,°C$; $T_{s0} = 900–1050\,°C$, $r_d = 1 - 2\,\mu m\,min^{-1}$

[a] T_s is the substrate temperature, T_{s0} is the source temperature, r_d is the rate of deposition.

sputter ion pumps backed by sorption pumps and assisted by a Ti sublimation pump. Cryopump systems using closed-cycle He-temperature probes are the new and clean, but more expensive, additions in the UHV field. It must be noted that each vacuum system has its own character from the point of view of pumping characteristics, vacuum, and residual gas composition. Further, each system may interact with a thin film condensation process in its own way (see Chopra[2]).

Besides a vapor source, one requires numerous other accessories in the vacuum system. These include shutters, substrate heaters, a planetary system (for uniform deposition over large areas), and monitors and/or controllers for deposition rate and film thickness. All the accessories must necessarily be made of materials compatible with UHV technology from the point of view of degassing and chemical reactions with the required vapors. The deposition rate is commonly monitored and controlled with the help of a quartz crystal oscillator, ion current in a nude ionization gauge and an appropriate mass spectrometer. The film thickness is monitored by integrating the rate monitor signal, or by other in-situ techniques such as an optical monitor (for nonmetallic films). Ideally, the vacuum deposition system should have facilities for structural and composition analysis (see Section 5.2.2.2). In view of their important role in obtaining films of desired properties, a detailed understanding of the deposition accessories is essential. We strongly recommend that the reader consult the standard references.[1–4]

5.2.2. *Epitaxial Deposition*

Thin films are inherently thermodynamically nonequilibrium structures as a result of the nature of the deposition process, in particular, the very high supersaturation, condensation kinetics, and rapid quenching of adsorbed atoms. Consequently, thin films are invariably fine-grained and have a frozen-in high concentration of structural defects. Solar cell applications require large oriented grains or mosaic monocrystals if single crystal films are not possible. Alternatively, techniques must be devised to passivate the grain boundaries so as not to affect the charge recombination processes.

Generally, oriented, large (~ 1 μm) grains are easily obtained by a judicious choice of deposition parameters for a given thin film material and the substrate. By selecting an appropriate single-crystal substrate, mosaic crystals are epitaxially grown in sizes up to tens of micrometers and small-angle grain boundaries. Although vacuum deposition is ideally suited for epitaxial deposition of films, the economic desirability of developing large-area, high-efficiency solar cells by a cheap process has limited investigations of epitaxially grown solar cell films to basic studies and studies for specialized applications such as high concentration and tandem cells. Some of the vacuum deposition techniques employed are described in the following sections.

5.2.2.1. *Hot Wall Epitaxy*

In this technique[8] one or two annular vapor sources are used and the vapors are transported through a heated cylindrical enclosure that is held at a temperature higher than the substrate. Thus, deposition of homogenized and equilibrated multicomponent vapors takes place on the substrate. This technique has been used to obtain good epitaxial films of several IV–VI and II–VI compounds.

5.2.2.2. *Molecular Beam Epitaxy*

Epitaxial growth onto a single crystal substrate obtained by the condensation of one or more directed beams of atoms and/or molecules (in some special cases) from an effusion source in an ultrahigh-vacuum (UHV) system is called molecular beam epitaxy. The low-density vapor beam is obtained from a high-vapor-pressure Knudsen type (effusion) source. Basically a slow evaporation technique, its significant advantages are best derived by making full use of the UHV analytical techniques for obtaining information on the structure, topography, and composition and its depth profile and the chemical state of the surface of the film during its growth. A practical molecular beam epitaxy (MBE) system should therefore include a quadrupole mass spectrometer for analyzing vapor species, sputter ion gun, secondary ion mass spectrometer (SIMS), scanning Auger microprobe (SAM), ESCA, LEED, and HEED. By keeping the deposition rates low (< 1 Å s^{-1}) onto ultraclean substrates, we can achieve epitaxial growth of high perfection at relatively lower substrate temperatures. In the case of multicomponent compounds, kinetics of condensation and thermodynamic reactions of the various ad-atoms can be monitored and controlled literally monolayer-by-monolayer.

Very slow deposition rates allow MBE to obtain multilayer structures of one or more material in a predetermined sequence with variable layer thicknesses ranging from about 10 Å to several micrometers, enabling the formation of quantum-well superstructures, heterostructures, heterojunctions, and graded composition/property structures. The surface of a film may be terminated with a desired material so that passivation, protection, or work-function alteration may be achieved. Further, three-dimensional geometrical structuring of thin films is possible by using appropriate masks/shutters, or simply by "writing" masked patterns utilizing fine vapor beams (~ 10 μm or less).

MBE is obviously a very sophisticated and expensive setup and has thus been used primarily for basic epitaxial growth studies and for specialized microelectronic applications. No doubt, the idea of using fine vapor beams under controlled condensation conditions is attractive for the study of novel structures and concepts in multijunction solar cells. MBE and its less sophisticated versions have been used extensively for the deposition of II–VI and III–V compounds, in particular, GaAs-based ternary (e.g., $Al_{1-x}Ga_xAs$ and $In_{1-x}Ga_xAs$, $GaSb_{1-y}As_y$) and quaternary ($Ga_xIn_{(1-x)}As_yP_{(1-y)}/InP$) heterojunction structures. The MBE technique has been used to deposit solar cell structures of the type $Ga_{1-x}Al_xAs/GaAs$. The technique and its applications have been reviewed by Esaki and Chang(9) and Arthur.(10)

5.2.2.3. *Graphoepitaxy*

Vapor-deposited films on amorphous substrates are generally polycrystalline. Partially textured or oriented films are obtained under suitable deposition conditions. Recently, Smith and co-workers(11,12) have shown that by depositing textured films (of Si in their case) on an amorphous substrate (SiO_2 in their case) having a relief pattern of grating lines with the right periodicity and profile, and subsequent crystallization by a laser (6 W) in a repeated raster pattern results in complete orientation of the film over the grating relief along the grating direction. The required grating spacing, according to this Lincoln laboratory group, should be less than the characteristic grain size of the film to be deposited. By creating a 3.8-micrometer grating of square-well profile through a photolithographic technique on Cr coated SiO_2 substrate, 5000-Å-thick CVD deposited Si films were oriented with (100)-plane parallel to the substrate and ⟨001⟩ directions parallel to the grating direction. A (111)-plane orientation would require a difficult-to-fabricate sawtooth grating profile. Although the technique continues to be an academic curiosity with doubts regarding the orientation processes, it nevertheless is an attractive possibility for growing large-area crystalline films from vapor deposited silicon on X-ray lithographically textured amorphous carbon or other conducting substrates for solar cell applications.

5.2.3. *Sputtering Techniques*

Vapor species may be created by kinetic ejection from the surface of a material (called target or cathode) by bombardment with energetic and nonreactive ions. The ejection process, known as sputtering, takes place as a result of momentum transfer between the impinging ions and the atoms of the target surface. The sputtered atoms are condensed on a substrate to form a film. The sputtering process has the following unique characteristics of interest in thin film technology: (1) In general, the sputtered species are predominantly neutral and atomic. A small ($<1\%$) percentage of the species is charged, both positively and negatively. The molecular or multiatom cluster content is also small and depends on sputtering parameters and the target material. (2) The sputtering yield, defined as the number of ejected atoms per incident ion, increases with the energy and mass of the ions. Typical yield variations with the ion energy for Cu, Ni, and Mo are shown in Figure 5.2. In most cases, the yield increases very slowly beyond ion energies of several thousand electron volts. (3) The yield depends on the angle of incidence of the ions (Figure 5.2) and increases as $(\cos\theta)^{-1}$, where θ is the angle between the normal to the target surface and the beam direction. Deviations from this relationship are observed as seen in Figure 5.2, which shows the sputter-etching rates of various materials. (4) The yield displays an undulatory behavior with periodicity corresponding to the groups of elements in the periodic chart. However, the maximum variation of the yield for various elements for a 1-keV Ar^+ ion is a factor of approximately 5. (5) The yield of a single crystal target increases with decreasing transparency of the crystal in the direction of the ion beam. (6) The energy of the ejected atoms shows a Maxwellian distribution with a long tail toward higher energies. The energy at which the peak of the distribution occurs shifts only slightly with increasing ion energy and, on the average, it is an order of magnitude higher than for thermally evaporated atoms at the same rate.

Since the number of sputtered atoms is proportional to the number of ions, the sputtering process provides a very simple and precise control on the rate of film deposition. However, the sputtering yield being low and the ion currents being limited, sputter deposition rates are invariably lower by one to two orders of magnitude compared to thermal evaporation under normal conditions. High ion current densities (~ 100 mA cm^{-2}) and hence high deposition rates (~ 100 Å s^{-1}) are achieved only in special sputtering geometries such as magnetron sputtering (discussed later). Note that the sputtering process is very inefficient from the energy point of view and most of the energy is converted to heat which becomes a serious limitation at high deposition rates. If the surface of a multicomponent (alloy, compound, or mixture) target does not change metallurgically by thermal diffusion, chemical reaction, or backsputtering processes, the sputtering process ensures layer-by-layer ejection and hence a homogeneous film of composition corresponding to that of the target. In case of multiple targets, the composition will be determined by the respective areas and yields of the target materials. Surface rearrangement and backsputtering can be significant in cases of low-melting-point and high-yield materials and, thus, in such cases, the composition of the film may be very different from that of the target. Finally, the high energy of the ejected particles and the attendant bombardment of the growing film (acting as anode) by the electrons and negative ions has considerable influence on the nucleation and growth of the films and, in particular, yields higher adhesion films. The sputtering rate of different materials is given in Table 5.3.

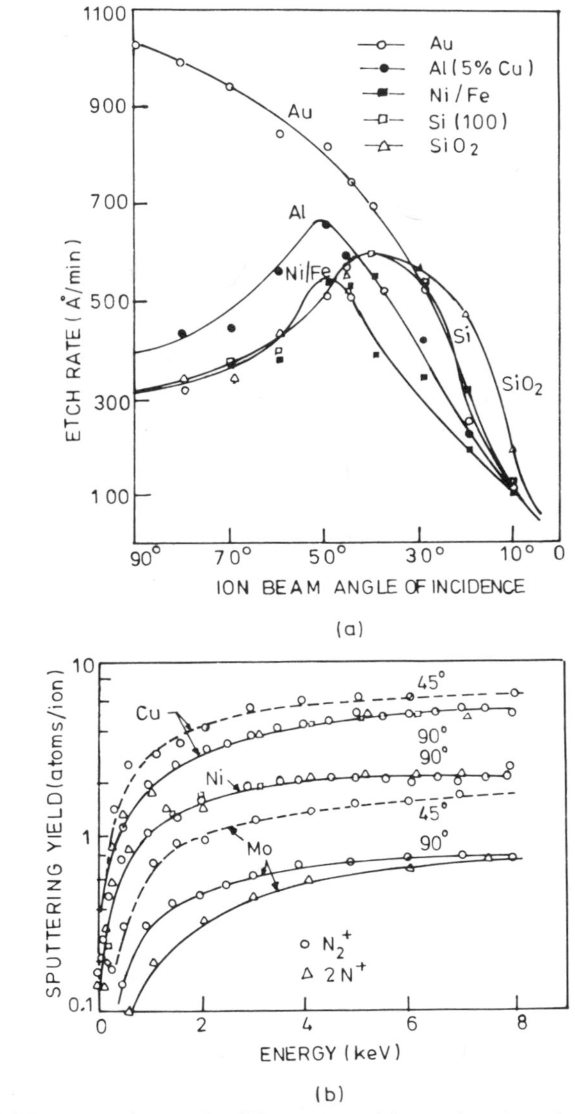

Figure 5.2. (a) Sputter etch rates for different materials as a function of angle of incidence and (b) sputtering yield vs. ion energy for different materials, ionic species, and angles of incidence.

Table 5.3. Sputtering Rate of Different Materials for 500-eV Ar Ions at 1 mA cm^{-2}

Target	Å min^{-1}	Target	Å min^{-1}	Target	Å min^{-1}
Ag	2000	Nb	390	W	340
Al	630	Ni	560	CdS	2100
Au	1500	Pb	2700	GaAs	1500
C	40	Pd	1100	GaP	1400
Cr	540	Pt	780	InSb	1400
Cu	870	Si	320	PbTe	3400
Fe	500	Sn	1500	SiC	320
Ge	920	Ta	380	SiO_2	400
Mo	470	Ti	340		

In spite of being energy-intensive, sputtering processes are best suited for depositing adherent films of multicomponent materials of any kind. Depending on the cathode geometry and mode of creation and transport of ions, a large number of sputtering variants have been developed over the years. These variants, their characteristics and those of the films obtained are discussed in detail in various books and reviews.[1–4] The processes described below are those of special interest to solar cell thin film materials.

5.2.3.1. *Glow Discharge Sputtering*

The simplest arrangement to produce ions is provided by a normal glow discharge created at a residual pressure of about 10^{-2} torr of the required gas (generally Ar) by applying 1 to 3 kV dc between a cathode (target) and an anode (on which the substrate is placed) separated by about 5 cm. The thickness d of the cathode dark space (across which most of the applied voltage drops) is inversely proportional to the gas pressure p (Paschen's law). The most commonly used sputtering gas is Ar for which the product $pd = 0.3$ torr cm. Owing to competing processes of increased creation of ions and scattering of ejected atoms by collisions with the gas atoms, the optimum pressure range for efficient sputter deposition of films is between 25 and 75 mtorr. Because of collisions with gas atoms, the sputtered atoms reach the substrate with randomized directions and energies. The diffuse nature of the transport of sputtered atoms causes deposition to occur at places not necessarily in the line of sight of the cathode. Crude analysis of the distribution of sputtered atoms shows that, under optimum conditions of deposition, uniformity of deposit extends to about half the area of the target when the cathode–anode distance is about twice the length of the cathode dark space.

Besides the more popular parallel plate diode configuration, wire, cylindrical, and concave cathodes may be used for particular applications. Also, multiple cathodes may be used for simultaneous or sequential sputtering for multicomponent or multilayer coatings.

Continuous bombardment of the depositing film with neutral and negatively charged Ar atoms, atoms of other reactive gases in the ambient, and electrons leads to a large concentration (up to several percent, depending on deposition conditions) of gas and impurity atoms being trapped in the film. The partial pressure of reactive gases can be decreased considerably in a getter-sputtering setup, which utilizes the gettering action of the sputtering material to purify the Ar before it is actually used to sputter-deposit a film. Lowering of the working pressure for glow discharge is also helpful in reducing the trapped gas content. This is achieved in a triode sputtering configuration in which a heated filament provides electrons that are accelerated to sustain the glow discharge at low pressures. The probability of ionization of Ar atoms by the electrons is further increased by using a magnetic field to increase the path of electrons prior to their collision with the anode. In this way, high sputter deposition rates (~ 1–10 Å s^{-1}) at approximately 10^{-4} torr are achieved. Trapped gases in the film can also be removed by negatively biasing the substrate relative to the anode so that the film is bombarded with ions throughout its growth.

5.2.3.2. *Magnetron Sputtering*

Arrangements in which the applied electric and magnetic fields are perpendicular to each other are called magnetron sputtering systems. In a planar cathode system, the magnetic field is applied parallel to the cathode to confine the primary electron motion to the vicinity of the cathode and thus increase the ionization efficiency and prevent the electron bombardment of the film. Permanent magnets are placed behind the cathode in various geometries in such a way that the cathode surface has at least one region where the locus of the magnetic field lines is parallel to the cathode surface in a closed path. The discharge plasma is constrained near the cathode surface by endless toroidal trapping regions bounded by a tunnel-shaped magnetic field. The tunnel shape and thus the electron paths depend on the magnet geometry and arrangement.

The distortion in plasma caused by $\mathbf{E} \times \mathbf{B}$ drift motion of electrons can be overcome by using cylindrical cathodes so that these drift currents close on themselves. This forms the basis of the cylindrical magnetron. The magnetron operation may be achieved with electron reflecting surfaces or with curved magnetic fields giving rise to a variety of cylindrical magnetron configurations.[4]

Magnetron sputtering makes it possible to utilize the cathode discharge power very efficiently (up to 60%) to generate high (up to ~ 50 mA cm^{-2}) current densities at relatively low (~ 500–1000 V) voltages to yield deposition rates that are at least one order of magnitude higher than those obtained in nonmagnetron systems. At a power density about 30 W cm^{-2}, a deposition rate as high as 25,000 Å min^{-1} has been achieved for Cu. The high deposition rates coupled with the fact that the film is not subjected to plasma and electron bombardment makes magnetron sputtering a very attractive large-area, low-temperature (deposition) process. The usefulness of this technique in depositing a number of solar cell materials such as TO,[13] ITO,[14] CdS,[15] and Cu_2S[15] has been demonstrated recently.

5.2.3.3. *rf Sputtering*

Sputtering at low pressures ($\sim 10^{-3}$ torr) is also possible by enhancing gas ionization with the help of an inductively coupled external rf field. If the cathode is an insulator material, dc sputtering is not possible owing to the building up of positive (Ar^+) surface charges. However, a high-frequency alternating potential may be used to neutralize the insulator surface periodically with plasma electrons, which have a much higher mobility than the positive ions. Whether or not the cathode surface develops a positive bias, which is responsible for sputtering, depends on the amplitude and frequency of rf and the geometry of the cathode. Typically, an rf power supply of 13.56 MHz (allowed by US Federal Communications Commission for commercial applications) and 1–2-kW power with about 2-kV peak-to-peak voltage is used to couple the cathode through a matching network. The rf technique can be used with any sputtering geometry in glow discharge or magnetron modes. It is an indispensable technique for deposition of thin films of semiconductors and insulators.

5.2.3.4. *Ion Beam Sputtering*

Sputter deposition under controlled high-vacuum conditions can be achieved by using an ion beam source. In the primary ion beam deposition process, the ions of the required material are produced and condensed on a surface to form a thin film. In the secondary ion beam deposition process, the Ar^+ ions from a beam source are used to sputter a target in vacuum and condense the sputtered species on a substrate. Both techniques have undergone major technology developments in the last decade or so and have now become standard but expensive tools for utilizing the benefits of a sputtering process under vacuum deposition conditions.

The two ion sources commonly employed for deposition are the duoplasmatron and the one developed by Kaufman. In the duoplasmatron source, the ions are created in a glow/arc discharge chamber and are then extracted through apertures into a second chamber at a much lower pressure ($\sim 10^{-4}$–10^{-5} torr). The Kaufman source employs chamber geometry and an applied magnetic field in such a way that the thermionically emitted electrons must travel long spiral paths to an anode cylinder located in the outer diameter of the discharge region. This results in high ionization efficiency as well as a uniform plasma. Applying a potential difference between a pair of grids with precisely aligned holes causes the ions to be extracted from the sheath around the grid holes and then accelerated by this potential difference. The grid optics focus the ions into a well-collimated beam, which can be neutralized by injecting low-energy electrons from a hot filament on the target side of the grids. Fully neutralized Ar beams up to 25 cm in diameter with current densities up to 50 mA cm^{-2} at 500–1000 eV have been obtained from the Kaufman source. This source is well suited for both etching (called ion beam milling) and sputter deposition of conducting and nonconducting materials. Ion milling is usefully employed to study spatial properties of multilayer junctions. Ion deposition has been used to deposit films of ITO[16] on Si for SIS solar cells. The sputter etch (and hence deposition) rates depend on the material and the angle of incidence of the ions as shown in Figure 5.2.

5.2.3.5. *Ion Plating*

A combination of thermal evaporation onto a substrate (cathode) which is simultaneously bombarded with positive ions (e.g., Ar^+) from a glow discharge or an ion source is called ion plating. The bombardment results in compactness and strong adherence of the films. The technique suffers from the disadvantage of sputter etching of films during growth and trapping of energetic gas ions in the film. A better and meaningful ion plating technique involves ionization of the vapor by bombardment with accelerated electrons from a thermal source and depositing the ions onto a substrate, with or without post-ionization acceleration.

5.2.3.6. *Reactive Sputtering*

The high chemical reactivity of ionic species and the atomic form of stable molecules, which are readily formed in a dc or rf glow discharge plasma, can be used very effectively to form thin film carbides, nitrides, oxides, hydrides, sulfides, arsenides, and phosphides. This is accomplished by introducing the reactant in gas form into the inert gas plasma. Whether the chemical reaction takes place on the cathode, in the plasma, or at the anode depends on the pressure and chemical activity of the reacting species under given surface and temperature conditions. The composition of the films obtained by this technique is determined by the prevailing plasma kinetics and thermodynamic conditions. Among the major applications of this technique is the preparation of controlled composition oxide films for MIS and SIS solar cells and for AR coatings. By introduction of specific halocarbons in the plasma that cause the reaction with the substrate to produce high-vapor-pressure halides, this technique has now been perfected to perform plasma or "dry" etching of a whole range of metallic and nonmetallic surfaces[4] on a selective or nonselective basis. It is thus possible to create patterned thin films for solar cell grid structures.

5.3. *Chemical Deposition Techniques*

5.3.1. *Spray Pyrolysis Process*

Spray pyrolysis was used as early as 1910 to obtain transparent oxide films.[17] In the 1960s, Chamberlin et al.[18–20] extended the technique to sulfide and selenide films. Spray pyrolysis involves a thermally stimulated reaction between clusters of liquid/vapor atoms of different chemical species. If we accept the strict definition that a thin film is an atom/molecule/ion-by-atom/molecule/ion condensation process, spray pyrolysis lies somewhere in the regime between a thin film and a thick film technique, depending on the atom cluster size.

The spray pyrolysis technique has been developed extensively by Chamberlin et al.[18–23] (now at Photon Power, Inc.), Bube et al.[24–30] at Stanford University, Savelli et al.[31–34] at Montpellier University, and Chopra et al.[35–42] at the Indian Institute of Technology, Delhi. Recently, Chopra et al.[43] published a review on the technique. In the following sections we shall examine in some detail the characteristic features of the spray process.

5.3.1.1. *Physical Aspects*

The spray technique involves spraying a solution, usually aqueous, containing soluble salts of the constituent atoms of the desired compound onto a substrate maintained at elevated temperatures. The sprayed droplet reaching the hot substrate surface undergoes pyrolytic (endothermic) decomposition and forms a single crystallite or a cluster of crystallites of the product. The other volatile by-products and the excess solvent escape in the vapor phase. The substrate provides the thermal energy for the thermal decomposition and subsequent recombination of the constituent species followed by sintering and recrystallization of the clusters of crystallites giving rise to a coherent film.

A schematic block diagram of a typical spray pyrolysis setup in operation in the authors' laboratory is shown in Figure 5.3. The atomization of the chemical solution into a spray of fine droplets is effected by the spray nozzle with the help of a filtered carrier gas which may (as in the case of SnO_x films) or may not (as for CdS films) be involved in the pyrolytic reaction. The carrier gas and the solution are fed into the spray nozzle at predetermined and constant pressure and flow rates. The substrate temperature is maintained with the help of a feedback circuit which controls a primary and auxiliary heater power supply. Large-area uniform converage of the substrate is effected by scanning either or both the spray head and the substrate, employing mechanical or electromechanical arrangements.

The geometry of the gas and liquid nozzle largely determine the spray pattern, the size distribution of droplets, and the spray rate. A wide variety of nozzles have been designed and employed for spraying on stationary and moving substrates. In Figure 5.4, we have shown schematically some nozzles and a cross-sectional view of a commercially available nozzle (from Spraying Systems Co., USA).

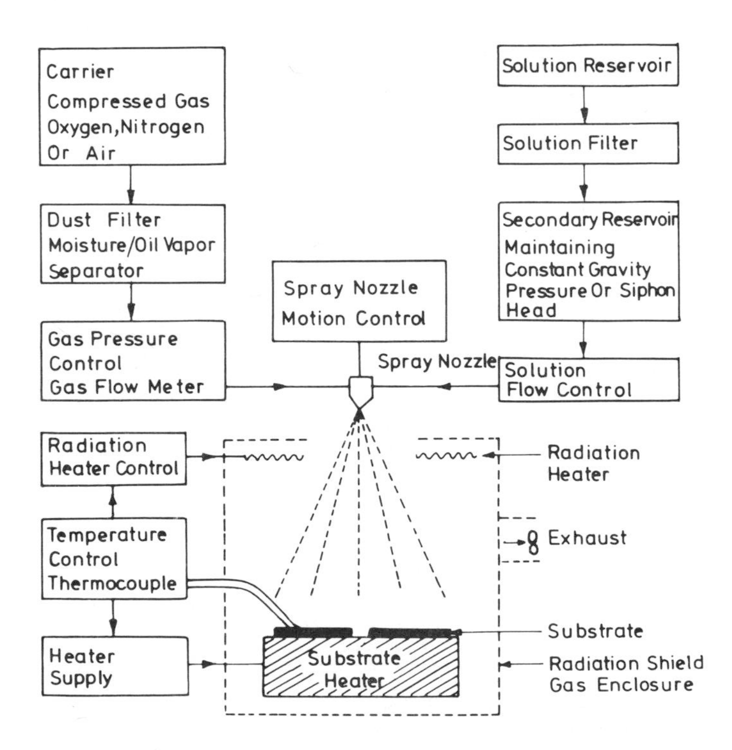

Figure 5.3. Schematic diagram of the spray process followed in the authors' laboratory for CdS thin film deposition.

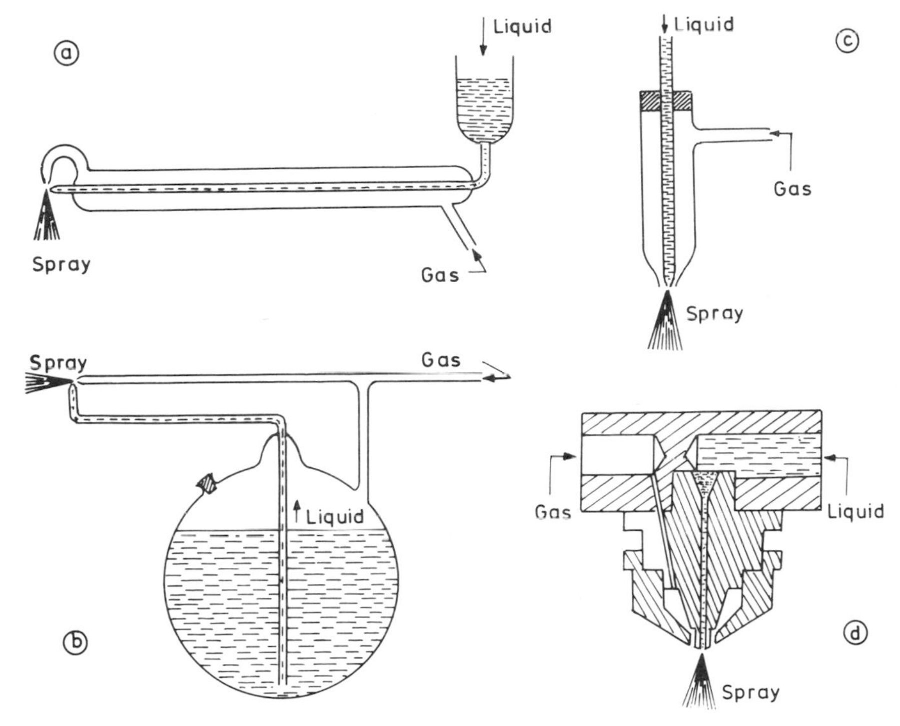

Figure 5.4. Schematic diagrams of some commonly employed nozzles (a, b, c) and the cross-section of a nozzle commercially available from Spraying Systems Co., USA (d).

5.3.1.2. *Growth Kinetics*

Lampkin[44] has studied the aerodynamics of the atomization and droplet impact processes and has correlated the dynamic features of the spray process with the kinetics of film growth and surface topography. When both the size and momentum of the spray droplets are uniform, optically good quality and smooth films are obtained in the case of CdS. In the presence of a field, the spray pattern becomes more defined and the field-induced higher droplet velocities and the coalescence kinetics of the droplets on the substrate surface are expected to have a considerable influence on the microstructure of the films.

According to Banerjee et al.,[37,39] the liquid droplet tends to flatten out into a disk on impact with the substrate surface. The disk geometry depends on the momentum and volume of the droplet, the substrate temperature, and the balance of the dynamical surface energy and thermal processes. The deposition process is a resultant of the following steps: (i) spreading of a drop into a disk, (ii) pyrolytic reaction between the decomposed reactants, (iii) evaporation of the solvent, and (iv) repetition of the preceding processes with succeeding droplets. Consequently, the film generally contains disks interspersed into each other. The lateral mobility of the droplets and coalescence and sintering kinetics of the superimposed disk crystallite clusters determine the growth kinetics and microstructural features of the spray deposited films. The following important characteristics are observed in this mode of growth: (i) the random disk-by-disk growth exposed to a continuous flow of pressurized liquid droplets eliminates microscopic and macroscopic voids and cavities in the growing film. Thus spray deposited films are coherent and pinhole-free even at very low (1000 Å) thicknesses, provided the substrate temperature is high enough to cause complete pyrolytic reaction, (ii) the microstructure of the film depends very sensitively on several deposition conditions, notably the spray head geometry, carrier gas and liquid flow pattern and rate, droplet velocities, sizes, and geometries, nature and temperature of the substrate, the kinetics and thermodynamics of the pyrolytic reactions, and the temperature profile during the deposition process.

5.3.1.3. *Chemical Aspects*

The chemicals used for spray pyrolysis have to satisfy the following conditions: (i) on thermal decomposition, the chemicals in solution form must provide the species/complexes that will undergo a thermally activated chemical reaction to yield the desired thin film material, and (ii) the remainder of the constituents of the chemicals, including the carrier liquid, should be volatile at the spray temperature. For a given thin film material, the above conditions can be met by a number of combinations of chemicals. However, different deposition parameters are required to obtain comparable quality (structurally) films.

5.3.1.3a. *Sulfides and Selenides.* To obtain CdS films, a dilute (0.001 M to 0.1 M) aqueous solution of a water-soluble cadmium salt and a sulfo-organic salt is most commonly used.[4,39,45,46] The commonly used chemicals $CdCl_2$ and thiourea yield CdS films according to the reaction:

$$CdCl_2 + (NH_2)_2CS + 2H_2O \rightarrow CdS + 2NH_4Cl + CO_2$$

Similar reactions occur using other cadmium salts such as $Cd(NO_3)_2$, $CdSO_4$, $Cd(CH_3COO)_2$, $Cd(CHO_2)_2$, and $Cd(C_3H_5O_2)$. Similarly, thiourea can be replaced by N-N-dimethyl thiourea [$N_2(CH_3)_2H_2CS$], allyl thiourea [$H_2NCSNHCH_2CH:CH_2$], thiolocetic acid [CH_3COSH], and ammonium thiocynate [NH_4CNS].

It should be noted that the various intermediate chemical reactions and products are quite complex. In most materials, very little is known about the reactions and they need to be investigated more fully since the quality of the films and the residual trapped impurities depend on these processes.

Selenide films are obtained by replacing thiourea by selenourea or other suitable selenium compounds such as N-N-dimethyl selenourea. The corresponding reaction is

$$CdCl_2 + (NH_4)_2CSe + 2H_2O \rightarrow CdSe + 2NH_4Cl + CO_2$$

Sulfide and selenide films of a number of other elements, such as Zn, Cu, In, Ag, Ga, Sb, Pb, and Sn have been obtained by using similar pyrolytic reactions. It has not been possible to obtain telluride films since the telluro-organic salts are extremely unstable and difficult to synthesize. However, it may be possible to increase their stability by appropriate and desired dopants.

5.3.1.3b. *Oxides.* An aqueous metal salt solution is sprayed onto a hot substrate in air to obtain the corresponding metal oxide films. Generally, metal chlorides, such as $SnCl_4$ for SnO_2,[38] $InCl_3$ for In_2O_3,[47] $AlCl_3$ for Al_2O_3,[48] $FeCl_3$ for Fe_2O_3,[49] $CoCl_3$ for Co_2O_3,[50] and $ZnCl_2$ for ZnO,[51] have been used. In addition nitrates, carbonates, acetates, and bromides have also been employed.[43]

Usually 0.07 to 0.1 *M* aqueous chloride solutions are used for good optical quality SnO_x films although concentrations as high as 2.85 *M*[52] have also been used. The optimum concentration depends on the desired optical and electrical quality of the film, the deposition rate, and the chemistry of the reaction. A typical chemical reaction for SnO_2 films is

$$SnCl_4 + 2H_2O \rightarrow SnO_2 + 4HCl$$

The choice of the anion in the metal salt depends on the thermodynamic driving forces. In the case of ZnO films, the heat of reaction for the anion A in the reaction

$$(ZnA)_{\text{aqueous}} + H_2O \rightarrow ZnO + 2HA$$

is 30 kcal/mole for chloride, -0.1 kcal/mole for acetate, and -10 kcal/mole for nitrate at room temperature.[43]

Organometallic compounds have also been used to obtain oxide films. Although more expensive, these compounds offer the advantage of low decomposition temperature, thereby reducing substrate/vapor/film interaction. Some of these compounds are: dibutyl tin diacetate $(C_4H_9)_2Sn(CH_3COO)_2$ for SnO_2[53] and indium acetylacetonate $In(C_5H_2O_2)$ for In_2O_3.[54,55] Kane et al.[56] have also used an indium chelate derived from dipivaloyl methane for In_2O_3.

5.3.1.4. *Characteristic Features of the Spray Pyrolysis Process*

5.3.1.4a. *Growth Rate.* The chemical and topographical nature and temperature of the substrate, the chemical nature and concentration of the spray solution and its additives, and the spray parameters largely determine the growth rate. In the case of SnO_x (TO) and In_xO_y : Sn (ITO) films, the thickness increases nearly linearly with time of spray, i.e., with the amount of sprayed solution. The growth rates can be as large as 1000 Å min^{-1} for oxide films and 500 Å min^{-1} for sulfide films. Figure 5.5 illustrates the substrate temperature dependence of the deposition rate for CdS films under different spray conditions.

5.3.1.4b. *Substrate Effects.* In general, the spray pyrolysis process affects the substrate surface. When it is not desirable for the substrate to

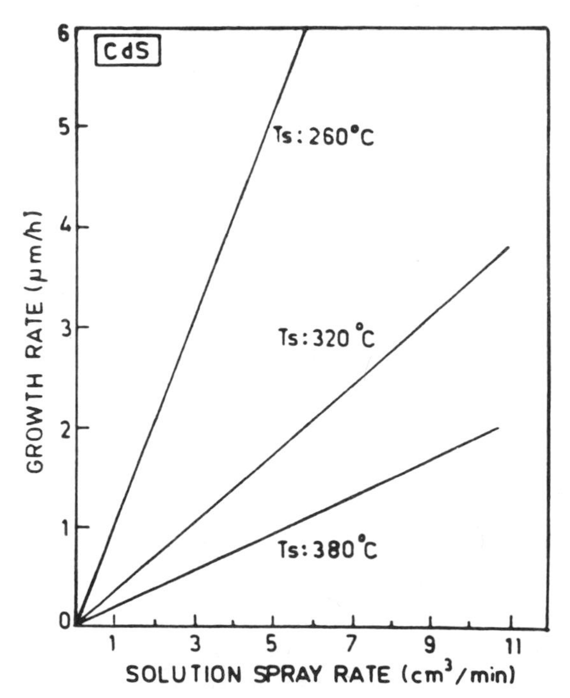

Figure 5.5. Dependence of the growth rate of spray deposited CdS films on solution spray rate and substrate temperature (T_s) (after Banerjee[39]).

take part in the pyrolytic reactions, neutral substrates such as glass, quartz, ceramics, or appropriate oxide/nitride/carbide coated substrates are employed. In the case of certain oxide films on Si, some desirable etching takes place during deposition.(57) Metallic substrates(39) have not been found suitable for this process.

Mobile alkali and other rare earth ions such as Li^+, Na^+, Ca^{2+}, Sr^{2+}, and Mg^{2+} contained in the chemically inactive substrates may be incorporated into the films(47) and the extent of inclusion would increase with the substrate temperature.

Generally, at lower substrate temperatures foggy and diffusely scattering films are obtained. High substrate temperatures yield thinner, continuous, hard, and specularly scattering films. Moreover, at higher temperatures, re-evaporation of anionic species may occur, leading to metal-rich deposits.

5.3.1.4c. *Film Composition.* The composition of the film is expected to depend on the kinetics of the spray process and the thermodynamics of the pyrolytic processes. Stoichiometric sulfide and selenide films and nearly stoichiometric oxide films have been obtained under appropriate conditions. The stoichiometry of sulfide films does not vary appreciably with the metal-to-sulfur ion ratio in the spray solution for ratios ranging from 1:1 to 1:5, but the microstructure of the films is strongly influenced by this ratio.(26,34,58)

At low enough temperatures, if the pyrolytic reactions have not been completed, some by-products or intermediate compounds will be trapped as impurities in the film. In the case of chloride salts, residual chlorine is often present(51) in films. As shown in Figure 5.6, the chlorine concentration in CdS films decreases with increasing substrate temperature during pyrolysis. Consequently, owing to the cooling effect at the growing film surface, a higher concentration of chlorine is observed at the surface. Moreover, the chlorine concentration(34,45) is sensitively dependent on the ratio of chloride- to sulfo-salt in the spray solution.

The stoichiometry of the oxide films is dependent on relatively more complex reactions.(47,59,60) The presence of oxygen ion vacancies (V_0^{2+}) in TO films is associated with the conversion of Sn^{4+} to Sn^{2+} according to the reactions:

$$O_0 \rightleftarrows \tfrac{1}{2}O_2 + 2e^- + V_0^{2+}$$

$$Sn^{4+} + 2e^- \rightleftarrows Sn^{2+}$$

The deviation from stoichiometry, i.e., the number of oxygen vacancies, is equal to the number of Sn^{4+} species reduced to Sn^{2+} ions, and this is controlled by the water and alcohol content in the spray solution. Water

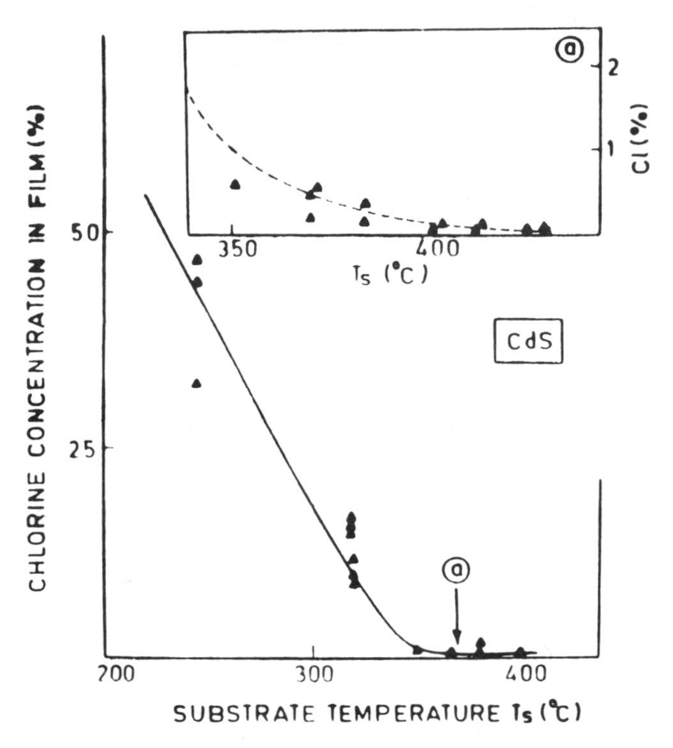

Figure 5.6. Chlorine concentration in spray deposited CdS films as a function of substrate temperature (after Coumar[34]).

molecules provide oxygen and alcohol acts as a reducing agent. The oxygen content in films is also influenced by the rate of cooling of the films after the spray is over, owing primarily to the adsorption of oxygen.

5.3.1.5. *Multicomponent Doping and Alloying*

Copyrolysis has been successfully utilized by a number of workers[20,28,30,36,37,39,40] to extend the technique to prepare doped and alloyed sulfide and selenide films of Cd, Zn, and Pb. Copyrolysis involves choosing appropriate salts and then spraying the common solution from one nozzle, or by using multiple nozzles to spray different solutions. Deshotels et al.[61,62] have used different carrier gases to dope CdS films with In and Ga. Pamplin and Fiegelson[63,64] have deposited a large variety of ternary, quaternary, and quinary stannite compounds, adamantine materials, and chalcopyrites.

It is important to note that the composition of the film is not simply related to the composition of the spray solution. Indeed, the correlation has to be established empirically for each system. $Cd_xZn_{1-x}S$ films having the same composition as that of the solution have been prepared[39] over the whole composition range by choosing an appropriate substrate temperature for each composition. Deviations from these conditions yield Zn- or Cd-rich films. In other systems, such as $SnO_x:F$, where one of the elements has a low vapor pressure, or forms a volatile gaseous product, the film composition can be drastically different from the solution composition. Typically, an F/Sn atomic ratio of 0.38 in solution yields less than a few percent of F in $SnO_x:F$.[65] On the other hand, in $SnO_x:Sb$ and $In_2O_3:Sn$ films, the Sb/Sn and Sn/In ratios are the same as in the solution. The composition is, however, affected by the nature of the substrates as shown in Figure 5.7 for ITO films deposited on Si and pyrex.[47]

Tin oxide films have been doped with cationic impurities of Sb,[38] In,[66] Cd,[67] Bi,[68] Mo,[69] B,[70] P,[69,71] Te,[43] and W,[72] and anionic impurities of F[65] and mixed F-Sb.[73] Similarly, indium oxide films have been doped with Sn,[47,74] Ti,[74] Sb,[74] F,[75] and Cl and mixed impurities of Sn and F.[76]

Copyrolysis has been used by some workers[58,77,78] to obtain a heterogeneous mixture of oxide and sulfide films. The authors have obtained CdS films with segregated Al_2O_3 formed at the grain boundaries by cospraying solutions of $CdCl_2$, $AlCl_3$, and thiourea.

An extension of the copyrolysis technique is to spray different solutions sequentially and thus employ sequential pyrolysis to obtain multilayer films of different materials, or films with gradient composition along the thickness. In the authors' laboratory, sequential pyrolysis has been utilized to obtain gradient-doped CdS : Al films for Cu_2S/CdS : Al solar cells.

In Table 5.4 we have given the solution compositions commonly used to prepare oxide thin films by the spray process.

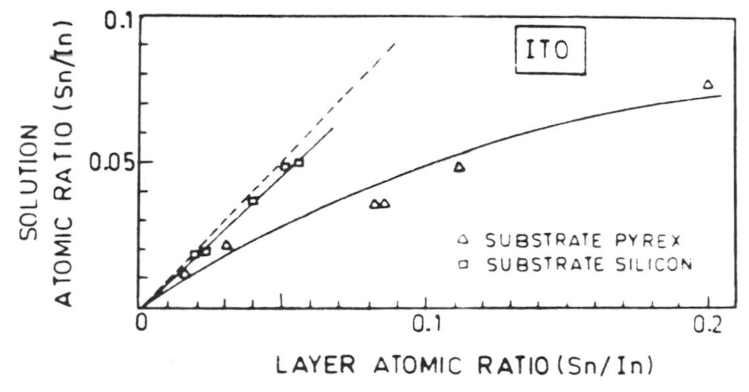

Figure 5.7. Dependence of the Sn/In atomic ratio in ITO films on the spray solution composition, for different substrates (after Manifacier et al.[47]).

Table 5.4. Some Commonly Used Spraying Solution Compositions (from Chopra et al.[43])

Oxide	Salt for matrix	Salt for dopant	Carrier solution	Additive
SnO_x:Sb	$SnCl_4.5H_2O$	$SbCl_3$ 0.135 g (1.4 m/o) Dissolved in conc. HCl	Isopropyl alcohol 150 cc	—
SnO_x : F	$SnCl_4 . 5H_2O$ 5 cc of 2.85 *M*	NH_4F 1.0 g (65.4 m/o)	Isopropyl alcohol 5 cc	—
In_xO_y : Sn	$InCl_3$ 8.2 g	$SnCl_4 . 5H_2O$ (2 at.%) 0.25 g	H_2O and C_2H_5OH 42 g each	7.5 g HCl
$CdSnO_3$	$CdCl_2 . 5H_2O$ 100 g of 1.67 *M* and $SnCl_4 . 5H_2O$ 10 g of 1 *M* (aq. solution)	—	H_2O (as existing in $CdCl_2$ and $SnCl_4$ solutions)	$InCl_3$ 3 g of 1 *M* 10 g HCl (conc.)
ZnO_x	$ZnCl_2$ 100 cc of 0.1 *M*	—	H_2O (as existing in $ZnCl_2$ aq. solution)	1.2 cc of H_2O_2

5.3.1.6. *Properties of Spray Deposited Films*

The properties of the spray deposited films are discussed in the next chapter in comparison with the properties of films obtained by other deposition techniques. However, we may note briefly here that, in general, spray deposited films are strongly adherent, mechanically hard, pinhole-free, and stable with time and temperature (up to the spray temperature). The surface topography of the films is rough with the roughness depending on the spray conditions and the substrate temperature. The microstructure ranges from amorphous to micropolycrystalline depending on the droplet mobilities and chemical reactivities of various constituents. Typical grain sizes of sulfide and selenide films range from 0.2 to 0.5 μm[39] and those of oxide films from 0.1 to 0.2 μm.[42] Postdeposition annealing of films generally affects the oxygen-dominated electrical properties significantly but not the microstructure.[41] At annealing temperatures above the spray temperature, or under some reactive environments, recrystallization increases the grain size and may produce some preferential orientation effects.[41]

5.3.2. *Solution Growth Process*

The solution growth technique was pioneered by the works of Bode and co-workers[(79–81)] at Santa Barbara Research Centre, G. A. Kitaev and co-workers[(82–87)] at Ural Polytechnic, U.S.S.R., and Chopra and co-workers[(88–91)] at the Indian Institute of Technology, Delhi. The technique itself was first used in 1946[(92)] to prepare PbS films for infrared applications. However, it is only recently[(88–91)] that large-area and large-scale applications of this technique to obtain doped and undoped multicomponent semiconductor films of usual, unusual, and metastable structures have necessitated an understanding of the physics and chemistry of the processes involved. This full-fledged technique, of great future promise, is the subject of this section.

5.3.2.1. *Chemical Aspects*

According to the solubility product principle, in a saturated solution of a weakly soluble compound, the product of the molar concentrations of its ions (each concentration term being raised to a power equal to the number of ions of that kind as shown by the formula for the compound), called the ionic product, is a constant at a given temperature. For example, $Cd(OH)_2$ when added to water will hydrolyze according to the reaction

$$Cd(OH)_2 \rightleftharpoons Cd^{2+} + 2(OH)^-$$

The ionic product IP is given by

$$[Cd^{2+}] \times [OH^-]^2 = \text{Constant (called the solubility product SP)}$$
$$= 2.2 \times 10^{-14} \text{ (at 25 °C)}$$

There is no equilibrium if this relationship is not satisfied. If the ionic product exceeds the solubility product, precipitation occurs. When IP < SP, the solid phase will dissolve until the above relation is satisfied.

It is necessary to eliminate spontaneous precipitation in order to form a thin film by a controlled ion-by-ion reaction. This can be achieved by using a fairly stable complex of the metal ions which provides a controlled number of the free ions according to an equilibrium reaction of the type

$$M(A)^{2+} \rightleftharpoons M^{2+} + A$$

The concentration of the free metal ions at a particular temperature is given by

$$\frac{[M^{2+}][A]}{[M(A)^{2+}]} = K_i$$

where K_i is termed the instability constant of the complex ion. By choosing an appropriate complexing agent, the concentration of the metal ions is controlled by the concentration of the complexing agent and the solution temperature. Table 5.5 lists the complexing agents for different ions.

If a high concentration of S^{2-} ions exists locally such that the solubility product is exceeded, localized spontaneous precipitation of a sulfide can occur. This problem can be overcome by generating chalcogen ions slowly and uniformly throughout the volume of the solution. This is achieved, for example, by having thiourea in an alkaline aqueous solution according to the reaction

$$(NH_2)_2CS + OH^- \rightarrow CH_2N_2 + H_2O + HS^-$$

$$HS^- + OH^- \rightarrow H_2O + S^{2-}$$

Allyl-thiourea or N–N dimethyl thiourea may be used in place of thiourea.

Selenide films[83,93] are obtained by replacing thiourea by selenourea or its other derivatives. Kainthla et al.[90] have generated Se^{2-} ions by dissolving inorganic sodium selenosulfate in an alkaline solution as given by the following reaction:

$$Na_2SeSO_3 + 2OH^- \rightarrow Na_2SO_4 + H_2O + Se^{2-}$$

This method has the advantage that sodium selenosulfate can be easily synthesized by dissolving Se in Na_2SO_3 solution. On the other hand, selenourea and its derivatives are difficult to synthesize and further, their aqueous solutions have to be stabilized using antioxidants like Na_2SO_3.

Table 5.5. Ions and Their Common Complexing Agents

Elements	Complexing agents
Ag	CN^-, NH_3, Cl^-
Cd	CN^-, NH_3, Cl^-, $C_6H_5O_7^{3-}$, $C_4H_4O_6^{2-}$, EDTA
Co	NH_3, CN^-, SCN^-, $C_6H_5O_7^{3-}$, $C_4H_4O_6^{2-}$
Cu	NH_3, Cl^-, CN^-, EDTA
Hg	NH_3, Cl^-, CN^-, EDTA
Mn	$C_2O_4^{2-}$, $C_6H_5O_7^{3-}$, $C_4H_6O_6^{2-}$, CN^-, EDTA
Ni	CN^-, SCN^-, EDTA, NH_3
Pb	EDTA, $C_6H_5O_7^{3-}$, $C_4H_6O_6^{2-}$, OH^-
Sn	$C_6H_5O_7^{3-}$, $C_4H_6O_6^{2-}$, $C_2O_4^{2-}$, OH^-
Zn	CN^-, NH_3, EDTA, $C_4H_6O_6^{2-}$, $C_6H_5O_7^{3-}$

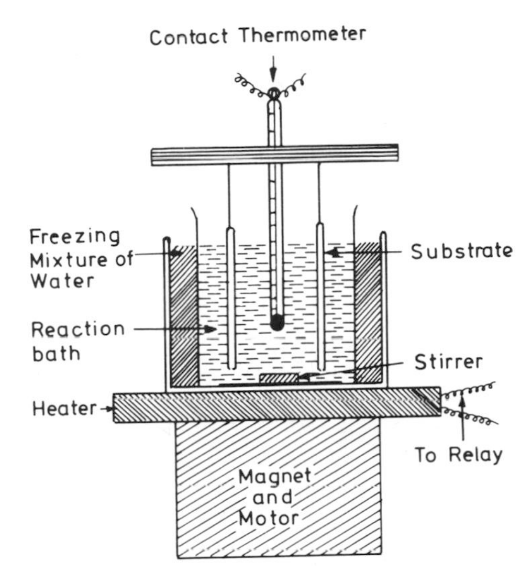

Figure 5.8. Experimental arrangement in the authors' laboratory for solution growth of semiconductor films.

Tellurium-bearing compounds are very unstable and cannot be synthesized. However, it may be possible to generate Te^{2-} ions in a solution by using an inorganic compound like sodium dithionate ($Na_2S_4O_6$) to dissolve Te.

The experimental setup to obtain film deposition is shown schematically in Figure 5.8. The substrates are immersed vertically in the reaction bath, which is stirred continuously with a magnetic stirrer. The temperature of the bath is monitored by a contact thermometer that forms a part of a feedback circuit controlling the heater to maintain a constant temperature. When the IP of the metal and chalcogen ions exceeds the SP of the corresponding chalcogenide, a metal chalcogenide film is formed on the substrate by an ion-by-ion condensation process.

The composition and temperature dependence of the various chemical reactions involved can be worked out thermodynamically. Such an analysis has been carried out for the following systems:

1. $Cd(NH_3)_4^{2+} - (NH_2)_2CS - OH^-$ (82,91)
2. $Cd(en)_3^{2+} - (NH_2)_2CS - OH^-$ (94)
3. $Cd(C_6H_5O_7)^- - (NH_2)_2CSe - OH^-$ (95)
4. $Cd(NH_3)_4^{2+} - SeSO_3^{2-} - OH^-$ (86,90)

The deposition conditions for CdS and CdSe films in Systems 1 and 4

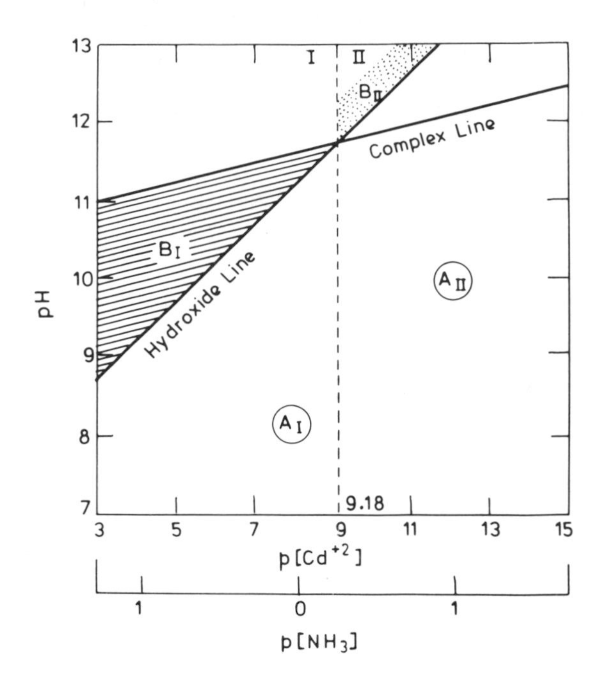

Figure 5.9. Plot of pH vs. $p[Cd^{+2}]$ and $p[NH_3]$ (after Kaur et al.[91]).

are determined by a graphical solution of the following equations:

$$Cd^{2+} + 2OH^- \rightleftharpoons Cd(OH)_2$$

with

$$[Cd^{2+}][OH^-]^2 = 2.2 \times 10^{-14}$$

and

$$Cd(NH_3)_4^{2+} \rightleftharpoons Cd^{2+} + 4NH_3$$

with

$$\frac{[Cd^{2+}][NH_3]^4}{[Cd(NH_3)_4^{2+}]} = 7.56 \times 10^{-8}$$

From the above two equations, a plot of pH against $p[Cd^{2+}]$ (at constant pC_{salt}) yields straight lines, called the hydroxide line and the complex line, as shown in Figure 5.9. The shaded regions B_I and B_{II} which lie above the hydroxide line correspond to the presence of $Cd(OH)_2$ in the solution. In these regions, thin (~600 Å), hard, physically coherent, and specularly reflecting CdS films having a wurtzite structure are obtained. In regions A_I and A_{II}, where no $Cd(OH)_2$ exists in the solution, powdery films of CdS with a sphalerite structure are obtained. For conditions corresponding to the points on the complex line in Region II, film formation takes place only above 45 °C when the IP becomes greater than the SP of the cadmium chalcogenide. The films so obtained are hard, physically coherent, and specularly reflecting with mixed sphalerite and wurtzite structures.[91]

CdSe films are obtained only in the Region B_{II} using Na_2SeSO_3 at room temperature.[86] These films have mixed sphalerite and wurtzite structures. CdSe films with a pure sphalerite structure can be obtained under conditions on the complex line at or above 45 °C.[90]

5.3.2.2. *Characteristic Features of the Solution Growth Process*

The kinetics of growth of a thin film in this process are determined by the ion-by-ion deposition of the chalcogenide on nucleating sites on the immersed surfaces. Initially, the film growth rate is negligible because an incubation period is required for the formation of critical nuclei from a homogeneous system onto a clean surface. Once nucleation occurs, the rate rises rapidly until the rate of deposition equals the rate of dissolution, i.e., IP = SP. Consequently, the film attains a terminal thickness as illustrated in Figure 5.10. On a presensitized substrate surface, no incubation period for nucleation is observed, since nucleation centers already exist on the substrate (see Figure 5.10). Also, when the substrates are suspended in the container before forming the complex in the solution, film thickness increases in a manner similar to that of the sensitized surface, thereby showing that the nuclei for the formation of the film are provided by the solution itself. This is shown in Figure 5.11 for the case of CdSe films with NH_3 concentration as a parameter.

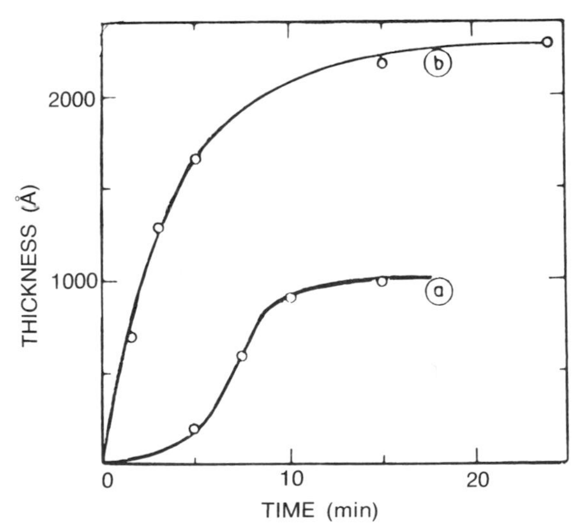

Figure 5.10. Typical growth kinetics for solution grown films deposited on: (a) unsensitized substrate and (b) sensitized substrate (after Chopra et al.[43]).

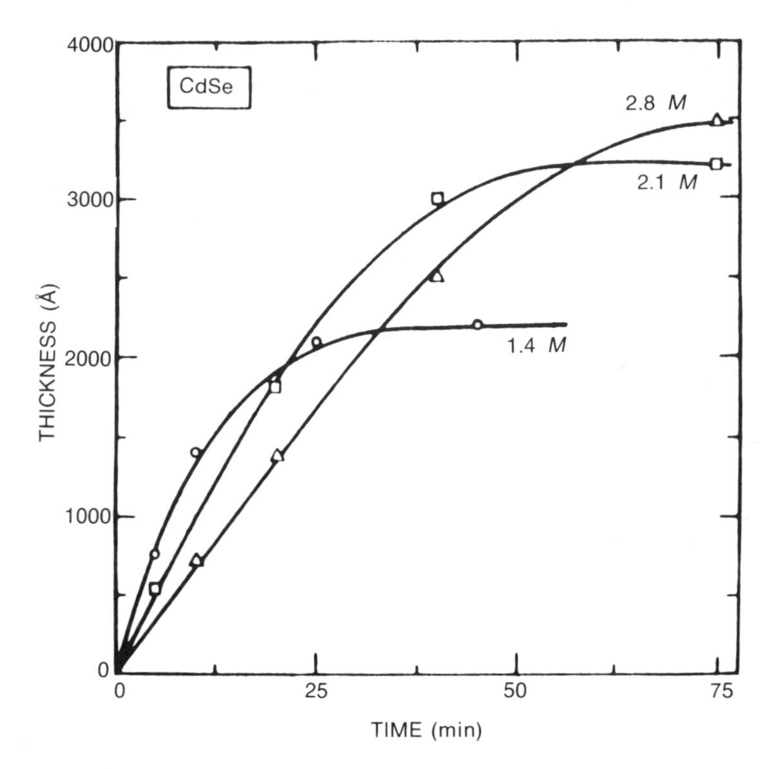

Figure 5.11. CdSe film thickness as a function of time for a sensitized surface with NH_3 concentration as a parameter (after Kainthla[95]).

The rate of deposition and the terminal thickness both depend on the number of nucleation centers, supersaturation of the solution (defined as the ratio of IP to SP), and stirring. The growth kinetics depend on the concentration of ions, their velocities, and nucleation and growth processes on the immersed surfaces. The effect of various deposition conditions on these parameters is discussed in the subsequent sections.

5.3.2.2a. *Nature of the Salt.* The growth kinetics depend on the salts/compounds used for metal and chalcogenide ions. The rate of deposition is expected to decrease and the terminal thickness to increase if metal sulfate is employed to deposit metal selenide films using sodium selenosulfate. Similar results are expected if $CdCl_2$ is used to deposit CdS and CdSe films. In the former case, the SO_4^{2-} ions, obtained from the metal sulfate, reduce the concentration of Se^{2-} ions, while in the latter case, the Cl^- ions formed by the dissolution of $CdCl_2$ reduce the concentration of Cd^{2+} ions by forming the complex $CdCl_4^{2-}$. In general, the rates and terminal thickness are higher for sulfide than for the corresponding selenide films under similar deposition conditions.

The deposition rate and terminal thickness initially increase with an increase in the chalcogen ion concentration. At high concentrations, however, precipitation becomes more significant, leading to decreased film thickness on the substrate. Figure 5.12 illustrates this effect for CdS films.

5.3.2.2b. *Complexing Agent.* The metal (M^{2+}) ion concentration decreases with increasing concentration of the complexing ions. Consequently, the rate of reaction and hence precipitation are reduced leading to a larger terminal thickness of the film. Such a behavior has been observed for CdSe, CdS, PbSe, and ZnS films and is illustrated in Figure 5.13 for CdSe films.

5.3.2.2c. *pH Value.* The addition of OH^-, i.e., increase in pH, makes the complex more stable, provided the OH^- ions take part in the complex formation (as in $Pb(OH)C_6H_5O_7^{2-}$). Thus, the free M^{2+} ion concentration is reduced, leading to a decrease in the deposition rate and an increase in the terminal thickness with increasing pH value. The dependence of the rate of deposition and terminal thickness on the pH value for PbSe films is shown in Figure 5.14.

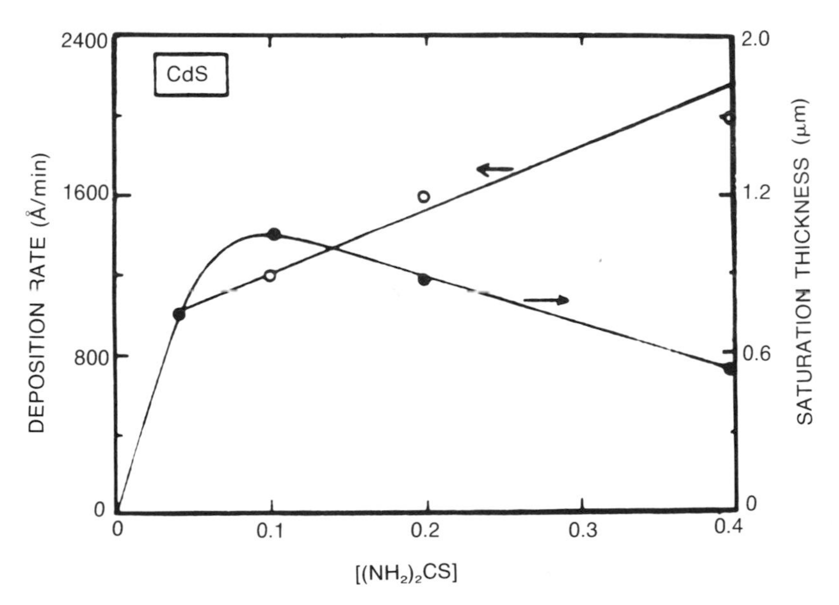

Figure 5.12. Deposition rate and saturation thickness of solution grown CdS films as a function of chalcogen ion concentration (after Kaur et al.[91]).

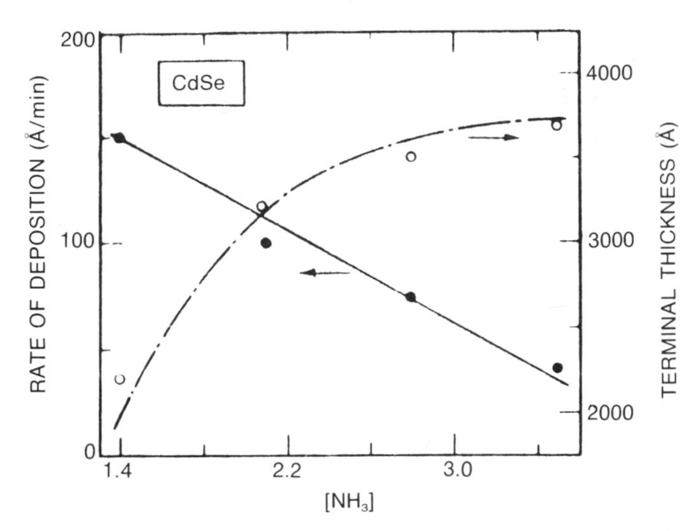

Figure 5.13. Terminal thickness and deposition rate of solution grown CdSe films as a function of complexing agent concentration (after Kainthla[95]).

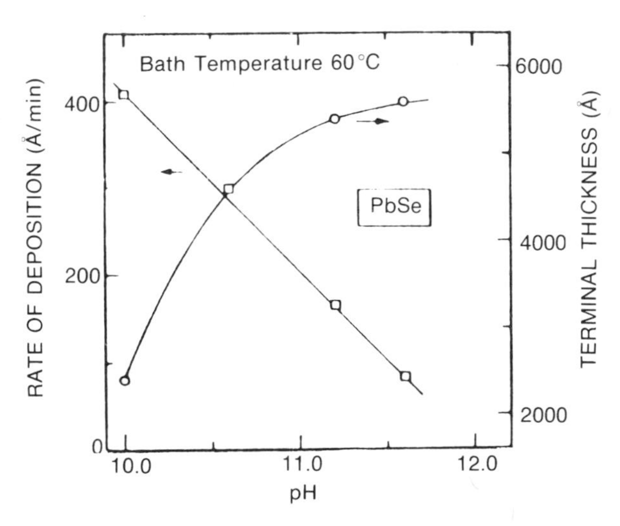

Figure 5.14. Effect of pH on the deposition rate and terminal thickness of solution grown PbSe films (after Chopra et al.[43]).

If the OH^- ions do not participate in complex formation [as in the case of $Cd(NH_3)_4^{2+}$], the addition of OH^- precipitates out the corresponding hydroxide. With increasing formation of $Cd(OH)_2$ in the bulk of the solution, most of the CdSe is precipitated on addition of Na_2SeSO_3, leading to lower terminal thicknesses at high pH values.

An important observation, due to Kitaev et al.,[86] is that the Cd:Se ratio in CdSe films is independent of the pH value.

5.3.2.2d. ***Substrate Effects.*** Higher deposition rates and terminal thicknesses are observed for those substrates whose lattices and lattice parameters match well with those of the deposited material. This effect is illustrated in Figure 5.15 for PbSe films deposited on glass, copper, polished single crystal Si, and Ge. Under similar conditions, higher rates and thicknesses have been observed on Ge rather than Si because of better matching of the lattice parameters of PbSe[95] with those of Ge.

5.3.2.2e. ***Bath/Substrate Temperature Effects.*** With increase in the solution temperature, the dissociation of the complex and the chal-

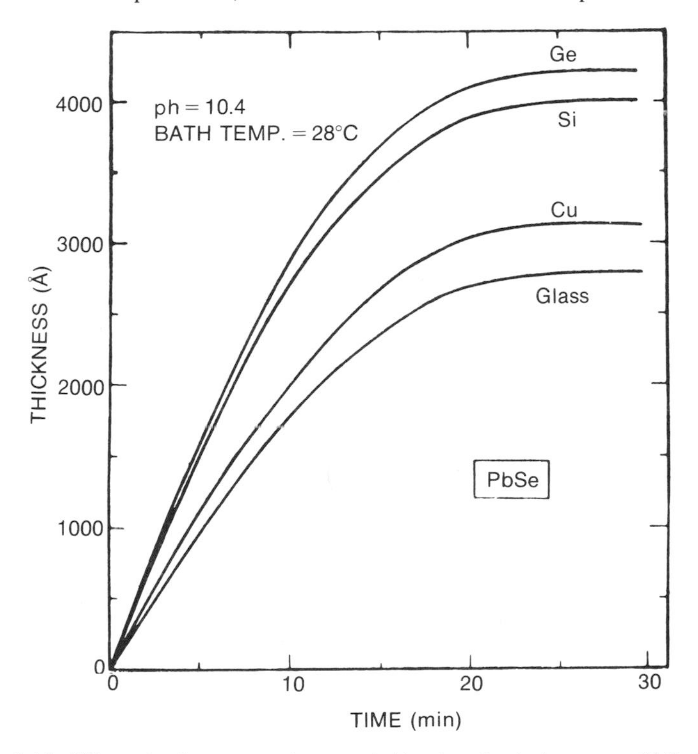

Figure 5.15. Effect of substrate on the growth kinetics of solution grown PbSe films (after Chopra et al.[43]).

cogen bearing compound increases. The increased concentration of metal and chalcogen ions, coupled with higher kinetic energy of the ions, results in increased interaction and yields a higher rate of deposition of the metal chalcogenide film. The terminal thickness may increase or decrease with increasing bath temperature, depending on the degree of supersaturation. The terminal thickness first increases with increasing supersaturation (owing to increased concentration of ions) and then it decreases at sufficiently high supersaturations at which precipitation dominates. The supersaturation may, however, be controlled by the bath temperature and also by the complexing agent concentration. Figure 5.16 shows the effect of bath temperature on the rate of deposition and terminal thickness for the case of CdSe films.

It should be noted that by using a larger surface area, it is possible to collect more chalcogenide on the surface in the form of film. Moreover, dipping the coated surface again in a fresh solution results in further deposition of material. Thus, thick and multilayer films can be obtained by sequential dippings.

5.3.2.3. *Doping*

Impurities in the starting chemicals can be incorporated into the films only if the impurities form insoluble chalcogenides under the same conditions of deposition and provided their corresponding IP is greater than the SP. Few dopants satisfy these conditions. An important consequence of this fact is that the degree of purity of the starting chemicals is

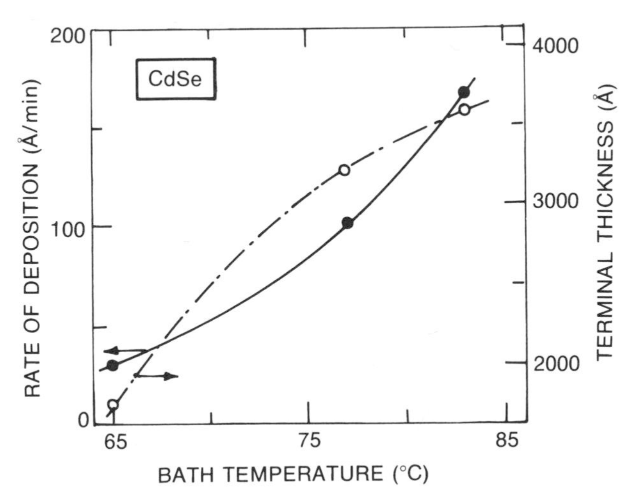

Figure 5.16. Effect of bath temperature on the deposition rate and terminal thickness of solution grown CdSe films (after Kainthla[49]).

not so important a factor in determining the purity of the resulting film if the impurity concentration is low and the corresponding IP $<$ SP.

5.3.2.4. *Multicomponent Films*

It is possible to form multicomponent chalcogenide films over a wide composition range. A variety of variable composition ternary alloys have been prepared in the authors' laboratory and these are listed in Table 5.6.

The alloy films in the nonisoelectronic systems have been prepared by reacting sodium-selenosulfate or thiourea with a mixture of different complexed ions. If two noninterfering, independent complexing agents are used for complexing the two cations, then the ions dissociate in an aqueous solution to give free metal ions according to the reactions

$$M(A)_n^{2+} \rightleftharpoons M^{2+} + nA$$

and

$$M'(B)_m^{2+} \rightleftharpoons M^{2+} + mB$$

The composition x of the films can be varied by controlling the initial salt concentration, complexing salt concentration, and temperature of the bath. An alternative technique to vary the composition is to change the concentration of the complexing agents without altering the ratio of the salt concentrations. In $Cd_{1-x}Pb_xSe$ films, the composition has been varied by adding different amounts of NH_4Cl solution to the reaction mixture.[95] On the other hand, variable composition $Cd_{1-x}Hg_xSe$, $Cd_{1-x}Zn_xS$, $Cd_{1-x}Hg_xS$, and $Pb_{1-x}Hg_xS$ films have been obtained by changing the relative amounts of the salts in the solution, using the same complexing agent for the two cations.[95–97]

Table 5.6. Alloys Formed by Solution Growth Technique (from Chopra et al.[43])

Alloy	Composition
$Pb_{1-x}Hg_xS$	$0 \leq x \leq 0.33$
$Pb_{1-x}Hg_xSe$	$0 \leq x \leq 0.35$
$Cd_{1-x}Hg_xS$	$0 \leq x \leq 0.30$
$Cd_{1-x}Hg_xSe$	$0 \leq x \leq 0.15$
$Cd_{1-x}Zn_xS$	$0 \leq x \leq 1.0$
$Cd_{1-x}Pb_xSe$	$0 \leq x \leq 1.0$
$CdSe_{1-x}S_x$	$0 \leq x \leq 1.0$
$PbSe_{1-x}S_x$	$0 \leq x \leq 1.0$

Since thiourea has a higher dissociation constant, the fraction of S^{2-} ions in the solution is expected to be more than the fraction of thiourea in the solution. Moreover, as the SP of sulfide and selenides do not differ much, $MSe_{1-x}S$ films are expected to be sulfur-rich, in each case relative to the fraction of thiourea in the solution. Thus, if the films are given the formula $CdSe_{1-y}S_y$ and x is the fraction of thiourea in the solution, then $y > x$.[95]

As is the case in an atom-by-atom deposition process, the solubility conditions of multicomponents in an ion-by-ion condensation process are relaxed. Also, one expects the stabilization of high-temperature and/or high-pressure polymorphs of the chalcogenide under certain deposition conditions. For example, $Pb_{1-x}Hg_xS$ $(0 \leq x \leq 0.33)$ and $Cd_{1-x}Pb_xSe$ $(0 \leq x \leq 1)$ alloy films have been prepared even though solid solubilities of the two chalcogenides in each case are not known to exist in bulk form.[95,96] Further, either the room-temperature-stable α-phase (2.0 eV bandgap, trigonal) or the high-temperature β-phase (0.1 eV bandgap, cubic, stable above 280 °C) of HgS can be stabilized yielding a uniphase fcc structure of α'- or β'-$Pb_xHg_{1-x}S$ films.[96]

5.3.2.5. *Oxide Films*

The deposition of oxide films is made possible by the fact that a large number of metal ions (M^{+n}) undergo hydrolysis in an aqueous solution precipitating a solid phase of hydrous oxide $(M_2O_n \, . \, xH_2O)$, which decomposes on heating to yield the corresponding oxide. For example,

$$Sn^{+4} \xrightarrow[\text{alkaline}]{} SnO_2 \, . \, xH_2O \xrightarrow{250\,°C} SnO_2$$

$$Zn^{+2} \xrightarrow[\text{alkaline}]{} ZnO \, . \, xH_2O \xrightarrow{120\,°C} ZnO$$

Thus oxide films of Mn, Fe, Zn, Sn, Pb, Cu, Cd, Cr, Ti, and Al can be obtained. The resistivity of as-deposited films of SnO_2 and ZnO varies from 10^2 to 10^8 Ω cm and transmission in the visible ranges from 10 to 70%, depending on deposition conditions. Under optimized conditions and by using suitable dopants (e.g., In for SnO_2), transmission of about 80% and resistivity of about 10^{-2} Ω cm have been obtained. This simple technique holds much promise of further development.

5.3.2.6. *General Properties of Solution Grown Films*

Transmission electron microscopy studies[90] of solution grown films have established that the film formation proceeds via nucleation and growth processes in a way similar to vapor deposited films. Usually, the films are micropolycrystalline, with grain sizes typically in the range from 300 to 1000 Å. The grain size depends on the composition and temperature of the bath and nature of the substrate. The grain size is larger at lower deposition rates (i.e., at lower supersaturations), higher bath temperatures, and for lattice-matched substrates.

When two or more chalcogenides are codeposited, a finer microstructure[95,98] is obtained, similar to the case of vapor deposited films. The microstructure in multicomponent films is dominated by the sizes of the various ions and interaction among them.

As reported by Sharma et al.,[99] under suitable deposition conditions, the ad-ion mobility is large enough to yield well-oriented, epitaxial films on single crystal substrates. The authors obtained the epitaxial growth of $Pb_{1-x}Hg_xS$ ($0 \leq x \leq 0.33$) films on single crystal Ge substrates below 20 °C bath temperature.

The complexing agent has a marked effect on the CdS film structure. CdS films obtained from $Cd(NH_3)_4^{2+}$ complex have sphalerite, wurtzite, or mixed structures depending on the deposition conditions. On the other hand, CdS films prepared from $Cd(CN)_4^{2-}$ and $Cd(en)_3^{2+}$ complexes always exhibit wurtzite structure with the c-axis perpendicular to the substrate.[91,94,97]

Postdeposition annealing at temperatures greater than 400 °C for a sufficiently long time leads to appreciable grain growth. Higher impurity-induced growth rates have been achieved at lower temperatures (~300 °C) using the embedding technique.[95,100] CdS and CdSe films have also been recrystallized[95] by depositing a thin layer of Cu or Ag on the film and heating at temperatures greater than 350 °C.

A detailed discussion of the electrical and optical properties of solution grown films and their relation to the microstructure is presented in the next chapter.

5.3.3. *Screen Printing*

Thick film technology offers a flexible inexpensive answer to the problems of grid pattern delineation on solar cells, array interconnections, and, in recent years, to preparation of active semiconductor layers and devices.[101–104] In this approach, pastes containing the desired material are screen-printed by conventional methods onto a suitable substrate to define conductor, resistor, and device patterns. Subsequently, the substrate is fired under appropriate conditions of time and temperature to yield rugged components bonded to the substrate. The device and/or circuit may be given a coating of resin for general protection. Improved protection may be effected through encapsulation, by transfer molding, or by hermetic sealing in metal or ceramic enclosures. The capital outlay and running costs are relatively low, highly qualified or skilled operators are not essential, and changes in device geometry or circuit design are readily and inexpensively implemented in production.

5.3.3.1. *Physical Aspects of Screen Printing*

5.3.3.1a. *Substrates.* For thick film applications, the substrates must possess the following characteristics: (i) a uniform smooth surface texture, allowing, however, adequate adhesion of the fired thick film layers, (ii) a minimum of distortion or bowing of the plate, (iii) capability of withstanding normal firing temperatures, usually in the range of 500 to 1000 °C, (iv) high mechanical strength, high thermal conductivity, and good general electrical properties, (v) chemical and physical compatibility with the thick film conductor, resistor, dielectric, and semiconductor paste compositions, and (vi) low cost in quantity production.

Many ceramic materials can, in principle, be used; for example, alumina, beryllia, magnesia, thoria, and zirconia.[5,105] Glass substrates[102,104] have also been used to prepare semiconductor layers. High-purity (96%) alumina exhibits the best combination of electrical, thermal, and mechanical properties. However, beryllia possesses a high value of thermal conductivity, thus enabling a circuit based upon it to dissipate greater power.

The surface topography plays an important role in the screen printing technique. An excessively smooth surface results in poor adhesion of the fired thick film layers, whereas a very rough surface leads to poor reproducibility of film thickness. A surface finish of 0.5 to 1 μm center line average (CLA) is observed to be the normal acceptable range in practice. The CLA height is defined as the arithmetical average value of the deviation of the whole profile above and below its center line throughout the prescribed meter cut-off. The trace profile may easily be obtained by a Talystep height measuring instrument.

In most thick film and screen printing applications, dimensional tolerances on length, width, and hole positioning are of the order of ±1 μm. Substrate sizes vary according to the requirements, normally in the range from 3 to 30 cm^2. Substrate thicknesses range from 0.05 to 0.1 cm with thickness tolerances of the order of ±25 to ±100 μm. The typical values to which the substrate may be bowed are within 40 to 50 $\mu m\, cm^{-1}$.

It is important to realize that close tolerances have to be set on substrate length, width, and thickness as well as on hole position and degree of bowing, in order to obtain precision in the subsequent printing operations. As noted previously, the substrate must be compatible with the thick film compositions. Normally, manufacturers of commercial thick film pastes ensure that the glaze compositions are compatible with standard substrate materials. However, for those preparing their own pastes for specific applications, a word of caution is appropriate. The expansion coefficients of the fully fired pastes must be as closely matched as possible with that of the substrate material, since poor expansion matching could produce strains and subsequent crazing, loss of adhesion, and degradation of properties.

5.3.3.1b. *Printing Procedures.* The screen printing process, shown schematically in Figure 5.17, involves positioning the substrate on a carriage, which is then brought beneath the screen so that the substrate is in accurate registration with the pattern on the screen. The pattern on the screen is photolithographically defined so that open mesh areas in the screen correspond to the configuration to be printed. The substrate, when in the printing position, is placed a short distance beneath the screen. The clearance between the screen and the substrate surface is termed the breakaway or snap-off distance. A small amount of the paste is dispensed onto the upper surface of the screen. A flexible wiper, called the squeegee, then moves across the screen surface, deflecting the screen vertically and bringing it into contact with the substrate and forcing the paste through the open mesh areas. On removal of the squeegee, the screen regains its original position by its natural tension, leaving behind the printed paste pattern on the substrate. The substrate carriage is then removed from beneath the screen and the substrate replaced and the process continues.

The screen printer contains the following basic functional components: (i) the screen and its mounting framework, (ii) the substrate carriage and the associated mechanical feed system, which may be manually or automatically controlled, (iii) the squeegee mechanism and pressurizing system, and (iv) the adjustment mechanism for precise positioning of the screen relative to the substrate.

The screen mounting is provided with micrometer screws to facilitate X, Y, and rotational adjustment of the screen relative to the substrate and

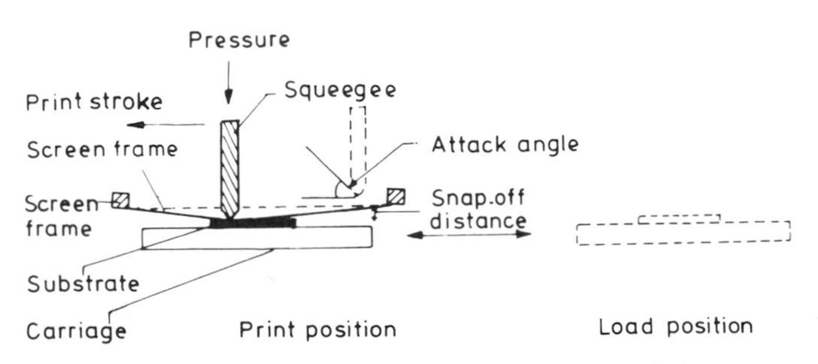

Figure 5.17. Schematic representation of the screen printing process.

to place it parallel to the substrate. The clearance between the screen and the substrate can also be adjusted. The snap-off distance critically determines the amount of ink delivered through the screen and the precision of the printed pattern.

The substrate holder comprises a platform with a recess in which the substrate is placed and held by a vacuum. The substrate holder is mounted on a moving carriage which travels repeatedly from a loading position to the print position.

Various microcontrols are incorporated into the squeegee mounting system to allow adjustment of the parallelism of the squeegee with the screen, the squeegee printing pressure, and the angle of attack of the squeegee. Generally, movement of the squeegee across the screen is automatic with speeds adjustable from 0.25 to 25 $cm\,s^{-1}$. The printing stroke may be unidirectional or bi-directional.

The paste is applied to the screen manually with a spatula; however, an automatic paste dispenser may be incorporated.

The screen serves a twofold purpose: it delineates the pattern to be printed and it acts as a metering device for the deposition of a controlled amount of the paste. The volume of paste held in each mesh opening depends on the mesh separation and diameter and also on the thickness of the material used to form the pattern. Four types of screens are commonly used: (i) Direct emulsion screen in which a photographic emulsion is spread uniformly over the mesh and dried. The pattern is then defined by exposing the emulsion to UV light through an appropriate photographic mask to harden the areas to be retained on the screen. The nonhardened (unexposed) areas of the pattern are removed in a developer solution. A small quantity of emulsion merely blocks the mesh openings and does not affect the screen thickness or the volume of paste delivered by each mesh opening. Larger quantities of emulsion increase both the effective screen thickness and the volume of paste delivered. (ii) Indirect emulsion screen utilizes a light-sensitive film backed by a gelatine-coated polyester sheet. Similar photographic exposure, development, and washing processes as used for direct emulsion screens are followed on the light-sensitive film/polyester sandwich, and the wet film is transferred to a screen to which the gelatine adheres. The polyester backing is removed after drying and the screen is then ready for use. The indirect emulsion increases the overall thickness of the screen.

The indirect type of screen gives better emulsion thickness uniformity and superior line definition as compared to the direct type. However, the adhesion of the indirect emulsion screen to the mesh is relatively poor. Moreover, the life of a direct screen is typically of the order of 20,000 prints while that of an indirect screen is only about 5000 prints.

Nylon and stainless steel are the most commonly used mesh materials, for both direct and indirect emulsion screens. Screen tension is dependent on the size of the screen frame and on the wire diameter. Table 5.7 gives the relationship between the screen mesh number, wire diameter, and deposited thickness. Under very controlled conditions, utilizing fine-line pastes and fine-mesh screens, the lower limit in line width and spacing is of the order of 100 μm. Metal masks used in conjunction with contact printing techniques enable patterns to be produced with line widths as small as 25 μm, spacings of about 75 μm, and line width tolerances of ±5 μm. (iii) Suspended etched metal screen comprises a metal foil 25 to 50 μm thick, chemically etched to the required pattern and bonded to a stainless steel mesh supported under tension on a screen frame. (iv) Solid metal etched mask consists of a bimetallic sheet in which the pattern to be printed is etched on the lower side and a grid of holes connecting to the pattern is etched on the upper side.

In comparison to the suspended metal screen (where the etched image is not precisely aligned with the wire mesh), the solid metal mask allows the attainment of lower line tolerances, since the patterns on the upper and lower sheets can be accurately registered under a microscope before etching. The metal masks have a longer life compared to the emulsion screens with the solid etched metal screen exhibiting the longest life.

Table 5.7. Relationship between Screen Mesh Number, Wire Diameter, and Deposited Thickness (from Hamer and Biggers[105])

Mesh number (meshes/in.)	Wire diameter (mils)	Deposited thickness[a] (wet, in mils)
105	3.0	4.44
120	2.6	3.85
145	2.2	3.25
165	2.0	2.92
165	1.9	2.82
200	2.1	2.20
200	1.6	2.35
250	1.6	2.12
250	1.4	1.99
270	1.6	2.02
270	1.4	1.91
325	1.4	1.75
325	1.1	1.55
400	1.1	1.37
400	1.0	1.32

[a] Assuming a 100% transfer to substrate from screen.

To obtain uniformity in transfer of paste through the screen, it is necessary to ensure that the printing edge of the squeegee remains parallel to the screen throughout the printing stroke. Another important consideration is that the squeegee blade should be made of a material inert to the commonly used solvents. Neoprene, polymethane, or PTFE are used. An attack angle of 45° is often recommended, although, under certain conditions, a blunt-edged, relatively soft blade gives better results.

The squeegee pressure should be just sufficient to wipe the print pattern areas free of paste and leave a thin film on the blocked areas of the screen. The printing force is decided by the screen tension and snap-off distance and is usually in a range of up to 12 lb. Excessive pressure causes irregular print edges and may permanently deform the screen over the edges of the substrate.

5.3.3.1c. *Print Quality.* The rheological characteristics of the paste, i.e., the viscosity, surface tension, and thixotropy, strongly influence the quality of the print. These factors affect the extent of wetting of both the substrate and the screen by the paste and thus determine the proportion of the paste volume in each mesh which is transferred to the substrate by the squeegee.

The important characteristics of the print are the film thickness and pattern definition. Thin prints result under the following conditions: (i) low viscosity of the paste, (ii) high squeegee pressure, (iii) high stroke speed, (iv) small snap-off distance, (v) small angle of attack, and (vi) hard squeegee blade. Relatively thicker films are obtained by the converse of each of these parameters. End-to-end nonuniformity results from: (i) lack of constancy in the stroke speed, (ii) lack of parallelism between screen and substrate, (iii) insufficient paste on the screen, and (iv) small snap-off distance. Side to side nonuniformity is caused by: (i) the substrate and screen not being parallel, (ii) insufficient paste on the screen, (iii) insufficient squeegee pressure, and (iv) a worn squeegee blade. A very high paste viscosity may also lead to nonuniform deposition.

The pattern definition may be marred by ragged edges, incomplete filling, or smearing. Ragged print edges result from: (i) a worn screen, (ii) a worn squeegee blade, (iii) insufficient squeegee pressure, and (iv) a rough substrate. Incomplete filling is caused by: (i) blocked meshes, (ii) high viscosity of the paste, (iii) low squeegee pressure, (iv) hard or worn squeegee blade, (v) low angle of attack, and (vi) excessive print speed. Smearing occurs if: (i) the viscosity of the paste is too low, (ii) the snap-off distance is too small, (iii) the print speed is too slow, (iv) the squeegee blade is worn, (v) the squeegee pressure is excessive, (vi) the screen is dirty, and (vii) the screen tension is too low.

5.3.3.1d. *Firing Procedures.* The process subsequent to printing can be separated into four operations, namely, ink coalescence, drying, organic binder removal, and high-temperature firing.

The pattern immediately after screen printing consists of a series of discrete ink spots each corresponding to a mesh opening in the screen. The substrates are then allowed to stand at ambient temperature for a few minutes to enable the ink/paste to coalesce sufficiently to form a coherent, level film. The time required for coalescence is determined by the nature of the paste composition. In the case of meshless metal masks, no time is needed for coalescence.

Temperatures of 70 to 150 °C are commonly employed, for periods ranging from 15 to 30 min, to dry the printed film and remove the more volatile ink components. Close control of the drying step is necessary for good results. Improper drying can lead to imperfections such as blisters, cracks, and crazing. Small ovens or infrared lamps are often used, although the best results are obtained utilizing a low-temperature tunnel kiln with resistance or radiant heating and belt feed. Adequate ventilation during the drying operation is a necessary condition.

The large amount of organic material left in the composite after drying is removed at relatively low temperatures ($\sim 400\,°C$) by carbonizing and oxidizing. Thus an oxidizing atmosphere is required. The binder removal is generally carried out as the first phase of the final firing process.

Firing is done in a multizone tunnel kiln. The temperature profile provides an adequate period at low temperatures so that all the organic binder is removed.

In the second stage of the firing process, the print is taken to the maximum firing temperature, which may be up to 1000 °C. The glass component of the ink melts to form a vitreous medium which consolidates the printed layer and promotes adhesion to the substrate. The important properties of the composite are determined by the chemical reactions that take place in the high-temperature zone of the kiln. For good results, the time–temperature profile must be accurately controlled.

5.3.3.2. *Chemical Aspects*

The chemical reactions that occur during firing can be expressed in the general form

$$\text{Reactants} \rightarrow \text{Intermediates} \rightarrow \text{Products}$$

The rate of the reaction is determined by the concentration of the reactants, intermediates, and products, the physical form of the reactants, the reaction environment, time, and temperature. For a specific paste composition, most of these variables are fixed. Time and temperature are the two variables the user can modily to control the reaction. The reaction rate can be expressed by the relation:

$$\text{Reaction rate} = A \exp(-B/T)$$

where A and B are constants and T is the absolute temperature.

The reactions that take place during firing can be classified as: (i) reactions between ink components, (ii) reactions between ink components and the substrate, (iii) reactions between ink components and the furnace atmosphere, and (iv) reactions between components of different inks in contact with each other.[105]

The reactions between ink components include reactions between (i) functional metals, (ii) functional metals and permanent binders, (iii) functional metals and temporary binders, (iv) permanent binder components, and (v) temporary and permanent binders.

The furnace atmosphere can react with the functional metals/compounds and permanent and temporary binders.

When multilayer coatings are made or when one coating (such as a conductor line) overlays another (such as a semiconductor/dielectric/resistor print), the possibility exists of interactions between components of the overlapping films during firing. A typical example is the diffusion of the metal phase of a conductor into a resistor.

5.3.3.3. *Paste Characteristics*

Thick film pastes normally consist of the following constituents: (i) a metallic/resistive/dielectric/semiconducting component in finely divided powder form, (ii) a bonding agent comprised of a finely divided glass frit, (iii) an organic suspension medium, and (iv) an organic diluent.

The choice of a particular paste composition is dictated by the desired characteristics of the final product, and a large number of conductor, resistor, and dielectric compositions are commercially available. Semiconductor pastes are not yet available commercially and their use has been limited. In Table 5.8 we have listed the important documented characteristics of conductor and semiconductor paste compositions.

5.3.4. *Chemical Vapor Deposition (CVD)*

The history of CVD techniques dates back to the early nineteenth century when the reduction of silicon tetrafluoride or silicon tetrachloride by sodium or potassium[106] was utilized to refine and deposit silicon. In the 1930s, increasing emphasis was laid on utilizing the CVD technique to prepare refractory compounds such as metal carbides, nitrides, silicides,

Table 5.8. Important Characteristics of Some (a) Conductor and (b) Semiconductor Paste Compositions

(a)

Characteristics	Au	Pt/Au	Pd/Au	Pd/Ag
Optimum firing temp., °C	850	850	850	850
Sheet resistivity, Ω/square	0.005–0.01	0.08–0.10	0.04–0.10	0.010–0.030
Solderability	Nonsolderable	Excellent	Excellent	Excellent
Bondability, wire	Excellent	Good	Good	Good
die	Excellent	Good	Good	Nonbondable
Line resolution, mils	5–15	5–15	5–15	5–15
Cost	High	Highest	High	Lowest

(b)

Characteristics	CdTe	CdS
Paste composition	CdTe(5N) + 1 wt.% $CdCl_2$ + propylene glycol	CdS(4N) + 5 wt.% $CdCl_2$ + 0.3 wt.% $GaCl_2$ (dopant) + propylene glycol
Optimum firing temp., °C	720	630
Resistivity, Ω cm	$\sim 10^3$	$\sim 10^{-2}$
Grain size, μm	~ 10	~ 10

borides, and oxides as well as sulfides, selenides, tellurides, intermetallic compounds, and alloys. Following the work on epitaxial growth of semiconductors by the technique, CVD gained wide acceptance as a means of growing thin layers. The process has since been extended to deposit a wide variety of films, including insulators, conductors, resistors, varistors, and ferrites.

Chemical vapor deposition involves, essentially, exposure of the substrate to one or several vaporized compounds or reagent gases, some or all of which contain constituents of the desired deposited substance. A chemical reaction is then initiated, at or near the substrate surface, producing the desired material as a solid-phase reaction product which condenses on the substrate. The chemical reaction may be activated by the application of heat, an rf field, light or X-rays, an electrical arc, a glow discharge, electron bombardment, or catalytic action of the substrate surface. It should be emphasized that the morphology of the deposited layer is strongly influenced by the nature of the chemical reaction and the activation mechanism. It is important to attain deposition conditions which enable the reaction to take place near or on the substrate surface (heterogeneous reaction) in order to avoid powdery deposits, which result when the reaction occurs in the gas phase (homogeneous reaction).

The film growth in the CVD process takes place by an atom/molecule-by-atom/molecule condensation process. The growth process is in many ways similar to that of physical vapor deposition processes such as evaporation and sputtering, since in every case the deposit is formed from a vapor phase. A clear distinction between chemical vapor deposition and physical vapor deposition (PVD) processes is that in CVD the formation of a film results from a heterogeneous chemical reaction without involving a mean free path of the gas molecules larger than or comparable with the dimensions of the deposition chamber as a necessary condition for the deposition process. However, CVD may be carried out at low pressures or in high vacuum, depending on the requirements.

The major advantages of the CVD technique are: (i) in general, no vacuum or pumping facilities are required and thus a relatively simple setup and fast recycle times are possible, (ii) high deposition rates are possible, (iii) it is possible to deposit compounds and control their stoichiometry easily, (iv) it is relatively easy to dope the deposits with controlled amounts of impurities, (v) it is possible to grow multicomponent alloys, (vi) refractory materials can be deposited at relatively lower temperatures compared to vacuum evaporation, (vii) epitaxial layers of high perfection and low impurity content can be grown, (viii) objects of complex shapes and geometries can be coated, and (ix) in-situ chemical vapor etching of the substrates prior to deposition is possible.

However, the technique suffers from several drawbacks, namely: (i) the thermodynamics and reaction kinetics involved in the deposition process are frequently very complex and poorly understood, (ii) usually, higher substrate temperatures are required than in the corresponding PVD technique, (iii) the reactive gases used for the deposition process and the reaction products are, in most cases, highly toxic, explosive, or corrosive, (iv) the corrosive vapors may attack the substrate, the deposited film, and materials of the deposition setup, and volatile products generated during the deposition process may lead to incorporation of impurities in the growing film, (v) the high temperatures may lead to diffusion, alloying, or chemical reaction on the substrate surface and thus the choice of substrates is limited, (vi) it is difficult to control the uniformity of the deposit, and (vii) masking of the substrate is usually difficult.

In the following sections, we discuss the salient characteristics of the CVD process. For a detailed treatment, the reader is referred to comprehensive reviews on the subject.[107,108]

5.3.4.1. *Chemical Aspects of CVD*

Any chemical reaction between one or several reactive vapors which yields a solid phase reaction product can be used for chemical vapor deposition. The substrate may, in some cases, take part in the reaction mechanism if the temperature is sufficiently high. For example, Si or Al substrates when exposed to an oxygen atmosphere grow SiO_2 or Al_2O_3 layers, respectively. The selection of a practical reaction, however, is dictated by the constraints imposed by the substrate and the reaction. Moreover, one has to be aware of and take into account the fact that the actual course of the reaction may be much more complex and involve formation of intermediate reaction species in accordance with the reaction kinetics. The reaction kinetics depend on several factors, notably flow rates, partial gas pressure, deposition temperature, temperature gradients, and nature and properties of the substrate surface.

The chemical reactions utilized in CVD processes can be classified as: (i) decomposition reactions, (ii) hydrogen or metal reduction of halogens, (iii) polymerization, and (iv) transport reactions. A brief summary of these reactions follows.

5.3.4.1a. *Decomposition.* If sufficiently high energy is supplied to the vaporized plating compound streaming over or being adsorbed on the substrate surface, it decomposes and a solid-phase reaction product condenses on the substrate. This reaction can be written as

$$AB\ (\text{gas}) \rightleftharpoons A\ (\text{solid}) + B\ (\text{gas})$$

Both organic and inorganic as well as polar and nonpolar plating compounds are suitable for this process. A typical example of a decomposition process is

$$SiH_4 \xrightarrow{800\text{–}1300\,^\circ\text{C}} Si + 2H_2$$

The decomposition process, also termed pyrolysis, can be classified as a high-temperature pyrolysis requiring a substrate temperature exceeding 600 °C or a low-temperature process at temperatures between room temperature and 600 °C. Primarily metal halides, in particular the iodides, undergo high-temperature decomposition. Compounds which decompose at low temperatures include metal hydrides, metal carbonyls and complex carbonyls, most organometallic compounds, metal borohydrides, and some of the more unstable metal halides and carbonyl halides.

For decomposition at low pressures, or with a large concentration of decomposition products, increased substrate temperature may be required. An increase in substrate temperature is beneficial in obtaining improved crystallinity, composition, purity, or adhesion of the deposited layer.

Despite the simplicity of the decomposition process, certain difficulties may arise, associated with the formation of more than one nonvolatile residual reaction product such as carbon from carbonyls and organometallic compounds, boron from borohydrides and oxides from oxygen-containing compounds. Electron bombardment, used in activated decompositions, converts various silicon oils adsorbed to a substrate surface to SiO or SiO_2 with incorporated organic constituents.[109]

5.3.4.1b. *Reduction.* A reduction process can be considered a decomposition process aided by the presence of a second reaction species. Consequently, deposition occurs at a temperature lower than that for pyrolysis of the first component. Hydrogen or metal vapors are employed as reducing agents while metal halides, carbonyl halides, oxyhalides, or other oxygen-containing compounds are used to obtain the material to be deposited. In some cases, addition of a reducing agent to a reactive vapor, from which the layer is formed mainly by pyrolysis, serves to prevent codeposition of undesired oxides or carbides. The strength of the reducing reaction in the chemical vapor deposition from metal halides is influenced by the reducing agent and increases in the sequence

$$H_2 — Cd — Zn — Mg — Na — K \qquad (T < 1000\,°C)$$

However, a very strong reducing reaction leads to premature reduction in the gas phase, yielding powdery deposits. A typical example of chemical vapor deposition by reduction is the preparation of Si from the corresponding halide vapors using H_2 or Zn as the reducing agents according to the reaction

$$SiCl_4 + 2H_2 \rightarrow Si + 4HCl$$

Hydrogen, although less strongly reducing, offers the advantage that it can be premixed with the metal halide without causing a premature reaction.

Metals, when used as a reductant, may contaminate the deposit. To overcome this problem, the metal is used in stoichiometric proportions and the process is carried out at a reduced pressure. If the reductant metal forms a halide in the reaction that is less volatile than the parent metal (and may, therefore, be codeposited), operating conditions have to be maintained such that the pressure of the halide of the reductant metal is lower than its saturation pressure at the deposition temperature. In this context, for the reductant metals mentioned above, the fluorides, the chlorides, and the bromides (except those of zinc), and the iodides (except those of zinc and magnesium) are less volatile than their parent metals with fluorides being the least volatile and the iodides the most volatile. Further, alkali metal halides are least volatile relative to the parent metal. Thus, iodides are preferable as deposition media and alkali metals are least suitable as reductants.

For thermally activated deposition, the reduction process may require high substrate temperatures. However, too high a temperature may lead to pyrolysis of a reaction product such as hydrogen halide leading to a reversal of the reaction or etching of the substrate.

5.3.4.1c. *Polymerization.* In the polymerization technique, organic and organic–inorganic composites form monomers whose molecules are linked together by one of the following activation processes: (i) electron or ion bombardment, (ii) irradiation with light, X-rays, or γ-rays, (iii) electrical discharge in the monomer vapor, and (iv) surface catalysis or surface recombination of monomers having free radicals. The films can be produced by (i) condensing monomer vapors on the substrate and subjecting them simultaneously or sequentially to the activation process, (ii) (ii) activating in the gas phase and allowing the polymerized product to deposit on the substrate, or (iii) depositing the monomer film of the substrate by other means and then activating the polymerization.

The polymerized films have electrical properties ranging from semiconducting to insulating and have certain desirable characteristics such as complete surface coverage, good adhesion, low stress, and high plasticity.

5.3.4.1d. *Transport Reactions.* The chemical transport technique involves the transfer of a relatively nonvolatile material from the source location to the substrate location utilizing a relatively highly volatile chemical vapor. The three basic steps involved are: (i) conversion of the source material into a volatile compound through a chemical reaction, (ii) transport of the vapor to the substrate, and (iii) decomposition of the chemical vapor over or on the substrate leading to deposition of the source material onto the substrate.

To achieve chemical transport, the reaction equilibrium is usually shifted to opposite sides in the source and substrate locations. Let us consider the example of an indirect distillation system:

$$\text{Ti (solid)} + 2\text{NaCl (gas)} \rightleftharpoons \text{TiCl}_2\text{ (gas)} + 2\text{Na (gas)}$$

The direction in which the reaction occurs is controlled by the temperature and pressure conditions. Thus, different conditions of temperature and pressure are maintained at the source and substrate regions. Another method to shift the reaction equilibrium is to introduce an additional chemical to perform either a reduction or an oxidation of the transport vapor near the substrate surface, leading to the deposition of a compound, for example, the transport of SiO_2 with the more volatile SiO as the transport medium, according to the reaction

$$\text{Si (solid)} + \text{SiO}_2 \text{ (solid)} \rightarrow 2\text{SiO (gas)}$$

$$2\text{SiO (gas)} + \text{O}_2 \text{ (gas)} \rightarrow 2\text{SiO}_2 \text{ (solid)}$$

Of the various reactions utilized for chemical transport, the disproportionation reactions, particularly the halide disproportionation reactions, are the most widely used. Transfer is accomplished by treating the nonvolatile metal with the vapor of its own higher-valent halide at a high temperature to yield a lower-valent, volatile halide which after transport into a cooler zone of the system disproportionates back into the higher-valent, volatile halide and the nonvolatile metal. The higher-valent halide is recycled by feeding back the hotter zone in which the source material is kept. Thus the system can operate as a closed system. A typical example is the transport of silicon in an iodine vapor atmosphere[110]:

$$\text{Si (solid)} + 2\text{I}_2 \text{ (gas)} \xrightarrow{1100\,^\circ\text{C}} \text{SiI}_4 \text{ (gas)}$$

$$\text{Si (solid)} + \text{SiI}_4 \text{ (gas)} \xrightarrow{1100\,^\circ\text{C}} 2\text{SiI}_2 \text{ (gas)}$$

$$2\text{SiI}_2 \text{ (gas)} \xrightarrow{900\,^\circ\text{C}} \text{Si (solid)} + \text{SiI}_4 \text{ (gas)}$$

An additional advantage of transport reactions is that refinement of the material also occurs in those cases, such as the three above, where the corresponding iodine vapors of the major contaminants of the source material have vapor pressures sufficiently different from those of the major transporting species.

Transport reactions exist where oxidation occurs on one side of the reaction equilibrium and reduction on the other, such as:

$$2\text{Ga (solid)} + \text{H}_2\text{O (gas)} \rightleftharpoons \text{Ga}_2\text{O (gas)} + \text{H}_2 \text{ (gas)}$$

$$\text{Ge (solid)} + \text{H}_2\text{O (gas)} \rightleftharpoons \text{GeO (gas)} + \text{H}_2 \text{ (gas)}$$

We may note here that transport reactions have been used for gaseous etching, e.g., Ge or Si with HCl[111,112] and Al_2O_3 with fluorinated hydrocarbons[98] to remove contaminants and surface damage prior to the deposition of epitaxial films and also to obtain a thin film by thinning (gaseous etching) a relatively thick deposit.[108]

5.3.4.2. *Physical Aspects of CVD*

5.3.4.2a. *Types of Systems.* A CVD system for depositing thin films combines the following functions: (i) generates reactive chemical vapors, (ii) transports, meters, and times the diluent and reactant gases entering the reactor, (iii) supplies activation energy to the reaction, which leads to the formation of the desired thin film, and (iv) removes and safely disposes of the reaction by-products.

CVD systems cover the range from extremely simple laboratory setups to highly sophisticated, completely automated, electronically controlled, and computerized industrial reactors. The system can be either a closed system, permitting complete recovery of the reagent species and recycling, or an open system requiring an external supply of source material and extraction of the reaction components.

The reactor is the most important part of a CVD system. The reactors can be classified as: (i) low-temperature CVD reactors for applications at less than 500 °C at normal pressure and (ii) high-temperature CVD reactors for use at more than 500 °C at normal or low pressures. The low-temperature reactors are further categorized into four main classes, according to their gas flow characteristics and principle of operation: (i) horizontal tube displacement flow reactors, (ii) rotary vertical batch-type reactors, (iii) continuous reactors employing premixed gas flow feed through an extended area slotted disperser plate, and (iv) continuous reactors employing separate nitrogen-diluted oxygen and hydride streams directed toward the substrate by laminar flow nozzles. The high-temperature reactors can be divided into (i) hot-wall reactors, used in systems where the deposition reaction is exothermic in nature and (ii) cold-wall reactors used when the deposition reaction is endothermic. Kern and Ban[114] have discussed various types of reactors in detail.

Obtaining high-quality deposits, reproducibly, requires stringent control of the CVD process, particularly of the following parameters: (i) the temperature in the reactor (single or multizone), (ii) the quantities and compositions of all gases or vapors entering the reactor, (iii) the time sequencing of the thermodynamic and chemical variables, and (iv) the pressure. The temperature is measured either by using a thermocouple or pyrometrically. The outputs of the sensing element are fed to temperature controllers of accuracy typically of the order of ± 1 to 5 °C. The gas flow rates are determined and controlled utilizing either rotameter type flowmeters or electronic mass controllers. Rotameters possess an accuracy of $\pm 5\%$ of the full scale compared to $\pm 2\%$ of full scale for the electronic flowmeters. Electronic flowmeters afford the possibility of automation and computer programming of CVD processes, thus allowing the user to synthesize complex structures with good control, since the amount and the duration of gaseous reactants determine the composition and the thickness of the deposited layer.

The most frequently employed method of supplying activation energy to the deposition reaction is thermal. Toward this end, the substrate in the

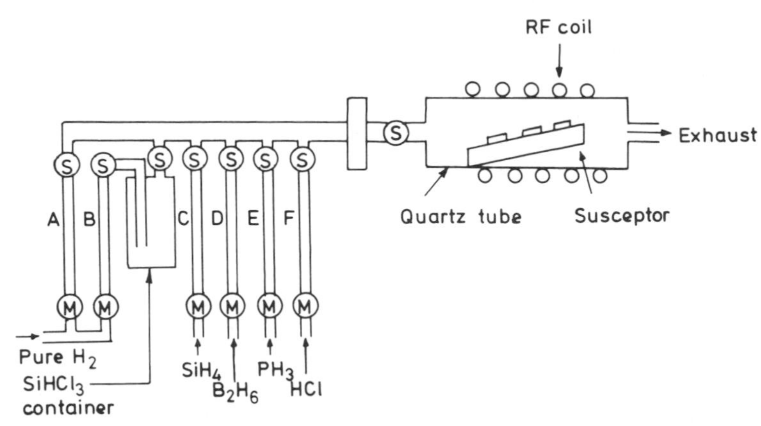

Figure 5.18. Schematic diagram of a CVD system used for fabricating thin film Si solar cells: S — shut-off values; M — metering values; A, B, C, D, E, and F — flowmeters (after Chu and Singh[115]).

reaction is heated by an external furnace or by internal heating. Internal heating may be accomplished by passing current through the substrate (if the substrate is conducting), by placing it on a resistive strip, by placing it on an rf inductively heated susceptor, by infrared or high-intensity radiation lamps, or by electron bombardment. Susceptors are usually made of high-density SiC-coated graphite. The reactor tubes are fabricated from fused silica.

Figure 5.18 illustrates, schematically, a CVD system used for the deposition of Si p/n junction structures.[115]

Kern and Ban,[114] while reviewing CVD processes for the deposition of inorganic thin films, have pointed out that despite considerable advance in reactor design and construction, significant improvements need to be realized to decrease the chemical quantity and power requirements, reduce heat losses, and increase reactant utilization.

5.3.4.2b. *Reaction Thermodynamics and Kinetics.* General process parameters can be established by thermodynamic calculations which give the theoretically obtainable amount of a deposit and partial pressures of all vapor species under specified experimental conditions. However, thermodynamics does not provide information on the rate of various CVD processes. Further, thermodynamic calculations assume the attainment of chemical equilibrium, which may not be valid. Nevertheless, such calculations are a useful guide to the feasibility of a particular CVD process.

The free energy of a chemical reaction ΔG_r^0 can be calculated according to

$$\Delta G_r^0 = \sum \Delta G_f^0\,(\text{products}) - \sum \Delta G_f^0\,(\text{reactants})$$

where ΔG_f^0 is the standard free energy of formation of a compound. ΔG_r^0 can be expressed by

$$-\Delta G_r^0 = 2.3RT \log k_p$$

where k_p is the equilibrium constant and is related to partial pressures in the system by the relation

$$k_p = \prod_{i=1}^{n} P_i \text{ (products)} \Big/ \prod_{i=1}^{n} P_i \text{ (reactants)}$$

In dealing with multicomponent and multiphase CVD systems we can adopt either the optimization[116] or the nonlinear equation[117] method to calculate the thermodynamic equilibrium. In the first, the free energy G of the whole system consisting of m gaseous species and solid phases is given by

$$G = \sum_{i=n_i}^{m} \left(n_i^g \Delta G_{f_{ig}}^0 + RT \ln P + 2T \ln \frac{n_{ig}}{N_g} \right) + \sum_{i=1}^{s} (n_i^s \Delta G_{f_{is}}^0)$$

where n_i^g and n_i^s are the number of moles of gaseous and solid species, respectively. N_g is the total number of moles of gaseous species, P is the total pressure and $\Delta G_{f_{ig}}^0$ and $\Delta G_{f_{is}}^0$ are the free energies of formation at CVD temperatures of gaseous and solid species, respectively. Optimization calculations aim to determine the set of n_i that minimizes G. A computer program[116] has been described to obtain the equilibrium compositions of gaseous and solid phases at some specified deposition parameters, such as temperature, pressure, and input concentrations. In the second method, a set of m simultaneous, generally nonlinear equations, specifying a quantitative relationship between partial pressures of species present, is established and solved (using the computer) to yield values of partial pressures of all species under specified parametric conditions.

The sequential events occurring in the usual heterogeneous processes are as follows: (i) diffusion of reactants to the surface, (ii) adsorption of reactants at the surface, (iii) chemical reaction, surface motion, and lattice incorporation, (iv) desorption of products from the surface, and (v) diffusion of products away from the surface. The slowest step in the above is the rate-determining one.

The following factors in CVD processes affect the deposition rate and the uniformity, composition, and properties of the films: (i) Substrate temperature — as illustrative examples, Figures 5.19a and b show the deposition rates of Si from various gaseous sources[118] as a function of the substrate temperature. It is evident that the temperature dependence is

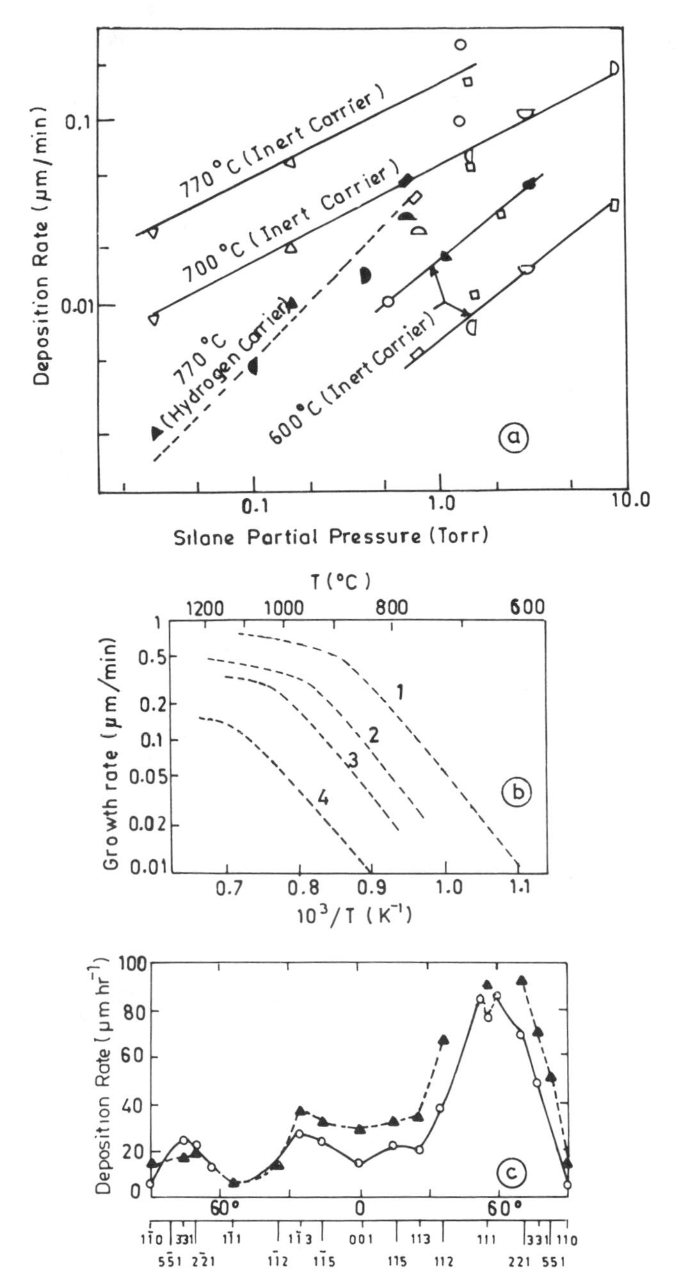

Figure 5.19. (a) Deposition rates of CVD grown Si films as a function of SiH_4 partial pressure for various gaseous carriers (after Eversteijn[118]); (b) deposition rate of CVD Si film as a function of substrate temperature for 1 — SiH_4, 2 — SiH_2Cl_2, 3 — $SiHCl_3$, and 4 — $SiCl_4$, (after Eversteijn[118]); and (c) GaAs deposition rate as a function of crystallographic orientation of the substrate for substrate temperatures of 750 °C (----) and 755 °C (————) (after Shaw[121]).

different in the lower and upper temperature ranges indicating that the rate-controlling step changes with temperature. The effect of temperature on the deposition rate has also been studied for compound semiconductors.[119,120] It should be emphasized that the temperature dependence shown here is not typical but differs from system to system and depends on other deposition parameters. (ii) Substrate orientation — the deposition rate may be strongly influenced by the crystallographic orientation of the substrate,[121] as shown in Figure 5.19c for GaAs film growth. The observed differences in growth rates can be due to the variations in the densities and geometric arrangements of surface sites, the number and nature of the surface bonds, the composition of various crystallographic surfaces, and the number and nature of surface features, such as steps, kinks, ledges, and vacancies. These factors can affect the adsorption, desorption, surface mobility, and reactivity. Further, substrate surface conditions such as adsorbed contaminants and substrate topography also affect the growth rate and structural quality of CVD films. (iii) Partial pressure of reactants — the molar ratios of different reactants play an important role in determining the deposition rate as well as the properties of the films. In addition, the effects of substrate temperature and molar ratios of reactants are closely interrelated.[114] (iv) Nature and flow rate of diluent gas — too high or too low a flow rate lowers the film deposition rate and causes nonuniformity. The nature of the gas can also significantly alter the deposition rates.[114] (v) Reactor geometry and wall temperature — the gas flow dynamics and hence the deposition rate and thickness uniformity of the films are affected by the exact shape and dimensions of the reaction chamber and the gas inlet/outlet.

Kern and Rossler[122] carried out a parametric investigation of the deposition rate and film properties of low-temperature CVD of phosphosilicate glass (PSG) and presented their results graphically, illustrating the effect of various CVD parameters on film growth.

5.3.4.2c. *Doping Techniques/Multicomponent Materials.* In the CVD process, since impurities can be implanted into the semiconductor film while it is being grown, desired and well-controlled impurity profiles can be achieved that are not limited by the laws of diffusion theory. Deposition of doped and multicomponent alloy[108,114] semiconductor layers can be accomplished by doping/alloying the source material with the desired impurity and chemically transporting both the host material and the impurity to the substrate in a closed system. However, this technique does not allow a given impurity concentration profile to be achieved or switching to a different impurity during the deposition process. In this respect, an open system allows greater flexibility by permitting the addition of impurities to the matrix reagent independently. The various techniques employed for this purpose are closely related to the methods utilized in connection with the chemical vapor deposition of compounds, mixtures, alloys, or solid solutions. These are discussed briefly below.

To deposit compounds, alloys, or doped materials, generation of closely controlled mixtures of source vapors is accomplished by simultaneous evaporation from the appropriate number of individually controlled vaporizers. Control of the mixing ratio is, however, difficult. A technique to achieve better control is to fill a long tube with the mixture of source materials (usually with quite different vapor pressures) and then move a hot zone gradually from the end nearest to the reaction zone across the tube, evaporating all the material.[108] Flash evaporation has also been used to obtain mixtures.

In yet another approach, evaporation of compound vapors, carrying elements in the proper ratio, occurs automatically if a small pool containing the source compound/mixture is heated to the evaporation temperature and the compound/mixture replenished by just the right amount to keep the pool size constant. This technique is called the constant-size pool method.[108]

Passing a halogen with (in case of Br_2 and I_2) or without (in case of Cl_2) carrier gas through a heated bed of a metal or a metal compound is found to be convenient in some cases. However, in using a chemical transport method, it is important to note that, though the exact conditions of temperature, gas flow, and surface area are not critical when the method is applied to the vaporization of univalent metals, the deposition parameters assume importance and need to be closely controlled in the case of multivalent metals in order to obtain an unambiguous oxidation state.

5.3.4.3. *Morphology of Films Prepared by CVD*

The microstructure of CVD-grown films range from very porous to impermeable, from amorphous to epitaxial single crystal, and from powdery to strongly adherent, depending on the nature of the substrate and the surface conditions, the deposition conditions prevailing during growth, and the nature and amount of impurities.

In general, in analogy with PVD, high substrate temperatures and low vapor concentrations lead to coarsely crystalline coatings whereas low substrate temperatures and high vapor concentrations yield amorphous or microcrystalline layers. It should be emphasized that this general principle is influenced to a great extent by the nature of the reacting vapors, the nature of the substrate, the vapor velocities, and the impurities in the system.

Nucleation processes occurring in the early growth stages largely determine the morphology of CVD films. Filby et al.[123] have pointed out that a relatively high nucleation rate yields a smooth layer whereas a high lateral growth rate leads to the formation of facets or steps. The smoothness of epitaxial films may also be influenced by the substrate orientation.[108]

The substrate temperature plays an important role in the microstructure of CVD films. At very low temperatures, microcrystalline, amorphous or porous layers are obtained, owing to slow or incomplete reactions. With increase in deposition temperature, the grain size and layer density increase. At very high temperatures, rapid grain growth occurs. However, the integrity of the coatings is generally inferior to that of low-temperature-deposited coatings.

The partial pressure of the reactants in the system influences the deposition rate and hence the grain size. With increase in deposition rate the deposited layer can vary from columnar (at low rates) to porous (at higher rates).[124]

5.3.4.4. *Films Prepared by CVD*

5.3.4.4a. *Semiconductors.* A large body of literature exists on various aspects of CVD of semiconductors. A review of this literature is outside the scope of this book, but we present in a tabular form (Table 5.9) the salient features of important CVD processes with particular reference to those semiconductors which are used in photovoltaic devices. A very brief discussion of some of the more important photovoltaic materials follows.

Si has been prepared primarily from silane and various chlorosilanes. Silane has the advantage that low deposition temperatures are required. However, chlorosilanes are relatively inexpensive, and allow higher-temperature deposition and thus higher growth rates. Si films have been obtained in epitaxial single crystal, polycrystalline, and amorphous forms. Typical substrates used for growing silicon layers range from a single crystal Si wafer to metallurgical grade polycrystalline silicon, graphite, stainless steel, sapphire, and quartz.

Group III–V semiconductors[114] such as GaAs, AlAs, GaSb, InP, InAs, GaP, and GaN have been obtained from CVD processes by the hydride method, the chloride method, and the organometallic method. Ternary alloys such as $Ga_xIn_{1-x}P$ or $GaAs_{1-x}P_x$ have been prepared with precisely tailored bandgaps determined by the compositional variable x. The CVD process has been extended to prepare quaternary alloys[114] such as $Ga_xIn_{1-x}As_yP_{1-y}$ with desired bandgap and lattice constants.

Films of II–IV compounds have been obtained by employing[114] the reaction of metal vapors with group VI hydrides, an open tube transport of II–VI compounds by HCl or HBr, and the reaction of organometallic diethyl or dimethyl compounds of metals with group VI hydrides.

Table 5.9. Input Materials for CVD of Various Solar Cell Materials (from Vossen and Kern[4])

Semiconductor	Input materials	Deposition temperature (°C)	Crystallinity[a]	Substrate
Si	$SiCl_4$, H_2	1200	E	Single crystal Si wafer
Si	$SiCl_3H$, H_2	1150	E	Single crystal Si wafer
Si	$SiCl_2H_2$, H_2	1100	E	Single crystal Si wafer
Si	SiH_4, H_2	1050	E	Single crystal Si wafer
Si	SiH_4, He	900	E	Single crystal Si wafer
Si	SiH_4, H_2	960	E	Sapphire
Si	SiH_4, H_2	960	E	Spinel
Si	SiH_4, N_2 or H_2	700–900	P	SiO_2
Si	$SiCl_3H$ in H_2	900–1000	P	Graphite metal
Si	SiH_4	600–700 (low pressure)	P	IC structures
Ge	$GeCl_4$, H_2	600–900	E	Sapphire, Spinel
Ge	$(CH_3)_4Ge$, H_2	700–1000	E	Sapphire, spinel
Ge	GeH_4, H_2	600	E	Sapphire, spinel
SiC	$SiCl_4$, toluene, H_2	1100	P	Single crystal Si wafer
SiC	$SiCl_4$, hexane, H_2	1850	E	Single crystal Si wafer
AlAs	Al, HCl, AsH_3, H_2	1000	E	GaAs
AlAs	$(CH_3)_3Al$, AsH_3, H_2	700	E	GaAs
CdSe	Cd, H_2Se, H_2	700	E	CdS, Sapphire
CdSe	$(CH_3)_2Cd$, H_2Se, H_2	600	E	Sapphire
GaAs	Ga, $AsCl_3$, H_2	760	E	GaAs
GaAs	Ga, HCl, AsH_3, H_2	725	E	GaAs, Ge
GaAs	$(CH_3)_3Ga$, AsH_3, H_2	700	E	Spinel
InP	In, PCl_3, H_2	650	E	InP
InAs	$(C_2H_5)_3In$, AsH_3, H_2	675	E	Sapphire
InAs	HCl, AsH_3, H_2	700	E	InAs
AlGaAs	$(CH_3)_3Al$, $(CH_3)_3Ga$, AsH_3, H_2	700	E	GaAs, sapphire
GaAsP	Ga, HCl, AsH_3, PH_3, H_2	750	E	Ge, GaAs
GaAsP	Ga, $AsCl_3$, PCl_3, H_2	800	E	GaAs
GaAsSb	Ga, HCl, AsH_3, SbH_2, H_2	700	E	GaAs

Table 5.9 (continued)

Semiconductor	Input materials	Deposition temperature (°C)	Crystallinity[a]	Substrate
GaInP	Ga, In, HCl, PH_3, H_2	700	E	GaAs
GaInAs	Ga, In, HCl, AsH_3, H_2	725	E	GaAs
GaInAs	$(CH_3)_3Ga$, $(C_2H_5)_3In$, AsH_3, H_2	600	E	GaAs
InAsP	In, HCl, AsH_3, PH_3, H_2	700	E	InAs
GaInAsP	Ga, In, HCl, PH_3, AsH_3, H_2	650	E	InP
ZnS	Zn, H_2S, H_2	825	E	GaAs, GaP
ZnS	$(C_2H_5)_2Zn$, H_2S, H_2	750	E	Sapphire
ZnS	ZnS, HCl, H_2	800	E	GaAs
ZnSe	Zn, H_2Se, H_2	890	E	GaAs, sapphire
ZnSe	ZnSe, HBr, H_2	550	E	GaAs, Ga
ZnSe	$(C_2H_5)_2Zn$, H_2Se, H_2	750	E	Sapphire, BeO
CdS	Cd, H_2S, H_2	690	E	GaAs, sapphire
CdS	$(CH_3)_2Cd$, H_2S, H_2	475	E	Sapphire
CdTe	Cd, Te, He	700–960	P	Carbon
CdTe	$(CH_3)_2Cd$, $(CH_3)_2Te$, H_2	500	E	Sapphire, Spinel, BeO
SnO_2	$(C_4H_9)_2Sn$, $(OOCH_3)_2$, O_2, H_2O, N_2	420	A	Si, quartz, glass
SnO_2 : Sb	$(C_4H_9)_2Sn$, $(OOCH_3)_2$, $SbCl_5$, O_2	450	A	Si, quartz, glass
In_2O_3 : Sb	In-chelate, $(C_4H_9)_2Sn$, $(OOCH_3)_2$, H_2O, O_2, N_2	500	A	Si, quartz, glass
V_2O_3, VO_2, V_2O_5	$(C_5H_7O_2)_4V$, N_2, O_2	400	P	Glass, sapphire, quartz
ZnO	$(C_3H_5O_2)_2Zn$	500–600	A	Metals
		650	SC	Metals
In_2O_3	$(C_5H_7O_2)_3In$	450–550	A	Metals
		630	SC	Metals

[a] A — amorphous, E — epitaxial, P — polycrystalline, SC — single crystal.

Transparent conducting oxide films, namely, SnO_2, In_2O_3, VO_2, V_2O_3, V_2O_5, SnO_2:Sb, and In_2O_3:Sn have been prepared either by hydrolysis of the metal chlorides or by pyrolysis of metal organic compounds.

5.3.4.4b. *Metals.* CVD of metals and metal alloys employs the following major processes: (i) thermal decomposition or pyrolysis of organometallic compounds, generally at low temperatures, (ii) hydrogen reduction of metal halides, oxyhalides, carbonyl halides, and other oxygen-containing compounds, (iii) reduction of metal hydrides with vapors of thermodynamically appropriate metals, and (iv) chemical transport reactions.

5.3.4.4c. *Insulators.* In many cases CVD produces higher-quality dielectric films than other deposition methods for such applications as optical coatings, surface passivation, insulation between multilayer arrangements, diffusion and photoetching masks, and corrosion-preventing coatings. Insulating films deposited by the CVD process include[108] silicon dioxide (SiO_2), silicon nitride (Si_3N_4), boron nitride (BN), alumina (Al_2O_3), aluminium nitride (AlN), tantalum pentoxide (Ta_2O_5), niobium pentoxide (Nb_2O_5), and titanium dioxide (TiO_2).

5.3.4.5. *Close-Spaced Vapor Transport (CSVT)*

The use of a close spacing in chemical transport systems for growing epitaxial layers of pure and compound semiconductors has been demonstrated by several workers.[125,126] In this CSVT technique, a temperature gradient is maintained between the closely spaced (~ 1 mm) source and substrate. A gas is used to react with the source to form a volatile compound that is subsequently decomposed at the surface of the substrate to form a thin film. As expected, the deposition rate is strongly dependent on the kinetics of transport of this reactive gas, the source and substrate temperatures, the substrate surface, and, of course, the thermodynamics of the reactions involved. The various processes occurring have been analyzed theoretically by Bailly et al.[127]

Using the CSVT technique, May[128] reported epitaxial deposition of Si by using a low-pressure I_2 atmosphere for transport of Si from a source held at a temperature lower than the substrate. Epitaxial films of CdS and GaAs grown at rates of approximately 1 μm min^{-1} have been obtained by Curtis and Brunner[129] by exploiting the displacement of the equilibrium of the reaction

$$\text{CdS (solid)} + \text{H}_2 \text{ (gas)} \rightleftharpoons \text{H}_2\text{S (gas)} + \text{Cd (gas)}$$

Epitaxial films of CdTe[130] and GaAs[131] have been obtained by this technique using water vapor as the transporting agent.

5.3.4.6. *Plasma Deposition*

This technique, also called the glow discharge deposition technique, is essentially a plasma-assisted CVD technique. Glow discharge plasma (dc or rf) is used to break the vapors up into different species that react to deposit as a film. For example, a reaction between the vapor species of SiH_4 and NH_3 yields Si_3N_4 films. Similarly, a reaction between SiH_4 and H_2O vapor species results in Si films. Decomposition of tetraethylorthosilicate vapors results in the formation of SiO_x films. The technique is well suited to the formation of complex polymer films.

The most exciting application of this technique is for the preparation of a-Si:H films for solar cells from pure, or diluted in Ar, SiH_4 glow discharge. In a simple setup, an rf electrodeless glow discharge generated by an outside coil at a frequency from 0.5 to 13.5 MHz is utilized. At SiH_4 pressures from 0.1 to 0.2 torr and flow rates from 0.2 to 5.0 standard $cm^3\,min^{-1}$ (sccm), deposition rates of about 100–1000 Å min^{-1} are obtained. The uniformity of the a-Si:H films is considerably improved by using a parallel-plate electrode geometry for a dc, or 13.56-MHz rf capacitive glow discharge in SiH_4. By varying the cathodic current density from 0.2 to 2.0 $mA\,cm^{-2}$ in about 1.0 torr of SiH_4, deposition rates of from 1 to 10 Å min^{-1} are obtained on the cathode in a dc discharge. In the case of rf discharge, deposition rates of about 500 Å min^{-1} are obtained for SiH_4 pressures of 5 to 250 mtorr and flow rates of 10 to 30 sccm. The composition of the films, in particular the hydrogen content, depends sensitively on glow discharge conditions, SiH_4 pressure, and substrate temperature. At high SiH_4 pressures, gas phase nucleation produces particulate films. The amorphous Si:H films are a Si/H_2 alloy with H_2 concentration varying up to 50 at.%. Films deposited at temperatures below about 200 °C contain dihydride and possibly trihydride groups. Above 200 °C, the H_2 appears to exist only in the monohydride form (SiH groups). There is no evidence of interstitial atomic or molecular hydrogen. It is possible to dope these films with boron or phosphorus by introducing B_2H_6 or PH_3 vapors in the glow discharge.

5.3.5. *Exchange Reactions*

Films of cuprous chalcogens can be prepared by a topotactial ion exchange reaction according to

$$CdY + 2CuX \rightarrow Cu_2Y + CdX_2$$

where X denotes a halogen (Cl, Br, or I) atom and Y a chalcogen (S, Se, or Te) atom. The process, also termed dipping or chemiplating, involves the displacement of one Cd ion by two Cu ions. The technique has been utilized to grow Cu_2S[132] and Cu_2Te[133] films. The reaction product CdX_2 is washed away in water or methanol.

The reaction is usually performed in aqueous solution at temperatures ranging from 90 to 100 °C. Salts such as NaCl or NH_4Cl are added to improve the CuX solubility. To prevent oxidation of Cu^+ ions, hydrazine or hydroxylamine is used as the reducing agent. The reaction is carried out in an inert atmosphere using an Ar or N_2 blanket over the bath. Some authors[134] include Cu filings in the bath to reduce the cupric ions. Use of an organic solution instead of an aqueous one has also been reported.[135]

A typical dipping solution used in our laboratory to fabricate Cu_2S/CdS solar cells is as follows:

CuCl (freshly prepared) : 4 g/l   NaCl : 6 g/l

pH : 3.0 (adjusted with HCl)   Bath temperature: 97 °C

The kinetics of the reaction, particularly in the case of Cu_2S, have been studied by several workers. The rate of the reaction is strongly influenced by the number of Cu ions present in the solution which, in turn, depends on the salt concentration in the solution, the concentration of NaCl, the pH, and the microstructure of the base. The thickness of the converted layer is a function of time and increases with time of dipping, all other parameters being constant. The conversion is rapid and the dipping time is generally of the order of 5 to 10 s to grow a 2000-Å-thick Cu_2S layer. The rate of deposition depends on the nature of the atom X and decreases with the sequence $Cl \rightarrow Br \rightarrow I$.[136]

Since the reaction is topotactial, the microstructure of the Cu_2S layer replicates the microstructure of the parent CdS layer. All the Cu chalcogens exist in several phases with different stoichiometries. The stoichiometry of the deposited layer depends on the rate of reaction and the oxidation state of the Cu ion in the solution. The electrical and optical properties are dominated by the stoichiometric effects.

In Figure 5.20, we have represented graphically the effects of the various deposition parameters on the growth kinetics of Cu_2S films prepared by dipping.[137]

Although the chemiplating technique has been widely used in the fabrication of Cu_2S/CdS solar cells, the process suffers from several limitations: (i) the necessity for a thick (~ 25 μm) CdS base layer to prevent shorting during the dipping process from the rapid diffusion of the highly mobile Cu ions through imperfections in the CdS layer, particularly the grain boundaries: (ii) an inhomogeneous pattern of growth of the Cu_xS

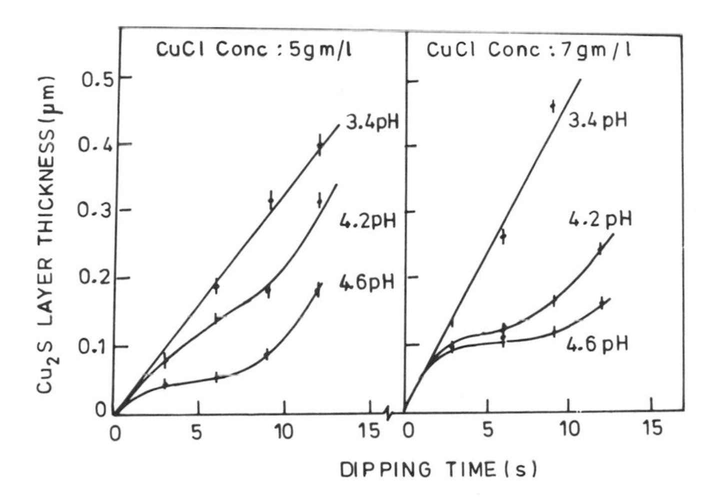

Figure 5.20. Effect of various deposition parameters on the growth kinetics of Cu_2S films prepared by dipping (after Salkalachen[137]).

layer owing to the sudden interruption of the reaction as soon as the desired thickness is reached; (iii) the presence of Cu and Cd gradients in the Cu_xS at the end of the reaction, and (iv) lack of control on the stoichiometry of the Cu_xS layer.

A solid state exchange reaction is often used to eliminate the above problems. The technique[138,139] involves the deposition of a thin film of CuCl by vacuum evaporation onto the CdS layer. The thickness of the CuCl layer is determined by the desired thickness of the Cu_2S layer. The CuCl/CdS film combination is heated to allow a solid state reaction between the two. Das et al.[139] have reported that stoichiometric Cu_2S films are obtained with both the wurtzite and sphalerite structures of CdS in the reaction temperature range of 200–250 °C. The growth is topotactial and the thickness of the Cu_2S layer is precisely controlled by the thickness of the CuCl layer. Since the solid state reaction is diffusion-controlled[140] and allowed to proceed to completion,[116] no Cd or Cu gradients are present in the layer. However, in order to obtain stoichiometric Cu_2S, any cupric ions present in CuCl must be reduced to the cuprous state prior to evaporation. Evaporation through a pad of Cu wool helps to reduce the cupric ions. Another technique is to first deposit a thin Cu layer on CdS and then evaporate the CuCl film.[140]

5.3.6. *Electrodeposition*

The occurrence of chemical changes owing to the passage of electric current through an electrolyte is termed electrolysis and the deposition of any substance on an electrode as a consequence of electrolysis is called electrodeposition.

The phenomenon of electrolysis is governed by the following two laws, first enunciated by Faraday in 1833: (i) the magnitude of chemical change occurring is proportional to the quantity of electricity passed and (ii) the masses of different species deposited at or dissolved from electrodes by the same quantity of electricity are in direct proportion to their chemical equivalent weights.

The two laws can be combined and expressed mathematically as

$$W = IEt/F$$

where W is the mass (in grams) of the substance deposited, I is the current (in amperes), E is the chemical equivalent weight (in grams), and t is the reaction time (in seconds). F is a constant, called the Faraday, equal to 96,500 C and is the amount of charge required to deposit one equivalent of any ion from a solution.

5.3.6.1. *Deposition Process*

When a metal electrode is dipped in a solution containing ions of that metal, a dynamic equilibrium is set up:

$$\mathrm{M} \rightleftharpoons \mathrm{M}^{+X} + Xe$$

where M denotes the metal atom. The resultant potential between the electrode and the electrolyte, in the absence of an external voltage, is called the electrode potential.

With the establishment of dynamic equilibrium, the electrode gains a certain charge on itself which attracts oppositely charged ions and water molecules, holding them at the electrode/electrolyte interface by electrostatic forces. Thus the so-called electrical double layer is formed; the inner layer is largely occupied by oriented water dipoles interposed by some preferentially adsorbed ions while the outer layer is composed of a file of ions of charge opposite to that of the electrode. During deposition, ions reach the electrode surface, move to stable positions on it, simultaneously releasing their ligands (water molecules or complexing agents) if solvated, release their charges, and, in the process, undergo the stipulated electrochemical reaction. The rapid depletion of the depositing ions from the double-layer region is compensated by a continuous supply of fresh ions from the bulk of the electrolyte. The transport of ions to the depleted region occurs because of the following: (i) diffusion owing to the concentration gradient, (ii) migration owing to the applied electric field, and (iii) convection currents in the electrolyte.

Various process variables critically determine the growth of the deposits: (i) Current density — defined as the total current divided by the electrode area — is one of the principal variables that determines the overall nature of the deposit, particularly the microstructure, efficiency of deposition, and the deposition rate. The optimum range of current densities can be experimentally determined for each individual deposition process to yield deposits suited to a particular need; (ii) Bath characteristics — the composition of the bath plays an important role in the deposition process. Usually the main constituent of the deposition bath is the electrolyte, which essentially serves to provide ions either in their simple form or in the form of a complex. Depending on whether the process is cathodic or anodic, the nature of the anions or cations can strongly influence the structure of the film deposited, particularly so if the ions are preferentially adsorbed at the electrode/electrolyte interface forming a part of the inner row of the double layer. For the same reason, certain organic molecules can also afford a similar control. In some cases suitable wetting agents may be used to promote the detachment of the hydrogen bubbles from the cathode surface and thus prevent pitting caused by hydrogen evolution in a cathodic deposition. If the electrolyte itself is not very conductive, acids, alkalis, or suitable salts that are highly ionizable are used to achieve the required conductivity and/or to control the pH value of the bath. The pH controls the overall conductivity of the electrolyte. However, a suitable optimum value has to be determined because too low a pH value may result only in hydrogen evolution, whereas too high a value may cause the inclusion of hydroxides in the deposits. The temperature of the bath controls the rate of diffusion of the ions, the convection currents in the bath, the nature and stability of any complex and the decomposition of additive, if any; (iii) Electrode shape — the distribution of current across the electrode and hence the uniformity of the deposited films is affected by the shape of the active electrode. Higher current densities at edges and projections, compared to crevices and hollows, lead to thicker deposits on the edges; (iv) Counter electrode — generally, the counter electrode is not actively functional but merely serves to complete the circuit. It may, in some cases, serve to replenish the electrolyte with the ion to be deposited; (v) Agitation — the convection currents in the bath are generated by externally provided agitation to reduce the possibility of a concentration overvoltage arising.

5.3.6.2. *Applications*

5.3.6.2a. *Electrodeposition of Metals.* Various metals like Cu, Ag, Au, Rh, and Pd may be electroplated for electrical contacts in solar cells. In view of its outstanding electrical conductivity, Ag is the best choice for

Table 5.10. Typical Bath Composition and Deposition Parameters for Depositing Metals and Alloy Films by Electroplating and Electroless Techniques

Metal or alloy	Bath composition: Constituents		Amount (g/l)	Temperature (°C)	Current density ($A\,dm^{-2}$): Cathode	Current density ($A\,dm^{-2}$): Anode	Plating voltage (V)	pH	Anode area/ cathode area	Ref.
Electroplating										
Cu	CuCN	Cyanide bath	15							
	NaCN or KCN		23	41–60	1.0–3.2	0.5–1.0	6	—	3:1	144
	Na_2CO_3		15							
	$CuSO_4 . 5H_2O$	Acid bath	188							
	Cu as metal		48	32–43	3–5	1.7	—	—	—	145
	Sulfuric acid		75							
	Cu^{+2}	Pyro-phosphate bath	22–38		10–80	20–40				
	$(P_2O_7)^{-4}$		150–250		($A\,ft^{-2}$)	($A\,ft^{-2}$)	1.4–4	8.1–8.4	1:1–2:1	146
	$(NO_3)^-$		5–10	50–60	0.5–2.5	0.5–1.5				
	Ammonia		1–3							
	Orthophosfate		< 10–130 ($oz\,gal^{-1}$)							
Ag	AgCN		30–55							
	Ag as metal		24–44							
	Total KCN		50–78	20–28	0.5–1.5					145
	Free KCN		35–50							
	K_2CO_3		15–90							
Au	$KAu(CN)_2$	Alkaline bath	6–23.5							
	gold as metal		4–6							
	KCN		15–90	25–70	0.2–1			6–8		145
	K_2CO_3		0–30							
	K_2HPO_4		0–45							
	KOH		0–30							
	$KAu(CN)_2$	Acid bath	3–23.5							
	Gold as metal		2–16							
	KHO_2PO_4		0–100	40–70	0.1–0.5			3–6		145
	Chelates		10–200							
	Secondary brightener		0–10							

60% Sn, 40% Pb alloy	Stannous (Tin)	60				
	Lead	25				
	HBF_4	100		3.2		144
	Peptone	5.0				
Pd	$Pd(NH_3)_4Br_2$	30	50	4	9.2	144
	NH_3Br	45				
Electroless						
Ni	$NiCl_2 . 6H_2O$ (Acid bath)	30				
	Sod. hydrophosfite	10				
	$(NaH_2PO_2 . H_2O)$		90			
	Sodium hydroxy	50			4–6	145
	acetate $(HOCH_2COONa)$					
	$NiCl_2 . 6H_2O$ (Alkaline bath)	30				
	$NaH_2PO_2 . H_2O$	10	90		8–10	145
	Sod. citrate					
	$Na_3C_6H_5O_7 . 2H_2O$	84				
	NH_4Cl	50				
Cu	$CuSO_4 . 5H_2O$	3.6				
	$KNaC_4H_4O_6 \cdot H_2O$	25	22			145
	(Sod. pot. tartrate)					
	NaOH	3.8				
	Formaldehyde	10				
Pd	$PdCl_2$	2				
	NH_4OH (27%)	160 ml/l				
	NH_4Cl	26	50			145
	$NaH_2PO_2 . H_2O$	10				
Au	KCN	13				
	$KAu(CN)_2$	5.8	75			145
	KOH	11.2				
	KBH_4	21.6				
Ag	AgCN	1.34				
	NaCN	1.49				145
	NaOH	0.75	55			
	Dimethylamine	2.0				
	borane					
	$(CH_3)_2NHBH_3$					
	Thiourea	0.0003				

contacts. However, owing to its tendency to get tarnished in atmosphere, suitable overlayers of Au or Rh are plated.

Electroplated In is often used as a *p*-type doping agent in Ge and GaAs transistors.[141–143]

5.3.6.2b. *Electrodeposition of Alloys.*[144] The ions of nobler metals deposit preferentially compared to the ions of baser metals. Thus for codepositing metals to form alloys, the concentration of the more noble metal at the cathode/electrolyte interface should be considerably reduced relative to the less noble metal. This is achieved by taking a bath concentration which is less rich in the noble metal than the concentration desired in the film. However, codeposition requires that the electrode potentials for the two ions should be close (260 mV) to each other.

In Table 5.10, we have listed the typical bath compositions and plating parameters used for plating various metals and alloys commonly used in solar cell fabrication.[144–146]

5.3.6.2c. *Electrodeposition of Semiconductors.* Cadmium selenide/cadmium telluride has been deposited on conducting substrates by codeposition of Cd and Se/Te. Hodes et al.[147] have used an acidic aqueous solution of $CdSO_4$ and SeO_2 to deposit CdSe on titanium substrates. Danaher and Lyons[148] have deposited CdTe galvanostatically in a two-compartment cell using $2M$ H_2SO_4 as an anolyte, a solution of $0.01M$ H_2TeO_3, $1.5M$ H_2SO_4, and $0.22M$ $CdSO_4$ as catholyte, Pt as anode, Ti as cathode, and a KCl-agar salt bridge to connect the two compartments.

Baranski and Fawcett[149] have used a new approach to deposit thin films of semiconducting metal chalcogenides like CdS, HgS, PbS, Tl_2S, Bi_2S_3, Cu_2S, NiS, CoS, and CdSe on various conducting substrates. They have used an electrolyte constituted of the corresponding metal salt and the elemental chalcogen, dissolved in a suitable nonaqueous solvent such as DMF (dimethyl formamide), DMSO (dimethyl sulfo-oxide), or ethylene glycol. The nascent metal formed at the cathode rapidly reacts with the dissolved chalcogen to form the required metal chalcogenide.

Nakayama et al.[150] have electroplated Cu_2S on CdS ceramic plates to form Cu_2S/CdS solar cells. This technique of electroconversion has been extended to CdS thin films by Saksena et al.,[151] who have used a simple electrolytic cell with Cu as the anode, CdS as the cathode, and $CuSO_4$ as the electrolyte. The authors have established the effect of pH, concentration, and temperature of the bath, and the cathode current density on the stoichiometry of the Cu_2S film. Figure 5.21 shows the growth kinetics and the dependence of the stoichiometry on temperature and current density for a typical bath. In the temperature range of 55 to 65 °C and current density range of 2.5 to 5 mA cm^{-2}, the most stoichiometric cuprous sulfide

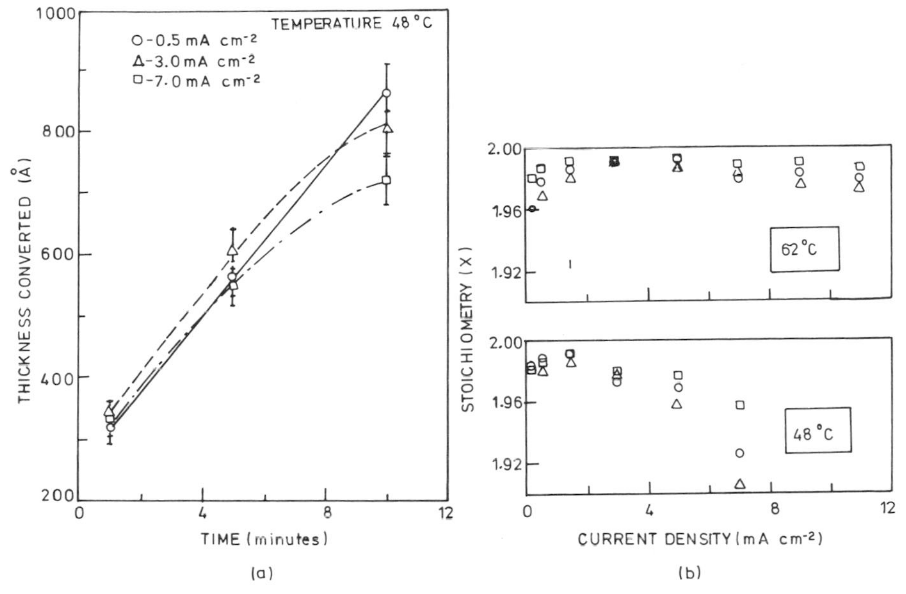

Figure 5.21. (a) Growth kinetics and (b) dependence of stoichiometry of electroplated Cu_xS films on temperature and current density for different plating times: □ — 1 min, ○ — 5 min, and △ — min (after Saksena et al.[151]).

films are obtained. A two-step competing reaction mechanism involving a chemical ion exchange and an electrochemical reduction has been proposed to explain the deposition process. An advantage of this technique over other techniques of preparing Cu_2S films is that unstable cuprous salts are avoided and the plating time is considerably longer, thus affording better control.

5.3.6.3. *Electroless or Autocatalytic Plating*

Similar to electrodeposition, the electroless technique also involves the reduction of metal ions to form the deposits, but no external power supply is required to provide electrons. Instead, a catalytic surface is used to initiate the deposition and then the metal itself must be catalytic to further the deposition. The electrons for the reduction process are provided by a chemical reducing agent in the solution. Since the metal being deposited has to be catalytic in nature, only a limited number of metals can be deposited by this technique. Table 5.10 gives the typical deposition parameters for various metals. The technique has the following advantages: (i) highly uniform deposits are obtained, with no excessive buildup on projections or edges, (ii) the deposits are less porous than electroplated films, and (iii) films can be grown on nonconducting substrates by pretreating the substrate surface to make it catalytic.

5.3.7. *Anodization*

Anodization is a field-assisted form of thermal growth.[152] The metal to be anodized is made an anode and immersed in an oxygen-containing electrolyte, which may be aqueous, nonaqueous, or fused salt. Growth may take place at constant voltage[153] or at constant current. In the former case, if the metal is left in the anodization bath for a sufficient time, a reproducible thickness will be obtained, depending on the applied voltage. We thus define a quantity called the anodization constant as the thickness of the layer grown per unit voltage. The anodization constant for different metals is listed in Table 5.11. A disadvantage of the constant voltage method is that in the initial stages of growth very high current densities are required. This difficulty is overcome in the constant current mode where the thickness is proportional to the time for which anodization is carried out. However, the voltage across the film has to be increased with increase in the oxide thickness. The limiting voltage and hence the limiting thickness is reached when breakdown of the film occurs. The limiting thickness for different metals is given in Table 5.11.

The pH of the electrolyte plays an important role in obtaining a coherent film. If the bath is too acidic or too alkaline, the film will dissolve during growth and a porous structure will be obtained.[154] With an optimum pH value, coherent films can be obtained by anodization on a variety of metals, including Al, Ta, Nb, Si, Ti, and Zr. The final thickness depends on the metal, the voltage applied on the metal, the temperature of the bath, and the time that the metal is immersed in the electrolyte. The limiting thickness in the case of constant current growth depends on, among other factors, the purity of the substrate and the purity of the electrolyte, since impurities cause breakdown at lower voltages.

It is now generally accepted that both metal and oxygen ions are mobile to equal extents in the anodization of Ta, Al, Nb, and Si. In the case of Zr, the growth is due to oxygen ion transport.[155]

The growth of the oxide occurs as continuous layers of amorphous materials and the oxide replicates the surface features underneath.

In semiconductors, the anodic behavior is of considerable importance in the production of devices, particularly in electroshaping, location of p/n junctions, preparation of stable surfaces for semiconductor devices, and making oxide films for MOS devices. Ge[156] and Si[157,158] have been anodized to form their respective oxides. Cd[159] and Bi have been anodized in a sulfide bath to prepare polycrystalline n-CdS and n-Bi_2S_3 films, respectively. The anodization can be accomplished galvanostatically as well as potentiostatically.

In Table 5.11, we list the bath composition and other relevant parameters for the anodization of various metals.

Table 5.11. Bath Composition and Deposition Parameters Used in the Anodization of Various Metals and Semiconductors

Material to be anodized	Product after anodization and type of film	Electrolyte used	Current density ($mA\ cm^{-2}$)	Maximum voltage (V)	Anodization constant ($Å\ V^{-1}$)	Breakdown field $V\ cm^{-1}$	Maximum thickness μm	Ref.
Al	Al_2O_3 (Porous)	15% H_2SO_4 (21 °C)	8–10	> 350	13.6	$1–5 \times 10^6$	1.5	160
	Nonporous	Boric acid — 52.5% by wt. Ethylene glycol — 43.4%. Ammoniacal glycol — 3.6% Ammonium Dihydrogen-orthophosphate — 0.8%	8–10	> 350	13.6	$1–5 \times 10^6$	1.5	162
Ta	Ta_2O_5	0.05% H_3PO_4 + 50% Ethylene glycol + water	1.0	300	16.0	5×10^6	1.1	160
Cd	CdO	25 g/l KOH	0.050	~ 10	—	—	—	161
	CdS	1F Na_2S (aq)	—	—	—	$1–2 \times 10^6$	0.25	159
Bi	Bi_2S_3	1F Na_2S + 0.05 FS	—	—	—	$1–2 \times 10^6$	—	159
Ge	GeO_2	Anhydrous sodium acetate in glacial acetic acid	0.010–0.03	5.00	—	$1–4 \times 10^6$	0.12	160
Si	SiO_2	0.04N KNO_3 + N-Methyl acetamide	7.0	560	3.5	$2–6 \times 10^7$	0.22	157 158, 160
Zr	ZrO_2	0.2N H_2SO_4 (O °C)	10.0	—	12–30	$2–4 \times 10^6$		160

5.3.8. *Electrophoresis*

In this technique, electrically charged particles suspended in a liquid medium are deposited on an electrode. The as-deposited films are usually loosely adherent coatings of powder. Further postdeposition treatment is required to produce an adherent, compact, and mechanically strong surface coating. The postdeposition treatment usually consists of pressurized compaction and a heat treatment to dry out traces of the suspension medium and sinter the particles within the film.

The electrophoretic technique is very versatile and well suited to the deposition of both conductors and nonconductors, including metals, alloys, salts, oxides, refractory compounds, polymers, and combinations of the various components. The particular advantages of the technique are: (i) practically any powdered material can be deposited on any conducting substrate, (ii) uniform thickness is obtained even on complicated shapes, (iii) very thick coatings can be deposited and the thickness can be very precisely controlled, (iv) codeposits of two or more materials can be prepared in the desired proportion, (v) the deposition period is very short, generally of the order of seconds to a few minutes, and (vi) no deposit material is wasted.

5.3.8.1. *Method*

The theoretical aspects of the deposition process have been discussed by several workers.[163–166] The technique basically involves the preparation of a colloidal suspension of the desired material. Colloidal particles are formed either by breaking larger particles up into smaller ones of colloidal size (10 to 5000 Å), or by effecting the merger of smaller particles, usually ions or molecules, to form larger aggregates of colloidal size. In the first process, ball-milling a liquid suspension of a powder (which has already been separated by sieves into a starting range of particle size) gives the desired particle size. Organic solvents such as acetone or methyl-, ethyl-, and propylalcohol are used as the suspension medium. After ball-milling, a suitable range of particles is selected by allowing the particles in the suspension to settle for a given time, calculated from Stoke's law. The powder of the desired size is separated from the liquid by centrifuging and, finally, the powder is resuspended in another liquid for electrophoretic deposition. The particles acquire their electric charge spontaneously when mixed with the suspending liquid.

In the second process, a colloidal precipitate is formed *in situ* in a suitable medium, using a chemical reaction. By this technique colloids of a large variety of meterials, including metals such as Cu, Ag, or Au, nonmetals such as S and Se, sulfides, hydroxides, and hydrous oxides, can be conveniently prepared.

In either case, surfactants may be added to the colloid to change the charge of the particles. Sometimes it may be necessary to add a small amount of a polymer to the suspension for stabilizing the colloid. Special solvents must be used when binders are codeposited with the powders. It may be noted that particles as large as 20 μm can be easily deposited electrophoretically.

In general, the concentrations of the suspensions used in electrophoretic deposition vary widely. We can choose a suitable concentration to allow the coating of an object of known area with a prescribed thickness by plating to depletion. Codeposits are formed from mixed suspensions, the particles depositing on the electrode usually in the same proportion as that existing in the suspension.

When depositing from aqueous solutions, voltages of the order of 15 V are used, whereas several hundred volts is required when depositing from organic solvents. Current densities are in the range of 1 mA cm^{-2}. The adhesion of the deposit can frequently be improved by adding small quantities of ionic additives such as NH_4OH, citric acid, and benzoic acid.[167] Periodic reversal of current has been used[168] to produce more uniform, nonporous deposits.

5.3.8.2. *Applications*

The technique has been used[169] to deposit alumina on tungsten wire for insulation purposes. Williams et al.[170] have prepared CdS films for Cu_2S/CdS solar cells by preparing the colloid *in situ* in a suitable medium. The suspension of CdS was prepared by passing H_2S through a solution of $Cd(CH_3COO)_2$ and water. $Zn(CH_3COO)_2$ was added in the solution to prepare $Zn_xCd_{1-x}S$ films. Deposition was carried out on a stainless steel anode after diluting the solution with alcohol using a current density of 2.56 mA cm^{-2} and a field of about 33 V cm^{-1}.

Several polymers have been coated on a variety of objects electrophoretically. These include polytetrafluoroethylene (PTFE), both alone[171,172] and in combination with ceramic,[173] polyethylene,[174] and polyvinylidene chloride, polyacrylates, and polymethylacrylates.[175,176]

5.4. *Liquid Deposition Techniques*

Although not "thin film" techniques, the following deposition techniques show great promise in preparing large-area thin layer materials in several-micrometer dimensions for solar cell applications.

5.4.1. *Liquid Phase Epitaxy (LPE)*

Basically, LPE involves the precipitation of a material from a cooling solution onto an underlying single crystal substrate. The solution and the substrate are kept apart and contact is achieved either by "tipping" the furnace with solution, or by dipping the substrate into the solution in a vertical furnace. The solution is saturated with the growth material at the desired growth temperature and then allowed to cool in contact with the substrate surface at a rate and for a time interval appropriate for the generation of the desired layer. Under optimum conditions, the layer grows as an extension of the single crystal substrate. It is obviously not easy to grow polycrystalline layers on grossly dissimilar substrates.

Since a solvent is needed for the material to be deposited in LPE, the usefulness of the process is limited to applications in which the solvent does not produce adverse effects. There is no problem in the case of III–V compounds since the normal constituent of the compound semiconductors can be employed as a solvent. In the case of Si and Ge, the useful solvent must also be a useful dopant (e.g., Sn in Ge). Note that the LPE technique is not easily adaptable for semiconductors of three or more components, largely because of different distribution coefficients of the constituent elements. Further, the presence of the solvent in the solution precludes easy adjustment of the stoichiometry of the grown layers.

The LPE technique has been used very successfully to prepare device-quality III–V compounds.[177] Its applicability for the production of large oriented grain structure layers of useful solar energy materials is still to be explored.

An interesting LPE technique using epitaxial nucleation in submicroscopic holes (called ENSH method) forming a linear parallel distribution of holes has been reported by Hadini and Thomas.[178] These authors have obtained oriented growth of triglycerine sulfate (TGS) from liquid (as well as vapor) phase on Mylar or photoresist-coated single crystal substrate. Holes of about 15 μm diameter and 100 μm separation were drilled by a laser in 4-μm-thick Mylar. Photoresist holes were created lithographically. This is certainly an interesting idea which needs further investigation with solar cell chalcogenide films.

5.4.2. *Melt Spinning Technique*

Melt spinning is a very promising technique to produce rapidly quenched ribbons at high speeds of several meters per second. By bringing the molten material in a nozzle into contact with a spinning wheel, the liquid drop is pulled out in the form of a ribbon. The rate of solidification of the melt and the ribbon thickness are dependent on the nozzle dimensions, detailed shape of the molten drop in contact with the spinning wheel surface, the linear velocity of the wheel, and the heat transfer processes. At high spinning speeds, this technique has been used extensively to obtain metallic glass ribbons of 10–100 μm thicknesses formed at quenching rates of about 10^6 K s^{-1}. The use of the technique to obtain rapidly solidified, large-grain-size ribbons of Si has been investigated at IBM and in our laboratory. *p/n* junction solar cells, on such ribbons, have also been demonstrated. In principle, by employing multiple as well as sequential nozzles, wide Si ribbons coated on conducting metallic substrates should be obtainable for solar cell applications. This simple technique, however, is dependent on a host of complex parameters that are currently being investigated.[179–180]

5.5. *Miscellaneous Techniques*

In this section we shall note some of the recent innovations and developments[181] in deposition techniques, motivated largely by the desire to find inexpensive and simple methods to coat large-area substrates with photovoltaic materials. Since the techniques are, as yet, specific to the materials, we shall discuss the processes with reference to the materials.

5.5.1. *Silicon (Si)*

1. Electrohydrodynamic (EHD) process: The high-temperature EHD method for deposition of Si involves the application of high electrostatic fields ($\simeq 10^5$ V cm^{-1}) to capillary nozzles to generate beams of charged Si droplets. Molten Si droplets, of diameters up to 150 μm, accelerated to a high velocity, impinge on heated substrates placed about 5 to 8 cm beneath the nozzle orifice.

Capillary flow replenishment of molten Si is sufficient to maintain a stable electrostatic spray of charged Si droplets. The flow rate is controlled by the magnitude of the voltage applied during the dispersion process.

Si droplet beams have been used to deposit Si films on graphite, mullite, single crystal Si, and vitreous C substrates. Film thickness ranges up to 500 μm. Deposits produced on cold or insufficiently heated substrates exhibit small grain sizes of the order of 2 μm. Columnar grains with length equivalent to the thickness of the quenched deposits ($\gtrsim 30$ μm) are obtained on substrates held at high temperatures.

2. Electrodeposition process: Polycrystalline Si can be electrolytically deposited on a variety of electrically conductive substrate materials by the electrodeposition method. Binary or ternary molten salt electrolytes, consisting of alkali metal fluoride solvents (KF, LiF) into which are dissolved potassium fluorosilicate (K_2SiF_6) or sand, are utilized. On application of a

voltage across the electrolyte, elemental polycrystalline Si is deposited on the cathode or negative electrode. Deposition parameters can be varied to produce desired crystal size and film thickness. The electrolytes may be doped intentionally to produce p- or n-type Si with desired resistivity.

Polycrystalline Si has been deposited on Ag, Ta, Mo, Ni, and graphite substrates by electrodeposition with film thicknesses ranging up to 250 μm and grain sizes up to 100 μm.

3. Thermal expansion shear separation (TESS) process: This technique utilizes reusable Mo substrates with high-pressure plasma (hpp) Si deposition and subsequent ribbon-to-ribbon (RTR) laser recrystallization. Mo substrates, used for shear separation, need periodic substrate resurfacing (after every seven cycles) for continued reliability of shear separation. The shear separation mechanism involves the formation of Kirkendall voids at the Si/temporary substrate interface as a result of predominant Si diffusion through the $MoSi_2$ interfacial layer. Shear stresses develop during cooldown leading to separation at the weakened (by voids) interface.

Semicontinuous plasma deposition has been investigated and the process characterized using on-line gas chromatographic analysis of the effluent gases from the reactor. The deposition efficiency and throughput rate are highest for SiH_2Cl_2 and lowest for $SiCl_4$. However, gas phase nucleation and powder growth problems limit the utility of SiH_2Cl_2 as the input gas. Further, $SiHCl_3$ formed during the deposition process when using $SiCl_4$ can be recycled and thus the use of $SiCl_4$ as the input gas is more economical.

The microcrystalline Si films produced by the TESS process and hpp are grain-enhanced by RTR laser recrystallization at a rate of 2.5 cm min^{-1}.

5.5.2. *Cadmium Sulfide (CdS)*

Chemical Spray Deposition: In the conventional process of CdS spray deposition using $CdCl_2$ and thiourea (Section 5.3.1), the chemical reaction of the spray solution on the substrate occurs quite rapidly. The lack of surface mobility of the ions at the relatively low substrate temperatures leads to structural defects in the films. Moreover, the complex residual by-products of the reaction have to be removed from the surface. To eliminate these drawbacks, a new spray deposition technique has been developed in which a film of CdO is first formed by spraying an appropriate Cd salt onto a heated substrate and subsequently completely converting the CdO into CdS by an anion exchange at an elevated temperature in a gaseous sulfide atmosphere. In this case, the initial oxide-forming reaction proceeds to its thermodynamic end-point. Further, the by-product of the sulfurizing reaction to form CdS is simply water and the sulfide lattice is contracted slowly during the ion exchange, under conditions close to equilibrium. The entire process, is, then, essentially a combination of two processes: (a) an initial spray deposition process, and (b) an ion-exchange CVD process.

The CdO films have been prepared by spraying solutions of various organic and inorganic Cd salts in aqueous or nonaqueous solvents onto glass substrates. The growth rates of the CdO films depend on the type of substrate, nature of the solvent, flow rate of the solution, and the substrate temperature. The choice of Cd salt and solvent determines the crystallinity and degree of preferred orientation. All salts do not result in the formation of CdO. Deposition rates as high as 0.2 $\mu m\, min^{-1}$ have been achieved.

The CdS films are obtained by annealing the CdO films at elevated temperatures in a flowing mixture of $H_2S + H_2$. The ion-exchange rate can be as high as 1–2 $\mu m\, h^{-1}$ and the process has an activation energy of 0.93 eV. Owing to the density difference, the fully converted CdS film is approximately double the thickness of the starting CdO film. The grain size of the CdS films is typically between 0.5 and 1 μm. The grain size can be increased up to 3 μm by increasing the H_2S flow rate at the end of the annealing process.

6

Properties of Thin Films for Solar Cells

6.1. *Introduction*

A comprehensive review of the literature on semiconducting thin films is beyond the scope of this book. Instead, we will, in this chapter, restrict ourselves to a discussion of those semiconducting thin films that have been prepared or suggested specifically for fabricating thin film solar cells. Further, we will consider only the properties relevant to photovoltaic operation. As the properties of semiconducting thin films are not general in nature but, rather, unique to the material and the technique of deposition, we have reviewed the individual properties with specific reference to a particular material and with special emphasis on comparisons among films prepared by different deposition techniques. In Sections 6.4 and 6.5 we give brief descriptions of the behavior of metal films and insulating films, respectively, as these films are integral parts of thin film solar cells, especially Schottky-barrier and CIS-type devices.

6.2. *Semiconducting Films*

6.2.1. *Silicon (Si)*

Thin films of Si have been deposited by several techniques, including chemical vapor deposition (CVD),[1-16] evaporation,[17-20] sputtering,[21-24] silicon-on-ceramic (SOC) technology,[25-27] thermal expansion shear separation (TESS),[28,29] electrodeposition,[30] and electrohydrodynamic (EHD)[31] processes.

6.2.1.1. *Structural Properties*

Adamczewska et al.[3] have observed that Si films deposited on amorphous substrates by CVD are amorphous at substrate temperatures less than 500 °C. At higher substrate temperatures (~ 550 °C), the films are randomly oriented polycrystalline. Films deposited in the temperature range from 600 to 700 °C are polycrystalline with a preferred orientation that changes from ⟨110⟩ through ⟨100⟩ to ⟨111⟩. At still higher temperatures (~ 750 °C), the films again become randomly oriented polycrystalline. The silane/nitrogen ratio does not influence the structure at substrate temperatures below 500 °C. At 600 °C and above, the silane/nitrogen ratio strongly influences the structure, the nature and magnitude of the effect depending on the substrate temperature. Hirose et al.[6] prepared Si films on quartz by thermal decomposition of silane and observed a crystallization temperature above 675 °C. Similar results have been reported by Kamins et al.,[11] who obtained amorphous films below 600 °C and polycrystalline films with a preferred ⟨110⟩ orientation at higher temperatures. However, polycrystalline Si films grown on alumina and recrystallized exhibit dendritic growth[15] with the dominant orientation being ⟨111⟩.

The microstructure of CVD films has been examined by several other workers.[7,10,12,13–15] Emmanuel and Pollock[12] have studied the effects of substrate temperature, type of carrier gas, and gas flow rate on the grain size of Si films deposited on Si_3N_4 and SiO_2 substrates. At substrate temperatures of 720 °C and above, the films are polycrystalline and under certain conditions uniform grain sizes of 300 Å are obtained. At high substrate temperatures (> 770 °C), the films are discontinuous. Chu et al. have deposited films on steel using borosilicate as a diffusion barrier,[13] graphite,[13] and recrystallized metallurgical grade Si.[10,14] The Si film on the borosilicate/steel substrate, deposited at a substrate temperature of 900 °C, is polycrystalline with a grain size of 1 to 5 μm. Some of the crystallites show well-developed faces. Graphite is more compatible with Si than steel, and Si deposited on graphite substrates at 1000 °C show considerably better microstructure. Si films deposited at 1150 °C onto recrystallized metallurgical grade Si are epitaxial with respect to the substrate and similar to the substrate in appearance. Saitoh et al.[15] have observed dendritic growth of about 0.3 mm wide and a few centimeters long in recrystallized Si films deposited on alumina/Ti.

Si films evaporated onto Ti-coated steel substrates,[19] kept at 525 °C or above, show well-developed crystal structure with oriented growth along the ⟨220⟩ direction. However, films grown on Al exhibit a ⟨111⟩ orientation. Moreover, the growth is columnar. Feldman et al.[20] have also observed columnar growth of Si. Under similar conditions, the average grain size of Si films on sapphire and glass is the same, in the range of 0.2 to 5 μm. Van

Zolingen et al.[17,18] deposited Si layers on sintered Si and single crystalline Si substrates coated with insulating Si_3N_4 and SiO_2. The as-deposited layers on sintered Si are columnar with column diameters of 10 to 15 μm, the columns being composed of fibrils of diameter of about 0.4 μm. On Si_3N_4-coated Si substrates, the layers are again columnar (diameter 2 to 8 μm) with fibrils of diameter 0.2 to 0.5 μm. The layers on SiO_2 show a very fine fibrous structure with fibril diameter of about 0.1 μm. Annealing of the layers at 1250 °C for 1 h in an Ar atmosphere improves the crystallinity dramatically. A microstructure composed of 1-μm crystallites is observed, but the columnar structure is destroyed.

rf sputtering[21] produces films with a very strong ⟨111⟩ orientation on ⟨0001⟩ sapphire when the substrate temperature is above 900 °C. Between 800 and 900 °C, the films have a polycrystalline structure. At still lower temperatures, the films are amorphous. Si films on mullite substrates grown by the SOC technique[25] exhibit a predominantly ⟨110⟩ texture with a ⟨211⟩ growth direction aligned in the direction of withdrawal of the substrate. The grain size is of the order of several millimeters.[25,26] Columnar growth is also obtained by the TESS process,[29] but in the direction of the thickness. However, the grain size is of the order of 1 to 5 μm. After laser crystallization, grain enhancement occurs in the Si ribbon resulting in grains a few millimeters wide and several centimeters long. Electrodeposited Si films[30] on Ag, Ta, Mo, Ni and graphite (thickness of the order of 200 μm) exhibit large grain diameters of up to 100 μm. Electrohydrodynamically deposited[31] Si deposits on graphite, mullite, and vitreous carbon substrates possess small grain sizes (~ 2 μm) at low substrate temperatures. High-temperature-deposited Si droplets show columnar grains with diameters of 30 μm.

6.2.1.2. *Electrical Properties*

Hirose et al.[6] have carried out a systematic investigation of the electrical properties of CVD Si films on quartz as a function of the deposition conditions. The dc conductivity at temperatures above 280 K exhibits an activation energy (0.53–0.61 eV) that is observed to be dependent on the deposition temperature. At lower temperatures, the observed activation energy is attributed to hopping conduction through deep traps near the Fermi level. Defect centers have been identified by ESR and have a density greater than or equal to 10^{18} cm^{-3}. The temperature dependence of the dc conductivity is shown in Figure 6.1. The temperature dependence of photoconductivity (Figure 6.2) reveals the presence of localized state bands both above and below the Fermi level. Hirose et al.[6] proposed that the localized bands are due to tailing that arises mainly from the grain boundary regions. Yoshihara et al.[8] have

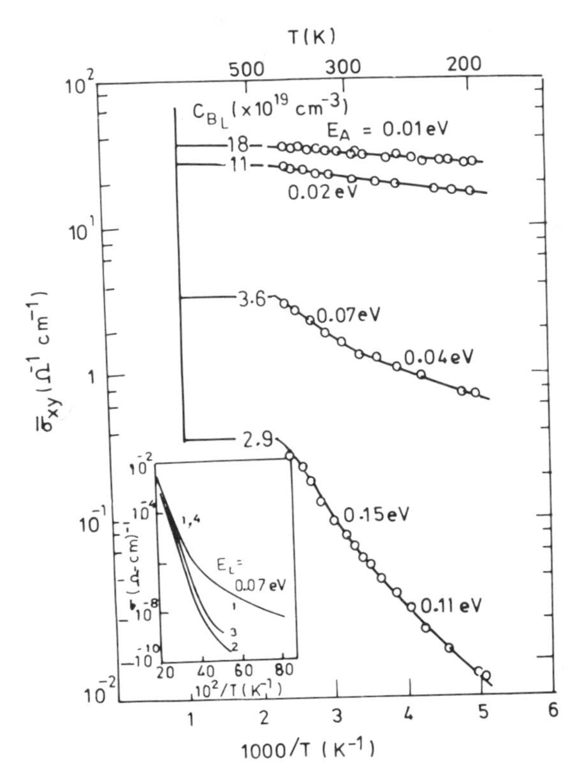

Figure 6.1. Temperature dependence of conductivity $\bar{\sigma}_{xy}$ of vacuum deposited and annealed Si films, for different doping concentrations C_{BL} (after Van Zolingen and Kipperman[18]). The inset shows the temperature dependence of the conductivity σ for polycrystalline, CVD Si films deposited at various substrate temperatures: 1 — 500 °C, 2 — 600 °C, 3 — 650 °C, and 4 — 700 °C (after Hirose et al.[6]).

found that the activation energy of the dc conductivity of B- and P-implanted polycrystalline Si films is inversely proportional to the doping concentration, as shown in Figure 6.3. They have interpreted their results on the basis of a grain boundary trapping model. Trapping state densities of 3.5×10^{12} cm^{-2} for B- and 5.2×10^{12} cm^{-2} for P-implanted films have been calculated. However, it should be noted that the trapping state density and, hence, the sheet resistivity depends on the deposition temperature and decreases with increasing temperature of deposition. This is illustrated in Figure 6.4.

The trapping state density decreases from 3.5×10^{12} to 2.0×10^{12} cm^{-2} for a deposition temperature range from 700 to 1050 °C, while the sheet resistance decreases from approximately 10^5 to 10^2 Ω/□. The trapping state density also decreases on annealing, but increases with an increase in O_2 incorporation. Dvurechensky et al.[2] have observed paramagnetic vacancy

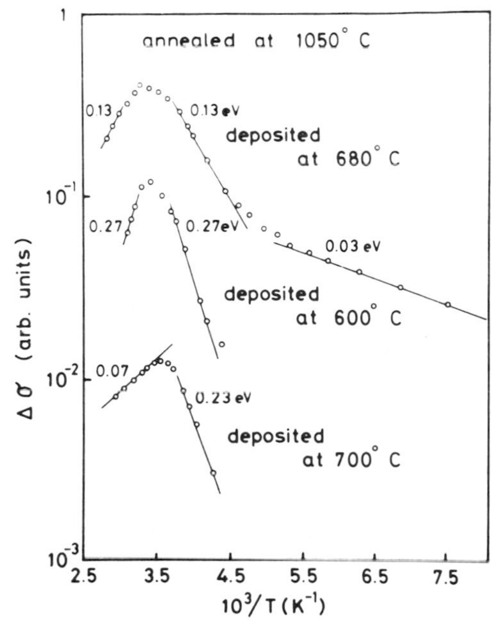

Figure 6.2. Temperature dependence of photoconductivity $\Delta\sigma$ (measured at $h\nu = 2.14$ eV) for polycrystalline, annealed, CVD Si films deposited at different substrate temperatures (after Hirose et al.[6]).

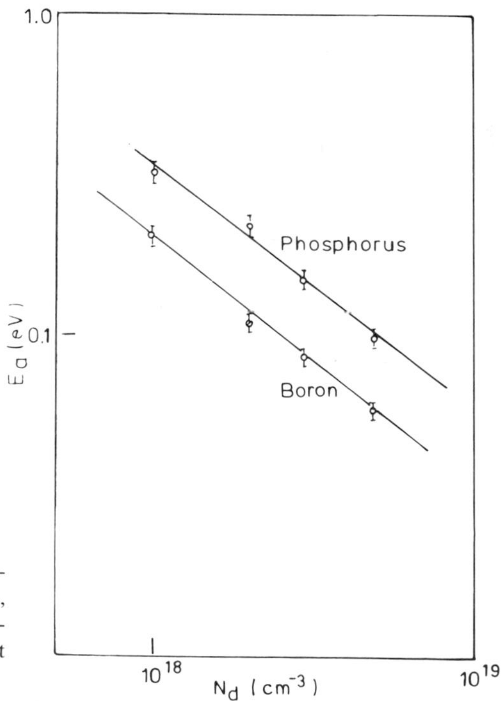

Figure 6.3. Dependence of the activation energy E_a of doped, polycrystalline, chemically deposited Si films on the doping concentration N_d (after Yoshihara et al.[8]).

types of defects in their films with densities in the range of 3×10^{14} to 4×10^{14} cm^{-2}.

As-deposited, evaporated layers[17] show a low conductivity ($\sim 10^{-3}$ Ω^{-1} cm^{-1}) and low mobility (0.5 cm^2 V^{-1} s^{-1}), almost independent of the doping concentration. The conductivity increases on annealing by several orders of magnitude. The annealed samples exhibit a doping-concentration-dependent conductivity, and as with CVD films, the activation energy of conductivity decreases with increasing doping concentration. Analysis of the mobility-vs.-temperature data (Figure 6.5) shows that at lower doping levels ($< 4 \times 10^{19}$ cm^{-3}) potential barriers between crystallites represent the dominant scattering mechanism, whereas at higher doping levels lattice scattering and impurity scattering predominate.

Haberle and Fröschle[23] have studied the dependence of resistivity, carrier concentration, and Hall mobility of sputtered Si films on the deposition, annealing, and diffusion parameters. Resistivities as low as

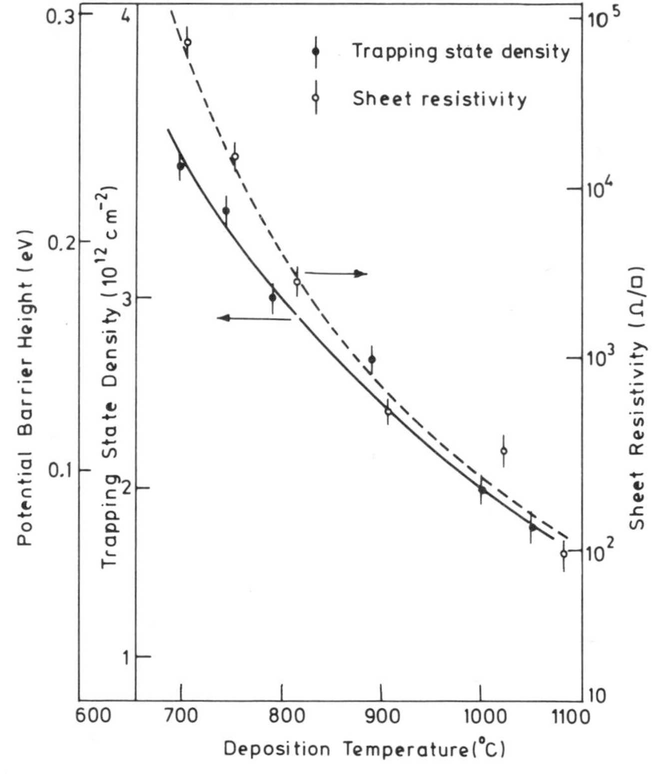

Figure 6.4. Sheet resistivity, potential barrier height, and trapping state density of B-implanted, polycrystalline CVD Si films as a function of deposition temperature (after Yoshihara et al.[8]).

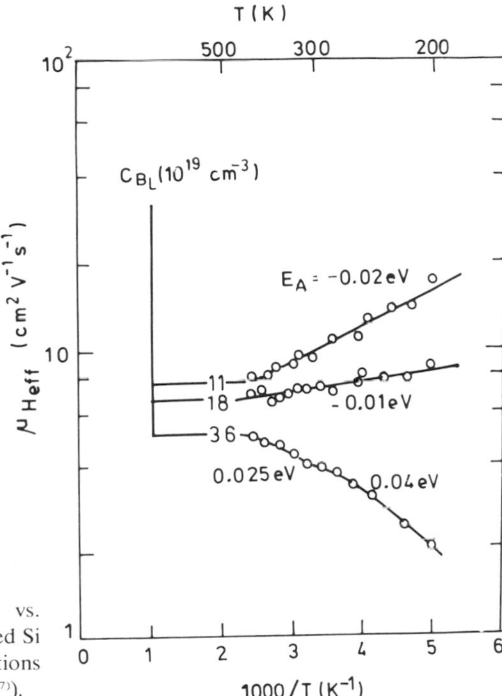

Figure 6.5. Effective Hall mobility vs. temperature plots for vacuum deposited Si films with different doping concentrations (after Van Zolingen and Kipperman[17]).

3×10^{-3} Ω cm and mobilities of 16 $cm^2 V^{-1} s^{-1}$ can be achieved by optimizing the preparation conditions. The carrier concentration is found to be about 10^{20} cm^{-3}. Harman et al.,[21] on the basis of measurements on strongly oriented rf sputtered films on sapphire, have concluded that: (i) the resistivity of the films ($\sim 10^{-1}$ Ω cm) is much smaller than that of the Si target (10^3 Ω cm), (ii) the type of conductivity is the same as that of the target, (iii) the electron mobility in the films (maximum 256 $cm^2 V^{-1} s^{-1}$) is smaller than that in the target (1100 $cm^2 V^{-1} s^{-1}$), and (iv) the carrier concentration in the films is high (10^{17} to 10^{18} electrons cm^{-3}). Hinneberg et al.[22] have also found that the conductivity type of the films is the same as that of the target. The room temperature mobility exhibits a strong dependence on the growth temperature and is in the range from 1 to 20 $cm^2 V^{-1} s^{-1}$, which is the same as that of the bulk material. The lower mobility in the thin (1-μm) films is attributed to additional scattering at crystal defects. At room temperature, the carrier densities in the epitaxial layers vary from 1×10^{16} to 2×10^{17} cm^{-3}. From the temperature dependence (Figure 6.6), acceptor levels at about 0.05 eV have been deduced, and these levels have been attributed to the doping element B present in

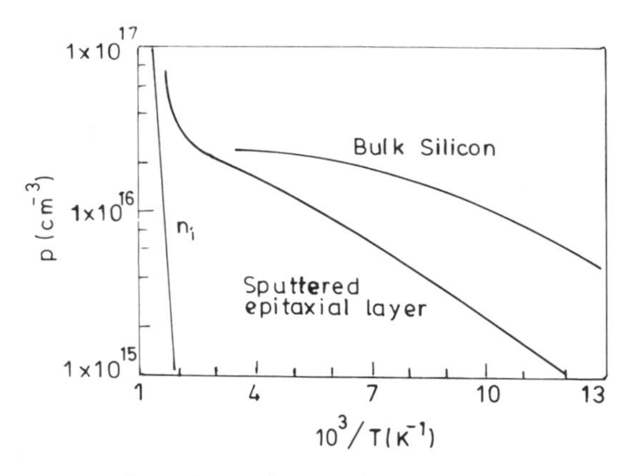

Figure 6.6. Temperature dependence of the carrier concentration p of a sputtered epitaxial Si layer (after Hinneberg et al.[22]).

the target. Baccarani et al.[24] have carried out a theoretical as well as an experimental treatment of the electronic properties of Si films.

The diffusion lengths of minority carriers in CVD Si layers have been measured on solar cells fabricated from the layers, and they usually lie in the range 1 to 25 μm.[9,15,16] Si layers on metallurgical-grade recrystallized Si substrates exhibit different diffusion lengths in the valley (15 to 20 μm) and face (30 to 40 μm) regions. Feldman et al.[20] have shown that lower-temperature-processed layers exhibit longer minority carrier diffusion lengths in solar cells fabricated from evaporated layers.

6.2.1.3. *Optical Properties*

Kühl et al.[1] have determined the optical absorption of CVD deposited Si films grown on sapphire as a function of thickness. They concluded that the optical absorption depends strongly upon both annealing and deposition temperature, with the film quality deteriorating toward the substrate region of the film. The aluminum silicate interface region together with the Al-rich region in the Si film contribute to the decrease in optical quality with decreasing film thickness. Under optimum deposition conditions for epitaxy, the film quality approaches the bulk value for thickness in excess of 0.3 μm.

Hirose et al.[6] have determined the refractive index and optical absorption coefficient from R and T measurements on CVD films. Figure 6.7 shows the refractive indices of amorphous and polycrystalline films. From their study of the change in refractive index with annealing, the authors conclude that macroscopic crystallization occurs at a temperature between 675 and 690 °C. The optical absorption spectra of the amorphous

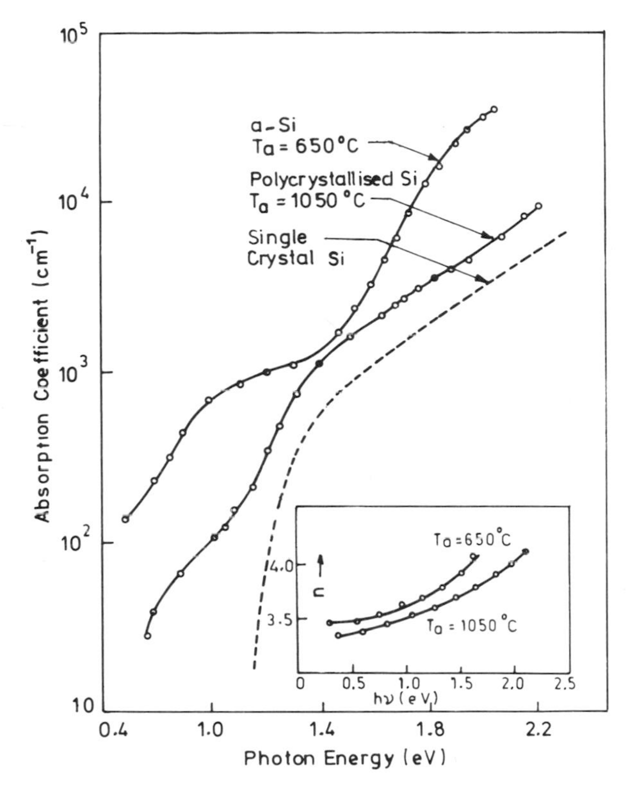

Figure 6.7. Optical absorption coefficient and refractive index n (see inset) of amorphous and polycrystallized CVD Si films as a function of photon energy (after Hirose et al.[6]).

and polycrystalline Si films are also shown in Figure 6.7. Hirose et al. found that the absorption coefficient of polycrystalline Si coincides with that of the monocrystalline Si within a factor of two above the bandgap energy. Below the bandgap, the appreciable optical absorption is due to transitions from filled deep levels to the conduction band or to its tailing states. These results are consistent with their measurements of dc conductivity, ESR, and photoconductivity in these films.

6.2.2. *Cadmium Selenide (CdSe)*

Thin films of CdSe have been deposited by evaporation,[32–37] sputtering,[38–40] spray pyrolysis,[41] anodization,[42] electrodeposition,[43,44] and solution growth[45,46] techniques. The structural and electrical properties of the films depend sensitively on the method of deposition and the deposition parameters.

6.2.2.1. *Structural Properties*

In bulk, CdSe exists in hexagonal (wurtzite) and cubic (sphalerite) structures. Wurtzite is the low-temperature phase which transforms to sphalerite at 700–800 °C.[47] In thin film form, either or both phases can be obtained at room temperature depending on deposition conditions. CdSe thin films deposited in vacuum onto unheated glass substrates consist mostly of the hexagonal phase,[34,48,49] with a (0001) texture orientation. At substrate temperatures between 200 and 500 °C, the films consist of a two-phase mixture of 60% hexagonal and 40% cubic CdSe.[50] With the vapor beam at normal incidence, (0001) or (111) growth textures are found. Epitaxial films of CdSe have been prepared by vacuum evaporation both in the sphalerite and the wurtzite structures.[51–55] Evaporated CdSe films on Al_2O_3 grow epitaxially with (0001) CdSe ∥ (0001) Al_2O_3 at a substrate temperature of about 580 °C,[53] while on Ge the epitaxial temperature varies between 300 and 450 °C, depending on the substrate orientation. Russak et al.[35] have also observed that coevaporated CdSe films deposited at 100 °C on Ti substrates are predominantly hexagonal with a (0001) growth texture.

Lehmann and Widner[38] have studied the growth of CdSe films by rf sputtering on sapphire and silica glass. At temperatures below 400 °C, the film properties on both substrates are similar. The films are polycrystalline with a large degree of preferred (0001) orientation. At temperatures above 400 °C, the films on glass become increasingly amorphous with increasing substrate temperature while those on single crystal sapphire become epitaxial with (0001) CdSe ∥ (0001) Al_2O_3. CdSe films prepared by spray pyrolysis also show[38] hexagonal structure while those prepared by a solution growth technique[45,46] exhibit either purely cubic or a mixture of cubic and hexagonal structures, depending on the deposition conditions. The films prepared by electrodeposition and anodization have not been characterized structurally.

The lattice parameters of CdS_xSe_{1-x} films are found to vary continuously from that of CdSe to CdS, indicating complete solid solubility throughout the range.[46,56,62]

6.2.2.2. *Electrical Properties*

The electrical properties of CdSe films depend on the method of preparation and deposition parameters as a consequence of the variations of stoichiometry and microstructure. CdSe films are always n-type owing to Se vacancies (Cd donors). The acceptor states resulting from Se provide a deep level ($\sim$0.6 eV from the top of the valence band) in the forbidden gap and, therefore, the preparation of p-type films is not feasible.

Glew[40] has observed that CdSe films, dc sputtered in a $H_2Se + Ar$ mixture, have higher dark resistivities of about $10^9\ \Omega\,cm$ measured through the film compared with about $10^7\ \Omega\,cm$ in the plane of the film. This anisotropy has been attributed to the existence of potential barriers resulting from stacking faults. At an illumination level of 3 $mW\,cm^{-2}$, the resistivity, both through and in the plane of the film, decreases to 10^5 to 10^6 $\Omega\,cm$. The films have a response time of less than 1 ms. The through-film gain (ratio of dark resistivity to light resistivity) is found to exhibit a maximum value of 1.8×10^3 at a thickness of 1.5 to 2 μm, whereas the in-plane photoconductive gain increases with increasing thickness from 10 to 80. The through-film photoconductive gain has a maximum value of 8×10^3 for films deposited at substrate temperatures of 150 °C compared to the in-plane maximum photoconductive gain of 60 for films deposited at 175 to 200 °C. These films exhibit current saturation at high fields.

Dhere et al.[34,57] have reported that vacuum evaporated, oriented CdSe films exhibiting a lower resistivity are characterized by a larger grain size and better ordering, whereas higher resistivity films contain amorphous regions and are less ordered. The as-deposited, low-resistivity films contain a small excess of Cd, which increases with heat treatment, resulting in higher carrier concentrations and still lower resistivities. Large increases in the resistivities of films deposited at high rates, when exposed to the atmosphere and heat-treated, are attributed to the depletion of the small individual grains. The deposition rate and the evaporant temperature strongly influence the microstructure and, hence, the electrical properties. Hamersky[36] has studied the effect of residual oxygen pressure and substrate temperature on the structure of evaporated CdSe films and has concluded that at high substrate temperatures special structures are obtained owing to incorporation of O_2 in the film in the form of complexes. The author has also observed that the resistivity of the films is strongly influenced by the O_2 partial pressure and to a lesser extent by the substrate temperature and the boat material. Resistivity as a function of deposition rate exhibits an anomalous minimum and the positions of the anomalous minima, for CdSe films prepared at different O_2 partial pressures, were observed to depend on the ratio of the density of impinging molecules of evaporated CdSe to that of impinging O_2 molecules. The magnitudes of the minima are found to be dependent on the O_2 partial pressure during evaporation.

Chan and Hill[33] have concluded from their electrical conductivity and Hall measurements of evaporated CdSe films (deposited under controlled deposition rates, source temperatures, and substrate temperatures), in the temperature range of -100 to 80 °C, that conduction above 0 °C is due to compensated donors with an ionization energy of 0.186 eV and a concentration of about 10^{20} cm^{-3}. In this temperature region, intercrystalline potential barrier scattering is the dominant scattering mechanism. Below 0 °C, impurity conduction processes dominate. Kutra et al.[58] have reported that the conductivity of *n*-CdSe layers implanted with Se ions converts to *p*-type on thermal treatment in an inert atmosphere. These layers show Hall hole mobilities of 0.4 to 0.6 $cm^2 V^{-1} S^{-1}$ and hole densities of 4.2×10^{14} to 9.8×10^{17} cm^{-3}.

As-deposited, solution grown CdSe films[46] exhibit a dark resistivity of 10^8 to 10^9 Ω cm, depending on the deposition conditions. On vacuum annealing at 300 °C, the resistivity decreases to 1 to 10 Ω cm. Kainthla[46] has studied the effects of annealing the films in different ambients and has found that O_2 increases the dark resistivity of the films. The annealing effects have been attributed to the adsorption and desorption of O_2, similar to the behavior of spray deposited CdS films. The electron concentration in the annealed CdSe films is found to be in the range from 5×10^{17} to 5×10^{18} cm^{-3} and mobility varies from 1 to 10 $cm^2 V^{-1} s^{-1}$. From the temperature dependence of the carrier concentration (Figure 6.8) an activation energy of 0.015 eV has been deduced, corresponding to Cd donors. The mobility is found to vary as $\mu \propto T^{3/2}$. The solution grown CdSe films,[59] after sensitizing in air, exhibit a photoconductive gain of 2×10^3 at 50 mW cm^{-2} white light intensity. The maximum peak of spectral sensitivity is found to occur at 0.7 μm (Figure 6.9), indicating an intrinsic photoconductivity mechanism. A trap density of from about 10^{17} to 10^{18} $cm^{-3} eV^{-1}$ in the energy range 0.21 to 0.26 eV has been estimated. Li and Cu doping are found to enhance the photoconductive gain.[46] Cu and Ag are observed to extend the spectral sensitivity to longer wavelengths.

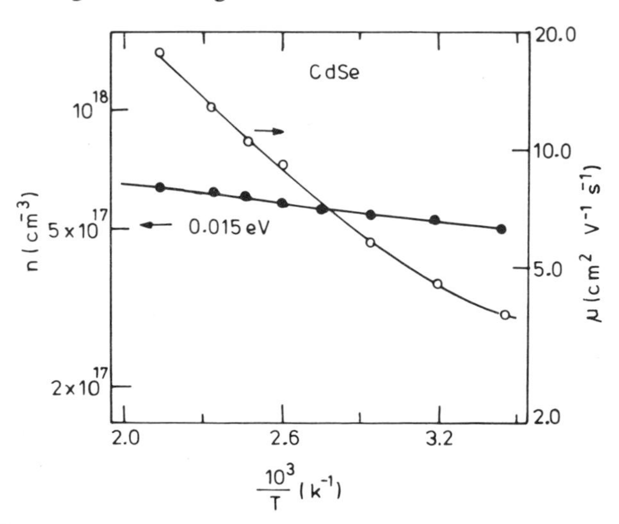

Figure 6.8. Temperature dependence of carrier concentration and mobility of annealed, solution grown CdSe films (after Kainthla[46]).

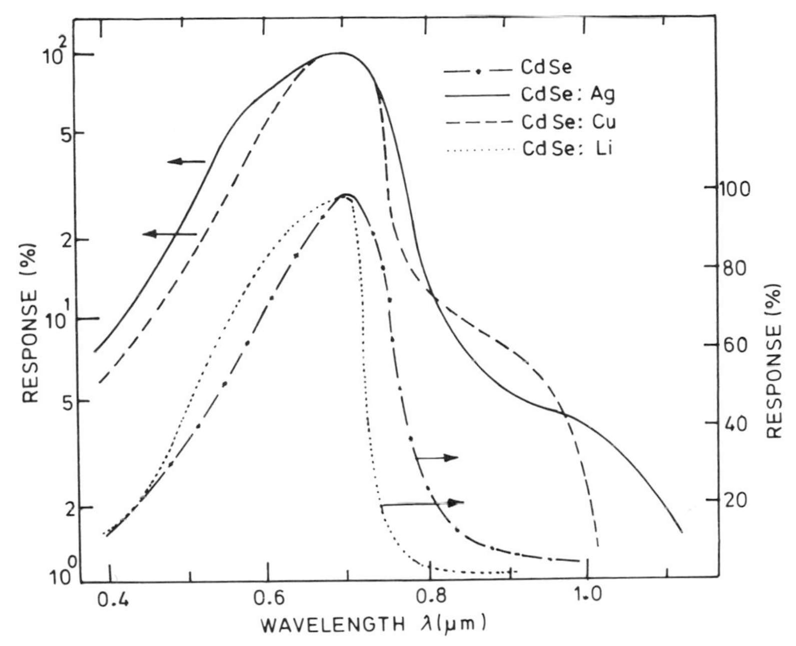

Figure 6.9. Spectral dependence of photoconductivity of solution grown, pure and doped CdSe films (after Kainthla[46]).

Alloy CdS_xSe_{1-x} solution grown films[46] show no change in the dark resistivity with composition. The annealing and photoconductivity behavior is similar to pure CdSe films. Svechnikov and Kaganovich[60] have prepared CdS_xSe_{1-x} films by triode sputtering and chemical deposition with a wide range of photoelectrical properties. However, Feigelson et al.[56] have observed that the resistivity of spray deposited CdS_xSe_{1-x} films does not vary appreciably with composition, although substrate temperature, spray rate, and cooling rate play an important role. Electron concentration in the alloy films is in the range of 10^{17} to 10^{18} cm^{-3} at all compositions.

6.2.2.3. *Optical Properties*

Pure solution grown CdSe films exhibit a direct energy gap of 1.74 eV. The absorption coefficient as a function of photon energy and the spectral dependence of n and α are shown in Figure 6.10. Rentzsch and Berger[61] have observed that the behavior of the fundamental absorption edge of untreated, evaporated CdSe films is dependent on the impurity concentration. Films deposited at high substrate temperatures or annealed at high temperatures exhibit exciton lines at the absorption edge.

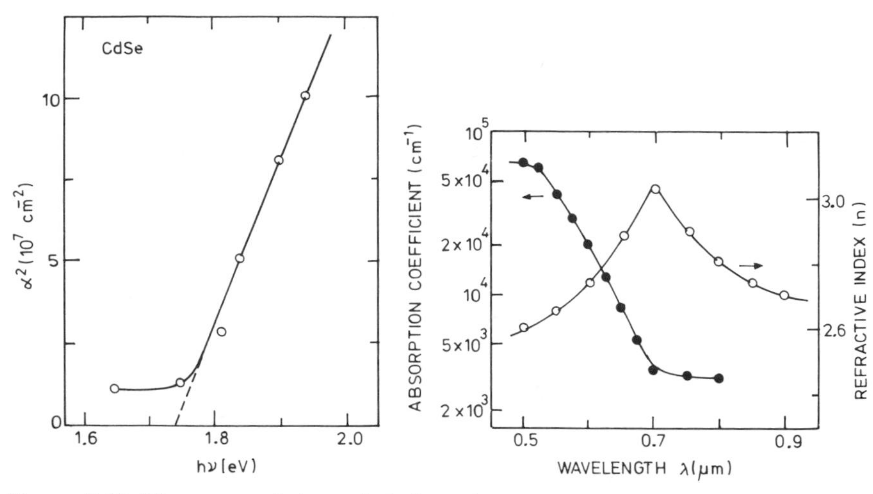

Figure 6.10. The square of the optical absorption coefficient α as a function of photon energy $h\nu$ and optical constants n and α plotted vs. wavelength λ for solution grown CdSe films (after Kainthla[46]).

Kainthla[46] observed that the bandgaps of solution grown CdS_xSe_{1-x} films vary continuously with composition from 1.74 eV for CdSe to 2.44 eV for CdS films. Other workers[56] have also observed a similar continuous variation in sprayed films. The bandgap is direct at all compositions.

6.2.3. *Cadmium Telluride (CdTe)*

CdTe thin films have been prepared by evaporation,[63,64] sputtering,[63,65,66] hot wall epitaxy,[67] spray pyrolysis,[63,65] chemical transport,[63,65,66] vapor transport,[63,68,69] close-spaced vapor transport,[63,65,70–72] screen printing,[73,74] and chemical deposition.[75]

6.2.3.1. *Structural Properties*

Room temperature evaporated CdTe films[64] exhibit a cubic structure and contain a small amount of free Te, as revealed by X-ray diffraction. Deposition at higher temperatures (150 and 250 °C) eliminates the free Te and yields a hexagonal structure. Annealing a room-temperature-deposited film at 300 °C yields diffraction patterns similar to those of a film deposited at 300 °C. Barbe et al.[66] have observed that sputtered CdTe films, obtained using a polycrystalline target, are polycrystalline with a preferential columnar growth direction perpendicular to the substrate surface. A multitwined structure at a 50-Å scale and alternating hcp and fcc stacked layers have been observed along the columns. For cosputtering from CdTe and Cd targets, it is observed that at low substrate temperatures

(20 °C) the film exists in the hexagonal form, whereas at higher substrate temperatures ($\geqslant$ 100 °C), a mixture of cubic and hexagonal forms is observed. At still higher substrate temperatures (350 °C), the film is purely cubic. However, at higher Cd target voltages, the hexagonal phase is enhanced, the films are less oriented, and the high-temperature (350 °C) phase is a different hexagonal phase. CdTe films deposited by vapor transport[68] are essentially polycrystalline and the average grain size increases with increasing substrate temperature. At 500 °C, grain sizes of 20 to 30 μm are obtained and at 600 °C, the grain size is 50 μm or larger. Mimila-Arroyo et al.[72] have observed that CdTe films deposited by CSVT onto CdTe single crystal substrates are always polycrystalline and oriented by epitaxy. The grain size is observed to be sensitively dependent on the surface state (chemical and thermal etches prior to deposition), on the substrate temperature, and on the growth rate. The effect of substrate orientation on the epitaxy of CSVT deposited CdTe films has been studied by Fahrenbruch et al.[63,71] If the c-axis of the CdS substrate is normal to the growth plane, the epitaxy is very strong and the $\langle 111 \rangle$ axis of the cubic CdTe is parallel to the CdS c-axis. The CdTe grains are observed to be truncated pyramids with bases of 50 μm and heights of about 5 μm. In general, the grain size increases with increase in substrate temperature. Thermal etching of the substrate at 750 °C prior to deposition favors strong epitaxy and yields grain sizes of up to 50 μm or more. The grain size is observed to be smaller ($\sim$ 20 μm), although epitaxial growth is retained, when the substrate temperature is maintained constant at the growth temperature before deposition. However, if the temperature is kept low before and during the initial stages of growth, smaller grain sizes ($\sim$ 5 μm) are obtained. Screen printed CdTe layers exist in the cubic phase[74] and exhibit grain sizes of about 10 μm. The free surface of the film exhibits a porous texture while the region near the substrate/film interface is denser. Electrochemically deposited CdTe layers[75] are found to be strongly adherent and well crystallized without any annealing.

6.2.3.2. *Electrical Properties*

The resistivity of sputtered CdTe films[66] deposited at low substrate temperatures is about 10^8 Ω cm, when no voltage is applied to the Cd target. The resistivity decreases to 1 Ω cm for a high Cd target bias. At high deposition temperatures, the resistivity is generally greater than 10^6 Ω cm. Films prepared from In doped targets also exhibit high resistivity, even if In is well transported in the layer, indicating that the dopant is inactive. n-CdTe films, doped with In, grown by a hot wall epitaxy technique[67] at substrate temperatures of 480 to 500 °C at growth rates of 1 to 3.7 μm h^{-1}, exhibit mobilities in the range of 54 to 69 $cm^2 V^{-1} s^{-1}$ and carrier concentrations of about 1×10^{17} cm^{-3}. For thermally etched substrates, the CdTe films exhibit substantially higher mobilities up to 240 $cm^2 V^{-1} s^{-1}$. The carrier concentrations are in the range of $(6–7) \times 10^{16}$ cm^{-3}. The films grown on thermally etched substrates exhibit a temperature dependence of mobility that is indicative of grain boundary scattering. CdTe films deposited by vapor transport on foreign substrates[68] exhibit a dark resistivity in excess of 10^5 Ω cm. n-CdTe films with carrier concentrations of about 3×10^{13} cm^{-3} can be obtained by using HI or CdI_2 as a dopant. However, the carrier concentration cannot be increased further by increasing the dopant concentration because of the etching of CdTe. The carrier mobility in these films is in the range of 20 to 30 $cm^2 V^{-1} s^{-1}$ and the effective intragrain hole diffusion length is found to be 0.8 to 1.5 μm. It should be noted that a carrier concentration of about 10^{16} cm^{-3} and mobilities of 10 to 15 $cm^2 V^{-1} s^{-1}$ have been obtained by carrying out the deposition in a He atmosphere, as a result of reduced etching of CdTe. p-CdTe films have been prepared using Sb or P as dopants. However, the p-type films exhibit high resistivity. Low-resistivity (10 Ω cm) n-CdTe films have been obtained by doping CdTe films simultaneously with I and Ga.[69] Epitaxial p-CdTe layers have been grown on In-doped CdTe single crystals, using As-doped CdTe sources, by CSVT.[72] Thermoelectric power measurements[71] on p-CdTe films deposited by CSVT on sapphire have indicated a hole concentration of 5×10^{15} cm^{-3} and a grain boundary limited mobility varying from 10^{-3} (at 270 K) to 1 $cm^2 V^{-1} s^{-1}$ (at 370 K). Epitaxial CdTe films on single crystal CdS substrates exhibit a 1000 times lower resistivity. Screen printed CdTe layers (10 μm thick) exhibit a resistivity between 10^{-1} to 1 Ω cm, close to the resistivity of the bulk CdTe source material.[73]

6.2.3.3. *Optical Properties*

The optical constants n and k of thin polycrystalline CdTe films have been determined by Thutupalli and Tomlin[64] over the wavelength range from 0.3 to 3.0 μm, from normal-incidence reflectance and transmittance data. Analysis of the absorption data reveals that CdTe films possess a direct bandgap of 1.50 eV and exhibit a spin-orbit splitting of the valence band of 0.88 eV. An indirect bandgap of 1.82 eV is also observed. The dispersion of n and k for CdTe films is shown in Figure 6.11.

6.2.4. *Indium Phosphide (InP)*

Thin layers of InP have been grown by metallorganic chemical vapor deposition (MOCVD),[76] CVD,[77–79] evaporation,[80] and planar reactive deposition.[81]

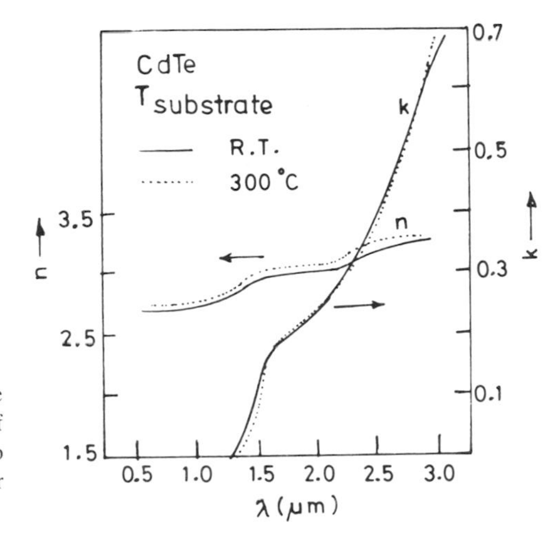

Figure 6.11. Wavelength dependence of the optical constants n and k of evaporated CdTe films deposited at two different substrate temperatures (after Thutupalli and Tomlin[64]).

6.2.4.1. *Structural Properties*

InP has a sphalerite structure at room temperature with a lattice constant a equal to 6.869 Å.[80] Single-source evaporation of InP results in films with elemental In and P, both crystalline and amorphous.[80] Low substrate temperatures and high deposition rates lead to excess elemental P and high substrate temperatures and low deposition rates yield films with excess elemental In. Two-source evaporation, involving elemental deposition of In and P, yields single-phase InP films with grain sizes exceeding 1 μm at a substrate temperature of 500 K. CVD grown InP layers are polycrystalline and randomly oriented.[77] InP films deposited on C substrates exhibit larger grains than those deposited on Mo substrates, although in both cases the grain size is comparable to the diffusion length in InP ($\leqslant 1$ μm). Saitoh et al.[79] have observed that the structures of n-InP films grown by CVD on Mo substrates vary with deposition temperature and growth time and not with reaction gas concentration. At a substrate temperature of 600 °C, the n-InP films are columnar and oriented predominantly in the ⟨110⟩ direction. At substrate temperatures lower than 550 °C or for films of thickness less than 5 μm, the grains are randomly oriented. p-type films exhibit random orientation, irrespective of deposition conditions. Moreover, with increasing dopant (Zn) concentration, InP whiskers tend to be formed. Be-doped, micrometer-thick InP films grown on recrystallized CdS films by planar reactive deposition[81] at a substrate temperature of 280 °C exhibit complete epitaxy and the lateral dimensions of the InP grains (~ 40 μm) replicate those of the substrate. At a substrate temperature of 380 °C, the epitaxy is poor owing to the formation of an intermediate In-Cd-S layer.

6.2.4.2. *Electrical Properties*

Undoped, *n*-type, epitaxial InP films grown on GaAs:Cr single crystal substrates[76] exhibit an increase in electron mobility with increase in electron density, for electron densities less than 10^{16} cm^{-3}. Small structural potential barriers significantly affect the mobility. Epitaxial InP films deposited on InP:Fe substrates exhibit an electron density greater that 10^{16} cm^{-3}. Electron mobilities at 77 and 296 K have been measured to be as high as 10,500 and 3100 $cm^2 V^{-1} s^{-1}$, respectively, for films on GaAs:Cr substrates and 16,600 and 3540 $cm^2 V^{-1} s^{-1}$, respectively, for films on InP:Fe substrates. Evaporated *p*-InP films[80] exhibit a temperature dependence of mobility that is indicative of grain boundary scattering, as shown in Figure 6.12. The grain boundary potential is determined to be 0.04 eV. The activation energy for the carrier concentration is found to be 0.19 eV. Bachmann et al.[77] have obtained a resistivity of 0.07 Ω cm, corresponding to an acceptor concentration of 9×10^{17} cm^{-3} and a mobility of 97 $cm^2 V^{-1} s^{-1}$, in Zn-doped, epitaxial InP layers grown by CVD on InP:Fe substrates. Saitoh et al.[79] have observed *n*-type conductivity in undoped, polycrystalline CVD films, with the electron concentration varying from 10^{15} to 10^{17} cm^{-3}, depending on the deposition temperature. The carrier concentration was observed to increase with increase in the deposition temperature. The electron mobility in these films was measured to be about 10 $cm^2 V^{-1} s^{-1}$ and the resistivity was observed to be in the range 10 to 100 Ω cm. The electrical properties of S-doped InP films was observed to depend strongly on the deposition conditions and the doping levels. Zn doped *p*-InP films exhibited a resistivity of 18 Ω cm, corresponding to a hole concentration of 4.6×10^{16} cm^{-3} and a mobility of 7.5 $cm^2 V^{-1} s^{-1}$. An electron diffusion length of 0.2 μm was estimated for the *p*-type films.

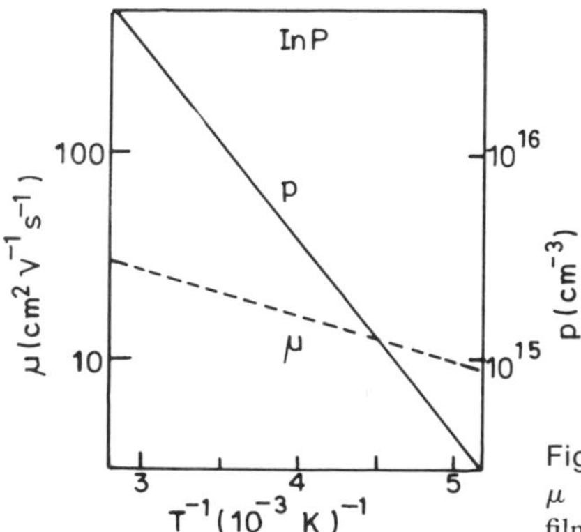

Figure 6.12. Temperature dependence of mobility μ and carrier concentration p for evaporated InP films (after Kazmerski et al.[80]).

6.2.5. *Zinc Phosphide (Zn_3P_2)*

Vacuum evaporation,[82,83] CVD,[68] and CSVT[84] have been successfully employed to deposit Zn_3P_2 thin films on a variety of substrates.

6.2.5.1. *Structural Properties*

In bulk, the Zn_3P_2 lattice has tetragonal symmetry with unit cell dimensions of $a = b = 8.097$ Å and $c = 11.45$ Å $= (2)^{1/2}a$. The properties of evaporated films are observed to depend strongly on substrate temperature, and to a lesser degree on source temperature.[83] At substrate temperatures greater than or equal to 220 °C, the films exhibit an increase in preferential orientation. The orientation coefficient, defined as the ratio of areas under the ⟨004⟩ and ⟨400⟩ diffraction peaks, increases with film thickness. For film thicknesses above 10 μm, nearly perfect basal plane orientation is obtained. The films exhibit grain sizes of 1 to 2 μm and the surface roughness is found to increase with film thickness. Zn_3P_2 films deposited by CSVT[84] in Ar and H_2 ambients show major effects from O_2 contamination of Ar. The growth rate in H_2 is found to be higher than in Ar.

6.2.5.2. *Electrical Properties*

Zn_3P_2 films deposited by vacuum evaporation under optimized conditions exhibit resistivities of the order of 10^5 Ω cm. The resistivity is found to decrease with increase in preferential orientation.[83] The films are p-type with hole mobilities in the range 10 to 40 $cm^2 V^{-1} s^{-1}$. The resistivity can be varied from 10 to 10^5 Ω cm by annealing at temperatures greater than or equal to 250 °C, under an appropriate partial pressure of Zn or P.[82] CSVT grown Zn_3P_2 films[84] exhibit an activation energy of conduction of 0.11 eV at higher temperatures and 0.04 eV at lower temperatures. CVD films[68] deposited at 600 °C have achieved a room temperature resistivity of 200 to 400 Ω cm, a hole mobility of 10 to 12 $cm^2 V^{-1} s^{-1}$, and an effective intragrain diffusion length of 0.5 μm. For comparison, diffusion lengths in single crystals, estimated from spectral response measurements, are in the range 5 to 10 μm.[85]

6.2.5.3. *Optical Properties*

The bulk absorption edge is found to be exponential in energy for (absorption coefficient) values less than 1500 cm^{-1}. The thin film absorption edge is also observed to be exponential in this range, although the slope is shallower. Fagen[83] has suggested that at higher values of absorption

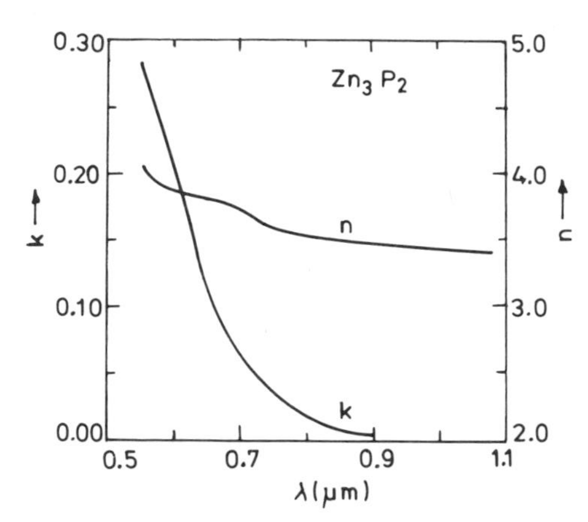

Figure 6.13. Wavelength dependence of the optical constants n and k of Zn_3P_2 films (after Fagen[83]).

coefficient the optical gap is probably direct and lies in the neighborhood of 1.6 eV. The near IR refractive index has been determined to be 3.3 ± 0.1, as shown in Figure 6.13.

6.2.6. *Gallium Arsenide (GaAs)*

GaAs thin films for solar cell applications have been grown almost exclusively by MOCVD[86–89] and CVD,[90–94] although sputtering,[95] MBE,[96–97] and vapor phase epitaxy[98] have also been employed. GaAs-based alloy films, such as GaAlAs, GaAsP, and GaInAs, have been deposited by a variety of techniques such as CVD,[94] MOCVD,[99–101] MBE,[96,97,102,103] VPE,[103,104] and LPE.[103,105]

6.2.6.1. *Structural Properties*

Chu et al.[93] have observed that CVD GaAs films deposited on W coated graphite are continuous at thicknesses less than 2 μm, under proper deposition conditions. The deposition parameters, particularly substrate temperature, composition and flow rates of the reactant gases, and hydrogen chloride concentration at the substrate surface, sensitively influence the microstructural characteristics of the films. The presence of hydrogen chloride reduces the rate of nucleation, promotes the growth of larger crystallites, and improves the bonding characteristics along the grain boundaries. The average grain size of GaAs films increases with increase in film thickness and is about 10 μm in 20- to 25-μm-thick films and about 5 μm in 8- to 10-μm-thick films. The CVD films do not exhibit any preferred orientations at high substrate temperatures. At a low arsine/hydrogen chloride molar ratio (0.5–1), the average crystallite size increases with increasing substrate temperature, attains a maximum at 750 to 775 °C, and decreases thereafter. At high arsine/hydrogen chloride molar ratio (e.g., 2), the average grain size is independent of substrate temperature in the range 725 to 775 °C and decreases at higher temperatures. At substrate temperatures of 725 to 750 °C, the grain size increases with increase in arsine concentration in the reactant mixture. Moreover, at low substrate temperatures, GaAs films exhibit a strong ⟨110⟩ preferred orientation at low arsine concentrations and ⟨111⟩ preferred orientation at high arsine concentrations.[92] Vernon et al.[88] have grown 5-μm-thick GaAs films by MOCVD with grain sizes of 0.2 to 0.5 mm after recrystallizing the films in an arsine flow.

6.2.6.2. *Electrical Properties*

GaAs films, grown by CVD on W/graphite substrates, without intentional doping[94] are n-type with carrier concentrations in the range of 5×10^{16} to 10^{17} cm^{-3}. With H_2S as a dopant, the carrier concentration can be increased to $(2–5) \times 10^{18}$ cm^{-3}. At a substrate temperature of 750 °C, the carrier concentration in GaAs films is essentially independent of the reactant composition.[92] However, at higher temperatures, the carrier concentration decreases with increasing arsine/hydrogen chloride molar ratio. The minority carrier diffusion lengths in CVD GaAs films have been estimated to be 0.2 to 0.4 μm.[92] GaAs films deposited on graphite substrates[91] exhibit diffusion lengths in the range 0.5 to 0.8 μm. GaAs films, formed by MOCVD,[87] have been observed to posses a carrier concentration of 6×10^{16} cm^{-3}, and a hole diffusion length in excess of 2 μm has been measured for single crystal films. Jang et al.[106] have measured the electrical properties of Se- and Zn-doped polycrystalline GaAs films deposited by MOCVD. The resistivity and mobility have been observed to be temperature-activated over a wide temperature range. The results have been interpreted in terms of grain boundary scattering.

We note here that the properties of GaAs-based alloys depend primarily on the composition.

6.2.7. *Cadmium Sulfide (CdS)*

CdS is one of the most extensively investigated semiconductors in thin film form and a large variety of deposition techniques have been utilized to obtain solar cell quality layers of CdS. These preparation techniques include evaporation,[107–130] spray pyrolysis,[109,131–143] sputtering,[144–148] MBE,[149] VPE,[150] CSVT,[114,151] CVD,[163] screen printing,[152,153] chemical deposition,[154–156] anodization,[157] and electrophoresis.[164]

6.2.7.1. *Structural Properties*

Evaporated CdS films grown for solar cell applications are usually 15 to 30 μm thick, deposited at substrate temperatures of 200 to 250 °C, source temperatures of 900 to 1050 °C, and deposition rates of 0.5 to 3 μm min^{-1}.[113,117,130] Under these conditions, the CdS films grow in the wurtzite structure and in a strongly preferred orientation with the (002) plane parallel to the substrate and the *c*-axis perpendicular to the substrate. Moreover, the microstructure is columnar in nature, as shown in Figure 7.2 (p. 356), each column being a single grain. The grain sizes of these films are generally observed to be in the range of 1 to 5 μm, although larger grain sizes up to 10 μm have also been obtained.[130] It should be emphasized that at lower film thicknesses, the grains are smaller and randomly oriented.[121] Also, the crystallographic structure and microstructure of the films are strongly dependent on the substrate temperature.

Vankar et al.[111] and Das[116] have investigated the structure of vacuum evaporated $Zn_xCd_{1-x}S$ films as a function of substrate temperature and observed that the crystallographic structure and the lattice parameter of the structures of CdS films exhibit strong dependence on the deposition temperature. CdS films prepared at substrate temperatures ranging from room temperature to 150 °C exhibit a sphalerite structure, while the wurtzite structure is found at 170 °C and higher substrate temperatures. At temperatures between 150 and 170 °C a two-phase structure, composed of sphalerite and wurtzite structures, is obtained. At 200 °C and above, preferred orientation effects are observed. The grain size of the CdS films increases with increase in temperature of deposition. However, the surface roughness of the CdS films first decreases with increase in temperature and then increases beyond 150 °C, probably because of reevaporation. Beyond 200 °C, the films exhibit blow holes.[116]

Fraas et al.[125] have recrystallized evaporated CdS films via a heat treatment in an H_2S flow to obtain grains of dimensions up to 100 to 800 μm. However, the *c*-axis orientation and the columnar nature of the microstructure are destroyed. Amith[113] has reported that the grain size, the preferential grain orientation, the preferential *c*-axis orientation toward the source, and the surface roughness all increase as the film thickness is increased. Hall[119] has pointed out that as-deposited CdS films normally have a *c*-axis that is tilted about 19° about the normal to the substrate. Moreover, the distribution of tilts about this average is very broad with a half-width at half-maximum height of 10 to 12°. Subsequent to high-pressure (100 psi) and high-temperature (190 °C) postdeposition treatments, the *c*-axis orientation has a distribution about the normal with a half-width at half-maximum height of 3°.

Tseng[121] has concluded on the basis of electron microscopy data that the grain boundaries in the top layer of wurtzite CdS films are of the tilt type and have misorientation angles in the range of 9 to 40°. Most of the grain boundaries are of the coincident-site type. Dhere and Parikh[108] have observed that the growth texture, crystallographic features, and faceting of CdS films are enhanced by using better vacuum conditions. Romeo et al.[107] have studied the effect of Cd:S ratio on the properties of CdS films deposited by a two-source technique. They have observed that although good-quality films can be grown over a wide range of Cd:S ratio, the best results are obtained at a Cd:S ratio of 1.5. Moreover, better crystallization has been observed in CdS films doped up to the limit of solubility of the impurity (In in their case).

The most significant parameter dominating the crystallographic and microstructure of sprayed films is the substrate temperature during deposition.[131,134,158] However, other factors such as type of salts,[136,159] ratio of cations to anions,[131–134,140,141,143] and dopants[134,140] also affect the grain size and any orientation, if present.

In Figure 6.14 we show the effect of Cd:S ratio, substrate temperature, film thickness, multilayers, and postdeposition annealing on the orientation of CdS films. We should point out that these results are by no means unique and different workers have reported different orientations under similar deposition conditions. If deposited from acetate solution, CdS films have a very small grain size.[159] Chloride solutions yield larger grains as well as c-axis orientation.[136,159] Typically, in sprayed CdS films, grain dimensions are in the range of 0.1 to 0.5 μm, but some workers[132,138] have reported grains up to 1 to 4 μm. Impurities such as In and Ag facilitate grain growth while insoluble impurities such as Al_2O_3[135,160,161] inhibit recrystallization and reduce the grain size of CdS films drastically as well as destroy preferred orientation, except at very small concentrations.

The segregated Al_2O_3 at the grain boundaries produces a characteristic serpentine structure on the surface of CdS films. The surface topography of pure CdS and CdS:Al_2O_3 films are shown in Figure 7.3 (p. 358). Ma and Bube[158] have observed the existence of sphalerite and wurtzite phases in sprayed CdS films deposited at low ($\leqslant 400\,°C$) and high ($> 400\,°C$) substrate temperatures, respectively. However, Banerjee et al.[139] have concluded that the crystallographic structure does not depend on the deposition temperature. Special features of the sprayed films are the strong adhesion of the film to the substrate and the coherence of the film even at low thicknesses.

Sputtered films exhibit a greater uniformity of orientation of the c-axis compared to evaporated films.[148] Moreover, for the same film thickness, sputtered films have a lower pin-hole density. However, the grain sizes in sputtered CdS films[145] are usually smaller ($\leqslant 1$ μm), although the films are columnar. The crystallographic structure of sputtered films is always

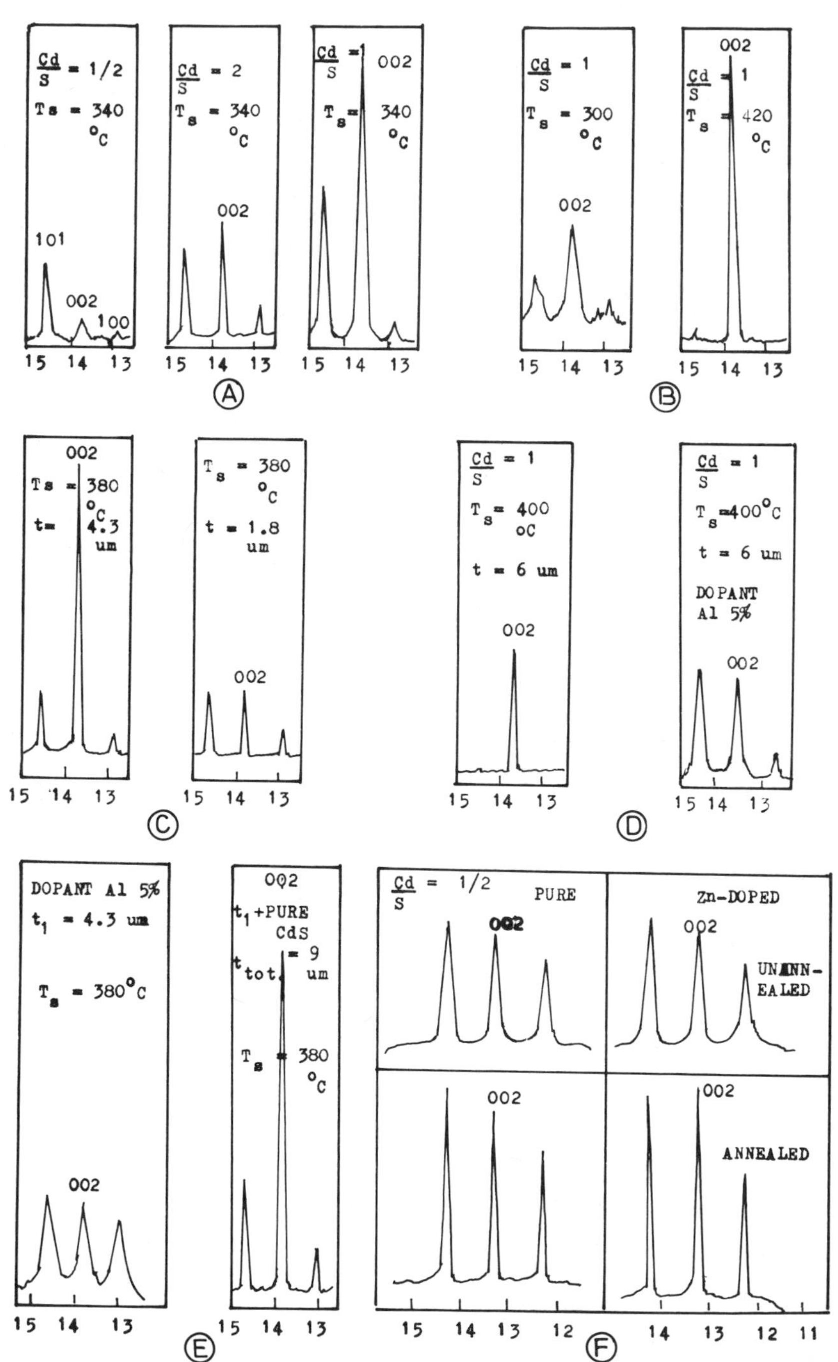

Figure 6.14. Effect of Cd:S ratio (A), substrate temperature (B), film thickness (C), doping (D), multilayers (E), and postdeposition annealing (F) on the orientation of CdS films.

hexagonal with a preferred orientation of the c-axis perpendicular to the substrate.[162] Piel and Murray[147] have pointed out that sputtered CdS layers contain ionized particles of the discharge gas trapped in the film. Mitchell et al.[114] have grown 1- to 3-μm-thick CdS films by CSVT with grain sizes of the same order (1 to 3 μm) and have observed no correlation between substrate temperature and grain size in the temperature range 325 to 500 °C. On the other hand, Yoshikawa and Sakai[151] have observed the surface morphology of CSVT films to be sensitive to the substrate temperature and have concluded that a high substrate temperature is necessary to obtain a smooth surface. However, at very high substrate temperatures whisker growth occurs. The c-axis of CSVT films is nearly perpendicular to the substrate.

Epitaxial growth of CdS[149–151,163] has been observed on GaAs, CdTe, Ge, spinel, Au, Al, and InP. Epitaxial CdS films grown by MBE on spinel and Au exhibit a wurtzite structure while those on Al exhibit a sphalerite structure.[149] Single crystal layers of hexagonal CdS have been deposited by VPE[150] on (111), (110), and (100) faces of InP in the following heteroepitaxial relationships: (0001) CdS ∥ (111) InP, (0113) CdS ∥ (110) InP, and (3034) CdS ∥ (100) InP.

Solution grown CdS films exhibit small grain sizes of up to 1000 Å. The grain size is larger at lower deposition rates and higher bath temperatures. The solution grown films can be obtained in different structures,[155] depending on the deposition conditions. The CdS films obtained from $Cd(NH_3)_4^{2+}$ complex has sphalerite, wurtzite, or mixed structure depending on the deposition conditions, while those obtained from $Cd(CN)_4^{2-}$ and $Cd(en)_3^{2+}$ complexes always exhibit wurtzite structure with the c-axis perpendicular to the substrate.[155,156]

6.2.7.2. *Electrical Properties*

The electrical properties of CdS thin films depend sensitively on the deposition condition. Evaporated CdS films prepared for solar cell applications usually have resistivities[116,118,130] in the range of 1 to 1000 Ω cm and carrier concentrations[80,108,116,117,126] in the range 10^{16} to 10^{18} cm^{-3}. The films are always n-type and the conductivity is dominated by the deviation from stoichiometry which leads to films with S vacancies or Cd excess. Mobilities[108,116,126] are in the range 0.1 to 10 cm^2 V^{-1} s^{-1}. Minority carrier diffusion lengths in the evaporated films have been measured to be in the range 0.1 to 0.3 μm.[116,120,165] CdS films grown at higher rates are observed to exhibit a higher carrier concentration[129] and the carrier concentration is found to increase with an increase in film thickness[113] accompanied by a corresponding decrease in resistivity.

The Cd:S ratio during evaporation sensitively influences the electrical properties as does doping.[107] For In-doped CdS films, the best electrical properties are obtained at a Cd:S ratio of 1.5 at which the films also exhibit the best structural characteristics. Resistivity values as low as 10^{-3} Ω cm and mobility of 90 $cm^2 V^{-1} s^{-1}$ have been obtained on CdS:In (1.5%) samples. The resistivity and mobility as a function of Cd:S ratio are shown in Figure 6.15 for two evaporated CdS films with different In concentrations. Weng[110] has reported that In doping increases the carrier concentration by almost three orders of magnitude and also significantly enhances the mobility up to a concentration of about 2% by weight. Thereafter, the carrier concentration does not increase and the mobility decreases somewhat. However, at low In concentrations, both the carrier concentration and the mobility decrease. For doped films (In ~ 2%), the carrier concentration and mobility remain relatively constant over a wide range of substrate temperatures, as shown in Figure 6.16. Cu doping, on the other hand, decreases the carrier concentration and increases the resistivity by several orders of magnitude.[126] The electron mobility is also decreased.

Several workers[108,122,126] have studied the transport mechanism in evaporated films. Deppe and Kassing[122] have attributed the electrical properties to one dominant deep impurity level caused by S vacancies. The energy level of the S vacancy is dependent on the concentration and, at low values ($< 10^{17}$ cm^{-3}), a discrete level about equal to 0.6 eV has been obtained. At higher values, an impurity band is reportedly formed. Dhere and Parikh[108] have obtained an electron energy level with an activation energy of 0.22 eV. Wu and Bube[126] have observed that the dark electron

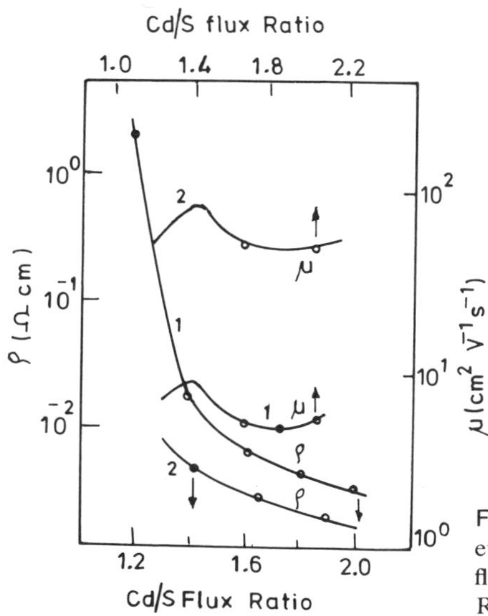

Figure 6.15. Resistivity ρ and mobility μ of evaporated CdS:In films as functions of Cd:S flux ratio: 1— 1% In, 2 — 1.5% In (after Romeo et al.[107]).

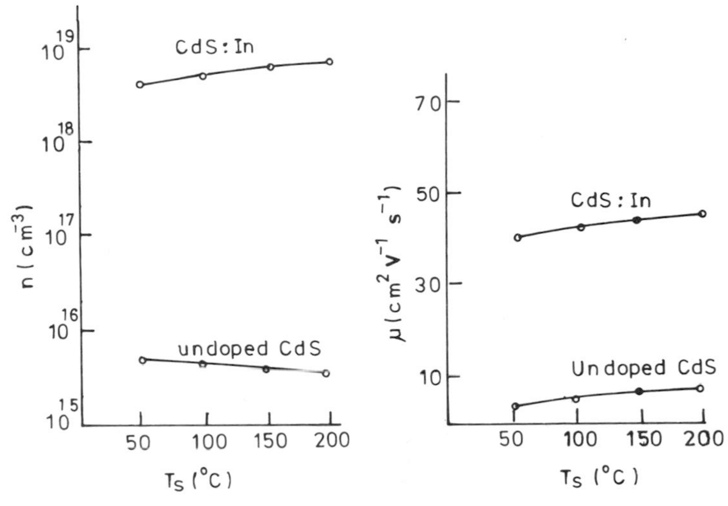

Figure 6.16. Carrier concentration n and mobility μ of undoped and In doped, flash evaporated CdS films as functions of substrate temperature (after Weng[110]).

densities of evaporated films are essentially independent of temperature between 200 and 330 K because of shallow donors. Activation energies for electron density are in the range 0 to 0.04 eV. For mobility, the activation energies are in the range 0.11 to 0.19 eV and the pre-exponential factors vary from 50 to 100 $cm^2 V^{-1} s^{-1}$. The structure and characteristics of intergrain boundaries dominate the transport processes in these films. As-evaporated CdS films are not photosensitive. However, Cu-diffused CdS films exhibit significant amounts of photoconductivity and the electron density is lower and the electron mobility is higher under high-intensity photoexcitation than without Cu diffusion.

In spray deposited CdS films, the electrical properties are dominated by the chemisorption of O_2 in these films at the grain boundaries, which reduces both the carrier concentration and the mobility.[136] The films are invariably n-type (owing to S vacancies) and the resistivity of these films can be varied over as much as 10^8.[136,166] Postdeposition annealing in air increases the resistivity of CdS films to about 10^7 Ω cm and makes them highly photoconducting. A photoconductive gain from 10^6 to 10^7 with response time of about 1 ms under 50 mW cm^{-2} illumination has been measured in our laboratory.[166] On vacuum annealing, the resistivity is decreased to 1 to 10 Ω cm and the photoconductivity is quenched, indicating the reversibility of chemisorption and desorption of oxygen processes.[136] The behavior of the resistivity of as-deposited CdS films with annealing is shown in Figure 6.17a.

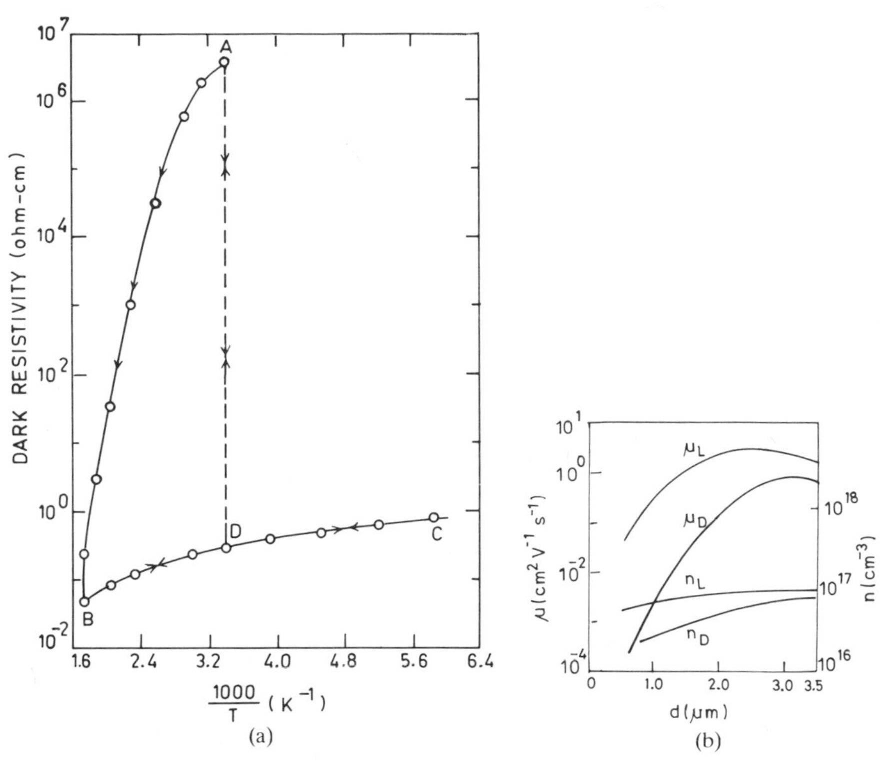

Figure 6.17. (a) Dark resistivity vs. temperature in vacuum and in the presence of different gases of sprayed CdS films. Point A corresponds to resistivity of as-sprayed films. Path A–B corresponds to effect of vacuum annealing. Path B–C corresponds to temperature dependence of resistivity of the annealed films in vacuum or in the presence of inert gases. Point D corresponds to the room temperature value of resistivity of the vacuum annealed films. Further annealing in air or oxygen results in the resistivity regaining the initial value A (after Banerjee(136)). (b) Light (subscript L) and dark (subscript D) mobilities and carrier concentrations as functions of film thickness for spray deposited CdS films (after Kwok and Siu(132)).

Detailed electron transport measurements on CdS films have been reported by several workers.(126,136,158,167) Ma and Bube(158) have reported an oscillatory behavior of electrical conductivity, carrier concentration, and mobility on the deposition temperature. The chemisorption kinetics are also affected by the rate of cooling of as-deposited films and, consequently, changes in the electron transport properties have been observed.(158) Kwok and Siu(132) have reported that the dark carrier concentration and mobility of sprayed CdS films increase with film thickness, accompanied by an increase in the grain size. The light and dark carrier concentrations and mobilities are plotted as functions of film thickness in Figure 6.17b. Photo-Hall effect and photothermoelectric power measurements[136,168] have shown that on illumination n or μ or both change. Which parameter changes more depends on the relative role of the microstructure (grain size) and postdeposition treatment (chemisorbed oxygen) in the electrical conduction processes. Hole diffusion lengths in spray deposited CdS films have been measured to be in the range 0.2 to 0.4 μm.[136]

As-deposited sputtered CdS films exhibit a high resistivity[145,148,162] of up to 10^8 Ω cm. Cosputtering with In yields films with resistivities of about 1 Ω cm and mobilities of about 40 $cm^2 V^{-1} s^{-1}$. Lichtensteiger[169] has reported p-type doped CdS films with hole mobilities of 6 to 15 $cm^2 V^{-1} s^{-1}$. Carrier concentration in In doped CdS films (1 at.%) have been measured to be about 7×10^{18} cm^{-3}.[144] The resistivity of the doped sputtered films is found to be relatively insensitive to the substrate temperature during deposition,[144] in contrast to the strong dependence of the resistivity of undoped films on deposition temperature. This behavior of the sputtered films is similar to that of evaporated films. Piel and Murray[147] have interpreted their high-electric-field conductivity data to be the result of a Poole–Frenkel mechanism. Hill[148] has measured an electron diffusion length of about 1 μm in high-mobility sputtered CdS films.

Chemically deposited CdS films[154,155] are n-type and have resistivities in the range 10^7 to 10^9 Ω cm. On annealing in vacuum, resistivities decrease to about 1 to 10 Ω cm. This reduction is attributed to the desorption of O_2 from the films, as in the case of spray deposited films. The original resistivity values can be recovered by subsequent heating in air/O_2. Pavaskar et al.[170] have reported a carrier concentration of about 10^{14} cm^{-3} and a mobility of about 5 $cm^2 V^{-1} s^{-1}$, measured under illumination. The chemically deposited films annealed in air exhibit a high photosensitivity.[154] Screen printed CdS films[152] also exhibit a good photosensitivity and a dark to light (100 m W cm^{-2}) resistivity ratio of about 10^4. Electrophoretically deposited CdS films[164] exhibit resistivities in the range 10^3 to 10^5 Ω cm.

Epitaxially grown CdS films exhibit very high mobilities. The electrical properties of CdS films grown epitaxially on GaAs by CSVT[151] show a strong dependence on growth conditions, especially substrate temperature. The carrier concentration is observed to increase exponentially with increase in substrate temperature. Electron mobility also increases with an increase in substrate temperature, and a maximum value of 241 $cm^2 V^{-1} s^{-1}$ has been obtained. The resistivity varies from 10^{-3} to 1 Ω cm as a function of the substrate temperature. Undoped, epitaxial CdS films grown by MBE[149] exhibit a resistivity of 10^5 Ω cm, which decreases exponentially with temperature with an activation energy of 1.6 eV. In-doped films possess a carrier concentration of 10^{18} cm^{-3} with Hall mobilities of 65 $cm^2 V^{-1} s^{-1}$. Epitaxial CVD films[163] exhibit as-deposited resistivities between 10 and 100 Ω cm. Annealing in H_2/Ar at 400 °C reduces the resistivity to values between 0.01 and 0.05 Ω cm. Mobilities of the higher resistivity films were found to be between 100 and 150 $cm^2 V^{-1} s^{-1}$.

6.2.7.3. *Optical Properties*

The optical properties of CdS films are determined to a large extent by the microstructure of the films and, hence, by the deposition conditions. Thin evaporated CdS films are smooth and specularly reflecting but the surface roughness increases with increase in thickness, leading to a large diffuse scattering component in thick films. Khwaja and Tomlin[124] have determined the optical constants *n* and *k* of thin evaporated films over the wavelength range of 0.25 to 2.0 μm by measuring the normal incidence reflectance and transmittance and taking into account the surface roughness.

Analysis of the data (Figure 6.18) yields direct transitions in the range of 2.42 to 2.82 eV and combined direct and indirect transitions beyond 2.82 eV. The *n* and *k* values are observed to be dependent on the substrate temperature during deposition. At high substrate temperatures, corresponding to increased grain size, the refractive index approaches that of single crystal material. Sputtered CdS films[162] exhibit a sharp optical transmission cut-off near 0.52 μm, corresponding to the bandgap of CdS. At longer wavelengths, the films are essentially transparent. In spray deposited films,[134–136,141] the bandgap and the fundamental optical absorption edge are not affected by the microstructure. The diffuse scattering and,

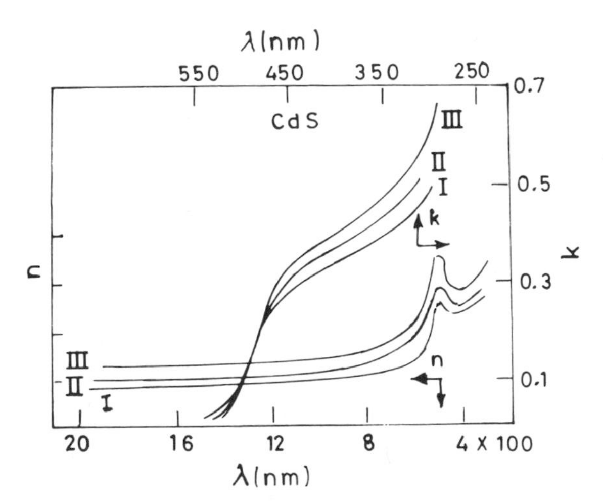

Figure 6.18. Refractive index *n* and extinction coefficient *k* dependence on wavelength for evaporated CdS films deposited at three different temperatures: I — room temperature, II — 140 °C, and III — 180 °C (after Khwaja and Tomlin[124]).

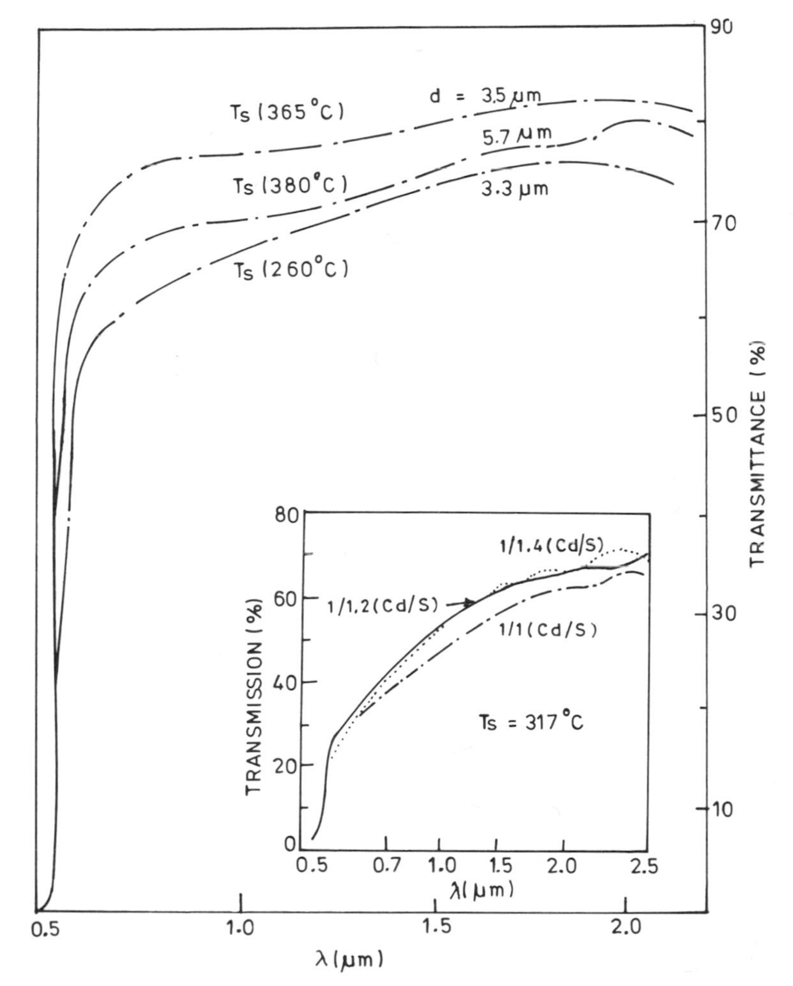

Figure 6.19. Transmittance of sprayed CdS films deposited under different conditions (after Oudeacoumar[134] and Bougnot et al.[141]).

hence, transmittance depend on film thickness, substrate temperature, and Cd:S ratio, as illustrated in Figure 6.19. While diffuse scattering increases with film thickness, it is reduced with increasing deposition temperature (owing to increased grain size and orientation effects). At very high temperatures (above 500 °C), the films become rough and translucent, presumably because of major changes in the growth kinetics of the films.

Berg et al.[135] have observed that the grain structure and morphology of spray deposited films (3–4 μm thick) produce extensive optical scattering and high effective absorption constants ($\sim 500\ \text{cm}^{-1}$) for wavelengths longer (~ 1.0 μm) than those corresponding to the CdS bandgap. The optical absorption edge of solution grown CdS films is the same as that of the corresponding bulk material.[154] However, the sharpness of the edge is considerably reduced owing largely to the diffuse scattering of light as a result of the fine-grained microstructure of the films.

6.2.7.4. $Zn_xCd_{1-x}S$ Alloy Films

The composition of the alloy films plays a dominant role in determining their structural, electronic, and optical properties. The alloy films have been prepared by evaporation,[111,112,115,116,127,130,171] spray pyrolysis,[136,137,139] and sputtering.[146] In general, CdS and ZnS form a solid solution over the the entire composition range and the alloy films exist in a single-phase wurtzite structure up to 60% ZnS concentration, irrespective of the deposition technique. Beyond 80% ZnS, the films exhibit a cubic structure. Between 60 and 80% ZnS concentration the films exist in both wurtzite and sphalerite structures. Evaporated films have a wurtzite structure up to 60% Zn concentration with the *c*-axis normal to the substrate.

Vankar et al.[111] have observed that the crystallographic structure and the lattice parameter of evaporated ZnCdS films exhibit a strong dependence on the temperature of deposition. The lattice parameter *a* (Figure 6.20a) varies continuously with composition. Kane et al.[128] have reported

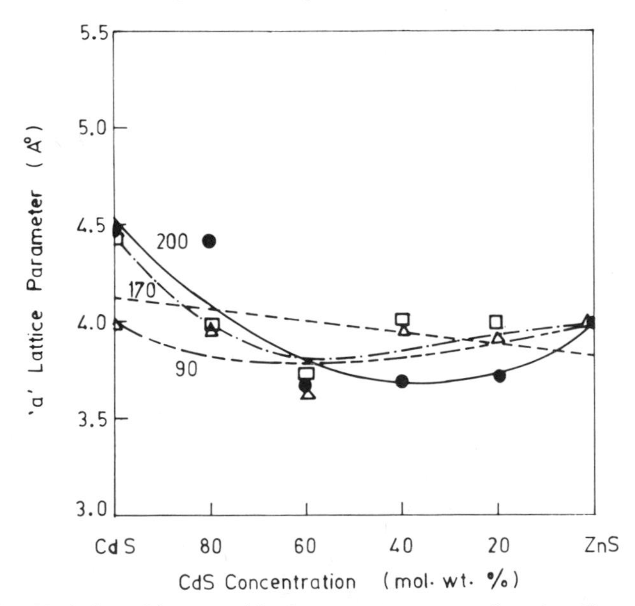

Figure 6.20a. Variation of hexagonal lattice parameter *a* as a function of composition for $Zn_xCd_{1-x}S$ films deposited at substrate temperature of 200 °C (———), 170 °C (—— · ——), and 90 °C (———–––———). The dashed line (– – –) shows the bulk lattice parameter values deduced using Vegard's law (after Vankar et al.[111]).

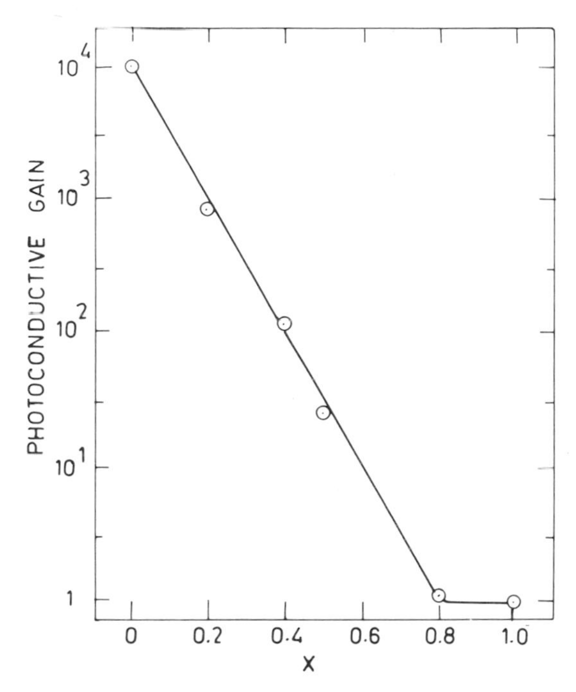

Figure 6.20b. Photoconductive gain as a function of x for spray deposited $Zn_xCd_{1-x}S$ films (after Banerjee et al.[139]).

that at compositions where both wurtzite and cubic structures are present, the basal plane spacing is the same for the hexagonal (002) and the cubic (111). Equivalent hexagonal lattice parameters can, therefore, be defined for the cubic form, and a and c calculated at all compositions. The variation in lattice parameters of the alloy films as a function of substrate temperature is qualitatively explained on the basis of nonstoichiometry arising from the presence of excess metal atoms.

Cadene et al.[171] have reported that crystallite size and orientation show a weak dependence on the nature of the substrate and ZnS concentration. A slight misorientation of the columnar crystallites appears and increases with an increase in ZnS concentration. Hall et al.[115] have reported a grain size of about 2 μm for evaporated ZnCdS films. Burton et al.[112] have observed that alloy films deposited from a single source have a spatial variation of composition along the thickness. Banerjee et al.[136,139] have reported that the lattice parameters of spray deposited $Zn_xCd_{1-x}S$ films vary continuously with composition. The spray deposited alloy films exist in a single phase (hexagonal or cubic, depending upon composition). In contrast to the evaporated films, the crystallographic structure of the spray deposited films does not depend on the deposition temperature.

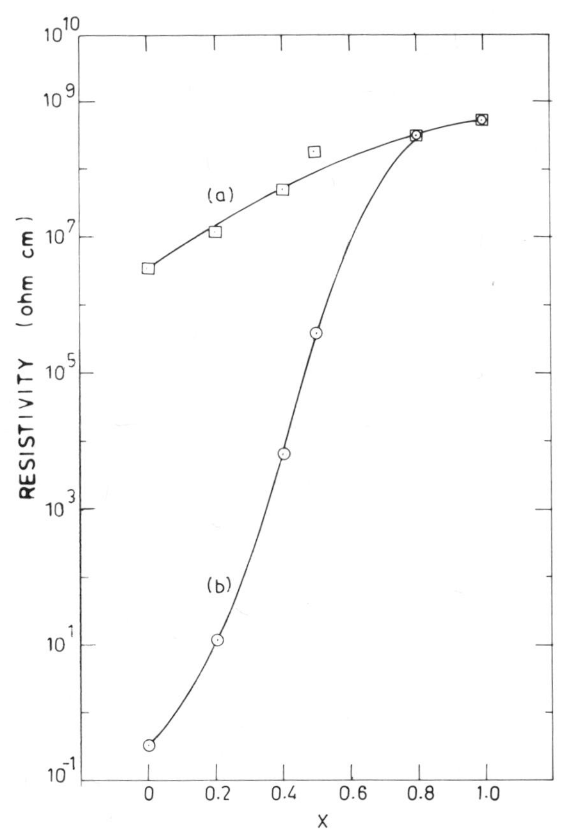

Figure 6.20c. Resistivity of (a) as-sprayed and (b) annealed $Zn_xCd_{1-x}S$ films as a function of x (after Banerjee et al.[139]).

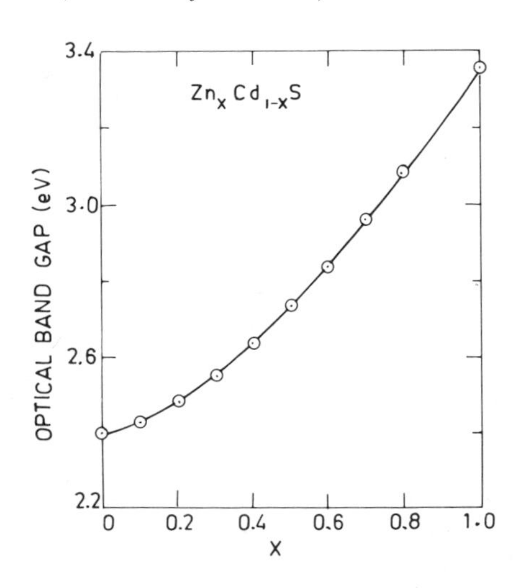

Figure 6.20d. Optical gap of $Zn_xCd_{1-x}S$ films as a function of x (after Das[116]).

Below 60% Zn concentration, the films are hexagonal and beyond 80% Zn concentration, the films exist in a cubic structure. For $0.6 \leq x \leq 0.8$, both hexagonal and cubic modifications are present. Singh and Jordan[137] have reported that 5- to 6-μm-thick spray deposited alloy films exhibit a strong preferential orientation along the c-axis and typical grain sizes are in the range from 1 to 2 μm. The average grain size changes slightly with Zn content. At x greater than 0.70, the films show signs of surface cracking.

In general, the electrical resistivity of the alloy films increases with increase in Zn content by several orders of magnitude. At low concentrations, however, low resistivity (1–20 Ω cm) alloy films can be obtained by evaporation.[115,116] Das[116] has observed that both the carrier concentration and the mobility decrease with increase in Zn content. Banerjee et al.[139] have observed that the photoconductive gain of the as-sprayed alloy films decreases from 10^4 for pure CdS films to 1 for pure ZnS films, as shown in Figure 6.20b. The dark resistivity of the as-sprayed films increases with Zn content. On annealing, the resistivity of the alloy films decreases, the decrease due to annealing being maximum for pure CdS films and negligible for films with $x \geq 0.8$, as illustrated in Figure 6.20c.

The optical properties of the alloy films vary smoothly with composition. The optical bandgap of these films is direct at all compositions[116] and varies sublinearly with composition from that of pure CdS to that of pure ZnS, as shown in Figure 6.20d. This increase in bandgap results in the increased V_{oc} measured on $Cu_2S/Zn_xCd_{1-x}S$ solar cells.

6.2.8. *Copper Selenide (Cu_2Se)*

It is only in recent years that Cu_2Se thin films have excited some interest for solar cell applications, and the literature on this material is, therefore, very sparse. Thin films of Cu_2Se have been prepared mainly by evaporation.[172,173,174] Cu_2Se exists in several phases and the properties of the films depend very strongly on the stoichiometry, which is governed, in turn, by the deposition conditions.

6.2.8.1. *Structural Properties*

Shafizade et al.[173] have observed that during heat treatment in vacuum at temperatures below 350 °C, CuSe undergoes consecutive phase transformations according to

$$\underset{}{\mathrm{CuSe}} \xrightarrow{100\text{–}300\,^\circ\mathrm{C}} \underset{\substack{\text{fcc}\\ a = 5.68\text{–}5.75\,\text{Å}}}{\mathrm{Cu_{2-x}Se}} \xrightarrow{300\,^\circ\mathrm{C}} \underset{\substack{\text{fcc}\\ a = 5.84\,\text{Å}}}{\mathrm{Cu_2Se}} \xrightarrow[\text{room temperature}]{\text{Cool to}} \underset{\substack{\text{low-temperature}\\ \text{modification}}}{\mathrm{Cu_2Se}}$$

The lattice parameter a of the fcc $Cu_{2-x}Se$ varies in the range 5.68 to 5.75 Å, corresponding to the composition range from $Cu_{1.4}Se$ to $Cu_{1.75}Se$. An ordered low-temperature phase of Cu_2Se exists in the range from $Cu_{1.75}Se$ to Cu_2Se. At temperatures above 350 °C, CuSe decomposes into two fcc phases with $a = 5.84$ and 5.65 Å. On cooling to room temperature, two phases are observed to coexist, a low-temperature modification of Cu_2Se and an independent fcc phase with $a = 5.63$ Å, corresponding to an approximate composition of $Cu_{1.15}Se$. Epitaxial films[(172)] of the low-temperature modification of Cu_2Se deposited onto NaCl single crystal substrates at 200 and 400 °C have been indexed on the basis of a hexagonal lattice with unit cell parameters $a = 7.07$ and $c = 6.68$ Å. Buldhaupt et al.[(174)] have reported single-phase fcc $Cu_{2-x}Se$ thin films in the substrate temperature range of 150 to 275 °C. At higher substrate temperatures, the film composition is observed to be less dependent on the Se deposition rate (provided that this rate is above a threshold value). The grain size of the films has been measured to be of the order of 1 μm.

6.2.8.2. *Electrical Properties*

Hall and resistivity measurements indicate that $Cu_{2-x}Se$ is p-type with a hole mobility of 10 $cm^2\,V^{-1}\,s^{-1}$ and a hole concentration of 2×10^{21} cm^{-3}, at a composition of $Cu_{1.8}Se$.[(174)]

6.2.8.3. *Optical Properties*

From the spectral dependence of the absorbance,[(174)] $Cu_{2-x}Se$ films have been observed to exhibit an indirect bandgap of 1.4 eV and a direct bandgap of 2.2 eV.

6.2.9. *Copper Sulfide (Cu_2S)*

Between the Cu-rich Cu_2S (chalcocite) and the S-rich CuS (covellite), Cu_xS in the bulk form exists in several other phases with either different crystallographic structures or different stoichiometry, or both. The phase diagram[(175)] of Cu_xS is quite complex as shown in Figure 6.21. In Table 6.1, we present the crystal structure and the lattice constants of the principal phases of Cu_xS.[(176)] It should be noted that in the crystallographic structure of these different phases of the Cu–S system, the sublattice of the chalcogen element forms the rigid armature of the crystal. The sublattice of the Cu cations varies since the Cu cations are very mobile and can position themselves in a large number of equivalent sites.

The electrical and optical properties of Cu_xS are almost exclusively governed by the Cu vacancies and, hence, the stoichiometry, whatever the

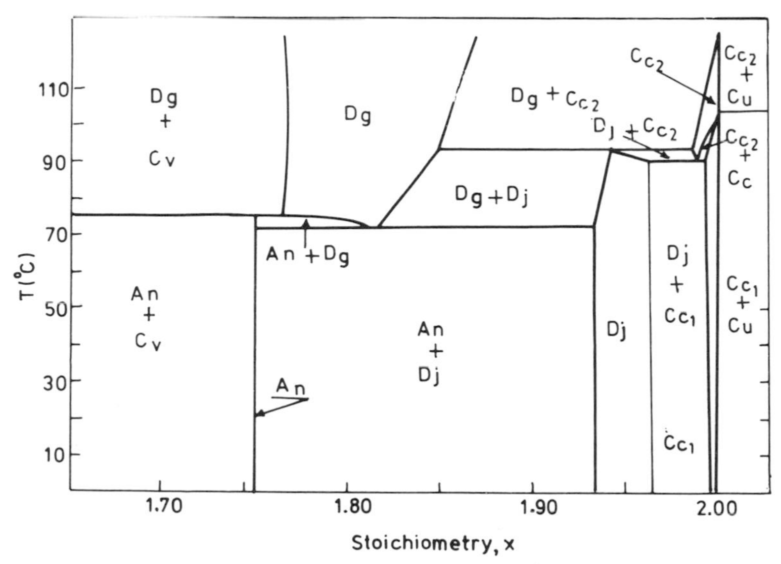

Figure 6.21. Phase diagram of Cu_xS for compositions involved in the Cu_xS/CdS solar cells. Dg — digenite, Cv — covellite, Dj — djurleite, Cc — chalcocite, An — anilite.

Table 6.1. Crystallographic Data for Different Phases of Cu_xS

	Hexagonal Cu_2S	Chalcocite orthorhombic	Chalcocite monoclinic	Djurleite orthorhombic	Hexagonal Cu_2S $1.96 > x > 1.80$	Hexagonal Cu_2S $Cu_{1.91}S$	Tetragonal $Cu_{1.96}S$	Digenite pseudo-cubic
a (Å)	3.961	11.848	15.246	15.71	15.475	11.355	4.008	5.56
b (Å)		27.330	11.884	13.56				
c (Å)	6.722	13.497	13.494	26.84	13.356	13.506	11.268	

crystallographic form. Cu_2S is always p-type. At high temperatures, in the range 60 to 300 °C, the temperature dependence of the conductivity, the Hall coefficient, and the thermoelectric coefficient exhibits two breakdowns as a result of phase changes. β-Cu_2S ($103.5\,°C < T < 435\,°C$) exhibits a large ionic conductivity, which is independent of the composition.[177,178] In the Cu vacancy density range of 3×10^{18} to 8.4×10^{19} cm^{-3} and temperature range of 150 to 400 °C, the ionic conductivity obeys the relation[178]

$$\sigma_{\text{ionic}} = (8.9 \times 10^{14}/T)\exp(-0.24\,e/kT)$$

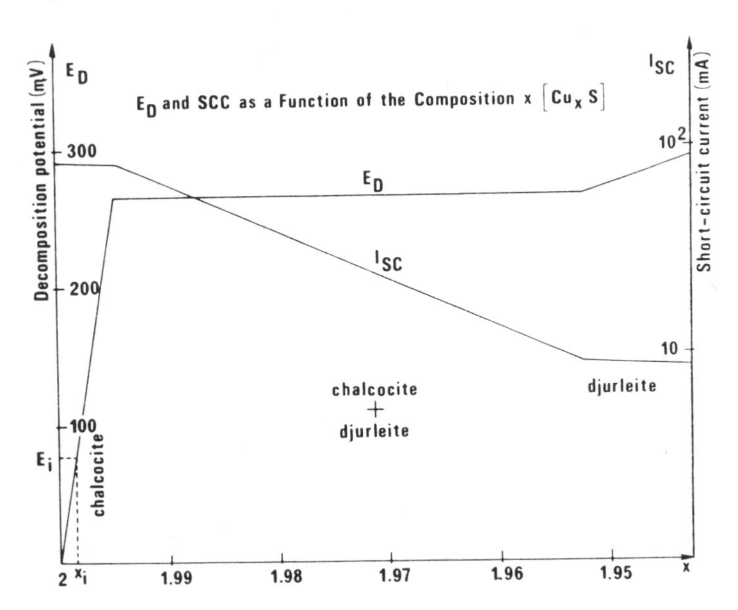

Figure 6.22. Electrodecomposition potential of Cu_xS as a function of stoichiometry. Also shown is the variation of short-circuit current of a Cu_xS/CdS solar cell (after Besson et al.[179]).

where e is the electronic charge, k is the Boltzmann constant, and T is the absolute temperature. The mobility of the cationic vacancies is observed to be inversely proportional to their density and independent of temperature.

The temperature dependence of chalcocite Cu_2S indicates a large number of carriers and imperfections. Thus, the stoichiometric composition Cu_2S is not really achieved. Indeed, the Cu_2S phase decomposes spontaneously to less stoichiometric phases owing to a very low electrodecomposition potential. The decomposition potential[179] vs. the stoichiometry is plotted in Figure 6.22. It is evident that to obtain a higher E_D value it is preferable to choose an initial x value of about 1.995. The mobility values are low (3 to 30 $cm^2 V^{-1} s^{-1}$). Large variations are noted in the carrier concentration values reported by different workers. Doping Cu_xS with In, Cd, or Zn[178,180] improves the stoichiometry and changes the electrical and optical properties to those of stoichiometric Cu_2S, as shown in Figure 6.23. The effective mass of holes has been determined to be about $1.7m_0$,[180] although both lower ($0.58m_0$) and higher ($3m_0$) values have also been reported in different studies.[175] The Fermi level, which is an extrinsic parameter, is observed to be situated in a region less than $3kT$ of the valence band edge because of the large number of free carriers. From the low-temperature variations of the conductivity and Hall coefficient, various workers have estimated activation energy values ranging from 0.007 to 0.6

eV.[175] The temperature dependence of the hole mobility has been studied by several workers to ascertain the scattering mechanism, and various mechanisms have been suggested in different temperature regimes.[175] The mobility is found to be linearly dependent on the hole concentration on a log–log scale, suggesting an impurity scattering process. The dependence of mobility on carrier concentration is depicted in Figure 6.24.[180]

Cu_2S in thin film form has been prepared by several deposition techniques. The most commonly employed process for Cu_2S/CdS solar cells has been the ion-exchange reaction in the wet (chemiplating)[181,182] and the dry (solid state reaction)[118,183] form. Other preparation methods include evaporation,[184,185] activated reactive evaporation,[186] sputtering,[144] electroplating,[187,188] and spray pyrolysis.[189] The properties of Cu_xS thin films are discussed below.

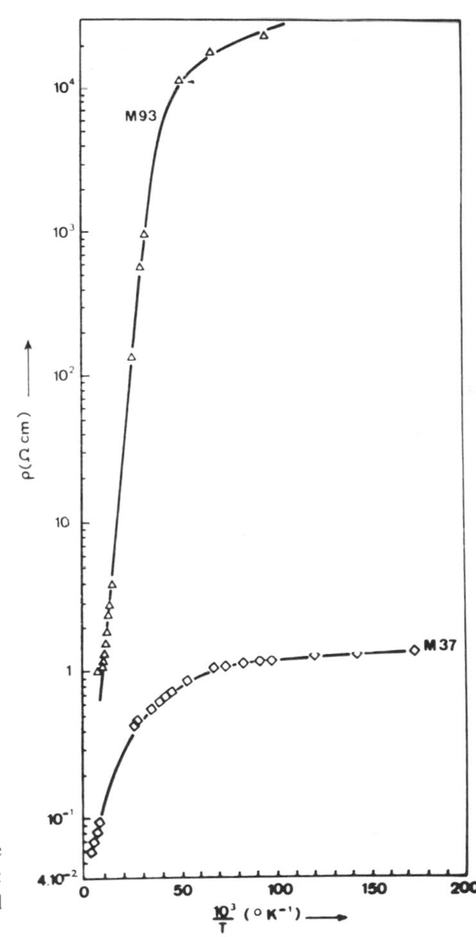

Figure 6.23. Temperature dependence of the resistivity of bulk Cu_2S samples: M-37 — undoped, M-93 — Cd doped (after Guastavino[180]).

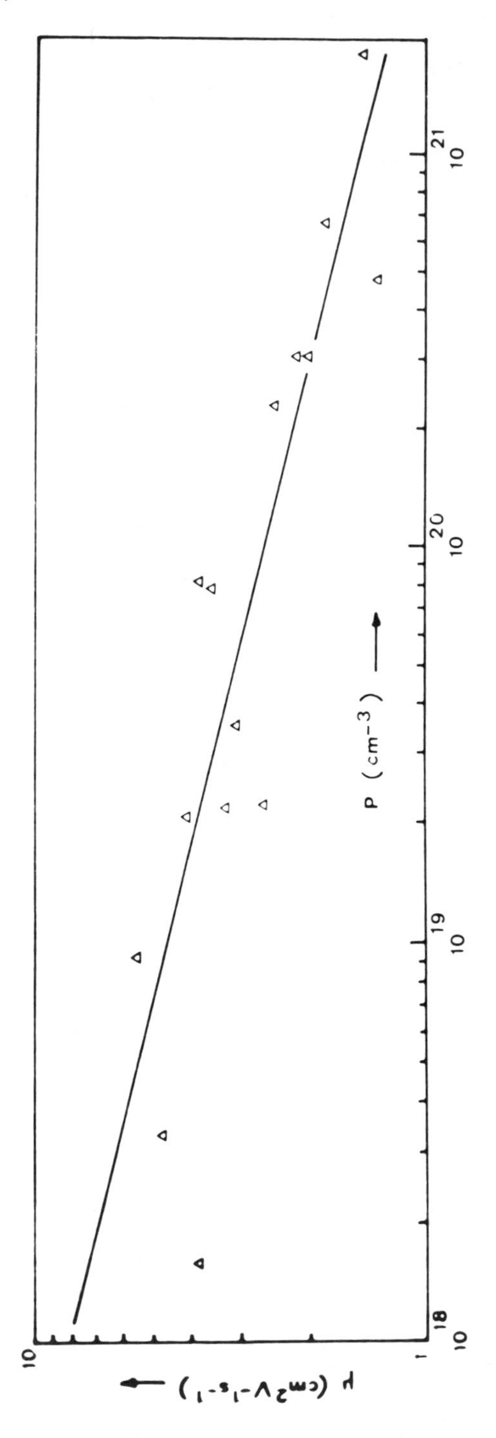

Figure 6.24. Variation of hole mobility as a function of hole concentration for bulk Cu_xS (after Guastavino[180]).

6.2.9.1. *Structural Properties*

The crystallographic structure of Cu_2S films is governed largely by the stoichiometry of the film. The microstructural features of Cu_2S films depend on the substrate, the mode of deposition, and the deposition conditions during growth. The chemiplated Cu_2S films, which are the result of a topotactial exchange reaction, possess a highly convoluted microstructure as shown in Figures 4.18 (p. 182) and 7.3 (p. 358). The surface topography of Cu_2S replicates that of the base CdS layer, in cases of both evaporated and sprayed CdS layers. Thus the grain size of the Cu_2S layer is the same as that of the base CdS layer. However, during the dipping process, rapid diffusion of Cu ions takes place through the grain boundaries, resulting in deep Cu_2S protrusions or fingers at the grain boundaries. The microstructure of the chemiplated Cu_2S layers is depicted schematically in Figure 7.4 (p. 359). Electroplated Cu_2S layers also exhibit deep penetration at the grain boundaries. In contrast to these techniques involving growth from liquids, the Cu_2S films grown by solid state reaction are planar. As in the chemiplated films, the dry reaction films also grow topotactially on the base CdS layer. A feature of the topotactial growth of Cu_2S by both chemiplating and dry reaction is that the preferred crystallographic orientations of the base CdS layer lead to strong orientation effects in the Cu_2S overlayer. Thus, Cu_2S films also exhibit a c-axis orientation perpendicular to the substrate.

6.2.9.2. *Electrical Properties*

As in the case of bulk Cu_2S, the electrical properties of thin Cu_xS films are governed primarily by the deviation from stoichiometry, which in turn is determined by the deposition process and conditions. In general, chemiplated layers exhibit resistivities in the range of 10^{-1} to 10^{-3} Ω cm, carrier concentrations in the range 10^{19} to 10^{21} cm^{-3}, and mobilities in the range of 1 to 5 cm^2 V^{-1} s^{-1}. Solid state reacted Cu_2S layers, possessing better stoichiometry, exhibit resistivities of 0.1 to 1 Ω cm, carrier concentrations of about 10^{19} cm^{-3}, and mobilities of 5 cm^2 V^{-1} s^{-1}. It should be noted that the Cu_2S films are p-type degenerate and the carrier concentration is determined by the number of Cu vacancies, i.e., stoichiometry. Couve et al.[185] have carried out a systematic study of the resistivity of evaporated Cu_xS layers as a function of composition. The composition of the films was found to vary with thickness and for $1.89 \leq x \leq 1.95$, the resistivity varied from 10^{-2} to 10^2 Ω cm. The temperature variation of resistivity is found to vary with composition and, hence, thickness, as illustrated in Figure 6.25. Annealing Cu_2S films in H_2 leads to higher resistivities and better stoichiometry while annealing in O_2 leads to the opposite effect. The minority carrier diffusion lengths in Cu_2S films prepared by different techniques lie in the range 0.05 to 0.5 μm. Dielman[190] has pointed out that the diffusion length of the minority carriers in the direction of the c-axis is larger than that perpendicular to the c-axis.

6.2.9.3. *Optical Properties*

The optical properties of Cu_2S are dominated by the stoichiometry. It is now generally agreed that the chalcocite phase has a direct bandgap at 1.2 eV and probably an indirect bandgap at about 1.8 eV. The optical bandgaps of the lower stoichiometry phases are higher. The transmission variations vs. wavelength for evaporated Cu_xS layers of different thicknesses and, consequently, different stoichiometries, are shown in Figure 6.26 and the corresponding α^2 vs. $h\nu$ plots are shown in Figure 6.27.

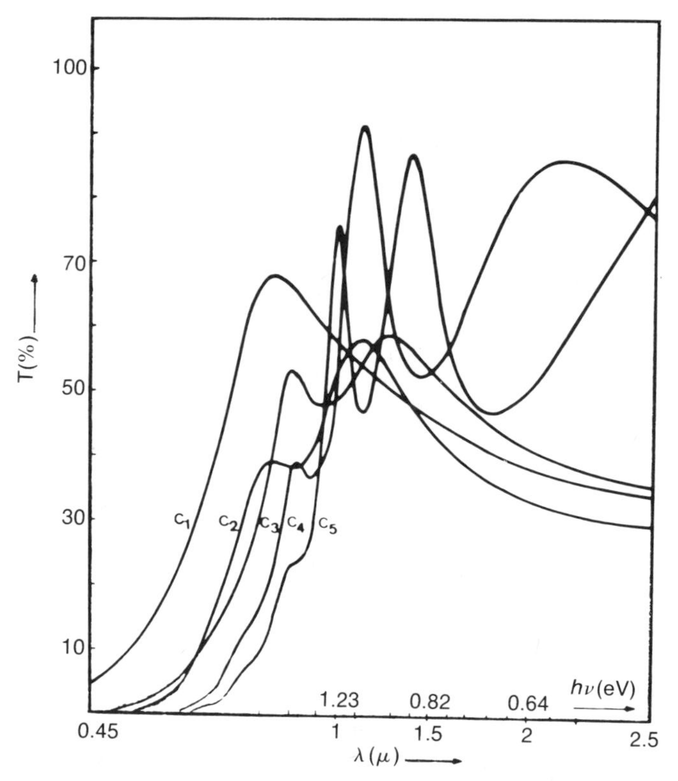

Figure 6.25. Resistivity variation vs. temperature for evaporated Cu_xS films with different thicknesses: (1) — 0.17 μm, (2) — 0.25 μm, (3) — 0.36 μm, and (4) — 0.45 μm (after Couve et al.[185]).

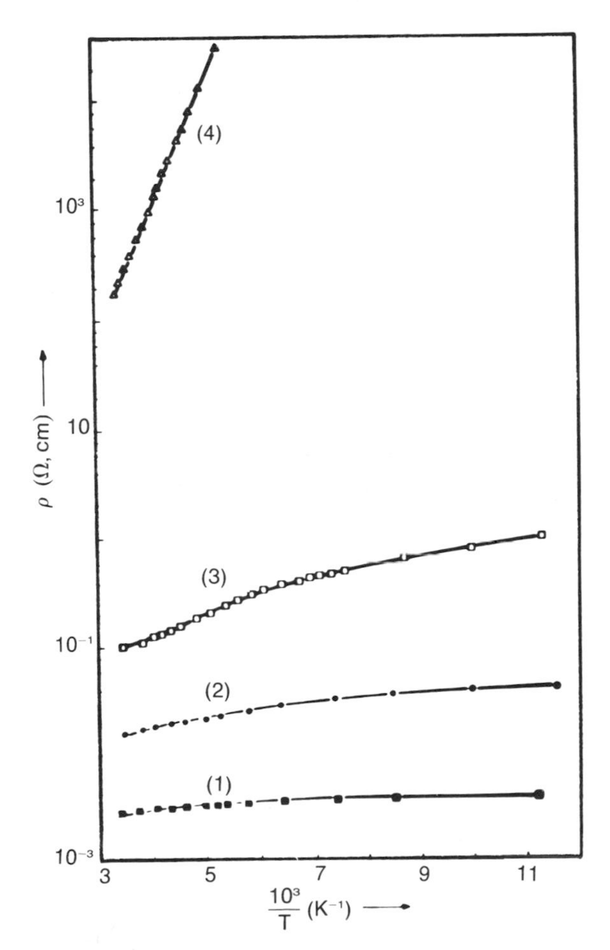

Figure 6.26. Transmission vs. wavelength for evaporated Cu_xS layers of different thicknesses: C_1 — 0.13 μm, C_2 — 0.25 μm, C_3 — 0.33 μm, C_4 — 0.45 μm, and C_5 — 0.6 μm (after Couve et al.[185]).

6.2.10. *Copper Indium Selenide ($CuInSe_2$)*

With a bandgap of 1.04 eV in bulk $CuInSe_2$, the $CuInSe_2$/CdS system has nearly optimum efficiency as a heterojunction solar cell[191] and has already shown promise as a thin film solar cell candidate. The lattice mismatch between the (0001) surface of CdS and the (112) surface of $CuInSe_2$ is 1.16%. Thin films of $CuInSe_2$ have been prepared by a variety of techniques including evaporation,[192,193,195,196] rf sputtering,[194] flash evaporation,[197,198] spray pyrolysis,[199] and MBE.[200]

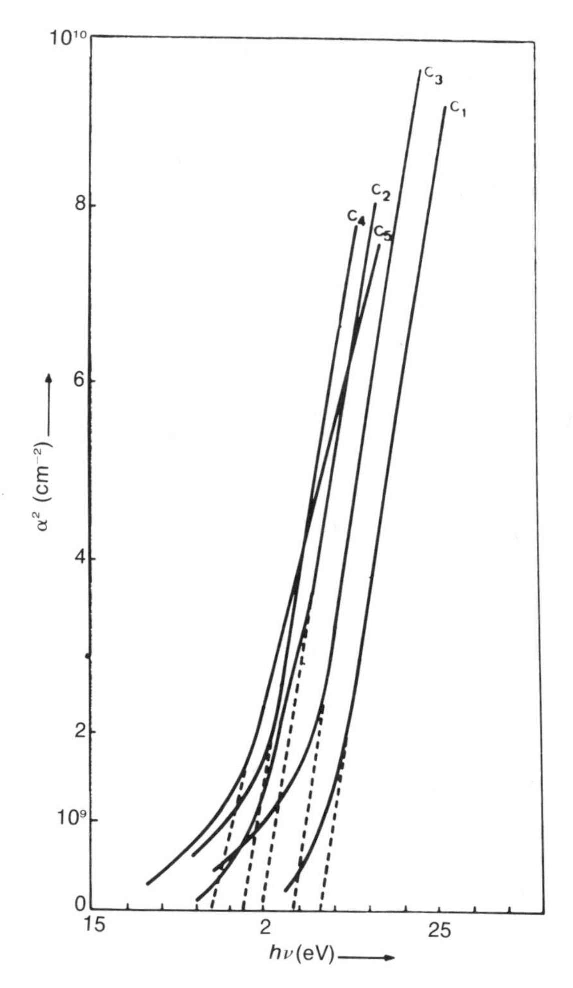

Figure 6.27. Plots of α^2 vs. $h\nu$ for the evaporated Cu_xS films of Figure 6.26 (after Couve et al.[185]).

6.2.10.1. *Structural Properties*

$CuInSe_2$ films prepared by evaporation from a single source onto substrates held at 500 K are single phase with preferred *c*-axis orientation. The grain size increases with increase in substrate temperature and at 500 K the as-deposited films exhibit a grain size of about 0.5 μm. Grain sizes greater than or equal to 1 μm have been obtained on postannealing. At lower substrate temperatures (350 K), Cu and In are also present. Durny et al.[197] have reported amorphous films of $CuInSe_2$ produced by flash evaporation onto glass substrates held at 77 K. Horig et al.[198] have obtained polycrystalline films with a grain size of about 0.1 μm by flash evaporation onto substrates held at 450 °C.

For sputtered films,[194] the particle size of the target material critically controls the quality of the films. Films sputtered from fine powder targets are Se-deficient and In-rich and consist of several phases, while films sputtered from coarse powder targets are stoichiometric and exhibit X-ray diffraction lines characteristic of the sphalerite or chalcopyrite phases. The substrate temperature is observed to influence crystal structure and grain size, but not the composition. At low substrate temperatures (20 °C), films sputtered from a coarse target are amorphous. At substrate temperatures between 50 and 300 °C, the sphalerite structure is observed. Between 450 and 505 °C, films exhibit a chalcopyrite structure with preferential orientation. Grain sizes in the high-temperature deposited films are about 1 μm.

Good-quality spray deposited films are obtained in the substrate temperature range of 250 to 450 °C. A purely sphalerite structure is obtained at 350 °C. Below 250 °C, the peaks decrease in size, while above 400 °C extra lines appear in the diffraction pattern.

6.2.10.2. *Electrical Properties*

Carrier transport in $CuInSe_2$ thin films is dominated by grain boundary scattering while carrier type and concentration are determined generally by deviation from stoichiometry. Excess Se in the films leads to p-type conductivity and the conductivity increases with increase in Se excess.[193,195] Se-deficient films are n-type. However, rf sputtered films[194] with Se deficiency have exhibited p-type conductivity. Kazmerski et al.[196] have reported that p-type films cannot be directly deposited by single-source evaporation. However, postdeposition annealing at low temperatures in an H_2Se atmosphere can yield p-type films. The dependence of carrier concentration and resistivity on excess Se content in the source for evaporated films[193] are shown in Figure 6.28. The dependence of resistivity of the same films on substrate temperature is shown in Figure 6.29. When deposited at 500 K, n-type evaporated films with an as-deposited room temperature mobility of 1 to 20 $cm^2\,V^{-1}\,s^{-1}$ exhibit grain boundary scattering.[196] Films prepared at lower substrate temperatures (480 K) also exhibit impurity scattering. On annealing, the n-type films retain the conductivity type but the mobility increases owing to grain growth.

6.2.10.3. *Optical Properties*

Flash evaporated, polycrystalline $CuInSe_2$ films exhibit a direct optical gap at 1.02 ± 0.01 eV,[198] and rf sputtered films have been reported to

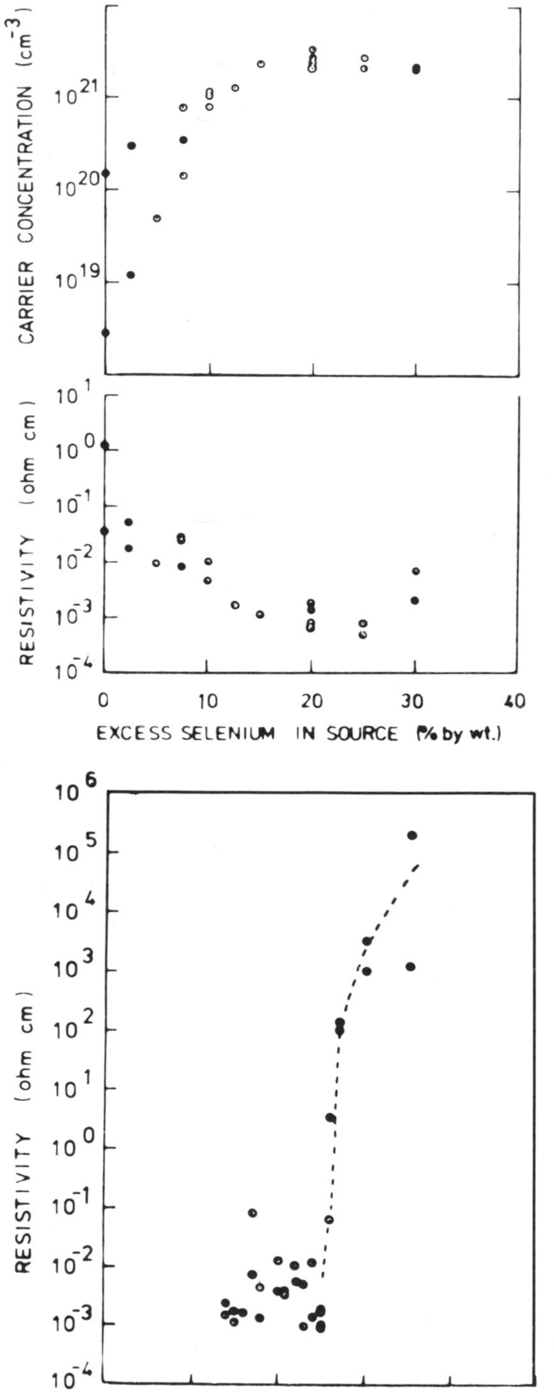

Figure 6.28. Carrier concentration and resistivity as functions of excess elemental Se (wt.%) for evaporated $CuInSe_2$ films (● — n-type and ○ — p-type films) deposited at 250 °C (after Fray and Lloyd[193]).

Figure 6.29. Resistivity of evaporated $CuInSe_2$ films as a function of substrate temperature. The source material has 20% excess Se (after Fray and Lloyd[193]).

possess bandgaps in the range of 0.86 to 1.00 eV.[194] In spray deposited films,[199] sharp absorption edges have been observed at about 0.95 eV in films deposited at substrate temperatures between 300 to 400 °C. Films deposited at 200 °C exhibit tailing of the absorption edge while films prepared at 400 °C exhibit an almost flat absorption curve.

In Table 6.2 we have compared the properties of semiconductor films deposited by different techniques.

6.3. *Transparent Conducting Oxides*

Transparent and conducting oxide thin films have figured prominently for many years in a wide variety of applications, such as heating elements on aircraft windows for deicing and defogging, antistatic coatings on instrument panels, and electrical contacts in liquid crystal, electrochromic, and electroluminescent displays. The high reflection in the infrared region, in conjunction with high transparency in the visible region, has been exploited to make heat reflecting mirrors. The renewed research interest in these films in recent years stems from the possibility of fabricating large-area, stable, and high-efficiency solar cells utilizing these films.

The desired properties of a transparent conducting film are a high bandgap, above about 3.0 eV, and a low effective mass and high carrier mobility. The most commonly used transparent conducting oxide films are those of SnO_2, In_2O_3, and Cd_2SnO_4. Recently, ZnO films possessing transparent electrode properties have been prepared. To compare the performance of different oxides, Fraser et al.[201] have defined a quantity, called the figure of merit, which relates the optical transmission and sheet resistance. A modified definition, more appropriate for solar cells, has been given by Haacke[202] and is expressed as

$$\phi_{TC} = T^{10}/R_{\square} = \sigma t \exp(-\alpha t)$$

where T is the transmittance, $R_{\square}$ is the sheet resistance, σ is the conductivity, α is the absorption coefficient, and t is the thickness of the coating. For a given sheet resistance, the figure of merit ϕ_{TC} can be related to basic material parameters by the expression

$$\phi_{TC} \sim \exp(-\text{constant} \cdot \alpha/\sigma)$$

A maximum value for ϕ_{TC} can be obtained only when α/σ is a minimum. α/σ is given by

$$\alpha/\sigma = e/\pi c n \nu^2 \mu^2 m^*$$

Table 6.2. Typical Characteristics of Several Semiconductor Films Employed for Fabricating Thin Film Solar Cells[a]

Material	Deposition technique	T_s (°C)	Crystallinity	Grian size (μm)	μ (cm^2 V^{-1} s^{-1})	n, p (cm^{-3})	ρ (Ω cm)	Remarks	Ref.
Si	CVD	600–750	P, oriented ⟨110⟩→⟨100⟩ →⟨111⟩	0.03			10^6	Orientation changes with temperature; conductivity, type dependent on doping	3, 7, 12, 13
		> 900	P	1–5				Epitaxial on recrystallized mg-Si at T_s = 1150 °C	10, 14
	Evap	⩾ 500	P, ⟨111⟩ directed ⊥ to substrate	0.5–5	0.3	10^{16}	10^3	Fibrous columnar grains	17–20
	Sputter	800–900	P		16	10^{20}	3.10^{-3}	Amorphous for T_s < 800 °C	21–24
		> 900	P, ⟨111⟩ oriented		256	10^{17}–10^{18}	~ 10^{-1}		
	SOC		P, ⟨110⟩ texture	~ 1000				On mullite	25–27
	TESS		P	1–5				On mullite; columnar grains	28, 29
	Elect.dep		P	~ 100				On Ag, Ta, Mo, or graphite	30
	EHD	Low temp.	P	~ 2				Grain size ~ 30 μm for high T_s	31
CdSe	Evap	200–500	P, (60% hexagonal, 40% cubic)			10^{20}	< 10^7	Hexagonal for T_s = room temperature; T_s (epitaxy) ~ 580 °C for CdSe (0001) ‖ Al_2O_3 (0001); ≃ 300–450 °C for CdSe/Ge	32–37
	Sputter	< 400	P, ⟨0001⟩ oriented				10^7–10^9	Epitaxial for T_s > 400 °C with (0001) CdSe ‖ (0001) alumina	38–40
	SP	200–300	P, hexagonal						41
	Sol. gr		P, cubic, (cubic + hexagonal)	0.05–0.2	1–10	5.10^{17}–5.10^{18}	10^8–10^9	High photoconductive gain	45, 46
CdTe	Evap	150–250	P, hexagonal					Cubic for T_s = room temperature	63, 64
	Sputter	~ 100	P, (cubic + hexagonal)				10^6–10^8	For two-source Cd and CdTe targets, hexagonal for T_s ~ 20 °C and cubic for	63, 65, 66

	CSVT		P	20–50	10^{-3}–1	5.10^{15}		Epitaxial on single crystal CdTe	63, 70–72
	Sc. ptg		P, cubic	~ 10			10^{-1}–1		73–74
	HWE	≃ 500	P		54–69	10^{17}			67
InP	Evap	≃ 225	P	> 1	≃ 25	2.10^{16}		Two-source	80
	CVD	< 550	P, randomly oriented	⩽ 1	~ 10	10^{15}–10^{17}	10–100	⟨110⟩ oriented for $T_s \sim 600\,°C$	77–79
	MOCVD	500–700	E		3100	> 10^{16}		On GaAs:Cr	76
	PRD	280	E	40				Poor epitaxy for $T_s \sim 380\,°C$	81
Zn_3P_2	Evap	~ 200	P, oriented	1–2	10–40		~ 10^5	Strongly oriented for $T_s > 220\,°C$	82, 83
	CVD	600			10–12		200–400		68
GaAs	CVD	≃ 700	P, oriented	5–10		5.10^{16}–10^{17}		Orientation nature depends on Arsine conc.	90–94
	MOCVD		P, sphalerite	200–500		6.10^{16}			86–89
CdS	Evap	200–250	P, oriented, c-axis ⊥ to substrate, wurtzite, columnar grains	1–5	0.1–10	10^{15}–10^{18}		Sphalerite for room temperature < T_s < 150 °C	107–130
	SP	300–350	P, oriented, wurtzite	0.1–1.0	1–5 (annealed)	up to 10^{18}	10^{-5}–10^{7} 10^{-1}–1 (annealed)	Sphalerite structure under certain conditions; photoconductive gain ~ 10^4–10^6; 4 μm grains by recrystallization	109, 131–143
	Sputter	300–500	P, oriented, columnar grains	⩽ 1			10^8		148
	Sol. gr		P, sphalerite, wurtzite or mixed	~ 0.1			10^7–10^9	Photoconductive gain ~ 10^4–10^6	154, 156

[a] CVD — chemical vapor deposition, Evap — vacuum evaporation, Sputter — sputtering, SOC — silicon on ceramic, TESS — Thermal evaporation shear stress, Elect.dep — electrodeposition, EHD — electrohydrodynamic process, SP — spray pyrolysis, VT — vapor transport, CSVT — closed-space vapor transport, Sc. ptg — screen printing, HWE — hot wall epitaxy, MOCVD — metallo-organic chemical vapor deposition, PRD — planar reactive deposition, Sol. gr — solution growth, Chem — chemiplating, Dry — solid state reaction, MBE — molecular beam epitaxy, P — polycrystalline, E — epitaxial, A — amorphous, T_s — substrate temperature, T_{so} — source temperature.

Table 6.2 (continued)

Material	Deposition technique	T_s (°C)	Crystallinity	Grian size (μm)	μ ($cm^2 V^{-1} s^{-1}$)	n, p (cm^{-3})	ρ (Ω cm)	Remarks	Ref.
	Sc. ptg		P	10			10^{-2}		73–74
	CSVT		E		241		10^{-3}–1	Electrical properties depend on T_s	161
	CVD		E		100–150		10–100		163
$Cu_{1.8}Se$	Evap	150–275	P, fcc	~1	10	2.10^{21}			174
Cu_2S	Chem	> 90 (solution temperature)	P, oriented, c-axis ⊥ to substrate, nonplanar	1–3	1–5	10^{19}–10^{20}	10^{-1}–10^{-2}	Grain size, orientation dependent on CdS grain size and orientation; variable stoichiometry	181, 182
	Dry	200 (reaction temperature)	P, oriented, c-axis ⊥ to substrate, planar	1–3	1–5	10^{19}	10^{-1}	Good stoichiometry	118, 183
	Evap	100–120	P				200–400	Stoichiometry factor is 2	184, 185
	Sputter	150	P				10–100		144
	SP	≃ 150	P	0.1–0.2				Poor stoichiometry	189
$CuInSe_2$	Evap (one-source)	220–250	P, ⟨001⟩ oriented for film thickness > 1500 Å	≃ 0.5	≃ 10	10^{17}–10^{19}	10^{-2}–10^{2}	ρ dependent on T_s; ρ, μ, p, depend on the excess Se. in source	193, 196
	Evap (two-source)	300–340	P	1–2		3–4.10^{16}	~ 10^2	$T_{so}(CuInSe_2) = 1150$ °C, $T_{so}(Se) = 200$ °C; ρ depends upon $T_{so}(Se)$	80, 195
	Flash evap	450	P	0.1				$T_{so} \simeq 1350$ °C	198
		200	A				2–8.10^{6}	$T_{so} \simeq 1300$ °C	197
	SP	~ 350	P, sphalerite				2	Structure depends on T_s	199
	Sputter		P, chalcopyrite	≃ 1	6		0.3–2		194
	MBE	300	$CuInSe_2(112) \parallel CdS(0001)$					$T_{so}(Cu) \simeq 1030$ °C, $T_{so}(In) \simeq 850$ °C, $T_{so}(Se) = 200$ °C	200

where $\mu = e\tau/m^*$ is the mobility of free electrons in the conduction band, m^* is the effective mass of free carriers, τ is the relaxation time, n is the refractive index, e is the electronic charge, and c is the velocity of light. It is thus clear that α/σ will be a minimum when μ is high and m^* is low, since μ is related to m^* by

$$\mu = (m^*)^{-x}$$

where $x \sim 1.35$ for many semiconductor materials.[203]

A variety of techniques have been employed to deposit transparent conducting oxides. These include dc,[201,204–209] rf,[210–215] and ion beam[207,216] sputtering, ion plating,[217] spray pyrolysis,[218–225] CVD,[226,227] electron beam,[228] flash,[229] reactive,[230–232] and activated reactive[233,234] evaporation, and anodization.[207] The various physical properties of interest for solar cell applications are discussed briefly in the following sections.

6.3.1. *Cadmium Oxide (CdO)*

6.3.1.1. *Structural Properties*

CdO films have an NaCl-type structure with a preferred ⟨100⟩ orientation normal to the substrate. The lattice parameter increases with an increase in carrier concentration, the change being as high as 0.88%.[235]

6.3.1.2. *Electrical Properties*

Carrier concentration and mobility in the range of 5×10^{16} to 10^{21} cm^{-3} and 2 to 120 $cm^2 V^{-1} s^{-1}$, respectively, have been obtained in CdO films.[235] The effective mass of electrons near the bottom of the conduction band has been estimated to be about 0.14[236] and is found to increase at higher carrier concentrations.[237] This has been explained on the basis of non-parabolicity of the conduction band and an increase in lattice parameter.[237] Resistivity as low as 5×10^{-4} Ωcm[238] has been obtained in these films.

6.3.1.3. *Optical Properties*

The absorption edge occurs in the visible region ($\sim$5000 Å). As the carrier concentration is high, plasma resonance occurs in the infrared region. Good agreement with Drude theory is obtained near the plasma edge.[237] The absorption edge shifts from 2.3 to 2.7 eV with increasing carrier concentration. The change in absorption edge has been attributed to a Moss–Burnstein shift.

6.3.2. Tin Oxide (SnO_2)

6.3.2.1. Structural Properties

SnO_x films have been the subjects of extensive studies. These films retain the bulk rutile structure,[221] and the crystallite orientation appears to be influenced by the deposition technique. Undoped and antimony doped SnO_2 films prepared by a spray pyrolysis technique exhibit ⟨200⟩ preferred orientation at lower substrate temperatures (623 to 723 K) and ⟨110⟩ orientation at higher substrate temperatures ($>$723 K). Fluorine doping leads to ⟨112⟩ and ⟨121⟩ preferred orientations at lower doping concentrations and ⟨121⟩ orientation at higher doping concentrations.[219,220] Films prepared by sputtering in O_2-rich ambient exhibit a ⟨110⟩ preferred orientation while films deposited at lower O_2 pressures exhibit a ⟨101⟩ preferred orientation.[206] An SnO phase has been detected in reactively sputtered SnO_2 films and its presence is found to be dependent upon several experimental parameters,[206] particularly the O_2 partial pressure during sputtering.[205]

6.3.2.2. Electrical Properties

The observed mobility of carriers in SnO_2 films is lower than the bulk values and is in the range of 10 to 50 $cm^2\,V^{-1}\,s^{-1}$ for carrier concentrations and resistivities varying from 10^{18} to 10^{21} cm^{-3} and 10^{-1} to 4×10^{-4} $\Omega\,cm$, respectively. Grain boundary scattering has been found to be the dominant scattering mechanism limiting the mobility.[219–221,223] The resistivity of F-doped SnO_2 films is lower than that of Sb-doped films.[219,220,239] The difference in the mobility variation with doping concentration in Sb- and F-doped films (Figure 6.30) is attributed to changes in the contribution from grain boundary scattering.[219,220] Annealing of these films in vacuum or H_2 atmosphere results in lower resistivity, while annealing in air leads to increased resistivity.[240] The annealing behavior has been shown to result from chemisorption or desorption of O_2, primarily from the grain boundaries.[240]

6.3.2.3. Optical Properties

A large Moss–Burstein shift, from 3.97 to 4.63 eV for a carrier concentration change from 10^{18} to 4.6×10^{20} cm^{-3}, has been observed in doped SnO_2 films. The plasma resonance in the infrared has been successfully interpreted in terms of the Drude theory. The effective mass, determined from plasma edge measurements, varies from 0.1 to $0.3m_e$ and is dependent upon the carrier concentration, indicating nonparabolicity of the conduction band.[219–221,223]

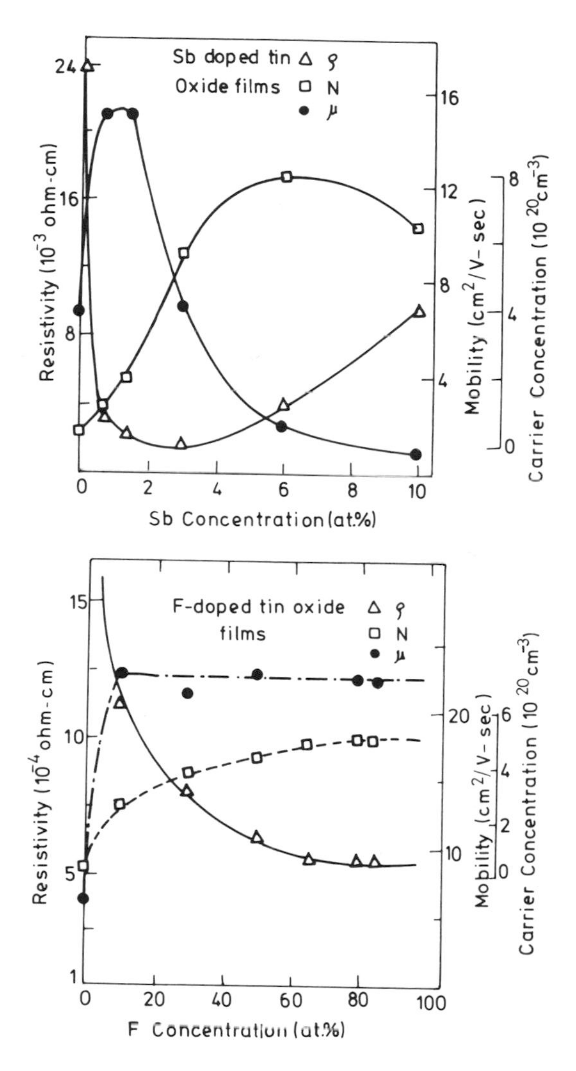

Figure 6.30. Variation in mobility, resistivity, and carrier concentration of SnO_x:Sb and SnO_x:F films with doping concentration (after Shanthi et al.[219,220]).

We should note in conclusion that SnO_2 films are stable, strongly adherent (when deposited on glass, quartz, etc.), and resistant to moisture and acids.

6.3.3. *Indium Oxide (In_2O_3)*

6.3.3.1. *Structural Properties*

In_2O_3 and ITO (indium tin oxide) films retain the bulk cubic bcc structure, exhibiting, in general, either ⟨111⟩ or ⟨100⟩ preferred orientations.[201,208,222,224] In dc sputtered films, the orientation is sensitively influenced by a number of deposition parameters, namely, substrate temperature, sputtering rate, and O_2 pressure.[213] There have been conflicting reports on the Sn-compound phase present in these films. Bosnell et al.[241] have detected Sn_3O_4 in sputtered films, while a SnO_2 phase has been observed in sprayed films for Sn concentrations exceeding 60 to 80%.[224] In contrast to these results, no Sn-compound phase could be detected by Manifacier et al.[222]

6.3.3.2. *Electrical Properties*

The resistivity, carrier concentration, and mobility of In_2O_3 films are typically in the range of 2×10^{-4} to 10^{-2} Ω cm, 10^{19} to 2×10^{21} cm^{-3}, and 15 to 70 cm^2 V^{-1} s^{-1}, respectively. For spray deposited films, high electrical conductivity and optical transmission values are achieved at an optimum Sn concentration of about 10 at%. Further increase in Sn concentration leads to lower conductivity[224] owing to the formation of $(SnO_2)_2$ clusters in In_2O_3 films.

The conduction mechanism in In_2O_3 films has not been investigated in great detail. Grain boundary scattering appears to be the dominant scattering mechanism in nondegenerate films prepared by reactive sputtering.[242] Several workers[208,214,243,244] have reported on the annealing behavior of the resistivity of sputtered films for different atmospheres. The changes in resistivity have been explained variously by grain growth,[208] reversible diffusion of O_2 from and into the film,[214] and grain boundaries.[243]

Cd doping of In_2O_3 films results in low resistivity ($\sim 3.3 \times 10^{-4}$ Ω cm), with carrier concentrations of about 3×10^{20} cm^{-3} and a mobility of about 6 cm^2 V^{-1} s^{-1}.[245] Very high mobilities are obtained on doping with Ti ($\mu \sim 120$ cm^2 V^{-1} s^{-1}, $n \sim 10^{20}$ cm^{-3}) and Zr ($\mu \sim 170$ cm^2 V^{-1} s^{-1}, $n \sim 8 \times 10^{19}$ cm^{-3}).

6.3.3.3. *Optical Properties*

In_2O_3 has a forbidden indirect bandgap at 2.6 eV[246] with the allowed direct transitions occurring at 3.6 to 3.85 eV.[214,242,244,246,247] The observed Moss–Burstein shift follows an $n^{2/3}$ dependence, indicating a parabolic nature of the conduction band.[214,224,247] The values of effective mass computed from a Burnstein shift vary from 0.25 to $0.45 m_e$[214,224,242,247] and the computed values of reduced mass are always higher than the electron effective mass, implying negative curvature in the valence band.[214,224,247] In_2O_3 films are readily etched by acids.

6.3.4. Cadmium Stannate (Cd_2SnO_4)

6.3.4.1. Structural Properties

Cd_2SnO_4 is the only conducting oxide in which the structure in the films (cubic spinel) is different from that of the bulk (orthorhombic). Haacke[210,249] has shown that rf sputtered films are generally multiphase, containing CdO, $CdSnO_3$, and Cd_2SnO_4 phases. Single-phase Cd_2SnO_4 films are obtained by annealing the as-deposited films in Ar/CdS atmosphere.

6.3.4.2. Electrical Properties

The carrier concentration and mobility in these films are in the range 10^{17} to 10^{21} cm^{-3} and 8 to 73 $cm^2 V^{-1} s^{-1}$, respectively.[210,212,248] The carrier concentration in annealed single-phase sputtered films is high ($>10^{20}$ cm^{-3}). Miyata et al.[212] have suitably controlled the deposition parameters during sputtering to obtain as-deposited films with a resistivity as low as 5×10^{-4} Ω cm.

6.3.4.3. Optical Properties

In amorphous sputtered films, the apparent bandgap shifts from 2.06 to 2.85 eV for a change in carrier concentration from 1.1×10^{17} to 1.2×10^{20} cm^{-3}. This bandgap change has been attributed to a Moss–Burstein shift.[248] Effective mass, calculated on the basis of optical measurements, is reported to be about $0.04m_e$ and is dependent on the carrier concentration. However, Haacke has questioned this interpretation, owing to the presence of multiple phases in these films.

6.3.5. Zinc Oxide (ZnO)

6.3.5.1. Structural Properties

Recently, high conductivity has been obtained in ZnO thin films.[218,223,250,251] Limited experimental investigations have been carried out on these films. A detailed analysis of the structural, electrical, and optical properties of spray deposited films has been made by Aranovich et al.[218] These films exhibit a wurtzite structure with the c-axis normal to the substrate. On doping with In, the ZnO films prepared by spray pyrolysis show random orientation of the 200- to 400-Å crystallites.[269] ZnO films prepared by MOCVD[270] are polycrystalline when deposited at a substrate temperature greater than 250 °C with columnar crystallites of about 100 Å. The crystallites are oriented with their c-axes perpendicular

Table 6.3. Comparison of the Performance of Various Transparent, Conducting Oxide Films Prepared by Different Techniques[a]

Films	Deposition technique	$R_\square$	Transmission (%) at different wavelengths (μm)										
			0.40	0.45	0.50	0.55	0.60	0.65	0.70	0.75	0.80	1.20	1.60
SnO_2	Sputter	500	0.73	0.93	0.94	0.85	0.82	0.85	0.92	0.92	—	—	—
SnO_2	SP	9.2	0.87	0.87	0.89	0.91	0.90	0.93	0.92	0.82	0.81	0.62	0.13
ITO	Sputter	5.5	0.66	0.87	0.92	0.93	0.92	0.90	0.94	0.9	0.9	—	—
ITO	SP	5.4	0.81	0.8	0.8	0.8	0.91	0.83	0.8	0.83	0.88	0.76	0.39
ITO	ARE	2.2	0.79	—	—	—	0.91	—	—	—	0.91	0.8	0.58
Cd_2SnO_4	Sputter	1.7	0.64	0.81	0.93	0.84	0.92	0.89	0.84	0.85	0.86	0.46	—
$CdSnO_3$	SP	17	0.56	0.78	0.85	0.80	0.88	0.80	0.89	0.89	0.85	—	—
ZnO	ARE	40	0.82	0.86	0.94	0.99	0.90	0.89	0.95	0.99	1.0	0.94	0.97

[a] Sputter — sputtering, SP — spray pyrolysis, ARE — activated reactive evaporation, $R_\square$ — sheet resistance (Ω/□

to the substrate and the crystallite size (50–200 Å) increases with increasing substrate temperature. Films of ZnO have also been deposited by rf reactive magnetron sputtering.[271]

6.3.5.2. *Electrical Properties*

The electrical properties of spray deposited films are strongly influenced by the substrate temperature and the rate of air flow. As-deposited films have a high resistivity of about 10^2 Ω cm, which decreases to about 10^{-3} Ω cm on annealing in vacuum[223] or H_2.[218] The change in resistivity is due mainly to the variation in mobility.[223] The temperature dependence of the thermoelectric power indicates predominantly ionized impurity scattering.[218] The photoconductive gain of 10^2 obtained in these films is less than the value obtained in bulk. The behavior of the transient change in conductivity on illumination is profoundly modified by the annealing history of the films prior to the onset of illumination, and is believed to be caused by the presence of chemisorbed oxygen on the surface and at the grain boundaries.

In recent developments, large-area ZnO films have been prepared by spray pyrolysis[269] with low resistivity ($\sim 8 \times 10^{-4}$ Ω cm), high electron concentration ($\sim 5 \times 10^{20}$ cm^{-3}), and an electron mobility of about 15 $cm^2 V^{-1} s^{-1}$. The mobility first decreases and then increases with carrier concentration. This dependence has been attributed to carrier trapping processes at the grain boundaries. MOCVD ZnO films,[270] deposited at substrate temperatures between 280 and 350 °C have a conductivity ranging from 10^{-2} to 50 $\Omega^{-1} cm^{-1}$. The carrier mobility is observed to be limited by both thermionic and thermal field emission at the grain boundaries. For the

Figure of Merit ($10^{-3}\ \Omega^{-1}$) at different wavelengths (μm)											
0.40	0.45	0.50	0.55	0.60	0.65	0.70	0.75	0.80	1.20	1.60	Ref.
0.09	0.97	1.1	0.39	0.27	0.39	0.7	0.87	—	—	—	215
16.9	27	33.9	42.3	37.9	52.6	47.2	25.5	13.2	0.91	—	219
2.9	45.2	79.0	88.0	79	63.1	97.9	63.4	63.4	—	—	214
22.5	19.9	19.9	19.9	72.1	28.7	19.9	28.7	51.6	11.9	0.02	225
43	—	—	—	177	—	—	—	177	54	2	233
6.8	71.5	284.7	102.9	255.55	183.4	102.9	115.8	130.2	0.25	—	252
0.18	4.9	11.6	6.3	16.5	6.3	18.3	18.3	11.6	—	—	253
3.4	5.5	13.5	22.6	8.7	7.8	15.0	22.6	25	13.5	18.4	234

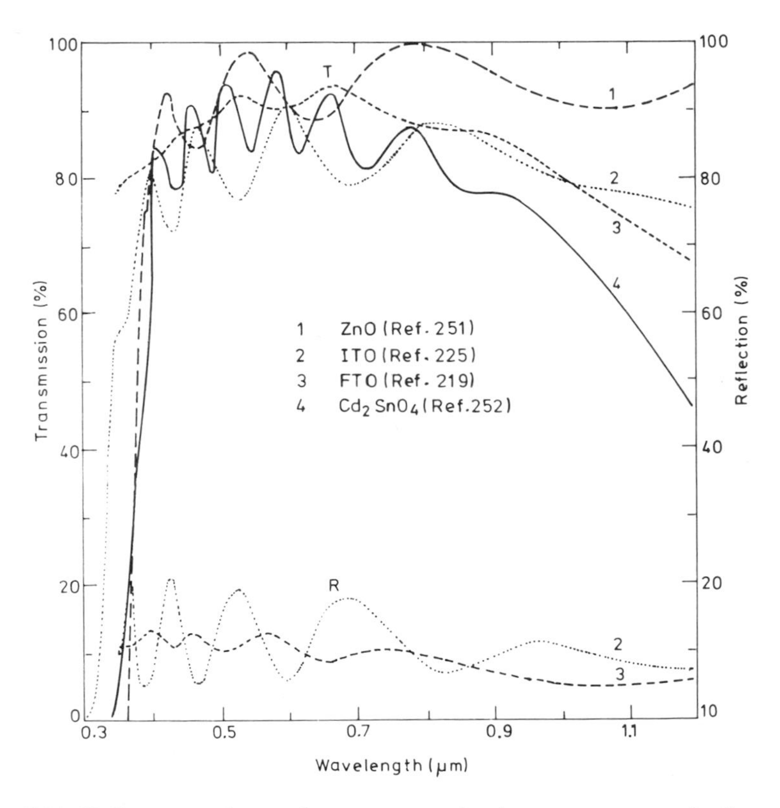

Figure 6.31. Reflectance and transmittance spectra of various transparent conducting oxide films prepared by different workers.

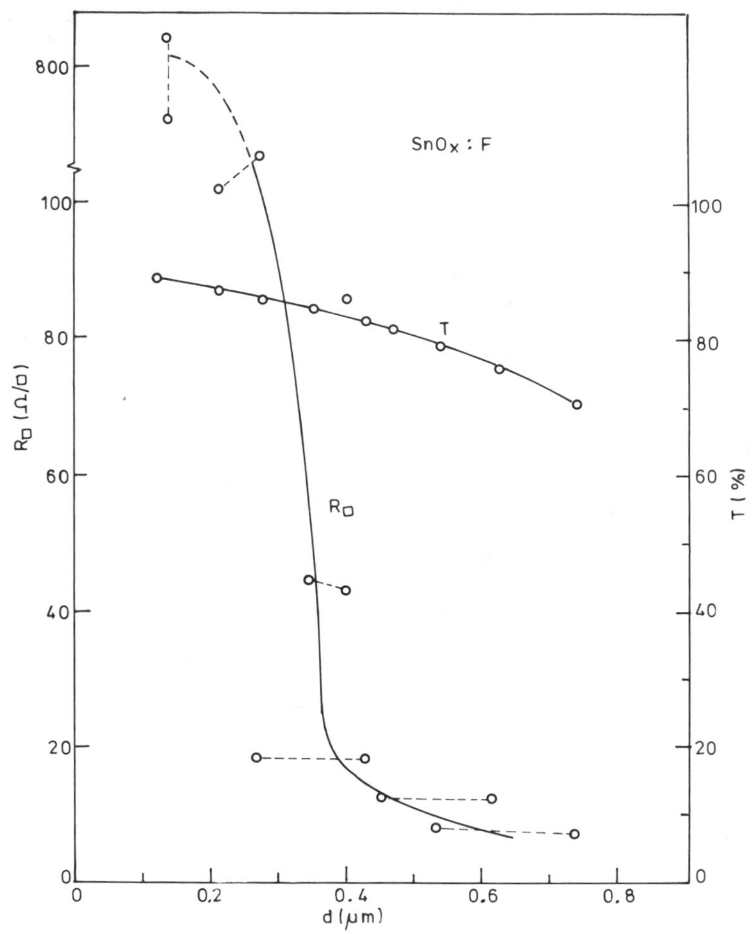

Figure 6.32. Sheet resistance $R_\square$ and transmittance T vs. thickness d for spray deposited SnO_x:F films.

rf reactive magnetron sputtering process, ZnO films can be prepared with conductivities ranging from about $10^{-8}\ \Omega^{-1}\,cm^{-1}$ to $5 \times 10^2\ \Omega^{-1}\,cm^{-1}$, depending on the sputter conditions.[271]

6.3.5.3. *Optical Properties*

ZnO films exhibit a direct optical bandgap at 3.3 eV,[218] similar to that observed in bulk. High optical transmission ($\sim 85\%$) in the solar spectrum has been obtained in spray deposited films with resistivities about $8 \times 10^{-4}\ \Omega cm$.[269] Films with sheet resistance of 85 Ω/□, deposited by magnetron sputtering,[271] exhibit a transmission of 90% in the spectral range from 4000 to 8000 Å.

In Table 6.3 we have compared the performance of several transparent, conducting oxide films prepared by different deposition techniques. In Figure 6.31, we have plotted the transmittance and reflectance spectra of various oxide films. In Figure 6.32 we have plotted the sheet resistance and transmittance data for SnO_x:F conducting oxide coatings as a function of film thickness.

6.4. *Transport Properties of Metal Films*

Metal films provide conducting and transparent barrier electrodes for Schottky barrier solar cells and grid structure and base electrodes for various types of solar cells. The relevant properties of metal films are reviewed briefly in a fairly general way in this section. For more thorough and extensive reviews of the subject, the reader is referred to standard texts[254,255] and reviews.[256–258]

6.4.1. *Electrical Properties*

Typical variations of the resistivity for a metal film as a function of thickness are shown in Figure 6.33. The variation has four regions: In

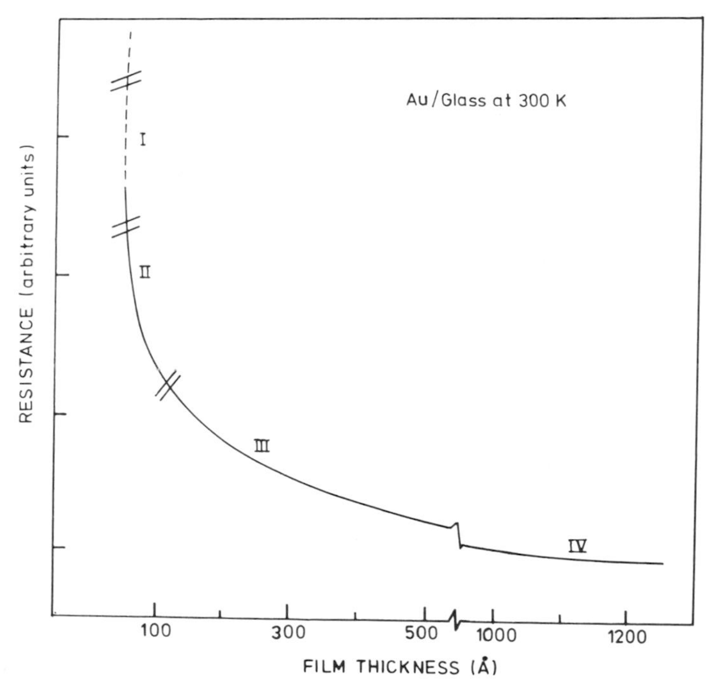

Figure 6.33. Typical variation of the resistance of a metal film with thickness. The various regions of thickness correspond to different conduction mechanisms.

Region I, the films are discontinuous and consist of discrete islands (granular). The conductivity is strongly dependent on the separation and size of the island and is thermally activated. Figure 6.34 shows the Arrhenius variation of the conductivity of ultrathin Pt films. The values of the activation energy are consistent with equation (2.202). The temperature coefficient of resistivity TCR is negative and its value is determined by the microstructure of the film.

In Region II, the film has a network structure that is electrically, but not physically, continuous. The resistivity and TCR here are determined by the volume fraction of the continuous regions and are the sum of the activated and nonactivated contributions.

The films in Region III are physically continuous and coherent, and the resistivity is the sum total of the various contributions, i.e., [equation (2.207)]:

$$\rho_F = \rho_B + \rho_S + \rho_{GB} + \rho_D + \rho_I$$

The surface scattering term is determined by the film thickness and the nature of the film surfaces. Epitaxially grown and large-grained, well-oriented films of metals deposited at high substrate temperatures exhibit nearly specular behavior for conduction electrons. In general, however, polycrystalline films prepared for most solar cell applications are expected to show "nearly" diffuse scattering behavior so that ρ_S is given by the size effect theories [equation (2.208)]. It should also be pointed out that the electron scattering property of a surface can be modified by the presence of

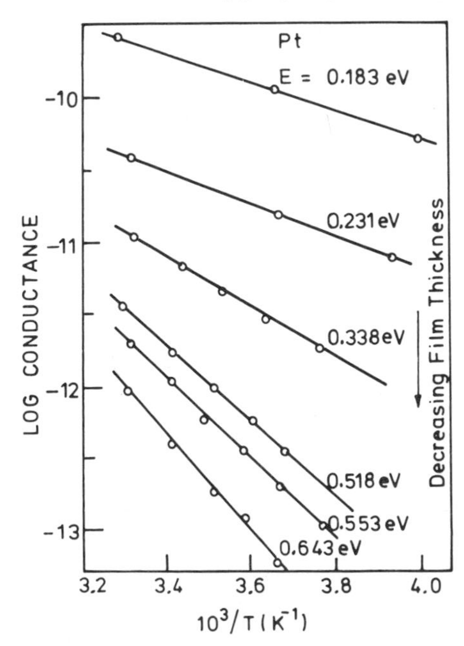

Figure 6.34. Arrhenius variation of conductance for discontinuous Pt films of various thicknesses. E is the calculated activation energy (from Chopra[254]).

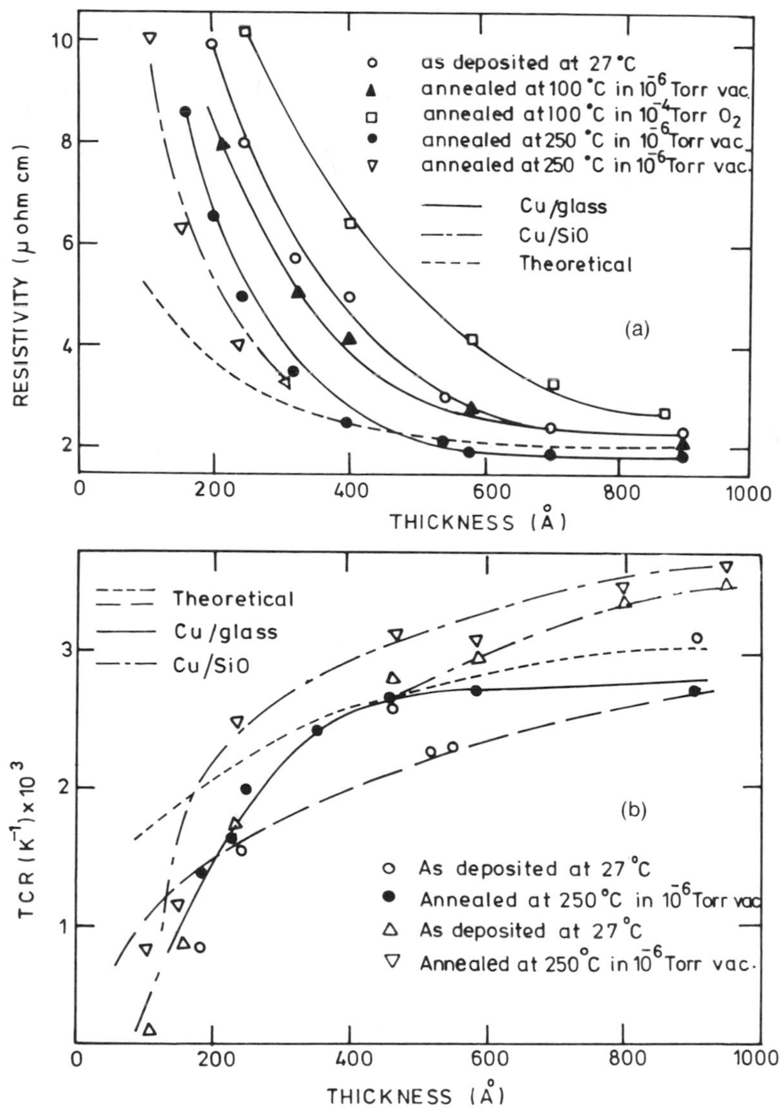

Figure 6.35. Thickness dependence of (a) the resistivity and (b) the TCR of Cu films deposited and annealed at different temperatures and deposited on different substrates (from Suri et al.[259]).

monolayers of another material, but the effect on resistivity is small ($\leqslant 20\%$). The ρ_{GB} and ρ_D contributions are sensitively dependent on the microstructure and hence on film thickness and deposition conditions as well as postdeposition treatment. This is best illustrated by the data (Figures 6.35a and b) on the thickness dependence of the resistivity and TCR of Cu films deposited and annealed at different temperatures.

Several important conclusions must be emphasized here: (i) Scattering of electrons from structural defects (primarily vacancies and vacancy clusters in the grains as well as the grain boundaries) provides the major contribution to resistivity at film thicknesses comparable to or greater than the mean free path l of conduction electrons. (ii) The geometrical scattering from grain boundaries is not significant because of their small thicknesses relative to the mean free path of conduction electrons in low-resistivity metals; however, this contribution becomes very significant if the grain boundaries are loaded with vacancies or their potentials are drastically modified by such processes as oxidation (in case of readily oxidizable metals, e.g., Cr, Ta, Ti, Nb). In the latter case of an active grain boundary, the resistivity increases rapidly and the TCR becomes negative. (iii) As expected, annealing out of vacancies is more effectively achieved by a higher deposition temperature rather than postdeposition annealing. With decreasing deposition temperature, the concentration of frozen-in structural defects of all varieties increases rapidly. The activation energy for diffusion of vacancies in noble metals being large (~ 1 eV), the high concentration of vacancies, if frozen-in, is maintained at or near room temperature, thereby making a substantial contribution to the resistivity. On increasing deposition (or annealing) temperature, migration and annihilation of vacancies result in a decrease in the resistivity. At higher temperatures, dislocations are removed followed by grain growth and recrystallization processes. These processes have only a marginal effect on resistivity and TCR, although a profound one on some other transport properties such as thermoelectric power.[259] (iv) Since different electron transport parameters depend on different aspects of electron dynamics within energy bands and, further, since size effects are only partly due to "geometrical" scattering at surfaces, the thickness dependence of different transport properties is not necessarily correlated in the simple fashion suggested by the free electron theory.

Finally, Region IV corresponds to a thick film ($t \gg l$) in which case the dominant contributions to resistivity are ρ_B and ρ_D. Any thickness dependence in this region will arise primarily because of the thickness dependence of the microstructure which, in turn, is dictated by the deposition process (see Chapter 1). Because of the higher value of ρ_F and the fact that only the ρ_B contribution is temperature-dependent, the TCR of films in Regions III and IV is invariably smaller than that of the bulk. The TCR of a film is modified by α_L (the linear expansion coefficient of the substrate) and can be changed considerably by the appropriate choice of α_L.

The resistivity of a metal film is increased by the addition of impurities, following Mathiessen's rule. If alloyed, intentionally or unintentionally (owing to interfacial diffusion), the resistivity of a disordered binary alloy film follows an $Ax(1-x)$ dependence (Mott–Jones), where A

is a constant and x is the concentration of one type of atom. If ordered alloys are formed at some compositions, the resistivity shows a minimum.

Now let us address ourselves to the question of what determines the position of the various regions in Figure 6.33. This is determined by the nucleation and growth processes of a particular film under given deposition conditions (Chapter 1), and, thus, these regions are not uniquely defined even for the same metal film deposited under different conditions. Generally speaking, the lower the deposition temperature the lower the ad-atom mobility, and the larger the nucleation centers, the smaller and narrower are Regions I–III. We may define a useful parameter called the critical thickness t_c (see Chopra[254]) at which a film becomes physically continuous. Figure 6.36 shows the dependence of t_c on deposition temperature, angle of oblique deposition, and electric field. The t_c can also be varied over a large range by modifying the nucleation process. As an example, the t_c of Cu films deposited on glass at 300 K can be decreased from 120 Å down to about 50 Å by coating glass with monolayers of SiO_x.

Among other electrical properties, the work function of metal films is of interest, and has been measured for a number of metal films by using the optical response of photoemission. Generally the work function of unannealed films is lower (by as much as 0.72 eV for Cu films) than the bulk value, and it increases on annealing out structural defects.

6.4.2. *Optical Properties*

Ultrathin discontinuous films exhibit nearly dielectric behavior with a high refractive index n and low extinction coefficient k. With increasing

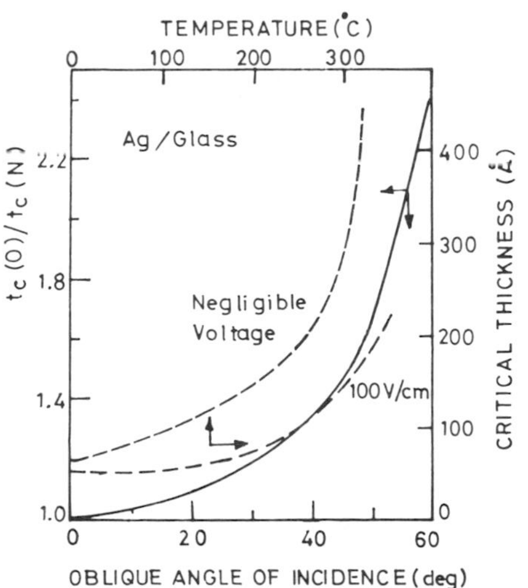

Figure 6.36. Dependence of the normalized critical thickness of Ag films, evaporated on glass at 25 °C and at a rate of about 1 Å s^{-1}, on the angle of incidence of the impinging vapor. O stands for oblique and N for normal incidence. The broken curves show the dependence of the critical thickness on the substrate temperature, with and without an applied electric field (after Chopra[254]).

thickness, n decreases and k increases as shown in Figure 6.37 for the case of Ag films. The observed variation of the optical constants is understandable in terms of the Maxwell–Garnett and other theories of granular films (see Section 2.4.1.3). A direct correlation between the optical constants of the film and the fractional volume occupied by the material, called the packing fraction q, exists, as is seen in Figure 6.37. Such granular films also exhibit anomalous absorption at a certain wavelength depending on the size of the island and if $k(\lambda) > n(\lambda)$. The occurrence of this absorption peak and corresponding maxima of n and k has been reviewed in the literature.[257] It is physically related to the fact that a system of metal particles acting as a damped oscillator resonates at a particular frequency.

The optical constants and hence R (reflectance) and T (transmittance) depend on the microstructure of the films. This is illustrated in Figures 6.38b and a for the case of Au films of different thicknesses deposited at the same substrate temperature and for films of the same thickness deposited at different temperatures, respectively. With decreasing deposition temperatures, the grain size decreases and the defect concentration increases. An additional absorption peak at 1 μm is due to the disordered structure.

The R and T data are also very sensitively dependent on the existence of surface roughness, anisotropy, inhomogeneity, and nonstoichiometry of films. Roughness gives rise to scattering and change in reflectance $\Delta R/R \sim (4\pi\sigma/\lambda)^2$ for small values of σ/λ, where σ is the rms height of the surface

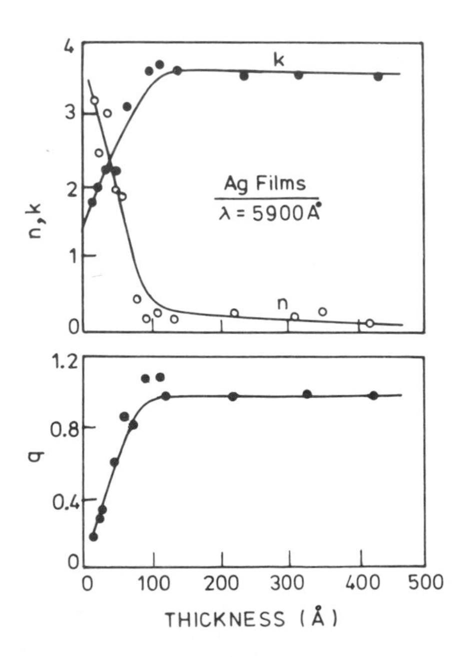

Figure 6.37. The optical constants n and k and packing fraction q of thin Ag films at a wavelength of 0.59 μm, as functions of film thickness (from Chopra[254]).

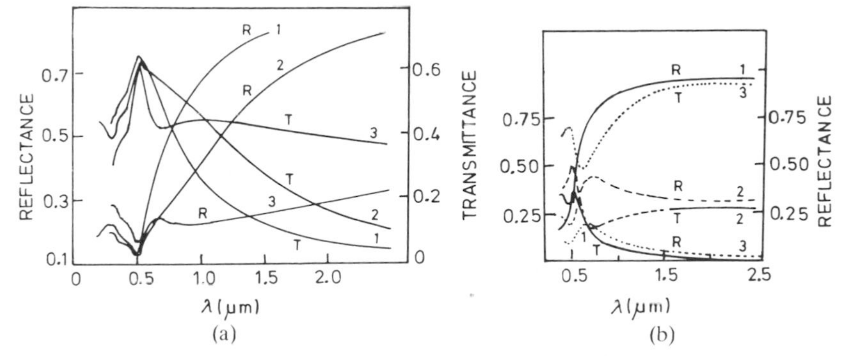

Figure 6.38. (a) Reflectance R and transmittance T for three gold films (1, 2, 3) of the same thickness (120 Å), but with very different crystalline structures. (b) R and T data for Au films of different thicknesses: 1 — 275 Å, 2 — 150 Å, and 3 — 56 Å (after Gadenne[262]).

irregularities. Loss in R from scattering in evaporated films is generally small ($< 1\%$), since $\sigma/\lambda \sim 10^{-2}$ is readily obtained. Anisotropic columnar structure obtained from oblique deposition results in large and anisotropic absorption with the direction of polarization related to the direction of columns. The inhomogeneity and nonstoichiometry problems are important in cases of alloy and semiconductor films.

Table 6.4. Some Useful Electrical Data for Various Metal Films (from Chopra[254] and Maissel and Glang[255])

Metal	l (Å)	ϕ_m (eV)	ρ (20 °C) (μΩ cm)	α (TCR at 20 °C) (K^{-1})
Al	169 (273 K)	4.28	2.65	0.00429
Ag	570 (300 K)	4.26	1.59	0.0041
Au	400 (293 K)	5.1	2.44	0.0034
Cu	450 (293 K)	4.65	1.67	0.0068
Ni	202	5.15	6.8	0.0069
Fe		4.5	9.7	0.0065
W		4.55	5.6	0.0045
Mo		4.6	5.7	0.004
Ta		4.25	2.5	0.001
Ti		4.33		
Cr		4.5		0.0025
Pt		5.65		
Pd		5.12		
Pb		4.25		
Mg		3.62		
Cd		3.94		

l — mfp, ϕ_m — work function

Tables 6.4 and 6.5 list some of the useful electrical and optical data for metallic films of interest. Structurally, the films are generally polycrystalline with grain sizes varying from about 100 Å (for high-melting-point materials) to about 5000 Å (for noble and average-melting-point materials) under normal deposition conditions. Besides affecting the electrical and optical properties, the grain size has a major effect on interdiffusion between superimposed layers because of the dominance of a grain boundary diffusion process.

Table 6.5A. Reflectance R and Transmittance T Spectral Data for Films of Different Thicknesses for Some Metal Films Used in Solar Cells (from Carlson et al.[263] and Beesley et al.[264])

Metal	t (Å)	$\lambda = 0.4\ \mu m$		$\lambda = 0.546\ \mu m$		$\lambda = 0.650\ \mu m$	
		R	T	R	T	R	T
Al on glass	40	0.125	0.65	0.33	0.51	0.38	0.42
	80	0.52	0.36	0.60	0.24	0.63	0.18
	120	0.70	0.19	0.74	0.12	0.75	0.09
	160	0.79	0.11	0.81	0.07	0.82	0.05
	200	0.84	0.05	0.85	0.03	0.85	0.02
	240	0.88	0.03	0.88	0.02	0.87	0.01
	280	0.90	0.019	0.89	0.01	0.88	0.008
	320	0.91	0.011	0.90	0.005	0.89	0.004
	360	0.91	0.006	0.90	0.004	0.90	0.003
	400	0.92	0.004	0.91	0.002	0.90	0.002
	500	0.92	<0.001	0.91	<0.001	0.90	<0.001
		$\lambda = 0.545\ \mu m$					
Cu	200	0.133	0.605				
	400	0.264	0.374				
	500	0.340	0.291				
	600	0.395	0.227				
	700	0.437	0.186				
	800	0.465	0.136				
	1000	0.513	0.073				
Cr	200	0.209	0.445				
	400	0.369	0.244				
	500	0.421	0.200				
	600	0.475	0.160				
	700	0.509	0.132				
	800	0.544	0.097				
	1000	0.614	0.053				

Table 6.5B. Optical Data for Some Thick Metal Films (after Barnes,[265] Hass and Waylonis,[266] and Schulz et al.[267,268])

Metal	λ (μm)	n	k	R	λ (μm)	n	k	R
Al	0.30	0.25	3.33	0.921	0.578	0.93	6.33	0.915
	0.32	0.28	3.56	0.922	0.650	1.30	7.11	0.907
	0.34	0.31	3.80	0.923	0.700	1.55	7.00	0.888
	0.36	0.34	4.01	0.924	0.750	1.80	7.12	0.877
	0.38	0.37	4.25	0.926	0.800	1.99	7.05	0.864
	0.40	0.40	4.45	0.926	0.85	2.08	7.15	0.863
	0.436	0.47	4.84	0.927	0.900	1.96	7.70	0.885
	0.450	0.51	5.00	0.925	0.950	1.75	8.50	0.912
	0.492	0.64	5.50	0.922	2.00	2.30	16.50	0.968
	0.546	0.82	5.99	0.916				
Ag	0.301	1.34	0.964	0.163	0.549	0.06	3.586	0.982
	0.311	1.13	0.616	0.806	0.582	0.05	3.850	0.987
	0.320	0.81	0.392	0.553	0.617	0.06	4.152	0.986
	0.331	0.17	0.829	0.669	0.660	0.05	4.483	0.990
	0.342	0.14	1.142	0.784	0.705	0.04	4.838	0.993
	0.354	0.10	1.419	0.875	0.756	0.03	5.242	0.995
	0.368	0.07	1:657	0.928	0.821	0.04	5.727	0.995
	0.381	0.05	1.864	0.956	0.892	0.04	6.312	0.996
	0.397	0.05	2.07	0.962	0.984	0.04	6.992	0.996
	0.413	0.05	2.275	0.968	1.088	0.04	7.795	0.997
	0.430	0.04	2.462	0.977	1.216	0.09	8.828	0.995
	0.451	0.04	2.657	0.980	1.394	0.13	10.100	0.994
	0.471	0.05	2.869	0.978	1.611	0.15	11.850	0.995
	0.496	0.05	3.093	0.981	1.938	0.24	14.08	0.995
	0.521	0.05	3.324	0.983				
Au	0.301	1.53	1.889	0.386	0.549	0.48	2.455	0.786
	0.311	1.53	1.893	0.387	0.582	0.29	2.863	0.882
	0.320	1.54	1.898	0.387	0.617	0.21	3.272	0.930
	0.331	1.48	1.883	0.389	0.660	0.14	3.697	0.962
	0.342	1.48	1.871	0.386	0.705	0.13	4.103	0.971
	0.354	1.50	1.866	0.383	0.756	0.14	4.542	0.974
	0.368	1.48	1.895	0.392	0.821	0.16	5.083	0.976
	0.381	1.46	1.933	0.403	0.892	0.17	5.663	0.979
	0.397	1.47	1.952	0.406	0.984	0.22	6.350	0.978
	0.413	1.46	1.958	0.409	1.088	0.27	7.15	0.979
	0.430	1.45	1.948	0.407	1.216	0.35	8.145	0.979
	0.451	1.38	1.914	0.408	1.394	0.43	9.519	0.981
	0.471	1.31	1.849	0.401	1.611	0.56	11.21	0.982
	0.496	1.04	1.833	0.446	1.938	0.92	13.78	0.980
	0.521	0.62	2.081	0.643				
Cu	0.284	1.45	1.638	0.330	0.521	1.18	2.608	0.591
	0.292	1.42	1.633	0.333	0.549	1.02	2.577	0.619

(*Continued*)

Table 6.5B (continued)

Metal	λ (μm)	n	k	R	λ (μm)	n	k	R
	0.301	1.40	1.679	0.347	0.582	0.70	2.704	0.725
	0.311	1.38	1.729	0.362	0.617	0.30	3.205	0.899
	0.320	1.38	1.783	0.375	0.660	0.22	3.747	0.943
	0.331	1.34	1.821	0.390	0.705	0.21	4.205	0.956
	0.342	1.36	1.864	0.398	0.756	0.24	4.665	0.958
	0.354	1.37	1.916	0.409	0.821	0.26	5.180	0.963
	0.368	1.36	1.975	0.425	0.892	0.30	5.768	0.965
	0.381	1.38	2.045	0.464	0.984	0.32	6.451	0.970
	0.397	1.32	2.116	0.464	1.088	0.36	7.217	0.973
	0.413	1.28	2.207	0.491	1.216	0.48	8.245	0.972
	0.430	1.25	2.305	0.518	1.394	0.60	9.439	0.973
	0.451	1.24	2.397	0.539	1.611	0.76	11.12	0.976
	0.471	1.25	2.483	0.554	1.938	1.09	13.43	0.976
	0.496	1.22	2.504	0.575				
Cr	0.301	1.53	2.34	0.486	0.549	3.18	3.33	0.554
	0.311	1.58	2.40	0.490	0.582	3.22	3.30	0.551
	0.320	1.65	2.47	0.497	0.617	3.17	3.30	0.551
	0.331	1.60	2.58	0.504	0.660	3.09	3.34	0.556
	0.342	1.76	2.58	0.506	0.705	3.05	3.39	0.562
	0.354	1.64	2.64	0.510	0.756	3.08	3.42	0.565
	0.368	1.87	2.69	0.516	0.821	3.20	3.48	0.569
	0.381	1.92	2.74	0.521	0.892	3.30	3.52	0.572
	0.397	2.0	2.83	0.529	0.984	3.41	3.57	0.576
	0.413	2.08	2.93	0.539	1.088	3.58	3.58	0.576
	0.430	2.19	3.04	0.548	1.216	3.67	3.60	0.577
	0.451	2.33	3.14	0.555	1.394	3.69	3.84	0.598
	0.471	2.51	3.24	0.559	1.611	3.66	4.31	0.636
	0.496	2.75	3.30	0.559	1.938	3.71	5.04	0.688
	0.521	2.94	3.39	0.558				
Ni	0.301	2.02	2.18	0.417	0.549	1.92	3.61	0.643
	0.311	2.01	2.18	0.417	0.582	1.96	3.80	0.662
	0.320	1.93	2.19	0.423	0.617	1.99	4.02	0.682
	0.331	1.84	2.22	0.433	0.660	1.99	4.26	0.706
	0.342	1.78	2.26	0.445	0.705	2.06	4.50	0.721
	0.354	1.74	2.32	0.460	0.756	2.13	4.73	0.735
	0.368	1.70	2.40	0.478	0.821	2.26	4.97	0.744
	0.381	1.72	2.48	0.492	0.892	2.40	5.23	0.753
	0.397	1.72	2.57	0.508	0.984	2.48	5.55	0.768
	0.413	1.70	2.69	0.531	1.068	2.65	5.93	0.781
	0.430	1.71	2.82	0.552	1.216	2.79	6.43	0.799
	0.451	1.73	2.95	0.571	1.394	2.96	7.08	0.820
	0.471	1.78	3.00	0.587	1.611	3.14	7.96	0.843
	4.496	1.82	3.25	0.606	1.938	3.47	9.09	0.864
	0.521	1.85	3.42	0.626				
Pt	0.256	1.169	1.929	0.445	0.999	3.419	6.299	0.769
	0.444	1.939	3.159	0.583	1.969	5.919	9.799	0.835

Table 6.5B (continued)

Metal	λ (μm)	n	k	R	λ (μm)	n	k	R
	0.588	2.629	3.539	0.590	3.289	7.499	12.199	0.864
	0.667	2.909	3.659	0.594	4.649	10.899	15.499	0.885
Pd	0.292	1.179	2.229	0.514	0.704	1.859	4.649	0.750
	0.310	1.209	2.349	0.534	0.756	1.949	4.889	0.760
	0.354	1.229	2.649	0.589	0.821	2.059	5.189	0.773
	0.381	1.259	2.829	0.615	0.892	2.229	5.499	0.780
	0.431	1.329	3.029	0.635	0.984	2.339	5.889	0.795
	0.450	1.409	3.259	0.656	1.087	2.519	6.329	0.807
	0.496	1.519	3.539	0.678	1.215	2.659	6.899	0.825
	0.548	1.639	3.839	0.697	1.393	2.799	7.649	0.846
	0.616	1.749	4.209	0.723	1.610	3.009	8.589	0.866
	0.659	1.799	4.419	0.737	1.937	3.339	9.889	0.885

6.5. *Dielectric Films*

As solar cell materials, dielectric films are used to form electrical and interfacial barriers between metals and semiconductors, for passivation and for antireflection (AR) coatings. The AR coatings and transparent conducting oxides are discussed in Appendix B and Section 6.3, respectively. A summary of the relevant properties of some of the useful dielectric films is given in the following and the appropriate data are presented in Table 6.6. For detailed reviews of the subject, the reader is again referred to standard texts.[254,255]

Table 6.6. Some Electrical and Optical Data for Dielectric Films Used in Thin Film Solar Cells[a]

Material	n (λ in μm)	E_g (eV)	ε	χ (eV)
SiO_2	1.44 (1.6)	~12	4	1.00
SiO	1.7 (6.0)		6	
Si_3N_4	2.05	4 (crystalline) 5 (amorphous)	9	
Ta_2O_5	2.2 (0.59)	4.3	25	4.4
TiO_2	2.2–2.7 (0.55)	3	30–40	4.3
ZnS	2.2 (2.0)	3.7	10	3.9
MgF_2	1.37 (0.59)		5.5	
Nb_2O_5		3.3	39	4.56
Cu_2O		2.0	12	
Sb_2O_3		4.20		4.22
Cr_2O_3		1.4	12	4.98
MgO	1.77 (0.36)	7.3	9.65	1.72

[a] n — refractive index, ε — dielectric constant, χ — electron affinity.

6.5.1. *Electronic Properties*

Theoretically, a structurally perfect dielectric film is expected to exhibit bulk-like dielectric and optical properties down to monolayer dimensions. Some solitary measurements on 24.6-Å-thick cadmium stearate films on cleaved single crystal $Bi_8Te_7S_5$ substrates appear to support this statement. However, vapor deposited ultrathin films are discontinuous and become continuous only at a thickness determined by the nucleation and growth processes, which, in case of amorphous films, may be as low as several angstroms. The dielectric and optical constants of discontinuous granular films are determined by the packing fraction of the material and the bulk constants, as discussed in Chapter 2.

Mean free path and surface state effects are not expected in the dielectric constants. The dielectric breakdown field of thin films of oxides is known to be comparable to the bulk value. If the breakdown mechanism is an avalanche type, one may expect the field to increase when the film thickness is comparable to or less than the mean free path of the avalanche electrons. This, in most materials, should occur for thicknesses less than 20 Å. However, thickness-dependent increase of breakdown field has been observed in some film materials at much greater thicknesses, and this may be attributed to microstructural effects.

In case of thick, physically continuous films, bulk-like dielectric properties are observed in single crystal and amorphous films (most oxide films belong to the latter category). The commonly occurring pinholes, spatial thickness variations, nonstoichiometry, inhomogeneity, anisotropy, and trapping centers associated with structural defects have a significant effect on the various dielectric and electron transport properties of the films. All these deviations from an ideal film are profoundly dependent on deposition process parameters in cases of multicomponent dielectric films. For example, x in evaporated SiO_x films can be varied from 1 to 2.[260] Films of SiO_x prepared by both PVD and CVD techniques invariably exhibit stoichiometry and composition gradient with a mixture of $Si + SiO_x$ in some cases. Various diagnostic techniques have indicated that the nonstoichiometric interfacial region for SiO_2 film grown on Si varies from a few atomic layers to about 12–13 Å. If composition gradients exist, the dielectric constant as well as the optical constants differ considerably from the bulk value and depend on the temperature and technique of deposition. This point is illustrated by the wavelength dependence of the refractive index of evaporated Al_2O_3 and anodized Ta_2O_5 films in Figure 6.39.

The loss factor of dielectric films is seriously affected by the microstructure and the nature and magnitude of stresses in the films. As a

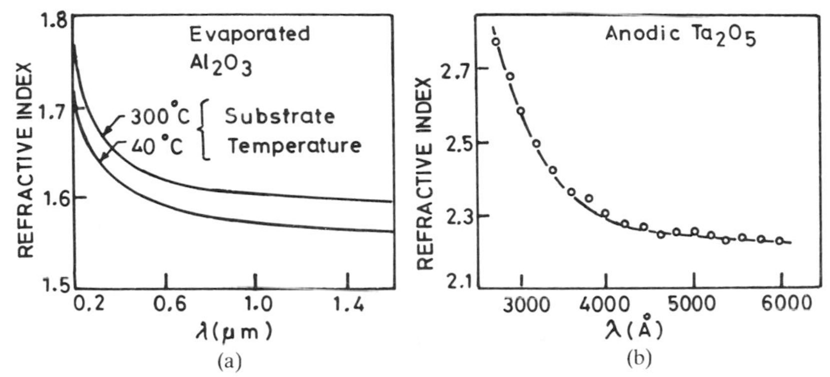

Figure 6.39. The dispersion of the refractive indices of (a) vacuum-evaporated films of Al_2O_3 deposited at 40 and 300 °C; (b) anodic films of Ta_2O_5 (from Chopra[254]).

consequence, the temperature coefficient of the dielectric constant, $\alpha_\varepsilon = (1/\varepsilon)(d\varepsilon/dT)$, is increased in view of the Gaever equation approximation

$$\alpha_\varepsilon = A \tan\delta - \alpha_L(1+\varepsilon)$$

where $\tan\delta$ is the loss factor, A is a constant, and α_L is the linear expansion coefficient.

The existence of pinholes, thickness variations, and trapping centers has considerable effect on electron transport through thin dielectric films. Although studied most extensively,[254,255] the electron transport processes remain controversial. Various possible conduction processes are described in Chapter 2, and some of these or combinations thereof have been observed in various film materials by different workers. The presence of different mechanisms in Si_3N_4 films is seen in Figure 6.40.

Major conclusions of numerous studies on films less than 100 Å thick sandwiched between metal electrodes are summarized here: (i) Under different field and temperature conditions, different conduction mechanisms dominate. Generally, for ultrathin ($\lesssim 30$ Å) films a tunneling process dominates. (ii) Quantitative comparison of the experimental data with the theoretical predictions is possible only by adjusting the barrier parameters and/or film thickness. Indeed, as expected, such parameters as barrier height, electron affinity, barrier shape, nonuniformity of film thickness, pinhole density, dielectric constant, interfacial diffusion, and interface and impurity states are essentially characteristics of a particular film.

Pinholes in films arise from the statistical nature of a vapor deposition process and from the presence of dust particles during the growth process. It is possible to reduce the density of pinholes of less than 1 μm to as low as

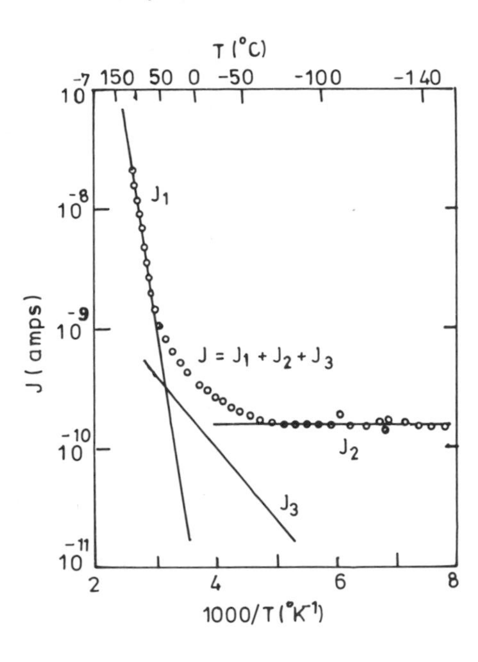

Figure 6.40. The temperature dependence of the current through an Au-Si_3N_4-Si diode at a field of 5.3×10^6 $V cm^{-1}$. The total current J exhibits three components: J_1 — Poole–Frenkel current, J_2 — tunnel current, and J_3 — ohmic conduction current (after Sze[261]).

about 10^3 cm^{-2}, which may not contribute significantly to the shorting of the tunnel current. Much larger spatial variations in the tunnel current would arise from the thickness variation because of its exponential dependence on thickness. In deposition conditions leading to an amorphous structure, thickness variation or roughness is given by a Poisson distribution and hence is proportional to $\sqrt{t}$ (t = film thickness). A much greater thickness variation is expected if the film has large and oriented grains.

The abundance of structural defects in micropolycrystalline and amorphous dielectric films and nonstoichiometry and inhomogeneity provide for high concentrations of traps, with energies distributed in the wide bandgap of these materials. Limited studies on the subject indicate that trap densities of about 10^{16} to 10^{20} cm^{-3} are entirely reasonable and are essential to explain drastic reduction in the tunnel current density.

In addition to the ionized traps, the dangling bonds at the surface and lattice mismatch at the dielectric/semiconductor interface give rise to a high density of surface states and surface charges. These states provide charge-storage and recombination–generation centers and tunneling paths. The role of such surface effects in the performance of SIS solar cells is discussed in Chapter 3. Finally, it should be noted that while considering the transport of hot electrons through a thin insulator, energy loss due to the creation of optical phonons must be considered as well (see Chopra[254]). This process results in considerable reduction of the electron–electron mean free path.

6.5.2. *Optical Properties*

Like the dielectric properties, the optical constants of dielectric films depend on the microstructure, as already illustrated in Figure 6.39. Since the short-range order in micropolycrystalline films is the same as that in corresponding bulk materials, the optical bandgap is not affected. However, the band edges may be smeared owing to structural defects and strains in the films.

7

Cu_2S Based Solar Cells

7.1. Introduction

Cu_2S possesses extremely favorable material characteristics, which make it suitable for large-scale photovoltaic applications, particularly in combination with CdS or ZnCdS. Moreover, Cu_2S and CdS are available in sufficient abundance. It is not surprising, therefore, that the Cu_2S/CdS thin film heterojunction solar cell has been the most extensively investigated thin film photovoltaic system.[1-5] The developmental effort has resulted in the fabrication of thin film Cu_2S/CdS solar cells with a conversion efficiency of 9.15%[6] and Cu_2S/ZnCdS cells with a conversion efficiency of 10.2%.[7] Theoretical analyses have revealed that the Cu_2S/CdS and Cu_2S/ZnCdS thin film photovoltaic systems are capable of achieving practical conversion efficiencies of 11 and 14–15%, respectively.[8]

The history of the Cu_2S/CdS solar cell spans more than a quarter of a century, beginning in 1954, when Reynolds et al.[10] first discovered a photovoltaic response in heat-treated Cu contacts on CdS at photon energies less than the CdS band edge. Early work on Cu_2S/CdS solar cells was carried out by research groups at Harshaw, Clevite, ARL, RCA, Stanford, SAT (France), AEG-Telefunken (Germany), and IRD (UK). During this phase, a number of fabrication techniques were developed for thin film solar cells and in the late sixties and early seventies, cell development had reached a stage where conversion efficiencies of about 7% could be achieved, though not reproducibly with a high yield, on small-area cells. On larger areas, efficiencies of 3–4% were common. Several problems associated with cell fabrication and operation were identified but remained unsolved. These included the problem of degradation, the low open-circuit voltage, control of the Cu_2S layer stoichiometry, and suitable encapsulation. Cell development was mostly empirical and not based on a clear understanding of the cell behavior. Consequently, cell performance and understanding of the cell operation was limited.

Subsequent development of the Cu_2S/CdS thin film device, through the seventies, was undertaken primarily at Delaware University, Stuttgart University, Stanford University, Photon Power, the CNRS Laboratories (France), and the Thin Film Laboratory, Indian Institute of Technology, Delhi. This later phase witnessed the development of high-efficiency cells, the fabrication of modules and panels, the emergence of the spray pyrolysis technique, and achievement of a fairly high degree of stability and reproducibility in cell fabrication and processing. The research and development effort was marked by refinement and modification of earlier processes, device design, and material properties. The earlier empirical approach was replaced by a more purposeful and cohesive program sustained and directed by extensive cell characterization, analysis of failure modes, and identification and minimization of photon and carrier loss mechanisms. It was established that the bulk of the photocurrent was contributed by Cu_2S. In recognition of the role of the Cu_2S layer, the familiar CdS cell has been renamed the Cu_2S cell. Based on the physical, electronic, and optoelectronic characteristics of the cell, several theoretical models were developed to explain the cell behavior. All these developments have led to a thin film $Cu_2S/ZnCdS$ cell of greater than 10% efficiency.

Apart from the practical possibility of large-scale terrestrial photovoltaic applications, the Cu_2S/CdS device is an extremely interesting heterojunction system from the purely academic point of view. This heterojunction exhibits a variety of effects and phenomena that would be expected in nonideal heterojunctions and that are associated with lattice mismatch, electron affinity mismatch, deep-level traps, and bias- and wavelength-dependent junction fields. Thus the Cu_2S/CdS cell provides an excellent case study for heterojunction phenomena. Indeed, most of the present knowledge and understanding of thin film heterojunction photovoltaic devices are based on the extensive experience gained with the Cu_2S/CdS thin film solar cell. Another unique feature of this cell is that it is perhaps the only photovoltaic system, to date, that has exhibited higher conversion efficiencies in the thin film form (9–10%) than in the single crystal form (7–8%).[9]

In keeping with the long history, a large volume of literature exists on the Cu_2S/CdS solar cell, and has been reviewed recently by several authors.[2–4] The review by Stanley,[2] who summarized the entire literature up to 1975, is particularly recommended to those readers interested in the early development of the cell. However, much of the information and data of the earlier studies have been rendered obsolete or have been superseded by the rapid development of the past few years. In this chapter, we shall, therefore, discuss only the more recent literature on Cu_2S/CdS thin film solar cells.

7.2. Fabrication Process

Cu_2S/CdS and $Cu_2S/Zn_xCd_{1-x}S$ thin film solar cells are currently being fabricated by a wide variety of techniques which include a combination of evaporation and chemiplating,[6,8,11–13] a combination of evaporation and solid state ion-exchange reaction,[14–22] all evaporation,[23] a combination of spray pyrolysis and chemiplating,[19,24–29] a complete spray pyrolysis process,[30] a combination of sputtering and chemiplating/solid-state reaction,[31–34] sequential sputtering,[31,33] a combination of screen printing and chemical deposition,[35] and a combination of electrophoresis and chemiplating/solid-state reaction/evaporation.[36] In our own laboratory, cells have also been fabricated by electroconversion of evaporated and sprayed CdS films. The evaporated/chemiplated, evaporated/solid-state reaction, and sprayed/chemiplated thin film $Cu_2S/Zn_xCd_{1-x}S$ cells have reached an advanced stage of development. The fabrication processes are described in greater detail in Sections 7.2.2 and 7.2.3.

7.2.1. Device Geometry

The Cu_2S/CdS thin film solar cell operates in two principal modes — front wall (FW) and back wall (BW). In the FW mode, light is incident on the absorber layer (Cu_2S) first, while in the BW mode, the illumination is through the window layer (CdS or $Zn_xCd_{1-x}S$). If the back contacts are highly reflecting, such that the light is forced to make a double pass or several passes through the cell, then the device operates in the front wall reflecting (FWR) or back wall reflecting (BWR) mode. The high-efficiency cells usually operate in the FWR or BWR mode and in these devices the reflectance of the back surface, the thickness of the Cu_2S layer, and any process parameters affecting these two quantities play a critical role in the performance of the device.[8]

A schematic diagram of a typical FWR Cu_2S/CdS solar cell is shown in Figure 7.1a. The cross-section of a back wall device is shown in Figure 7.1b. Most Cu_2S/CdS thin film cells fabricated from evaporated CdS layers are of the FW type, while sprayed cells are usually of BW configuration. However, evaporated BW cells have also been fabricated.[18,37] Sputtered and all-evaporated cells have generally been fabricated in the FW configuration. The BW cells possess the advantage that they are naturally encapsulated by the back metal contact and do not require further elaborate encapsulation as do the FW cells. Das et al.[15] have fabricated FW cells in a novel configuration designed to protect the Cu_2S layer from exposure to ambients. The geometry of this cell is shown in Figure 7.1c. The metal/CdS layer, deposited on top of the initially prepared Cu_2S layer

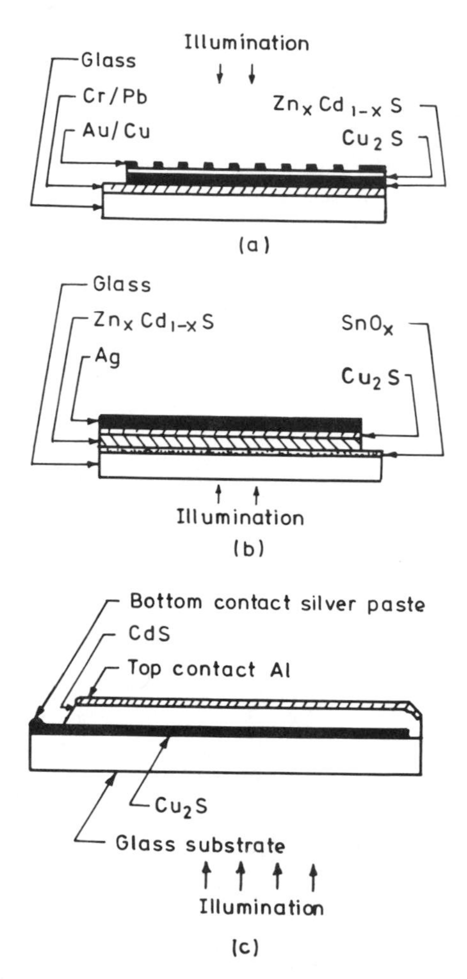

Figure 7.1. Schematic diagrams of: (a) FWR, (b) BWR, and (c) novel geometry Cu_2S/CdS thin film solar cells.

on glass (with a predeposited grid pattern), provides the natural encapsulation.

A large variety of substrates have been successfully utilized for the fabrication of thin film Cu_2S cells. These include Zn coated Kapton,[38] Zn plated Cu,[6–8,13] Cd_2SnO_4/silica,[37] Cr/Ag,[11,12] Cr/Pb,[19] Cr/Au,[31], Ag,[32] and Nb[33,34] coated glass, Fe and Au plated polyamides,[36] and Fe[36] substrates for FW devices. BW devices have been fabricated on SnO_x,[22,36] SnO_x:Sb,[18,19] SnO_2/Sb_2O_3,[32] ZnO,[39] and ITO[39,41] coated glass and Cd_2SnO_4/silica[37] substrates. The best cells, to date, have been fabricated on Zn coated Cu substrates[6–8] and operate in the FWR mode.

7.2.2. Evaporated Cells

The fabrication process for high-efficiency[8] evaporated CdS/chemiplated Cu_2S cells consists of the following steps: a flexible, Cu foil about 35 μm thick is electroplated on the topographically rougher side (8–10 μm granular topography) with a 800-Å-thick Zn layer. The Zn layer provides an ohmic contact to the subsequent CdS layer and also inhibits direct Cu_2S formation with the incoming S_2 molecules during CdS evaporation. In addition, the Zn and Cu interdiffuse during the deposition of CdS resulting in the formation of a highly reflecting, low-Zn-content brass. A 30-μm-thick, 1 to 10 Ω cm, *n*-type, columnar, oriented (*c*-axis perpendicular to the substrate) CdS layer is deposited on the Zn/Cu substrate at a substrate temperature of about 220 °C from an indirectly heated, CdS powder-filled graphite source bottle (described in Section 5.2.1.4) maintained at about 1000 °C. The deposition rate is about 1 to 2 μm min^{-1}. The CdS layer is polycrystalline, with a grain size of about 5 μm. Next, a 300-Å-thick Cu_2S layer is topotactially formed by reacting the CdS film with a cuprous ion solution, containing 2 g of NaCl and 6 g of CuCl per liter of H_2O, maintained at 90 °C (pH ~ 3–4). However, prior to the chemiplating process, the CdS surface is textured, using a 50% HCl etch at 60 °C for 2 s, to promote light trapping. The freshly prepared Cu_2S/CdS junction is heat-treated in a reducing atmosphere, typically for 16 h at 170 °C in CO, to adjust the Cu_2S stoichiometry, cause further brassification and hence reduce substrate absorption at the Zn/Cu interface, and promote the formation of a junction field in the CdS. In the subsequent step, a linear grid of evaporated Au lines 2 to 4 μm thick, with 32 lines cm^{-1} and 95% transmission, is applied to the Cu_2S. Finally, a 700-Å-thick SiO AR layer is evaporated onto the device. The AR coating/textured CdS/rough Zn-Cu substrate/high-transmission grid combination achieves a high efficiency of light trapping and captures about 92% of the incoming radiation.

An essentially similar fabrication process has been reported by Szedon et al.[13] However, the top contact is a Au plated Cu grid laminated onto Cu_2S and the heat treatment takes the form of a 2-min soak in a circulating air oven maintained at 250 °C. Arndt et al.[11,12] reported a similar junction fabrication process for their large-area (7×7 cm^2), stable, 7.1% efficiency (AM1) Cu_2S/CdS thin film solar cells on Cr/Ag coated glass substrate. The front Cu_2S contacting grid is fabricated directly on the glass sheet that is used for final lamination and encapsulation. The post-junction-formation treatment consists of either the evaporation of a thin (50–100 Å) Cu layer onto Cu_2S or the application of a H_2 glow-discharge to the cells, followed by subsequent heat treatment in air or in O_2-containing atmosphere at about 220 °C for a few minutes. Bhat et al.[18] and Banerjee et al.[19] have observed that their Cu_2S/CdS cells, prepared by the solid state reaction technique, require a heat treatment at about 200 °C for 5 min in a vacuum of about 10^{-1} torr for optimum performance.

The fabrication of $Cu_2S/Zn_xCd_{1-x}S$ solar cells[7] is identical in all respects except that $Zn_xCd_{1-x}S$ films are vacuum evaporated from a dual-chamber concentric graphite source (Section 5.2.1.4) instead of CdS. The grain size of the $Zn_xCd_{1-x}S$ films, in the range $0.1 \leq x \leq 0.2$, averages about 2 μm. The 10%-efficiency $Cu_2S/Zn_xCd_{1-x}S$ thin film devices were provided with a 700-Å Ta_2O_5 AR coating and optimized by heat treatment at 170 °C in a flowing H_2 atmosphere. Single-chamber sources[19,40] have also been employed for $Zn_xCd_{1-x}S$ film deposition, but with less successful results.

7.2.3. *Sprayed Cells*

Backwall Cu_2S/ZnCdS cells[19,26] are fabricated by spray pyrolytically depositing a transparent, conducting SnO_x:Sb layer onto chemically cleaned glass slides at about 400 to 450 °C. The SnO_x:Sb coatings possess a sheet resistance of 5 to 10 Ω/□ and a transmittance of about 90% over the solar spectrum in the visible region. A layer of $Zn_xCd_{1-x}S$ of 3 to 4 μm is deposited on the conducting glass by spraying an aqueous solution of $CdCl_2$, $ZnCl_2$, and $CS(NH_2)_2$, mixed in the desired proportion. The substrate temperature is maintained at about 300 °C. The $Zn_xCd_{1-x}S$ films are annealed at about 300 °C for 30 min in a vacuum or at 200 °C in a H_2 flow[27] to decrease the resistivity of the as-deposited films. Subsequently, the Cu_2S film is formed by chemiplating. The junction forming treatment consists of a vacuum anneal, the temperature and duration of which is determined by the composition of the film. Typically, for $x = 0$, the temperature is 250 °C and the duration is about 5 min. The top contact is provided by complete metallization of the Cu_2S surface with Ag paste. Singh[29] has employed vacuum evaporated multilayer (Au/Cu/Pb) metal electrodes to make ohmic contact to Cu_2S. Cu treatment has been employed by some workers[39,41] to improve the characteristics of sprayed cells.

Martinuzzi et al.[27] have varied the S/Cd ratio (1 to 3) in the spray solution and observed that the CdS layers deposited at 380 °C from a spray solution containing S/Cd in the ratio 1:1 possess crystallite sizes of 0.5 μm, exhibit a strong ⟨002⟩ orientation, are transparent at $\lambda > 5200$ Å, and are, therefore, suitable for solar cells. In our laboratory, however, we have observed a S/Cd ratio of 3:1 to produce the best results. It should be noted that the stoichiometry of the CdS is not changed by changing the S/Cd ratio. Only the film microstructure is affected. Bodhraj et al.,[26] and Martinuzzi et al.[27] and Jordan[30] have utilized gradient doped CdS:Al films, grown by varying the $AlCl_3$ concentration, the source of Al in the spray solution, continuously, for fabricating Cu_2S/CdS solar cells.

All-spray Cu_2S/CdS thin film solar cells have been produced by Jordan[30] on SnO_x coated glass substrates by first spray depositing a 2-μm-thick, gradient doped CdS layer and subsequently growing a Cu_xS layer by spraying a solution of copper acetate, N-N-dimethyl thiourea, and other ingredients onto CdS at approximately 150 °C.

A common feature of all the fabrication processes for Cu_2S/CdS solar cells is the postjunction formation heat treatment. This treatment is one of the most critical process steps and will be discussed in more detail in Section 7.5.1.

7.3. *Physical Model*

It is now known that the microstructure of each component layer of the cell (back contact, CdS, and Cu_2S) and the morphology of the Cu_2S/CdS interface strongly influence the cell performance. A clear picture of the physical structure of the cell has emerged from several electron microscopy studies of Cu_2S/CdS thin films cells. A variety of sample preparation techniques, described in Section 4.4.1.2, have been developed to examine the cross-section of the cell as well as the Cu_2S/CdS interface. These structural and morphology studies, coupled with composition and depth profile analyses, have been instrumental in clarifying many aspects of junction formation and cell operation. The results of these studies are presented in this section.

7.3.1. *Microstructural*

The microstructure and morphology of the evaporated cells are in marked variance with those of spray deposited cells. We discuss the evaporated cells first.

The microstructure and CdS topography are determined by the topography of the underlying substrate. In general, CdS films, deposited at temperatures of 200 °C or more are columnar, with grain sizes of the order of 3 to 5 μm at the top of the columns, and strongly oriented. The initial CdS layer, next to the substrate, is composed of randomly oriented smaller grains, as revealed by SEM and TEM micrographs of a Cu_2S/CdS cell cross-section.[42] The surface of CdS films grown on Zn plated Cu substrates has been reported to replicate the substrate surface.[8] Etching of the CdS in HCl prior to junction formation provides a textured CdS surface consisting of pyramids.[13,42–46] However, the exact shape and size of the pyramids depend sensitively on the etchant concentration, the temperature, and the etch time.[13,44,45] The surface topography has to be optimally etched to ensure proper junction formation and, subsequently, good performance.[13,44,45]

The Cu_2S layer formed topotactially on the polycrystalline CdS film by chemiplating is extremely nonplanar and nonconformal. During the chemiplating process, the reaction penetrates deep down the grain boundaries and cracks and the resultant Cu_2S morphology is composed of a thin superficial conformal skin, produced by volume diffusion of the reactants, and an interconnected network of predominantly vertical intrusions produced by intergranular surface diffusion.[13,43–46] These vertical fingers may constitute from 20 to 80% of the total mass of Cu_2S and may exceed several micrometers in length, depending on the grain size and topography of the original CdS substrate and the etching and chemiplating process parameters. In Figures 7.2a–c, we have shown SEM micrographs of the surface

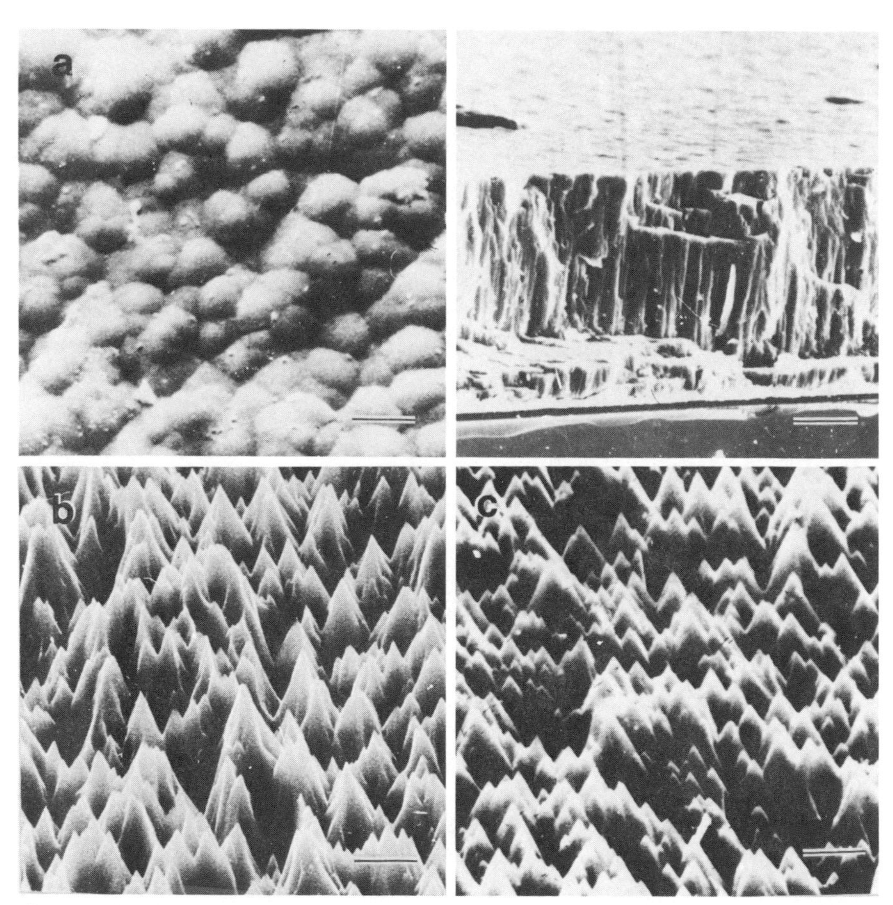

Figure 7.2. Scanning electron micrographs of: (a) surface topography and fracture surface of an as-deposited, evaporated CdS film, (b) CdS surface after etching, and (c) Cu_2S surface topography formed on etched CdS. Scale bars: (a) 6 μm (left) and 21 μm (right), (b) 6 μm, (c) 2 μm.

topography and fractured surface of an as-deposited evaporated CdS film, the surface of the CdS after texturizing, and the surface topography of the Cu_2S formed on the etched CdS by chemiplating. In Figures 4.18a–c (p. 182), SEM micrographs of the Cu_2S/CdS interface (obtained after leaching the Cu_2S in KCN), a free-standing Cu_2S layer as viewed from the CdS side, and the cross-section of a Cu_2S/CdS cell are presented.

In marked contrast, Cu_2S layers formed by the solid state reaction are substantially conformal and free of grain boundary intrusions.(14–16,21) Casperd and Hill(21) have observed that grain boundary penetrations in solid state reaction cells is confined to depths of less than 1 μm.

The morphology of the spray deposited/chemiplated CdS/Cu_2S solar cells arises from the vastly different microstructure of the sprayed CdS film.(26,41,47,48) SEM micrographs of pure and Al-doped sprayed CdS films are shown in Figure 7.3. Pure sprayed CdS films are oriented when prepared at substrate temperatures of 380 °C.(49) Doping, in general, destroys the preferred orientation. However, at low dopant concentrations the films are better oriented. Pure CdS films exhibit a nodular surface topography, whereas CdS:Al films possess a predominantly serpentine structure.(26,47–49) Gradient doped CdS:Al films or double-layered CdS:Al/CdS films also exhibit serpentine surface topography.(26,41,48,49) The modules are essentially hollow hemispheres(47,50) with a thin outer crust of material consisting of microcrystallites with grain sizes in the range of 0.1 to 0.5 μm. It should be noted that Al does not dope the CdS but forms Al_2O_3 and gives rise to the serpentine structure by segregating at the internodular regions.(26,30,48)

Cu_2S prepared by chemiplating sprayed CdS films is composed of a three-dimensional network structure and forms on the crust of each nodule.(39,47) The gradient doped CdS:Al films are denser and relatively free of faults and voids (characteristic of pure sprayed CdS films) owing to the presence of Al_2O_3. Thus the penetration of the Cu_2S is limited in the gradient doped films(26,30,49) and this allows the use of thinner (3–5 μm) CdS layers. An SEM micrograph of the CdS/Cu_2S interface of a spray deposited cell, revealed after etching away the Cu_2S layer in KCN, is shown in Figure 7.3c. On comparing with Figure 7.3a, we note the presence of a network structure on the nodules in the former. These are the grain boundaries where Cu_2S penetrated deep inside. In Figure 7.3d, we show the micrograph of a Cu_2S layer separated from a sprayed cell and viewed from the CdS side.

Sputtered CdS films(32) are more coherent than evaporated films, even at very low thicknesses. Moreover, sputtering produces oriented films, as in the case of evaporated CdS. Hill et al.(32) have reported that Cu_2S formed on sputtered CdS by solid state reaction is very planar and grain boundary penetration is even less than for evaporated CdS films.

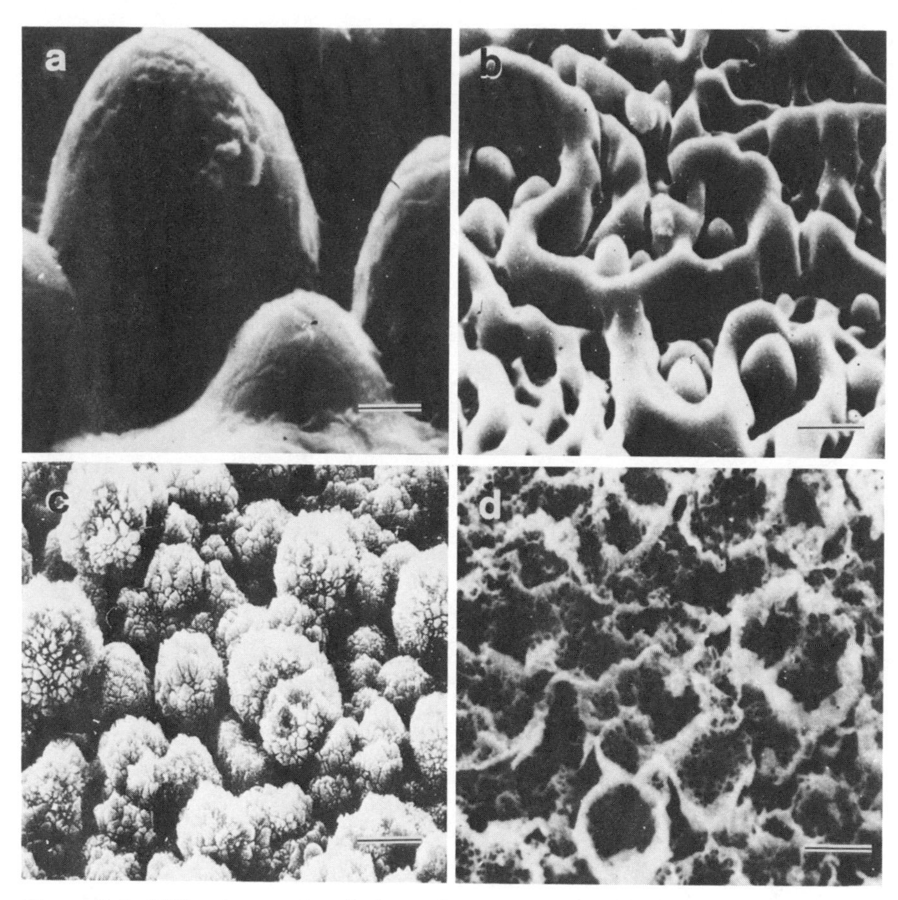

Figure 7.3. SEM micrographs of: (a) surface topography of pure, sprayed CdS film, (b) surface topography of Al doped CdS film, (c) Cu_2S/CdS interface of a chemiplated/sprayed cell, revealed after etching away Cu_2S in KCN, and (d) free-standing Cu_2S film separated from a chemiplated/sprayed Cu_2S/CdS thin film solar cell. Scale bars: (a) 2 μm, (b) 6 μm, (c) 9 μm, (d) 6 μm.

Based on the results of microstructural investigations, physical models of the Cu_2S/CdS thin film solar cells prepared by different techniques have been evolved and these are schematically depicted in Figures 7.4a–c, which show the cross-sections of an evaporated CdS/chemiplated Cu_2S, an evaporated CdS/solid state reaction Cu_2S, and a sprayed CdS/chemiplated Cu_2S thin film solar cells.

7.3.2. *Compositional Analysis*

Limited studies of compositional characteristics of Cu_2S/CdS solar cells have been reported in the literature. A fluorescence spectrometer scan across several grain boundaries has been utilized to map the lateral spatial profiles of Cu K_α and Cd K_α radiations.[51] The two profiles have been found to be almost perfectly anticorrelated, indicating strong preferential segregation of Cu_2S at the grain boundaries, in conformity with the structural studies. Similar results have been obtained by Mukherjee et al.[46] on bevelled Cu_2S/CdS samples using cathodoluminescence techniques. Tseng and Greenfield[52] have reported, on the basis of AES depth profile data, that the surface layer of a chemiplated Cu_2S/CdS cell contains

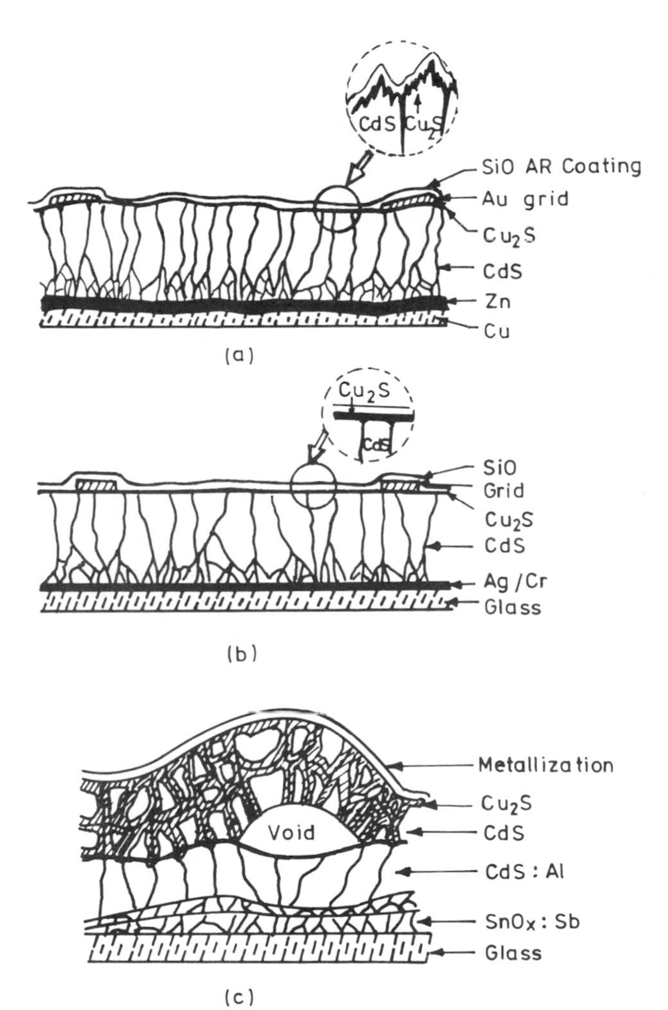

Figure 7.4. Schematic representation of the microstructures of: (a) chemiplated/evaporated, (b) solid state reaction/evaporated, and (c) chemiplated/sprayed Cu_2S/CdS thin film solar cells.

significant amounts of O, C, Cl, and Cd, in addition to S and Cu. The O and C were suggested to originate from exposure to atmosphere and Cl and Cd were thought to be present due to the dipping process. Further, cells with good fill factors were observed to contain a large percentage of chalcocite in the Cu_xS layer, with the remaining portion being composed of djurleite. Poor cells were found to be deficient of a significant amount of chalcocite.

The thickness of the chalcocite layer was observed to decrease with time, attended by Cu diffusion into CdS and to the surface followed by subsequent oxidation. Chemiplated single crystal CdS also exhibits similar characteristics, as revealed by ESCA and AES studies.[13] Both heat treated (in air) as well as freshly prepared, unheated junctions were found to contain Cu, Cd, S, Cl, O, and C on the surface. The heat treated sample exhibited a much higher Cd/Cu concentration ratio than the unheated sample. Moreover, S was found to exist in two different valence states (S and SO_4^{2-}) in the heat treated sample. Cd was observed to be present to a considerable depth in the Cu_2S layer in heat treated samples, while in unheated samples, Cd was present only on the surface. Cu nodules were detected as the Cu_2S layer was sputtered off. Pfisterer et al.[53] have reported AES and ESCA measurements on efficient, Cu treated chemiplated Cu_2S/CdS solar cells. The salient features of their data are the presence of Cu mainly in a single valence state, the existence of S as S^{2-}, the presence of an oxide layer at the outer surface of Cu_2S, and the existence of O in the O^{2-} state, indicating that the oxide layer of efficient solar cells consists of Cu_2O and not of CuO. No indication of $CuSO_4$ was present.

Satkiewicz and Charles[54] carried out an SIMS analysis of chemiplated Cu_2S/CdS samples and reported that different cell processing steps produce differences in the constituent (Cu, Cd, and S) profiles. They attributed the observed changes in the composition profiles to the possible formation of $Cu_2SO_3 \cdot H_2O$ on the surface, CuS or CuO at the Cu_2S/CdS interface, and/or CdO on the CdS surface prior to barrier formation. Several impurities such as Fe, Mn, Li, Na, Ca, K, and Si were detected in both CdS and Cu_2S layers. The concentration of $CuOH^+$ was observed to increase with time, indicating the possible introduction of H_2O as a result of prolonged exposure to the atmosphere.

Auger depth profiling of spray deposited CdS:Al films[26] indicated the absence of O in pure Al films and the presence of Cl in both doped and undoped films. Incorporation of residual Cl from $CdCl_2$ in the spray solution is thought to be responsible for the Cl presence. The position and shape of the Al peak in doped films have established that Al is present in the form of Al_2O_3.

teVelde[14] has observed that the concentration of Cd in the solid state reacted Cu_2S layer is below the detectability limit of the analysis equipment, indicating that the Cu_2S layer and, consequently, the junction, formed by the dry technique are relatively cleaner. Similar results have been reported by other workers,[21,39] although Cd presence at the surface of Cu_2S and Cu diffusion in CdS have been observed.

Martinuzzi et al.[22] have reported the presence of a slight transversal Zn gradient, with enriching within the free surface, for $Zn_xCd_{1-x}S$ layers prepared by evaporation from a single source. Coevaporation yields films that do not exhibit any gradient. AES and angle-resolved ESCA studies[55] have revealed that in $Cu_2S/Zn_xCd_{1-x}S$ cells, prepared by solid state reaction, the Zn and Cd profiles are distinctly different through the outer surface, up to the bulk of the base material. As the profile proceeds into the Cu_2S, an anomalously high Zn concentration, roughly proportional to the Zn content of the base layer, is observed. The Zn/Cd ratio is observed to be apparently larger than unity over much of the bulk of the Cu_2S layer. From the ESCA spectra, Cu^+, Cd^{2+}, and Zn^{2+} have been identified. S is present mostly as sulfide, with some sulfate (SO_4^{2-}) present on the outer surface. The results indicate that Zn^{2+} diffuses out more slowly, compared to Cd^{2+}, during the ion-exchange reaction. This conclusion is supported by the observed reduced rate of Cu_2S formation (in chemiplated cells) with increased Zn content.

A typical AES depth profile of a Cu_2S/CdS thin film solar cell is given in Figure 4.19 (p. 184).

7.4. Photovoltaic Performance

We shall first present some representative performance characteristics of $Cu_2S/ZnCdS$ thin film solar cells fabricated by different techniques. Later, we will discuss the effect of various process and material parameters on the performance of these cells.

7.4.1. *I–V* Characteristics

Chemiplated Cu_2S/evaporated $Zn_xCd_{1-x}S$, AR coated, thin film cells on Zn/Cu substrates[7] have exhibited the highest efficiencies to date. The photovoltaic parameters are: $\eta = 10.2\%$, $V_{oc} = 0.599$ V, $I_{sc} = 18.5$ mA cm^{-2}, and FF $= 0.748$ for a cell area of 0.98 cm^2 at an outdoor illumination intensity of 81.2 mW cm^{-2}. The Zn concentration x is 0.16. A similar cell with a lower Zn concentration ($x = 0.10$) has yielded comparable results, viz., $\eta = 10.1\%$, $I_{sc} = 21.1$ mA cm^{-2}, $V_{oc} = 0.561$ V, and FF $= 0.697$ at an intensity of 82 mW cm^{-2}. Conversion efficiencies up to 15% are expected at higher (0.30–0.35) Zn concentrations owing to an expected increase in V_{oc}.

AR coated, Cu_2S/CdS thin film cells, fabricated by chemiplating an evaporated CdS layer,[6,8] have achieved conversion efficiencies of up to 9.15% with $V_{oc} = 0.516$ V, $I_{sc} = 21.8$ mA cm^{-2}, FF = 0.714, and cell area = 0.884 cm^2, when tested under natural insolation of 87.5 mW cm^{-2}. The above results on Cu_2S/ZnCdS and Cu_2S/CdS cells have been obtained at Delaware University by Hall et al.[6–8]. Figure 7.5 shows the *I–V* characteristics of chemiplated/evaporated, FWR, AR coated, high-efficiency Cu_2S/ZnCdS and Cu_2S/CdS thin film solar cells.

In our laboratory, solid state reaction/evaporated BW cells have been fabricated exhibiting high I_{sc} values and efficiencies of about 10%.[18] However, some contribution to the I_{sc} has been detected from outside the designed active area. Thus the actual efficiency is slightly lower.

Hewig et al.[43] have reported at AM2 (85 mW cm^{-2}) an efficiency of 7.3% for encapsulated, 7 × 7-cm^2 area, chemiplated Cu_2S/evaporated CdS thin film cells on Ag/Cr coated glass substrates, corresponding to a $V_{oc} = 0.53$ V, $I_{sc} = 19$ mA cm^{-2} (for a Cu_2S layer area of 42 cm^2), and FF = 0.62. Photovoltaic generators composed of eight discrete 7 × 7-cm^2 cells connected by an integrated grid, single glass cover exhibit efficiencies up to 4.3%.[11]

Conversion efficiencies up to 6% have been achieved by Szedon et al.[13] on chemiplated Cu_2S/evaporated CdS cells on Zn/Cu substrates with $V_{oc} = 0.480$ V, $I_{sc} = 20$ mA cm^{-2}, and FF = 0.63, under 100 mW cm^{-2} W simulator illumination. Banerjee et al.[19] have fabricated AR coated,

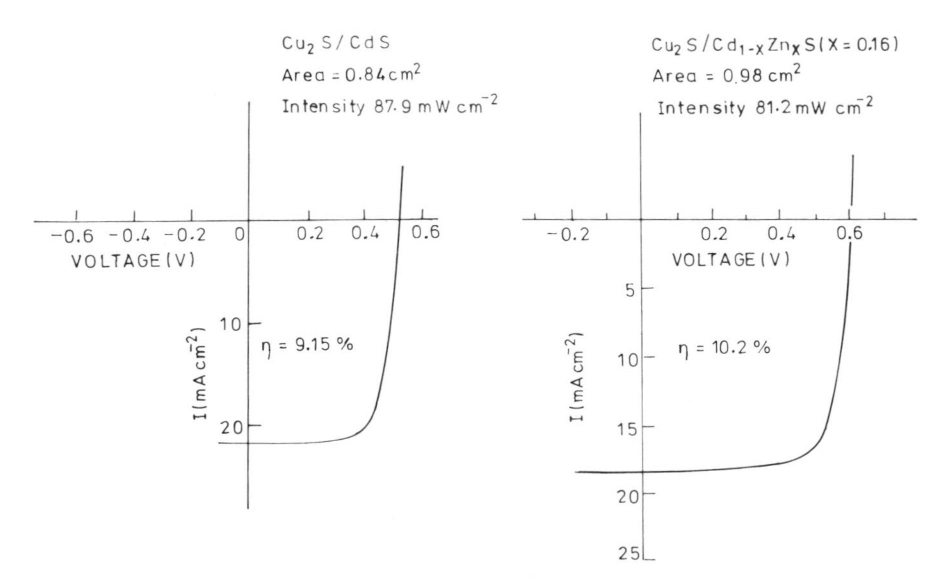

Figure 7.5. *I–V* characteristics of high-efficiency Cu_2S/ZnCdS and Cu_2S/CdS chemiplated/evaporated thin film solar cells (after Barnett et al.[6] and Hall et al.[7]).

$Cu_2S/Zn_xCd_{1-x}S$ thin film solar cells on Pb/Cr coated glass substrates by solid state reaction. The cells exhibited an efficiency of 6.5%, with $V_{oc} = 0.54$ V, $I_{sc} = 19$ mA cm^{-2}, and FF = 0.67, at an intensity of 100 mW cm^{-2}, on areas of 1 cm^2, for pure CdS films.

BWR chemiplated Cu_2S/evaporated CdS solar cells on Cd_2SnO_4 coated silica substrates have yielded efficiencies up to 5.2% under AM1 illumination over areas of 1.7 cm^2, with $V_{oc} = 0.48$ V, $I_{sc} = 17.1$ mA cm^{-2}, and FF = 0.634. Bhat et al.[18] have also obtained $V_{oc} = 0.47$–0.49 V, under 100 mW cm^{-2} W illumination on 1-cm^2 area, BW Cu_2S/CdS cells fabricated on SnO_x:Sb coated glass substrates by solid state reaction. Das et al.[15] have obtained $V_{oc} = 0.45$ V, $I_{sc} = 9$ mA cm^{-2}, and FF = 0.5, corresponding to $\eta = 2\%$ at 100 mW cm^{-2} illumination for their inverted geometry cells with Cu_2S below CdS and illumination through Cu_2S. High series resistance and low short-circuit current, resulting from unoptimized Cu_2S thickness and grid geometry, are the causes of the low efficiencies. Thin film Cu_2S/CdS cells fabricated by electroconversion in our laboratory have yielded typical efficiencies in the range of 2 to 3%. However, the cell design has not been optimized.

Spray deposited $Cu_2S/Zn_xCd_{1-x}S$ thin film cells have, in general, exhibited lower efficiencies than the corresponding evaporated cells. Martinuzzi et al.[56] and Bougnot et al.[41] have reported 1-cm^2 area, chemiplated Cu_2S/sprayed CdS BW solar cells with efficiencies up to 4.5%, accompanied by $V_{oc} = 0.425$ V, $I_{sc} = 19$ mA cm^{-2}, and FF = 0.56, under 100 mW cm^{-2} illumination. A high series resistance of 4 Ω degraded the FF. Singh[29] has prepared spray deposited cells with efficiencies up to 5.3% over a 1-cm^2 area. Jordan[30] has achieved efficiencies up to 4.92% on BW, all spray-deposited, Cu_2S/CdS solar cells on SnO_x coated glass substrates. Banerjee et al.[19] have achieved efficiencies of 5.6% on 1-cm^2, chemiplated Cu_2S/sprayed $Zn_{0.1}Cd_{0.9}S$ cells on SnO_x:Sb coated glass substrates, with $V_{oc} = 0.45$ V, $I_{sc} = 19$ mA cm^{-2}, and FF = 0.65, under 100 mW cm^{-2} illumination. The parameters for pure CdS base films are $V_{oc} = 0.4$ V, $I_{sc} = 15$ mA cm^{-2}, and FF = 0.7, corresponding to an efficiency of 4%.[24] The main factors responsible for the relatively lower efficiencies in spray deposited cells are the poor V_{oc} values (0.37–0.45) and the high series resistance and concomitantly low FF values. In unpublished communications, Singh has reported efficiencies of about 7% in sprayed cells. Martinuzzi has recently communicated an efficiency of 8% in spray deposited cells. In our laboratory, we have obtained high-efficiency BW sprayed cells with I_{sc} values exceeding 30 mA cm^{-2}, using an optimized design incorporating a back surface field. However, the short-circuit current is found to decay rapidly to lower values of about 20 mA cm^{-2} within a short time.

Reactively sputtered Cu_2S/evaporated CdS thin film solar cells on Au coated glass substrates[57] have exhibited a best efficiency of about 4% when tested in 82 mW cm^{-2} sunlight; considerably lower efficiencies have been obtained on all sputter-deposited Cu_2S/CdS cells.[31,58] Anderson and Jonath[58] have obtained $V_{oc} = 0.43$ V, $I_{sc} = 3.4$ mA cm^{-2}, and FF = 0.41, yielding an efficiency of 0.58% for an incident intensity of 107 mW cm^{-2} and cell area of 2.4 cm^2. Müller et al.[31] have obtained efficiencies of up to 1%. Hybrid cells consisting of sputter deposited CdS and solid state reaction grown Cu_2S solar cells[33,34,58] have exhibited $I_{sc} = 6.9$ mA cm^{-2} (AM1 illumination), $V_{oc} = 0.4$ V, and FF = 0.43, corresponding to an efficiency of about 1.2%. A low junction collection efficiency, owing to a low junction field in sputtered CdS, is concluded to be mainly responsible for the poor I_{sc} values obtained on these cells.

Screen printed Cu_2S/CdS solar cells[35] have been tested under 100 mW cm^{-2} illumination. $V_{oc} = 0.45$, FF ~ 0.5, and $\eta \sim 2\%$ have been observed. Efficiencies up to 4.7% have been observed on 0.01-cm^2-area, evaporated Cu_2S/electrophoretic CdS solar cells. On larger areas (1 cm^2), efficiencies up to 2% have been obtained. Chemiplated Cu_2S/electrophoretic CdS cells of 1-cm^2 area have yielded efficiencies in the range of 1 to 2%.

A characteristic feature of all thin film Cu_2S/CdS solar cells, irrespective of the fabrication process, is the crossover between the dark and light characteristics. This effect is due to the existence of deep levels in the compensated region of CdS near the junction interface. However, if the dark I–V characteristics are determined point by point and sufficient time is allowed at each point for the cell to attain equilibrium, then the resultant I–V plot, which is the true curve, does not exhibit a crossover with the light characteristics. This phenomenon is discussed in greater depth later.

Closely related to the crossover phenomenon is the spectral dependence of the I–V characteristics.[59] Although the light generated current and V_{oc} remain the same, I_{sc} and FF are lower for red-rich light compared to blue-rich light, as shown in Figure 4.3f (p. 158) for identical photon fluxes. Wavelength sensitive inhomogeneities have been observed on large cells by optical scanning techniques.[73]

Several workers have investigated the effect of illumination intensity and temperature on the light I–V characteristics of Cu_2S/CdS solar cells.[29,36,56,60–63] In an early work, Shirland[62] measured the characteristics of chemiplated/evaporated Cu_2S/CdS thin film solar cells over the illumination range 25 to 200 mW cm^{-2} and observed a linear variation of I_{sc} with intensity. The V_{oc} was found to rise, as expected, with higher light levels and saturated at very high intensities. The conversion efficiency, however, remained the same because of a decrease in the FF values from 0.71 to 0.66 with increase in intensity. The shunt resistance decrease from 100 to 7 Ω and series resistance decrease from 0.16 to 0.04 Ω were attributed to a photoconductive effect in the CdS. The maximum power point voltage V_m

increased with increasing intensity. In contrast to the above results, Bryant and Glew[61] have observed that the efficiency of Cu_2S/CdS cells increases very slowly as the intensity is decreased until a region is reached in which the series resistance is most sensitive to intensity changes. Below this intensity level, the efficiency decreases rapidly with decrease in intensity, as shown in Figure 7.6a. The shunt resistance R_{sh} decreases steadily with intensity, while the series resistance R_s increases with increasing intensity. The light generated current I_L increases linearly with intensity, as shown in Figure 7.6b. Also shown in the same figure are the variations of the reverse saturation current I_s and the diode factor n. The dependence of I_s and n on intensity have been suggested to be indicative of the influence of intensity on the barrier height at the Cu_2S/CdS interface.

In general, Cu_2S/CdS solar cells exhibit a linear dependence of I_{sc} and a logarithmic dependence of V_{oc} on intensity.[36,56,60,63] However, some workers[63] have reported a sublinear variation of I_{sc} with intensity at low

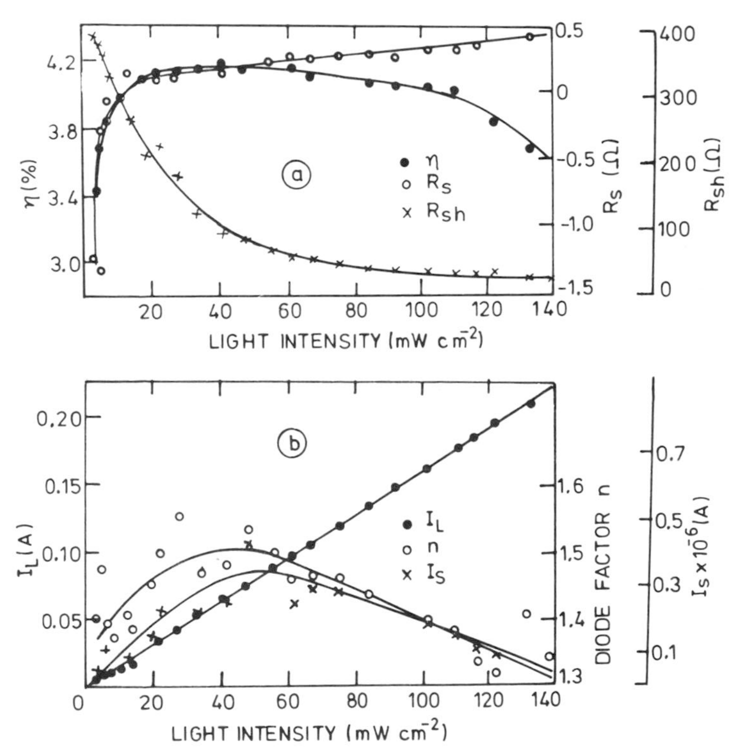

Figure 7.6. (a) Efficiency of a Cu_2S/CdS solar cell as a function of illumination intensity. The variations of R_s and R_{sh} are also shown. (b) I_L, I_s, and n vs. illumination intensity (after Bryant and Glew[61]).

intensities and superlinear variation at high intensities for long-wavelength monochromatic illumination (~9000 Å). Other workers[28] have observed a nonlogarithmic variation of V_{oc} with intensity for Cu_2S/ZnCdS, spray deposited solar cells. The variation of V_{oc} is shown in Figure 7.7. Also shown in the figure is the more typical logarithmic behavior of V_{oc}, exhibited by most Cu_2S/CdS thin film cells.

The effect of temperature on V_{oc}, I_{sc}, and η has been investigated by Shirland[62] in the temperature range −180 to 150 °C. The V_{oc} drops over the entire range. The decrease is initially slow up to −50 °C and rapid and linear thereafter up to 150 °C, with a temperature coefficient of −1.6 mV/°C. I_{sc} is relatively insensitive to temperature up to 100 °C, with a maximum at about −40 °C. Beyond 100 °C, I_{sc} falls rapidly. The efficiency rises rapidly at very low temperatures, reaching a maximum at −80 °C. At higher temperatures, η falls rapidly. Singh[29] has obtained a value of -1.3×10^{-3} V K^{-1} for the slope of the V_{oc} vs. T curve.

7.4.2. *Spectral Response*

A pronounced hysteresis is a common feature of the spectral response of all Cu_2S/CdS solar cells, with the response markedly lower while traversing from longer wavelengths to shorter wavelengths and higher while scanning from shorter wavelengths to longer wavelengths.[13,19,62,63] Moreover, the response decreases for $\lambda \geq 6500$ Å.[37] These effects occur owing to quenching phenomena related to the wavelength and intensity

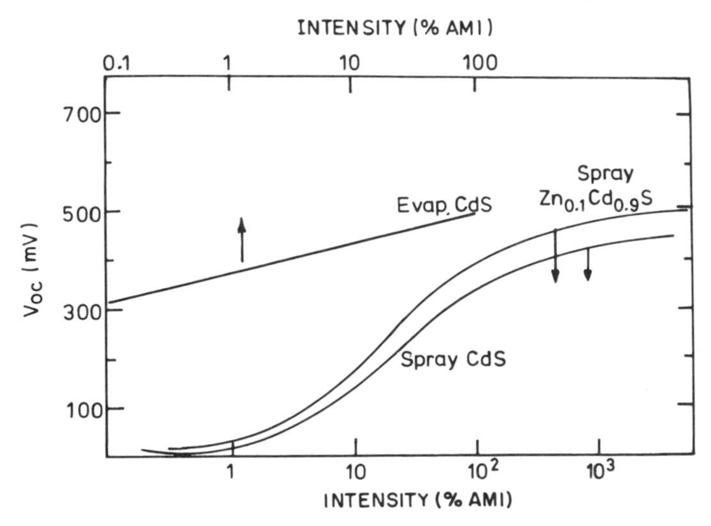

Figure 7.7. Dependence of V_{oc} on illumination intensity for spray deposited Cu_2S/ZnCdS cells. Also shown is the generally observed logarithmic V_{oc} variation in typical evaporated Cu_2S/CdS solar cells (after Martinuzzi et al.[28]).

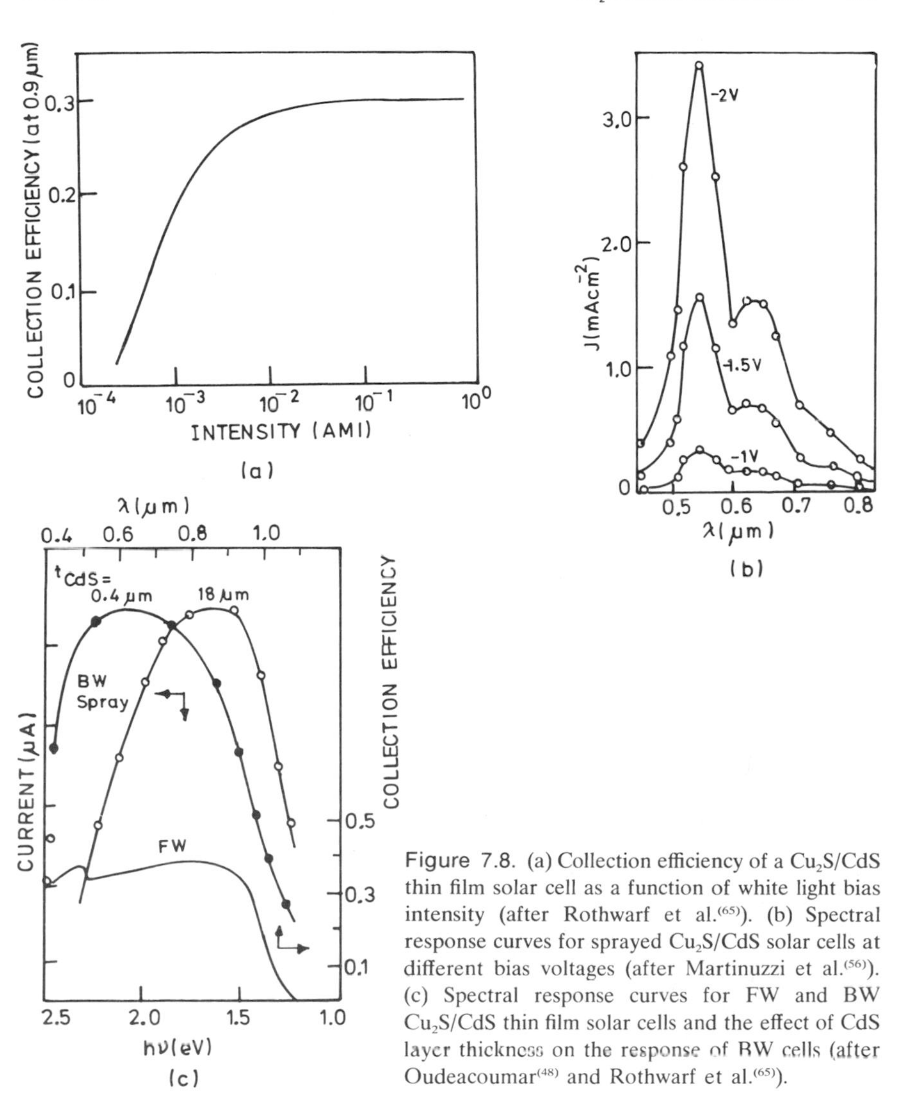

Figure 7.8. (a) Collection efficiency of a Cu_2S/CdS thin film solar cell as a function of white light bias intensity (after Rothwarf et al.[65]). (b) Spectral response curves for sprayed Cu_2S/CdS solar cells at different bias voltages (after Martinuzzi et al.[56]). (c) Spectral response curves for FW and BW Cu_2S/CdS thin film solar cells and the effect of CdS layer thickness on the response of BW cells (after Oudeacoumar[48] and Rothwarf et al.[65]).

dependence of the junction field.[59] However, these effects can be removed by simultaneous illumination with shorter-wavelength light or by white light.[13,19,37] Thus, the monochromatic response of Cu_2S/CdS cells measured under white light bias does not exhibit any hysteresis and the photoresponse over the entire wavelength range of interest is higher and more uniform.[64] The presence of the white light bias maintains a high field at the junction, improving the collection efficiency. In Figure 7.8a, we show the effect of bias light intensity on the collection efficiency of a high-efficiency Cu_2S/CdS thin film solar cell.[64,65]

The spectral response of Cu_2S/CdS solar cells is strongly influenced by an electric bias,[56] as illustrated in Figure 7.8b. The photocurrent increases with applied reverse voltage, approximately as $I_L \simeq V^4$. This voltage-bias-dependent photocurrent is attributed to the voltage-dependent component of the junction field and, hence, bias-dependent collection efficiency.[59]

Although the collection efficiency of BW cells is, in theory, higher, the photocurrent values of FW cells are, in general, relatively higher because of the more extended wavelength response of FW cells. Figure 7.8c shows the spectral response curves (with bias light) for a high-efficiency FWR[65] and BW spray deposited[48] Cu_2S/CdS solar cells. Also shown is the effect of the CdS layer thickness on the response of the BW cells.[48] Salient features of the curves are: (i) significant response for $\lambda \leqslant 5000$ Å for the FWR cell, resulting from minority carrier generation in and collection from both the Cu_2S and the CdS, (ii) cut-off at $\lambda \sim 5000$ Å for BW cells because of complete absorption of light of $\lambda \leqslant 5000$ Å in CdS, (iii) long-wavelength cut-off at $\lambda \sim 1$ μm, corresponding to the absorption edge of Cu_2S, and (iv) dip in the spectral response of FW cells in the region 5100 to 5500 Å.[19,38,42,58,64,66] The origin of this dip is controversial. While some workers[66] have invoked the formation of a photoconductive layer in CdS and a conduction band spike at the Cu_2S/CdS junction to explain the effect, other workers[42] have proposed insufficient absorption in the thin Cu_2S top layer and absorption in the bulk of the base CdS layer at this wavelength (and consequently lower contribution to the photocurrent) to be responsible for this effect. Indeed, the calculated spectral response of FW Cu_2S/CdS solar cells for different Cu_2S layer thicknesses show that the dip becomes more pronounced at lower Cu_2S thicknesses and disappears at very large thicknesses.[42]

7.4.3. *Minority Carrier Diffusion Lengths*

Diffusion lengths of minority carriers in Cu_2S/CdS solar cells have been estimated from spectral response[53,58,67] and EBIC[19,20,24,68–71] measurements and from laser scanning of tapered Cu_2S/CdS solar cells.[13,44] Banerjee et al.[18] have obtained minority carrier diffusion lengths of about 0.13 μm in CdS and 0.18 μm in Cu_2S for solid state reaction/evaporated Cu_2S/CdS thin film solar cells and about 0.41 μm in CdS and 0.27 μm in Cu_2S for spray deposited cells from EBIC measurements. Partain et al.[70] have reported diffusion length values in the ranges 0.11 to 0.57 μm in Cu_xS and 0.10 to 0.31 μm in CdS for chemiplated/evaporated cells. Gill and Bube[72] measured diffusion length values in the range of 0.1 to 0.4 μm in Cu_xS layers, using a light microprobe. However in CdS, which was a single crystal, they observed diffusion lengths of 3 to 7 μm.

From spectral response measurements, Moses and Wasserman[67] have estimated diffusion lengths in the range 0.08 to 0.26 μm in Cu_xS layers of thicknesses in the range 0.2 to 0.3 μm. Interestingly, the workers did not observe any significant changes in diffusion length after heat treatment of the cells in air and hydrogen, although large changes occurred in I_{sc} values. It was concluded that heat treatments led to changes in the junction electric field. Anderson and Jonath[58] have estimated diffusion lengths of about 0.1 μm in Cu_2S layers of all-sputter-deposited Cu_2S/CdS cells from spectral response measurements. Pfisterer et al.[53,68] and Schock[69] have obtained diffusion lengths of about 0.3 μm in Cu_xS from EBIC and spectral response measurements. Surface recombination velocities are found to be of the order of 10^4 cm s^{-1}. These results are for optimally heat treated cells, for which the Cu_2S stoichiometry is 1.995.

In contrast to the results of Moses and Wasserman,[67] Pfisterer et al.[53,68] have pointed out that heat treatment leads to drastic changes in the diffusion length and surface recombination velocities. Freshly dipped unheated cells, with Cu_xS stoichiometry of 1.98, yield low values of diffusion length ($\sim$0.1 μm) and high values of surface recombination velocity (10^6–10^7 cm s^{-1}). From EBIC measurements, Oakes et al.[71] also concluded that close-to-stoichiometric Cu_2S yields diffusion length values that are higher ($\geqslant$0.4 μm) than diffusion length values ($\sim$0.1 μm) obtained on less stoichiometric layers. Szedon et al.[13,44] have estimated diffusion lengths of approximately 0.2 μm in Cu_2S from laser scanning experiments.

7.4.4. *Capacitance Measurements*

Extensive capacitance measurements have been conducted on Cu_2S/CdS thin film solar cells to elicit information about the heterojunction.[20,24,29,42,56,58,59,64,68,74–78] In general, for unheated cells, the C^{-2}-vs.-V plots are straight lines, characteristic of an abrupt heterojunction, and the voltage intercept yields the value of the diffusion voltage V_D. Heat treatments have a profound effect on the C–V characteristics, consistent with the observed effects on the I–V characteristics and spectral response (discussed later). Heat treated cells invariably exhibit C^{-2}-vs.-V plots with two linear regions, a flat region extending from a few tenths of a volt in the forward bias to far reverse bias and a steep region in the extended forward bias region. The steeper region intersects the voltage axis. A typical C^{-2}-vs.-V plot for a Cu_2S/CdS solar cell is shown in Figure 7.9a.

Illumination of the solar cell causes an increase in the zero bias capacitance, associated with a decrease in the depletion layer width. The space charge density is observed to increase under illumination. Spectral dependence of photocapacitance reveals quenching and enhancement of capacitance at about the same wavelengths at which quenching and enhancement effects occur for the photocurrent.

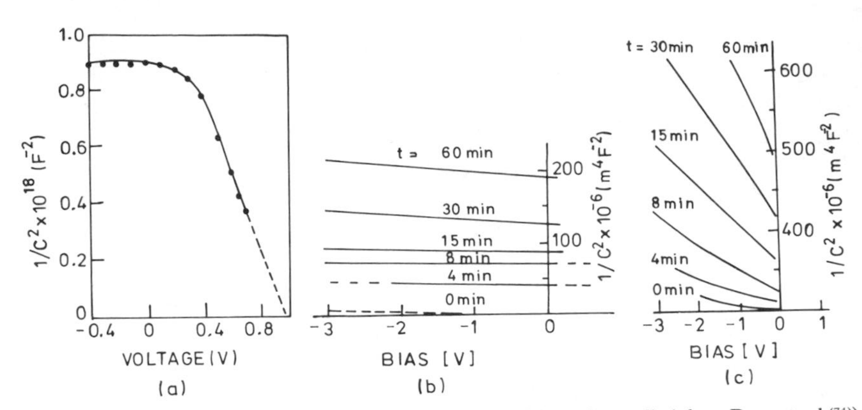

Figure 7.9. (a) C^{-2}-vs.-V plot for a sprayed Cu_2S/CdS solar cell (after Das et al.[74]). (b) C^{-2}-vs.-V plots of unetched Cu_2S/CdS evaporated cells for different heat treatment durations. (c) C^{-2}-vs.-V plots of etched Cu_2S/CdS evaporated cells for different heat treatment durations (after Pfisterer[76]).

From C^{-2}-vs.-V plots, Das et al.[74] have obtained, in sprayed cells, a value of approximately 1.5×10^{15} cm^{-3} for the carrier concentration near the junction and a value of about 10^{18} cm^{-3} in the bulk of the CdS. The depletion layer width was calculated to be about 0.9 μm. In evaporated cells, the carrier concentrations were determined to be about 3.5×10^{16} cm^{-3} near the junction and 3×10^{17} cm^{-3} in the bulk. The depletion width was estimated at 0.12 μm. Values of 0.96 and 0.87 V were obtained for the diffusion voltage in sprayed and evaporated cells, respectively. Singh[29] reported that his sprayed cells exhibited depletion layer widths of 0.74 μm in dark and 0.23 μm under AM1 illumination. Space-charge density was observed to change from 1.4×10^{15} cm^{-3} in the dark to 1.2×10^{16} cm^{-3} in the light. V_D was estimated to be 0.8 V. On additional heat treatment, the space-charge density in the dark increased to 2.4×10^{15} cm^{-3}. Hall and Singh[75] have measured space-charge densities of 8.9×10^{19} cm^{-3} in a 45-s heat treated (at 240 °C in air) and 2.5×10^{19} cm^{-3} in a 300-s heat treated chemiplated/evaporated thin film cell. They have observed that for heat treatment time of less than 900 s, the extrapolated intercept on the voltage axis increases with increasing annealing time. For $t > 900$ s, the capacitance becomes independent of the voltage and the space-charge density cannot be determined.

Pfisterer[76] and Pfisterer et al.[68] have obtained a doping concentration of $(1–15) \times 10^{17}$ cm^{-3} in planar (unetched) junction, unheated, chemiplated/evaporated Cu_2S/CdS solar cells. With increasing heat treatment, the C^{-2}-vs.-V curve shows a shift and a small increase in slope, as shown in Figure 7.9b. For nonplanar etched cells, these effects are much more pronounced, as shown in Figure 7.9c. Anderson and Jonath[58] have reported a saturation of the space-charge layer width in the vicinity of 0.48 μm, which is in good agreement with the thickness (0.5 μm) of the undoped, high-resistivity CdS layer in their Cu_2S/CdS/CdS:In sputter deposited solar cell. Martinuzzi et al.[56] have obtained a depletion layer width of 2 μm in spray deposited cells, with bulk donor densities in the range of 10^{14} to 10^{15} cm^{-3} and surface densities (at the interface) of 10^{15} to 10^{16} cm^{-3}.

Similar results have been reported by workers at Delaware.[42,64] Space-charge densities of 5×10^{15} cm^{-3} and of the order of 10^{17} cm^{-3} near the interface and in the bulk, respectively, were estimated. Under illumination, the space-charge density near the interface was observed to increase to 1.4×10^{16} cm^{-3}. On heat treatment, the space-charge density near the interface decreased to a value somewhat under 2×10^{15} cm^{-3} (under illumination) and the space charge layer width increased to about 0.5 μm from 0.1 μm for unheated cells. Quenching and enhancement effects have been observed in the photocapacitance behavior,[79] as illustrated in Figure 7.10a. The variation of capacitance with illumination intensity is shown in Figure 7.10b. The capacitance varies monotonically with intensity and at low illumination levels the capacitance values approach the dark value.[64]

The C–V characteristics have been interpreted by several workers on the basis of a compensated region formed near the interface from diffusion of Cu into CdS during heat treatment. Unheated cells show a linear C^{-2}-vs.-V plot characteristic of an abrupt heterojunction, and as the Cu_2S layer is nearly degenerate, the entire voltage is assumed to be supported by the depletion width in the CdS. With successive heat treatment, and, consequently, increasing compensation, a parallel shift occurs in the C^{-2}-vs.-V curve and the depletion width increases. The capacitance can be expressed[68] as a sum of a voltage-independent term and a voltage-dependent term, according to equation (4.13) (p. 174).

The doping concentration N_D can be determined from the voltage-dependent term, and the voltage-independent term, which causes the parallel shift, yields values of the product $N_A d^2$, where d is the distance from the interface to the further edge of the compensated layer and N_A is the acceptor concentration. If N_A is too high, the junction changes into a CdS p/n homojunction and for very low values of N_A, the depletion layer width is confined to the compensated region. In these two extreme cases the C^{-2}-vs.-V curves do not show a parallel shift and, if linearly extrapolated, intersect the voltage axis at $V = V_D$. Also, if the junction is nonplanar then an area factor has to be introduced into the capacitance. For an etched junction, with a larger area factor,[76] in addition to the shift in the C^{-2}-vs.-V curves, a change of slope also occurs. This change of slope has been attributed to a change in the area factor as a result of the smoothing of the edge of the space-charge region.[75,76] The area factor decreases with both heat treatment time and reverse bias voltage,[76] as shown in Figure 7.9c. The observations of a large charge density at the interface and a low intercept voltage can be explained by postulating a sheet of charge of large density at the junction as a result of the appreciable lattice mismatch between Cu_2S and CdS.[29]

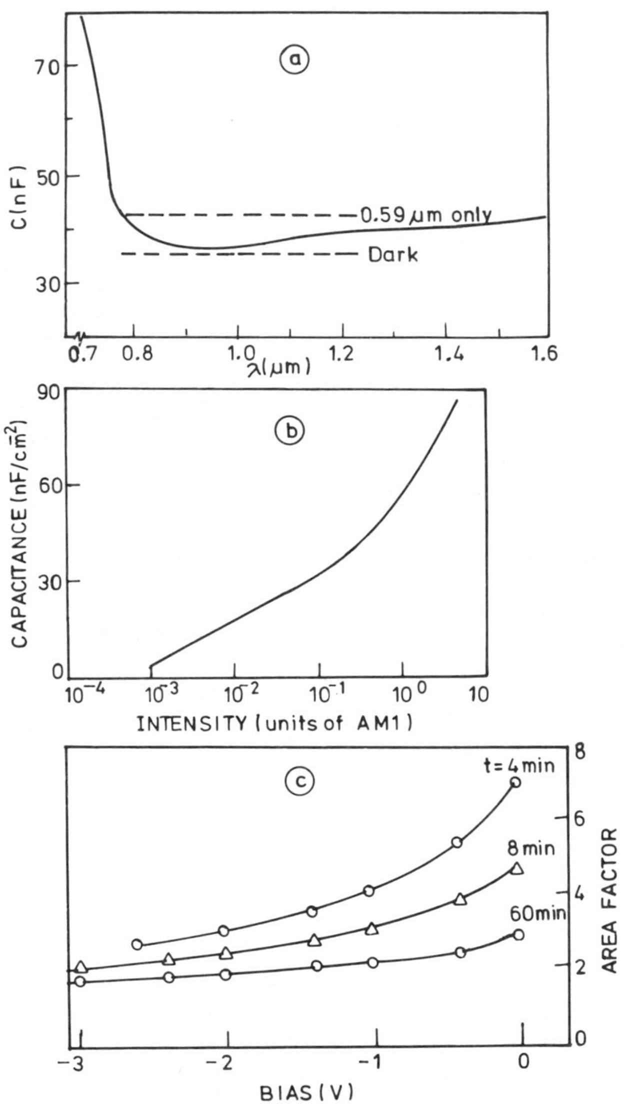

Figure 7.10. (a) Spectral dependence and (b) illumination intensity dependence of the capacitance of a Cu_2S/CdS thin film solar cell (after Rothwarf et al.[65] and Delaware Report[79]). (c) Reduction in the area factor of a Cu_2S/CdS thin film solar cell with heat treatment time and reverse bias (after Pfisterer[76]).

An alternate explanation for the capacitance behavior in dark and light has been offered by Rothwarf,[42] who has postulated the existence of deep levels in the CdS. Evidence for the existence of deep levels in the CdS of Cu_2S/CdS solar cells has come from DLTS measurements (Chapter 4). Starting from a forward bias corresponding to no space charge, as the voltage is decreased, electrons in a region w are swept out in order to support the voltage drop $V_D - V$ needed in the CdS. If $N_D^* = N_D - N_A$ is the net donor density, then

$$w = [2(V_D - V)\varepsilon_s / qN_D^*]^{1/2}$$

and

$$C^{-2} = 2q(V_D - V)/\varepsilon_S N_D^*$$

where ε_S is the dielectric permittivity of the semiconductor and q is the electronic charge. With further decrease in voltage, the Fermi level near the junction falls below the point where deep levels are uncovered. Thereafter, the additional voltage drop is provided by the charge removed from these deep levels rather than by widening of the space charge region. Thus, with further decrease in voltage, the width of the space-charge region changes very little and a low-slope region is observed.

The effects of heat treatment are viewed as an increase in the density of deep levels, rather than as an increase in the compensated region width. The increased density leads to a more compensated region and hence a lower value of N_D^*. In this case, while lowering the voltage from forward bias, a wider space-charge region is traversed before the deep levels are encountered, accounting for the observed increase in the depletion width.

The results of photocapacitance measurements have been explained in this model in terms of the ionization of deep levels. It should be borne in mind that light can ionize much deeper levels. Under illumination, the ionized deep levels provide the space charge needed to sustain the voltage drop, thereby reducing the width of the region supplied by the shallow donors, which is reflected in the higher capacitance values observed. The spectral dependence (see Figure 7.10) and, hence, the quenching and enhancement effects are then obviously related to the ionization of deep levels at some wavelengths and filling of deep levels at others.

7.4.5. *Analysis of I–V Characteristics*

Das et al.[74] have analyzed the temperature dependence of the $I-V$ characteristics of solid state reaction/evaporated Cu_2S/CdS thin film solar cells and observed two distinct slopes in the curves of $\log(I' + I_L)$ vs. V at all temperatures. The higher voltage region corresponds to a diode with diode factor $n = 1$ and the lower voltage region corresponds to a diode factor $n = 2$. The barrier height calculated for the $n = 1$ diode, assuming interface recombination, was found to be 0.98 eV. The reverse saturation current density J_s was determined to be about 7×10^{-9} mA cm^{-2}. A value of approximately 8×10^8 mA cm^{-2} was obtained for $J_{so} \simeq qN_cS_I$, from which a value of about 2×10^6 cm s^{-1} was determined for S_I, where S_I is the interface recombination velocity and N_c is the effective density of states in CdS. For the lower-voltage region generation–recombination in the depletion region was suggested to be the transport mechanism. Martinuzzi and Mallem[80] have reported that the dark forward and reverse currents, J_f and J_r, respectively, in chemiplated/evaporated cells could be expressed by $J_f = C \exp(\beta V + \gamma T)$ and $J_r = -CV \exp[-\lambda(V_d - V)^{-1/2}]$, where C is a constant and β and γ are practically independent of T and V, respectively. λ is a function of T. The observed exponential dependence of J_f and J_r on T, the linear variation of $\ln(J_r)$ vs. $(V_d - V)^{-1/2}$, and the weak dependence of $d(\ln J_f)/fV$ on T were suggested to be indicative of multistep tunneling mechanisms, with the number of tunneling steps being 50 to 80 for J_f and 10^3 to 5×10^3 for J_r. The linear variation of $\ln(J_r)$ with T was attributed to a linear variation of the bandgap of CdS with temperature. Hadley and Phillips[60] have measured the intensity dependence of V_{oc} and the temperature dependence of the I–V characteristics in chemiplated/evaporated cells and have observed two-diode behavior with the high-voltage region diode factor equal to unity and the low-voltage region diode factor equal to 2. The $n = 1$ diode was observed in all good cells and at higher intensities (10 to 100 mW cm^{-2}). The $n = 2$ diode was found to be active at lower intensities and led to reduced fill factors. The reverse current density J_s corresponding to the $n = 1$ diode was observed to vary exponentially with temperature, characteristic of a barrier height of about 0.9 eV. Barrier heights of about 0.4 eV were determined for the $n = 2$ diodes. Storti and Culik[78] have obtained barrier heights of 0.86 to 0.94 eV and J_{s0} values of 3.2×10^7 to 3.7×10^8 mA cm^{-2} for chemiplated/evaporated thin film cells.

For spray deposited cells, Das et al.[74] have observed a two-diode behavior in the temperature range 300 to 333 K and a barrier height of 0.86 eV, corresponding to the $n = 1$ diode and $J_s \simeq 10^{-9}$ mA cm^{-2}. Similar observations have been reported by Singh,[29] who reported that the $\log(J + J_L)$-vs.-V curves can be broken into two distinct linear regions, a high region, whose slope is practically temperature-independent, and a low region, whose slope is very weakly temperature-dependent. The breakpoint voltage between these two regions is observed to be dependent on the temperature and lies below the maximum power point. The dominant forward current mechanism in light is determined to be multistep tunneling–recombination, in the temperature range from -66 to 25 °C.

Martinuzzi et al.[56] have studied the I–V characteristics of chemiplated/sprayed Cu_2S/gradient doped CdS thin film solar cells and have concluded that multistep tunneling is the dominant current mechanism.

In contrast to the results on evaporated and spray deposited cells, Anderson and Jonath[58] have reported that the forward I–V characteristics of sputter deposited Cu_2S/CdS thin film solar cells appear to be described by a space-charge-limited current flow mechanism.

7.5. *Effect of Various Process and Material Parameters*

In this section we examine the effects of various process and material parameters on the performance of Cu_2S/CdS thin film solar cells, including the effects of heat treatments,[3,42,53,78,81–87] Cu_xS stoichiometry,[52,53,67,70,88–91] CdS resistivity,[93] doping,[3,79,89,91,94] $Zn_xCd_{1-x}S$ alloy composition,[7,19,22,28,38,40,59,74,79,95–97] and microstructure.[3,13,26,27,42,44,45,48,63,64,78,90,99–101]

7.5.1. *Heat Treatments*

We have noted previously that some sort of heat treatment seems to be necessary to obtain optimum performance in Cu_2S/CdS solar cells. This heat treatment takes the form of a vacuum heat treatment at 190 to 200 °C followed by a longer-duration heat treatment in H_2, or a glow discharge in H_2 followed by heat treatment in air, or deposition of a thin Cu layer on Cu_xS followed by heat treatment in air, or a simple heat treatment in air. Initially, the cells exhibit an almost ohmic behavior with very low V_{oc}, I_{sc}, and FF values. The baking process improves the photovoltaic behavior and the rectifying nature of the junction, as shown in Figure 7.11.

Various reasons have been advanced for the observed improvements. Rothwarf[42] suggested that freshly prepared cells consist of uncompensated CdS and nearly stoichiometric Cu_2S. The space-charge region is thus narrow[102] and ionization of deep levels near the junction on illumination causes a further narrowing. Tunneling to interface states occurs, lowering the effective barrier height, and, hence, V_{oc}. Heat treatment allows O_2 and/or Cu (from Cu_2S) to reach the space-charge region, forming compensating acceptor states in the CdS and thereby widening the space-charge region. The widened space-charge region restricts tunneling, thereby increasing the effective barrier, and, consequently, V_{oc}. The increase in photocurrent can be attributed to a decrease in S_I, or the formation of Cu_2O or CuO on the Cu_xS surface leading to a reduced surface recombination velocity. Overheating causes a reduction in the photocurrent (leaving V_{oc} unchanged) either because of a reduced field at the interface from

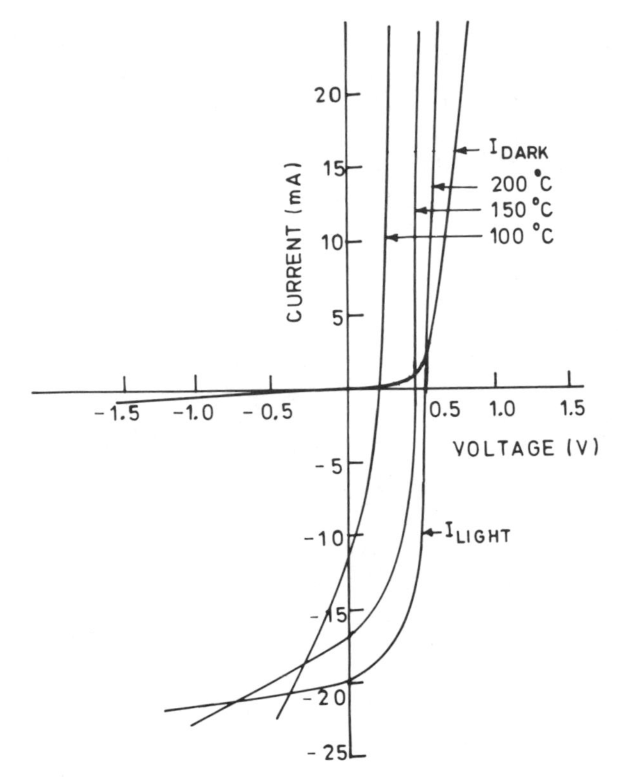

Figure 7.11. Evolution of the $I-V$ characteristics of a thin film Cu_2S/CdS solar cell with heat treatment.

overcompensation and a very wide space-charge region or because of changes in the stoichiometry of Cu_xS to Cu-deficient phases. Electronic states, caused by the mismatch between Cu_2S and CdS lattices, are associated with short-range stresses and long-range elastic stresses. It has been suggested that the long-range stress fields are eliminated by nucleation of prismatic dislocations in the interface region on heat treatment.

Devaney et al.[87] suggested that a heat treatment of 1 h at 190 °C in about 0.2 torr and 2 h at 170 °C in an H_2 atmosphere during grid lamination and subsequent reducing heat treatments have the first-order effect of reducing Cu oxides present, thereby releasing Cu, which enhances the stoichiometry of the Cu_xS layer. The higher stoichiometry is believed to be accompanied by longer minority carrier diffusion lengths,[78] and, hence, higher quantum efficiencies. Another factor contributing to the increased photocurrent is the alloying of Zn and Cu on heat treatment, leading to a highly reflecting α-brass layer, which causes the light to make a second pass through the Cu_2S layer. The increase in photocurrent from this effect is found to be of the order of 15%.

Pfisterer et al.[53] stated that the oxide layer consists of Cu_2O and not CuO. Further, they found no indication of $CuSO_4$. The Cu_2O layer, it is suggested, can be formed only from metallic Cu, for which either deposition of Cu or reduction of Cu_xS in H_2 glow discharge is necessary. These workers[53,83] have observed that the characteristic features of their heat treatment are: (i) poor stoichiometry of the Cu_xS layer and pronounced surface recombination losses directly after chemiplating, (ii) improved stoichiometry by Cu diffusion, but low efficiency owing to a Cu/Cu_2S Schottky barrier at the surface after Cu deposition or H_2 glow discharge, (iii) formation of a Cu_2O window layer, leading to reduced surface recombination, and an optimized Cu_xS stoichiometry, providing large electron diffusion lengths, after optimal heat treatment in air, and (iv) formation of Cu_xO $(1 < x < 2)$, accompanied by increased surface recombination losses and low electron diffusion lengths owing to degradation in Cu_xS stoichiometry on overheating or prolonged exposure to O_2.

Loferski et al.[85] suggested the formation of a wide bandgap $Cu_xS_yO_{1-y}$ window layer on Cu_xS on heat treatment. Akramov et al.[84] have obtained improvement in I_{sc} and V_{oc} with heat treatment in air. The rectification factors for dark $I-V$ characteristics were found to increase from 100 to 1000 with heat treatment. The shunt resistance was observed to increase by a factor of 2 to 5. These workers have suggested that microclusters of Cu present in the initially formed devices tend to short the p/n heterojunction. Thermal treatments lead to the diffusion of Cu from these clusters and to the breaking of the shunting channels. Amith[102] has, however, suggested that ohmic shunting is caused by thin layers of Cu_2S formed along the grain boundaries. Heat treatment causes the thin layers of Cu_2S to disappear from the grain boundaries, thus eliminating the shunting paths. Akramov et al.,[84] Caswell and Woods,[86] and Amith[102] have observed widening of the space-charge region in the CdS after heat treatment. Akramov et al.[84] opined that the widened compensated region in CdS serves as a more efficient absorber of the incident radiation and a better collector of the minority carriers thus leading to an enhanced photocurrent. The long-wavelength response increases dramatically.

Hall[81] has pointed out that a general effect of the pressure and heat treatments, perpetrated on the cell during lamination and subsequent fabrication processes, is to enhance the normal c-axis orientation and decrease the distribution span of c-axis tilts about this normal. He observed that after chemiplating and before laminating the c-axis of most of the grains is oriented about 19° from the normal and the distribution of tilts about this average is very broad, with a half-width at half-maximum height of 10 to 12°. After the lamination treatment, the c-axis distribution assumes an orientation normal to the substrate. Extended heat treatment after lamination sharpens the distribution about the normal, reducing the half-width at half-maximum height to about 3°.

7.5.2. Cu_xS Stoichiometry

Experimental results have revealed that the sheet resistance of the Cu_xS layer and I_{sc} depend very sensitively on the stoichiometry of the Cu_xS layer. The most direct evidence of this comes from the measurement of I_{sc} as a function of the stoichiometry[88,89] in the range $1.9 < x < 2$ shown in Figure 7.12a. Palz et al.[89] have reported that the highest quantum efficiency is obtained with the orthorhombic chalcocite phase of Cu_2S. For Cu-deficient phases, the quantum yield decreases, leading to a decrease in I_{sc}. Nakayama et al.[103] have observed that the relative efficiencies of Cu_xS/CdS cells varies with stoichiometry and are 100% for Cu_2S, 60% for $Cu_{1.96}S$, 10% for $Cu_{1.8}S$, and 5% for CuS. Similar results have been obtained by other workers.[53,91]

Wyeth and Catalano[92] have measured the spectral response of I_{sc} of Cu_xS/CdS cells as a function of x over the range $1.91 < x < 2$. Their results are shown in Figure 7.12b. Comparison with theoretical calculations have led to the conclusion that the decrease in the response with decreasing value of x is due primarily to the shortening of the minority carrier diffusion length in Cu_xS. The nearly parallel shift downward in the 6000 to 8000 Å range has been attributed to a possible decrease in the absorption coefficient in this range. Dielman[90] has reported that the minority carrier diffusion lengths in the various Cu_xS phases are below 50 Å, except for chalcocite, in which $L_e \simeq 350$ Å perpendicular to the c-axis and $L_e \simeq 900$ Å

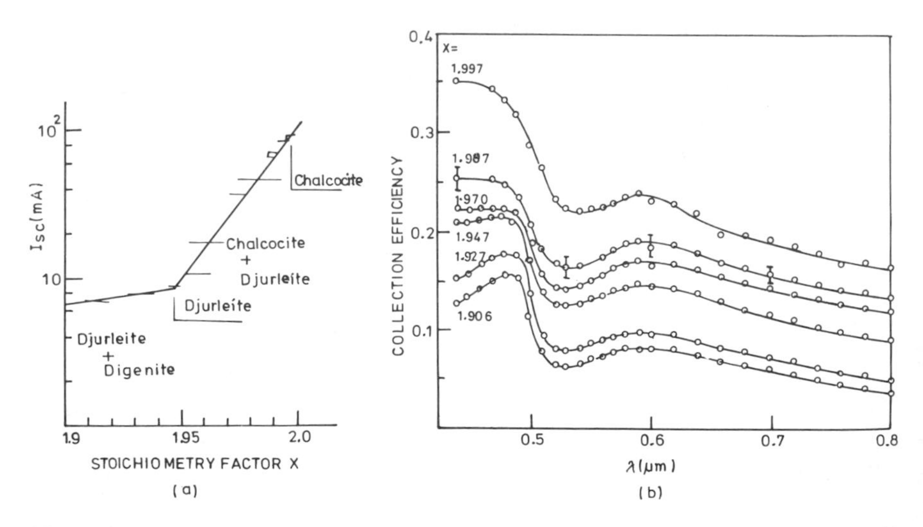

Figure 7.12. (a) I_{sc} and (b) spectral response of Cu_xS/CdS solar cells as a function of Cu_xS stoichiometry (after Palz et al.[89] and Wyeth and Catalano[92]).

at an angle of 35° to the c-axis. He concluded that only chalcocite has reasonable values of L_e, and L_e is highest parallel to the c-axis. A further observation is that chalcocite is the most efficiently absorbing phase and maximum absorption occurs for c-axis perpendicular to the substrate, i.e., along the c-axis. Other phases have a lower absorption coefficient.

Some workers[42] have explained the dependence of the electron diffusion length in Cu_xS on stoichiometry through the hole concentration, which is observed to be equal to the Cu vacancy density. If Auger recombination is assumed to be the dominant recombination mechanisms then the electron recombination lifetime $\tau \propto p^{-1}$, where p is the hole concentration. The diffusion length is given by

$$L_e = (\mu kT\tau/q)^{1/2} \propto (\mu kT/pq)^{1/2}$$

The sheet resistance depends on p as

$$R_\square = \rho/t = (q\mu_p pt)^{-1}$$

where t is the thickness of the Cu_xS layer. Thus if $I_{sc} \propto L_e$ (for $L_e < t$), then $I_{sc} \propto R_\square^{1/2}$. This relation is experimentally observed to hold over a considerable range.

In contrast to the above results, Moses and Wasserman[67] have not found any correlation between the measured L_e and the stoichiometry of the Cu_xS layer. They have attributed the changes in the I–V characteristics and spectral response with heat treatments to changes in the junction electric field. In our laboratory we have also observed that, for cells in the efficiency range up to 3 to 4%, the efficiency and other photovoltaic parameters are not very sensitive to the Cu_xS stoichiometry and no definite correlation between I_{sc} and x can be established.

7.5.3. *CdS Resistivity*

Marek[93] has measured various junction parameters of Cu_2S/CdS cells as a function of the base (CdS) carrier concentration in the range 3×10^{16} to 4×10^{17} cm^{-3}. The carrier concentration of the substrate is influenced by the density of grown-in donor-type defects. These defects also govern the extent of Cu diffusion into the CdS, and, hence, the depletion width. Junctions with higher carrier densities in CdS exhibit a higher degree of Cu compensation. Recombination has been observed to be the dominant transport mechanism. At low carrier concentrations, tunneling via recombination centers is found to be the transport mechanism in light. These results are corroborated by the decrease in J_s and increase in $\alpha = (1/nkT)$ values with increase in carrier concentration.

Rothwarf[42] has pointed out that high-resistivity CdS can lead to a reduced field at the junction, thereby adversely affecting carrier collection.

7.5.4. *Doping*

Bougnot et al.[104] have investigated the effect of doping Cu_xS with Cd, Zn, and In and observed an increase in stoichiometry directly proportional to the number of dopant atoms added during preparation. The improved stoichiometry is accompanied by an improvement in the electrical properties.

Experimental evidence suggests that Cu diffusion from Cu_xS into CdS after fabrication strongly influences cell degradation by causing the chalcocite phase to change to Cu-deficient phases. Partain and Birchenall[91] suggested that stability can be achieved through doping of the CdS to inhibit Cu diffusion. Vecht[105] has shown that doping with Cl increases Cu diffusion. Woodbury[106] found that heavy In doping reduces the diffusivity of Ag into CdS.

In addition to degradation of I_{sc}, Palz et al.[89] have observed a degradation in FF when the diode is kept under bias for extended periods of time. The origin of this degradation is the low shunt resistance caused by the electrodecomposition of Cu_2S under bias or at V_{oc} and subsequent diffusion and precipitation of Cu in CdS. Doping CdS with unspecified impurities has been found to inhibit Cu precipitation. Doped CdS cells show no degradation in performance after accelerated tests. For example, the shunt resistance of undoped cells decreases from 5000 Ω before test to 100 Ω after accelerated tests, while for doped cells, R_{sh} remains unchanged. The $I-V$ characteristics of the doped cells also do not exhibit any change after accelerated tests, while undoped cells show a marked degradation. Doping is found to yield a smaller depletion width, a lower resistivity, and a slightly higher spectral response. Luquet et al.[107] have obtained similar results on doped CdS/Cu_2S cells. They have observed that the space-charge width is divided by 4, the donor concentration near the interface is multiplied by 16, and the donor concentration in the bulk of CdS is multiplied by 50 in doped cells relative to undoped cells, as revealed by $C-V$ measurements. The ratio of the reverse currents of doped and undoped cells is observed to be as large as the ratio of the respective conductivities.

Umarov et al.[94] have investigated the effect of dopant (In and Ga) concentration in CdS, in the range 10^{12} to 10^{17} cm^{-3}, on the photovoltaic parameters of thin film Cu_2S/CdS heterojunctions. The V_{oc} and spectral sensitivity were observed to depend on the degree of doping. The variation of the efficiency and V_{oc} with impurity concentration have been explained in terms of changes in R_s, R_{sh}, and I_s. An optimum doping concentration for maximum efficiency was determined to be 10^{15} to 10^{16} cm^{-3}. However, their observation of a $V_{oc} \sim 0.8$ V is at sharp variance with the results of other workers who have obtained V_{oc} in the range of 0.5 to 0.55 V for pure CdS films. Hill[3] has pointed out that although In doping leads to low-resistivity CdS, segregation of In at the grain boundaries, particularly in films of low thicknesses (< 5 μm), causes low shunt resistances. Gradient doped CdS layers with low resistivity, $CdCl_2$ doped layer below, and high resistivity, pure layer above lead to an improved back contact and higher photogenerated currents.[3] Bodhraj et al.[26] have reported improved photovoltaic performance and greater junction stability with gradient doped CdS:Al cells fabricated by a spray process.

7.5.5. $Zn_xCd_{1-x}S$ Alloy Composition

On increasing the Zn content in $Zn_xCd_{1-x}S/Cu_2S$ solar cells, in general, the open-circuit voltage increases. However, the increase in V_{oc} is accompanied by a decrease in J_{sc} and η. Burton et al.[96] have observed that solid state reaction/evaporated $Cu_2S/Zn_xCd_{1-x}S$ thin film cells exhibit V_{oc} consistently in the 0.67 to 0.68 V range for Zn content above about 10%. Chemiplated cells exhibit such large voltages only at a Zn content of 40%. A maximum V_{oc} of 0.72 V has been obtained. The J_{sc} in both types of cells was found to decrease drastically from about 15 mA cm^{-2} for pure CdS cells to about 1 mA cm^{-2} for a Zn content greater than 35%. The V_{oc} was observed to vary linearly with the measured barrier height.

In a separate paper, Burton et al.[40] pointed out that $Zn_xCd_{1-x}S$ films deposited by vacuum evaporation from a single source are nonuniform in a direction normal to the substrate, owing to the preferential sublimation of Cd from the $Zn_xCd_{1-x}S$ powder charge. They have advocated the use of a concentric, dual-chamber source for obtaining greater uniformity and for controlling composition.[7] Burton and Hench[95] suggested nonoptimized Cu_xS thickness, poor Cu_xS stoichiometry, and high reflection from the unetched cell surface as probable causes for the poor J_{sc} values obtained on $Zn_xCd_{1-x}S$ cells. Series resistance was not found to be a limitation for $x < 0.2$. Subsequent experiments[79] confirmed that the Cu_xS thicknesses were indeed lower in $Zn_xCd_{1-x}S$ films, compared to those on CdS films for identical processing conditions. It was found that Cu_xS thickness was strongly dependent on the Zn content, as shown in Figure 7.13. However, even for Cu_xS thicknesses equal to those produced on CdS, the J_{sc} values were lower for $Zn_xCd_{1-x}S$ cells, although the stoichiometry was found to be good. This led to the suspicion that the collection factor in $Zn_xCd_{1-x}S$ cells was lower owing to a reduced field as a result of the high resistivity of these layers. Doping of $Zn_xCd_{1-x}S$ with In to reduce the resistivity did not yield the expected increase in current. Recently, Hall et al.[7] have obtained

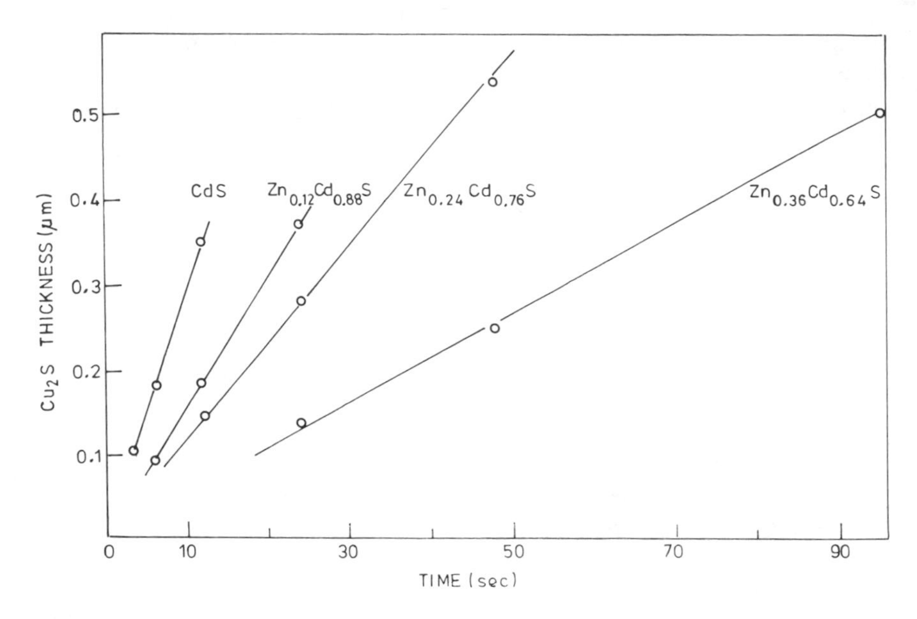

Figure 7.13. Cu_2S thickness as a function of chemiplating time for different composition $Zn_xCd_{1-x}S$ films (after Delaware Report[79]).

high current densities (~22–26 mA cm^{-2}, prorated to 100 mW cm^{-2}) in $Zn_xCd_{1-x}S$ solar cells, comparable to the best achieved in pure CdS cells for identical cell design. However, the improvements in cell processing leading to these current values have not been specified. Some workers[59] have reported that the open-circuit voltage of $Zn_xCd_{1-x}S/Cu_2S$ cells exhibit a time-dependent behavior. However, steady state values can be achieved after waiting for a sufficient period of time (~30 min). These effects have been attributed to deep levels in the depletion width region. The capacitance of these cells is observed to increase with Zn content, indicating that Cu diffusion is less in Zn-rich material.

Martinuzzi et al.[38] have observed an increase in V_{oc} from 0.48 to 0.60 V and a decrease in J_{sc} from 12 to 7 mA cm^{-2} on increasing Zn content from 0 to 15%. However, by using a bifilm (CdS–CdZnS) layer, J_{sc} values of 10.5 mA cm^{-2} could be obtained for a Zn content of 15%. A penalty has to be paid in the open-circuit voltage which is only 0.54 V. For solid state reacted cells, Martinuzzi et al.[22] observed that J_{sc} values of 12 mA cm^{-2} could be conserved up to a Zn content of 8%, corresponding to a V_{oc} value of 0.63 V.

Banerjee et al.[19] have carried out an extensive investigation of the performance of $Cu_2S/Zn_xCd_{1-x}S$ cells fabricated by vacuum evaporation and spray pyrolysis. They have reported highest conversion efficiencies of

6.5%, corresponding to $x = 0$, for evaporated cells and 5.6%, corresponding to $x = 0.1$, for sprayed cells. The variation of V_{oc}, I_{sc}, FF, and η with ZnS concentration for both evaporated and sprayed cells is shown in Figure 7.14. While the I_{sc} decreases with increasing x for evaporated cells, a peak in the I_{sc} variation is observed at $x = 0.1$ for sprayed cells. For the backwall sprayed cells, the short-wavelength cut-off in the spectral response is observed to shift toward shorter wavelengths on addition of Zn.

Singh and Jordan[97] have reported sprayed $Zn_xCd_{1-x}S/Cu_2S$ solar cells with stable open-circuit voltages up to 0.784 V at $x = 0.55$. The depletion layer width (in light) in $Zn_xCd_{1-x}S$ films was observed to increase with x, from 0.16 to 3.1 μm. J_{sc} values decreased from 18.7 to 1.9 mA cm^{-2} for a variation in x from 0 to 0.55. Martinuzzi et al.[28] reported an increase in V_{oc} from 0.48 to 0.58 V and a decrease in J_{sc} from 14.5 to 11 mA cm^{-2} in sprayed cells for a change of Zn content from 0 to 5%.

Das et al.,[74] on the basis of I–V measurements, reported that $Zn_xCd_{1-x}S/Cu_2S$ cells exhibit double-diode behavior at all x values. The higher-voltage region corresponds to a diode factor of $n = 1$ and the lower-voltage region to a diode factor of $n = 2$. For sprayed cells, the barrier height for reverse current ϕ_B increases from 0.86 eV at $x = 0$ to 0.96 eV at $x = 0.2$. For the evaporated cells, ϕ_B increases from 0.98 eV at $x = 0$ to 1.14 eV at $x = 0.3$. The reverse saturation current I_s for both types of cells decreases with increasing Zn content at all temperatures. For evaporated cells, the increase in V_{oc} with increasing Zn content is due to the reduction in I_s, caused primarily by the increase in ϕ_B. The authors have observed that the density of interface states is not significantly affected by the lattice parameter mismatch of the two materials. For

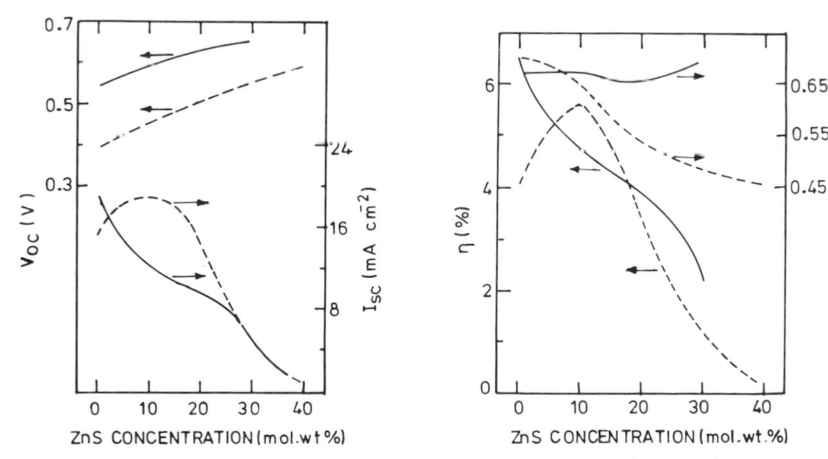

Figure 7.14. Variations of V_{oc}, I_{sc}, FF, and η of evaporated (———) and sprayed (-----) $Cu_2S/Zn_xCd_{1-x}S$ thin film solar cells with ZnS concentration (after Banerjee et al.[19]).

sprayed cells, the increase in V_{oc} is related to the increase in the barrier height which, in turn, is related to the increase in the bandgap of $Zn_xCd_{1-x}S$ with increasing x value, i.e., $\Delta V_{oc} \sim \Delta\phi_B \sim \Delta E_g$. For both evaporated and sprayed cells ΔE_c, the electron affinity mismatch, is found to decrease with increasing x. Martinuzzi et al.[22] also stated that the increase in V_{oc} in their cells is directly related to the increase of barrier height, and, consequently, to the decrease of ΔE_c.

7.5.6. *Microstructure*

The microstructure of the Cu_2S and CdS films strongly influences the junction characteristics, and, hence, the photovoltaic parameters of the Cu_2S/CdS heterojunction. Several workers have attempted to correlate the microstructural features with the observed photovoltaic performance in a quantitative way. Although the bulk of the photoresponse is contributed by Cu_2S, the collection takes place as a result of the field in the CdS. Thus the microstructure of the CdS plays a direct and important role. Moreover, since the growth of Cu_2S by the ion exchange reaction is topotactial, the structural features of Cu_2S are strongly influenced by the structure of the underlying CdS film. Theoretical calculations show that the photocurrent increases with increase in grain size.[100] Further, if the grains are columnar, recombination along the length of the columns would be less and higher photocurrents would result. Therefore, columnar films give better performance and the performance improves with increasing grain size. The columnar growth in evaporated films is accompanied by a strong preferred orientation of the c-axis perpendicular to the substrate. However, at low thickness, the grains are not oriented and are very small. Thus for low thicknesses of CdS films, the performance is expected to be poor. Amith[101] has investigated the thickness dependence of the photovoltaic parameters of Cu_2S/CdS solar cells. He found that the grain size, the preferential grain orientation, and the surface roughness increased as the CdS film thickness increased. The donor concentration and the quality of the diode improve with the film thickness while the voltage-dependent shunting paths diminish.

We noted previously that chalcocite with c-axis perpendicular to the substrate is found to be necessary for optimum performance. The topotactial growth of Cu_2S requires that the base CdS layer be oriented for the Cu_2S to be oriented. Some workers[63] have reported that evaporation of CdS at lower source temperatures ($\sim 950\,°C$) results in cells having higher V_{oc} values, other process parameters remaining the same. I_{sc} is not significantly affected. The lower-source-temperature CdS cells exhibit more spectral hysteresis than cells made from CdS deposited at higher source temperatures (1050 °C). The cells fabricated from higher-source-temperature deposited layers contribute strongly to the collection efficiency from the CdS for $\lambda < 5000$ Å for thin Cu_2S layers. It has been found that low evaporation temperatures produce higher resistivity CdS layers. Storti and Culik[78] have observed that CdS films grown at high rates exhibit high donor concentrations, larger positive space charge in the dark, and a higher field at the interface. The field in the dark is sufficiently high to sweep all carriers arriving at the interface past the junction without recombining.

Oudeacoumar[48] has reported that the spectral response of backwall sprayed CdS/Cu_2S cells is influenced strongly by the CdS film thickness, as illustrated in Figure 7.8c. Martinuzzi et al.[27] have observed that the crystallinity of sprayed CdS films depends on the S/Cd ratio in the spray solution. For a S/Cd ratio of unity, crystallite size is about 0.5 μm, the ⟨002⟩ orientation is pronounced, and the CdS layers are highly transparent ($\lambda > 5200$ Å). These films yield the best cells. Bodhraj et al.[26] have reported improved cell characteristics and greater stability with gradient doped CdS:Al films which exhibit a dense serpentine microstructure.

As noted earlier, in evaporated cells, Cu_2S forms down grain boundaries in the CdS, when Cu_2S is grown by the chemiplating technique. This effect produces a large increase in the effective area of the junction ($\gtrsim$ factor of 10) and a consequent decrease in V_{oc}. Moreover, etching of the CdS surface (to minimize reflection losses) accentuates this effect.[100] Hewig and Bloss[98] reported that the Cu_2S morphology is composed of horizontal and vertical structures and, for grain size of 0.5 to 1 μm, the area of the vertical part of the junction is larger by a factor of 10 compared to the horizontal part. Owing to the absorption of the light, the vertical junctions are strongly illuminated at the top, while at the lower part of the junction the intensity of illumination is nearly zero. The lower parts of the vertical junction, therefore, exhibit a behavior comparable to the dark junction, and, hence, reduced losses. For a 2000-Å horizontal layer thickness and 2000-Å width of the vertical layers, the calculated current densities in the horizontal and vertical junctions are found to be 20 and 10 mA cm^{-2}, respectively. The open-circuit voltage is found to be lower by 0.1 V for such cells compared to single crystal cells. The dependence of V_{oc}, I_{sc}, and η on crystallite size for such cells has been calculated by Hewig and Bloss.[98]

For planar Cu_2S/CdS cells formed by the dry technique, the peak in efficiency lies between a Cu_2S thickness of from 1000 to 1500 Å. For chemiplated cells, some workers have observed a strong correlation between cell performance and the Cu_2S morphology.[13,44,45] Higher output cells have been observed to be characterized by a high density of deep etch pits, extensive pyramiding of the tops of the CdS crystallites, and deep penetration of the Cu_2S film. Examination of Cu_2S structures of cells of varying outputs has led to the conclusion that denser, closer packed Cu_2S penetrations absorb more photons leading to higher output currents. The photoresponse at the grain boundaries has been observed to be 1.5 times the midgrain response. It has been calculated that 40% more photocurrent would be collected in a cell with grain diameters of 1 μm than in a cell with grain diameters of 2 μm. Rothwarf[100] has reported that optimally etched CdS surfaces exhibit a pyramidal structure with a 0.2 to 0.4 μm base and a 0.5 to 2 μm height.

An important observation is that polarity of the basal plane of CdS has an important influence on the growth and structure of Cu_2S film. Caswell et al.[99] have observed that the conversion of CdS to Cu_2S proceeds 1.5 times faster on sulfur faces and the photovoltages of unheated cells formed on sulfur faces are about 20% higher than for cells in which Cu_2S is formed on the Cd faces.

Although some workers[24] have reported a columnar structure in sprayed CdS films, the microstructure of the sprayed cells is markedly different from that of evaporated cells. Yet, efficiencies in the range of 7 to 8% have been obtained. These observations lead to the conclusion that the columnar grains microstructure thought to be necessary for attaining high efficiencies in evaporated cells is not an essential condition for attaining good performance in sprayed cells. Indeed, the disadvantage of smaller grain size and noncolumnar growth observed in sprayed cells can be compensated by providing back surface fields and drift fields in the CdS film, utilizing gradient doped films. It is, therefore, important to realize that the optimized design and microstructure of a sprayed cell is bound to be different from that of an evaporated cell of comparable efficiency.

A comparison of the performance characteristics of Cu_2S/ZnCdS thin film solar cells prepared by different techniques is given in Table 7.1.

7.6. *Band Diagram And Loss Mechanisms*

Several models have been proposed for the Cu_2S/CdS heterojunction.[1–4] Rothwarf[42] has synthesized all the previous models and developed the theory of interface recombination. An energy band diagram based on the above theory was proposed by Das et al.[74] and is shown in Figure 7.15. The values of the various diode parameters have been experimentally determined. In the model of interface recombination, for the usual values of carrier concentrations ($p \gtrsim 10^{19}$ cm^{-3} in Cu_2S and $n \gtrsim 10^{17}$ cm^{-3} in CdS), the space-charge region exists almost entirely in the CdS layer. In the frontwall mode, the solar spectrum greater than $h\nu = 1.2$ eV (bandgap of Cu_2S) is almost entirely absorbed in the 1000- to 3000-Å-thick Cu_2S layer creating electron–hole pairs. In the backwall mode, about 20% of the

Table 7.1. Performance Characteristics of Thin Film $Cu_2S/Zn_xCd_{1-x}S$ Solar Cells Fabricated by Different Techniques[a]

Device configuration	Deposition technique	Mode	Area (cm^2)	Int. ($mW\,cm^{-2}$)	V_{oc} (V)	J_{sc} ($mA\,cm^{-2}$)	FF	η (%)	Remarks	Ref.
p-Cu_2S/n-$Zn_{0.16}Cd_{0.84}S$	chem/evap	FW	0.98	81.2	0.599	18.5	0.748	10.2	With AR	7
p-Cu_2S/n-$Zn_{0.1}Cd_{0.9}S$	chem/evap	FW		82	0.561	21.1	0.697	10.1	With AR	7
p-Cu_2S/n-CdS	chem/evap	FW	0.884	87.5	0.516	21.8	0.714	9.5	With AR	6, 8
	chem/evap	FW	42	85	0.53	19.0	0.62	7.3	Encapsulated	43
	chem/evap	FW		100	0.48	20.0	0.63	6.0	With AR	13
	dry/evap	FW	1	100	0.54	19.0	0.67	6.5		19
	chem/evap	BW	1.7	AM1	0.48	17.1	0.634	5.2		18
	dry/evap	BW	1	100	0.47	37.1	0.60	10.4	Contribution from edges, see text	18
	dry/evap			100	0.45	9	0.5	2	Inverted geometry	15
	chem/sp	BW	1	100	0.425	19	0.56	4.5		41
	chem/sp	BW	1					5.3		29
	sp/sp	BW	1					4.5		30
p-Cu_2S/n-$Zn_{0.1}Cd_{0.9}S$	chem/sp	BW	1	100	0.43	19	0.65	5.6		19
p-Cu_2S/n-CdS	sputter/evap			82				4		57
	sputter/sputter		2.4	107	0.43	3.4	0.41	0.58		58
	dry/sputter			AM1	0.4	6.9	0.43	1.2		33, 34, 58
	chem.ptg			100	0.45		0.5	2		35
	evap/elph	FW	0.01					4.7		36

[a] evap — vacuum evaporation; chem — chemiplating; dry — solid state reaction; sp — spray pyrolysis; sputter — sputtering; chem.ptg — chemical printing; elph — electrophoretic; FW — front wall; BW — back wall.

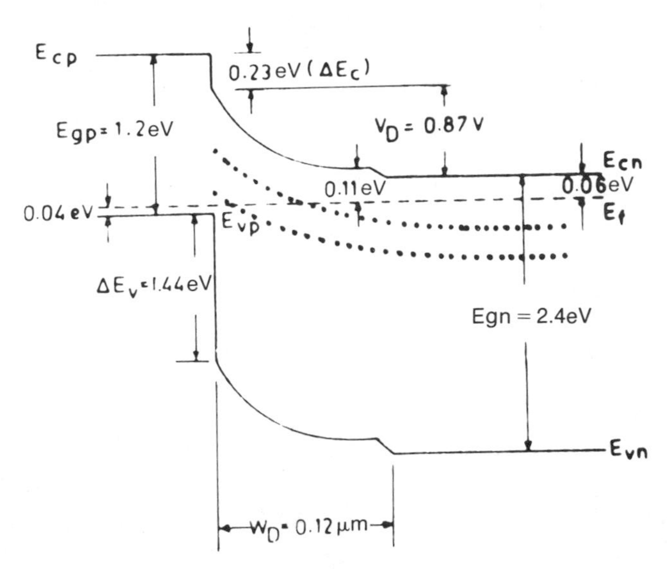

Figure 7.15. Energy band diagram of a Cu_2S/CdS heterojunction solar cell.

incident light is absorbed in the CdS layer. The electron–hole pairs in Cu_2S and/or CdS diffuse to the interface, where the electrons cross into the CdS and are either swept into the CdS by the field F_2 at the junction or are trapped by interface states and ultimately recombine with holes in Cu_2S. The field at the junction depends on the voltage across the cell, the photon flux, the wavelength of light, and the distribution of donors and acceptors near the junction in CdS. The field at the junction, and consequently the factors controlling the field, influence the short-circuit current and fill factor. The open-circuit voltage is determined by the interface recombination process, as the reverse saturation current flows by this mechanism.

The short-circuit current for a FW cell is given by $I_{sc} = A_{\perp} j_{sc}$, where $A_{\perp}$ is the geometrical area of the cell and

$$j_{sc} = \frac{\mu_2 F_2(\phi')}{S_I + \mu_2 F_2(\phi')} \int_{\lambda=0}^{\lambda_g} \Phi_0(\lambda) T_g [1 - R(\lambda)] \times [1 - A(\lambda)] \eta_{coll}(\alpha, L, d_1, F_1, r, S) d\lambda \tag{7.1}$$

where λ_g corresponds to the Cu_2S bandgap, μ_2 is the mobility in CdS, S_I is the interface recombination velocity, $\Phi_0(\lambda)$ is the incident photon flux density, T_g is the transmission of the grid contact, $R(\lambda)$ is the effective reflection coefficient of the cell, $A(\lambda)$ is the effective absorption loss coefficient of the cell, and η_{coll} is the collection efficiency of the Cu_2S layer. η_{coll} depends on the following properties of the Cu_2S layer: absorption coefficient $\alpha(\lambda)$, diffusion length L, effective thickness d_1, grain size r, drift field F_1, and surface recombination velocity S. The mode of operation (FW, BW, FWR, or BWR) also influences η_{coll}. Based on this model, the various loss mechanisms and their relative contributions to the short-circuit current in high-efficiency Cu_2S/CdS cells are: absorption/reflection (5–8%), grid shading (5–10%), surface recombination (1–2%), bulk recombination (10–15%), interface recombination (5%), grain boundary recombination (1%), and back surface + bulk CdS light loss (2%), yielding a total loss of about 25 to 35% and an AM1 short-circuit current density of 22 to 26 mA cm^{-2}.

Assuming that interface recombination is the dominant process and that the cell has a sufficiently large shunt resistance, we have

$$qV_{oc} = E_{g1} - \Delta E_C + kT \ln j_{sc} - kT \ln qN_{C2}S_I - kT \ln(A_j/A_\perp) \quad (7.2)$$

Here ΔE_C is the electron affinity mismatch between Cu_2S and CdS, N_{C2} is the effective density of states at the conduction band edge of CdS, q is the electronic charge, and A_j is the junction area. The various terms in equation (7.2) have been evaluated to be $(E_{g1} - \Delta E_C)_{max} \simeq 1$ eV, $kT \ln j_{sc} = -0.1$ eV, $kT \ln qN_{C2}S_I = -0.33$ eV, and $-kT \ln(A_j/A_\perp) = -0.03$ to -0.06 eV (depending on the junction area), yielding a maximum V_{oc} of 0.54 V in a planar cell. Indeed, in solid state reacted Cu_2S/CdS cells formed on unetched CdS, V_{oc} values of 0.54 V have been achieved.[19] Significant improvement in the V_{oc} can be achieved by changing ΔE_C and S_I. However, since ΔE_C and S_I are fundamental material properties, they can only be modified significantly by modifying/changing materials. $Zn_xCd_{1-x}S/Cu_2S$ cells are expected to yield $E_{g1} - \Delta E_c = 1.2$ eV and a $V_{oc} \sim 0.74$ V. Experimentally observed values of V_{oc} for such cells are in the range of 0.65 to 0.71 V. The increase in V_{oc} is mainly due to an increase in the bandgap of $Zn_xCd_{1-x}S$ (relative to CdS), which is accompanied by a decrease in ΔE_C and an increase in $E_{g1} - \Delta E_C$.

In an ideal diode, for a $V_{oc} \sim 0.51$ V, the fill factor is expected to be 0.80. In Cu_2S/CdS thin film cells, the fill factor is reduced by 0.04 to 0.02 owing to series resistance and 0.03 from the field dependence of the short-circuit current. The best Cu_2S/CdS cells to date exhibit a fill factor of 0.68 to 0.72.

Based on the above loss analysis, the highest efficiency expected in Cu_2S/CdS cells is about 11%. For $Cu_2S/Zn_xCd_{1-x}S$ cells, efficiencies of 15% are expected. It should be noted that if bulk recombination in Cu_2S becomes the dominant process (instead of interface recombination), then the possible room temperature value of V_{oc} increases to 0.86 V and the value of the fill factor increases to 0.86, yielding a value of 26% for lossless efficiency.

7.7. Conclusions

Efficiencies exceeding 10% have been demonstrated on small-area $Cu_2S/Zn_xCd_{1-x}S$ thin film solar cells prepared by chemiplating evaporated CdS films. Higher efficiencies are expected with optimized $Zn_xCd_{1-x}S$ composition and better control of the growth and formation of Cu_2S on $Zn_xCd_{1-x}S$. High efficiencies have also been obtained on sprayed cells. However, the optimum microstructure in sprayed cells is vastly different from that of evaporated cells. On large-area, production-scale devices, efficiencies of 6 to 7% have been achieved. Major efforts are needed to obtain, reproducibly, efficiencies exceeding 10% on large-area devices on a pilot plant basis with high yield and stability. The control of degradation, stabilization, and encapsulation of Cu_2S/CdS cells remain major challenges and are discussed in Appendix F.

8

Polycrystalline Thin Film Silicon Solar Cells

8.1. *Introduction*

Silicon, in its bulk single crystal and polycrystalline form, is the most extensively characterized and best understood semiconductor material. The highly developed and fast moving Si technology continues to form the backbone of the semiconductor industry and the dominance of Si in electronic applications is almost total. In addition, Si finds extensive applications in optoelectronics, integrated optics, and computer technology.

In photovoltaic solar energy conversion, the only photovoltaic system to find wide-scale applicability to date (as, for example, in power generators in artificial satellites) is based on single crystal Si solar cells. Thus, extensive experience and technical expertise exist on Si, which make it possible for the industry to take up a technology based on Si thin films and scale it up to suitable proportions to maintain a viable photovoltaic industry. Add to this the high conversion efficiencies obtained on bulk Si solar cells ($\eta \sim 16$–19%) and the abundance of Si and we have all the ingredients of a successful candidate for thin film photovoltaic applications. It is, therefore, surprising that, until recently, little research effort has been expended on developing solar cells based on Si thin films. Perhaps the lack of success in fabricating good-quality devices in the early years[1] discouraged further vigorous developmental activity on Si and, consequently, other II–VI (CdS, CdTe) and III–V (GaAs) semiconductors were more intensively investigated as potential thin film solar cell materials.

Recent studies in several laboratories have resulted in dramatic improvements in the performance of thin film Si solar cells. Large-area (30 to 50 cm^2), 8 to 9% efficient cells have been fabricated with Si films deposited on recrystallized metallurgical-grade Si (mg-Si) substrates by CVD processes.[2] Conversion efficiencies of 12% (area 2×2 cm^2) and

9.93% (area 28 cm^2) have been obtained on RTR (ribbon-to-ribbon) laser recrystallized, TESS (thermal expansion shear stress) separated Si films grown on Mo substrates by high-pressure plasma deposition.[3] Several innovative deposition processes have been developed in an effort to grow large-area, good-quality Si thin films suitable for photovoltaic devices. In Chapter 6 we discussed the properties of polycrystalline Si thin films prepared by various techniques. In this chapter, we confine ourselves to a description of the fabrication processes and performance characteristics of polycrystalline Si thin film solar cells. Amorphous Si solar cells are discussed in Chapter 10.

8.2. *Current Status of Bulk Silicon Solar Cells*

To provide a relevant background against which to view the development of thin film Si solar cells, we shall briefly review the current status of bulk single crystal and polycrystalline solar cells. The performance characteristics of these cells are tabulated in Table 3.3 (p. 150).

In the last decade, several developments in Si solar cell design led to improvements in efficiency to the range of 16 to 19%. Based on the calculations of Wolf,[4] Lindmayer and Allison[5] obtained a great improvement in the short-wavelength response and fill factor by introducing a shallow junction coupled with a change in grid structure. Named the violet cell, it had an efficiency around 16%. Reduction in front surface reflection losses was achieved by replacing evaporated SiO_x as an AR coating by Ta_2O_5, CeO_2, and TiO_2, all of which, possessing refractive indices between those of Si and glass, showed least reflectance in conjunction with a cover glass. By etching the ⟨100⟩ Si surface with a suitable etchant (e.g., KOH, NaOH, or hydrazine hydrate), a textured surface, composed of pyramids having ⟨111⟩ facets, was obtained, which helped to trap the incident light by multiple reflections and thus reduce reflection losses to very low values. The textured cells, also called the CNR (Comsat nonreflecting) cells achieved efficiencies in the range of 17 to 18%. Modification of the back surface ohmic contact, by including a lower resistivity layer of the same type as the base material at the back surface, provided an aiding built-in field at the back surface of a conventional n^+/p or p^+/n solar cell. The presence of the back surface field resulted in an unusually high V_{oc} (~ 0.61 V) accompanied by a high fill factor (~ 0.82), an improved J_{sc} (~ 49.6 mA cm^{-2}, AMO), an enhanced long-wavelength response, and a lower-resistance ohmic contact for minority carriers.[6–8] The overall effect was an AM1 efficiency around 19%. A combination of all these approaches has enabled the realization of photocurrents that are 90% of the theoretical limit and V_{oc} that is within 75 to 80% of the expected value.[8]

Dendritic web grown ribbon solar cells[9,10] have exhibited AM1 efficiencies of about 14% with typical efficiencies of about 12%, while edge-defined film-fed growth (EFG) ribbons[11,12] have been used to fabricate solar cells with a maximum AM1 efficiency of about 12% and typical efficiencies of around 10%. Spray deposited ITO/polycrystalline n-Si SIS solar cells 1 cm^2 in area have exhibited efficiencies of 11.2%.[13] Ion beam sputtered ITO/polycrystalline p-Si SIS solar cells[13] have achieved efficiencies of 11.5% over areas of 11.5 cm^2 and 9% over areas of 18.5 cm^2. MIS inversion layer cells[13] on polycrystalline p-Si have been fabricated with efficiencies of 8% over a 1-cm^2 area.

8.3. *Fabrication Technology*

Thin film Si solar cells have been fabricated by CVD,[2,14–22] e-beam evaporation,[23] SOC,[24–26] and TESS[3] processes. The highest efficiencies (9–12%) have been obtained on epitaxial Si films deposited by CVD[2,14,15,18,19] and on recrystallized TESS separated Si films.[3] A wide variety of substrates have been employed including steel with a borosilicate diffusion barrier,[20] graphite,[20] a purified mg-Si wafer,[17,21] and Ti coated alumina[22] for CVD processes; sapphire[23] for evaporation; carbon[26] and C coated mullite[24] for SOC techniques; Mo for the TESS method[3]; and recrystallized mg-Si,[2,14,16,18,19] and EFG ribbons[15] for epitaxial growth.

Although a few studies of MIS[25,26] devices have been reported, most thin film Si solar cells have been fabricated in the AR coating/grid/n^+-Si/p^+-Si/substrate type of configuration. A typical process[2,16] for fabricating high-efficiency cells consists of the following steps in order: (i) pulverization of mg-Si and purification by repeated leaching in aqua regia, (ii) unidirectional solidification of mg-Si on graphite plates to produce low-resistivity (0.01 Ω cm) p^+-type mg-Si sheets with relatively large crystallites on graphite plates to serve as substrates, (iii) deposition of about a 25 μm, 0.1- to 1-Ω cm, p-type epitaxial Si layer and about a 10 μm, gradient doped, n^+-type Si film sequentially, by CVD technique (thermal reduction of trichlorosilane with H_2 containing appropriate dopants) at a substrate temperature of about 1150 °C and an average deposition rate of 1 μm/min, (iv) evaporation of a Ti/Ag grid structure through a metal mask, (v) application of an SnO_2 AR coating, employing oxidation of tetramethyltin at 400 °C in an Ar atmosphere, and (vi) annealing of the deposited structure in a He atmosphere to promote diffusion of impurities to grain boundaries. The graphite plate acts as an ohmic contact to the p^+-region and the low-resistivity p^+ mg-Si substrate provides a back-surface field at the p-Si/p^+-Si interface. The top n^+-Si layer is graded to provide a drift field.

Evaporated,[23] SOC,[24] and TESS[3] deposited Si films have been fabricated into devices using conventional diffusion techniques to dope the films and form *p*/*n* junctions.

8.4. *Photovoltaic Performance*

For ready reference, we have listed the performance characteristics of several Si thin film solar cells fabricated by different techniques in Table 8.1. In the following sections, we shall examine the factors affecting the performance of these cells.

8.4.1. *Photovoltaic Characteristics*

Si_3N_4 antireflection layer coated solar cells fabricated from RTR laser recrystallized, TESS deposited Si films have exhibited AM1 conversion efficiencies of up to 12% with $V_{oc} = 0.582$ V, $J_{sc} = 28.3$ mA cm^{-2}, and FF = 0.73 on areas of 2 cm × 2 cm. Lower efficiencies are observed on large-area (28 cm^2) cells, with maximum reduction resulting from a lowering of J_{sc}. The performance figures for the large-area cells are

Table 8.1. Performance Characteristics of Thin Film Si Solar Cells Fabricated by Different Deposition Processes

Deposition technique	Type	AR coating	Area (cm^2)	Intensity (mW cm^{-2})	V_{oc} (V)	J_{sc} (mA cm^{-2})	FF	η (%)	Ref.
TESS	Homo	Si_3N_4	4	AM1	0.582	28.3	0.73	12	19
TESS	Homo	Si_3N_4	28	AM1	0.595	23.1	0.72	9.93	19
CVD	Homo	SnO_2	30	AM1	0.56–0.58	19–22	0.75	9	2
CVD	Homo	SiO	8.3	100	0.58	15.2	0.67	5.9 (no AR) (7.3 with AR)	17
CVD	Homo	—	4.4	80	<0.1	6	—	0.05	20
CVD	Schottky	—	6.25	AM0	0.33	13	0.44	1.4	20
SOC		SiO	1.05	100	0.53	26.8	0.675	9.6	24
SOC	Homo	TiO_2	—	AM1	0.495	23	0.55	6.2	26
SOC	MIS	TiO_2	—	AM1	0.38	21.6	0.48	3.9	26
SOC	Hetero	ITO	—	AM1	0.28	25	0.23	1.6	26
Evap.	Homo	—	—	75	~0.2	12	0.6	1.9	23

$V_{oc} = 0.595$ V, $J_{sc} = 23.1$ mA cm^{-2}, FF = 0.72, and $\eta = 9.93\%$. Large dislocation densities (locally up to about 10^6 cm^{-2}) in the grain enhanced films are thought to be the factor limiting the performance of TESS solar cells.

AM1 efficiencies of up to 12%[19] have been obtained on epitaxial Si (epi-Si) films grown on 0.05 Ω cm mg-Si substrates purified by the heat exchanger method (HEM). The cell performance is enhanced by the naturally occurring high–low junction behind the epitaxial layer. Robinson et al.[18] reported efficiencies exceeding 10% for solar cells fabricated on epitaxial CVD thin films on upgraded polycrystalline mg-Si. Large-area epi-Si/mg-Si cells have been developed by Chu and co-workers[2] with AM1 efficiencies of about 9% (30-cm^2 area) and about 8% (50-cm^2 area) with V_{oc}, J_{sc}, and FF in the ranges of 0.56 to 0.58 V, 19 to 22 mA cm^{-2}, and 0.70 to 0.72, respectively. The large-area cells have lower V_{oc} and J_{sc} values than the small-area cells. The effective minority carrier diffusion length in these cells has been measured to be 15 to 25 μm.

Warabisako et al.[17] have fabricated polycrystalline Si solar cells on mg-Si substrates by CVD and obtained an AM1 efficiency of 7.3% with a cell configuration of a 25-μm-thick p-type active layer, a 0.5-μm-thick n^+-type surface layer, and a SiO AR coating for a relatively large area of 8.3 cm^2. The low J_{sc} (13.1 to 15.2 mA cm^{-2}, without AR coating) observed is attributed to the unsatisfactory diffusion length of the photocarrier in the active layer, which was measured to be in the range of 7 to 11 μm. Grain boundaries were found to play a relatively minor role in the photocurrent degradation. Similar results were obtained by the same workers on dendritic Si thin film solar cells on alumina substrates.[22] From spectral response measurements they estimated an electron diffusion length in the p-type layers of about 1 μm. From their laser scanning studies, they have inferred that the low diffusion length value is not determined by the presence of defects but by deep-level impurities, based on their observation of a variation of only approximately 10% in the photocurrent at irregular grain boundaries.

The microstructure of the active Si layer plays a dominant role in the performance of these thin film solar cells. The film microstructure depends sensitively on the deposition conditions and the quality and nature of the substrate. Schottky barrier and p/n junction polycrystalline CVD Si solar cells on steel substrates have yielded low V_{oc} ($\lesssim 0.1$ V) and η ($\sim 0.05\%$) because of the poor microstructure of the Si film (average grain size is 2.5 μm) and stress in the junction region from thermal expansion coefficient mismatch between Si and steel.[20] CVD films on graphite, with considerably better microstructure, have exhibited V_{oc}, J_{sc}, and FF of 0.33 V, 13 mA cm^{-2}, and 0.44, respectively, corresponding to an AM0 conversion efficiency of 1.4%,[20] without AR coating. Grain boundary limited AM0 efficiencies of about 2.8% (without AR coating) were improved to about 3.5% by increasing the dopant concentration in the p-layer of an n^+-Si/p-Si/p^+-Si/mg-Si CVD solar cell.[21] Doping reduced the grain boundary effects and resulted in a higher V_{oc} and FF and lower series resistance. From their spectral response measurements on CVD Si/mg-Si solar cells, Chu et al.[16] estimated the effective minority carrier diffusion lengths to be about 15 to 20 μm in the valley regions and 30 to 40 μm in the face regions, corresponding to a 15 to 20% lower photoresponse in the valley region. It should be noted that the valleys are usually high-angle grain boundaries.

AR coated large-grained (up to 4 cm × 0.4 cm) SOC solar cells of 1 cm^2 active area have exhibited a conversion efficiency of 9.6% with J_{sc} as high as 26.8 mA cm^{-2} at 100 mW cm^{-2}, V_{oc} equal to 0.53 V, and an FF value of 0.675. As the cells were fabricated with both the contacts on the same side, the series resistance was high, thus limiting the performance. It was found from light and electron beam induced current (LBIC and EBIC) measurements that grain boundaries detracted from device performance, while twin boundaries did not significantly influence performance. The carrier diffusion lengths were found to vary from 6 to 25 μm.

Fabre and Baudet[26] observed an enhancement in minority carrier diffusion length from 6 to 22 μm on increasing the photon flux from about 10^{13} to 3×10^{17} photons cm^{-2} s^{-1} in their SOC cells. The cells exhibit an AM1 efficiency of 6.2% with $V_{oc} = 0.495$ V, $J_{sc} = 23$ mA cm^{-2}, and FF = 0.55. The diffusion length enhancement has been attributed to the light saturation of traps present in the material. The photocurrent is observed to follow a supralinear ($J_{sc} = K\phi^{1.05}$, where ϕ is the photon flux and K is a constant) dependence on the light flux. Light enhancement of minority carrier diffusion length has been observed in EFG ribbon solar cells also.[27,28] A trap filling phenomenon at various depths in the bulk neutral region of the EFG cell has been found to be consistent with the observed strong dependence of the diffusion lengths on the volume generation rate as well as on the wavelength of the superimposed light.

MIS type devices on Si films have, to date, not exhibited high efficiencies. A Cr-MIS process[25,29] has been used to fabricate 6 to 8% efficient cells on SOC films. AR (TiO_2) coated Ti-MIS and ITO coated SIS diodes on Si films[26] have exhibited efficiencies of 3.9 and 1.6%, respectively, with corresponding J_{sc} values of 21.6 mA cm^{-2} and 25 mA cm^{-2} under AM1 illumination.

8.4.2. *Junction Analysis*

Diode characteristics of thin film Si junctions have been investigated by several workers[16,17,20–23,30] in an effort to elucidate the current transport mechanisms and reveal the material parameters affecting cell performance.

From I–V measurements on a series of evaporated Si solar cells of different grain sizes, prepared by varying the deposition temperature, Feldman et al.[23] have found that the dark characteristics could be analyzed by a two-exponential diode equation with one of the preexponential factors, J_{s1}, and diode factor, $n_1 = 1$, corresponding to diffusion-limited operation and the other preexponential factor, J_{s2}, and diode factor, $n_2 = 2$, corresponding to a recombination-dominated mechanism. Independent measurements of diffusion length and grain size show a decrease in J_{s1} with increase in diffusion length, as expected theoretically, and a decrease in J_{s2} with increasing grain size, implying decreased recombination sites with larger grain diameters.

The dependence of J_{s1} and J_{s2} on diffusion length and grain size, respectively, is shown in Figure 8.1. V_{oc} and J_{sc} as a function of grain size is plotted in Figure 8.2a and η vs. grain size is depicted in Figure 8.2b. Also shown in Figure 8.2b are the theoretical estimates of several researchers.[31–34] Extrapolating their experimental results, Feldman et al.[23] inferred that films containing 30-μm grain diameters would yield 10% efficient cells. They did, however, point out that owing to columnar growth of evaporated films, 50-μm-thick films would be needed to obtain 30-μm grain diameters. Also, based on the reported efficiency of only 1.5% in CVD Si films with 20 to 30 μm grains deposited on graphite,[35] compared to this efficiency value in evaporated films with 5 μm diameter grains, the role of lattice mismatch, purity of grain surfaces, and doping concentration has been emphasized.

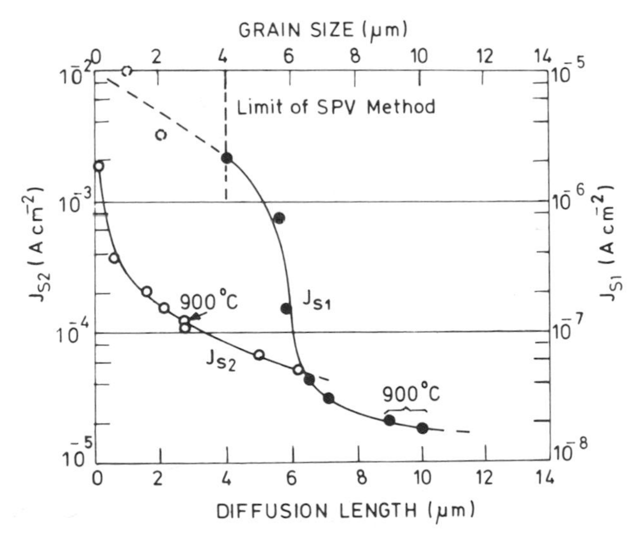

Figure 8.1. Saturation current densities J_{s1} and J_{s2} as a function of diffusion length and grain size, respectively, for an evaporated Si solar cell (after Feldman et al.[23]).

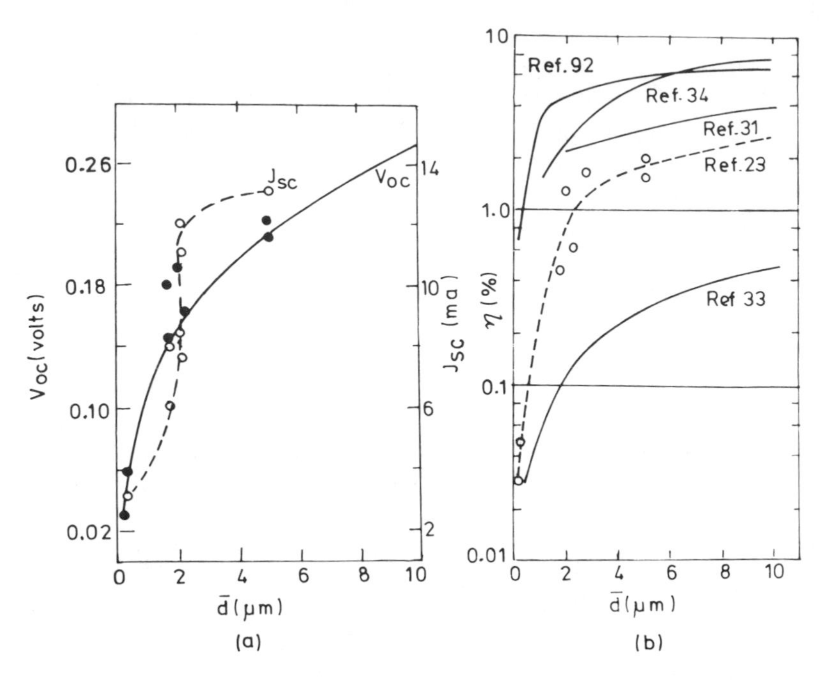

Figure 8.2. (a) V_{oc} and J_{sc} as a function of grain size, for evaporated Si cells. Solid curve represents a theoretical plot. (b) Dependence of η on grain size. Dotted curve represents experimental data (after Feldman et al.[23]).

Two-exponential diode characteristics have also been observed in CVD Si solar cells. Warabisako et al.[17] have determined that $J_{s1} \leq 10^{-11}$ A cm^{-2} and $J_{s2} \sim 10^{-7}$ A cm^{-2}, with $n_1 = 1$ and $n_2 = 2$.

Chu et al.[20] observed poor rectification properties of Schottky barrier and p/n junction solar cells fabricated on CVD-Si film on steel. The diode factor calculated from the forward characteristics was found to be 3.9 and 2.8 for the p/n junction and the Schottky barrier, respectively, compared to 1.8 for single crystalline p/n junctions and 1 for ideal Schottky barriers. The reverse current density in the thin film p/n junction was found to be several orders of magnitude higher than that of single crystal cells with similar resistivity profiles. The reverse current density of the Schottky cells was considerably higher than that of the p/n junction. The high reverse current in these films deposited on steel was attributed to the small grain size in the film and stress in the junction region. Better rectification properties were obtained on Si films deposited on graphite.[20] A diode factor of 1.9, comparable to the single crystal value, was determined from the forward characteristics, indicating that grain boundaries in these films do not significantly influence current conduction. The reverse currents were observed to be more than an order of magnitude less than that of films on steel.

Chu and Singh[21] found that passivation of grain boundaries by heavy doping improved the dark *I–V* characteristics of CVD-Si solar cells on mg-Si substrates.[21] They also observed a considerable improvement in the forward and reverse characteristics of an n^+-Si/p^+-Si/mg-Si cell over those of an n^+-Si/p-Si/mg-Si cell. The improved diode characteristics were reflected in an enhanced photovoltaic performance. Seager et al.[30] reported better performance of polycrystalline solar cells after hydrogenation of grain boundaries, as revealed by EBIC and *I–V* characteristics. They observed a marked improvement in the dark *I–V* characteristics after hydrogenation, with the improvement in reverse current being quite dramatic, as shown in Figure 8.3. In a separate study, Seager and Ginley[36] have optimized the hydrogenation process to the point where all measurable grain boundary potential barriers are removed to depths of 0.2 to 0.5 mm in Si. The geometric configuration and the density of the hydrogen plasma and the temperature and surface preparation of the polycrystalline Si were observed to be important parameters affecting the passivation process.

In an elegant experiment, Chu et al.[16] demonstrated the effect of film microstructure on the photovoltaic performance of epi-Si/mg-Si solar cells by measuring the dark *I–V* characteristics of a set of mesa diodes fabricated at regions exhibiting different microstructural features. The

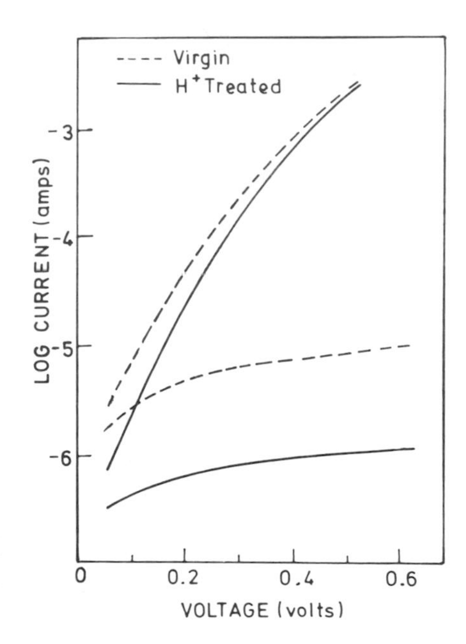

Figure 8.3. Dark *I–V* characteristics before and after hydrogenation (for 16 h at 350 °C in 2 torr H^+ plasma) of an n/p polycrystalline Si solar cell (after Seager et al.[30]).

diodes exhibited a two-exponential diode behavior with $n_1 = 1$ and $n_2 > 1.5$. The J_{s1} values of diodes fabricated at the face and valley regions were observed to be similar to those of single crystalline epitaxial diodes with similar resistivity and thickness of n- and p-regions. The J_{s2} values of face diodes were similar to those of the single crystalline diodes. However, the J_{s2} values of valley diodes, containing randomly oriented crystallites, were found to be appreciably higher, indicating high density of recombination centers at grain boundaries. From the temperature dependence of J_{s1} and J_{s2} (Figure 8.4), the corresponding barrier heights ϕ_{B1} and ϕ_{B2} were deduced and the values were found to be as follows: ϕ_{B1} is 1.17 to 1.24 eV for valley diodes, ϕ_{B1} is 1.19 to 1.24 eV for face diodes, ϕ_{B2} is 0.53 to 0.63 for valley diodes, and ϕ_{B2} is 0.58 to 0.66 eV for face diodes; these values are in agreement with expected values, i.e., E_g for J_{s1} and $E_g/2$ for J_{s2}.

In contrast to the above results, the diode characteristics of dendritic Si thin film solar cells on alumina[22] show space-charge-limited conduction, with the current increasing with voltage as a quadratic function. Deep-level impurities possibly play an important role.

8.4.3. *Loss Mechanisms*

From the preceding discussion it is clear that grain boundaries play a dominant role in the performance of thin film Si solar cells. The two most important solar cell degradation mechanisms associated with the polycrystalline nature of the films are a low collection efficiency owing to increased minority carrier recombination at grain boundaries and a reduction in the p/n junction quality. The estimates of some workers[23] on the optimum

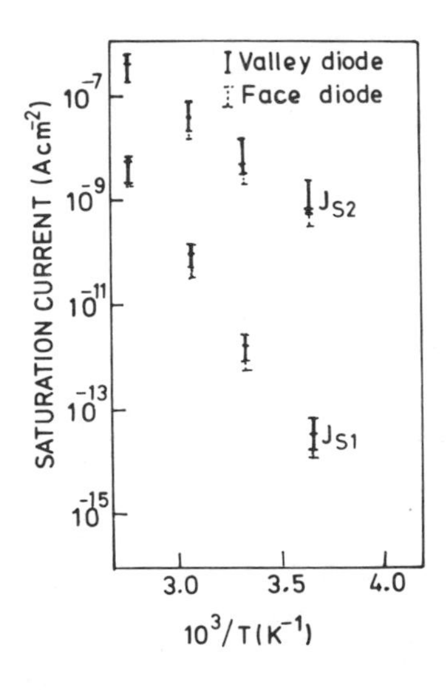

Figure 8.4. Saturation current densities J_{s1} and J_{s2} of several face and valley diodes as a function of temperature (after Chu et al.[16]).

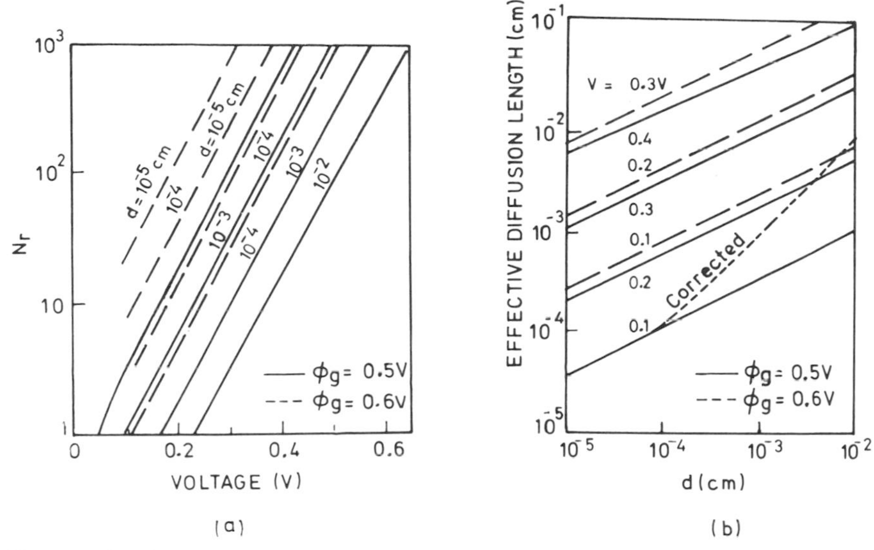

Figure 8.5. (a) Theoretical plot of the number of grain boundaries contributing to recombination N_r as a function of bias voltage. (b) Dependence of minority carrier diffusion length on grain size and bias voltage (after Card and Hwang[37]).

grain size to achieve conversion efficiencies in excess of 10% need to be treated with circumspection, as other film properties such as deep-level impurities, doping concentration, microstructural defects like dislocations and vacancies, and stresses may also influence minority carrier transport and thus modify the relative contribution of grain boundaries.

Card and Hwang[37] have carried out a theoretical investigation of Si thin film Schottky barriers taking into account the doping concentration, the grain size, the grain boundary potential, and the interface state density at grain boundaries. In a separate paper, Card and Yang[33] discussed the dependence of grain boundary potentials on illumination. An important result for the Schottky barrier cells on thin silicon films is that the number of grain boundaries contributing to recombination is dependent on the bias voltage, as shown in Figure 8.5a.[37] The bias-voltage-dependent effective diffusion length as a function of grain size is plotted in Figure 8.5b for a range of bias voltages and for two different values of the grain boundary potential. Wu et al.[38] have fabricated Al–poly-Si Schottky barrier solar cells and observed that low-angle grain boundaries have little effect on the I–V characteristics. The diode factor of a device containing twin and low-angle boundaries was measured to be 1.17. However, high-angle grain boundaries significantly alter both the I–V and low-frequency C–V characteristics, introducing recombination centers and traps, resulting in an increase in recombination current and a reduction of the effective mobility.

Recombination losses represent carrier losses. Thus, while the dependence of carrier losses on microstructural properties, and hence on deposition conditions, has been studied, there has been no investigation of photon losses. Si, because of its indirect bandgap and low absorption coefficient, requires a large thickness for adequate absorption. The thin film Si cells, made of layers 30 to 50 μm thick, must, therefore, be losing a substantial amount of the incident photon flux as a result of insufficient absorption. Moreover, reflection losses also cause a decrease in short-circuit current. Although most solar cells are AR coated, it is not quite clear if the AR layers are optimized for that particular device design.

8.5. *Future Directions*

Epitaxial Si thin film solar cells seem to be at a fairly advanced stage of development. The large-area cells need to be assembled into panels and the modules subjected to field tests and accelerated life tests to ascertain the stability of these cells against environmental effects. Although experience with bulk Si devices indicates that Si is a stable material, it should be remembered that thin films, with their large surface-to-volume ratio, high density of defects, and peculiar microstructural properties, may exhibit enhanced chemical reactivity and/or diffusion. It would, therefore, be prudent to establish the degradation characteristics of these thin film Si cells. Further, the devices need to be produced on a pilot plant basis to demonstrate the cost viability of the processes, after reproducing the high efficiencies on larger areas.

Further improvements in device efficiencies would require systematic and extensive material-related studies, primarily: (i) A complete understanding of the deposition process and its effect on the microstructure of the deposited film. (ii) Development of the deposition process, based on the above study, to prepare large-area, large-grain, high-quality, defect-free films with low concentrations of deep-level impurities and microstructural defects like dislocations and vacancies. (iii) Investigatioh of the nature and type of the grain boundaries. (iv) Passivation of grain boundaries by hydrogenation and/or doping. (vi) Proper design of the device geometry based on carrier loss and photon loss analysis. The last would require optimization of the active and doped layer thickness, the grid structure, the AR coating, the back contact reflection, and the doping profile of the top layer. In this regard, the possibility of increasing carrier collection by decreasing the absorber layer thickness, while maintaining adequate photon absorption by causing a double pass of the light using a highly reflecting back contact, as suggested by Redfield[39] needs to be investigated.

9

Emerging Solar Cells

9.1. *Introduction*

There are in addition to polycrystalline Si, amorphous Si, and Cu_2S/CdS thin film solar cells, several other semiconductors that have exhibited potential as high-efficiency photovoltaic materials in thin film form: GaAs, CdTe, InP, Zn_3P_2, CdSe, Cu_2Se, $CuInSe_2$, $ZnIn_2Se_4$, and Cu_2O. The interest in GaAs and CdTe dates back to the 1960s, but doubts about the adequate availability of these materials kept interest at a low level and confined to only a few laboratories. Recent years have witnessed a revival of research and development efforts on these two materials. The interest in the other materials is of more recent origin and, in the case of InP and $CuInSe_2$, has been inspired by the observation of high (~ 15 and 12.5%, respectively) conversion efficiencies when they are fabricated into single crystal heterojunctions with CdS as the window material. Zn_3P_2 is attractive from the point of view of availability. For the same reason, and also because of very low cost, organic materials have excited interest in certain quarters. Of all the above-mentioned materials, only two, $CuInSe_2$ and CdTe, have reached an advanced stage of development in thin film form, exhibiting conversion efficiencies of about 10 and 8%, respectively. Interestingly, both $CuInSe_2$ and CdTe high-efficiency thin film solar cells employ CdS as the heterojunction window. The progress in the development of the other materials is encouraging, though not spectacular.

The properties of GaAs, CdTe, InP, Zn_3P_2, CdSe, Cu_2Se, $ZnIn_2Se_4$, $CuInSe_2$ and other Cu-ternaries, and Cu_2O thin films were discussed in Chapter 6. The photovoltaic performance of these materials as thin film solar cells forms the subject of this chapter.

9.2. *Gallium Arsenide (GaAs)*

The best studied bulk GaAs solar cell has been the *p*/*n* homojunction. Typically, GaAs *p*/*n* cells were made on polished single crystal wafers, doped n-type to about 10^{17} cm^{-3}. Zn or Cd acceptors were diffused to a depth of about 1 μm to form the p/n junction.[1] Some workers have fabricated cells with a shallower (0.25 to 0.5 μm) junction and the best terrestrial efficiencies obtained were about 12%.[1] Spectral response measurements revealed two major loss mechanisms: (i) small minority carrier diffusion length in the n-type GaAs and (ii) high surface recombination velocity and poor diffusion length in the p-type GaAs, both leading to loss in photocurrent.

In recent years, GaAs solar cells have been developed in the $Ga_{1-x}Al_xAs/p$-GaAs/n-GaAs heteroface[2] and the p-$Ga_{1-x}Al_xAs/n$-GaAs heterojunction[3] structures with vastly improved efficiencies of 15% AM0 and 19% AM1, respectively. p-GaAlAs/p-GaAs/n-GaAs/n^+-GaAs concentrator solar cells[4] have exhibited efficiencies of 23% at a concentration ratio of 9.9 and temperature of 30 °C. The $Ga_{1-x}Al_xAs$ layer reduces the surface recombination losses by confining the electrons generated in the p-GaAs layer and also provides a transparent, low-resistance contact to the GaAs, thereby decreasing the series resistance.

GaAs solar cells have also been fabricated as Schottky barrier and MIS devices. Stirn and Yeh have reported conversion efficiencies of 15% for single crystal Au/GaAs MIS solar cells.[5]

With this outline of single crystal GaAs solar cells, we shall now discuss the thin film cells.

9.2.1. *Fabrication Process*

Thin film GaAs solar cells have been produced using chemical vapor deposition (CVD)[6–10] and metallorganic chemical vapor deposition (MOCVD)[11–14] techniques, in homojunction,[6] heteroface junction,[11] Schottky barrier,[7,9,12–14] and MIS[7,10] configurations. Zn doped p^+-, ⟨100⟩ oriented, single crystal GaAs,[6] ⟨100⟩ oriented, single crystal GaAS : Si,[11] graphite,[8] W coated graphite,[7,9,10] and Mo[12–14] have been employed as substrates. The processes employing single crystal substrates are expensive and, therefore, are merely of academic interest.

A typical thin film GaAs MOS configuration of the type Au/oxide/n-GaAs/n^+-GaAs/W/graphite is fabricated[7] by depositing GaAs films on W coated graphite substrates by the CVD reaction of Ga, HCl, and AsH_3 in a H_2 flow at a substrate temperature of 750 to 775 °C and source (Ga) temperature of 800 to 900 °C. The flow rates of HCl, AsH_3, and H_2 are 45 ml/min, 90 ml/min, and 1 liter/min, respectively. The n-GaAs layer is about 20 μm thick with a carrier concentration in the range of 5×10^{16} to 10^{17} cm^{-3}. The 3- to 5-μm thick n^+-GaAs doped with sulfur, using H_2S gas in the reaction mixture, has a carrier concentration of (2 to 5) $\times 10^{18}$ cm^{-3} and provides a low ($\sim 10^{-3}$ Ω) resistance contact to the substrate. The oxide layer on GaAs is formed *in situ* immediately after the deposition process by using O_2 or an O_2/Ar mixture at 200 °C for 0.5 to 1 h followed by oxidation for 8 to 12 h at 30 °C using O_2 saturated with water vapor.[8]

To improve cell performance, a 1-μm-thick, graded composition $GaAs_{1-x}P_x$ layer is deposited on GaAs by the reaction of Ga, HCl, AsH_3, and PH_3, prior to oxidation.[7] The Au layer (60 to 100 Å thick) is evaporated at a pressure less than 10^{-6} torr and the grid contact is formed by evaporating Ag through a metal mask. A TiO_2 layer 600 to 700 Å thick deposited at 80 to 100 °C by the hydrolysis of tetraisopropyl titanate in an Ar atmosphere has been employed as an AR coating on these devices.[8]

9.2.2. *Photovoltaic Performance*

Bozler and Fan[6] reported V_{oc} of 0.91 V, I_{sc} of 10.3 mA (at 100 mW cm^{-2}), and FF of 0.82, giving a measured AM1 efficiency of 15.3% for their antireflection layer coated, CVD, homojunction GaAs solar cells fabricated on single crystal substrates. Recently, Fan et al.[6] reported improved efficiencies of 17% for GaAs thin film solar cells fabricated by the CLEFT process. MOCVD deposited GaAlAs/*p*-GaAs/*n*-GaAs heteroface solar cells on GaAs:Si crystal substrates have exhibited V_{oc} of 0.99 V, I_{sc} of 24.5 mA cm^{-2} (128 mW cm^{-2}, AM0), FF of 0.74, and η of 12.8%, without AR coating.[11]

Schottky barrier devices with CVD GaAs thin films on W/graphite substrates[9] have yielded efficiencies of about 3% over areas as large as 8 cm^2. The performance parameters, under AM1 conditions, are V_{oc} of 0.456 V, I_{sc} of 10.4 mA cm^{-2}, and FF of 0.59. However, these devices have not been optimized. Vernon et al.[13] have prepared Schottky cells with MOCVD films on Mo and obtained AM1 efficiencies of 4.3% with V_{oc} of 0.39 V, I_{sc} of 19 mA cm^{-2}, and FF of 0.58. Grain boundary edge passivated (anodized), MOCVD GaAs films on Mo substrates, in the Schottky barrier configuration, have exhibited superior performance.[14] Without AR coating, the AM1 efficiency is 5.45%, V_{oc} is 0.49 V, I_{sc} is 20.6 mA cm^{-2}, and FF is 0.54.

Chu et al.[7,10] reported AM1 efficiencies of up to 4.4% over areas as large as 9 cm^2, for MOS devices fabricated from CVD *n*-GaAs/n^+-GaAs films on W/graphite. These devices exhibit V_{oc} of 0.45 to 0.52 V, I_{sc} of 11 to 13.4 mA cm^{-2}, and FF of 0.55 to 0.68 without AR coating. Deposition of a thin $GaAs_{1-x}P_x$ layer on top of the *n*-GaAs layer enhances the V_{oc} to 0.63 V and I_{sc} to 13.3 mA cm^{-2}, corresponding to an improved efficiency of 5.0%. Incorporation of an AR coating is expected to increase the efficiency to 7.5%. AR coated, MOS devices with CVD films on graphite substrates, fabricated by the same workers, have yielded V_{oc} of 0.5 V, I_{sc} of 18 mA cm^{-2}, and FF of 0.675 on 9-cm^2-area cells, corresponding to an AM1 efficiency of 6.1%.

Recently,[41] thin film polycrystalline GaAs solar cells have been fabricated using 10 μm thick GaAs films (average grain size ~5 μm) that have exhibited efficiencies up to 8.5% with V_{oc} of 0.56 V, I_{sc} of 22.7 mA cm^{-2}, and FF of 0.67, over areas of 9 cm^2.

Solar cells have been fabricated on 8-μm-thick, single crystal GaAs films grown on a GaAs single crystal substrate.[41] After fabrication, the cell has been bonded to glass, separated from the substrate, and metallized on the back surface. The performance parameters of such a cell have been measured to be $V_{oc} = 0.92$ V, $I_{sc} = 22.0$ mA cm^{-2}, FF = 0.73, and $\eta = 15\%$, over a total cell area of 0.4 cm^2.

Spectral response measurements on thin film GaAs MOS devices show a peak at 0.6 to 0.7 μm with an external quantum efficiency of 60 to 70%.[8] Effective diffusion lengths, deduced from spectral response measurements, are in the range of 0.5 to 0.8 μm.

9.2.3. *Junction Analysis*

Very limited studies exist on the electronic and structural quality of thin film GaAs photovoltaic junctions. Chu et al.[8] have measured a diode factor of 2.3 and barrier height of 1.0 to 1.02 eV on MOS devices on graphite, from I–V and C–V characteristics, respectively. Reverse saturation current densities of about 10^{-7} mA cm^{-2}, observed in MOS devices on W/graphite[7] have been attributed to carrier recombination at grain boundaries. However, the grain boundaries do not seriously degrade the photocurrent. Grain boundary shunting effects are ameliorated by Zn diffusion into the grain boundaries to compensate the donor levels present at the grain edges.[10] Improvement in the V_{oc} of $GaAs_{1-x}P_x/n$-GaAs/n^+-GaAs MOS solar cells have been attributed to an increase in the barrier height. The barrier height, from C–V measurements, was determined to be 1.2 eV, about 0.2 eV higher than n-GaAs MOS devices. The saturation current density is an order of magnitude lower for the $GaAs_{1-x}P_x$ device.

Essentially similar results have been obtained on Schottky cells by Ghandhi et al.,[12,14] who have observed diode factors of 2.7 at low forward bias and 1.3 at high forward bias. The saturation current density corresponding to the $n = 1.3$ diode was found to be in the range of 10^{-6} to 10^{-8} A cm^{-2}. From the temperature dependence of the I–V characteristics, a barrier height of 0.85 eV has been determined. Reverse C–V characteristics have yielded a carrier concentration of 6×10^{16} cm^{-3}. Grain boundary recombination has been proposed to explain the dark I–V characteristics. Photocurrent collection is, however, not affected by grain boundaries, since these are heavily doped during the film growth process. Grain boundary shunting effects, leading to high saturation current densities, have been reduced by selective anodization of the grain boundaries. These workers have also observed bias dependent collection of photocurrents and suggested recombination at the interface states as a probable mechanism.

9.3. *Cadmium Telluride (CdTe)*

With a very favorable direct bandgap of 1.44 eV, CdTe is a potentially high-efficiency photovoltaic material. Moreover, it is one of the few II–VI semiconductors that can be prepared in both conductivity types. The solar conversion efficiency for an *n*-CdS/*p*-CdTe heterojunction has been theoretically estimated to be about 17%. Therefore, CdTe has been the subject of continued interest. Photovoltaic devices utilizing CdTe as the active semiconductor can be classified into three categories: (i) single crystal or bulk polycrystalline CdTe wafers used as the base substrate for forming homojunctions[15,16] or heterojunctions with thin film window layers such as CdS,[17–20] ZnCdS,[17,18] ITO,[17,18,21,22] SnO_x,[23] ZnO,[18] and ZnSe,[18] (ii) epitaxial layers of CdTe grown on single crystal CdS or ZnTe substrates to form photovoltaic heterojunctions,[24,25] and (iii) all thin film CdTe/CdS and CdTe/Cu_xTe heterojunction solar cells[26–28] and CdTe homojunction solar cells.[29] The solar cells of the first type are essentially bulk devices and of interest insofar as they represent the best performance obtained on CdTe. The second class of devices using CdTe films on single crystal substrates, though not low cost, serve as useful guideposts for the all-thin-film solar cells and, therefore, junction studies on these devices are of considerable interest. From the point of view of large-scale applications, only the third class of CdTe devices are important and will be discussed in detail.

CdS/CdTe heterojunction solar cells prepared by vapor phase epitaxy of CdS on *p*-CdTe wafers[19] have yielded the highest efficiency to date of 10.5% at AM1.3 (68 mW cm^{-2}), with V_{oc} of 0.67 V, I_{sc} of 20.1 mA cm^{-2}, and FF of 0.59. If only the active area is considered, the cell efficiency is 11.7%. Solar cells fabricated by vacuum deposition of CdS onto *p*-CdTe wafers[18] have exhibited an active area efficiency of 7.9%, with V_{oc} of 0.63 V, I_{sc} of 16.1 mA cm^{-2}, and FF of 0.66. Sputtered ITO/*p*-CdTe junctions[22] have been produced with solar efficiencies of up to 8%, V_{oc} of 0.82 V, I_{sc} of 14.5 mA cm^{-2}, and FF of 0.55. The spectral response of the vapor phase CdS/*p*-CdTe heterojunction[19] extends from about 0.52 to 0.86 μm with a uniform response between 0.58 and 0.81 μm. The short-wavelength and long-wavelength cut-offs correspond to the bandgaps of CdS and CdTe, respectively, and are independent of bias voltage, although the collection efficiency and, hence, the photocurrent, are bias-dependent. Another interesting feature of this cell is that the electrically active photovoltaic junction does not exist at the metallurgical junction between CdS and CdTe but at a point a few micrometers deep inside CdTe, because of an n-CdTe layer formed as a result of diffusion of In from CdS into CdTe. Similarly, in the ITO/p-CdTe junction,[22] the photovoltaically active junction exists inside the CdTe crystal and not at the ITO/CdTe interface, because of the formation of an n-type CdTe layer during the sputtering process.

CdS/p-CdTe and ZnCdS/p-CdTe junctions prepared by spray pyrolysis of CdS and ZnCdS on CdTe wafers[18] have exhibited V_{oc} values of 0.74 and 0.80 V, respectively, and efficiencies of 6.0 and 7.8%. CdTe homojunctions have exhibited very low efficiencies of 3.25%, with V_{oc} of 0.62 V and I_{sc} of 12 mA cm^{-2}.

p-CdTe/n-CdS solar cells fabricated by closed space vapor transport (CSVT) epitaxial growth of CdTe onto single crystal CdS window substrates have yielded a V_{oc} of 0.61 V, quantum efficiency of 0.85, and η of 4.0%.

We shall now describe the fabrication process and photovoltaic characteristics of various thin film solar cells.

9.3.1. *Fabrication Process*

Thin film CdTe solar cells have been fabricated using a combination of screen printing and chemiplating techniques,[26,27] a combination of vapor transport and chemiplating,[28] a combination of evaporation and flash evaporation,[30] a combination of evaporation and chemiplating,[31] and a combination of gas phase deposition and evaporation.[32] In all these processes, the first-named technique has been employed to deposit CdTe layers and the second deposition method has been used to grow another semiconductor layer to form a heterojunction. The highest efficiencies ($\sim 8\%$) have been reported for the screen printed and chemiplated p^+-Cu_2Te/n-CdTe/n-CdS/In_2O_3/glass heterojunction structure.[27] The details of this process are given below.

The schematic representation of the cross-section of a screen printed cell is shown in Figure 9.1a. A borosilicate glass plate (Corning Glass 7059) coated with a transparent and conducting In_2O_3 layer is used as the substrate. A CdS film 20 μm thick, 0.2 Ω cm is deposited onto the substrate by screen printing from a CdS + $CdCl_2$ + $GaCl_2$ + propylene glycol paste in a N_2 atmosphere at about 630 °C. Subsequently, an n-CdTe layer 10 μm thick, 10^{-1} to 1 Ω cm is screen printed in a N_2 atmosphere at 500 to 800 °C from a CdTe:In + $CdCl_2$ + propylene glycol paste. The CdTe layer is dipped for several seconds in a hot cuprous ion solution to convert a top surface layer to p^+-Cu_2Te. Finally, the cells are heat-treated for 10 min at 200 °C and a Ag paste is applied to Cu_2Te and In–Ga alloy solder to CdS film to form the contacts. In recent 4 cm × 4 cm cells, which have been used

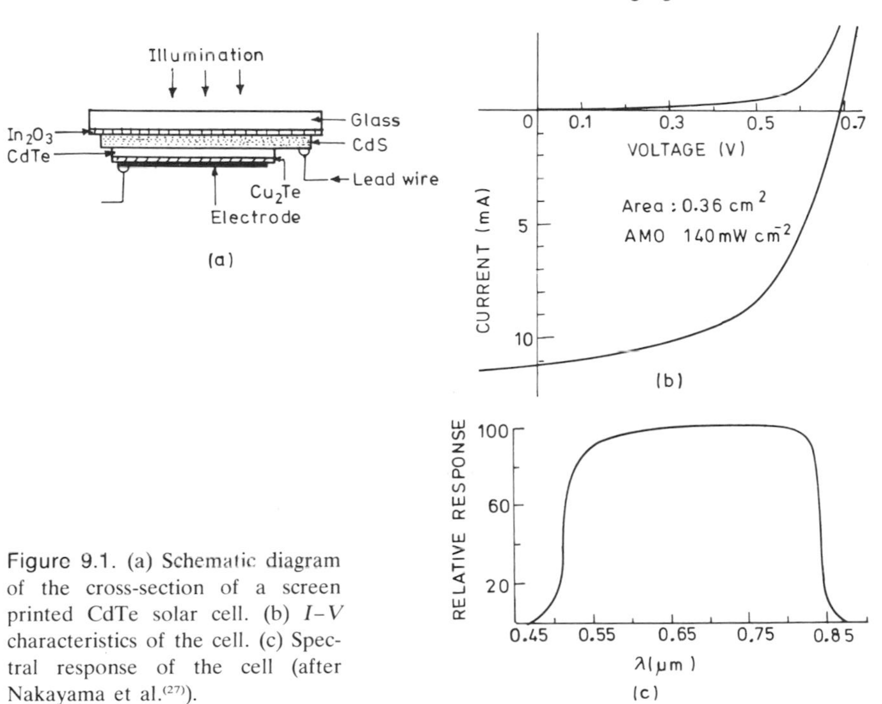

Figure 9.1. (a) Schematic diagram of the cross-section of a screen printed CdTe solar cell. (b) I–V characteristics of the cell. (c) Spectral response of the cell (after Nakayama et al.[27]).

to produce 1-W modules consisting of 25 elemental cells, the expensive In_2O_3 film has been replaced by a low-resistivity CdS screen printed film and the Cu_2Te has been replaced by a more stable screen printed C electrode. Moreover, undoped, weakly p-type CdTe has been used since a photovoltaic effect is observed to originate at the n-CdS/p-CdTe, rather than the p-Cu_2Te/n-CdTe junction.

9.3.2. *Photovoltaic Performance*

The first CdTe homojunction thin film solar cells, fabricated by Vodakov et al.,[33] exhibited a conversion efficiency of 4% in sunlight. Further development[29] yielded a V_{oc} of 0.75 V, I_{sc} of 9.8 mA cm^{-2}, and FF of 0.63 giving an efficiency of 6% for an illumination intensity of 77.2 mW cm^{-2}. Cu_2Te/CdTe heterojunction cells produced by Cusano[28] exhibited V_{oc} of 0.7 V for single crystal CdTe and 0.5 V for thin film CdTe. The I_{sc} values were higher in the thin film cells, but the FF values were lower, about 0.45 to 0.6, in the thin film cells, compared to about 0.7 for single crystal cells. Thin film cells achieved efficiencies of 6% as against 7% for single crystal cells. Large area (50 cm^2) thin film cells exhibited efficiencies of about 5%, the performance being limited by series resistance, which gave poor FF values. Justi et al.[31] obtained V_{oc} of 0.6 V and η up to 4.1%

on $Cu_xTe/CdTe$ thin film solar cells. High series resistance and low shunt resistance degraded the FF. High-temperature-stable (up to 200 °C) CdTe/CdS thin film solar cells fabricated by Bonnet and Rabenhorst[32] achieved V_{oc} of 0.5 V, I_{sc} of 15 mA cm^{-2} (for 50 mW cm^{-2} intensity), FF of 0.45, and η of 5 to 6%.

The $I-V$ characteristics of a screen printed CdTe thin film solar cell,[27] described in 9.3.1, is shown in Figure 9.1b. A small-area cell (0.21 cm^2) exhibits an efficiency of 8.2% at an illumination level of 72 mW cm^{-2}, with V_{oc} of 0.67 V, I_{sc} of 14.2 mA cm^{-2}, and FF of 0.58. The same cell, with a C electrode in place of the Cu_2Te and an area of 10 cm^2, yields an efficiency of 6.3%, V_{oc} of 0.73 V, I_{sc} of 11.4 mA cm^{-2}, and FF of 0.51.[26] A 1-W module, constructed from the C coated cells, has an efficiency of 2.9%, with V_{oc} of 3.43 V, I_{sc} of 711 mA, and FF of 0.51 at 90 mW cm^{-2} intensity.

The spectral response of a screen printed CdTe thin film solar cell, with Cu_2Te electrode, is shown in Figure 9.1c. The curve exhibits well defined cut-off at the CdS and CdTe absorption edges. Cu_2Te is observed to be optoelectronically inactive. Cusano[28] has observed similar behavior in $Cu_xTe/CdTe$ thin film cells.

In unpublished reports, Kodak has claimed a conversion efficiency of 10% for CSVT deposited CdS/CdTe thin film solar cells.

9.3.3. *Junction Analysis*

The $Cu_xTe/CdTe$ thin film solar cells prepared by Cusano[28] showed a poor diode behavior with a diode factor of about 2.7 and a rectification factor of only a few hundred. Diode behavior of screen printed CdTe cells has not been reported. However, X-ray microanalysis and EBIC studies of the junction have been performed to elucidate information on the location and electronic nature of the junction. The elemental distribution, as obtained from X-ray microanalysis, and EBIC distribution for a heat treated $Cu_xTe/CdTe/CdS$ screen printed solar cell is shown in Figure 9.2. From their studies on these cells, Nakayama et al.[26] have concluded that in optimally heat treated high-efficiency cells, the photovoltaic junction (as determined by EBIC) occurs at the metallurgical junction between CdS and CdTe owing to the change of n-CdTe to p-CdTe as a result of diffusion of Cu from Cu_2Te into n-CdTe. In very heavily heat treated cells, the Cu also diffuses into the CdS and causes a high series resistance, thereby causing a degradation in performance. In C coated CdTe/CdS screen printed cells, heat treatment does not cause significant diffusion of acceptor impurities into CdS, although impurities diffuse into the CdTe forming a p-CdTe/n-CdS photovoltaically active junction at the CdTe/CdS metallurgical interface. The suppression of acceptor impurity diffusion into CdS lends greater stability to the C coated cell.

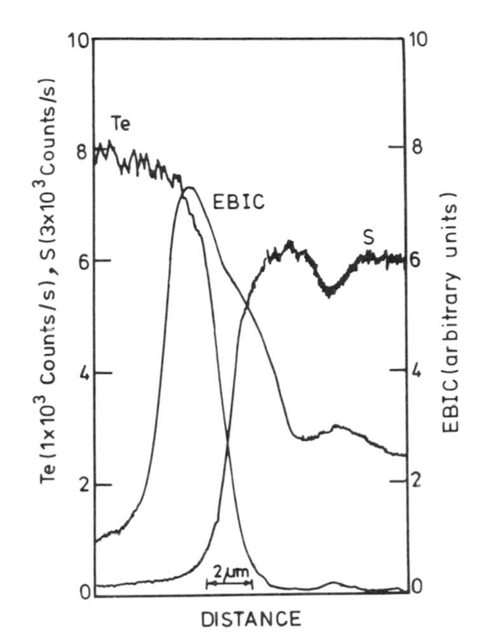

Figure 9.2. X-ray microanalysis elemental distribution and EBIC distribution for an optimally heat treated $Cu_xTe/CdTe/CdS$ thin film cell (after Nakayama et al.[26]).

9.4. CdSe, $Cu_{2-x}Se$, $ZnIn_2Se_4$

Limited studies have been carried out on photovoltaic devices of CdSe, $Cu_{2-x}Se$, and InSe thin films. n-CdSe/p-CdTe and n-CdSe/p-ZnTe solar cells, fabricated by CSVT growth of n-CdSe on single crystal substrates, have exhibited[25] V_{oc} of 0.61 V and 0.56 V, respectively, and I_{sc} values (at 85 mW cm^{-2}) of 0.75 and 1.89 mA cm^{-2}. $Cu_{2-x}Se$/Si junctions,[34] prepared by vacuum evaporating a $Cu_{2-x}Se$ layer onto n-Si single crystal substrates, show a V_{oc} of 0.45 V, I_{sc} of about 23 mA cm^{-2}, FF of 0.62 to 0.65, and η of 8.8%, under 75 mW cm^{-2} illumination. The electrical and photovoltaic properties of the junction are governed mainly by the corresponding properties of the base Si material.

Thin film $Cu_{2-x}Se$/CdS solar cells have been fabricated on glass substrates in the front-wall as well as back-wall geometries.[35] The back-wall cells consist of Mo/Au (3000 Å/5000 Å) grid pattern vacuum deposited onto glass substrates, followed sequentially by a 10-μm-thick In doped or undoped evaporated CdS layer and a 5-μm-thick evaporated $Cu_{2-x}Se$ layer. A 5000-Å-thick Au layer on top of the $Cu_{2-x}Se$ layer completes the cell and provides the back contact. For an undoped CdS cell, V_{oc} is in the range of 0.3 to 0.45 V and I_{sc} is of the order of 2 to 4.5 mA cm^{-2} for AM1 illumination. Efficiencies of 0.51% have been achieved. It has been suggested that reducing the CdS layer resistivity would improve the V_{oc} and optimizing the $Cu_{2-x}Se$ stoichiometry would increase I_{sc}. The poor FF (~ 0.36) is attributed to a poor metal/CdS contact. The band diagram of the heterojunction has been modeled on the lines of the Cu_2S/CdS solar cell.

Recent improvements[41] in the fabrication of $Cu_{2-x}Se$/CdS thin film cells have led to I_{sc} of 11.6 mA cm^{-2}, V_{oc} of 0.46 V, and FF of 0.62, yielding an AM1 efficiency of 3.3%. It has been observed that deposition of $Cu_{2-x}Se$ at low substrate temperatures ($\sim 160\,°C$) and texturing of the CdS film by etching in HCl, prior to evaporation of $Cu_{2-x}Se$, are important steps in achieving good performance.

Thin film MIS CdSe solar cells[36] have been fabricated by evaporating a 2-μm-thick CdSe layer onto a 1000-Å-thick Cr coated glass substrate, at a substrate temperature of 425 °C and deposition rate of about 10 Å s^{-1}. The CdSe layers are polycrystalline and have their c-axis perpendicular to the substrate. The film is highly stoichiometric and exhibits a carrier concentration of 10^{14} cm^{-3} and mobility of about 10 to 30 $cm^2\,V^{-1}\,s^{-1}$. A thin (40 to 50 Å) insulator layer of ZnSe or Sb_2Se_3 is evaporated onto the CdSe film, prior to the deposition of a 200-Å-thick Au layer. The transparency of the Au layer is 30 to 40%. Finally, the completed device is heat treated in N_2 at temperatures below 200 °C for 10 to 30 min.

The MIS CdSe thin film solar cells yield a V_{oc} between 0.55 and 0.6 V, an I_{sc} of about 20 mA cm^{-2} (for 100 mW cm^{-2} intensity), an FF of 0.4 to 0.5, and an efficiency greater than 5%. The spectral response of these cells is dominated by reflection from the metal layer. For the thickness of the Au film used, the minimum in the reflection and the maximum in the spectral response occur at 0.5 μm. Use of an AR coating can significantly alter this behavior. The collection efficiency at shorter wavelengths is observed to be relatively higher than that at longer wavelengths, indicating a low density of recombination centers at the junction interface but a small diffusion length in the bulk of the film. An enlarged space-charge layer, generated by a suitable doping profile, has been suggested to improve collection.

From $C-V$ measurements, a diffusion voltage of 0.75 V has been deduced. Considerably higher diffusion potentials are expected from an increase in the doping concentration. However, doping with Cd led to reduced space-charge widths and, consequently, poorer collection efficiency. Compensation of Cd donors by diffusion of Se at the surface yielded a V_{oc} up to 0.7 V. From $I-V$ characteristics, a diode factor of about 2, a reverse saturation current density of 6×10^{-8} A cm^{-2}, and a barrier height of 0.85 eV have been determined.

With an optical gap of 1.7 eV, the V_{oc} in CdSe solar cells is expected to be about 0.8 to 0.9 V. Improved doping is suggested to realize these values. Further, on decreasing the reflection losses with an AR coating and increasing the FF by improving minority carrier diffusion length, an efficiency of 10% is expected to be achieved.

García and Tomar[37] have prepared n-CdS/p-$ZnIn_2Se_4$ thin film solar cells on Zn metallized glass substrates. $ZnIn_2Se_4$ films, 7 to 11 μm thick are evaporated on the substrates at 425 K in a vacuum of 10^{-6} torr. Subsequently, a 3- to 7-μm thick CdS layer is deposited over the $ZnIn_2Se_4$ layer. The $ZnIn_2Se_4$ and CdS layers are codeposited with Se and In, respectively. Evaporated In grids form the top contact. Under 100 mW cm^{-2} illumination, the cells exhibit, typically, a V_{oc} of 0.27 V, an I_{sc} of 16 mA cm^{-2}, and an FF of 0.31, yielding an efficiency of about 1.5%.

9.5. *Zinc Phosphide (Zn_3P_2)*

In recent years Zn_3P_2 has emerged as a very promising solar cell material. The optical gap lies in the range of 1.55 to 1.60 eV and the absorption data suggest a direct bandgap.[38] Schottky diodes made from p-type films exhibit barrier heights of 1.0 to 1.4 eV with Mg and 0.75 eV with Al. The electron affinity has been determined to be 3.6 eV.[39] Minority carrier diffusion lengths[39,40] measured on single crystal p-Zn_3P_2 Schottky diode solar cells are in the range of 5 to 10 μm, while those measured on thin film diodes are approximately in the range of 3 to 4 μm. Based on the optical and electronic data, it has been concluded that Zn_3P_2 thin films are suitable for use in photovoltaic cells, provided AR coatings are employed. One of the major advantages of this material is that it is available in sufficient abundance to fulfill the requirements of large-scale power generation.

Zn_3P_2 thin films, prepared by evaporation of the bulk material from graphite crucibles[39] at source temperatures of 700 to 800 °C and substrate temperatures between 140 to 200 °C, exhibit p-type conductivity with hole mobilities in the range of 10 to 40 $cm^2V^{-1}s^{-1}$. The resistivity of the films can be varied from 10 to 10^5 Ω cm by annealing. The films make ohmic contact with Fe, which is observed to be a suitable substrate from the lattice and thermal expansion point of view. Thin film Schottky barrier Mg/Zn_3P_2 solar cells[41] have exhibited efficiencies of 2.5%, with V_{oc} of 0.37 V, I_{sc} of 12.0 mA cm^{-2}, and FF of 0.46 at an illumination intensity of 83 mW cm^{-2}.

9.6. *Indium Phosphide (InP)*

The direct, 1.34-eV optical bandgap of InP is well matched to the solar spectrum. The high absorption coefficient implies low absorption lengths and, consequently, less critical minority carrier diffusion length requirements. Thus InP is eminently suitable for photovoltaic applications. The lattice parameter of the zincblende InP is $a_Z = 5.869$ Å, while the corresponding parameter of wurtzite CdS is $\sqrt{2}\, a_W = 5.950$ Å. Thus, with a mismatch of only 0.32% between the ⟨111⟩ plane of InP and the basal plane of hexagonal CdS, the InP/CdS forms an ideal heterojunction, as the electron affinities are also well matched so there are no interfacial spikes in the conduction band edge.

Single crystal p-InP/n-CdS solar cells, provided with an evaporated SiO AR coating, have been prepared[42] by evaporating 5- to 10-μm-thick CdS from a coaxial isothermal source, containing elemental Cd and S, maintained at 350°C, onto chemically polished oriented ⟨111⟩ Czocharalski grown, Cd doped, p-type InP wafers ($\sim 0.4\ \Omega$ cm resitivity) maintained at 200 to 250 °C at a rate of about 0.15 μm/min. Electrodeposited Au/Zn/Au formed the contact to InP, and In or In–Ga alloy grid was deposited onto the CdS. The cells exhibited an efficiency of 12.5% with V_{oc} of 0.62 V, I_{sc} of 15 mA cm^{-2}, and FF of 0.71 for illumination intensity of 53 mW cm^{-2}. The spectral response was uniform between 0.55 and 0.91 μm, the short- and long-wavelength cut-offs corresponding to the CdS and InP bandgaps, respectively. The cells exhibited very low reverse saturation currents of about 10^{-8} A cm^{-2}. The short-circuit current and the collection efficiency are observed to be governed solely by the properties of the bulk InP and the CdS window[43]; the role of the heterojunction interface is insignificant. A diffusion length value of $L_n \sim 1.3$ μm has been determined for InP from spectral response measurements. The performance has been observed to be essentially the same when CdS is grown on ⟨100⟩ and ⟨110⟩ faces of InP.

The preparation of single crystal p-InP/n-CdS solar cells by CVD growth of CdS on ⟨111⟩ InP, in H_2S/H_2 flow, at source temperatures of 700 °C and substrate temperatures of 620 °C, have led to AM2 efficiencies of 15% for SiO AR coated cells (and 12.8% for uncoated cells),[43] with V_{oc} of 0.79 V, I_{sc} of 18.7 (16.0) mA cm^{-2}, and FF of 0.75. The InP crystal was Cd doped and possessed a hole concentration of approximately 2×10^{17} cm^{-3} and mobilities of about 80 cm^{2}V^{-1}s^{-1}. The CdS films had an electron concentration of $(2\text{–}10) \times 10^{18}$ cm^{-3} and mobilities of 100 to 150 cm^2 V^{-1} s^{-1}. Preliminary accelerated life tests on single crystal InP/CdS cells have revealed that the unencapsulated cells are insensitive to the ambient, and temperatures greater than 400 °C must be reached before the cell performance deteriorates. For higher conductivity CdS films, the degradation data indicate a lifetime of 1 month at 485 °C.

Ito and Ohsawa[44] reported an efficiency of 4.1% on InP/CdS single crystal solar cells prepared by CVD growth of CdS on ⟨111⟩ face of InP. The cells were not coated with an AR layer and the long-wavelength response was found to decrease sharply. This decrease was attributed to the formation of an InP–CdS mixed layer at the interface, as revealed by electron probe microanalysis.

The high efficiencies and predicted stability of the single crystal InP/CdS solar cells inspired several workers to fabricate thin film solar cells of this system. The performance of the thin film devices is discussed below.

9.6.1. *Fabrication Process*

Thin film *p*-InP/*n*-CdS solar cells have been fabricated by evaporation,[45] CVD,[43] and a combination of CVD and evaporation.[46,47] Mo,[46,47] C,[47] and heavily Zn doped, *p*-type, 25-μm-thick GaAs coated C[43] have been used as substrates. The best results[43] have been obtained on devices made by all CVD processes on GaAs coated C substrates The GaAs layer makes excellent electrical contact to the C and provides a low-resistance ohmic contact to the InP layer, in contrast to the C and Mo substrates which yield poor back contacts and thus lower efficiencies.[45,47]

The fabrication[43] of a typical InP/CdS thin film solar cell of efficiency exceeding 5% consists of depositing a 50-μm-thick, polycrystalline InP layer on GaAs coated C substrates by PCl_3 CVD,[47] followed by a 25-μm CdS film, using the H_2S–H_2 transport process. The *p*-InP has a doping of 0.83×10^{17} cm^{-3}, as estimated from C–V measurements. The CdS film is more heavily doped ($\sim 10^{19}$ cm^{-3}). SiO AR coatings are deposited by evaporation at substrate temperatures of about 100 °C. Soldered In provides the contact to CdS. An In–Au grid pattern yields higher FFs of 0.75.

We should note that evaporated InP layers do not exhibit a single phase but consist of elemental In and P. Two-source coevaporation of In and P is required to produce single-phase InP films with grain sizes exceeding 1 μm.[45]

9.6.2. *Photovoltaic Performance*

All-CVD, 0.03-cm^2-area, InP/CdS thin film solar cells[43] have exhibited V_{oc} in the range of 0.40 to 0.46 V, I_{sc} up to 13.3 mA cm^{-2} for uncoated and up to 15.4 mA cm^{-2} for SiO coated cells, and FF of up to 0.68, under AM2 illumination, leading to efficiencies of up to 4.9% for uncoated cells and 5.7% for AR coated cells. AM1 efficiencies of 2.0%, with V_{oc} of 0.37, I_{sc} of 18 mA cm^{-2}, and FF of 0.30, have been achieved on CVD InP/evaporated CdS cells on Mo substrates.[46] The substrate temperature during CdS deposition was observed to be critical in the performance of the cell. Also, annealing at 500 °C for 10 min in air improved the cell characteristics. A rectifying contact at the InP/Mo interface, as evidenced by a kink in the light I–V characteristics, was suggested to be the cause of poor FF. A high series resistance was also noted in the all-evaporated InP/CdS cells[45] on Mo, which exhibited efficiencies of 2.1% with V_{oc} of 0.51 V and FF of 0.51. The somewhat better FF indicated that the back contact is less problematic than in the case of CVD InP/evaporated CdS thin film cells on graphite,[47] which exhibit η of 2.8%, V_{oc} of 0.40 V, and FF of 0.31, under AM1 conditions.

Shay et al.[43] reported that the AM2 short-circuit current in the thin film cells is lower by only 18% compared to the single crystal cells and the absolute quantum efficiency is essentially the same in both the cells. On the basis of this observation, they have concluded that recombination at the interface does not significantly affect the quantum efficiency even in thin film cells. The diffusion length in the InP layer of these cells has been estimated to be about 0.6 μm. However, these workers have observed that the response of their cells extends below the bandgap of bulk InP, indicating the existence of band tails in the InP grains. Moreover, the V_{oc} decreases as the amount of band tailing increases. These workers have opined that the band tails, attributable to defects within the InP grains, do not affect the photocurrent, but reduce the voltage and, hence, the conversion efficiency by about a factor of 2. It has been suggested that improvement in the quality of the InP layer would increase the efficiency.

In contrast to the above results, Kazmerski et al.[45] observed a short-wavelength cut-off owing to absorption by CdS and a long-wavelength cut-off owing to the bandgap of InP in their thin film cells prepared by evaporation. No tailing of the response below the bandgap of InP was observed (Figure 9.3). Moreover, the quantum efficiency was

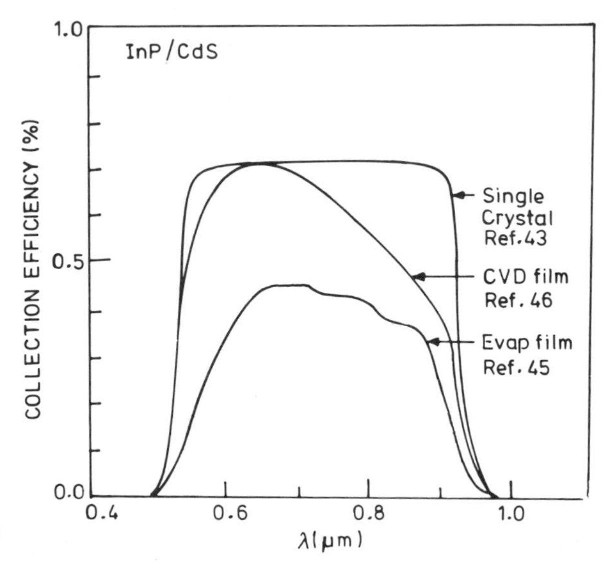

Figure 9.3. Spectral response of several InP/CdS thin film solar cells (after Shay et al.,[43] Kazmerski et al.,[45] Saitoh et al.[46]).

found to be considerably lower than that in the single crystal cell, leading to lower I_{sc} values. The lower quantum efficiency was attributed to recombination in the high-defect density InP films and at the grain boundaries. Saitoh et al.[46] obtained a collection efficiency of close to 70% at 0.6 to 0.7 μm in CVD InP/evaporated CdS thin film solar cells on Mo. They deduced an electron diffusion length of 0.2 μm in p-InP films. This low diffusion length was found to decrease the collection efficiency at longer wavelengths (Figure 9.3).

Kazmerski et al.[45] reported an exponential variation of the dark forward current with voltage. The diode factor was determined to be 2.34, indicative of generation–recombination processes.

CVD InP/evaporated Cu_xSe thin film solar cells have also been fabricated on Mo substrates by Saitoh et al.[46] AM1 efficiencies of 1.7% have been achieved with a V_{oc} of 0.47 V, an I_{sc} of 11 mA cm^{-2}, and an FF of 0.34. A collection efficiency of 40% was obtained at 0.65 μm. Analysis of the spectral response leads to the conclusion that only the photons absorbed in the depletion layer contribute to the light-generated current.

9.7. Copper Indium Selenide ($CuInSe_2$)

$CuInSe_2$ possesses certain exceptional material characteristics which are particularly suited to photovoltaic heterojunction applications. The semiconductor has a direct bandgap, which minimizes the requirements for minority carrier diffusion length. Thin films of $CuInSe_2$ can be prepared easily in both n- and p-type conductivity and, therefore, both homojunction and heterojunction potential exists for this material. The bandgap of 1.04 eV is near the optimal value for terrestrial conditions. $CuInSe_2$ forms an ideal heterojunction with CdS, since the lattice mismatch between chalcopyrite $CuInSe_2$ and hexagonal CdS is only about 1.2%. No interfacial spikes are formed in the conduction band as the electron affinities of the two materials are very close.

The excellent photovoltaic potential of the $CuInSe_2$/CdS heterojunction has indeed been realized on single crystal cells.[48–50] Efficiencies of 12%, with V_{oc} of 0.5 V, I_{sc} of 38 mA cm^{-2}, and FF of 0.60, have been obtained on 0.79-mm^2 areas at 92 mW cm^{-2} solar intensity. The cells were fabricated on chemically polished p-$CuInSe_2$ single crystals by depositing a 5- to 10-μm-thick n-CdS layer from a coaxial, isothermal, dual (Cd + S) source at a source temperature of 350 °C, a substrate temperature between 130 and 210 °C, and a deposition rate of about 0.15 μm/min. An evaporated SiO layer provides the AR coating. A high (~70 to 80%), uniform quantum efficiency is observed between 0.55 and 1.25 μm. The short- and long-wavelength cut-offs correspond to the bandgaps of CdS and $CuInSe_2$, respectively. Larger-area cells exhibited lower efficiencies, mainly because of a lower V_{oc}, as a result of microcracks present in the $CuInSe_2$ crystal.

The preparation of $CuInSe_2$ films[51,52] led to the development of all-thin-film, all-evaporated p-$CuInSe_2$/n-CdS solar cells. These cells are described below.

9.7.1. *Fabrication Process*

Thin film solar cells of p-$CuInSe_2$/n-CdS have been fabricated in the front-wall mode (illumination through $CuInSe_2$) and the back-wall mode (illumination through CdS) on metallized substrates.[53] In the front-wall mode, a 6- to 8-μm-thick, 10-Ω cm CdS layer is grown on the substrate by vacuum evaporation at a substrate temperature of 500 K. Subsequent to removal from the vacuum system, the CdS surface is lightly etched in 10% HCl solution. Thereafter, a thin (0.25 to 0.50 μm) p-$CuInSe_2$ layer is grown on the CdS by vacuum evaporation from a two-source ($CuInSe_2$ + Se) boat at a substrate temperature of 525 K. The Se is used to control carrier type. The etching is necessary to obtain photovoltaic response. The grain sizes of both the films are in the range of 0.8 to 1.2 μm. In an alternative process, the entire cell is fabricated *in situ*, in which the entire process is accomplished without breaking the vacuum. In the back-wall mode, a p-$CuInSe_2$ film is deposited on Au-metallized glass substrates, prior to *in situ* deposition of the CdS (5–6 μm). The carrier concentrations in the p-$CuInSe_2$ films are in the range of $(3–4) \times 10^{16}$ cm^{-3}. The films exhibit a preferential ⟨112⟩ growth, i.e., ⟨221⟩-axis perpendicular to the substrate. A grid pattern of Al on CdS or Au–Ag paste of $CuInSe_2$ provides the top contact. Finally, the devices are annealed at 450 K in about 10^{-1} torr for 15 to 20 min. These cells have efficiencies up to 5.7%.

Recently, Buldhaupt et al.[54] described an AR coated, back-wall, gradient doped n-CdS/p-$CuInSe_2$ thin film solar cell with AM1 efficiency of 9.53% for a 1-cm^2-area cell. The cell configuration is schematically

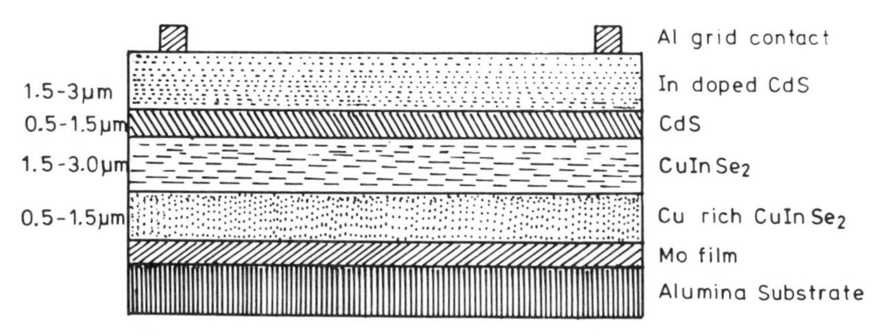

Figure 9.4. Schematic representation of a high-efficiency p-$CuInSe_2$/n-CdS gradient doped thin film solar cell (after Buldhaupt et al.[54]).

represented in Figure 9.4. The Cu doped $CuInSe_2$ layer is deposited by vacuum evaporation onto Mo coated alumina substrates, using a three-source (Cu + In + Se) technique. The CdS film is doped with In. An Al-grid pattern provides the top contact. An SiO_x layer is used as the AR coating. The devices are heat-treated at 200 °C in H_2/Ar.

$CuInSe_2$/CdS thin film cells have also been fabricated by spray pyrolysis[55] and by a combination of sputtering and evaporation.[56] The efficiencies have been low, typically below 2%.

9.7.2. *Photovoltaic Performance*

The *I–V* characteristics of a high-efficiency gradient doped p^+-$CuInSe_2$:Cu/p-$CuInSe_2$/n-CdS/n^+-CdS:In back-wall thin film solar cell of 1 cm^2 area is shown in Figure 9.5, with an AR coating. The performance parameters are V_{oc} of 0.396 V, I_{sc} of 35 mA cm^{-2} (for simulated AM1

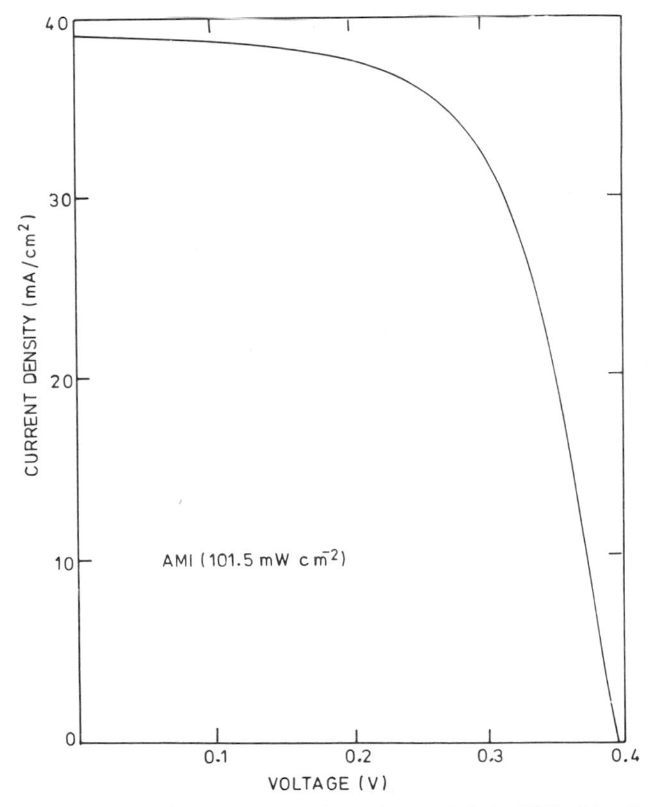

Figure 9.5. *I–V* characteristics of a gradient doped $CuInSe_2$/CdS thin film cell (after Buldhaupt et al.[54]).

illumination of 101.5 mW cm^{-2}), FF of 0.64, and η of 8.72% for an uncoated cell and V_{oc} of 0.396 V, I_{sc} of 39 mA cm^{-2}, FF of 0.63, and η of 9.53% for an SiO_x AR coated cell.

Kazmerski et al.[53] have obtained an efficiency of 4.4% for their front-wall *in situ* fabricated $CuInSe_2$/CdS cells. The etched cells yielded lower efficiencies (2.46%) probably because of a higher series resistance arising from surface states developed during the atmosphere exposure–etching process. The photovoltaic parameters for the *in situ* and etched devices (area 1.2 cm^2) at 100 mW cm^{-2} are $V_{oc} = 0.41$ V, $I_{sc} = 19.9$ mA cm^{-2}, and FF = 0.5 and $V_{oc} = 0.34$ V, $I_{sc} = 12$ mA cm^{-2}, and FF of 0.36, respectively. The back-wall devices exhibited much better characteristics with a V_{oc} of 0.44 V, I_{sc} of 22.5 mA cm^{-2}, FF of 0.59, and η of 5.7%. The better performance of the back-wall cell has been attributed to better absorption of carriers via the CdS window and improved junction characteristics. Also, Auger electron spectroscopy combined with depth profiling indicated a more abrupt junction in the case of the back-wall cell.[57] An interfacial oxide layer has been detected in the etched cells.

The spectral dependence of the quantum efficiency for a 9.53% gradient doped $CuInSe_2$/CdS thin film cell[54] is shown in Figure 9.6. The quantum efficiency is uniform and high (> 0.85) over the wavelength range of 0.60 to 0.95 μm and extends from 0.5 to 1.1 μm. The response for the 5.7% cell fabricated by Kazmerski et al.[53] is flat over the wavelength range $0.58 < \lambda < 1.3$ μm but the quantum efficiency is lower (~0.50 to 0.55).

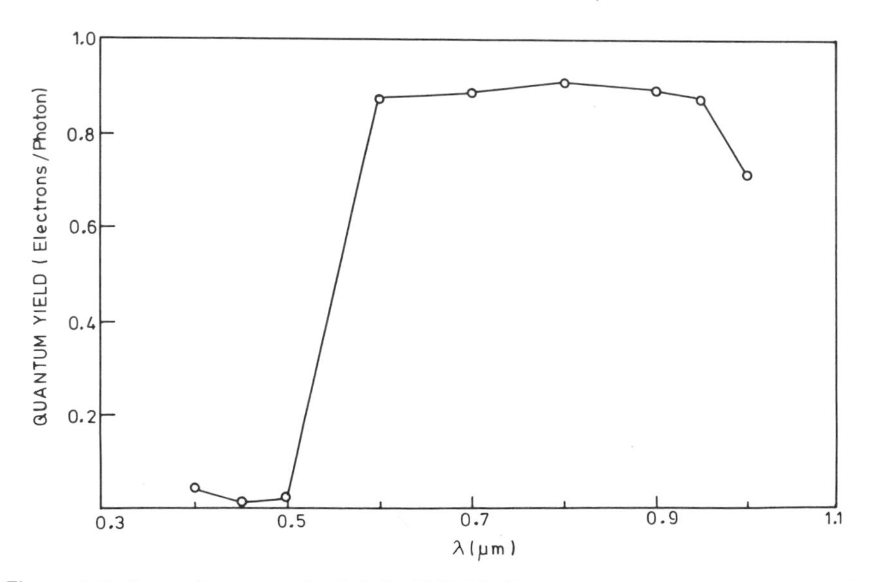

Figure 9.6. Spectral response of a $CuInSe_2$/CdS thin film solar cell (after Buldhaupt et al.[54]).

Buldhaupt et al.[54] have carried out detailed studies on the effect of heat treatment at 200 °C in different ambients on cell performance. They have observed that the cells show an improvement in I_{sc} with annealing which is insensitive to the ambient. However, V_{oc} and FF are found to be sensitive to the ambient, improving in H_2/Ar and air but degrading in pure H_2 or vacuum.

Kazmerski et al.[57] reported Auger studies on *in situ* fabricated $CuInSe_2$/CdS thin film cells annealed at 600 K for 2 h. Cd is observed to diffuse into the ternary film. A diffusion coefficient of 10^6 $cm^2 s^{-1}$ and activation energy of 1.5 eV have been determined for Cd diffusion from CdS into $CuInSe_2$. Cu diffusion into CdS is minimal and occurs only at temperatures exceeding 575 K. It has been proposed that, since only a single phase of $CuInSe_2$ exists at low temperatures and the Cu atoms are more tightly bound in the chalcopyrite lattice, $CuInSe_2$/CdS devices would be more reliable than Cu_2S/CdS cells.[53] Although no detailed investigations of the stability have been undertaken, thin film $CuInSe_2$/CdS cells have been observed to be relatively stable with no significant degradation in V_{oc} or I_{sc} over a time period of a few weeks.

Detailed junction analyses of these devices have not been performed. Limited data[53] suggest that the forward current follows an exponential behavior with the diode factor being in the range of 1.9 to 2.1. Recombination–generation processes are, therefore, dominant in these devices. Buldhaupt et al.[54] have carried out a loss analysis for their high-efficiency, gradient doped cells. They have obtained a series resistance of 1.2 Ω, R_{sh} of 10^5 Ω, J_s of 1.8×10^{-7} A cm^{-2}, and diode factor of 1.3. They have proposed a reduction in the series resistance to 0.5 Ω to achieve 10.6% efficiency. Investigation of cells with CdS replaced by ZnCdS have led to thin film $CuInSe_2$/ZnCdS cells of 1-cm^2 area exhibiting efficiencies of 10%.[58]

Before concluding this section, we should note that several other Cu-ternary compounds and penternary compounds have been developed for fabricating solar cells. *p*-$CuInTe_2$/*n*-CdS and *p*-$CuInS_2$/*n*-CdS thin film solar cells, fabricated by vacuum evaporation, have exhibited photovoltaic behavior.[45] The $CuInS_2$ cell achieved efficiencies of 2.55%. Thin film CdS/$CuGaSe_{0.90}Te_{1.10}$ solar cells,[56] prepared by a combination of evaporation and sputtering, have yielded a V_{oc} of 0.50 V, an I_{sc} of about 0.9 mA cm^{-2}, an FF of 0.35, and an η value of 0.1%. On large-grained sintered material, cells yielded V_{oc} of 0.43 V, $I_{sc} \sim 26$ mA cm^{-2}, and FF of 0.40. Quaternary Cu–In–S–Se and penternary Cu–Ag–In–S–Se alloys have been prepared by spray pyrolysis with a view toward developing materials lattice-matched to CdS and possessing suitable bandgaps. However, solar cells have not yet been fabricated with these materials.

9.8. Copper Oxide (Cu_2O)

With a direct bandgap of 1.95 eV, Cu_2O is not ideally matched to the solar spectrum. However, interest in this material stems from the possibility of extremely low production costs, albeit at low conversion efficiencies. The low cost process is possible because fairly well oriented films of Cu_2O can be formed by heating a Cu substrate in air at 1050 °C. The junction between the base Cu and Cu_2O constitutes a Schottky barrier.[59] An ohmic contact applied to Cu_2O completes the back-wall Cu/Cu_2O Schottky barrier cell. Front-wall cells can be formed by reducing the surface of a Cu_2O sample to form a Cu/Cu_2O junction. Alternatively, a metallic layer can be deposited onto the Cu_2O to form a Schottky device.

Olsen and Bohara[59] have fabricated back-wall Cu/Cu_2O Schottky barrier devices by placing a clean Cu substrate in a furnace for a few minutes at 1050 °C to form Cu_2O and subsequently quenching the oxidized sheets in distilled water. The CuO layer formed during the quenching process is then removed by etching in a 20% NaCN solution. The cells are rinsed and finally coated with vapor deposited Cu or Ni to form ohmic contacts.

The back-wall cells are in effect MIS devices. The I–V characteristics correspond to a diode factor $\geqslant 2$. C–V data have yielded values of diffusion voltage in reasonable agreement with calculated values (assuming an MIS model). Barrier heights determined from I–V measurements and photoresponse measurements are in the range of 0.7 to 0.8 eV. Collection efficiencies of 0.25 and 0.65 (at $\lambda = 0.65\ \mu m$) have been obtained for Cu_2O thicknesses of 161 and 37 μm, respectively. Olsen and Bohara[59] have suggested, on the basis of their calculations, that with a controlled thickness of the interfacial layer and reduced series resistance, efficiencies exceeding 10% could be achieved. Actual Cu/Cu_2O devices have exhibited, to date, a best performance of $V_{oc} = 0.37$ V, $I_{sc} = 7.7$ mA cm^{-2}, FF = 0.57, and $\eta = 1.6\%$ for 100 mW cm^{-2} illumination.[41]

9.9. Organic Semiconductors

The organic semiconductors are very attractive from the viewpoint of cost as well as large-area production. However, because of the intrinsic properties of these materials, the solar conversion efficiencies are very low. In an early work, Fang[60] analyzed the conversion efficiency of organic semicondcutor solar cells and predicted conversion efficiencies of about 10^{-2}% if one of the two electrodes to the organic semiconductor could be made ohmic. Ghosh et al.[61,62] studied the photovoltaic and rectification properties of Al/Mg-phthalocyanine/Ag Schottky barrier cells and

Al/tetracene/Au sandwich cells. They observed that in the Mg-phthalocyanine cell, the short-circuit photocurrent J_{sc} was proportional to $F^{0.5}$, where F is the light intensity. V_{oc} was proportional to $\ln F$, as for a Schottky or a p/n junction. A large series resistance was observed and from the spectral response of the device an electron diffusion length of 1.5×10^{-6} cm was deduced. The spectral response was influenced by the direction of illumination. From $C-V$ measurements, a diffusion voltage of 0.6 eV and a depletion width of 2.5×10^{-6} cm were estimated. The quantum efficiency for carrier generation was observed to be 1.5×10^{-3}. At 0.69 μm, with light incident on the Al side, a photovoltaic efficiency of about 0.01% was measured. For the tetracene sandwich cells,[62] the forward $I-V$ characteristics were found to be space-charge-limited and attributed to hole injection from the Au side. I_{sc} was found to be proportional to F^n, with n varying between 0.6 and 1. The spectral response was sensitive to the light direction, similar to the results on phthalocyanine cells.

Cr/chlorophyll-a/Hg cells,[63] prepared by electrodeposition of micro-crystalline chlorophyll-a films, showed a rectification ratio of 10^3. The forward bias dark current was observed to obey the Shockley equation, with a diode factor of 1.6 and J_s was 2×10^{-11} A for a cell of area 0.08 cm^{-2}. The power conversion efficiency at 0.745 μm was measured to be 10^{-2}% for an incident power of 6×10^{-5} W. The V_{oc} was about 0.32 V and observed to be nearly independent of the light intensity at this light level, but the I_{sc} was linearly dependent on the incident intensity. A quantum efficiency of 0.007 was measured for this cell.

Schottky barrier cells,[64] fabricated by dispersing metal-free phthalocyanines in a binder polymer and sandwiching between SnO_2:Sb and Al electrodes, have yielded a power conversion efficiency exceeding 6% at 0.67 μm, for low power densities (0.06 W m^{-2}). At peak solar power density (1400 W m^{-2}), the extrapolated efficiency at 0.67 μm decreases to 0.01%. This decrease is attributed to space-charge effects. V_{oc} values as high as 1.1 V have been achieved. However, the device performance is limited by a field-dependent quantum efficiency.

Significant improvements in the photovoltaic characteristics of organic solar cells have been achieved by Morel et al.[65] Devices based on merocyanine dyes have exhibited conversion efficiencies of up to 0.7% for 1 cm^2 area under sunlight (78 mW cm^{-2}) illumination, with I_{sc} of 1.8 mA cm^{-2} and FF of 0.25. V_{oc} values as high as 1.2 V have been obtained, although the device performance is limited by field-dependent, energy-dependent quantum efficiency. Quantum efficiencies approach 100% at high photon energies but are about 35% at the main absorption peak in the visible range. These results are for a cell with a 50% transparent Al contact. With a correction for 100% transmission, a conversion efficiency in excess of 1% is predicted.

Table 9.1. Performance Characteristics of Thin Film Solar Cells of Emerging Materials

Active material	Device configuration	Type	Deposition technique[a]	Area (cm^2)
GaAs	n^+-GaAs/p-GaAs/p^+-GaAs	Homo	CVD	0.51
	p-GaAs/n-GaAs	Homo	CVD	0.5
	p-GaAlAs/p-GaAs/n-GaAs	Heteroface	MOCVD	0.29
	n-GaAs/n^+-GaAs/W(or graphite)	Schottky	CVD	8
	GaAs/Mo	Schottky	MOCVD	
	n-GaAs/n^+-GaAs/Mo	Schottky	MOCVD	
	n-GaAs/n^+-GaAs/W/graphite	MOS	CVD	9
	n-GaAs/n^+-GaAs/graphite	MOS	CVD	9
	n-GaAs/n^+-GaAs/graphite	MOS	CVD	9
	GaAs/GaAs(single crystal)	Homo	CVD	0.4
CdTe	p-CdTe/n-CdTe	Homo		
	Metal/CdTe	Schottky	Sp.	0.034
	p-Cu_2Te/n-CdTe	Hetero	Chem/VT	
	p-Cu_xTe/n-CdTe	Hetero	Flash evap.	
	p-CdTe/n-CdS	Hetero		
	n-CdS/p-CdTe	Hetero	Sc. ptg.	0.21
	n-CdS/p-CdTe	Hetero	Sc. ptg.	10
	n-CdS/p-CdTE	Hetero	VT	—
CdSe	n-CdSe/p-CdTe(single crystal)	Hetero	CSVT	
	n-CdSe/p-ZnTe(single crystal)	Hetero	CSVT	
	Au/Sb_2Se_3/n-CdSe	MIS	Evap.	
$Cu_{2-x}Se$	p-$Cu_{2-x}Se$/n-Si(single crystal)	Hetero	Evap.	
	p-$Cu_{2-x}Se$/n-CdS	Hetero	Evap.	
	p-$Cu_{2-x}Se$/n-CdS	Hetero	Evap.	
$ZnIn_2Se_4$	n-CdS/p-$ZnIn_2Se_4$	Hetero	Evap.	
Zn_3P_2	Mg/p-Zn_3P_2	Schottky	Evap.	
InP	p-InP/n-CdS	Hetero	CVD	
	p-InP/n-CdS	Hetero	CVD	
	p-InP/n-CdS/Mo	Hetero	CVD/Evap.	
	p-InP/n-CdS/Mo	Hetero	Evap.	
	p-InP/n-CdS/C	Hetero	CVD/Evap.	0.52×10^{-2}
	n-InP/p-Cu_xSe/Mo	Hetero	CVD/Evap.	
$CuInSe_2$	p^+-$CuInSe_2$:Cu/ p-$CuInSe_2$/n-CdS/n^+-CdS:In	Hetero	Evap.	1
	p^+-$CuInSe_2$:Cu/p-$CuInSe_2$/ n-CdS/n^+-CdS:In	Hetero	Evap.	1
	p-$CuInSe_2$/n-CdS	Hetero	Evap.	1.2
	p-$CuInSe_2$/n-CdS	Hetero	Evap.	1.2
	p-$CuInSe_2$/n-CdS	Hetero	Evap.	1.2
	p-$CuInSe_2$/n-ZnCdS	Hetero	Evap.	1
Cu_2O	Cu/p-Cu_2O	Schottky		
Merocyanine dyes				1
Chlorophyll-a	Cr/chlorophyll-a/Hg	MIM	Electro dep.	0.08

[a] The first mentioned deposition technique refers to the first semiconductor: Sc.ptg. — screen printing; Electro dep. — electrodeposition; Chem. — chemiplated; VT — vapor transport; Sp. — spray pyrolysis.

Intensity ($mW cm^{-2}$)	V_{oc} (V)	J_{sc} ($mA cm^{-2}$)	FF	η (%)	Remarks	Ref.
				17	AR	6
AM1	0.91	10.3	0.82	15.3	SiO + MgF_2 AR	6
128	0.99	24.5	0.74	12.8	No AR	11
AM1	0.456	10.4	0.59	3		9
AM1	0.39	19	0.58	4.3		13
AM1	0.49	20.6	0.54	5.45	No AR	14
AM1	0.45–0.52	11–13.4	0.55–0.68	4.4	No AR	7, 10
AM1	0.5	18	0.675	6.1	No AR	7, 10
—	0.56	22.7	0.67	8.5		41
	0.92	22	0.73	15		41
77.2	0.75	9.8	0.63	6		29
—	0.36	13.4	0.42	2		41
	0.5		0.45–0.6	6		28
	0.6			4.1		31
50	0.5	15	0.45	5–6		32
71	0.67	14.2	0.58	8.2		26
71	0.73	11.4	0.51	6.3		26
—	—	—	—	10	—	Kodak (unpublished)
85	0.61	0.75				25
85	0.56	1.89				
AM1	0.55–0.6	20	0.4–0.5	>5		36
75	0.45	23	0.62–0.65	8.8		34
AM1	0.3–0.45	2–4.5		0.51		35
AM1	0.46	11.6	0.62	3.3		41
AM1	0.27	16	0.31	1.5		37
83	0.37	12	0.46	2.5		41
AM2	0.4–0.46	13.3	0.68	4.9	No AR	43
AM2	0.4–0.46	15.4	0.68	5.7	SiO AR	43
AM1	0.37	18	0.30	2	With AR	46
	0.51		0.51	2.1		45
AM1	0.40	—	0.31	2.8		47
AM1	0.47	11	0.34	1.7		46
101.5	0.396	35	0.64	8.72	No AR	54
AM1	0.396	39	0.63	9.53	SiO_x AR	54
AM1	0.41	19.9	0.50	4.4	Front wall; *in situ*	53
AM1	0.34	12	0.36	2.46	Front wall; etched	53
AM1	0.44	22.5	0.59	5.7	Back wall	53
	0.418	36.3	0.65	10		58
AM1	0.37	7.7	0.57	1.6		41
78	1.2	1.8	0.25	0.7		65
	0.32					63

9.10. *Future Directions*

The performance characteristics of thin film solar cells discussed in this chapter are summarized in Table 9.1.

Considerable developmental work and device analysis are necessary to enhance the efficiencies of InP, GaAs, Zn_3P_2, CdSe, and $Cu_{2-x}Se$ thin film solar cells to acceptable limits. These studies should take the form of device configuration optimization, process parameter optimization to achieve reproducibly high-quality films of desired optoelectronic properties, and loss analyses to determine factors which could be eliminated or improved to reduce photon and carrier losses.

Cells based on organic materials should be investigated in relation to the effect of dyes, dye concentration, transport mechanisms, and device geometry to establish the feasibility of fabricating solar cells of reasonable efficiencies. At present, however, these cells remain an academic curiosity.

CdTe/CdS and $CuInSe_2$/CdS thin film cells present the best possibility, as of today, for fabricating all thin film solar cell modules which could compete favorably with thin film arrays based on Si or Cu_2S/CdS. To achieve this goal, the fabrication processes for the CdTe and $CuInSe_2$ cells must be developed further to be capable of reproducibly producing large-area cells with a high yield. In this connection, the spray pyrolysis technique for fabricating multicomponent, gradient composition cells such as $CuInSe_2$ should be pursued more vigorously. Finally, the long-term stability of these devices has to be unambiguously established on the basis of accelerated life tests. As with all thin film solar cells, suitable encapsulant materials and encapsulation processes have yet to be developed.

Besides the proven and being-proved (emerging) materials for solar cells, a host of other exotic materials offer the possibility of an efficient photovoltaic effect. Schoijet[66] has listed these exotic materials under various categories of optical gap and electron affinities. Among them, the interesting ones for fabricating junctions are: β-Zn_4Sb_3 (1.2), Cd_4Sb_3 (1.25), β-ZnP (1.33), $CdSiAs_2$(1.55), $ZnSiAs_2$(1.75), Bi_2S_3(1.3), Sb_2S_3(1.7), $PbSnS_2$(1.05), and WSe_2(1.35). The number within the bracket is the optical gap in eV.

10

Amorphous Silicon Solar Cells

10.1. *Introduction*

In 1969, Chittick et al.[1] reported preliminary results on hydrogenated amorphous Si (a-Si:H) and substitutional doping of a-Si:H by P. In 1976, the first *p*/*n* junction in a-Si:H was reported by Spear et al.[2] following a detailed and extensive study[3] of substitutionally doped a-Si:H films. This was followed by the fabrication of a-Si:H photovoltaic cells at RCA laboratories.[4,5] The photovoltaic effect was observed in several types of device structure such as *p*/*n*, *p*/*i*/*n*, and Schottky barrier junctions as well as heterojunctions.

The demonstration of photovoltaic energy conversion in a-Si:H generated worldwide interest in this material, as can be seen by the explosion of literature,[6–35] reviews[36–39] and conferences[40] on the subject. The reasons are obvious. The technology of a-Si:H solar cells fabricated by the glow discharge process promises low-cost fabrication of large-area solar arrays on inexpensive substrates on an in-line continuous basis.[41] The total material costs are very low in view of the fact that very low thicknesses ($\sim 1\ \mu m$) are sufficient for adequate device performance and the material in thin film form is deposited directly from the basic raw material, silane, thus eliminating intermediate process steps of converting the raw chemicals to Si ingots or Si powder. Further, Si being abundant, adequate availability is no problem. Moreover, the bandgap of a-Si is about 1.6 eV[42] and a better match to the solar spectrum than crystalline Si. The absorption coefficient is much greater[42,46] and has characteristics typical of a direct bandgap material. A 1-μm-thick a-Si film absorbs up to 70% of the AM1 incident radiation greater than 1.6 eV. Theoretical estimates place the maximum possible conversion efficiency of a-Si:H solar cells at 15%.[5,30]

While large-scale photovoltaic applications and attendant device-related studies, such as device configuration, junction analysis, and loss analysis, provide the main motivation for research in a-Si:H, the poorly understood transport properties, the effect and nature of doping, the optical properties, the microstructural characteristics, and problems associated with preparation of requisite quality a-Si:H films of desired properties continue to provide a rich pasture for a variety of detailed and intensive experimental and theoretical studies on the basic material itself.

Although a-Si films exhibit certain properties similar to its crystalline bulk and polycrystalline thin film counterpart, and a-Si solar cells, in general, follow the same operational principles as devices made of conventional materials, the peculiar characteristics of the a-Si:H films lend to the corresponding devices certain unique features. Also, to do justice to the considerable literature on the subject of a-Si:H solar cells, we feel that amorphous materials and devices merit a detailed and separate treatment. This chapter is, therefore, devoted to the properties of amorphous materials and photovoltaic devices fabricated from such materials, primarily a-Si:H.

10.2. *Deposition of a-Si:H*

The electronic properties of a-semiconductors depend strongly on the method of preparation. For instance, in vacuum evaporated samples, the observed transport and optical properties may be completely dominated by the presence of structural defects which lead to localized states in the mobility gap of the material. Indeed, the high density of defects present in the earlier evaporated and sputtered a-Si films led to the belief that amorphous semiconductor films could not be doped. The rationale for this generalization, based on experimental evidence, was that defects involving dangling bonds are present in the traditional (i.e., evaporated) forms of a-Si in sufficient quantities to produce a high density of states in the gap ($\sim 10^{20}$ cm^{-3}), pinning the Fermi level. In this light, the role of H_2 in a-Si:H films prepared by the glow discharge method is seen to be critical.

It had been shown by several workers[44,45] that the dangling bonds in a-Ge could be saturated by the introduction of metallic impurities in concentrations up to 40 at.%. H_2 present during deposition plays precisely the same role by preventing the formation of dangling bonds as these provide favorable bonding sites for H_2 atoms. The removal of dangling bond states by H_2 is manifested in the drastic reduction in spin density[50] and hopping conductivity,[51] the increase in photoconductivity[32] and photoluminescence,[33,52] and the shift in the optical absorption edge to higher energies.[53] Several studies[28,46–49] have shown that many glow discharge a-Si:H films contain atomic proportions of H_2 as high as 5 to 50%. In this sense, a-Si:H is basically an alloy. In recent years, some workers[6,14] have investigated a-Si:F:H alloys for photovoltaic applications.

The incorporation of H or F into a-Si leads to a relatively low density of states in the gap ($< 10^{17}$ cm^{-3}), thereby facilitating doping of the material in either type.

By incorporating H_2 during the deposition, it has been possible to prepare a-Si:H films by rf sputtering[24,54] as well as pyrolysis.[38] Photovoltaic response[55] has been observed in devices made by rf sputtering. As a matter of fact, the role of hydrogenation in reducing gap states in tetrahedral semiconductors was revealed in studies on sputtered a-films.[56] a-Si films have also been prepared by evaporation,[57] ion plating,[58] arc discharge,[23] corona discharge,[23] magnetron sputtering,[40] and ion implantation.[40]

a-Si:H films have a high photoconductivity and a high resistivity and can be doped either *p* or *n* type, and these properties are considered to be necessary and beneficial for good photovoltaic performance. In this regard, the best device quality films have been prepared by glow discharge of SiH_4 and the solar cells made thereof have exhibited efficiencies in the range 5 to 10%.[37–39,14] Although rf sputtered films have not achieved the same photovoltaic performance, there are certain features connected with the sputtering process that make it attractive for film preparation. We shall, therefore, briefly enumerate the salient deposition conditions of glow discharge and rf sputtering techniques.

10.2.1. *Glow Discharge Deposition Conditions*

In the electrodeless glow discharge, used in the early works, an rf coil external to the discharge chamber generates the discharge.[1] The operating frequency range is usually 0.5 to 13.5 MHz at typical SiH_4 (silane) pressures of from about 0.1 to 2.0 torr. Flow rates ranging between 0.2 and 5.0 standard cm^3 min^{-1} (sccm) yield deposition rates of 100 to 1000 Å min^{-1}. However, the small dimensions of the discharge chamber tend to produce films with poor uniformity.

Capacitatively coupled rf glow discharge systems[59] yield films with better uniformity. The parallel-plate electrodes inside the discharge chamber operate at a frequency of 13.5 MHz. Silane pressure of usually 5 to 250 mtorr and flow rates ranging from 10 to 30 sccm yield deposition rates of about 500 Å min^{-1}. When using capacitative coupling with a superimposed dc bias, the properties of the films deposited on the two electrodes (anode and cathode) may differ.[60]

In a dc glow discharge in SiH_4, a-Si:H films can be deposited on substrates acting as the cathode at rates ranging from 1000 to 10,000 Å min^{-1} by varying the cathodic current density from 0.2 to 2.0 mA cm^{-2}, at a SiH_4 pressure of about 1.0 torr. On anodic substrates, the deposition rate is an order of magnitude less.

Most of the a-Si:H films possessing good electronic properties are prepared at substrate temperatures in the range of 200 to 400 °C. We should note that under conditions of low pressure, surface generated reactions are promoted, causing deposition of hydrogenated a-Si on any local surface.[37–39] In the high-pressure regime, polymerization is favored and, in such cases, the a-Si:H films appear hazy owing to the precipitation of disilane (Si_2H_6) and trisilane (Si_3H_8) clusters from the discharge.[27] It is evident from the characterization studies that film quality is strongly influenced by substrate temperature and discharge power. The role of other deposition parameters such as substrate position, reaction vessel geometry, pressure, and gas flow rates are not so clear. There is evidence that device quality is adversely affected by contaminants such as O_2 and N_2.[39]

10.2.2. rf *Sputtering*

There are several advantages of sputtering, namely the possibility of avoiding toxic gases, the greater degree of control, and less contamination from O_2 and N_2.

a-Si:H is produced by sputtering a Si target in a partial pressure of hydrogen. The source is undoped if intrinsic a-Si is required, or doped with P or B to prepare *n*- or *p*-type a-Si:H. Doping can also be effected by introducing hydrides of P or B into the chamber during the sputtering process. Anderson et al.[32,50–53] have carried out an extensive and systematic study of rf sputtered a-Si:H films and have obtained optimum values for two of the deposition parameters, deposition temperature (~ 200 °C) and H_2 partial pressure (5 mtorr). However, the magnitude of photoconductivity and photoluminescence intensity indicate that this optimization does not reduce the gap state density to the low level typically found in glow discharge a-Si:H. Thus the performance of devices fabricated from sputtered films is inferior as compared to devices made from glow discharge films.

The higher density of gap states in sputtered a-Si:H films is thought to result from defects introduced in the growing film owing to bombardment by energetic Si atoms during the deposition process,[22] as sputtering pressures (typically at 5 mtorr) are lower and the accelerating voltage (~ 1.5 kV) and rf power levels (~ 1.6 W cm^{-2}) are generally higher in the sputtering process as compared to the silane glow discharge process (applied voltage 650–800 V). Although hydrogenation reduces the gap state density, additional means to eliminate bombardment are considered desirable. Paesler et al.[61] have incorporated new compensating impurity atoms in an attempt to minimize the dangling bond states. In another approach,[62] some workers[22,24,62] have observed a reduction in bombardment induced defects in a-Si:H films prepared under a higher partial pressure of Ar (~ 30 mtorr). Significant improvements have been obtained in the photoconductive, photoluminescent, optical, and transport properties of these a-Si:H films, comparable to those of glow discharge deposited films. However, there is difference of opinion on the mechanism of bombardment reduction.

10.2.3. *Pyrolysis of Silane*

In this process, temperature and pressure conditions govern the state of the product, whether gaseous or solid. The composition of the solid product varies continuously during film formation with the ratio of hydrogen decreasing with the progress of reaction time, forming an initial film with an empirical formula SiH_2 and a-Si:H film later on.

10.2.4. *Discharge Kinetics*

Relatively little information exists in the literature on the discharge kinetics or discharge chemistry. Mass spectroscopy results[63,64] indicate that dihydrides (SiH_2^+) and trihydrides (SiH_3^+) are relatively abundant in SiH_4 discharges and higher silanes (Si_xHy, $x > 2$) are created in varying amounts depending on deposition conditions. Although there is controversy regarding the most desirable species for good-quality a-Si:H films, study of the properties of a-Si:H films prepared under diverse conditions leads to the conclusion that monohydrides (SiH^+) are favorable.

Films prepared at substrate temperatures below 200 °C, containing dihydrides and possibly trihydrides,[46,65,66] do not exhibit good electronic properties, while films prepared at substrate temperatures above about 200 °C exist only in the monohydride form[46,65,66] and exhibit the requisite solar cell quality. Films deposited near room temperature are more like polysilanes. Plasma diagnostics, utilizing emission spectroscopy, mass spectroscopy, and electrical measurements, carried out by Bauer and Bilger[23] reveal that the relative abundance of different species depends strongly on several deposition parameters. They have devised methods for controlled variation of plasma parameters to enable preparation of a-Si:H films with highly reproducible properties which are strongly dependent on the plasma parameters. A comparison of the plasma characteristics of different gas discharges used in the production of a-Si:H films, made by the above workers, is presented in Table 10.1. The properties of the corresponding films are also tabulated therein. Figure 10.1 shows the relative densities of Si and SiH species in rf sputtering and dc glow discharge systems as a function of applied voltage and current density. Recently, however, a-Si:H films have been prepared by planar rf magnetron sputtering on substrates held at room temperature[122] with properties comparable to films prepared by glow discharge and rf sputtering on substrates held at elevated temperatures.

Table 10.1. Plasma Characteristics for Different Gas Discharges (after Bauer and Bilger[23])

Feature / Discharge	Neutral density (cm^{-3})	Neutral gas composition	Gas temperature	Electron energy	Degree of ionization	Cathode material	Cathode mechanism
rf sputtering	10^{13}–10^{15}	H_2 (+rare gas)	300 K thermal non-equilibrium	10 eV	$< 10^{-4}$	Si	rf sputtering by high-energy ions
dc glow discharge	10^{19}–10^{14}	SiH_4/H_2 (+rare gas)	≤ 3000 K thermal non-equilibrium	2–5 eV	$< 10^{-4}$	Si	dc sputtering dependent on current density
Corona discharge	10^{19}–10^{18}	SiH_4/H_2 (+rare gas)	300 K thermal non-equilibrium	6–8 eV	$< 10^{-7}$	No cathode erosion effect	—
Arc discharge	10^{18}–10^{16}	H_2 (+rare gas)	(3000–6000) K localized thermal equilibrium	—	$\leq 10^{-3}$	Si	Thermal evaporation

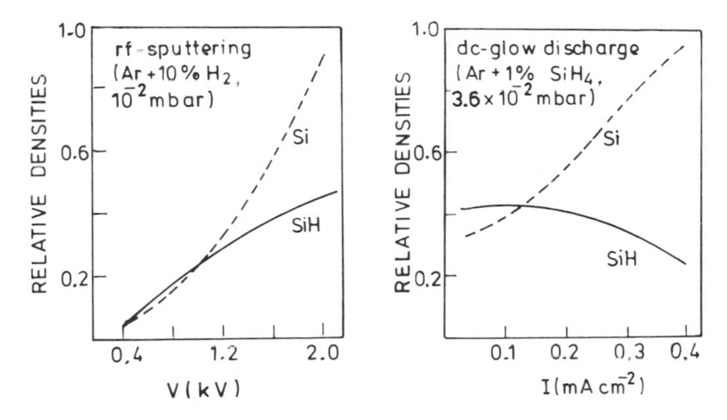

Figure 10.1. Relative densities of Si and SiH in rf sputtering and dc glow discharge systems as a function of applied voltage and current density, respectively (after Bauer and Bilger[23]).

10.2.5. *Substrate Effects*

Substrate materials such as Cr, Ti, V, Nb, Ta, and Mo do not influence the properties of a-Si:H films significantly, as the diffusion coefficient of these materials in a-Si:H, at temperatures up to 400 °C, is negligible ($\leq 10^{-18}$ cm^2 s^{-1} at 450 °C for Mo) as revealed by AES studies. However, Fe and Al, while suitable at temperatures lower than about 300 °C, have diffusion coefficients (2–3) $\times 10^{-15}$ cm^2 s^{-1} at approximately 400 °C. In the case of Al, the presence of an oxide layer can cause variations in the contact resistance or influence silicide formation. Silicides may be formed even at low temperatures and interdiffusion occurs with certain metals like Au and Cu, while Ag is unsuitable owing to the poor adhesion of the a-Si:H films. Alkali containing glasses may dope the film with alkali impurities that act as donors in a-Si:H. Fused silica is suitable in this regard. In all cases, especially metal substrates, chemical cleaning or sputter cleaning of the substrates just prior to deposition of the a-Si:H film is desirable.

10.3. *Properties of a-Si:H Films*

The properties of a-Si:H films are strongly dependent on the deposition technique and the conditions prevailing during growth. In this section, we shall discuss the structural, compositional, electronic, and optical properties of a-Si:H films deposited by different techniques.

10.3.1. *Structural Properties*

The idealized structural model of pure a-Si is the random network structure proposed by Polk and others.[67] The a-Si has the same tetrahedral coordination as in crystalline Si, but very limited short-range order extending to only the nearest neighbors. Barna et al.[68] have shown from electron diffraction studies that the radial distribution function (RDF) of a-Si is not drastically perturbed by the introduction of H_2, except for an additional peak at 5 Å for glow discharge produced a-Si:H films. The existence of this peak, corresponding to a dihedral angle of 45°, led to the conclusion that a-Si:H films possess a greater degree of short-range tetrahedral ordering than evaporated a-Si films. Similar results have been reported by Mosseri et al.,[13] who have attributed the observed peak at 4.9 Å in their glow discharge a-Si:H films, corresponding to a dihedral angle between 35 and 45°, to the formation of seven-membered Si rings or larger ones in the Si–H bond vicinity. They have deduced the coordination number to be 3.5 for a-Si:H (instead of 4 for pure a-Si), corresponding to a 33% H content in the films.

Knights et al.[69–77] have made significant contributions in the area of preparation–structure–composition relations of glow discharge deposited a-Si:H films. They have established a picture in which the microstructure consists of columns of about 100 Å diameter of more or less intrinsic a-Si, perhaps with point defects of monohydride substituting for Si. The columns are nucleated at the substrate film interface and grow in the direction of the impinging vapor stream, with lower-density regions separating the individual columns. The lower-density regions are thought to be comprised of polysilane chains, $(SiH_2)_n$, along with monohydride saturation of the dangling bonds. The low-density regions are minimal and monohydride arrangements are predominant in a-Si:H films of good device quality. Moreover, such materials are more resistant to postdeposition oxidation and show no noticeable microstructure at the 1000-Å level, though the 100-Å columns are still present. The microstructural features strongly affect the physical properties. It should be noted that the microstructural characteristics are influenced by the deposition parameters. The densities of a-Si:H films given in Table 10.2 indicate the variability of hydrogenated Si and its dependence upon deposition conditions.

Raman scattering data[78] suggest that doped a-Si:F:H films are somewhat polycrystalline. However, these films may still provide good contact layers for cells.

Void, or low-density, network structures, similar to those found for glow discharge a-Si:H, have been observed in pure sputtered a-Si films.[22,53,65] Recent studies[79–81] have shown that similar void networks exist in sputtered a-Si:H also. Bombardment processes[24] are thought to be

Table 10.2. The Dependence of the Density ρ and Atomic Fraction f_H of H in a-Si Samples on Silane Pressure P and Substrate Temperature T (after Wilson et al.[38])

Sample	T (°C)	P (torr)	$\rho_{a\text{-}Si}/\rho_{cryst.\ Si}$	f_H
78	25	1.0	0.63	0.35
79	25	0.1	0.73	0.25
76	250	0.1	0.98	0.18
77	250	1.0	0.86	0.14

responsible for this void structure. Similar to the glow discharge films, sputtered a-Si:H films have a columnar microstructure, with the columns aligned in the direction of the impinging vapor stream. The microstructures of sputtered a-Si and a-Si:H films are observed to be qualitatively similar. The a-Si:H films can be anisotropically etched in a manner similar to a-Si[24] and the resulting microstructure is a dense array of columns with a spacing of the order of 1000 to 3000 Å. The total surface reflectance is found to decrease from about 40% (for a smooth, as-deposited film) to about 3% on etching.

As-deposited boron doped evaporated a-Si films have also been reported to generally consist of a high density of microvoids and microcrystallites.

10.3.2. Compositional Properties

The deposition conditions during glow discharge strongly influence the H_2 concentration in a-Si:H films. Several investigations have shown that glow discharge deposited a-Si:H films contain anywhere between 5 to 50 at.% bonded H_2. In general, the H_2 content decreases with increasing substrate temperature.[46,65,82] For an rf electrodeless discharge, the H_2 content increases with power and decreases with increase in SiH_4 pressure. Similar results are obtained for glow discharge deposited a-Si:F:H films, where F content decreases with an increase in SiF_4/H_2 ratio.[6] However, several other deposition parameters affect hydrogen content. Solar cells have been made with a-Si:H films containing 10 to 15 at.% H_2.[82] Thus, it would seem that no specific concentration of H_2 is required for good electronic properties. However, as noted earlier, the monohydride form is favorable for better-quality material and, consequently, more efficient devices.

We have stated previously that the H_2 is pictured to be incorporated in the form of compensated microvoids such as monovacancies containing four H atoms, or divacancies containing six H atoms. These compensated microvoids do not produce localized states in the gap,[83] and their size and concentration depend strongly on the deposition conditions. On the basis of density measurements,[84] it is suggested that the average microvoid in the films investigated is slightly larger than a divacancy and that the changing H_2 content results primarily from a change in concentration of these compensated microvoids.

Messier and Tsong[24] have prepared a-Si:H films, containing a predominance of monohydride units, with no noticeable postdeposition oxidation. They have measured the H, Ar, and O contents on a series of A-Si:H films, sputter deposited in an Ar/H_2 mixture. The H concentration ranged from 2 to 25 at.% with the best-quality films (at 5 mtorr) containing 17 at.% H. The Ar content increases with decreasing sputtering gas pressure while the O content increases with increasing pressure. This indicates that the postdeposition oxidation is greater at the higher gas pressures. This behavior can be correlated to the void or low-density structures present in these materials, since at low gas pressures the beneficial effect of Ar bombardment, which increases with decreasing sputtering gas pressure,[24] leads to a-Si:H films with high monohydride content, high Ar content, and absence of microstructure.

10.3.3. *Optical Properties*

The optical absorption coefficient α of a-Si:H is more than an order of magnitude larger than that of crystalline Si over most of the visible light range.[46] The dependence of α on the substrate temperature is shown in Figure 10.2 for glow discharge produced films. The optical bandgap E_{opt}, determined by plotting $(\alpha h\nu)^{\frac{1}{2}}$ as a function of $h\nu$, is shown as a function of substrate temperature for two types of glow discharge films[37] in the inset. The increase in E_{opt} with decreasing substrate temperature is attributed to the increase in the H_2 content of the films.[46] It has been shown from photoelectron emission studies[66] that the addition of H_2 to the a-Si structure causes the top of the valence band to move down, hence causing an increase in E_{opt}. Any other parametric variation causing a change in H_2 content, such as discharge power or pressure, will also affect α.

Optical absorption observed for photon energies less than E_{opt} is not directly related to the H_2 content of a-Si:H but appears to be due to absorption by defect states in the energy gap and is a minimum at a substrate temperature around 300 °C,[37] similar to the behavior of photoluminescence and spin density.

In agreement with a decreasing gap, the refractive index of a-Si:H films, which is relatively close to that of crystalline Si, increases (by only ~5%) as the substrate temperature increases from about 195 to 420 °C, corresponding to a decrease in the H_2 content from about 40 to about 10

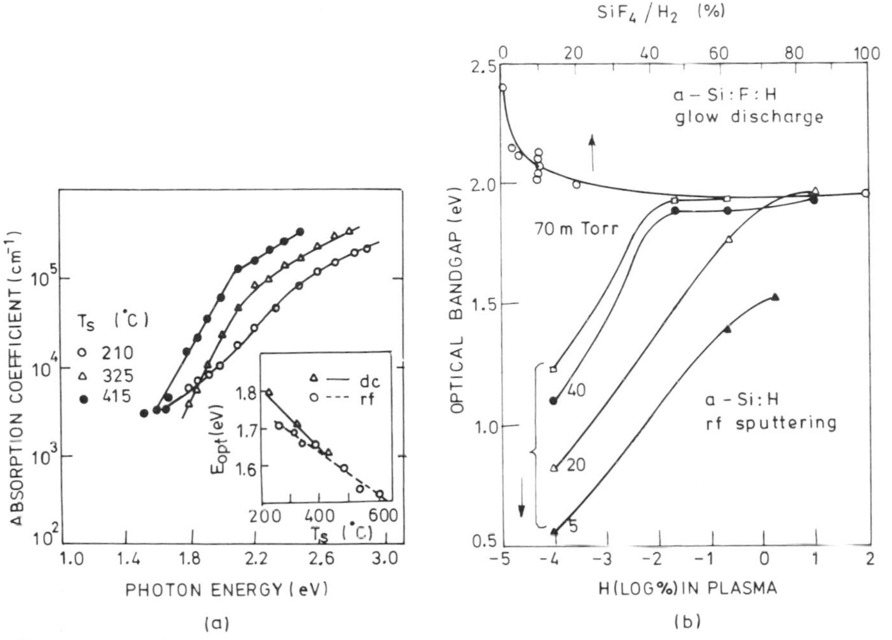

Figure 10.2. (a) Optical absorption coefficient of glow discharge a-Si:H films as a function of photon energy for different substrate temperatures. Inset shows the optical bandgap E_{opt} vs. substrate temperature T_s for dc proximity and rf glow discharge films (after Carlson and Wronski[37]). (b) Optical bandgap of a-Si:F:H glow discharge deposited films as a function of SiF_4/H_2 ratio and rf sputtered a-Si:H films as a function of H content in plasma for different total pressures (after Madan et al.[6] and Messier and Tsong[24]).

at.%.[46] Brodsky and Leary[11] investigated the temperature dependence of intrinsic and doped a-Si:H films prepared by glow discharge and observed quantitative spectral similarity with crystalline Si, with small differences on doping.

The optical gap of a-Si:F:H glow discharge deposited films decreases with increase in the SiF_4/H_2 ratio as shown in Figure 10.2b. With the greater inclusion of F and H in the films at lower SiF_4/H_2 ratios,[6] strong Si–F and Si–H bonds are formed, which are stronger than the Si–Si bond. Thus, E_{opt} increases.

The rf sputtered a-Si:H films also exhibit a similar optical behavior. The absorption edge shifts to higher energies with increasing H_2 content as well as with increase in the total pressure, as shown in Figure 10.2b.

10.3.4. *Electrical Properties*

The electronic properties are directly related to the photovoltaic performance of a device. There have been several detailed investigations of both intrinsic and doped a-Si:H films.[4,20,59,86,87] However, a-Si:H films exhibit large reversible photostructural and photothermal changes.[88,89] We shall, therefore, discuss the electronic properties of a-Si:H films which have been well-annealed and exposed and thus exhibit reproducible behavior.[46,90]

10.3.4.1. *Conductivity*

The dark resistivity of glow discharge deposited, annealed, undoped a-Si:H films varies from about 10^{11} Ω cm at a substrate temperature of 100 °C to something under 10^5 Ω cm for a substrate temperature of about 550 °C.[1,91] With increase in substrate temperature, the undoped films exhibit strongly n-type conductivity. It should be noted that the resistivity and photoconductivity are also influenced by other deposition parameters, indicating thereby that the density of gap states is sensitive to the deposition conditions.

The temperature dependence of the dark conductivity of undoped films, expressed by $\sigma = \sigma_0 \exp(-E_a/kT)$, where E_a is an activation energy, yields values of E_a in the range from about 0.2 to 0.8 eV.[1,91] The behavior of the resistivity and photoconductivity of undoped a-Si:H films can be analyzed in terms of free carrier conduction over a wide temperature range. Some workers[37] have reported that a single activation energy (0.68 eV) fits the dark conductivity data over a temperature range from about 180 to 330 K. However, a transition in carrier transport at about 250 K has been reported by other authors,[91] who have suggested hopping through localized states in the gap to be the transport mechanism below 250 K. These conflicting results may be attributed to different deposition conditions and, consequently, films with different properties.

Both doped and undoped a-Si:H films follow the Meyer–Neldel rule,[92] where the pre-exponential term σ_0 obeys the relation $\sigma_0 = \sigma_{00} \exp(E_a/kT_0)$, where σ_{00} and kT_0 are constants.[37]

At average electric fields exceeding about 10^4 V cm^{-1}, a-Si:H films exhibit nonohmic behavior. Temperature and film thickness dependence of the high-field effect have identified Poole–Frenkel conduction as the likely mechanism.[37,93] It may be noted that this field ionization of carriers out of traps may be the cause of the high collection efficiency of photogenerated carriers in the space-charge region of a-Si:H solar cells.

We have already noted that a-Si:H films can be doped in either conductivity type. Resistivities as low as 10^2 Ω cm have been obtained. The room temperature conductivity of n- and p-type a-Si:H films produced by glow discharge is shown in Figure 10.3a as a function of gas mixture composition.

Similar to the dark conductivity, the photoconductivity of undoped

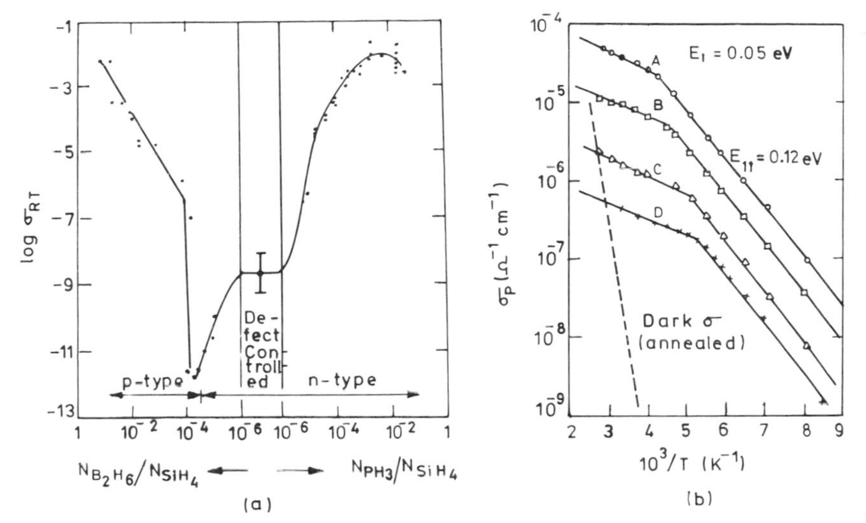

Figure 10.3. (a) Room temperature conductivity of doped a-Si:H films as a function of gas mixture composition (after Gibbon et al.[43]). (b) Temperature dependence of photoconductivity for an undoped a-Si:H film at $\lambda = 0.61$ μm for different photon fluxes: $A = 2 \times 10^{15}$, $B = 2 \times 10^{14}$, $C = 2 \times 10^{13}$, and $D = 2 \times 10^{12}$ $cm^{-2} s^{-1}$. Dark conductivity in the annealed state is also shown (after Carlson and Wronski[37]).

a-Si:H is sensitively influenced by the deposition conditions and exhibits a wide range of recombination kinetics, reflecting the variety of distributions of deep centers in a-Si:H. The photoconductivity σ_p has a power dependence on the illumination intensity F of the form $\sigma_p \propto F^{\gamma}$, where γ is a constant. Values of γ of 0.7 to 0.75 over four orders of magnitude preclude monomolecular ($\gamma = 1$) or bimolecular ($\gamma = 0.5$) recombination, but suggest an exponential or quasi-exponential distribution of recombination centers.

The temperature dependence of the photoconductivity of undoped a-Si:H films is shown in Figure 10.3b for various illumination levels. The results indicate that free carrier transport occurs even at temperatures as low as 120 K. The transition from one temperature region ($E_a = 0.05$ eV and $\gamma = 0.7$) to the other ($E_a = 120$ eV and $\gamma = 0.5$) occurs at temperatures that depend on the intensity of illumination. This behavior is consistent with the change in the recombination kinetics. Spear et al.,[86] however, observed the transition at temperatures below about 250 K in their rf glow discharge deposited films and interpreted the behavior in terms of a transition from free carrier transport to hopping conduction in localized states. The absence of hopping conduction at lower temperatures is significant in the sense that efficient solar cell operation can then be achieved even at low temperatures.

Madan et al.[6] have prepared a-Si:F:H films by glow discharge from a gas ratio of SiF_4:H_2 = 10:1 possessing a localized density of states less than 10^{17} $cm^{-3}\,eV^{-1}$. These films are highly photoconductive, devoid of photostructural changes and are mechanically stable. The low density of states facilitates *n*- and *p*-type doping, and very high conductivities (~ 5 $\Omega^{-1}\,cm^{-1}$) and low activation energies ($\Delta E \simeq 0.05$ eV) have been achieved with very small amounts of dopant. The dark conductivity, the photoconductivity, and the activation energy are all strongly dependent on the SiF_4/H_2 ratio, as shown in Figure 10.4.

The photoconductivity of sputtered a-Si:H films is shown in Figure 10.5 as a function of Ar partial pressure.[22] The photoconductivity increases with increase in partial pressure of Ar by several orders of magnitude. Temperature dependence of the dark resistivity in these films indicates that increasing Ar partial pressure from 5 to 7.5 mtorr does not affect the dark conductivity, but further increase causes a higher conductivity accompanied by a decrease in the activation energy. The photoconductivity data as a function of H_2 partial pressure show a similar behavior. These results and changes in photoconductivity and dark conductivity with partial pressure of Ar and/or H_2 have been explained on the basis of changes in the density of gap states following microstructural changes resulting from variation in deposition conditions.

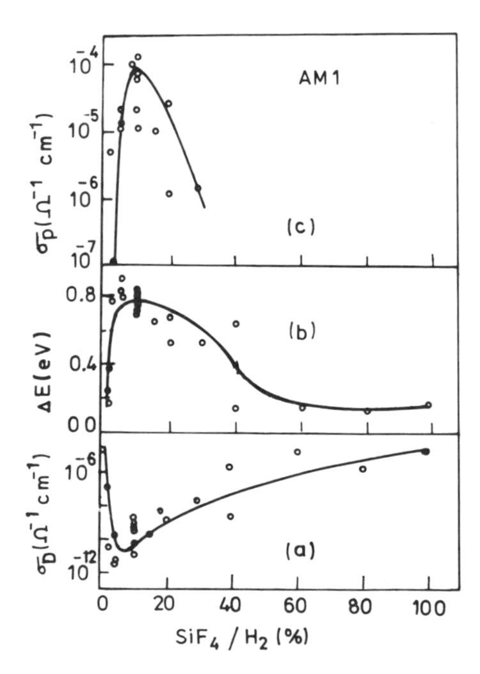

Figure 10.4. (a) Dark conductivity, (b) activation energy, and (c) photoconductivity of glow discharge a-Si:F:H films as a function of SiF_4/H_2 ratio (after Madan et al.[6]).

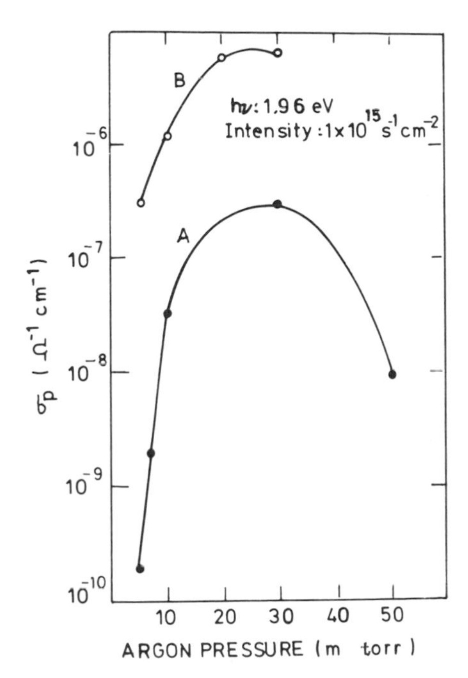

Figure 10.5. Room temperature photoconductivity of sputtered a-Si:H films as a function of Ar pressure for different values of H_2 partial pressure: $A = 2 \times 10^{-4}$ torr and $B = 6 \times 10^{-4}$ torr (after Anderson et al.[22]).

10.3.4.2. *Mobility and Lifetime*

The drift mobilities μ_d of both carriers, measured between 400 and 200 K, in rf and dc glow discharge produced a-Si:H films are trap-controlled. The values of μ_{dn} and μ_{dp}, at room temperature, are found to be $(2\text{–}5) \times 10^{-2}$ and $(5\text{–}6) \times 10^{-4}$ $cm^2 V^{-1} s^{-1}$, respectively, with corresponding activation energies of 0.19 and 0.35 eV.[34] The data[34] are depicted in Figure 10.6. In both cases, the mobilities correspond to majority carriers. The free carrier mobilities have been estimated to be about 1 $cm^2 V^{-1} s^{-1}$.[91,94]

Electron lifetimes deduced from photoconductivity measurements[46] range from 10^{-3} to 10^{-7} s in undoped a-Si:H films. The lifetime depends on the illumination level, reflecting the displacement of the electron quasi-Fermi level and the change in carrier recombination kinetics.[46] Although minority carrier lifetimes have not yet been determined in a-Si:H films, analysis of a-Si:H solar cells performance yield hole diffusion length estimates up to approximately 0.2 μm.[31] If the extended state mobility is about 1 $cm^2 V^{-1} s^{-1}$, hole lifetimes may be as high as 2×10^{-8} s.

10.3.4.3. *Density of States in the Gap*

The energies, densities, and types of states present in the gap of a-Si:H play an important role in device performance. The density and nature of

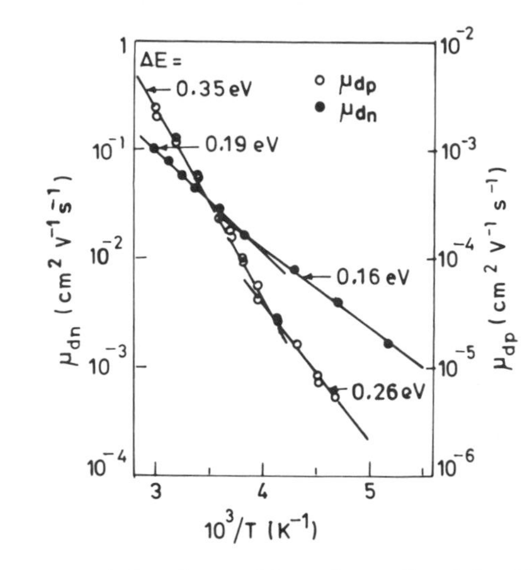

Figure 10.6. Temperature dependence of the drift mobilities of majority carriers in a-Si:H films (after Moore[34]).

gap states depend largely on the deposition conditions. Measurements on a variety of films[95,96] have yielded estimates of density of deep centers to be in the range of 10^{15} to 10^{16} cm^{-3} in a-Si:H films produced by glow discharge. The density of gap states decreases with increase in the substrate temperature[95,96] and goes through a minimum at a substrate temperature around 330 °C.[37,48] At higher substrate temperatures, the density of defects increases as a result of out-diffusion of H_2.[48,97] a-Si:H films in the light-soaked state[37] exhibit a high density of states near the midgap, owing to the creation of deep centers on exposure to light.

The density of gap states in glow discharge produced a-Si:F:H and rf sputtered a-Si:H films are determined to be in the range of 10^{17} $cm^{-3}\,eV^{-1}$.

10.3.4.4. *Decomposition of a-Si:H Films*

a-Si:H films treated to temperatures greater than the deposition temperature (or $\gtrsim 350$ °C) decompose accompanied by evolution of H_2 gas.[9,10,12,98] The loss of H_2 increases the density of dangling bonds.[98] H_2 loss occurs by elimination of molecular H_2 from sites in which pairs of H_2 atoms are in close proximity at low temperatures.[10] The relative concentrations and distribution of the SiH and SiH_2 groups within the film critically determine the kinetics of the decomposition mechanism. The out-diffusion of H_2 may limit the operational life of an a-Si:H solar cell. However, diffusion studies[49] of a-Si:D (D-deuterium) with fields of about 3×10^4 $V\,cm^{-1}$ and illumination levels of approximately 100 $mW\,cm^{-2}$ at temperatures near to 300 °C indicate that degradation of a-Si:H solar cells will be negligible for out-diffusion of H_2 at 100 °C up to 10^4 years.

In Table 10.3 we have tabulated the various properties of doped and undoped a-Si:H (a-Si:F:H) films prepared by different techniques.

Table 10.3. Properties of a-Si:H and a-Si:H:F Films Deposited by Different Techniques[a]

Material	Deposition technique	N_g ($cm^{-3} eV^{-1}$)	σ_d ($\Omega^{-1} cm^{-1}$)	σ_p ($\Omega^{-1} cm^{-1}$)	E_g (eV)	α vis. (cm^{-1})	α IR (cm^{-1})	μ ($cm^2 V^{-1} s^{-1}$)	τ (μs)	Remarks	Ref.
a-Si:H	E. beam	—	$\sim 10^{-8}$	6×10^{-7}	—	40	300	—	—	3.5 at.% H_2	42
a-Si:H	Glo. Dis.	10^{15}–10^{16}	$\sim 10^{-10}$	$\sim 10^{-5}$	1.6–1.8	10^4–10^5	—	$\mu_n = 2$–5.10^{-2} $\mu_p = 5$–6.10^{-4}	—	$T_{sub.}$ 200–400 °C	37
a-Si:H	Glo. Dis.	—	10^{-8}–10^{-9}	10^{-6}–10^{-7}	1.6–1.9	10^4–10^5	—	—	—	—	23
a-Si:H	Glo. Dis.	—	2×10^{-10}	4×10^{-7}	—	2×10^4	1400	—	0.1–0.3	14 at.% H_2	42
a-Si:H	Glo. Dis.	—	5×10^{-9}	2×10^{-7}	—	10^3–10^4	—	—	—	25 at.% H_2	42
a-Si:H	Sputter.	—	10^{-9}	10^{-5}	1.6–1.8	10^4	—	—	—	—	23
a-Si:F:H	Glo. Dis.	10^{17}	10^{-9}	10^{-4}	—	10^4	—	—	—	$T_{sub.} \sim 380$ °C, $\gamma = SiF_4/H_2 = 10$	6

[a] N_g — density of states in the gap, σ_d — dark conductivity (300 K), σ_p — photoconductivity (300 K), E_g — band gap, α — absorption coefficient, μ — mobility, τ — minority carrier lifetime, Glo. Dis. — glow discharge, sputter. — sputtering.

10.4. *a-Si:H Solar Cells*

Solar cells have been fabricated using both glow discharge and rf sputter deposited a-Si:H films. However, the best results to date have been obtained on glow discharge a-Si:H and a-Si:F:H films.[14,37–39] The efficiencies range from 4 to 10%. In comparison, the best efficiency obtained with sputtered a-Si:H films is about 2%. As we discussed in the preceding sections, sputter deposited films have not exhibited the microstructural and electronic properties that are required for good-quality devices. Recent studies[22,24] on a-Si:H films deposited in the presence of high Ar partial pressures show that sputtered films of structural and electronic quality comparable to glow-discharge deposited films can be obtained under controlled deposition conditions. However, results on solar cells prepared with these films have not yet been reported.

The performance history of a-Si:H solar cells has been reviewed by several authors.[37,39,42] We shall, therefore, confine ourselves to a discussion of the performance characteristics of present state-of-the-art solar cells and examine the loss mechanisms based on our present level of understanding of cell operation. The data presented will be primarily those obtained on cells fabricated using glow discharge deposited a-Si:H (a-Si:F:H) films.

10.4.1. *Device Configuration*

At present, conversion efficiencies exceeding 5% have been obtained on a-Si:H solar cells only with the three configurations shown in Figure 10.7a–c.

10.4.1.1. *Schottky Barrier Cells*

The Schottky barrier cell (Figure 10.7a) is the simplest to fabricate and is, therefore, frequently employed as a diagnostic tool. The main factor limiting the performance of Schottky barrier cells is the relatively low V_{oc} ($\lesssim 600$ mV). Schottky cells can be fabricated by depositing undoped a-Si:H on Mo and subsequently evaporating about 50 Å of Pd onto the top of the a-Si:H film. An initial thin layer (~ 200 Å) of P doped a-Si:H helps to improve the photovoltaic and diode characteristics.[39]

10.4.1.2. *MIS Cells*

In the MIS device (Figure 10.7b), a thin insulating layer (~ 20–30 Å) is deposited on the undoped a-Si:H prior to deposition of a high work function metal such as Pt. The metal film is necessarily thin (~ 50 Å) to ensure adequate optical transmission into the semiconductor. An antireflection layer (e.g., ~ 450 Å of ZrO_2) helps to reduce reflection losses.[30]

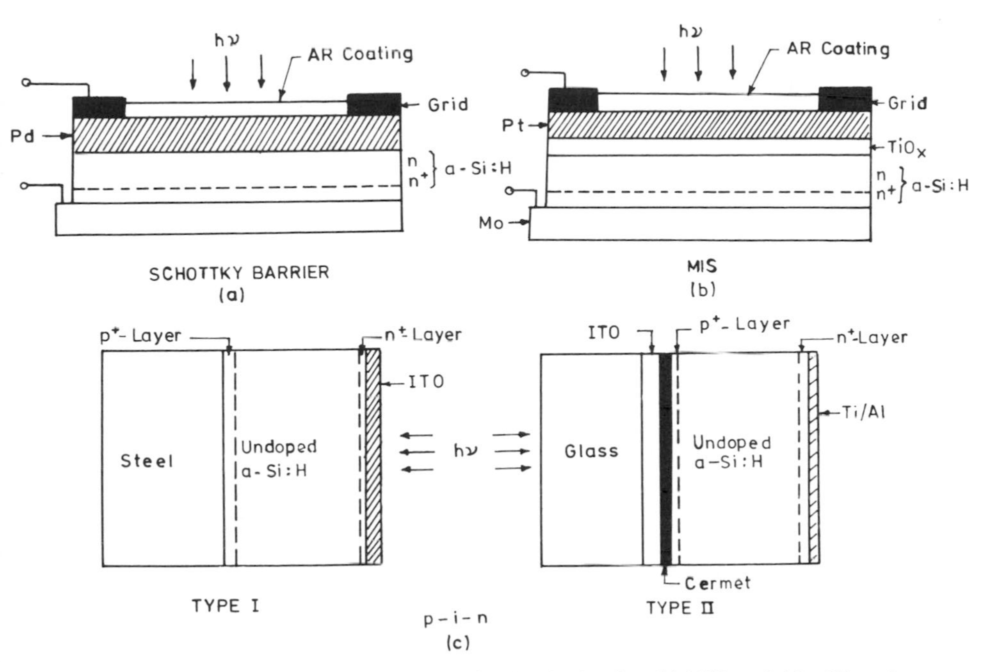

Figure 10.7. Schematic representation of (a) Schottky barrier, (b) MIS, and (c) $p/i/n$ solar cells.

10.4.1.3. *p/i/n Cells*

The $p/i/n$ devices shown in Figure 10.7c exhibit good performance and can be fabricated in two types of structures. Type-I device is fabricated by first depositing boron doped a-Si:H layer about 200 Å thick on steel, followed by an undoped a-Si:H layer about 5000 Å thick and finally a top P doped a-Si:H layer about 80 Å thick.[39] A 70-Å-thick ITO layer deposited on the top n-type layer acts as the top contact as well as an antireflection coating.[39]

In the Type-II[99] configuration, the device is illuminated through the glass substrate which is coated with about 600 Å of ITO and about 100 Å of a cermet ($Pt–SiO_2$). The cermet ensures good electrical contact to the thin 80-Å p-type layer. The undoped layer is 6000 to 8000 Å thick. A 1000-Å-thick Ti or Ti/Al layer forms the back contact to the 200-Å-thick n-type top layer.

10.4.2. *Photovoltaic Performance*

The $p/i/n$ a-Si:H solar cells, of area 1.19 cm^2, fabricated at RCA Laboratories[39] have exhibited conversion efficiencies greater than 5%.

Although the highest V_{oc} (910 mV) has been obtained with Type-II devices, the best efficiency (6.1%) and I_{sc} (~ 12 mA cm^{-2} under AM1) have been obtained in Type-I cells illuminated through the top n-layer. The best FF observed in $p/i/n$ cells is about 0.61. Schottky barrier cells have exhibited FF values as high as 0.674.

The spectral response for a Type-I $p/i/n$ cell is shown in Figure 10.8. The observed decrease in the collection efficiency at short wavelengths is due primarily to absorption losses in the top doped layer, while the decrease at long wavelengths is due to the decreasing absorption coefficient of undoped a-Si:H.[100] Based on this spectral response, the calculated I_{sc} under AM1 is 9.27 mA cm^{-2}. However, the measured J_{sc} in sunlight is typically 15 to 20% higher than the calculated values, owing to a contribution from the diffused sunlight.

Under illumination, $p/i/n$ cells exhibit diode factors of about 1.1,[39] indicating low recombination at the $p/i/n$ junction. Similar values have been obtained on Schottky barrier cells.[101] Although the behavior of a-Si:H cells under illumination is close to that of an ideal diode, the dark

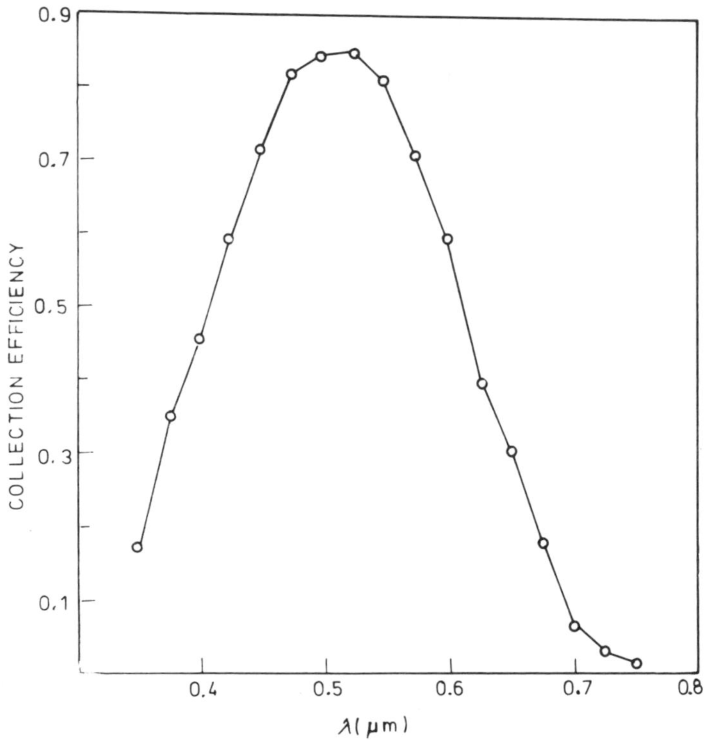

Figure 10.8. Collection efficiency as a function of wavelength for a Type-I $p/i/n$ solar cell (after Carlson[39]).

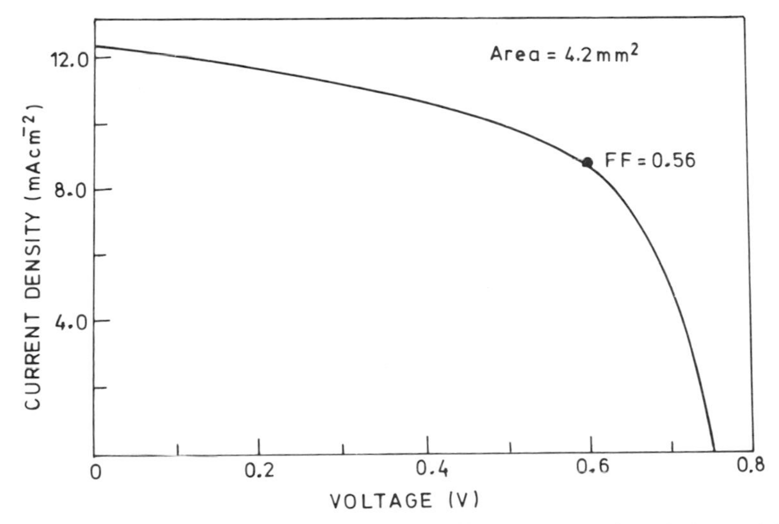

Figure 10.9. I–V characteristics of an a-Si:F:H MIS solar cell (after Madan et al.[14]).

I–V characteristics are poor. The data can be explained in some cases by space-charge-limited currents,[5] while in other cases the Poole–Frenkel mechanism appears to be dominant.[39] In yet other cells, relaxation semiconduction[102] may be responsible for the poor rectification in the dark. It should be noted that all types of a-Si:H cells exhibit a crossover between the dark and light I–V characteristics as is also the case with CdS/Cu_2S solar cells.

The recent $p/i/n$ cells are fully depleted with an electric field of about 2×10^4 V cm^{-1} extending across the undoped layer under short-circuit conditions.[39,103]

The highest conversion efficiencies reported to date in MIS-type photovoltaic devices using a-SiF:H films produced by glow discharge[14] is about 6.3% under AM1 illumination in devices of area 4.2 mm^2. A typical structure for these cells consists of an 800-Å-thick highly conducting n^+-layer deposited onto a reflecting Mo bottom contact, followed by about 5000 Å of active photoconductive a-Si:F:H. The undoped layer possesses a relatively low density of states[6] and the photoconductivity under AM1 is typically in the range 10^{-4} to 10^{-3} Ω^{-1} cm^{-1}, providing for a low series resistance in operation. A 20-Å-thick Nb_2O_5 layer serves as the insulator. Finally, a 70-Å-thick high-work-function Au/Pd (90:10) metal film and a 350-Å ZnS antireflection coating complete the device structure.

The illuminated I–V characteristics curve measured at 83 mW cm^{-2} is shown in Figure 10.9. The device characteristics are $V_{oc} = 0.75$ V, $J_{sc} =$ 12.24 mA cm^{-2}, FF = 0.56, and $\eta = 6.2\%$. Further improvements in the

MIS cell require optimization of the antireflection coating (to minimize reflection losses over the entire range from UV to $\lambda = 6000$ Å), the contacts, and fine tuning of the intrinsic and the n^+-layer thicknesses.[6]

The rectification ratio in the dark for the MIS devices at 0.5 V is about 10^5. The diode factor is 1.12 and the departure from ideality ($n = 1$) is explained on the basis of the oxide layer.[104] Analysis of the dark characteristics based on diffusion theory yields a value for the barrier height ϕ_B of 1.0 eV for a measured J_s (reverse saturation current) of 10^{-12} mA cm^{-2}.[6] A similar value of ϕ_B has been determined from the temperature dependence of J_s. The diode factor under illumination is equal to unity and is temperature-independent over the range 180 to 300 K.

Wilson et al.[15] have observed efficiencies of 4.8% in 60 mW cm^{-2} simulated sunlight of a-Si:H MIS devices with a Ni–TiO_x contact on top of the amorphous films. The a-Si:H film was undoped throughout the major portion of its depth (~ 1 μm), except for a narrow n^+-region adjacent to the stainless steel substrate to provide an ohmic contact. The device was not coated with an AR layer. The estimated efficiency with an AR coating is 6.3%. Devices with 50% transparent contacts ($\sim 7 \times 10^{-2}$ cm^2 in area) yielded (AM1 illumination) $J_{sc} > 9.6$ mA cm^{-2}, $V_{oc} > 0.6$ V, and FF ~ 0.51. Schottky barrier diodes prepared by the Wilson group[16] show that there is virtually no carrier collection outside the space-charge region and a drift field is desirable. High substrate temperatures during a-Si:H deposition increase the long-wavelength response but give poorer diodes.

Efficient a-Si:H devices have only been fabricated in the substrate temperature range between 200 to 400 °C. As discussed in the preceding sections on the properties of a-Si:H films, the defect density is a minimum in this temperature range. The dependence of V_{oc} and J_{sc} on substrate temperature (T_s) for a Pt Schottky barrier cell is shown in Figure 10.10. The increase in J_{sc} with T_s is due to the decrease in E_{opt} and increase in α and also because of enhancement in carrier recombination lifetime. V_{oc} decreases with T_s as a result of the decrease in ϕ_B with decrease in E_{opt}. The temperature dependence of V_{oc} in the case of MIS and $p/i/n$ cells is more complicated.

The effect of various impurities on the photovoltaic parameters of a-Si:H solar cells is given in Table 10.4.[37] These a-Si:H films contain significant amounts of the impurities normally found in the glow discharge environment. However, the photovoltaic properties are not strongly impaired by most impurities, indicating that a-Si:H films can accommodate relatively large impurity contents without deleterious effects. Some impurities, though, can severely restrict the performance of the cells. These impurities are H_2O, H_2S, GeH_4, and PH_3. Films made from SiH_2Cl_2/H_2 mixtures also yield cells with poor performance.

Several large-area cells (~ 100 cm^2) have been fabricated at RCA

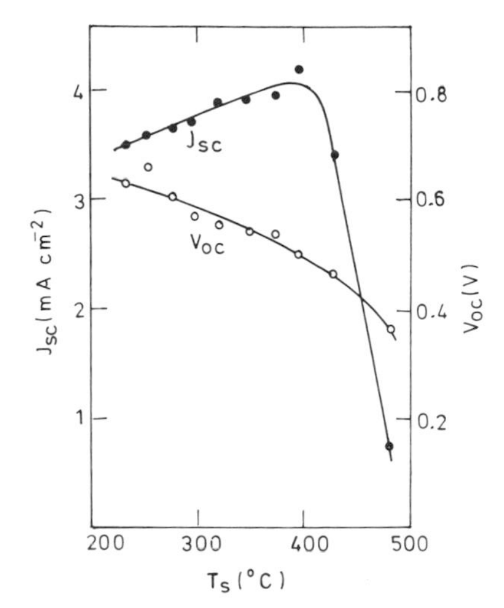

Figure 10.10. J_{sc} and V_{oc} of a Pt Schottky barrier cell as a function of substrate temperature (after Carlson and Wronski[37]).

Table 10.4. Effect of Impurities on the Photovoltaic Properties of a-Si:H Solar Cells (after Carlson and Wronski[37])

Impurity gas (vol. % of total atmosphere)	Type of discharge	V_{oc} (mV)	J_{sc} (mA cm^{-2})	FF	Impurity content of film
None (control)	dc(P)	700	6.0	0.54	(H/Si $\simeq$ 0.20)[a]
2.3% H_2O	dc(P)	400	0.6	0.23	O/Si $\simeq$ 0.037
2.3% H_2S	dc(P)	425	3.0	0.21	—
1.0% GeH_4	rf(E)	370	3.0	0.27	Ge/Si $\simeq$ 0.025
10% CH_4	rf(E)	662	4.0	0.53	C/Si $\simeq$ 0.07
30% CH_4	rf(E)	230	0.02	0.18	C/Si $\simeq$ 0.21
10% N_2	rf(C)	595	6.0	0.55	N/Si $\simeq$ 0.008
0.06% PH_3	dc(P)	130	1.5	0.42	P/Si $\simeq$ 0.0004
50% SiH_2Cl_2 + 50% H_2	dc(C)	321	0.05	0.25	Cl/Si $\simeq$ 0.07

[a] All films contained from 10 to 30 at.% hydrogen. P — proximity, E — external, C — capacitative.

Laboratories with efficiencies exceeding 3%.[25] Sanyo and Fuji of Japan have fabricated large-area (100–144 cm^2) cells with 5–6% efficiencies. In Table 10.5, we have tabulated the performance characteristics of a wide variety of a-Si:H (a-Si:F:H) solar cells prepared by different techniques.

10.4.3. *Loss Mechanisms*

Most of the discussion in this section will be based on the characteristics of *p/i/n* cells. However, the conclusions are generally applicable to other types of a-Si:H solar cells as well.

Table 10.5. Comparison of the Performance of Amorphous Si Solar Cells Fabricated by Various Workers

Material	Type	Deposition techinque[a]	Area (cm^2)	Intensity ($mW\,cm^{-2}$)	V_{oc} (V)	J_{sc} ($mA\,cm^{-2}$)	FF	η (%)	Remarks	Ref.
a-Si:H	*p/i/n*	Glo. Dis.	1.19	99.58 (AM1)	~0.85	12	0.61	6.1	ITO contact, AR coating	39
a-Si	Al/Si/Al	Sputter.	0.385	—	0.037	0.006	—	0.006	—	42
a-Si:H	M-S	Sputter.	—	—	0.080	0.001	0.25	2×10^{-5}	Pt-barrier	42
a-Si:H	M-S	Glo. Dis.	0.002	—	0.88	12	0.58	6	Pt-barrier steel substrate	30
a-Si:H	MIS	Glo. Dis.	0.072	60	0.60	9.6	0.51	4.8	Ni barrier, No AR, TiO_x-*i* layer	15
a-Si:H	MIS	Glo. Dis.	0.042	83	0.75	12.24	0.56	6.2	Au–Pd barrier, ZnS AR, Nb_2O_5-*i* layer	14
a-Si:H	SIS	Glo. Dis.	—	100	0.43	10.0	0.28	1.2	ITO barrier	30
a-Si:H	9 cells — series connected inverted hetero-junctions	Glo. Dis.	41.2	—	6.0	6.1	0.34	1.38	High series resistance	42
a-Si	22 cells — series connected	Glo. Dis.	12.5 × 17.5	—	11.5	—	—	0.33	High series resistance	42
a-Si:H	Hetero-junction a-SiC:H/a-Si:H	Glo. Dis.	0.033	—	0.909	13.45	0.617	7.55	—	120
a-Si:Sn	Heterojn. *a*-SiC/*a*-Si:Sn	Glo. Dis.	—	—	—	—	—	8.7	Si_3N_2 AR	121
a-Si:H:F								9.2		ECD (private commun.)
a-Si:H	—	—	—	—	—	—	—	10	—	RCA

[a] Glo. Dis. — glow discharge, Sputter. — sputtering.

The major limitation in a-Si:H solar cells, indicated by early work on Schottky barrier cells, is the minority carrier diffusion length,[30,31] which in undoped a-Si:H is in the range of approximately 0.03 to 0.2 μm[39,105] for holes. The hole lifetime in undoped a-Si:H has been estimated to be about 0.34 μs, while the electron lifetime is about 1.2 μs.[39] Saell et al. (in Carlson[39]) have shown from their recent studies that hole lifetimes at current densities of about 10 mA cm^{-2} are in the range from 10 to 20 μs. Longer hole lifetimes ($\sim$1 ms) have been reported by some workers.[39]

Several types of defects may limit the carrier lifetimes. Recombination centers can be created by dangling bonds (as a result of H_2 out-diffusion) in a-Si:H,[106] by polymer chains or groups,[70] and by certain impurities such as O_2, N_2, and P.[39] $p/i/n$ cells of 5 to 6% conversion efficiency typically[39] have the following impurity contents: O — 10^{20} cm^{-3}, C — 5×10^{19} cm^{-3}, N — 7×10^{18} cm^{-3}, and 10–14 at.% H_2. Geminate recombination, expected in any low-mobility material where the photogenerated electron–hole pairs have difficulty overcoming their mutual coulombic attraction, may be responsible for limiting performance in some a-Si:H solar cells.[107,108] It has been estimated that geminate recombination accounts for a 56% loss of photogenerated carriers in a quasi-neutral region for $\lambda \simeq 0.633$ μm. This recombination loss is about 30% when there is a field of about 10^4 V cm^{-1}. Further, geminate recombination losses decrease as the photon energy increases as the photogenerated carriers have a higher initial kinetic energy and thus a higher probability of escaping their mutual coulombic attraction. Experimental studies indicate that geminate recombination in *a*-Si:H may be related to the defect density of the film and may be negligible in the best quality material. The loss in J_{sc} due to geminate recombination has been estimated to be less than or equal to 15%.

Whatever be the nature of recombination in a-Si:H films, the space-charge field plays a critical role in the operation of a-Si:H solar cells.[31] The higher efficiency $p/i/n$ and MIS a-Si:H devices are depleted over the entire thickness of the undoped layer. Modeling[39] of $p/i/n$ devices yields an FF value of about 0.60 for a drift length of approximately 3.5 μm at zero bias, corresponding to a hole diffusion length of about 0.2 μm. Thus, the lifetime (or mobility) needs to be improved for better fill factors.

Based on the results of his computer model of a $p/i/n$ device, Swartz[109] has opined that the performance is being severely limited by the quality of the doped layers. That doping increases the defect density in a-Si:H films has been shown by several investigations.[5,35,39,110,111] At the optimum doping levels for $p/i/n$ cells [$\sim$1 vol% B_2H_6 in SiH_4 for the p-layer and about 1 to 2 vol% PH_3 in SiH_4 for the n-layer[30,31]], the lifetime of the minority carriers is very short and the doped layers act as dead regions and only serve as filters.[100] The collection efficiency as a function of thickness of the top doped layer[112] is shown in Figure 10.11a. It is clear that the short-wavelength response is considerably affected by the top doped layer. Since the absorption coefficient of a-Si:H is about 8×10^{-5} cm^{-1} at $\lambda = 0.4$ μm, the thickness of the top layer must be less than 100 Å to ensure adequate collection at shorter wavelengths. Figure 10.11b shows the dependence of V_{oc} and J_{sc} on the thickness of the n-layer for a $p/i/n$

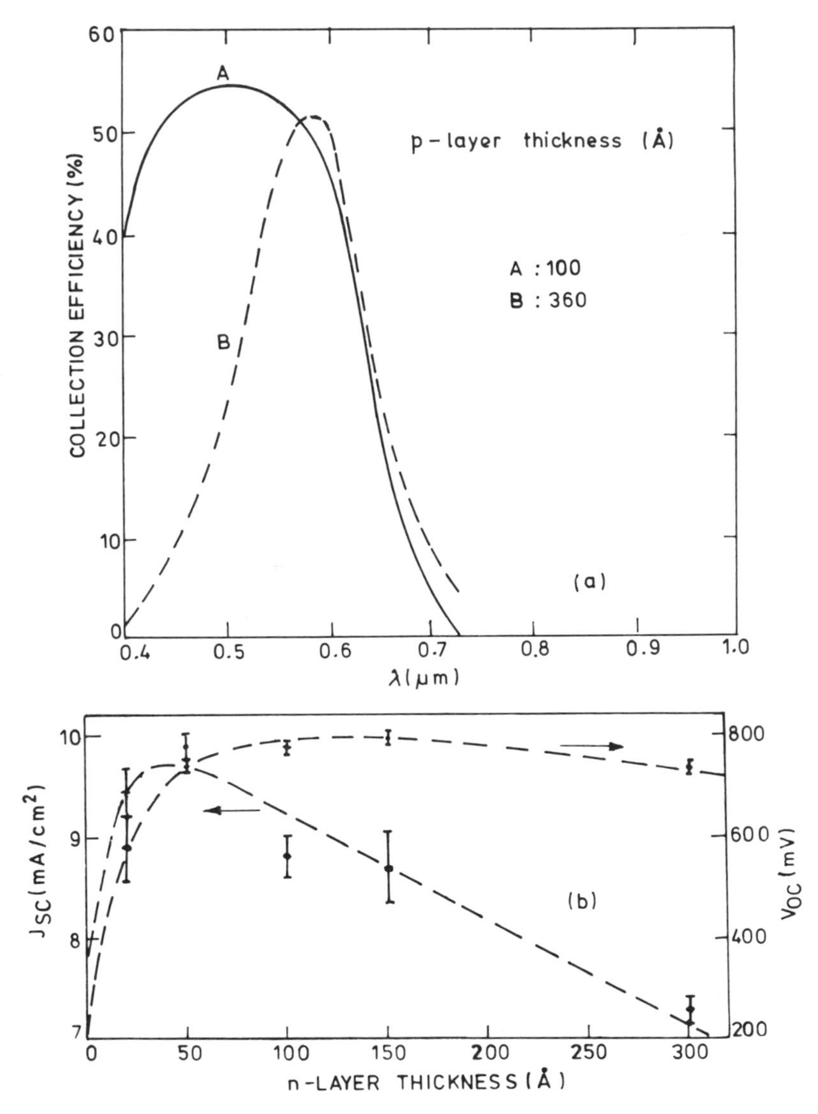

Figure 10.11. (a) Collection efficiency of a $p/i/n$ cell as a function of the thickness of the top doped layer (after Hanak et al.[112]). (b) Dependence of V_{oc} and J_{sc} on the n-layer thickness for a $p/i/n$ cell illuminated through the n-layer (after Carlson[39]).

cell illuminated through the n-layer.[39] The decrease in collection efficiency at shorter wavelengths with thickness of the n-layer is reflected in the decrease in J_{sc} with thickness. The initial increase in J_{sc} with thickness (at small thicknesses) can be attributed to the increase in the built-in potential, as evidenced by the rapid increase in V_{oc}.

Although the optical gap in undoped a-Si:H is roughly from 1.65 to 1.70 eV, the built-in potential V_D in $p/i/n$ cells is of the order of 1.1 V.[101,113] From electrical measurements, the Fermi level in P doped a-Si:H films is deduced to be about 0.2 eV below the conduction band,[114] while it is about 0.5 eV above the valence band[111] in B doped specimens. Fermi levels closer to the respective band edges would result in increased V_D. In this regard, the preparation of highly conductive ($\geq 5\ \Omega^{-1}\,cm^{-1}$) a-Si:F:H films, with the Fermi level about 0.05 eV below the conduction band,[6] is particularly encouraging. Some workers[25] have prepared highly conductive n^+- and p^+-a-Si:H films. As with the a-Si:F:H films, the highly doped a-Si:H films are found to be microcrystalline (grain size ~ 100 Å).

Improving the conductivity of the doped layers, consistent with defect-free material, would also lead to a reduction in the contact resistance in $p/i/n$ cells (~ 2–$5\ \Omega\,cm^2$ at present) and a concomitant increase in the fill factor.

Presence of surface states at the ITO/doped a-Si:H interface may cause a reduction in V_D, as the top doped layer will be fully depleted if it is very thin (~ 100 Å). Moreover, a resistive oxide layer, if formed at the ITO/doped a-Si:H interface, would contribute to the series resistance, lowering the fill factor.

10.4.4. *Stability*

It appears that a-Si:H solar cells are quite stable. Degradation noted in some Schottky barrier and MIS cells on interaction with water vapor can be checked by proper encapsulation.[98] However, processing at elevated temperatures (~ 1 hr at 350 °C) or prolonged exposure to light can change the characteristics of some a-Si:H films[39,98,89] and, hence, cause degradation in device performance. Recent studies indicate serious degradation in high-efficiency (7–8%) cells.

10.5. *Recent Advances*

Research in a-semiconductors is presently in a dynamic phase and new materials, discoveries, and concepts are evolving rapidly. In this section we shall mention some of the recent developments, the results of which are exciting though preliminary.

10.5.1. *New Materials*

Several semiconductor materials in the amorphous form, other than a-Si:H, are being evaluated for photovoltaic applications. These include a-B:H,[58] II–IV–V tetrahedral chalcogenide glasses,[40] and amorphous organic semiconductors.[115]

A glow-discharge process has been used to produce a-B:H films with controlled optical bandgaps close to the theoretical optimum. *n*-type doping has been achieved by both Si and C. Alloying with F (an analogy with a-Si:F:H) appears to markedly influence the optical and electrical properties. a-B/a-Si:H heterojunctions[25] have exhibited a short-circuit density of 13 $mA\,cm^{-2}$.

Amorphous organic semiconductor films have exhibited photovoltaic effect, and efficiencies of about 1% have been achieved recently.[115]

10.5.2. *New Processes*

A variety of new processes are being actively investigated for large-area deposition of a-Si:H films. a-Si:H films have been prepared by electrodeposition[40] and doped *n*-type with Li and *p*-type with Ga and B by codeposition. H_2 content measurements have revealed that films deposited at 35 °C have 30 at.% H, while deposits made at 70 °C contain less than 1 at.% H. Annealing the films in the temperature range 350 to 460 °C has proved useful for activating the dopants.

Magnetron sputtering with reactive gases and/or doped cathodes has been employed by some workers[40] to achieve reproducible and controlled fabrication of a-Si solar cells. Both planar and cylindrical magnetron sputtering systems have been investigated. Alloying and doping have been achieved by ion implantation and both *n*- and *p*-type a-Si films have been deposited by dc magnetron sputtering, in pure Ar, from doped single crystal Si targets.

H_2 ion implantation of vapor deposited Si films[40] is also being investigated for fabrication of a-Si:H solar cells. Other workers[58] have used ion plating to prepare a-Si:H thin film, codepositing Ga to dope the films and change the electrical resistivity over a wide range.

10.5.3. *Novel Structures*

Stacked junction cells with *p*/*i*/*n* junctions stacked on top of one another[112,116] have been proposed for obtaining higher conversion efficiencies. For optimum performance, the optical bandgap of the various junctions must be suitably tailored, with the widest bandgap on top and the bandgap decreasing for each subsequent junction. Wide bandgap alloys such as a-Si:C:H[20,117] and a-Si:O:H[118] have been prepared. Recently Ohnishi et al.[121] have reported an integrated, series-connected Si_3N_2/*a*-SiC:H/*a*-Si:Sn solar cell exhibiting high voltages and conversion efficiencies of 8.7%. In an unpublished, private communication ECD, Troy (Michigan) claimed a conversion efficiency of 9.2% for their glow discharge produced *a*-Si:H:F cells. RCA has recently announced the development of a 10% efficient a-Si:H thin film solar cell. Solar cells made of narrow bandgap alloys of a-Si:Ge:H have shown degradation in performance with increase in Ge content.[112]

A recent innovation, the series connected, tandem junction a-Si:H cells have exhibited high voltages.[25] Problems remain with the short-circuit current, which is low.

10.6. *Scope of Future Work*

The impressive studies made in the field of a-Si solar cells give rise to a feeling of "cautious optimism". However, considerable experimental and theoretical research and development effort is needed in a variety of interrelated disciplines before a-Si solar cells become a viable techno-economic reality. The discussion in the preceding sections on the performance of these devices indicates areas where further research efforts are required. These areas are enumerated below.

The glow discharge deposition process needs to be developed further, based on a better understanding of the plasma processes and surface reaction chemistry. Important gas phase and growth parameters must be identified and controlled to improve film quality. The effects of impurities, modifiers, and dopants on film quality and device performance should be studied in greater detail. The improvement in quality of doped layers on alloying with F in the case of a-Si:F:H films suggest that other discharge atmospheres, deposition conditions, and dopants must be investigated.

Although ITO/a-Si:H and SnO_2/a-Si:H heterojunctions have been fabricated, the observed V_{oc} is relatively low[30,119] in these devices. New wide bandgap, highly conductive films, such as polycrystalline SiC or GaN, deposited by low cost, thin film processes, could be used to form suitable heterojunctions with a-Si:H and need to be developed.

The stacked junction and series connected tandem junction devices have to be optimized to realize better performance. This would require preparation of variable bandgap alloys.

The limited studies on the stability of a-Si:H cells must be augmented by field trials, under actual operating conditions, of amorphous submodules. Accelerated tests must be performed to establish the long-term stability of these materials over a period of 15 to 20 years. If necessary, the thermal stability of a-Si:H may have to be enhanced, possibly by modifying the material. In this regard, the work of Tanaka et al.,[120] showing suppression of H_2 evolution on incorporation of Ar into the a-Si:H structure, is particularly relevant.

The above studies are expected to provide answers to several questions and problems regarding correlation of optical gap to current density, improvement of fill factor, optimization of doped layers, impurity levels affecting device performance, factors affecting ohmic contacts, surface and bulk defects in a-Si, effects of local structure on transport properties, mechanisms at the plasma/substrate interface affecting film quality, and suitability of other elements or compounds as substitutes for H_2. A comprehensive picture of a-Si solar cell operation and its dependence on material parameters would then emerge and pave the way for high-efficiency, large-area cells.

11

Photoelectrochemical Cells

11.1. *Introduction*

The change in the electrode potential (on open circuit) or in the current flowing in the external circuit (under short-circuit conditions) of an electrode/electrolyte system on irradiation is termed a photoelectrochemical effect.[1] Photoelectrochemical (PEC) cells employing a semiconductor/electrolyte junction have gained popularity in recent years for solar energy conversion in view of some potential advantages over conventional solid state solar cells. These are: (i) PEC devices can both store energy in the form of conventional fuels and convert light directly to electrical energy. (ii) The devices can be fabricated easily and the band-bending characteristics of the semiconductor can be suitably modified by proper choice of electrolyte and cell variables. (iii) Problems associated with differential thermal expansion of solid–solid junctions are not present. (iv) Capability for *in situ* storage exists. (v) It is possible to fabricate hybrid (photovoltaic and photothermal) systems.

11.2. *Theoretical Considerations*

The photoresponse and stability characteristics of the semiconductor critically determine the performance of a PEC system. The principles governing the choice of a suitable PEC electrode system, for high conversion efficiency, are intimately related to basic concepts in semiconductor electrochemistry. We will now examine some of the electrochemical concepts.

11.2.1. *The Semiconductor/Electrolyte Interface*

Gerischer,[2–6] Green,[7] Myamlin and Pleskov,[8] and Rajeshwar et al.[9] have extensively reviewed the theoretical treatments of the semiconductor/electrolyte interface. In this section, we consider the behavior of n-type and p-type semiconductor photoelectrodes in contact with an electrolyte.

When a semiconductor is brought into contact with an electrolyte containing a redox system, equilibrium is attained by electron exchange at the interface. At equilibrium, the Fermi levels are adjusted by the formation of an electric double layer between the semiconductor and the electrolyte. The structure of this double layer depends mainly on the concentration of the mobile and immobile charge carriers on both sides of the interface.[3,8,10] Since, in PEC cells, concentrated electrolyte solutions are used to minimize internal resistance, ions are present in concentrations of the order of 10^{21} cm^{-3} (about 1 M).

To effect charge separation during illumination, the electric field in the space-charge layer below the semiconductor surface has to be extended, which means that the carrier density in the semiconductor must be kept much lower than that in the electrolyte. In the absence of any specific interaction between charged species and the surface, a simple double-layer structure occurs, with a diffuse space charge on the side of the semiconductor and a condensed ionic countercharge on the electrolyte side. However, presence of electronic surface states on the semiconductor or chemical bonding between ionic components of the electrolyte and the semiconductor surface atoms can lead to specific interactions. The situation in an n-type semiconductor with positive excess charge, such that a depletion layer is formed with a constant density of positive charge of ionized donors in the bulk of the space-charge region, is shown in Figure 11.1.

The Fermi level $E_{F,\text{redox}}$ of the electrolyte is equivalent to the equilibrium redox potential U_0 of electrochemistry and the relation between the two is given by

$$E_{F,\text{redox}} = \text{constant} - U_0 e \tag{11.1}$$

When U_0 is measured against the standard H_2 electrode, the constant has a value of about 4.5 eV.[11] We should point out that the Fermi level is here referred to the vacuum level of the electron, as is usual in solid state physics.

If a semiconductor is in contact with an electrolyte, a depletion layer develops as a consequence of the mismatch of the Fermi level E_F^0 of the semiconductor and the Fermi level of the electrolyte $E_{F,\text{redox}}$. At equilibrium, the band bending is energetically equal to the initial difference in the Fermi levels. At the particular redox potential of the electrolyte when $E_{F,\text{redox}} = E_F^0$, there is no excess charge in the semiconductor and, hence, no band bending. The bands are thus flat from the bulk to the surface and the corresponding redox potential is termed the flat-band potential V_{FB}. The

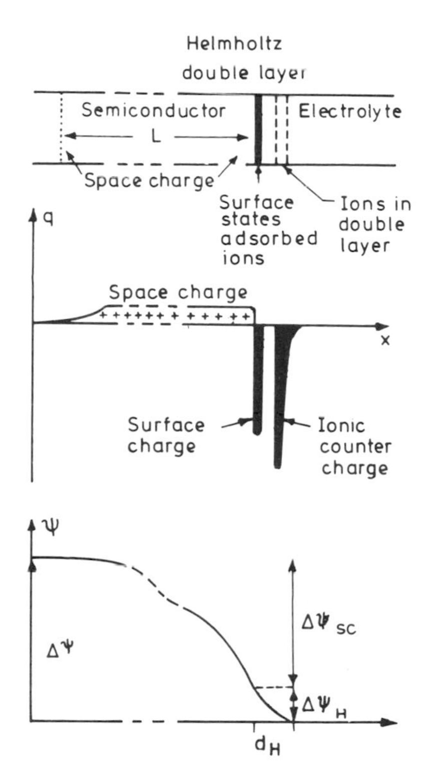

Figure 11.1. The electric double layer at an n-semiconductor/electrolyte interface and distribution of space charge q and potential ψ (after Gerischer[4]).

value of V_{FB} depends on the properties of the semiconductor, since the chemical nature and the doping concentrations control E_F^0, and on the composition of the electrolyte, because ionic double layers or oriented dipole layers are formed by specific interaction with ions on polar molecules,[2,8] which changes the work for electron transfer into the electrolyte.

A favorable equilibrium situation for the generation of photocurrents and/or photovoltages occurs whenever the excess charge and the respective band bending causes a depletion of the majority carriers below the interface. Some of the favorable band-bending situations are shown in Figure 11.2 for n- and p-type semiconductors. The equivalent solid state situation is a Schottky barrier between a semiconductor and a metal.[12] Here, the redox electrolyte performs the function of the metal, with the counter electrode (Fermi level denoted by $E_{F\,ce}$) acting as the grid. In principle, all semiconducting materials, both n- and p-types, can be used as photoelectrodes once a suitable redox system is found which would induce the formation of a depletion space-charge layer on contact with the particular semiconductor.

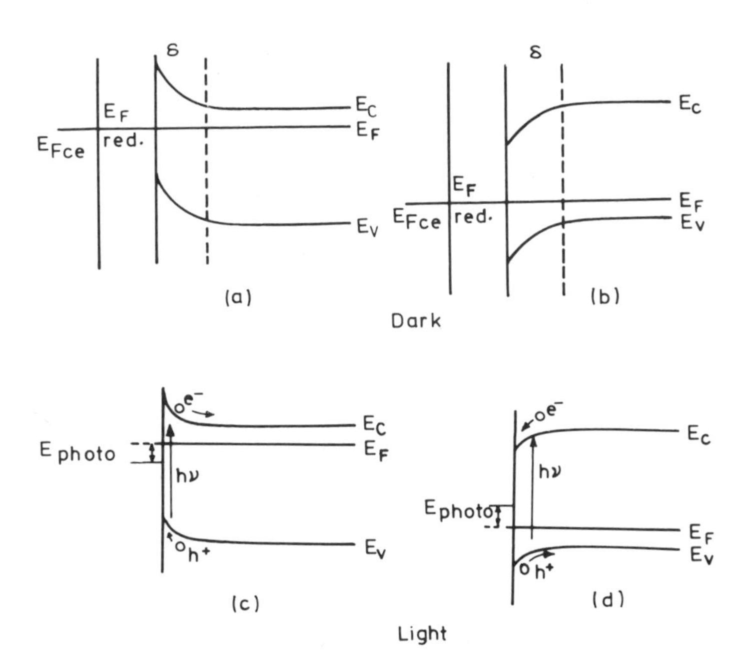

Figure 11.2. Band bending at the semiconductor/electrolyte interface favorable for observation of photocurrent for (a) n-type and (b) p-type semiconductor in dark. (c) and (d) are the band diagrams under illumination.

Electron–hole pairs are generated when light of energy $h\nu \geq E_g$ (bandgap of the semiconductor) falls onto the semiconductor surface after crossing the transparent electrolyte layer. If carriers are generated in the space-charge layer (scl), they move in opposite directions as a result of the electric field. If no recombination occurs, the majority carriers migrate to the bulk and the minority carriers to the surface. Thus, n-type semiconductors act as photoanodes and dark cathodes, while p-type materials act as photocathodes and dark anodes. For n-type semiconductor electrodes, in an ideal case, the holes react at the interface exclusively with the electrolyte, oxidizing the electron donors (Red) of the redox system. The following equations describe the various steps taking place:

Light absorption $\quad h\nu + \text{scl} \rightarrow h^{+}_{\text{scl}} + e^{-}_{\text{scl}}$ (11.2a)

Recombination $\quad h^{+}_{\text{scl}} + e^{-}_{\text{scl}} \rightarrow \text{heat}$ (11.2b)

Charge separation $\quad e^{-}_{\text{scl}} \rightarrow e^{-}_{\text{bulk}}$ (11.2c)

$h^{+}_{\text{scl}} \rightarrow h^{+}_{\text{surface}}$ (11.2d)

$$\text{Interfacial reactions} \qquad h^+_{\text{surface}} + \underset{\text{(reduced species)}}{\text{Red}} \longrightarrow \text{O}_x^+ \tag{11.2e}$$

$$e^-_{\text{surface}} + \underset{\text{(oxidized species)}}{\text{O}_x^+} \longrightarrow \text{Red} \tag{11.2f}$$

Reaction (11.2f), which occurs to the extent that electrons can reach the surface against the force of the electric field in the space-charge layer, is important for attainment of equilibrium at the electrode which needs the accumulation of a positive charge on the semiconductor.

On illumination, a photovoltage is obtained (at open circuit), which can be measured versus any reference electrode in contact with the electrolyte. Photogeneration of charge carriers causes a deviation from the equilibrium charge distribution and a reduction in band bending, as shown in Figures 11.2c and 11.2d. A steady state is achieved when the charge generation by reactions (11.2a), (11.2c), and (11.2d) is balanced by the steps (11.2b) and (11.2f).

The concentration of the excess light-generated minority carriers on the surface of a semiconductor electrode is determined by the incident spectrum, the incident intensity, the absorption coefficient of the semiconductor, the quantum yield for electron–hole pair generation, the recombination rates in the space-charge layer, and the transport phenomena. At low illumination levels, the surface concentration of minority carriers varies linearly with light intensity. At high intensities, nonlinear processes occur and the photocurrent saturates. At saturation, the band bending in the space-charge layer is drastically reduced and the potential distribution is close to that of the flat band situation.

11.2.2. *Design of a Photoelectrochemical Cell*

To draw power from a semiconductor/electrolyte junction, the semiconductor has to be combined with a suitable counter electrode in a complete galvanic cell. The natural counter electrode for a semiconductor/redox electrolyte system is a reversible redox electrode for the same redox system. The same redox potential then controls the equilibrium at both the electrodes, the Fermi levels being equal in both the solids and in the electrolyte. The counter electrode provides the further advantage that it guarantees a rapid attainment of equilibrium at the semiconductor electrode after bringing both electrodes into electronic contact, as the time constant in this case is controlled by the double-layer capacity and the external resistance instead of the interfacial electron-transfer (11.2f) resistance.

Under illumination, the photovoltage drives the electrons from the semiconductor to the counter electrode, while the holes react with the electrolyte. As a result, the oxidation of Red [reaction (11.2e)] at the semiconductor electrode is compensated by the reduction of O_x^+ at the counter electrode [reaction (11.2f)]. In this ideal case, no net chemical change occurs.

The maximum open-circuit voltage that can be generated is equal to the amount of band bending and is, therefore, controlled by the electrolyte Fermi level for a given semiconductor. The operational efficiency of the PEC cell depends on the relative rates of electron–hole recombination and electron transfer.[17] However, as in the case of solid junctions, several energy loss mechanisms may be operative. These include: (i) $[e\Delta\psi_{sc}]$ — energy loss while the charge carriers cross the space-charge layer, (ii) $[e(I.R)]$ — ohmic loss with $R = R_{int} + R_{ext}$, (iii) $[E_{F,\,redox} - E_V + e\eta_{sc}]$ — energy loss connected with the reaction $h^+ + \text{Red} \rightarrow O_x^+$, where η_{sc} is the overvoltage at the semiconductor electrode caused by slow transfer of holes at the semiconductor/redox interface, and (iv) $[e\eta_{ce}]$ — energy loss related to the reaction $O_x^+ + e^- \rightarrow \text{Red}$ at the counter electrode, where η_{ce} represents the overvoltage at the counter electrode. To achieve better performance, one has to minimize not only the energy losses in the cell circuit but also the light losses caused by absorption outside the space-charge layer, in the electrolyte, and by reflection.

For efficient collection of photogenerated carriers, the depletion region δ_{sc} should correspond to the reciprocal of the absorption coefficient $\langle\alpha\rangle$, averaged over the effective part of the absorbed light. Assuming that band bending of the order of 0.25 V is necessary for efficient charge separation, we have

$$\delta_{sc} = \left(\frac{2\varepsilon\varepsilon_0}{eN}\,\Delta\Psi_{sc}\right)^{1/2} = \left(\frac{2\varepsilon\varepsilon_0}{eN}\,0.25\right)^{1/2} \simeq \frac{1}{\langle\alpha\rangle} \qquad (11.3)$$

where N is the donor or acceptor concentration, ε is the dielectric constant of the semiconductor, and $\Delta\Psi_{sc}$ is the amount of band bending. For $\langle\alpha\rangle = 3\times10^4\ \text{cm}^{-1}$, N/ε is approximately $10^{15}\ \text{cm}^{-3}$, which corresponds to a relatively low donor concentration (unless ε is very large). Thus, one must try to find materials where the lifetime of the minority carriers is sufficiently large, so that they can also reach the space-charge layer by diffusion.

Photodecomposition of the semiconductor (see Section 11.2.5) imposes yet another restriction on the extent of band bending. Prevention of photodecomposition requires that the Fermi level of the redox system in the electrolyte not exceed the decomposition Fermi level of the semiconductor for the particular reaction with the minority carrier. The farther the redox Fermi level of the electrolyte is below the anodic decomposition Fermi level for n-type material (or above the cathodic decomposition Fermi level for p-type material), the greater is the risk in using the device.

Electrolysis at the electrodes causes local changes in chemical composition of the electrolyte next to the electrode. Equilibration of the electrolyte composition is achieved by diffusion and convection. To minimize concentration differences, the distance between the two electrodes has to be as short as possible. Alternatively, stirring may be employed to achieve efficient rates by convection. Further, since electrolytic conductivity is much lower than the conductivity of good electronic conductors, the electrolyte layer between the photoelectrode and the counter electrode should be kept thin.

The electrolyte containing the redox couple must be transparent to the useful wavelength of the incident light, so that maximal light intensity is incident at the depletion layer from the front. Moreover, the counter electrode employed should be stable in the electrolyte and should be able to effectively exchange majority carriers with the electrolyte. Finally, the semiconductor photoelectrode, the electrolyte, and the counter electrode should be arranged in a configuration optimum for uniform current distribution in the cell.

11.2.3. *Photocurrent and Photovoltage*

The PEC cell can be modeled as a Schottky barrier.[13-15] The photogenerated current in the semiconductor (assuming n-type behavior) consists of two basic components: (i) the current owing to carriers generated within the depletion region and (ii) the current arising from carriers generated in the bulk that diffuse into the depletion region. The first component is given by

$$J_{\text{dep}} = e\int_0^L g(x)dx = -eI_0[\exp(-\alpha L)-1] \tag{11.4}$$

where $g(x)$ is the function describing the generation of electron–hole pairs, I_0 is the photon flux, α is the optical absorption coefficient, x is the distance into the semiconductor, and L is the depletion layer width. The diffusion component of the photocurrent is given by

$$J_{\text{diff}} = eI_0\frac{\alpha L_p}{1+\alpha L_p}\exp(-\alpha L) + eI_0\frac{D_p}{L_p} \tag{11.5}$$

where L_p and D_p are the diffusion length and diffusion constant, respectively, of holes. For wide-bandgap semiconductors, the last term in equation (11.5) can be neglected[3] so that the total current density J, given by the sum of equations (11.4) and (11.5), can be expressed as

$$J = J_{\text{dep}} + J_{\text{diff}} = eI_0 \left[1 - \frac{\exp(-\alpha L)}{1+\alpha L_p}\right] \tag{11.6}$$

Ghosh and Maruska[14] and Ghosh et al.[16] have developed an expression similar to equation (11.6) with the inclusion of a factor G related to the quantum yield in the depletion region and to the charge transfer across the electrolyte/electrode interface, namely

$$J = eI_0G\left\{1 - \exp(-\alpha L) + \frac{\alpha}{\alpha+\beta}\exp(\beta L)[\exp(-\alpha-\beta)L - \exp(-\alpha-\beta)t]\right\} \tag{11.7}$$

where $\beta = 1/L_p$ and t is the sample thickness. The first part gives J_{dep} and the second part is related to $J_{\text{diff.}}$. Ghosh and Maruska[14] have generated spectral response curves predicted by equation (11.7) for a variety of combinations of L and L_p and constant sample thickness. According to these calculations, the total response is rather insensitive to the depletion layer width. However, the response is strongly affected by L_p.

11.2.4. *Determination of Flat-Band Potential*

The flat-band potential V_{FB} is an important parameter in a semiconductor/electrolyte system since it determines the relative Fermi levels of the electrolyte and the semiconductor and the amount of band bending at the interface. The flat-band potential can be determined either from capacitance–voltage (C–V) measurements or from current–voltage (J–V) measurements.

1. C–V measurements: In an ideal depletion layer, at equilibrium, in the absence of a charge variation in surface states, the measured capacity depends only on the potential drop in the space-charge layer according to the Mott–Schottky relation

$$\frac{1}{C^2} = \frac{V - V_{\text{FB}} - (kT/e)}{\frac{1}{2}\varepsilon\varepsilon_0 e N_D} = \frac{1.41\times 10^{32}}{\varepsilon N_D}[V - V_{\text{FB}} - (kT/e)] \ [\text{in F}^{-2}] \tag{11.8}$$

where V is the applied voltage, ε_0 is the permittivity of free space, and N_D is the concentration of donors per cubic centimeter. From a plot of $1/C^2$ vs. V, the flat-band potential V_{FB} can be determined from the intercept on the V-axis.

2. J–V measurements: From equation (11.6), expressing the depletion layer $L = L_0\,(V - V_{\text{FB}})^{\frac{1}{2}}$, where L_0 is the depletion layer width for an

applied potential of 1 V across it and V is the applied voltage, we obtain

$$J = eI_0\left[1 - \frac{\exp - \alpha L_0(V - V_{FB})^{\frac{1}{2}}}{1 + \alpha L_p}\right] \tag{11.9}$$

for

$$\alpha L_p \ll 1 \quad \text{and} \quad \alpha L_0(V - V_{FB})^{\frac{1}{2}} \ll 1$$

$$V - V_{FB} \sim \left(\frac{J}{\alpha L_0 e I_0}\right)^2 \tag{11.10}$$

A plot of J^2 vs. V yields a straight line and the intercept on the potential axis gives a measure of V_{FB}.

11.2.5. *Photodecomposition of Semiconductor Electrodes*

Thus far we have assumed that the electrodes are inert and play the role of electron donors or acceptors. In practice, however, the electrodes in contact with the electrolyte are susceptible to decomposition, either in the dark or under illumination, the decomposition resulting in an ionic oxidation or reduction product of the semiconductor. The electrolytic oxidation of semiconductors is always connected with the holes of the valence band as electronic reactants, while the electrolytic reduction of semiconductors is connected with the electrons of the conduction band.[18] The simplest types of decomposition reactions involving a binary compound semiconductor MX and an electrolyte AY can be formulated as follows

$$\mathrm{MX} + zh^+ + z\,\mathrm{Y}^-_{\mathrm{solv.}} + \mathrm{solv.} \rightarrow \mathrm{M}^{z+}_{\mathrm{solv.}} + z\,\mathrm{Y}^-_{\mathrm{solv.}} + \mathrm{X} \tag{11.11}$$

$$\mathrm{MX} + ze^- + \mathrm{solv.} + \mathrm{A}^{z+}_{\mathrm{solv.}} \rightarrow \mathrm{M} + \mathrm{X}^{z-}_{\mathrm{solv.}} + \mathrm{A}^{z+}_{\mathrm{solv.}} \tag{11.12}$$

where z is the number of holes or electrons. More complex reactions involve higher oxidation or reduction states with additional hole or electron consumption, respectively.

Reactions (11.11) and (11.12), representing redox reactions, can be treated as reversible and a thermodynamic redox potential can be assigned to them.[19] Since redox potentials are usually measured against an H_2 reference electrode and thermodynamic quantities are generally defined only for electroneutral reactions, the corresponding electrode reaction of the reference system has to be added to the reaction in question to obtain an electroneutral cell reaction for which thermodynamic data can be found in standard tables. Thus, for an H_2 reference electrode

$$z\,\mathrm{H}^+_{\mathrm{solv.}} \rightleftharpoons \tfrac{1}{2}z\,\mathrm{H}_2 + \mathrm{solv.} + zh^+ \tag{11.13}$$

$$z\,\mathrm{H}^+_{\mathrm{solv.}} + ze^- \rightleftharpoons \tfrac{1}{2}z\,\mathrm{H}_2 + \mathrm{solv.} \tag{11.14}$$

Combining reactions (11.11) and (11.13), we obtain

$$MX + z\,H^{+}_{solv.} + z\,Y^{-}_{solv.} \rightleftharpoons M^{z+}_{solv.} + z\,Y^{-}_{solv.} + H_2 \qquad (11.15)$$

with a free-energy difference of ΔG_1. Denoting the standard redox potential of reaction (11.11) vs. the H_2 electrode as the thermodynamic decomposition potential for oxidation by holes ${}_pV_{decomp.}$, we have

$${}_pV_{decomp.} = \Delta G_1/zF \qquad (11.16)$$

where F is the Faraday constant. Similarly, combining reactions (11.12) and (11.14), we obtain the thermodynamic decomposition potential for reduction by electrons ${}_nV_{decomp.}$ as

$${}_nV_{decomp.} = -\Delta G_2/zF \qquad (11.17)$$

The Fermi energies ${}_nE_F$ and ${}_pE_F$, of electrons and holes, respectively, for nondegenerate semiconductors can be related to the electrode potential V measured vs. a reference electrode by the relation

$$E_F = -eV + \text{constant (ref.)} \qquad (11.18)$$

Similarly, ${}_pV_{decomp.}$ and ${}_nV_{decomp.}$ correspond to definite positions of the Fermi levels of holes or electrons and they can be transformed into critical Fermi energies for decomposition of the semiconductor by holes (${}_pE_{decomp.}$) or by electrons (${}_nE_{decomp.}$) on the scale of electronic energies representing the properties of the solids. ${}_pE_{decomp.}$ and ${}_nE_{decomp.}$ are termed decomposition Fermi levels and their position relative to the band edges at the interface,[19] which are particularly instructive, can be obtained from the relation

$$E_{decomp.} = E_{FB} - e(V_{decomp.} - V_{FB}) \qquad (11.19)$$

The criteria for thermodynamic stability of a semiconductor in a particular electrolyte, at a given redox potential, are:

For stability:

$${}_pV_{decomp.} > V_{redox} > {}_nV_{decomp.} \equiv {}_pE_{decomp.} < E_{redox} < {}_nE_{decomp.} \qquad (11.20)$$

For instability:

$$V_{redox} > {}_pV_{decomp.} \quad \text{or} \quad V_{redox} < {}_nV_{decomp.} \equiv E_{redox} < {}_pE_{decomp.}$$
$$\text{or} \quad E_{redox} > {}_nE_{decomp.} \qquad (11.21)$$

Thus, if both the decomposition Fermi levels are known, or can be derived

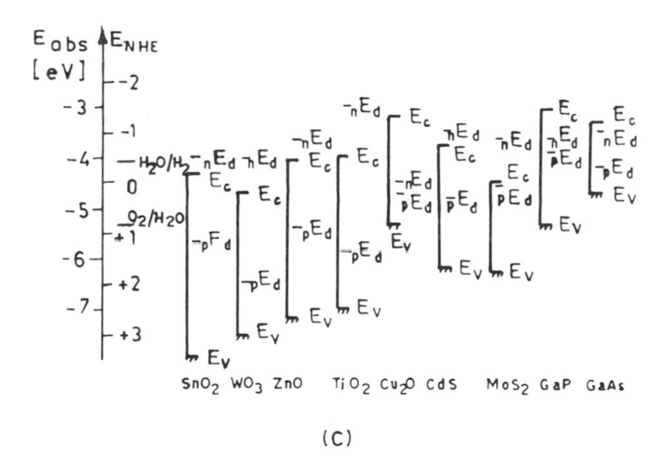

Figure 11.3. (a) Thermodynamics of photodecomposition of semiconductors: (1) stable, (2) anodically and cathodically unstable, (3) cathodically stable, and (4) anodically stable. (b) Ranges of thermodynamic and kinetic protection of semiconductors against photodecomposition by redox systems. E^c_{dec} are the critical Fermi levels with respect to kinetics. (c) Decomposition Fermi levels E_d for several semiconductors (after Gerischer[4]).

from thermodynamic data, one can predict the stability of a particular semiconductor/electrolyte system.

For an illuminated semiconductor, the quasi-Fermi levels in the steady state, ${}_nE_F^*$ and ${}_pE_F^*$, have to be considered instead of the real equilibrium values. However, the conditions for anodic or cathodic decomposition remain the same as above. The maximum splitting of the two quasi-Fermi levels cannot exceed the bandgap and, hence, the range of the quasi-Fermi levels that can be reached under illumination is limited by the position of the band edges at the semiconductor/electrolyte interface. Thus, the position of the decomposition Fermi levels in relation to the position of the band edges decisively governs the stability of a semiconductor electrode.

On the basis of the preceding discussion, four different cases of photodecomposition behavior of semiconductor electrodes arise. These are represented in Figure 11.3a. Known semiconductors do not fall into category (1) of Figure 11.3a. Instead, they belong to either category (2) or (3) and are, therefore, at least susceptible to anodic photodecomposition. A semiconductor belonging to category (2) can decompose anodically and cathodically simultaneously under illumination.

The actual occurrence of photodecomposition depends largely on the kinetics of the reactions. If high activation barriers exist, a semiconductor can be stable enough for practical use even under conditions that are thermodynamically unfavorable. The kinetics can help to prevent decomposition from effects connected with the mechanism of crystal decomposition.[20] Additionally, photodecomposition can also be prevented by using redox systems in which the photogenerated minority carriers are transferred rapidly across the interface. There are usually one-electron (hole) transfer reactions (if no kinetic complications are involved) with rate constants much higher than those for electrochemical decomposition reactions. Figure 11.3b indicates schematically the critical conditions for such competitive reactions which help to prevent photodecomposition of semiconductor electrodes. In Figure 11.3c we have shown the decomposition Fermi levels for various semiconductors.

11.3. *Photoelectrolysis of Water*

The production of H_2 as a fuel by photodecomposition of water at semiconductor electrodes has attracted considerable attention since the first publication of Fujishima and Honda.[21]

11.3.1. *Energetic Conditions*

As discussed earlier, electrons and holes reach different redox potentials in an illuminated electrode. The photodecomposition of water at the

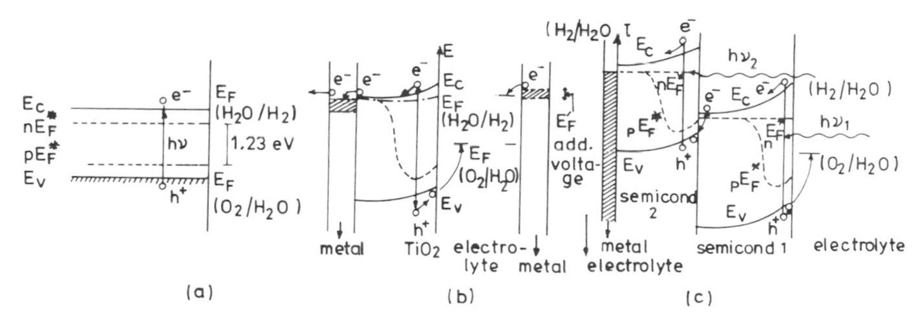

Figure 11.4. (a) Positions of energy levels for water decomposition. (b) Electron energy correlations in photoassisted electrolysis of H_2O, using n-TiO_2 electrode. (c) Energy correlations necessary for photoelectrolysis of water, using a two-layer n-type semiconductor electrode forming two Schottky barriers in series (after Gerischer[4]).

same semiconductor surface requires that the difference between the redox potential of electrons and holes exceed the decomposition voltage of water, which is 1.23 V.[22] Using the H_2 and O_2 electrode reactions and taking into account the different stoichiometry of these reactions, we obtain the following equations defining the position of the Fermi energy of the respective redox reactions:

$$E_{F(H_2O/H_2)} = E^0_{F(H_2O/H_2)} - kT\ln(a^+_H/a^0_{H^+}) + \tfrac{1}{2}kT\ln(p_{H_2}/p^0_{H_2}) \qquad (11.22)$$

and

$$E_{F(O_2/H_2O)} = E^0_{F(O_2/H_2O)} - kT\ln(a^+_H/a^0_{H^+}) + \tfrac{1}{4}kT\ln(p_{O_2}/p^0_{O_2}) \qquad (11.23)$$

where $a^0_{H^+}$ denotes the standard activity of protons in water and p^0 is the standard pressure of gases. The decomposition energy at room temperature has the value

$$\Delta E_{\text{decomp.}} = E^0_{F(H_2O/H_2)} - E^0_{F(O_2/H_2O)} = 1.23\ \text{eV} \qquad \text{at } 25\,^\circ\text{C} \qquad (11.24)$$

The condition for decomposition of water at the semiconductor surface is

$$_nE^*_F > E_{F(H_2O/H_2)} \quad \text{and} \quad _pE^*_F < E_{F(O_2/H_2O)} \qquad (11.25)$$

The above condition is graphically represented in Figure 11.4a.

Manassen et al.[24] have analyzed the energy losses in a working corrosion-free PEC cell. They have divided the energy balance into the following constituent units:

$$h\nu = E_g = (E_{\text{redox}} - E_v) + (E_F - E_{F,\text{photo}}) + (E_c - E_F) + \imath R + \eta_{\text{ce}} + \eta_{\text{sc}} + V_{\text{ph}} \qquad (11.26)$$

where IR denotes the resistance losses in the system and η_{sc} and η_{ce} are overpotentials at the semiconductor and counter electrode, respectively. The estimated values are $E_{redox} - E_v \sim 0.4$ eV, $E_F - E_{F,\,photo} \sim 0.1$ eV, $E_c - E_F + IR + \eta_{ce} \sim 0.1$ eV, and $\eta_{sc} \sim 0.4$ eV. Thus, from equation (11.26), the obtainable photopotential, V_{ph}, is given by

$$E_g - V_{ph} = 1 \text{ eV} \tag{11.27}$$

For $V_{ph} \geq 1.23$ eV, E_g has to be 2.23 eV or more. Such a large optical bandgap results in the loss of a considerable portion of the solar spectrum ($\lambda > 0.55$ μm).

Nozik[25] has taken into account the potential barrier at the semiconductor/electrolyte junction, Ψ_{sc}, and the drop across the Helmholtz layer (double layer) in the electrolyte, Ψ_H, and has derived an energy balance equation (for n-type TiO_2) of the form

$$E_g + E_b = \Psi_{sc} + \Delta G/2F + \Psi_H + IR + E_c - E_F + \eta_{ce} + \eta_{sc} \tag{11.28}$$

where E_b is the external bias and $\Delta G/2F$ is the free energy per electron for water decomposition (1.23 eV). Nozik estimates $\Psi_{sc} \sim 0.8$ eV, $\Psi_H \sim 0.05$ eV, $IR \sim 0.05$ eV, $E_c - E_F \sim 0.2$ eV, and $\eta_{sc} \sim 0.9$–1.1 eV. Thus,

$$E_g + E_b = 3.3\text{–}3.5 \text{ eV} \tag{11.29}$$

11.3.2. *Experimental Results*

TiO_2 has been most extensively employed as a photoelectrode in a PEC system in single crystal, polycrystalline, and thin film form. Fujishima and Honda[21] employed single crystal n-type TiO_2 [(001) face] as the photoanode and Pt black as the counter electrode in an electrolyte consisting of a buffered solution of pH = 6.7 and obtained an open-circuit voltage of 0.5 V, corresponding to a quantum efficiency of about 0.1. Fujishima et al.[26] used reduced single crystal n-TiO_2 in a heterogeneous PEC consisting of TiO_2 in 5 M NaOH and Pt in 0.5 M H_2SO_4 (corresponding to a pH differential of 13) and obtained an open-circuit voltage of about 0.64 V with quantum efficiency of 0.3. Wrighton et al.[27] obtained optical conversion efficiencies of about 1–1.4% using single crystal TiO_2 wafers in a homogeneous cell. Mavroides et al.[28] reported a maximum quantum efficiency of about 0.6 at 0.3 μm using polycrystalline disks of reduced n-TiO_2 in a buffered electrolyte of pH = 8. Thin films of TiO_2 have also been employed in PEC cells for the photoassisted decomposition of water.[28–34] In all cases, an additional voltage of 0.25 to 0.5 V is required for simultaneous O_2 evolution at the TiO_2 electrode and H_2 evolution at the Pt counter electrode.

Several workers[25,28,35,36] have investigated the effect of light intensity on the performance of TiO_2/electrolyte PEC cells. Nozik,[25] Mavroides et al.,[28] and Carey and Oliver[35] have observed photocurrent saturation at light intensities of 30 mW cm^{-2} and higher, while Bocarsly et al.[36] have not detected any current saturation for intensities up to 380 mW cm^{-2}.

Various authors[21,26,27] have demonstrated the stability of TiO_2 photoanodes. However, evidence exists that these electrodes undergo slow dissolution in acidic solutions.[37,38] Addition of 1.0 *M* $CoSO_4$ or 1.0 *M* $Co(ClO_4)_2$ to 0.5 *M* H_2SO_4 suppresses the anodic photodissolution.

Since the bandgap of TiO_2 is 3.0 eV, the electrode absorbs only a small part of the solar spectrum in the UV. Thus the efficiency achieved is very low. Another reason for the low efficiency for photoelectrolysis of water using TiO_2 electrodes is the unfavorable position of the band edges (Figure 11.4b). The conduction band is only slightly above the Fermi level for water reduction. The quasi-Fermi level of the electrons cannot create a large enough driving force for H_2 evolution. This requires the imposition of an external voltage. On the other hand, the valence band edge of TiO_2 is far below the Fermi level for water oxidation. The quasi-Fermi level of holes will, therefore, reach and surpass this energy even at very weak illumination, as long as $h\nu > E_g$. Thus, even at low intensities, this system has a much higher driving force for the oxidation of water than necessary and a lot of excess energy is dissipated.

More suitable semiconductors for photoelectrolysis of water should have a somewhat higher position of the conduction band edge than TiO_2 and a position of the valence band edge much closer to the redox Fermi level for water oxidation. The first condition is fulfilled by some other oxides like $SrTiO_3$,[39–41] $BaTiO_3$,[42–44] $KTaO_3$,[45] Nb_2O_5,[46] and SnO_2.[47,48] However, these oxides have even wider bandgaps than TiO_2 and, therefore, absorb less of the solar spectrum.

Doping of TiO_2 photoelectrodes has been attempted to shift the photoresponse of the PEC cells to longer wavelengths.[49] Ghosh and Maruska[14] reported the spectral sensitization of TiO_2 on doping with Al^{3+}. Doping apparently increases the minority carrier diffusion length and, consequently, the collection efficiency of photogenerated carriers. The presence of Cr^{3+} in the rutile lattice is also observed to extend optical absorption in TiO_2 from its fundamental band edge at 0.415 μm to 0.55 μm. Augustynski et al.[50] have studied thin films of mixed oxides such as TiO_2–M_xO_y (where M = Al, Sr, Ga, Eu, or B) as photoanodes in PEC cells. After activation by heat treatment in O_2, these electrodes exhibit quite different current–voltage characteristics under illumination, although only slight changes were observed in the spectral response, as compared with a TiO_2 electrode. This indicates that the bandgap of TiO_2 is apparently not modified by doping with Al^{3+}, Sr^{2+}, Ga^{3+}, Eu^{3+}, and B^{3+}.

A systematic study of the suitability of various *n*-type oxide semiconductors has been executed by Kung et al.,[51] including many oxides with similar or smaller bandgaps than TiO_2 that have been investigated by others [e.g., WO_3,[52,53] Bi_2O_3,[54] $YFeO_3$,[55] and Fe_2O_3[54,56–60]]. This was done with the hope of increasing the efficiency by absorbing more sunlight. However, the flat-band potential of all these oxides is below the Fermi level for H_2 evolution, and their use, therefore, requires external voltage.

From the above discussion, it seems obvious that water photoelectrolysis with a semiconducting oxide as the only power generator has little chance of performing at reasonable efficiency. Two alternatives have been proposed. The first alternative studied by Nozik[59] and Yoneyama et al.[60] is to upgrade the driving force of the photoelectrolytic cell by combining an *n*-type and a *p*-type semiconductor electrode, both illuminated simultaneously. Half the sunlight will be lost in such a cell, but since two photovoltages act together, each cell has to generate only part of the photovoltage needed for the cell reaction. The advantage is that electrodes with smaller bandgaps can be used, since the Fermi level of the majority carriers in the illuminated material does not have to reach the energy needed to drive the reaction at the counter electrode. On the other hand, when two semiconductor electrodes are used in one electrolytic cell, the photogenerated currents have to be well balanced in order to avoid additional losses. Moreover, the stability problem is doubled in such a device, since both semiconductors must be stable in the same aqueous electrolyte, in dark and under illumination. Since the redox reactions are fixed to the water decomposition Fermi levels, this imposes a serious limitation on the semiconductors that can be used. The Fermi levels at the flat-band situation should be in the middle between the redox Fermi levels for water decomposition. The band edges of the minority carriers must be sufficiently above or below these redox Fermi levels to provide for the overvoltage in the decomposition reactions.

In the experimental studies made with a combination of *n*-TiO_2 and *p*-GaP,[59,60] the quantum yield was totally controlled by the TiO_2 electrode because of its excessively large bandgap. So far, this system has shown no improvement in efficiency. Combinations of *n*-type $SrTiO_3$ with *p*-type CdTe or GaP can do no better because the wider bandgap of $SrTiO_3$ decreases the quantum yield.[61] No combined systems with smaller bandgaps have been studied yet, because of the difficulties in adjustment of the band edges between two different semiconductors and the stability problem.

Another alternative[4,5] is a combination of two junctions in series. The first junction must be transparent to light of longer wavelength in order to allow it to be absorbed in the second junction. The first semiconductor must, therefore, have the wider band gap. Figure 11.4c shows the energy correlations necessary for photoelectrolysis of water with a two-layer *n*-type semiconductor electrode forming two Schottky barriers in series. Such an arrangement has, theoretically, advantages for solar energy conversion, but has not been studied experimentally.

The deep position of the valence band in oxides is related to their stability against photodecomposition in aqueous solutions. Nonoxidic semiconductors are more susceptible to photochemical deterioration. Many nonoxidic semiconductors, namely CdS,[5,62] CdSe,[63] CdTe,[64,65] GaAs,[66] GaP,[64,65,67] and Si[68,69] have been used for photoassisted decomposition of water in PEC systems. The unfortunate obstacle for the use of these materials is their instability against photodecomposition. All efforts to prevent decomposition by covering the surface of unstable semiconductors with coatings of stable oxides have so far either failed to protect the surface, or have blocked the photocurrent of the semiconducting substrate.[69–71] The main difficulty with the use of protective coatings is that associated with solid/solid junctions. In this respect, the main advantage in the use of a PEC cell over conventional photovoltaic devices, namely absence of solid junctions and substantial reduction in fabrication costs, is considerably offset by the use of these coatings.

11.4. *Photoelectrochemical Cell*

The use of wide-bandgap semiconductors in PEC systems is hampered by the fact that these materials utilize only a small portion of the solar spectrum. This has led several workers to concentrate their efforts on the achievement of good stability with low-bandgap semiconductors, like CdX (X = S, Se, and Te), GaY (Y = P and As), WX_2 (X = S and Se), and MoX_2 (X = S and Se), by incorporation of suitable redox couples in the electrolyte. These redox couples have been shown to compete effectively with the anodic dissolution of the semiconductor electrode, suppressing the dissolution process completely in many cases. The presence of an electrochemically active electrolyte stabilizes the photoanodes in such a manner that its redox chemistry occurs at the expense of electrode decomposition. Thus, rather than attempting to convert optical energy to chemical energy (in the form of electrolytic products such as H_2), redox couple stabilization of low-bandgap semiconductor materials has led to PEC systems for direct conversion of light to electricity.

11.4.1. *Cadmium Chalcogenide PEC Cells*

So far, only *n*-type semiconductors have been used for the generation of electricity with a PEC cell. The first cell studied in the regenerative mode consisted of single crystal CdS in contact with the redox system $Fe(CN)_6^{3-}/Fe(CN)_6^{4-}$ in a concentrated KCl solution.[72] The system is theoretically capable of giving a photovoltage of about 1.4 V, which has indeed been observed.[5] This system has shown an efficiency of approximately 6% under AM1 solar radiation. But the system suffers from the problem of instability of the photoelectrode. This is caused by the decomposition of CdS into Cd^{2+} ions and molecular S, if holes are accumulated in the surface at too high a concentration, according to the reaction

$$CdS + 2h^+ + \text{solv.} \rightarrow S + Cd^{2+}\text{solv.} \tag{11.30}$$

To prevent decomposition, one needs a hole scavenger that reacts very fast with holes at the interface. S^{2-} ions are very efficient in this function, since these interact strongly with a sulfide surface and the concentration can be made very high. The difference between the redox potential of the S^{2-}/S_2^{2-} redox system and the flat-band potential of the CdS electrode in sulfide solution is smaller than for the $Fe(CN)_6^{3-}/Fe(CN)_6^{4-}$ redox system, which results in a maximum photovoltage of about 0.7 V only. Ellis et al.[73] have shown that the CdS electrode is very stable in polysulfide solution, in dark as well as under illumination. Heller et al.[74] have obtained an efficiency of about 1.3% for single crystal CdS in polysulfide solution. Polycrystalline thin films of CdS prepared by anodization,[75] evaporation,[76] and spray pyrolysis[76,77] have also been used in PEC cells. However, the reported efficiencies are much less than 1%.

The efficiency in CdS–polysulfide electrochemical cells is limited because of the wide bandgap of CdS (2.4 eV) and the strong absorption by the electrolyte, which absorbs all radiation below 0.46 μm. To absorb more sunlight, other Cd chalcogenides, namely CdSe ($E_g = 1.7$ eV) and CdTe ($E_g = 1.48$ EV), have been used in PEC cells. It has been shown that CdTe is unstable in polysulfide solutions as a photoanode at the current densities expected under normal solar irradiance.[78] However, when used with polyselenide solution, single crystal CdTe has shown an efficiency of about 8.4%.[79] Rockwell International[80] has obtained efficiencies approaching 16% for single crystal n-CdTe in aqueous electrolytes containing CN^- and $Fe(CN)_6^{3-}/Fe(CN)_6^{4-}$ redox couple. The outstanding feature of the ferro-/ferricyanide redox couple is the high photovoltage in the neighborhood of 1.2 V. Both systems, however, exhibit instability owing to formation of CdSe/Te on the surface of the semiconducting electrode which blocks the flow of holes to the surface.

Electrodeposited thin films of CdTe have also been used in PEC cells.[81] However, the efficiency obtained is quite low (0.4%).

CdSe is the most widely studied material for PEC cells, both in bulk and in thin film form. In bulk, the material has been used as single crystals[73,74,80,82–85] and large-grained (10–20 μm) pressure sintered pellets.[86] Thin films of CdSe used in PEC cells have been prepared by evaporation,[87] coevaporation,[87] anodization,[86,88] spray pyrolysis,[89] solution growth,[90,91] and electrodeposition.[92–94] Such a wide interest in CdSe stems from the observation[86] that it is the only material which when used in polycrystalline and thin film forms retains about 75% of the efficiency obtained in single crystals. As we would like to use polycrystalline material and thin films for economic reasons and large-area applications, CdSe is best suited for the purpose.

Miller et al.[86] reported efficiencies of about 8.4 and 5.1% for single crystal and pressure sintered pellets, respectively, using polysulfide electrolyte. An efficiency of about 5% has also been obtained with thin films prepared by spray pyrolysis[89] and coevaporation of Cd and Se onto Ti substrate.[87] Tomkiewich et al. (see Deb et al.[80]) have recently reported efficiencies approaching 6% with electrodeposited CdSe films in polysulfide solution. The use of anodized and solution grown films in PEC cells has resulted in efficiencies of about 0.5[86] and 3.0%,[91] respectively.

The salient features of CdSe based PEC cells are: (i) The efficiency is increased when the electrodes are etched in appropriate etching solutions (Figure 11.5a). The cell characteristics (V_{oc}, I_{sc} and FF) are sensitive to etching time and preparation of the etchant. It has been shown by Heller et al.[74,85] that the improvement in efficiency is due to removal of imperfections from the surface of the semiconductor, an increase in the surface area, and prevention of reflection losses owing to production of hillocks of micrometer or submicrometer size. (ii) CdSe films have to be annealed at temperatures higher than 300 °C to improve adhesion, crystallinity, and stoichiometry of the films. (iii) The current saturates at high illumination intensities, probably because of an insufficient number of reduced ions in

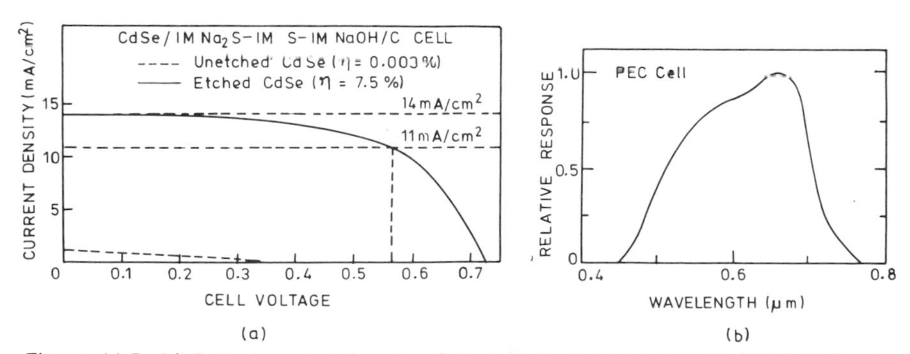

Figure 11.5. (a) I–V characteristics of n-CdSe/1 M Na_2S–1 M S–1 M NaOH/C PEC cell with etched and unetched CdSe electrodes (after Heller and Miller[82]). (b) Spectral response of an n-CdSe/1 M Na_2S–1 M S–1 M NaOH/C PEC cell (after Kainthla[91]).

the electrolyte.[87] Liu and Wang[89] have attributed the saturation to the high series resistance of the SnO_2 back electrode. (iv) The spectral response shows (Figure 11.5b) a lower-wavelength cut-off because of absorption of light by the electrolyte and a higher wavelength cut-off corresponding to the bandgap of the semiconductor. (v) The cell characteristics are improved by stirring the solution (see Figure 11.7b), owing to higher mobility of ions and the removal of concentration gradients in the electrolyte.

The CdSe/Na_2S–S–NaOH/C PEC cells are stable at low current densities (low illumination intensities). However, if the rate of transport of holes from the semiconductor to the redox couple falls below the rate at which holes arrive at the interface, a process of photocorrosion-induced ion exchange with the solution may take place.[82,85,95,96] Thus, for the case of *n*-CdSe in polysulfide solution

$$\mathrm{CdSe} + 2h^{+} \rightarrow \mathrm{Cd}^{2+} + \mathrm{Se}^{0} \tag{11.31}$$

$$\mathrm{Se}^{0} + \mathrm{S}^{2-} \rightarrow \mathrm{SeS}^{2-} \tag{11.32}$$

$$\mathrm{Cd}^{2+} + \mathrm{S}^{2-} \rightarrow \mathrm{CdS} \tag{11.33}$$

Holes arriving at the surface of the semiconductor initiate the photocorrosion reaction (11.31). The Se^0 produced dissolves by reacting with sulfide or some polysulfide species (11.32), while the Cd^{2+} ions reprecipitate on the surface as CdS (11.33). The presence of CdS at the surface of CdSe photoanode has been observed by many workers.[95,96]

The *n*-CdS layer on the surface introduces an added heterojunction (Figure 11.6), which blocks the flow of holes to the electrolyte.[96] Such blocking is expected in any heterojunction if the valence band of the exchanged layer is below that of the unexchanged semiconductor and if the exchanged layer exceeds in its thickness the tunneling distance of the carriers.

The ion exchange reaction is a self-accelerating process. Once part of the *n*-CdSe surface is converted to CdS, the current density at the remaining available area, and thus the rate of the induced ion-exchange reaction, also increases, leading to the spread of the inhibiting layer. Figure 11.7a shows the decline of the short-circuit current in an *n*-CdSe/1 *M* Na_2S–1 *M* S–1 *M* NaOH/C cell as a function of the level of illumination (initial photocurrent) and crystal face.[96]

Various methods have been suggested to improve the stability of *n*-CdSe polysulfide PEC cells. Since at adequately high light intensities, the kinetics of hole transport to the redox couple become mass-transport-dependent, the photocurrent stability can be improved by stirring. Figure 11.7b shows the effect of stirring on the stability of photocurrent.[82]

The problem of ion exchange can be suppressed by having common

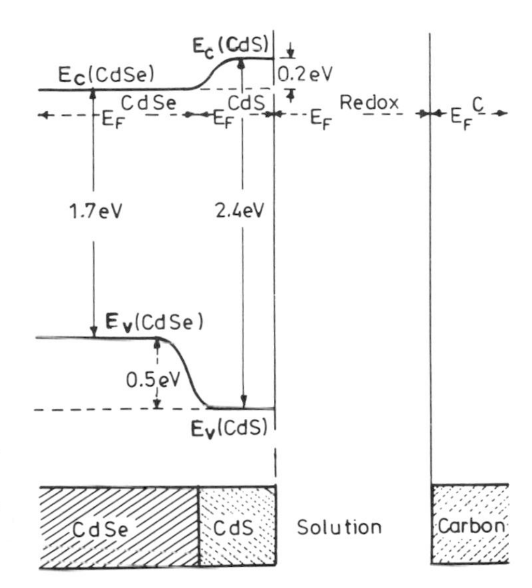

Figure 11.6. Barrier for hole transport from an n-CdSe photoanode to a redox couple solution owing to an n-CdS surface layer (after Heller and Miller[85]).

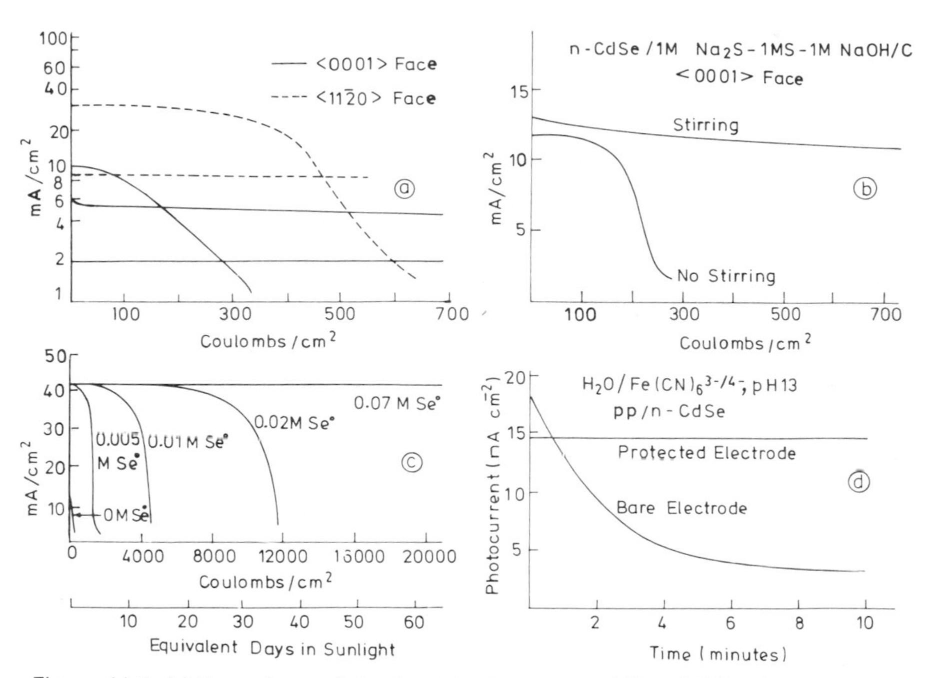

Figure 11.7. (a) Dependence of the short-circuit current stability of CdSe photoanodes in the n-CdSe/1 M Na_2S–1 M S–1 M NaOH/C PEC cell on current density (light intensity) and on the crystallographic face, with and without stirring. (b) Effect of electrolyte stirring on the short-circuit current stability. (c) Effect of Se addition on the short-circuit current stability of an n-CdSe $\langle 11\bar{2}0 \rangle$ face photoanode in a 1 M Na_2S–1 M S–1 M NaOH solution (after Heller and Miller[82]). (d) Photocurrent-vs.-time plots for polypyrrole coated and bare n-CdSe photoanodes in an aqueous Fe $(CN)_6^{3-/4-}$ redox system (after Deb et al.[80]).

ions in the solution and in the semiconductor lattice. In the case of n-CdSe in polysulfide solutions, this is achieved by adding elemental Se to the solution [Figure 11.7c]. Elemental Se dissolves as SeS_x^{2-}. In the presence of such species, the Cd^{2+} ions formed in the photocorrosion reaction (11.33) reprecipitate to form a CdSe-rich surface by the following reaction:

$$Cd^{2+} + SeS_x^{2-} \rightarrow CdSe + x\,S^0 \tag{11.34}$$

Such a surface, even if it contains some CdS, no longer blocks the flow of holes to the interface, because its valence band is adequately close to that of CdSe itself. Excess S is removed according to the reaction

$$x\,S^0 + x\,S^{2-} \rightarrow x\,S_2^{2-} \tag{11.35}$$

The addition of Se^0 causes the solution to absorb a larger fraction of the solar spectrum, and, thus, leads to a decline in the photocurrent. Nevertheless, there is little or no loss in the solar conversion efficiency as elimination of the current blocking n-CdS layer improves the operating voltage and the fill factor.[82]

The photodegradation of semiconductor electrode has also been overcome by the placement of an electrically conducting barrier to ion/solvent transport. Noufi et al.[97] have deposited polypyrrole films electrochemically from acetonitrile solution containing pyrrole on semiconductor electrodes. Semiconductor photoanodes coated with the polypyrrole polymer films exhibit stable photocurrent in acidic and basic electrolytes. Figure 11.7d shows the effect of depositing the polypyrrole film on a CdSe photoanode.

An improvement in stability with the use of nonaqueous electrolytes has also been reported.[80] Thus, under constant AM1 illumination, n-CdSe (single crystal) in a methanol electrolyte containing the ferro-/ferricyanide redox couple has been stable for 29 days.

As noted earlier, CdTe is unstable in polysulfide solution. Single crystal CdTe in polyselenide solution has given efficiencies of about 8.4%, which is limited owing to the high absorption of polyselenide solution up to 0.6 μm. To broaden the spectral response of CdSe and CdTe PEC cells, Hodes et al.[98,99] have used polycrystalline painted layers of the alloy $CdSe_{0.65}Te_{0.35}$ (having a minimum bandgap for CdSe–CdTe system), in a polysulfide electrolyte with sulfide of brass gauze as the counter electrode.[100] By suitably photoetching the electrodes and treating with an aqueous K_2CrO_4 solution, a solar conversion efficiency of about 8.0% has been obtained. The electrodes, operated at 20 mA cm^{-2} current density, have remained stable for more than 40 hr. Russak et al.[87] have deposited films of $CdSe_{0.65}Te_{0.35}$ by coevaporation and achieved efficiencies approaching 5.0% with polysulfide solution and a C counter electrode.

11.4.2. *Gallium Arsenide PEC Cells*

GaAs is one of the best materials for one-step solar energy conversion because of its favorable bandgap of about 1.4 eV. Single crystals of GaAs have been used in photoelectrochemical cells by various workers with ferro-/ferricyanide,[97] polyselenide,[82,85,101–105] and polytelluride[106] redox systems and efficiencies of about 10.5, 12, and 3.8%, respectively, have been achieved. The most outstanding feature of GaAs in ferro-/ferricyanide aqueous solution is the observation of an open-circuit voltage of 1.37 V, which is almost equal to the bandgap of the material. However, the GaAs photoanode corrodes according to the reaction

$$GaAs + 6h^+ + 8OH^- \rightarrow Ga(OH)_4^- + AsO_2^- + 2H_2O \qquad (11.36)$$

The corrosion has been reduced by depositing polypyrrole polymer film on the GaAs electrode and using a nonaqueous methanol-based electrolyte.[97] Such a system has been stable for 100 h without detectable degradation under continuous illumination.

The GaAs electrode is stable in the selenide/polyselenide redox system because of the very negative redox potential of the redox system ($V^0_{NHE} = -0.7$ eV), which is above the decomposition potential of GaAs. This, however, reduces the open-circuit voltage to about 0.7 V. In spite of this, however, a conversion efficiency for sunlight of nearly 12% has been obtained in this cell. This efficiency has been achieved by carefully etching the photoelectrode to give a matte surface, which reduces reflection losses from the film, and treating the electrode with a solution containing Ru^{3+} ions. The improvement in cell characteristics on treating the GaAs photoelectrode with Ru^{3+} ions has been explained by Heller and Miller[85] on the basis of interaction of the adsorbed Ru^{3+} ions with the surface states. In summary, Ru^{3+} incorporation in n-GaAs improves cell performance by stabilizing a new surface composition that produces a shift of surface state energies so as to defeat the power loss mechanisms. Essentially, the GaAs surface is made to closely approach the ideal of noninterfering surface states in the bandgap by deliberate chemical modification.

Polycrystalline layers of n-GaAs[107] in a polyselenide redox system have also been used in PEC cells and efficiencies approaching 7.3% have been achieved. The lower efficiency values in polycrystalline films have been attributed by Heller and Miller[82] to film imperfections as well as to grain size.

11.4.3. *Layered Semiconductor PEC Cells*

Photocells with semiconducting layer crystals, which are supposed to be particularly stable against photodecomposition for kinetic reasons, have been studied by various workers.[80,108–113] Ames Laboratory (see Deb et al.[80]) has succeeded in obtaining efficiencies of 10.2 and 9.4% in sunlight for single crystal n-WSe_2 and n-$MoSe_2$ photoanodes, respectively, using an aqueous I_3^-/I^- electrolyte. Stability has been monitored for single crystal $MoSe_2$ in an aqueous Br_3^-/Br^- electrolyte for 70 days of continuous quartz–halogen (100 $mW\,cm^{-2}$) illumination under constant current conditions with no degradation.[80]

11.4.4. *Other n-Type Materials for PEC Cells*

Heller et al.[75,114] have used anodized Bi_2S_3 films and single crystals of $CuInS_2$ in polysulfide redox system PEC cells. The systems are stable because the lattice and the solution have the common anion S. The efficiency of the $CuInS_2$-based cell at 70 °C is 6%, while the Bi_2S_3-based cell shows an efficiency approaching 1%.

11.4.5. *PEC Cells with p-Type Materials*

Regenerative cells with p-type semiconductors have not yet been tested in detail. The reason is that materials that can be made p-type (like GaP, GaAs, and CdTe) have a rather high position of the valence band edge. Consequently, redox couples with very negative redox potentials are needed to form a depletion layer with great enough band bending at equilibrium in the dark. Such redox systems are not stable in aqueous solution as they react with water.[4] Experiments with p-GaAs and p-GaP were reported by Memming.[115] Butler[116] also found p-GaP unstable in both basic and acidic mediums.

11.5. *Concluding Remarks*

It is clear that in recent years extensive investigations have been undertaken in the field of photoelectrochemical cells (see Table 11.1). Because of higher conversion efficiencies, the regenerative-type PEC cells are gaining more attention than the cells for photoelectrolysis of water. Although efficiencies of about 12 and 7% have been achieved with single crystal and polycrystalline materials (GaAs/polyselenide system), these values are quite low compared with the solid state photovoltaic devices. Also, while the short-term stability of some of the photoelectrodes has been demonstrated, the stability for long-term use remains to be seen. The development of viable PEC cells requires: (i) systematic study of various processes at the semiconductor/electrolyte interfaces such as charge-transfer phenomena, corrosion behavior, and role of surface states,

Table 11.1. Cell Characteristics of Various PEC Cells

Semiconductor (morphology) electrolyte	Preparation technique	V_{oc} (V)	I_{sc} (mA cm^{-2})	FF	η (%)	Ref.
n-GaAs (SC) 0.8 M Se^{2-}, 0.1 M Se_2^{2-}, 1 M OH^-		0.72	24	0.70	12	(102)
n-GaAs (P) 0.8 M Se^{2-}, 0.1 M Se_2^{2-}, 1 M OH^-	CVD	0.57–0.62	22	0.42	4.8	(82)
n-CdS (SC) 0.2 M $Fe(CN)_6^{4-}$, 0.1 M $Fe(CN)_6^{3-}$, 0.4 M KCl		0.95	6.2	0.68	5.5	(72)
n-CdTe (SC) 1 M Se^{2-}, 0.1 M Se_2^{2-}, 1 M OH^-		0.81	18.1	0.4	8.4	(117)
n-CdSe (SC) 1 M S^{2-}, 1 M S, 1 M OH^-		0.72	14.0	0.60	8.4	(86)
n-CdSe (P) 1 M S^{2-}, 1 M S, 1 M OH^-	Pressure sintering	0.67	12.0	0.45	5.1	(86)
n-CdSe (P) 2.5 M S^{2-}, 1 M S, 1 M OH^-	Coevaporated thin film	0.51	8.2	0.61	5.0	(87)
n-CdSe (P) 1 M S^{2-}, 1 M S, 1 M OH^-	Electro-deposition	0.52	9.2	0.55	3.1	(118)
n-CdSe (P) 1 M S^{2-}, 1 M S, 1 M OH^-	Anodization	0.41	4.6	0.25	0.6	(86)
n-WSe_2 (SC) 0.5 M H_2SO_4, 0.5 M Na_2SO_4 1.0 M NaI, 0.025 M I_2		0.71	65.0	0.46	14.0	(113)
n-Bi_2S_3 (P) 1 M Na_2S, 0.05 M S	Anodization	0.31	0.22	—	—	(75)
n-CdS (P) 1 M Na_2S	Anodization	0.60	2.0	—	—	(75)
n-CdS (P) 1.25 M OH^-, 0.2 M S^-	Spray pyrolysis	0.60	1.8×10^{-3}	0.21	0.038	(76)
n-CdS (P) 1.25 M OH^-, 0.2 M S^{2-}	Evaporated	0.71	1.1×10^{-4}	0.25	0.054	(76)
n-$CdSe_{0.65}Te_{0.35}$ (P) 1 M S^{2-}, 1 M S, 1 M OH^-	Screen printing	0.74	19	0.555	7.9	(98)
n-CdSe (P) 1 M S^{2-}, 1 M S, 1 M OH^-	Spray pyrolysis	0.63	11.8	0.35	5.2	(89)
n-CdSe (P) 1 M S^{2-}, 1 M S, 1 M OH^-	Solution growth	0.41–0.62	12.0	0.45	2.25	(91)

through experimental and theoretical studies, (ii) development of low-cost, high-efficiency amorphous and polycrystalline thin film semiconductor photoelectrodes and characterization of aqueous and nonaqueous electrolytes containing various redox systems, and (iii) development and characterization of electrochemical devices with proper sealing, encapsulation, and optimum cell configuration.

12

Novel Concepts in Design of High-Efficiency Solar Cells

12.1. *Introduction*

The most important factor in a discussion of the economic aspects of a particular photovoltaic system is the cost per peak watt output of the device. In earlier chapters we discussed the various thin film systems that are being actively investigated to meet the cost goals. In the last couple of years, a novel approach, that of concentrator solar cells, has been pursued vigorously in several laboratories. In this scheme, the high cost of the solar cell is taken as an accepted fact, but the dollar per peak watt index is reduced by increasing the cell output power using concentrators. It is assumed that the concentrating lens and tracking system would add only marginally to the cost of the total system.

From the bar chart of energy distribution (Figure 3.25, p. 137), we saw that the combined losses from excess photon energy and energy from photons incapable of ionizing charge carriers account for more than half of the energy incident on a conventional solar cell. It is thus clear that the conventional solar cells with their associated high-energy losses are not quite suitable for operation under concentration, particularly for illumination levels greater than 10 suns ($1\ \mathrm{W\,cm^{-2}}$). Further, at high illumination intensities, high-injection and high-temperature effects begin to dominate cell performance, lowering the conversion efficiency. These problems have necessitated new designs for conventional devices that can maintain high values of conversion efficiency at large values of concentration ratio and elevated temperatures.

The thermodynamics for conversion of direct solar radiation into work impose a limit of 93% on the efficiency, as noted in Chapter 1. In recent years, radically new concepts have emerged involving multiple cells/junctions made of different semiconductor materials, acting in tandem in a cascade configuration, thereby achieving efficiencies as high as 40%, far in excess of the theoretical limits of conventional homo-, hetero-, or Schottky-junction devices employing a single semiconductor absorber layer and/or only one junction.

The subsequent discussion in this chapter has been split into three sections. In Section 12.2, we enumerate the effects of high-intensity illumination on the operation of solar cells. In Section 12.3, we describe new designs for conventional solar cells that enable satisfactory performance, within the theoretical limits, with a single absorber layer. Finally, in Section 12.4, we discuss ultrahigh-efficiency cascade solar cells. Concentrators are discussed in Appendix E.

12.2. *High-Intensity Effects*

For an ideal diode solar cell with saturation current $I_s \ll I_L$, negligible series resistance, and infinite shunt resistance, $I_{sc} \propto P_i$ and $V_{oc} \propto \ln P_i$, where P_i is the input illumination intensity. Under the assumption that FF remains constant with increasing P_i, we can expect, in the ideal case, an increase in efficiency with illumination intensity by an amount directly proportional to the increment in photovoltage. However, it is well known that for low-injection conditions, the photovoltage saturates at a value very close to the diffusion potential V_d.[1] Thus an unlimited increase in conversion efficiency is not obtained. Moreover, since the base resistivity of most solar cells is higher than the junction layer, high-level injection conditions would be effective in the base before the condition $V_{oc} = V_d$ is achieved. To maintain charge neutrality in the base, an increase in majority carrier concentration is required to match the new value of minority carrier concentration. The implications are that the quasi-Fermi levels in the base are no longer flat and that a potential drop exists in this region even on open circuit, and thus a reduction occurs in photovoltage below that of V_d.[2] Moreover, under high-injection conditions, the current–voltage relationship takes the form $I = I_s \exp(qV/2kT)$. The saturation dark current value in this case is greater than that for low-injection levels.[3] This increase in I_s leads to a reduction in FF.

In real solar cells, the presence of a finite series resistance leads to a reduction in efficiency (owing to a lowering in FF) even at illumination levels below that at which high-injection conditions are reached. Series resistance has a deleterious effect in solar cells even at normal intensities but it is particularly damaging under concentration when current levels are high. Pulfrey[4] has calculated FF as a function of concentration ratio for a Si solar cell with series resistance as a parameter (Figure 12.1) and has shown that for satisfactory operation at 100 suns the series resistance

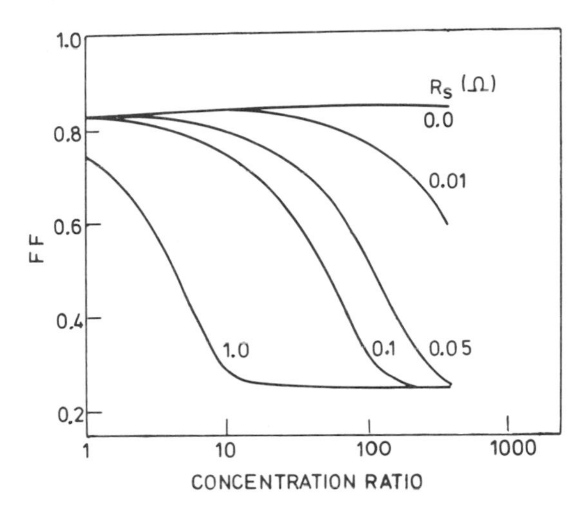

Figure 12.1. Calculated variation of FF of a Si cell with sunlight concentration ratio for different values of series resistance. It has been assumed that $n = 1.5$, $J_s = 3.3 \times 10^{-12}$ A cm^{-2}, $T = 300$ K, cell area = 1 cm^2, and $I_L \propto$ concentration ratio (after Pulfrey[4]).

should be less than a few hundredths of an ohm. Cells with lower photocurrent response, e.g., GaAs, can tolerate somewhat higher values of series resistance.

It is generally assumed that the photocurrent increases linearly with the input intensity. Dhariwal et al.[5] suggested photocurrent saturation ($\sim V_d/R_s$) at high illumination levels. Although this effect would alleviate to some extent the falloff in FF, the overall conversion efficiency would not improve because of the reduced photocurrent. Also, under high-injection conditions, the increase in majority carrier concentration would modulate the semiconductor resistivity and the series resistance would decrease with increase in the input power density. This would again lead to a reduction in the degradation of FF. However, as a result of increased dark currents, the V_{oc} would be reduced and the conversion efficiency would not increase, although the fall in the efficiency with input power may be less rapid than in the corresponding low-injection case. Further, it is likely that the initial value of efficiency would be higher in the low-injection conditions. Fossum and Burgess[6] suggested the use of a back surface field arrangement to combine the desirable features of conductivity modulation with a high initial value of V_{oc} in a single device. In a GaAs solar cell, it may be noted, the short minority carrier diffusion lengths do not permit the use of a back surface field. But, on the other hand, since doping levels in GaAs cells are higher than in Si cells, high-injection-level effects are not important in GaAs except at very high values of the concentration ratio.

We noted earlier that the maximum conversion efficiency for any single junction device is about 25%. This implies that at high concentration ratios there is a considerable amount of thermal energy at the solar cell which could lead to a high operating temperature for the device unless some cooling method is devised. A large increase in temperature would affect the minority carrier properties and intrinsic carrier concentrations in the cell and, to a lesser extent, the photon absorption properties from bandgap variations. Since minority carrier mobilities decrease somewhat in the case of Si and change very slightly for GaAs (for typical doping densities), the minority carrier diffusion coefficient is nearly temperature-independent for the former and increases monotonically with temperature for the latter. Consequently, minority carrier diffusion lengths increase with temperature, particularly for GaAs, as the minority carrier lifetime increases with temperature as a result of increased thermal velocity. Thus higher temperature would imply large short-circuit currents resulting from the improved diffusion lengths. On the other hand, V_{oc}, which would be expected to increase with an increase in I_{sc}, is strongly dependent on the intrinsic carrier concentration n_i, which increases exponentially with T. The saturation current increases as n_i^2 and n_i for emission–diffusion and generation–recombination mechanisms, respectively. Thus the V_{oc} and FF would decrease with T in cases where the above mechanisms are operative. Further, the slight decrease in bandgap with temperature would also lead to an increase in the saturation dark current, negating the beneficial effect of improved long-wavelength absorption.

Overall, the conversion efficiency would be reduced if the temperature is allowed to greatly exceed the room temperature. As the degradation in V_{oc} seems to be the principal mechanism leading to efficiency reduction at high temperatures, a high initial value of V_{oc} is required. The incorporation of a back surface field is effective in performance improvement at high temperatures, as shown in Table 12.1. The improvement stems in part from the increased minority carrier diffusion lengths and in part from the dark current suppression qualities of the back surface field. Moreover, the temperature coefficient of the contribution to the output voltage by the p/p^+ junction is positive. This leads to a reduction in the degradation of V_{oc} with temperature.

GaAs cells are potentially superior high-temperature devices than Si cells. The higher bandgap and dominance of recombination–generation mechanisms in the dark current ensures a higher V_{oc} and less degradation in V_{oc} with temperature. Also, since minority carrier diffusion lengths are short in GaAs, the improvements in L_D owing to temperature lead to a higher increase in J_{sc} than in the case of Si. Since GaAs has a high absorption coefficient, the decrease in bandgap with temperature may lead to an attenuation in the high-energy photons reaching the junctions.

Table 12.1. Calculated Performances of 10 Ω cm n^+/p and $n^+/p/p^+$ Solar Cells at 1 and 40 suns AM0 and 27 and 100 °C (from Pulfrey[4])

T (°C)	X (suns)	BSF	J_{sc} (mA cm^{-2})	V_{oc} (V)	FF	η (%)
27	1	No	49.9	0.541	0.808	16.1
27	1	Yes	52.0	0.606	0.809	18.8
27	40	No	2130.0	0.605	0.715	17.0
27	40	Yes	2130.0	0.721	0.755	21.4
100	1	No	50.4	0.350	0.702	9.2
100	1	Yes	52.6	0.438	0.717	12.2
100	40	No	2110.0	0.432	0.625	10.5
100	40	Yes	2110.0	0.580	0.714	16.1

However, the use of a shallow junction can minimize this effect to a certain extent.

GaAs solar cells, used under high concentrations, employ a high conductivity surface layer of $Ga_{(1-x)}Al_xAs$, which also serves to reduce the surface recombination. The thickness of the surface layer is 0.3 to 5 μm and the compositional factor x is about 0.8. TiO_2[7] and Si_3N_4[8] are used as AR coatings. With a grid pattern (shading loss 10%), the series resistance of a 1.27-cm-diameter cell can be made as low as 0.027 Ω. At 200 °C, the efficiency is 14% at a concentration ratio of 312.[8] The conversion efficiencies at room temperature for concentration ratios of 10 and 1735 are 23% and 19.1%, respectively.[8]

Si solar cells do not employ a high conductivity heteroface surface layer and thus have to rely on improved grid patterns to reduce series resistance to acceptable values for high-concentration operation. Various grid designs[6,9] have been employed. Operation at 100 suns appears possible with negligible degradation. Presently, devices with back surface field have been operated at efficiencies of 15.5%[9] and 12.2%[6] at concentration ratios of 23 and 60, respectively. The above results refer to working temperatures of 20 to 30 °C. Small-area (0.16 cm × 0.16 cm) $p^+/n/n^+$ devices[10] with a base resistivity of 30 Ω cm have been operated at 120 suns at a conversion efficiency of 8%. The base layer thickness in this case was 50 μm and some improvement is expected with increased base thickness and consequent enhanced photon absorption.

12.3. *Conventional Cells–New Designs*

12.3.1. *Single Absorber Layer Multijunction Solar Cell*

Multijunction devices using Si have been investigated for operation at high concentration ratios.[11–14] A particularly effective design is an interdigited structure with a high-lifetime base material and a transparent SiO_2 top layer to reduce surface recombination. Carrier separation occurs at the back surface. A high-resistivity base material facilitates conductivity modulation and a series resistance value of 0.022 Ω has been obtained.[15] At 15 °C and 220 suns, an efficiency of 16.5% is reported.

A more advanced form of the multijunction device, the V-groove multijunction (VGMJ) solar cell, has been recently described by Chappell.[16] All the elements of the cell are formed simultaneously from a single Si wafer by V-groove etching. Figure 12.2 illustrates the VGMJ solar cell, which consists of several $p^+/i/n^+$ diode elements connected in series. The trapezoidal-shaped diode elements are obtained by anisotropically etching ⟨100⟩ Si through a thermally grown SiO_2 layer. Thus the fabrication of the VGMJ cell requires only one photomasking step — that used to define the SiO_2 V-groove etching mask. The n^+ and p^+ junction regions are created by ion implanting the dopant atoms at such an angle that either the n^+ or p^+ faces of the structure are implanted at one time. The structure is annealed following ion implantation to activate the implanted dopant atoms and remove implantation damage. Evaporation of a metal layer connects the individual elements in series; the poor metal coverage over the edges of the photolithographically defined oxide strips, undercut during the V-groove etching step, prevents shorting of the n^+ and p^+ regions of the same element. The oxide strips also help to protect the narrow, flat regions they cover from being ion implanted. The metal or top of the oxide serves as a reflector, increasing the light trapping characteristics of the cell. The Si is mounted on a glass (Corning-type 7070) substrate. The cell is illuminated from the glass side.

Most of the photogenerated carriers in the VGMJ cell are produced near its illuminated surface. However, these carriers are not constrained to

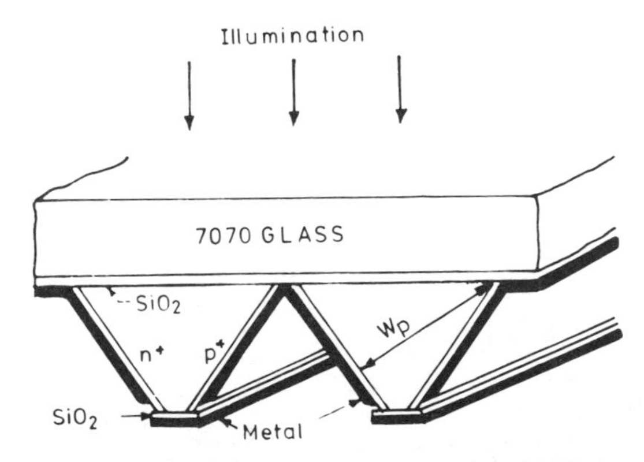

Figure 12.2. Schematic diagram of a V-groove multijunction (VGMJ) solar cell (after Chappell[16]).

move only parallel to the illuminated surface and need not travel further than the distance W_p to be collected. Further, some of the carriers are generated substantially below the illuminated surface and, hence, closer to the junction. In addition, multiple internal reflection of photons from the metal covered regions on the back of the cell or from the metallized junction regions give the VGMJ cell a fundamental collection efficiency which can be greater than that of a conventional planar cell having a thickness equal to the height of the trapezoidal elements. The fundamental collection efficiency here is defined as the ratio of the number of photons absorbed to the number that enter the cell. Thus, the VGMJ structure has an effective optical thickness many times its actual thickness. An average effective optical thickness of 267 μm has been calculated[16] for 50-μm-high elements, resulting in a fundamental collection efficiency exceeding 93%. The high fundamental collection efficiency using thin Si ameliorates the need for a long bulk carrier lifetime in obtaining a high total carrier collection efficiency.

The internal collection efficiency of the VGMJ cell, defined as the ratio of the number of electron–hole pairs collected by the cell to the number generated within it, is controlled primarily by bulk recombination. The calculated variation of the internal collection efficiency as a function of sunlight concentration ratio is shown in Figure 12.3a. An internal collection

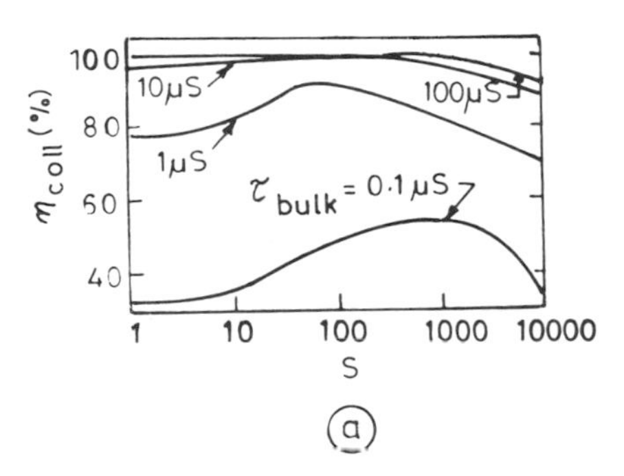

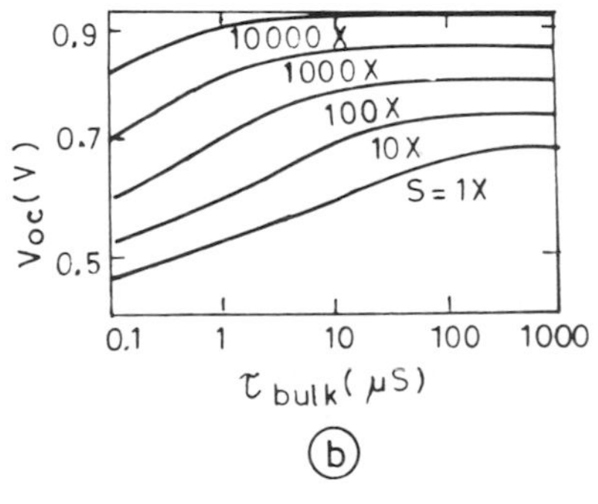

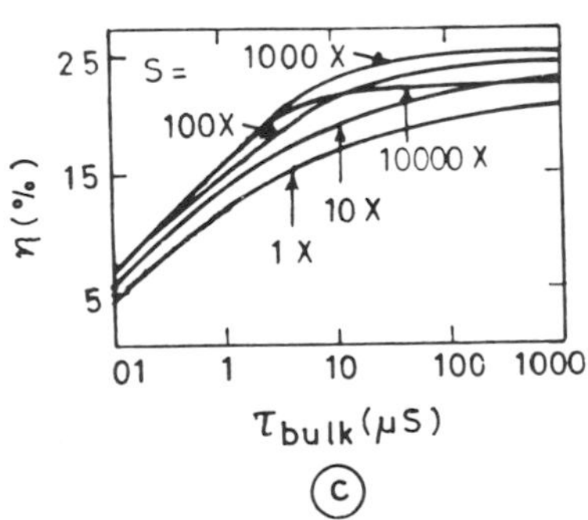

Figure 12.3. Performance characteristics (calculated) of a VGMJ Si solar cell: (a) Internal collection efficiency η_{coll} as a function of sunlight concentration factor S for different bulk carrier lifetimes τ. (b) V_{oc} as a function of τ for different values of S. (c) Conversion efficiency η as a function of τ for different values of S (after Chappell[16]).

efficiency of 95% in the VGMJ cell can be obtained with a bulk carrier lifetime somewhat less than 10 μs, in the concentration range 1 to 1000 suns. A 100-μm-thick interdigited solar cell would require a lifetime of 90 μs for a comparable internal collection efficiency. It should be noted that the observed rise in internal collection efficiency with concentration ratio can be attributed to a field enhancement effect caused by the electric field associated with the appreciable ohmic voltage drop in the bulk region of the cell at high current densities. The decrease in internal collection efficiency at still higher concentration ratios results from a decrease in the diffusion length owing to carrier–carrier scattering and to a reduction in the effectiveness of the field enhancement effect as a result of conductivity modulation and reduced carrier mobilities.[16]

The VGMJ has a larger ratio of illuminated area to cell volume than the planar device and, thus, the photogenerated carriers in the VGMJ cell are confined in a relatively smaller volume, resulting in a higher excess carrier concentration and a correspondingly greater V_{oc}. A 50-μm-thick Si VGMJ cell with oxide strips 10 μm wide at the back has an effective optical thickness of 267 μm and an illuminated area to cell volume ratio 10 times that of a 267-μm-thick planar device. As a result, for a 100-μs bulk carrier lifetime, the V_{oc} of the VGMJ cell shows an increase of 73 mV at 1 sun, 51 mV at 10 suns, 31 mV at 100 suns, 30 mV at 1000 suns, and 30 mV at 10,000 suns over that of the planar cell. The computed V_{oc} of the VGMJ cell as a function of bulk carrier lifetime is shown in Figure 12.3b for sunlight concentration ratios from 1 to 10,000.

Calculations[16] show that FF values of 0.8 or higher are possible at concentration ratios of 1 to 1000 suns with bulk carrier lifetimes exceeding 50 μs. For a 50-μs lifetime, the FF is 0.81 at 300 suns. The internal series resistance associated ohmic power loss of the VGMJ cell, at concentration ratios up to 1000 suns, is quite low by virtue of the small dimensions of the elements and the significant conductivity modulation of the bulk region in these elements at high concentrations. However, the power loss at 10,000 suns is significant and the FF degrades to 0.74 for bulk carrier lifetimes equal to or slightly exceeding 10 μs.

The room temperature conversion efficiency of an optimized VGMJ cell as a function of bulk carrier lifetime is shown in Figure 12.3c for different concentration ratios. The bulk carrier lifetime needed to obtain a high conversion efficiency decreases with increasing concentration, a result of the improvement in V_{oc} with intensity. For a 20% conversion efficiency, a lifetime of 229 μs is required at 1 sun, while at 100 suns a lifetime of 5 μs is adequate.

VGMJ Si solar cells fabricated with 43 series connected diode elements have exhibited a V_{oc} of 30.2 V, an I_{sc} of 44.3 mA, and an FF of 0.63 at 300 suns and 300 K, corresponding to a conversion efficiency of

12.2% against a designed value of 24.5%. Low bulk carrier lifetimes (0.8 μs) are responsible for the poor FF. The designed efficiency of 24.5% is based on a bulk lifetime of 50 μs and an AR coating to reduce reflection losses to 7% (observed reflection losses are 23%).

It is evident from the preceding discussion that the VGMJ Si solar cell enjoys several advantages over its conventional counterpart, which make it particularly suited to concentrator applications. These include: (i) capability for greater than 20% conversion efficiency using existing technology, (ii) high internal collection efficiency (>95%) with only modest bulk carrier lifetimes, (iii) higher V_{oc} than in planar cells of equivalent optical thickness, (iv) very low series resitance, permitting efficient cell operation at concentrations exceeding 1000 suns, (v) elimination of grid on the illuminated surface, (vi) excellent environmental protection provided by the glass front surface, and (vii) a relatively simple fabrication procedure.

12.3.2. *Fluorescent Wavelength Shifting*

Fluorescent wavelength shifting has been used to enhance the spectral response and increase, by 0.5 to 2 percentage points, the AM0 conversion efficiency of several types of cells.[17] We shall now examine the concept of wavelength shifting and its effect on cell operation.

A fluorescent plastic, glass, or crystalline sheet is placed in optical series with a solar cell, as shown in Figure 12.4. The fluorescent material absorbs short-wavelength light and re-emits it at longer wavelengths. Thus the use of these materials can enhance the short-wavelength response of solar cells that have a high collection efficiency at longer wavelengths but low quantum efficiency at short wavelengths (poor blue response).

The important properties of the fluorescent-material–solar-cell combination that need to be considered are the absorption and emission characteristics of the fluorescent material and the optical efficiency, defined

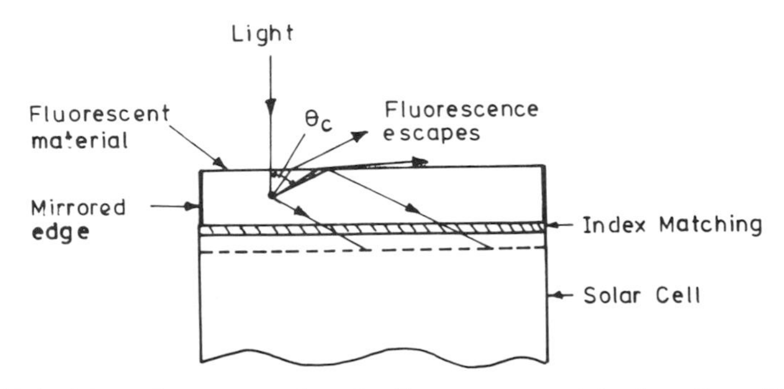

Figure 12.4. Schematic representation of a fluorescent material–solar cell combination.

as the number of photons transmitted into the solar cell divided by the number of photons incident on the upper surface.

Many fluorescent materials exhibit high quantum efficiencies for transforming short-wavelength light to longer-wavelength light. The internal quantum efficiencies (defined as the number of photons emitted per incident absorbed photon) of many fluorescent materials approach 100% for low dye concentrations and in the range of 70 to 90% for practical dye concentrations, where the absorption coefficient is high enough to absorb most of the incident light.

The absorption and emission spectra of ruby, containing 0.7% Cr by weight, are shown in Figure 12.5. Also shown are the spectral responses of a commercial Si cell with and without a ruby sheet (no AR coating on ruby). The absorption band is fairly broad (with two peaks) and extends into the UV, while the emission band is narrow and centered in the visible region at about 1.75 eV. The Stokes shift (i.e., the wavelength separation between the two nearest absorption and emission peaks) is large and there is almost no overlap between the absorption and emission bands.

The fluorescent material and the solar cell are optically coupled by a refractive index matching medium, such as cover glass adhesive or an optical fluid. Longer wavelengths pass through the fluorescent sheet without being absorbed while shorter wavelengths are absorbed and re-emitted at longer wavelengths. The optical efficiency is determined by reflection losses from the top surface of the fluorescent material, the

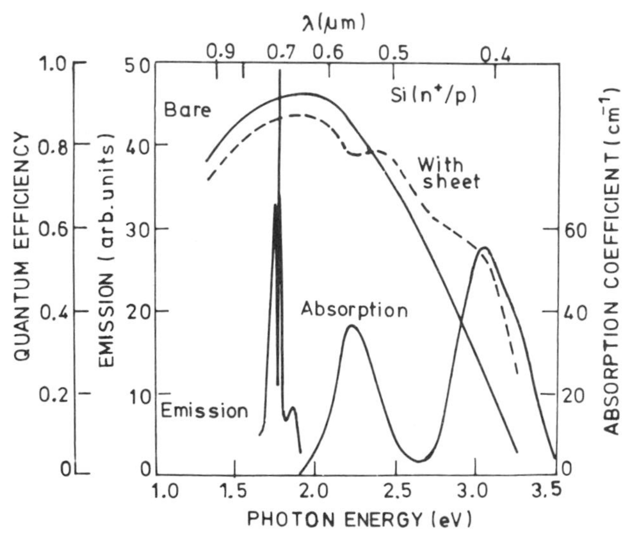

Figure 12.5. Absorption and emission spectra of ruby (0.7% Cr by weight) and the spectral responses of an n^+/p Si solar cell, with and without a ruby sheet (after Hovel et al.[17]).

Table 12.2. Optical Efficiency for Some Fluorescent Materials (from Hovel et al.[17])

Region	Characteristic	Optical efficiency
Nonabsorbing		
	Reflectance loss	5–10%
	Overall efficiency	90–95%
Absorbing		
	Reflectance loss	4% plastics, 7.6% ruby
	Escape loss	12.2% plastics, 8.2% ruby
	Overlap loss	5% plastics, 0% ruby
	Quantum efficiency	70–90% plastics, 100% ruby
	Overall efficiency	56–72% plastics, 85% ruby

internal fluorescence quantum efficiency, the escape of some of the fluorescent light through the upper surface, self-absorption losses from absorption–emission overlap, and, if the refractive index of the fluorescent material is not matched to that of the solar cell, reflection losses at the lower boundary owing to light incident at angles greater than the critical angle for total internal reflection. The overall optical efficiency in both the absorbing and nonabsorbing regions is summarized in Table 12.2 for plastic and ruby fluorescent sheets.

It is clear that the desired properties of a good fluorescent material are a high absorption coefficient at short wavelengths, a high internal quantum efficiency, a large Stokes shift with a low absorption–emission overlap, and a high optical efficiency.

Two types of fluorescent materials have been studied: plastic sheets doped with organic dyes and ruby crystals. For a given optical density of the fluorescent sheet, better results were obtained with thin, more heavily doped sheets rather than thicker sheets, since thin sheets minimize the losses from the edges. Plastic fluorescent materials have a small Stokes shift and work best with solar cells that have a sharp cut-off in quantum efficiency, such as GaAlAs/GaAs cells, Cu_2S/CdS cells, and a-Si cells.

GaAlAs cells with a commercially available Roehm–Haas 2154 fluorescent sheet exhibited improved spectral response at high energies with a corresponding increase in cell efficiency from 14 to about 15%, and in some cases the efficiency improved from 11.5 to 13.5%.[17] For a $p/i/n$ a-Si solar cell, the addition of a Roehm–Haas 2154 sheet resulted in a considerable improvement in the quantum efficiency at high energies and a decrease in the peak reponse, and it thus exhibited an essentially unchanged AM0 efficiency. The spectral responses of a back-wall Cu_2S/CdS solar cell on Cd_2SnO_4 coated quartz substrate, with and without fluorescent

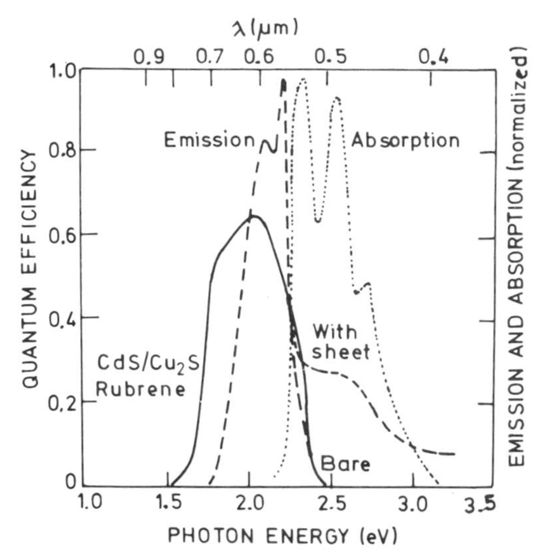

Figure 12.6. Normalized emission and absorption characteristics of rubrene and the spectral responses of a back-wall Cu_2S/CdS cell, with and without rubrene doped fluorescent sheets (after Hovel et al.[17]).

sheets doped with rubrene, are shown in Figure 12.6. The fluorescent sheet substantially improves the high-energy response without degrading the low-energy response. Although rubrene has a relatively small Stokes shift, as shown in Figure 12.6, the emission peak lies almost exactly at the response peak of the cell, resulting in an increase in the AM0 efficiency from 3 to about 3.5%.

The spectral response of a Si solar cell with a ruby sheet (see Figure 12.5) shows a local minimum owing to the absorption peak in ruby at 2.2 eV. At low energies, the lowering of the response is due to reflection losses as a result of nonoptimized layers. At higher energies the spectral response shows considerable enhancement and the AM0 efficiency of the ruby sheet/solar cell combination is expected to improve by 1 to 2 percentage points with proper index matching.

Diffused GaAs solar cells have shown an improvement from 9 to 9.3% in the AM0 efficiency with the addition of ruby fluorescent sheets. Elimination of reflection losses should further increase the conversion efficiency.

12.4. *Ultrahigh-Efficiency Cascade Solar Cells*

We saw in Chapter 3 that a conventional solar cell employing a single semiconductor utilizes only a limited portion of the incident solar spectrum. Moreover, the V_{oc} of these devices is limited by the bandgap of the semiconductor material (the lower of the two bandgaps in the case of a heterojunction). The above two effects lead to high internal losses in conventional cells and a conversion efficiency limited to about 25% for an optimum bandgap cell. The concept of cascade solar cells[18–35] envisages two or more solar cells made of different semiconductor materials with suitably separated bandgap values, operating in tandem to produce a high V_{oc}. The use of different semiconductors with widely spaced bandgaps ensures efficient utilization of the incident spectrum and low internal losses. The overall efficiency of the cascade combinations can be very high (exceeding 30%) with proper choice of materials.

Two approaches have been suggested to realize the above concepts: The first is the optical filter-mirror/multiple-cells combination and the second is the tandem junction solar cells (TJSC). In the first approach the incident spectrum is split into parts and each part is directed toward a separate cell with characteristics designed to match the spectral distribution of the incident part.

The idea of the tandem arrangement, which is the second type of cascade solar cell, is to place solar cells made of materials with different energy gaps one behind the other, such that the largest-gap material faces the incident radiation first. The high-energy photons are absorbed by the first material and the rest of the solar spectrum is allowed to be incident on the second solar cell, which now absorbs the higher-energy photons of the transmitted spectrum, allowing the rest of the spectrum to go through to the third cell. This selective absorption process continues down to the cell with the lowest energy gap. Obviously, this arrangement ensures a much better match to the solar spectrum, and higher total conversion efficiencies are achieved.

Further logical development of the tandem concept envisages a monolithic integrated tandem junction solar cell (ITSC) system wherein the different solar cells are fabricated on a common base and are internally electrically interconnected (in contrast to a series of independent solar cells in tandem). Although the fabrication of an ITSC system may be more involved, the aim is to eliminate the process costs associated with fabricating two or more solar cells for a tandem system and avoid the problems of electrical interconnections, optical coupling, and packaging.

We describe the operation of the filter-mirror/multiple-cells cascade solar cells and the integrated tandem junction solar cells in greater detail below.

12.4.1. *Optical Filter-Mirror and Multiple Cells*

The schematic diagram of a filter-mirror, concentrator, and two cells cascade arrangement is shown in Figure 12.7. The Fresnel lens serves as the

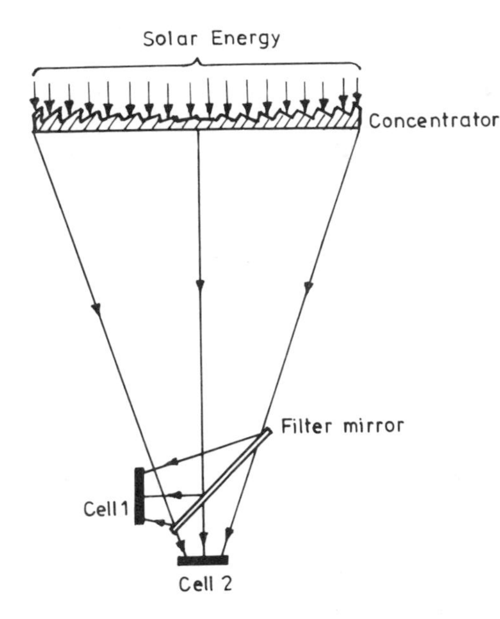

Figure 12.7. Filter-mirror, concentrator, and two cells in a cascade arrangement.

concentrator. The plane surface optical filter-mirror is placed at an angle of 45° to the axis of the concentrated beam. Thus, the reflected part of the beam, directed toward cell 1, has the same geometrical characteristics as the transmitted part, directed toward cell 2. Both the transmitted and reflected beams can be suitably focused to give the optimum concentration required for each cell. Additional cells can be included in this arrangement by placing additional filter-mirrors in the light path, each reflecting a certain part of the spectrum to a target cell.

Under the assumptions that: (i) the spectral distribution $s(\lambda)$ of the radiation is unaltered by the concentrator, (ii) the filter-mirror is lossless with a sharp (infinite slope) reflectance cut-off at wavelength λ_c, corresponding to E_{g_1}, i.e., reflectance is 1.0 for $\lambda < \lambda_c$ and 0 for $\lambda > \lambda_c$, and (iii) cells have no reflection and shading losses, no ohmic losses, and unity collection efficiency, Masden and Backus[24] have analyzed the limit efficiency of a two-cell system. The limit efficiency for an N-cell system under concentration can be expressed by

$$\eta_{\text{lim}} = \frac{\sum_{m=1}^{N} (\text{VF})_m (\text{FF})_m E_{gm} J_m}{C \int_0^\infty s(\lambda) d\lambda} \tag{12.1}$$

where $(\text{VF})_m$ is voltage factor of cell m:

$$(\text{VF})_m = \frac{qV_{\text{oc}}}{E_{gm}} = \frac{(kT/q)\ln[(J_m/J_{sm})+1]}{E_{gm}} \tag{12.2}$$

and the fill factor of cell m is given by

$$(\mathrm{FF})_m = \left[1 - \frac{nkT}{qV_{\mathrm{oc}}} \ln\left(1 + \frac{qV_{mp}}{nkT}\right)\right]\left[1 - \frac{J_{sm}}{J_m} \exp\left(\frac{qV_{mp}}{nkT} - 1\right)\right] \quad (12.3)$$

V_{mp} is obtained from the equation

$$\left(1 + \frac{qV_{mp}}{nkT}\right) \exp\left(\frac{qV_{mp}}{nkT}\right) = \left(\frac{J_m}{J_{sm}} + 1\right) \quad (12.4)$$

where

$$J_{sm} = qn_i^2 \left[\frac{1}{N_{Dm}}\left(\frac{\mu_{pm}}{\tau_{pm}}\right)^{1/2} + \frac{1}{N_{Am}}\left(\frac{\mu_{nm}}{\tau_{nm}}\right)^{1/2}\right] \quad (12.5)$$

$$n_i^2 = \begin{cases} 2.57 \times 10^{29}\, T^3 \exp(-E_{gm}/kT), & \text{for } E_{gm} \geq 1.29 \text{ eV} \\ 4.81 \times 10^{3}\, T^3 \exp(-E_{gm}/kT), & \text{for } E_{gm} < 1.29 \text{ eV} \end{cases}$$

where E_{gm} is the bandgap of cell m. The photocurrent for cell m, J_m, is given by

$$J_m = \frac{Cq}{hc} \int_{\lambda_{gm-1}}^{\lambda_{gm}} s(\lambda)\lambda\, d\lambda \quad (12.6)$$

where $\lambda_{gm} = 1.24/E_{gm}$, $\lambda_{g1} = 0.0$, and C is the concentration ratio.

Assuming values for μ_n, μ_p, τ_n, τ_p, N_A, and N_D, Masden and Backus[24] obtained constant limit efficiency contours as a function of E_{g1} and E_{g2} for a two-cell system at a concentration of 100 AM2 suns. The efficiency contours are shown in Figure 12.8. For a Si–Ge system, the conversion efficiency is 34%. For GaAs–Si and GaAs–Ge combinations, the efficiencies are 35 and 37%, respectively. The conversion efficiency of a combination of a Si cell and a cell (X) of bandgap 1.75 eV is 41%. An important feature of this spectrum splitting method is the significant reduction in the internal losses in each cell over corresponding losses occurring in a single cell under the same illumination. This is shown graphically in Figure 12.9, which gives the distribution of energy components in two independent single cells (GaAs and Ge) operating at 125 AM2 and a two-cell combination (of the same two cells) operating at 125 AM2 suns.

Cape et al.[25] have done theoretical and experimental investigations of two-cell systems. For a GaAs–GaSb combination and a filter-mirror with a cut-off at 0.88 μm, the predicted efficiency exceeds 30% at 300 suns. The experimental efficiencies obtained with a GaAs–Si combination and a filter-mirror with a cut-off at 0.73 μm operated at 340 and 800 suns are in good agreement with predictions. Apparently, the GaAs–Si combination is no better than a single GaAs cell. The authors have pointed out that the success of the filter-mirror multiple cell concept depends critically on optimal splitting of the spectrum and high mirror efficiency. Also, in a split spectrum application, a reduced total current, roughly in proportion to the relative splitting of the photon flux, flows through each cell. The series resistance losses are, therefore, significantly reduced at very high concentrations by spectral splitting.

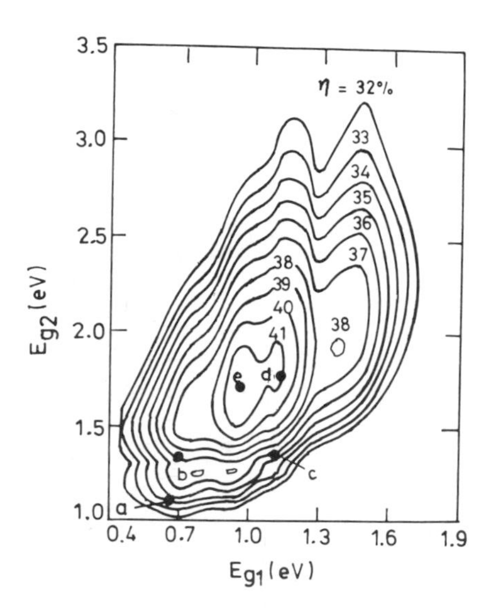

Figure 12.8. Limit conversion efficiency contours for a two-cell/filter-mirror/concentrator system at 100 AM2 suns. a–Si/Ge (34%), b–GaAs/Ge (37%), c–GaAs/Si (35%), d–X/Si (41%), e–optimum (42%) (after Masden and Backus[24]).

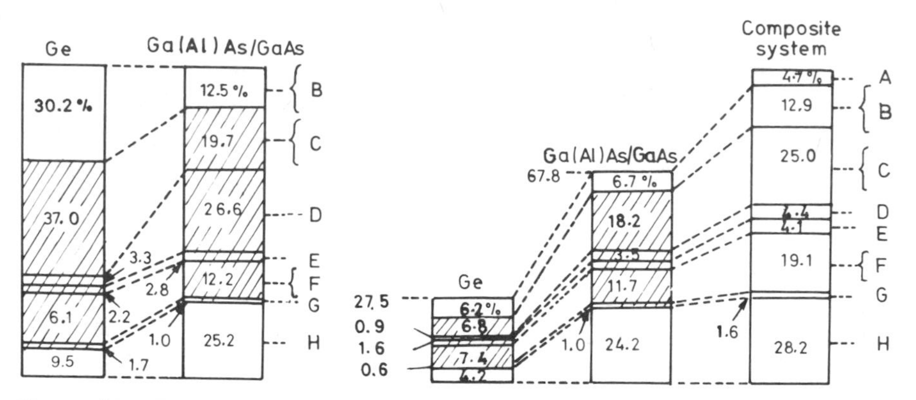

Figure 12.9. Distribution of energy components for single Ge and GaAlAs/GaAs cells and a two-cell cascade combination of Ge and GaAlAs/GaAs cells at 125 AM2 suns. Shaded areas denote energy components that cause heating in the cell. A — filter loss, B — reflection and shading loss, C — loss from excess photon energy, D — loss from weak photons, E — collection loss, F — junction losses, G — I^2R loss, H — power output (after Masden and Backus[24]).

While the limit efficiency analysis yields information on the selection of materials, one must remember that several assumptions are involved in the calculations. To predict the expected performance of an actual system, the various loss factors must be taken into account. Following the theoretical analyses and conclusions of several researchers[26,27] that 1.69–1.1 eV and 1.69–1.43 eV represent two very efficient bandgap combinations, corresponding to $Ga_{0.8}Al_{0.2}As$ (1.69 eV), GaAs (1.43 eV), and Si (1.1 eV) solar cells, Fanetti et al.[28] suggested that experimental efficiencies exceeding 25% can be expected for the 1.69–1.43 eV combination, even if the optical losses of a spectrum splitting mirror are included. Efficiencies exceeding 30% can be realistically expected for a three-cell combination composed of Si (1.1 eV), GaAs (1.43 eV), and $Ga_{0.63}Al_{0.37}As$ (1.92 eV) cells. Accordingly, Fanetti et al.[28] have fabricated, by a modified LPE technique, GaAs and $Ga_{0.8}Al_{0.2}As$ cells with efficiencies of 21.1 and 19.2%, respectively, measured at 210 AM1.5 suns.

Experimental evidence of high-efficiency operation has been given by Vander Plas et al.[29] for the Si (1.1 eV)–$Ga_{0.83}Al_{0.17}As$ (1.65 eV) system with a spectrum splitting factor. Overall efficiencies of 27% at 113 suns and 26% at 489 suns have been measured.

12.4.2. *Integrated Tandem Junction Solar Cells*

A schematic representation of an integrated tandem junction solar cells (ITSC) system, consisting of three p/n homojunctions is shown in Figure 12.10a. We note here that other variations of this basic structure can be conceived with heterojunctions and Schottky junctions. The unique feature of this integrated structure is the system of built-in electrical interconnections, which is equivalent to a permanent series connection of the various cells. Figure 12.10b shows the equivalent circuit of the device.

In presenting a theoretical model of an ITSC device, we shall follow the treatment of Vecchi[30] who has analyzed the structure in terms of the equivalent circuit of Figure 12.10b. Neglecting contact and leakage resistance, we can formulate the following relationships:

$$J = J_{Li} - J_{si}[\exp(\beta V_i) - 1] \tag{12.7}$$

and

$$V = \sum_{i=1}^{N} V_i \tag{12.8}$$

where $i = 1, 2, \ldots, N$ corresponds to the various cells in the ITSC device, J_{si} is the reverse saturation current density of the ith p/n junction and $\beta = q/nkT$. If the junctions are considered to be ideal Shockley diodes, $n = 1$.

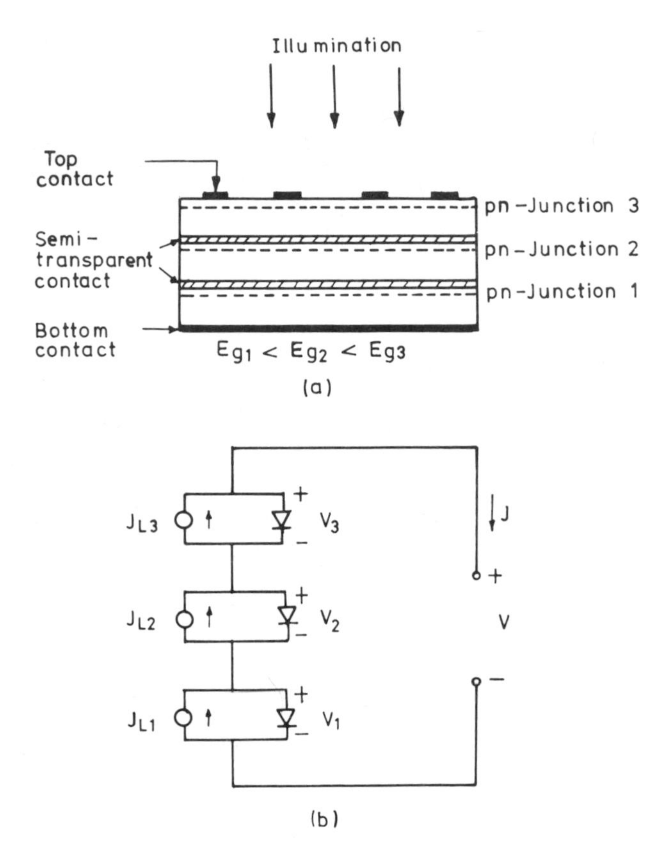

Figure 12.10. (a) Schematic cross-section of an ITSC device composed of three p/n homojunctions. (b) Equivalent circuit of (a).

The J–V characteristics of the ITSC device is given by the general solution of equation (12.7) and can be expressed as

$$\exp(\beta V) = \prod_{i=1}^{N} \frac{(J_{Li} + J_{si} - J)}{J_{si}} \tag{12.9}$$

The V_{oc} is given by [from equation (12.9)]

$$V_{oc} = \frac{1}{3} \sum_{i=1}^{N} \ln \left[\frac{J_{Li} + J_{si}}{J_{si}} \right] \tag{12.10}$$

In general, under short-circuit conditions, $V = 0$, the voltage across the individual cells, V_i, will not be zero. Only when all the J_{Li} have the same magnitude will all the $V_i = 0$ at $V = 0$. In this condition (all J_{Li} same), the ITSC is referred to as a balanced device. However, regardless of whether the ITSC is balanced or not, J_{sc} is given by (since $J_{si} \ll J_{Li}$)

$$J_{sc} = \text{smallest of } \{J_{Li}\} \tag{12.11}$$

The reverse saturation current J_{si} is obtained from the expression for an ideal Shockley p/n junction, viz.,

$$J_s(E_g) = K \exp(-E_g/kT) \tag{12.12}$$

where $K = 4.7 \times 10^9$ mA cm^{-2} based upon known values of J_s for Si. The photocurrent J_{Li} in a single solar cell is given by

$$J_{Li}(E_g) = q \int_{\hbar\omega = E_g}^{\infty} n_{\text{photons}}(\hbar\omega) d(\hbar\omega) \tag{12.13}$$

where the symbols have their usual meanings. The above equation assumes a unity collection efficiency. For two (or more) solar cells in tandem,

$$J_{L1}(E_{g1}, E_{g2}) = J_{L1}(E_{g1}) - J_{L2}(E_{g2}) \tag{12.14}$$

where $J_{L1}(E_{g1}, E_{g2})$ is the current flowing in the cell of bandgap E_{g1} of the combination, placed behind the cell of bandgap E_{g2}. $J_{L1}(E_{g1})$ and $J_{L2}(E_{g2})$ are the photocurrents that would flow in the cells had they been operating as single solar cells. The above expression assumes no internal reflection losses and it can obviously be extended to more cells.

From equations (12.10) to (12.14), we can calculate the theoretical optimum limit for ITSC devices made of p/n homojunction solar cells. However, to calculate the efficiency of an ITSC device made from materials of known parameters, suitable corrections need to be incorporated in the calculated values of J_{Li} and J_{si} and contact and/or leakage resistances included. Based on the calculated performance of several p/n homojunctions, Vecchi[30] has derived the performance characteristics for several two-cell and three-cell ITSC devices and compared the performance with the corresponding independent tandem systems (cells not internally electrically connected). He has shown that the efficiencies of balanced ITSC devices are the same as those of the corresponding independent tandem systems. Unbalanced ITSC devices exhibit much lower efficiencies than the corresponding independent tandem systems and in some cases lower than a single solar cell. For a two-cell device with $E_{g1} = 1.10$ eV and $E_{g2} = 1.68$ eV, the ITSC is balanced and exhibits an AM1 efficiency of 33%, the same as that of the independent tandem system

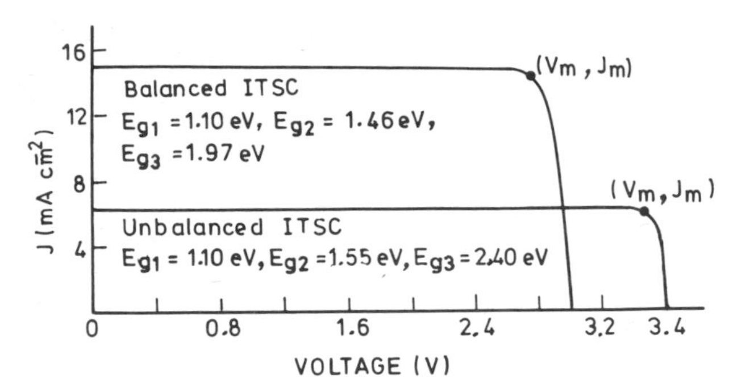

Figure 12.11. *J–V* characteristics (calculated) of balanced and unbalanced ITSC devices composed of three elements (after Vecchi[30]).

with two cells of bandgap 1.10 eV and 1.68 eV. However, for an unbalanced ITSC with $E_{g1} = 1.10$ eV and $E_{g2} = 2.00$ eV, the efficiency is only 26.6%, compared to an efficiency value of 33.2% for the corresponding independent tandem system. The *J–V* characteristics for a three-element balanced ITSC device, with $E_{g1} = 1.10$ eV, $E_{g2} = 1.46$ eV, and $E_{g3} = 1.97$ eV, is shown in Figure 12.11. The efficiency of this device is 37.6%, the same as that of the corresponding independent tandem system. Also shown in the same figure are the *I–V* characteristics of a three-element unbalanced ITSC device ($E_{g1} = 1.10$ eV, $E_{g2} = 1.55$ eV, and $E_{g3} = 2.40$ eV), whose calculated efficiency is only 20.4%, compared to 37.4% for the corresponding independent system. Thus, it is of paramount importance to select the values of the energy gaps properly to obtain a balanced device. It should also be pointed out that the increment in total efficiency that is obtained in going from a single-cell to a two-cell ITSC device is greater than that obtained in going from a two-cell to a three-cell ITSC device. Therefore, because of the diminishing returns with additional elements, it is unlikely that ITSC systems with more than three cells will prove to be economical.

GaAs and GaAs-based alloys have received both theoretical and experimental attention for cascade solar cell applications. Fraas and Knechtli[31] have considered the specific case of a monolithic AlGaAs/GaAs/Ge dual junction and calculated a conversion efficiency of 25.6% at AM0. The photocurrents available to both the cells at AM0 are nearly equal. For a three-junction stacked combination of Ge/$Ga_{1-x}In_xAs$/$Ga_{1-x}In_xP$/$Cd_{1-x}Zn_xS$/ITO with bandgaps of 0.66, 1.25, 1.77, 2.8, and 3.1 eV, respectively, an AM1.5 efficiency of 33% is projected. For operation at 300 AM1.5 suns, the efficiency is calculated to be 40%.

Lamorte and Abbott[23,32] have optimized theoretically the designs of two-junction AlGaAs–GaInAs cascade solar cells for operation at 290 K and under AM0 and AM1.5 spectral conditions. The energy band diagram of the device is shown in Figure 12.12. The design consists of the wide (AlGaAs) and narrow (GaInAs) bandgap junctions joined electrically by an n^+/p^+ tunnel diode formed as an integral part of the monolithic structure. The AM0 efficiency of this device exceeds 30%. An extensive study of this cascade solar cell has been carried out to determine the performance under AM0 to AM5 spectral conditions and at a concentration ratio of 10^3 over a temperature range from 290 to 600 K. The results of these studies show that while the optimized wide bandgap cell parameters under AM0 and AM1.5 are somewhat different, the optimized narrow bandgap cell parameters are identical. The theoretical efficiency has been found to increase with increasing concentration ratio, attaining a value of 40% at a concentration ratio of 10^3.

Experimental results on ITSC structures have been obtained in the AlGaAs/GaAs dual junction system fabricated by LPE techniques on a GaAs substrate.[33] The two cells are monolithically connected in series by means of a low resistance p^+/n^+ junction. A V_{oc} of about 2.0 V was observed for an illumination of 1 sun. The best cells exhibited an efficiency approaching 9% with J_{sc} of about 7 mA cm^{-2} and FF of 0.7 to 0.8. The poor

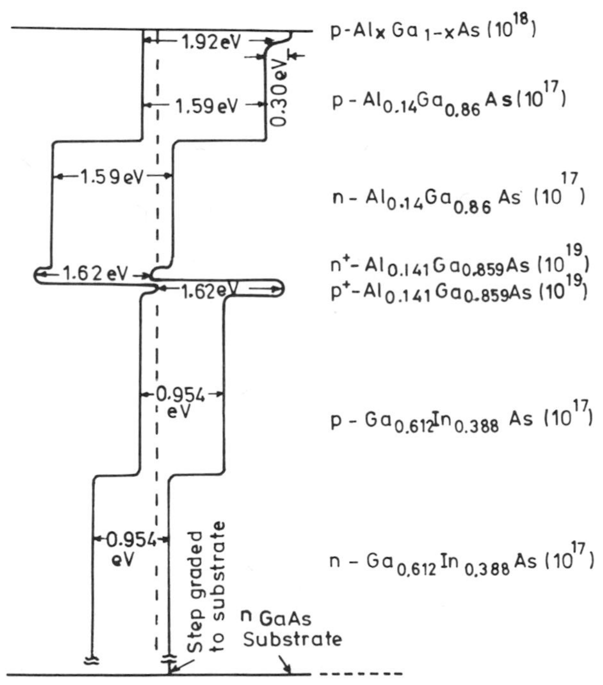

Figure 12.12. Energy band diagram of an idealized, two-junction, AlGaAs–GaInAs ITSC device, optimized for AM0 operation (after Lamorte and Abbott[32]).

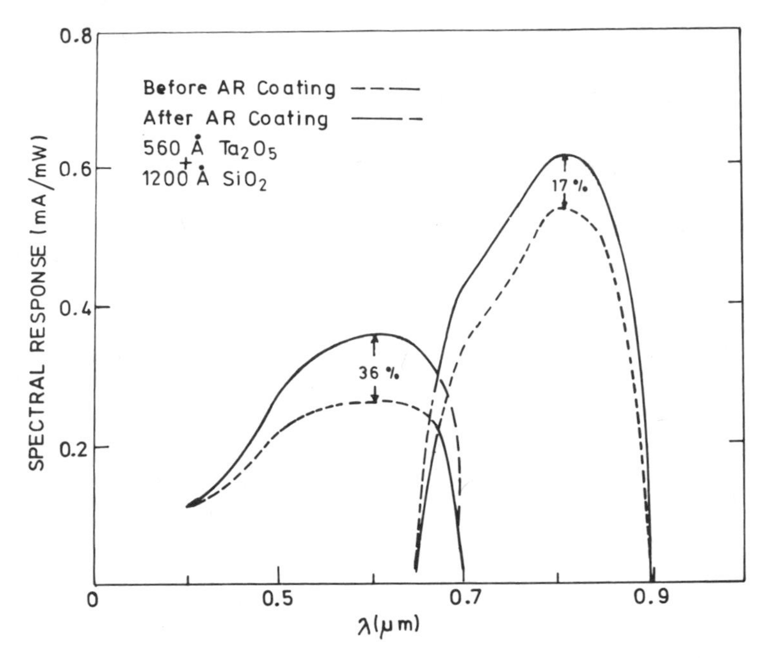

Figure 12.13. Spectral responses of a GaAlAs–GaAs ITSC device, with and without AR coating (after Hutchby et al.[34]).

short-circuit current and low efficiency are attributed to nonoptimized bandgap combinations, nonoptimized layer thicknesses, and the absence of a window layer for the upper cell.

A recent improved version of the AlGaAs/GaAs cascade solar cell[34] has exhibited values of J_{sc} in the range of 10 to 13 mA cm^{-2} without an AR coating, V_{oc} between 1.5 to 2.15 V, and FF ranging from 0.72 to 0.84, giving efficiencies of 10 to 15% and 12 to 16% at AM0 and AM1.5 respectively. The high-resistance front-surface contact grid on the high-bandgap AlGaAs window is held responsible for the poor fill factors. Improved front-surface contacts, using a thin intermediate p^+-GaAs layer between the metal and the top window layer, resulted in an R_sA value of approximately 5×10^{-3} Ω cm^2, compared to an R_sA value of about 0.2 Ω cm^2 for the earlier contact. Direct metallization of the AlGaAs surface, following a series of etching and oxidation steps, with an e-beam evaporated multilayer structure consisting of Mg/Ti/Pd/Ag/Al also led to low R_sA values of about 1×10^{-2} Ω cm^2. A two-layer AR coating composed of Ta_2O_5 560 Å thick and SiO_2 800 to 1000 Å thick was observed to yield typically a 30% increase in J_{sc} for cascade cells exhibiting a good current match. The spectral responses of a GaAlAs/GaAs ITSC device, with and without AR coating, are shown in Figure 12.13.

Timmons et al.[35] have investigated the AlGaAsSb–GaAsSb ITSC structure. They have successfully fabricated diffused junctions in both top (GaAlAsSb) and bottom (GaAsSb) cell materials. The 1.2-eV GaAsSb bottom cell with a 1.7-eV GaAlAsSb window layer exhibited a V_{oc} of 0.5 V, J_{sc} of about 20.4 mA cm^{-2}, and FF of 0.72 under AM0 illumination. The GaAlAsSb (1.55-eV) top cell with a window and AR coating yielded a V_{oc} of 0.57 V, J_{sc} of 8 mA cm^{-2}, and FF of 0.68. However, cascade cells using these materials have not yet been fabricated.

Sakai and Umeno[18] proposed a cascade structure based on InP and $In_{0.58}Ga_{0.42}As_{0.84}P_{0.16}$ (bandgap 0.827 eV) solar cells that can be fabricated by the usual epitaxial growth techniques. Theoretical analysis of the structure yields an AM0 efficiency of 19.5%, which can be improved to 22.2% with the addition of a thick CdS window layer on the InP surface. If the thickness of CdS is about 0.2 μm, an efficiency of 27% is predicted. The workers have fabricated a 23.5% GaAlAs–GaAs cascade cell, comparable to a conventional heteroface or graded-bandgap solar cells.

Chiang et al.[20] reported tandem junction solar cells of Si with junctions on both the illuminated and most of the nonilluminated sides. The Si thickness was 100 μm. Efficiencies approaching 15% were obtained. Milnes[36] has discussed the concept of tandem junction solar cells with reference to thin film cells grown by a rheotaxial method and has suggested growth processes and structures for dual cells with potential efficiencies of 25 to 30%.

A potentially high-efficiency ITSC device consisting of two heterojunctions on a common wide bandgap semiconductor base has been discussed by Ariezo and Loferski.[19] The semiconductor on which sunlight is incident has a bandgap E_{g1} greater than that of the third semiconductor E_{g3}. The intermediate semiconductor has a bandgap $E_{g2} \geqslant E_{g1} \geqslant E_{g3}$. A device with the two photovoltaically active semiconductors having energy gaps $E_{g1} = 2.0$ eV and $E_{g3} = 1.2$ eV should have a conversion efficiency in excess of 30%. The authors proposed a structure of the type $CuInSe_2$ (1.1 eV)–CdS (2.4 eV)–$AgInS_2$ (1.9 eV) to achieve the desired performance. It should be noted that both the active semiconductors are well matched to CdS, with the difference in lattice parameters being less than 1%.

Henry[37] has calculated the limiting efficiencies of ideal single solar cells and tandem cells of 2, 3, and 36 junctions to be 37 and 50, 56, and 72%, respectively, at 1000 suns and 300 K. The author has pointed out that the efficiency value of 72% is less than the thermodynamic limit of 93% owing to emission of light from the forward biased p/n junctions.

APPENDIX A

Spectral Distribution of Solar Radiation

The magnitude of the light-generated current, and hence the performance of a solar cell, depends critically on the intensity and spectral distribution of the incident solar radiation. For outer space, the solar spectrum is designated as air mass zero (AM0) and the radiation intensity is 1353 $W\,m^{-2}$.[1] For terrestrial cases, the intensity and spectral distribution of the solar radiation reaching the earth's surface depends on the local atmospheric conditions, such as the total atmospheric mass, the H_2O, O_3, O_2, CO_2, and dust content, etc.[1] The designation used in practice for terrestrial insolation is AMn, where $n = 1/\cos\theta$, θ being the angle between the sun's position and the zenith. The various solar spectra ($n = 0$, 1, 1.5, and 2) are plotted in Figures 1.3a and b (p. 4), in the form of irradiance ($W\,m^{-2}\,\mu m^{-1}$) and photon flux (number of photons $cm^{-2}\,s^{-1}\,\mu m^{-1}$) vs. wavelength, respectively. The corresponding data are tabulated in Tables A.1a, b. The data for the spectral distribution of the irradiance E_λ for various air masses have been taken from References 2 to 4. The photon flux N_{ph} in the wavelength interval between the corresponding wavelength and the one preceding has been calculated using the average wavelength and irradiance for each wavelength interval.

Table A.1a. Energy Density E_λ and Photon Flux N_{ph} Spectra for AM0 and AM1 Conditions

	AM0		AM1	
λ (μm)	E_λ (W m^{-2} μm^{-1})	N_{ph} (cm^{-2} s^{-1} μm^{-1})	E_λ (W m^{-2} μm^{-1})	N_{ph} (cm^{-2} s^{-1} μm^{-1})
0.295	584	—	0.0	—
0.300	514	8.2074×10^{16}	9.1	6.76×10^{14}
0.305	603.0	8.4898×10^{16}	11.7	1.559×10^{15}
0.310	689.0	9.9872×10^{16}	30.5	3.239×10^{15}
0.315	764.0	11.4143×10^{16}	79.4	8.633×10^{15}
0.320	830.0	12.7223×10^{16}	202.6	2.2508×10^{16}
0.325	975.0	14.6332×10^{16}	269.5	3.8273×10^{16}
0.330	1054.0	16.7043×10^{16}	331.6	4.9454×10^{16}
0.335	1081.0	17.8453×10^{16}	383.4	5.9763×10^{16}
0.340	1074.0	18.2834×10^{16}	431.3	6.8583×10^{16}
0.345	1069.0	18.5856×10^{16}	449.2	7.5809×10^{16}
0.350	1093.0	18.8862×10^{16}	480.5	8.1214×10^{16}
0.355	1083.0	19.2821×10^{16}	498.0	8.6707×10^{16}
0.360	1068.0	19.3309×10^{16}	513.7	9.0921×10^{16}
0.365	1132.0	20.0478×10^{16}	561.3	9.7961×10^{16}
0.370	1181.0	21.3682×10^{16}	603.5	10.7608×10^{16}
0.375	1157.0	21.8930×10^{16}	609.4	11.3576×10^{16}
0.380	1120.0	21.6080×10^{16}	608.0	11.5528×10^{16}
0.385	1098.0	21.3269×10^{16}	609.8	11.7096×10^{16}
0.390	1098.0	21.3914×10^{16}	623.4	12.0127×10^{16}
0.395	1189.0	22.5653×10^{16}	691.2	12.9709×10^{16}
0.400	1429.0	26.1603×10^{16}	849.9	15.3994×10^{16}
0.405	1644.0	31.0931×10^{16}	992.8	18.6447×10^{16}
0.410	1751.0	34.7778×10^{16}	1073.7	21.1689×10^{16}
0.415	1774.0	36.5526×10^{16}	1104.5	22.5869×10^{16}
0.420	1747.0	36.9537×10^{16}	1104.3	23.1819×10^{16}
0.425	1693.0	36.5359×10^{16}	1086.5	23.2651×10^{16}
0.430	1639.0	35.8077×10^{16}	1067.9	23.1525×10^{16}
0.435	1663.0	35.9003×10^{16}	1109.1	23.6690×10^{16}
0.440	1810.0	38.1960×10^{16}	1215.5	25.5659×10^{16}
0.445	1922.0	41.5136×10^{16}	1310.4	28.0973×10^{16}
0.450	2001.0	44.1875×10^{16}	1388.4	30.3598×10^{16}
0.455	2057.0	46.2169×10^{16}	1434.8	32.1141×10^{16}
0.460	2066.0	47.4176×10^{16}	1452.2	32.8922×10^{16}
0.465	2058.0	47.9475×10^{16}	1450.7	33.7504×10^{16}
0.470	2038.0	48.0780×10^{16}	1451.2	34.1035×10^{16}
0.475	2044.0	48.4260×10^{16}	1470.3	34.7011×10^{16}
0.480	2074.0	49.4305×10^{16}	1503.4	35.6949×10^{16}
0.485	1976.0	48.7594×10^{16}	1443.3	35.7411×10^{16}
0.490	1950.0	48.1127×10^{16}	1435.2	35.2757×10^{16}
0.495	1960.0	48.4081×10^{16}	1453.6	35.7651×10^{16}
0.500	1942.0	48.7995×10^{16}	1451.2	36.3283×10^{16}
0.505	1920.0	48.7846×10^{16}	1440.1	36.5228×10^{16}

Table A.1a (continued)

	AM0		AM1	
λ (μm)	E_λ (W m^{-2} μm^{-1})	N_{ph} (cm^{-2} s^{-1} μm^{-1})	E_λ (W m^{-2} μm^{-1})	N_{ph} (cm^{-2} s^{-1} μm^{-1})
0.510	1882.0	48.5047×10^{16}	1416.8	36.4474×10^{16}
0.515	1833.0	47.8617×10^{16}	1384.9	36.0953×10^{16}
0.520	1833.0	47.6912×10^{16}	1390.0	36.0988×10^{16}
0.525	1852.0	48.4015×10^{16}	1409.5	36.7707×10^{16}
0.530	1842.0	48.9840×10^{16}	1406.9	37.3467×10^{16}
0.535	1818.0	48.9932×10^{16}	1393.6	37.4878×10^{16}
0.540	1783.0	48.6560×10^{16}	1371.7	37.3588×10^{16}
0.545	1754.0	48.2359×10^{16}	1354.2	37.1745×10^{16}
0.550	1725.0	47.8822×10^{16}	1336.6	37.0340×10^{16}
0.555	1720.0	47.8472×10^{16}	1335.7	37.1132×10^{16}
0.560	1695.0	47.8598×10^{16}	1319.2	37.2073×10^{16}
0.565	1705.0	48.0769×10^{16}	1330.0	37.4604×10^{16}
0.570	1712.0	48.7468×10^{16}	1338.4	38.0673×10^{16}
0.575	1714.0	49.3058×10^{16}	1346.9	38.6459×10^{16}
0.580	1715.0	49.7800×10^{16}	1346.7	39.1039×10^{16}
0.585	1712.0	50.2075×10^{16}	1347.3	39.4483×10^{16}
0.590	1700.0	50.3909×10^{16}	1340.7	39.6983×10^{16}
0.595	1682.0	50.3729×10^{16}	1329.4	39.7696×10^{16}
0.600	1666.0	50.2873×10^{16}	1319.6	39.7883×10^{16}
0.605	1647.0	50.1780×10^{16}	1311.0	39.8425×10^{16}
0.610	1635.0	50.1210×10^{16}	1307.9	39.9945×10^{16}
0.620	1602.0	50.0441×10^{16}	1294.2	40.2285×10^{16}
0.630	1570.0	49.8366×10^{16}	1280.9	40.4585×10^{16}
0.640	1544.0	49.7018×10^{16}	1272.1	40.7530×10^{16}
0.650	1511.0	49.5343×10^{16}	1257.1	41.0089×10^{16}
0.660	1486.0	49.3473×10^{16}	1244.2	41.1853×10^{16}
0.670	1456.0	49.1812×10^{16}	1226.8	41.3076×10^{16}
0.680	1427.0	48.9197×10^{16}	1204.0	41.2466×10^{16}
0.690	1402.0	48.7146×10^{16}	1196.2	41.3307×10^{16}
0.698	1374.0	48.4300×10^{16}	1010.3	38.4910×10^{16}
0.700	1369.0	48.1990×10^{16}	1175.3	38.4046×10^{16}
0.710	1344.0	48.0811×10^{16}	1157.4	41.3412×10^{16}
0.720	1314.0	47.7745×10^{16}	1135.1	41.2051×10^{16}
0.728	1295.5	47.9846×10^{16}	1003.1	38.9118×10^{16}
0.730	1290.0	47.3813×10^{16}	1117.8	38.8672×10^{16}
0.740	1260.0	47.1154×10^{16}	1095.1	40.8869×10^{16}
0.750	1235.0	46.4128×10^{16}	1076.6	40.6716×10^{16}
0.762	1205.0	46.3710×10^{16}	796.0	35.5879×10^{16}
0.770	1185.0	46.0216×10^{16}	1039.2	35.3384×10^{16}
0.780	1159.0	45.6662×10^{16}	1019.4	40.1060×10^{16}
0.790	1134.0	45.2490×10^{16}	1000.3	39.8558×10^{16}
0.800	1109.0	44.8262×10^{16}	981.2	39.6601×10^{16}
0.806	1095.1	44.4900×10^{16}	874.4	37.4572×10^{16}
0.825	1048.0	43.9320×10^{16}	931.6	37.0235×10^{16}

(*Continued*)

Table A.1a (continued)

λ (μm)	AM0 E_λ (W m^{-2} μm^{-1})	AM0 N_{ph} (cm^{-2} s^{-1} μm^{-1})	AM1 E_λ (W m^{-2} μm^{-1})	AM1 N_{ph} (cm^{-2} s^{-1} μm^{-1})
0.830	1036.0	43.3512×10^{16}	921.8	38.5543×10^{16}
0.835	1024.5	43.1213×10^{16}	912.4	38.3854×10^{16}
0.846	998.1	42.7350×10^{16}	476.2	29.3393×10^{16}
0.860	968.0	42.1590×10^{16}	506.4	21.0698×10^{16}
0.870	947.0	41.6409×10^{16}	453.8	20.8792×10^{16}
0.875	936.0	41.3001×10^{16}	449.2	19.8056×10^{16}
0.887	912.5	40.9384×10^{16}	448.6	19.8834×10^{16}
0.900	891.0	40.5085×10^{16}	448.9	20.1588×10^{16}
0.907	882.8	40.2873×10^{16}	455.2	20.5343×10^{16}
0.915	874.5	40.2438×10^{16}	461.5	20.9933×10^{16}
0.925	863.5	40.1951×10^{16}	279.9	17.1465×10^{16}
0.930	858.0	40.1380×10^{16}	221.8	11.6975×10^{16}
0.940	847.0	40.0748×10^{16}	313.4	12.5795×10^{16}
0.950	837.0	40.0045×10^{16}	296.5	14.4886×10^{16}
0.955	828.5	39.8791×10^{16}	321.1	14.7879×10^{16}
0.965	811.5	39.5777×10^{16}	344.4	16.0603×10^{16}
0.975	794.0	39.1487×10^{16}	576.9	22.4651×10^{16}
0.985	776.0	38.6777×10^{16}	544.6	27.6287×10^{16}
1.018	719.2	37.6431×10^{16}	617.5	29.2570×10^{16}
1.082	620.0	35.3484×10^{16}	512.9	29.8371×10^{16}
1.094	602.0	33.4222×10^{16}	464.1	26.8937×10^{16}
1.098	596.0	33.0067×10^{16}	503.7	26.6644×10^{16}
1.101	591.8	32.8302×10^{16}	504.8	27.8745×10^{16}
1.128	560.5	32.2835×10^{16}	135.1	17.9278×10^{16}
1.131	557.0	31.7299×10^{16}	152.2	8.1575×10^{16}
1.137	550.1	31.5599×10^{16}	143.1	8.4181×10^{16}
1.144	542.0	31.3107×10^{16}	191.2	9.5844×10^{16}
1.147	538.5	31.1139×10^{16}	174.5	10.5307×10^{16}
1.178	507.0	30.5529×10^{16}	399.3	16.7683×10^{16}
1.189	496.0	29.8404×10^{16}	402.2	23.8455×10^{16}
1.193	492.0	29.5804×10^{16}	424.0	24.7362×10^{16}
1.222	464.3	29.0280×10^{16}	391.8	24.7632×10^{16}
1.236	451.2	28.2843×10^{16}	340.8	23.7570×10^{16}
1.264	426.5	27.5798×10^{16}	324.2	22.8962×10^{16}
1.276	416.7	26.9197×10^{16}	342.6	21.2880×10^{16}
1.288	406.8	26.5391×10^{16}	347.3	22.2336×10^{16}
1.314	386.1	25.9317×10^{16}	298.3	21.1143×10^{16}
1.335	369.7	25.1648×10^{16}	190.0	16.2782×10^{16}
1.384	343.7	24.3808×10^{16}	5.7	6.7086×10^{16}
1.432	321.0	23.5268×10^{16}	44.6	1.7804×10^{16}
1.457	308.6	22.9413×10^{16}	85.4	4.7369×10^{16}
1.472	301.4	22.4571×10^{16}	77.4	5.9935×10^{16}
1.542	270.4	21.5323×10^{16}	237.3	11.8507×10^{16}
1.572	257.3	20.6453×10^{16}	222.6	18.0789×10^{16}

Table A.1a (continued)

	AM0		AM1	
λ (μm)	E_λ (W m^{-2} μm^{-1})	N_{ph} (cm^{-2} s^{-1} μm^{-1})	E_λ (W m^{-2} μm^{-1})	N_{ph} (cm^{-2} s^{-1} μm^{-1})
1.599	245.0	20.0200×10^{16}	216.0	17.4812×10^{16}
1.608	241.5	19.6104×10^{16}	208.5	17.1113×10^{16}
1.626	233.6	19.3121×10^{16}	206.7	16.8773×10^{16}
1.644	225.6	18.8736×10^{16}	197.9	16.6295×10^{16}
1.650	223.0	18.5733×10^{16}	195.7	16.2754×10^{16}
1.676	212.1	18.1893×10^{16}	181.9	15.7856×10^{16}
1.732	187.9	17.1342×10^{16}	161.5	14.7097×10^{16}
1.782	166.6	15.6575×10^{16}	136.7	13.1709×10^{16}
1.862	138.2	13.9604×10^{16}	4.0	6.444×10^{16}
1.955	112.9	12.0469×10^{16}	42.7	2.2405×10^{16}
2.008	102.0	10.7045×10^{16}	69.4	5.5839×10^{16}
2.014	101.2	10.2724×10^{16}	74.7	7.2847×10^{16}
2.057	95.6	10.0700×10^{16}	69.5	7.3786×10^{16}
2.124	87.4	9.6169×10^{16}	70.0	7.3309×10^{16}
2.156	83.8	9.2099×10^{16}	66.0	7.3162×10^{16}
2.201	78.9	8.9101×10^{16}	66.1	7.2343×10^{16}
2.266	72.4	8.8810×10^{16}	61.6	7.1099×10^{16}
2.320	67.6	8.0699×10^{16}	57.2	6.8479×10^{16}
2.338	66.3	7.8394×10^{16}	54.7	6.5514×10^{16}
2.356	65.1	7.7525×10^{16}	52.0	6.2952×10^{16}
2.388	62.8	7.6264×10^{16}	36.0	5.2473×10^{16}
2.415	61.0	7.4737×10^{16}	32.5	4.1353×10^{16}
2.453	58.3	7.2996×10^{16}	29.6	3.7997×10^{16}
2.494	55.4	7.0698×10^{16}	20.3	3.1028×10^{16}

Table A.1b. Energy Density E_λ and Photon Flux N_{ph} Spectra for AM1.5 and AM2 Conditions

λ (μm)	AM1.5 E_λ (W m^{-2} μm^{-1})	AM1.5 N_{ph} (cm^{-2} s^{-1} μm^{-1})	AM2 E_λ (W m^{-2} μm^{-1})	AM2 N_{ph} (cm^{-2} s^{-1} μm^{-1})
0.295	0.00	—	0.00	—
0.305	1.32	9.979×10^{13}	0.00	—
0.315	20.96	1.740×10^{15}	8.00	7.08×10^{14}
0.325	113.48	1.084×10^{16}	75.00	10.16×10^{15}
0.335	182.23	2.459×10^{16}	138.00	2.04×10^{16}
0.345	243.43	3.569×10^{16}	193.00	3.18×10^{16}
0.355	286.01	4.590×10^{16}	236.00	4.02×10^{16}
0.365	355.88	5.823×10^{16}	288.00	4.96×10^{16}
0.375	386.80	6.924×10^{16}	334.00	6.14×10^{16}
0.385	381.78	7.359×10^{16}	353.00	6.72×10^{16}
0.395	492.18	8.589×10^{16}	421.00	7.84×10^{16}
0.405	751.72	12.593×10^{16}	630.00	10.76×10^{16}
0.415	822.45	16.264×10^{16}	725.00	14.74×10^{16}
0.425	842.26	17.619×10^{16}	737.00	15.70×10^{16}
0.435	890.55	18.777×10^{16}	771.00	16.44×10^{16}
0.445	1077.07	21.817×10^{16}	949.00	20.20×10^{16}
0.455	1162.43	25.396×10^{16}	1066.00	23.80×10^{16}
0.465	1180.61	27.161×10^{16}	1097.00	25.40×10^{16}
0.475	1212.72	28.347×10^{16}	1131.00	26.60×10^{16}
0.485	1180.43	28.948×10^{16}	1130.00	27.80×10^{16}
0.495	1253.83	30.058×10^{16}	1157.00	28.40×10^{16}
0.505	1247.28	31.451×10^{16}	1161.00	29.40×10^{16}
0.515	1211.01	31.530×10^{16}	1127.00	29.36×10^{16}
0.525	1244.87	32.182×10^{16}	1156.00	30.16×10^{16}
0.535	1299.51	33.983×10^{16}	1153.00	31.02×10^{16}
0.545	1273.47	35.013×10^{16}	1130.00	30.99×10^{16}
0.555	1276.14	35.338×10^{16}	1122.00	31.22×10^{16}
0.565	1277.74	36.040×10^{16}	1124.00	31.68×10^{16}
0.575	1292.51	36.919×10^{16}	1144.00	32.86×10^{16}
0.585	1284.55	37.666×10^{16}	1150.00	33.70×10^{16}
0.595	1262.61	37.870×10^{16}	1141.00	34.18×10^{16}
0.605	1261.79	38.169×10^{16}	1134.00	34.46×10^{16}
0.615	1255.43	38.695×10^{16}	1137.00	34.82×10^{16}
0.625	1240.19	38.992×10^{16}	1137.00	35.44×10^{16}
0.635	1243.79	39.436×10^{16}	1139.00	36.04×10^{16}
0.645	1233.96	39.664×10^{16}	1140.50	36.70×10^{16}
0.655	1188.32	39.677×10^{16}	1138.00	37.23×10^{16}
0.665	1228.40	40.195×10^{16}	1131.50	37.65×10^{16}
0.675	1210.08	41.171×10^{16}	1123.00	37.97×10^{16}
0.685	1200.72	41.210×10^{16}	1116.50	38.28×10^{16}
0.6983	973.53	37.755×10^{16}	887.00	34.83×10^{16}
0.700	1173.31	37.731×10^{16}	1102.00	35.00×10^{16}
0.710	1152.70	41.324×10^{16}	1089.00	38.90×10^{16}

Table A.1b (continued)

	AM1.5		AM2	
λ (μm)	E_λ ($W\,m^{-2}\,\mu m^{-1}$)	N_{ph} ($cm^{-2}\,s^{-1}\,\mu m^{-1}$)	E_λ ($W\,m^{-2}\,\mu m^{-1}$)	N_{ph} ($cm^{-2}\,s^{-1}\,\mu m^{-1}$)
0.720	1133.83	41.199×10^{16}	1071.00	38.90×10^{16}
0.7277	974.30	38.455×10^{16}	906.00	35.97×10^{16}
0.730	1110.93	38.299×10^{16}	1058.00	36.04×10^{16}
0.740	1086.44	40.700×10^{16}	1040.00	38.80×10^{16}
0.750	1070.44	40.493×10^{16}	1026.00	38.80×10^{16}
0.7621	733.08	34.361×10^{16}	656.00	32.04×10^{16}
0.770	1036.01	34.152×10^{16}	997.00	31.90×10^{16}
0.780	1018.42	40.103×10^{16}	981.00	38.54×10^{16}
0.790	1003.58	39.850×10^{16}	965.00	38.50×10^{16}
0.800	988.11	39.804×10^{16}	950.00	38.4×10^{16}
0.8059	860.28	37.402×10^{16}	812.00	35.59×10^{16}
0.825	932.74	36.846×10^{16}	906.00	35.29×10^{16}
0.830	923.87	38.716×10^{16}	898.00	37.60×10^{16}
0.855	914.95	38.576×10^{16}	889.00	37.40×10^{16}
0.8165	407.11	28.010×10^{16}	359.00	26.43×10^{16}
0.860	857.46	27.190×10^{16}	397.00	16.22×10^{16}
0.870	843.02	37.067×10^{16}	344.00	16.20×10^{16}
0.875	835.10	36.896×10^{16}	340.00	15.04×10^{16}
0.8875	817.12	36.692×10^{16}	344.00	15.26×10^{16}
0.900	807.83	36.598×10^{16}	348.00	15.58×10^{16}
0.9075	793.87	36.477×10^{16}	357.00	16.05×10^{16}
0.915	778.97	36.117×10^{16}	288.00	16.57×10^{16}
0.925	217.12	23.093×10^{16}	180.00	12.63×10^{16}
0.930	163.72	8.880×10^{16}	131.00	7.26×10^{16}
0.940	249.12	9.727×10^{16}	214.00	8.13×10^{16}
0.950	231.30	11.441×10^{16}	199.00	9.84×10^{16}
0.955	255.61	11.68×10^{16}	224.00	10.14×10^{16}
0.965	279.69	12.950×10^{16}	250.00	11.47×10^{16}
0.975	529.64	19.783×10^{16}	523.00	18.89×10^{16}
0.985	496.64	25.345×10^{16}	487.00	24.90×10^{16}
1.018	583.03	27.299×10^{16}	580.00	26.94×10^{16}
1.082	486.20	28.345×10^{16}	464.00	27.66×10^{16}
1.094	448.74	25.634×10^{16}	434.00	24.58×10^{16}
1.098	486.72	25.837×10^{16}	465.00	24.83×10^{16}
1.101	500.57	27.355×10^{16}	491.00	26.50×10^{16}
1.128	100.86	16.891×10^{16}	78.00	15.96×10^{16}
1.131	116.87	6.197×10^{16}	93.00	4.87×10^{16}
1.137	108.68	6.446×10^{16}	85.00	5.08×10^{16}
1.144	155.44	7.591×10^{16}	129.00	6.15×10^{16}
1.147	139.19	8.505×10^{16}	114.00	7.00×10^{16}
1.178	374.29	15.042×10^{16}	343.00	13.39×10^{16}
1.189	383.37	22.596×10^{16}	356.00	20.82×10^{16}
1.193	424.85	24.257×10^{16}	415.00	23.14×10^{16}
1.222	382.57	24.569×10^{16}	361.00	23.62×10^{16}

(*Continued*)

Table A.1b (continued)

λ	AM1.5		AM2	
(μm)	E_λ ($Wm^{-2}\mu m^{-1}$)	N_{ph} ($cm^{-2}s^{-1}\mu m^{-1}$)	E_λ ($W m^{-2}\mu m^{-1}$)	N_{ph} ($cm^{-2}s^{-1}\mu m^{-1}$)
1.236	383.81	23.735×10^{16}	369.00	22.57×10^{16}
1.264	323.88	22.292×10^{16}	305.00	21.21×10^{16}
1.276	344.11	21.378×10^{16}	328.00	20.25×10^{16}
1.288	345.69	22.285×10^{16}	323.00	21.00×10^{16}
1.314	284.24	20.652×10^{16}	251.00	18.81×10^{16}
1.335	175.28	15.337×10^{16}	151.00	13.43×10^{16}
1.384	2.42	6.088×10^{16}	1.00	5.20×10^{16}
1.432	30.06	1.152×10^{16}	21.00	7.81×10^{16}
1.457	67.14	3.538×10^{16}	52.00	2.66×10^{16}
1.472	59.89	4.688×10^{16}	46.00	3.62×10^{16}
1.542	240.85	11.421×10^{16}	230.00	10.49×10^{16}
1.572	226.14	18.323×10^{16}	219.00	17.63×10^{16}
1.599	220.46	17.843×10^{16}	214.00	17.29×10^{16}
1.608	211.76	17.466×10^{16}	205.00	16.89×10^{16}
1.626	211.26	17.237×10^{16}	205.00	16.72×10^{16}
1.644	201.85	17.021×10^{16}	196.00	16.50×10^{16}
1.650	199.68	16.665×10^{16}	193.00	16.15×10^{16}
1.676	180.50	15.932×10^{16}	169.00	15.15×10^{16}
1.732	161.59	14.689×10^{16}	150.00	13.22×10^{16}
1.782	136.65	13.205×10^{16}	121.00	11.98×10^{16}
1.862	2.01	6.367×10^{16}	1.00	5.60×10^{16}
1.955	39.43	1.993×10^{16}	30.00	1.47×10^{16}
2.008	72.58	5.593×10^{16}	59.00	4.45×10^{16}
2.014	80.01	7.733×10^{16}	67.00	6.38×10^{16}
2.057	72.57	7.827×10^{16}	62.00	6.60×10^{16}
2.124	70.29	7.526×10^{16}	60.00	6.43×10^{16}
2.156	64.76	7.283×10^{16}	56.00	6.25×10^{16}
2.201	68.29	7.304×10^{16}	64.00	6.58×10^{16}
2.266	62.52	7.362×10^{16}	60.00	6.98×10^{16}
2.320	57.03	6.908×10^{16}	55.00	6.65×10^{16}
2.338	53.57	6.491×10^{16}	52.00	6.27×10^{16}
2.356	50.01	6.126×10^{16}	49.00	5.94×10^{16}
2.388	31.93	4.898×10^{16}	30.00	4.72×10^{16}
2.415	28.10	3.633×10^{16}	26.00	3.39×10^{16}
2.453	24.96	3.016×10^{16}	23.00	3.00×10^{16}
2.494	15.82	2.542×10^{16}	14.00	2.31×10^{16}

APPENDIX B

Antireflection Coatings

B.1. *Introduction*

Photon loss analysis (see Chapter 3) shows that reflection losses in solar cells can result in a large fraction (20–35%) of the incident radiation being sent back to the incident medium without effecting carrier generation, particularly in cells made of high refractive index materials like Si, GaAs, and Cu_2S and in M-S or MIS cells having a top metallic layer. To reduce the reflectance and, consequently, increase the photovoltaic output, we need to introduce additional layers of appropriate thickness and refractive index (called antireflection or AR coatings) such that the effective optical admittance of the surface is reduced to a value close to that of the incident medium. Physically, this means that the waves reflected at the top of the layer(s) and those reflected at the bottom of the layer(s) interfere destructively on emerging in the incident medium. Since a destructive interference condition requires a phase difference of π or a path difference of $\lambda/2$, the condition for interference is satisfied only over a limited range of wavelength decided by the thickness and refractive index of the layer(s).

The basic criterion for an antireflection coating in most solar cell applications is that the effective optical impedance of the reflecting surface should match the refractive index of the incident medium, air, or the encapsulant material. Several designs to achieve this are possible.[1–3] The choice of a particular design depends on the availability of optical materials of required refractive indices and the wavelength range over which the minimum reflectance condition is required. The various designs[4–8] of AR coatings are the subject of this Appendix.

B.2. *Single Layer AR Coatings*

(i) n_s real

Consider a layer of a material of refractive index n_f on a substrate of refractive index n_s. If λ_0 is the wavelength at which the layer is a quarter wavelength in optical thickness, i.e., $n_f t_f = \lambda_0/4$ (where t_f is the thickness of the layer), then the normal reflectance at λ_0 is given by

$$R_{\lambda_0} = \left(\frac{n_0 n_s - n_f^2}{n_0 n_s + n_f^2}\right)^2 \qquad \text{for } n_f t_f = (2m+1)\lambda_0/4 \tag{B.1a}$$

$$= R_{\min}, \qquad \text{if } n_s > n_f > n_0 \tag{B.1b}$$

$$= R_{\max}, \qquad \text{if } n_s < n_f > n_0 \tag{B.1c}$$

where n_0 is the refractive index of the incident medium.

Clearly, the condition for the minimum reflectance to be zero follows from equation (B.1a) as

$$n_f^2 = n_0 n_s \tag{B.2}$$

For Si, $n_s = 3.5$ at $\lambda = 1.7$ μm and, therefore, the refractive index for an AR coating is $n_f = (3.5)^{\frac{1}{2}} = 1.87$ and the thickness $t_f = 2272$ Å. Coatings of SiO ($n_f = 2.0$), TiO_2 ($n_f = 2.2$), and Ta_2O_5 ($n_f = 1.98$) have been used as AR coatings on Si in the infrared-wavelength region.

(ii) n_s complex

If the substrate is absorbing in the wavelength region of interest, then the extinction coefficient $k_s \neq 0$. The conditions for minimum reflectance to be zero are now given by

$$n_f = \left[\frac{n_s(n_s - n_0) + k_s^2 n_0}{n_s - n_0}\right]^{1/2} \tag{B.3a}$$

and

$$t_f = \frac{\lambda}{2\pi n_f} \tan^{-1}\left[\frac{n_f(n_s - n_0)}{n_0 k_s}\right] \tag{B.3b}$$

Since n_f has to be real, the optical constants of the substrate should be such that $n_s > n_0$ or $k_s^2 < (n_s/n_0)(n_0 - n_s)$.

Yeh et al.[6] have determined, by ellipsometric techniques, that the complex refractive index for an Au/GaAs Schottky barrier solar cell is $n_s = 2.89 - i2.23$. For $n_0 = 1.0$ (air), the AR condition at 6000 Å is satisfied by $n_f = 2.35$ and $t_f = 449$ Å. A Ta_2O_5 610 Å thick coating with refractive index ($n_f = 1.975$) slightly less than the optimum gives a near optimum reflectance as shown in Figure B.1a.

In the case of Si solar cells, k_s is very small ($\simeq 0.1$) at wavelengths near the peak of the solar spectrum and the refractive index of the AR coating is nearly equal to $n_s^{1/2}$. n_s is 4.00 at $\lambda = 5600$ Å. Thus, a film of refractive index 2.00 and thickness 693 Å acts as an AR coating for Si. The materials that have been successfully used for this purpose are SiO, TiO_2,[5] a mixture of the two and SnO_2.[8] The computed reflectance[9] of Si with a single-layer coating is shown in Figure B.1b.

For Cu_2S films ($n_s = 3.46 - i0.84$), used in Cu_2S/CdS solar cells, the AR condition at 5500 Å is satisfied by a film with refractive index 1.94 and thickness 630 Å.

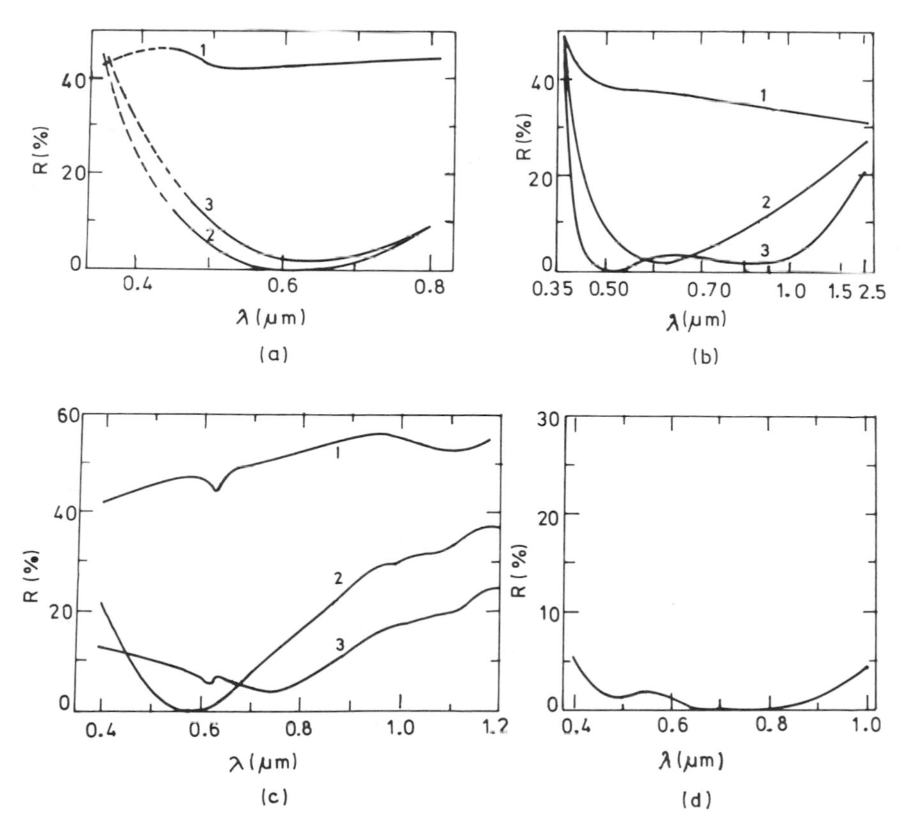

Figure B.1. Reflectance spectra of (a) 60-Å Au/GaAs solar cell: curve 1 — uncoated, curve 2 — with ideal AR coating ($n = 2.35$ and $t = 449$ Å) at 0.6 μm, and curve 3 — with Ta_2O_5 AR coating ($n = 1.975$ and $t = 610$ Å) (after Yeh et al.[6]); (b) Si: curve 1 — uncoated, curve 2 — with single-layer AR coating, and curve 3 — with two-layer (MgF_2 : $n = 1.38$, $t = 119$ ÅR, and ZnS : $n = 2.30$ and $t = 61$ Å) AR coating (after Apfel[9]); (c) 100-Å Ag/Si solar cell: curve 1 — uncoated, curve 2 — with single-layer (ZnS : 420 Å) AR coating, and curve 3 — with two-layer (ZnS : 420 Å and cryolite : 790 Å) AR coating (after Gandham et al.[7]); and (d) Cu_2S/CdS solar cell with a two-layer (ZnS : 600 Å and MgF_2 : 1000 Å) AR coating (after Rothwarf et al.[10]).

B.3. *Double Layer AR Coatings*

(i) Quarter-wavelength-thick coatings

The optical admittance at normal incidence for two quarter-wavelength ($\lambda/4$) coatings with refractive indices n_{f1} and n_{f2} is given by

$$n_{\text{eff}} = n_{f1}^2 n_s / n_{f2}^2 \tag{B.4}$$

n_{eff} should be equal to n_0 for zero reflectance. Thus,

$$n_{f2} = \left(\frac{n_{f1}^2 n_s}{n_0}\right)^{1/2} \tag{B.5}$$

The condition for low reflectance is satisfied only at wavelengths near λ_0. To increase the range of low reflectance, the refractive indices should be such that

$$\frac{n_{f1}}{n_0} = \frac{n_{f2}}{n_{f1}} = \frac{n_s}{n_{f2}} \tag{B.6}$$

Then, for $\lambda = \frac{3}{4}\lambda_0$ and $\lambda = \frac{3}{2}\lambda_0$ (λ_0 is the wavelength at which the layers are quarter wave) the reflectance is zero. At $\lambda = \lambda_0$,

$$R = \left[\frac{1-(n_s/n_0)^{1/3}}{1+(n_s/n_0)^{1/3}}\right]^2 \tag{B.7}$$

For an air/Si ($n_s = 3.5$) interface, n_{f1} and n_{f2} can be calculated to be 1.51 and 2.30, respectively. The reflectance at λ_0 is 4%. The theoretical curve for the stack is shown in Figure B.1b.

(ii) Non-quarter-wavelength coatings

Conditions for zero reflectance in non-quarter-wavelength double-layer coatings can be easily determined by the matrix method (see Chapter 2). The effective optical admittance is given by $Y = Y1/Y2$, where $Y1$ and $Y2$ are defined by

$$\begin{pmatrix} Y1 \\ Y2 \end{pmatrix} = \begin{pmatrix} \cos\delta_2 & i\sin\delta_2/n_{f2} \\ in_{f2}\sin\delta_2 & \cos\delta_2 \end{pmatrix} \begin{pmatrix} \cos\delta_1 & i\sin\delta_1/n_{f1} \\ in_{f1}\sin\delta_1 & \cos\delta_1 \end{pmatrix} \begin{pmatrix} 1 \\ n_s \end{pmatrix} \tag{B.8}$$

The reflectance will be zero if $Y = n_0$. Therefore, for n_s real

$$\tan^2\delta_1 = \frac{(n_s - n_0)(n_{f1}^2 - n_0 n_s)n_{f2}^2}{(n_{f2}^2 n_s - n_0 n_{f1}^2)(n_0 n_s - n_{f2}^2)} \tag{B.9a}$$

and

$$\tan^2 \delta_2 = \frac{(n_s - n_0)(n_0 n_s - n_{f2}^2)n_{f1}^2}{(n_f^2 n_s - n_0 n_{f1}^2)(n_{f1} - n_0 n_s)} \tag{B.9b}$$

The values of δ_1 and δ_2 as obtained from these equations are so chosen that

$$\tan \delta_1 \tan \delta_2 = \frac{n_{f1} n_{f2}(n_s - n_0)}{n_{f2}^2 n_s - n_0 n_{f1}^2} \tag{B.10a}$$

and

$$\frac{\tan \delta_2}{\tan \delta_1} = \frac{n_{f1}(n_0 n_s - n_{f2}^2)}{n_{f2}(n_{f1}^2 - n_0 n_s)} \tag{B.10b}$$

For real solutions of equations (B.10a) and (B.10b), the expressions $(n_{f1}^2 - n_0 n_s)$, $(n_{f2}^2 n_s - n_0 n_{f1}^2)$ and $(n_0 n_s - n_{f2}^2)$ should either all be positive, or only two of them should be negative. This gives one possible condition, namely,

$$n_{f2} \geqslant (n_0 n_s)^{\frac{1}{2}} \geqslant n_{f1} \tag{B.11}$$

which means that n_{f2} can be chosen conveniently.

Gandham et al.[7] have designed single- and double-layer AR coatings for Ag/Si Schottky barrier solar cells, for which n_s is complex. The reflectance curve for a 100-Å Ag on Si with a 420-Å ZnS and 790-Å cryolite stack is shown in Figure B.1c. A two-layer (600-Å ZnS + 1000-Å MgF_2) AR coating on planar junction Cu_2S/CdS solar cells has been reported by Rothwarf et al.[10] The reflectance curve of this stack is shown in Figure B.1d.

We should point out that the double-layer coatings described above are restricted to near normal incidence. Different designs have to be developed for applications at higher angles of incidence.[1]

B.4. *Multilayer AR Coatings*

A multilayer AR coating can be designed using the equivalent film concept of Berning.[11] Here, a symmetric combination of three films can be equated to a layer whose equivalent refractive index and thickness can be determined using the matrix method. A combination can thus be obtained for which the AR condition is satisfied.[2] Though the equivalence is valid for a limited range of wavelengths, it has been shown to be sufficiently good over a wide range.

B.5. *Inhomogeneous AR Coatings*

If the optical admittance of the coating is graded in such a manner that it continuously decreases from the refractive index of the incident medium to that of the substrate, a minimum reflectance will be achieved over a wide range of wavelengths. In the limit, the condition for a smoothly varying optical admittance or gradient index is achieved if the number of layers tends to infinity.

The lower- and higher-wavelength limits for a multilayer with n layers will be

$$\lambda_s = \left(\frac{n+1}{n}\right)\frac{\lambda_0}{2} \tag{B.12a}$$

and

$$\lambda_L = (n+1)\frac{\lambda_0}{2} \tag{B.12b}$$

where λ_0 is the wavelength at which the layers are quarter wave. For wavelengths longer than $2\lambda_L$, the AR condition cannot be achieved. If n is increased to infinity, keeping the total thickness of the multilayer $[T = (n\lambda_0)/4]$ finite, λ_s goes to zero and λ_L to $2T$. For all wavelengths within these limits the reflectance will be zero, and thus a perfect AR coating will be formed. However, since in practice the inhomogeneous layer has to be terminated at an MgF_2 layer (lowest refractive index material), the reflectance of such an inhomogeneous stack will be the reflectance for MgF_2/air interface, i.e., 2.2%.

B.6. *Moths'-Eye Coating*

If the substrate surface topography consists of cones of specific dimensions, light of wavelengths longer than the cone diameter d is scattered or diffracted. For wavelengths greater than d, the surface will appear to have gradient index varying continuously from that of the incident medium at the top of the cones to that of the substrate at the bottom of the cones. The effective refractive index is then simply the average $(n_s + n_0)/2$ and the optical thickness is $(n_s + n_0)(d/2)$. The film will then act as an AR coating for $n_s = 1.5$ from $\lambda = d$ to $2.5d$. Clapham and Hutley[12] prepared such cones (with a diameter of 3000 Å) of photoresist on glass and obtained low reflectance in the visible region. A similar surface prepared by Hewig et al.[13] on Cu_2S/CdS solar cells reduced the reflectance from 30% to 2%.

APPENDIX C

Grid Design

C.1. *Introduction*

We noted in Chapter 3 that an excessive series resistance has a deleterious effect on the fill factor and, consequently, on the conversion efficiency of a solar cell. One of the main components of the series resistance is the lateral (sheet) resistance of the top semiconductor layer. In order to reduce this contribution of the top layer to the series resistance, it is necessary to deposit a grid pattern, which serves the dual purpose of providing a low-resistance ohmic contact to the top layer for collecting the current and distributing it to the external circuit and at the same time reducing the series resistance of the top layer, maintaining at the same time a high transmission of the incident light to the active layers below.

The grid adds its own contribution to the series resistance in the form of grid metal resistance and surface contact resistance between metal grid and semiconductor. Thus, in designing a grid pattern, it is imperative that the metal be sufficiently thick that its resistance is negligible and it has to be appropriate to the top layer semiconductor to form a good ohmic contact with negligible contact resistance.

Qualitatively, the more closely the grid lines are spaced the lower the series resistance owing to the top layer will be. However, the necessity for maintaining high transmission requires that the width of the grid lines be small. The lower limit of the grid line width and intergrid spacing is imposed by the deposition/fabrication technique. With screen printing techniques, grid lines about 50 μm wide can be deposited. Evaporation through photolithographically prepared masks yield grid line widths down to about 20 μm. The sheet resistance component of the series resistance, R_g, owing to a given grid pattern and the transmission T_g represent the design goals while designing a suitable grid structure. It is necessary, therefore, to formulate design equations relating R_g and T_g to the physical dimensions and geometry of the grid structure and the sheet resistance of the top semiconductor layer.

C.2. Grid Design

The sheet resistance component of the series resistance has been analyzed by several authors[1–7] for different grid geometries. Wyeth[8] has derived quantitative expressions for the series resistance under the following assumptions: (a) current generation is uniform over the area of the cell, (b) the thickness of the upper semiconductor layer is very much smaller than the lateral dimensions of the cell, (c) the resistance of the grid electrode is much less than the sheet resistance of the top semiconductor layer, and (d) the current flow in the top layer is ohmic. These assumptions are usually valid in most solar cells. The various grid geometries and the corresponding series resistances are discussed below. It is convenient to analyze the grid geometries by selecting a unit field, which on repetition generates the entire grid pattern. The expression for the effective series resistance of each unit is presented. The series resistance of the whole cell can be calculated by adding the unit resistances in parallel.

C.2.1. Parallel Grid Lines

The grid pattern and the basic unit for this grid (as well as for the subsequent grid geometries) is shown in Figure C.1a. No cross bars or bus bars touch the semiconductor layer. The effective unit series resistance R_{gA} is given by

$$R_{gA} = \frac{1}{12}\frac{a}{b}\frac{\rho}{t} \tag{C.1}$$

where ρ is the resistivity and t is the thickness of the top semiconductor layer, i.e., ρ/t is the sheet resistance of the top layer. It can be shown that each grid line collects the current from exactly one half of the area and for each of these half units

$$R = \frac{1}{3}\frac{(a/2)}{b}\frac{\rho}{t} \tag{C.2}$$

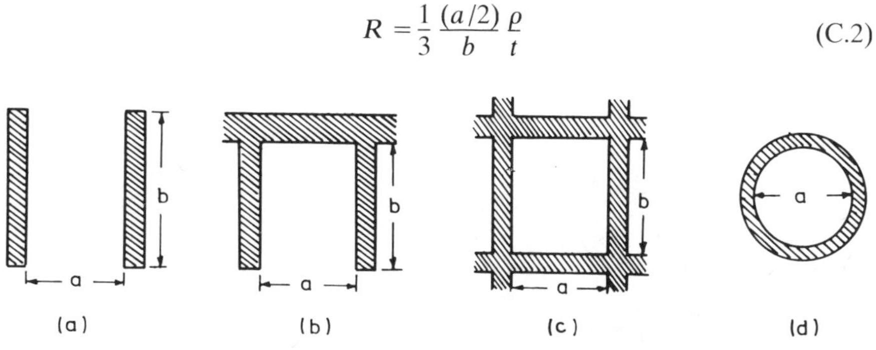

Figure C.1. Units of (a) parallel grid lines, (b) parallel grid lines with one bus bar, (c) rectangular mesh grid, and (d) circular grid along the circumference of a circular solar cell.

C.2.2. *Parallel Grid Lines with Bus Bar Along One Side*

The basic unit for this grid geometry is shown in Figure C.1b. The effective unit series resistance R_{gB} is expressed by

$$R_{gB} = \left[\frac{1}{12}\frac{a}{b} - \left(\frac{a}{b}\right)^2 \frac{8}{\pi^5} \sum_{m=0}^{\infty} (2m+1)^{-5} \tanh(2m+1)\pi \frac{b}{a}\right] \frac{\rho}{t} \quad \text{(C.3)}$$

R_{gB} is plotted as a function of a/b in Figure C.2. It approaches R_{gA} for $(a/b) \ll 1$ and $(1/3)(b/a)\rho/t$ for $(a/b) \gg 1$. At $(a/b) = 2$, R_{gB} is a maximum as given by

$$R_{gB_{max}} \simeq 7.029 \times 10^{-2} \frac{\rho}{t} \quad \text{(C.4)}$$

C.2.3. *Rectangular Mesh Grid*

The unit field of a rectangular mesh grid is shown in Figure C.1c. The effective unit series resistance R_{gC} is

$$R_{gC} = \left[\frac{1}{12}\frac{a}{b} - \left(\frac{a}{b}\right)^2 \frac{16}{\pi^5} \sum_{m=0}^{\infty} (2m+1)^{-5} \tanh(2m+1) \frac{\pi}{2}\frac{b}{a}\right] \frac{\rho}{t} \quad \text{(C.5)}$$

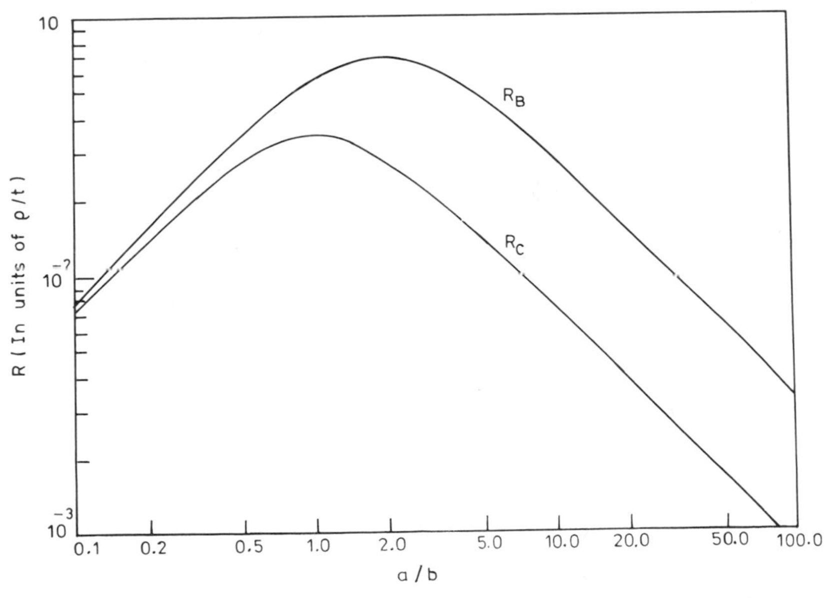

Figure C.2. Plots of R_{gB} and R_{gC} as a function of a/b (after Wyeth[8]).

R_{gC} is plotted as a function of a/b in Figure C.2 and exhibits asymptotes of $(1/12)(a/b)\rho/t$ and $(1/12)(b/a)\rho/t$. The maximum occurs at $(a/b)=1$ and is

$$R_{gC_{max}} \simeq 3.514 \times 10^{-2} \frac{\rho}{t} \tag{C.6}$$

C.2.4. Circular Grids

We consider the case of circular solar cells. For a circular grid (Figure C.1d) along the entire circumference, the effective series resistance R_{gD} is

$$R_{gD} = \frac{1}{8\pi} \frac{\rho}{t} \tag{C.7}$$

For a circular grid with an electrode along one diameter and the assumption that the electrode width is infinitesimal, the effective series resistance is given by

$$R_{gD} \simeq 7.328 \times 10^{-2} \frac{\rho}{t} \tag{C.8}$$

R_{gA}, R_{gB}, and R_{gC} can be expressed as a product of R_{gA} and a correction factor G, namely,

$$R_{g(A,B,C)} = \frac{1}{12} \frac{a}{b} \frac{\rho}{t} G_{A,B,C} \tag{C.9}$$

where $G_A \equiv 1$ and G_B and G_C are functions of a/b. Values of G_B and G_C for different values of a/b are tabulated in Table C.1.

Table C.1. Values of Correction Factor G_B and G_C for Different Values of a/b[8]

a/b	G_B (parallel lines and bus bar)	G_C (mesh grid)
1	0.6860	0.4217
0.5	0.8424	0.6860
0.2	0.9370	0.8740
0.1	0.9685	0.9370
0.05	0.9842	0.9685
0.02	0.9937	0.9874
0.01	0.9968	0.9934

For a cell consisting of N units, the total effective series resistance due to sheet resistance is given by

$$R_g = \frac{R_{g(A,B,C)}}{N} \tag{C.10}$$

If we neglect the grid line width, the area A of the cells described in Sections C.2.1 to C.2.3 is

$$A = Nab \tag{C.11}$$

The product of the series resistance and area is then given by

$$R_g A = \frac{1}{12} a^2(\rho/t)G_{A,B,C} \tag{C.12}$$

The above quantity is a measure of the current collection efficiency of a particular grid pattern and is extremely useful in directly comparing grids on cells of different areas.

Calculation of T_g is fairly straightforward. For a parallel line grid,

$$T_g = \frac{a}{a+\delta} \tag{C.13}$$

where δ is the grid line width. To obtain an expression for the grid spacing that gives the highest efficiency, we maximize the product $T_g \times \mathrm{FF}$. Thus,[9]

$$a = \left\{\left(\frac{6V_{oc}A}{CI_{sc}}\right)\left(\frac{t}{\rho}\right)\delta\left[\mathrm{FF}_0 - C\left(\frac{I_{sc}}{V_{oc}}\right)R_2\right]\right\}^{1/3} \tag{C.14}$$

where R_2 is the contribution to the series resistance from causes other than the sheet resistance of the top layer. The other terms in equation (C.14) have already been defined in Chapter 3. R_2 is primarily due to the base semiconductor layer (assuming base contact resistance and base metal resistance to be negligible) and is expressed by

$$R_2 A = \rho^* d \tag{C.15}$$

where ρ^* is the resistivity and d the thickness of the base layer.

Equations (C.1) to (C.15) represent the design equations, using which one can design a suitable grid pattern for a given device to achieve the desired values of R_g and T_g. The value of R_g is fixed by the maximum series resistance that can be tolerated to obtain a predetermined value of FF.

C.3. Grid Fabrication

After a suitable grid pattern has been designed, the grid has to be deposited on the device. This can be effected in several ways. The pattern can be directly generated on the device by screen printing techniques, which is particularly suitable when large areas need to be covered. However, resolution limitations make this technique unsuitable for patterns where the grid line spacing or grid line width is exceedingly fine ($\sim$10–20 μm), as in the case of concentrator cells or high-efficiency Cu_2S/CdS cells. In such cases, evaporation through a metal stencil mask is employed to deposit the grid pattern.[10,11] Another approach is to integrate the gridding and lamination process by laminating a metal grid onto the device using a cover glass coated with a transparent epoxy.[12] Some workers[13] have used conducting epoxy to bond the metal grid to the device. In yet another technique for grid fabrication, the entire cell surface is coated with a suitable metal film by evaporation and the grid pattern is generated by etching away the appropriate regions of the metal film, using photolithographic techniques.[14] This method is the best in terms of reproducibility, resolution, and accuracy but suffers from the disadvantage that the etchants used for the metal film may react or damage the cell surface.

All the above methods of grid deposition utilize photolithographic techniques, whether it be for grid fabrication or for stencil mask fabrication for evaporation. A brief description of the photolithographic technique for pattern generation is, therefore, appropriate.

C.3.1. Photolithography[15,16]

The lithographic technique involves the following basic steps: artwork generation and photoreduction; high-resolution photoresist application; precision exposure, developing, and etching.

C.3.1.1. Artwork

Photographic masks for contact printing are precision images of the pattern to be etched. In view of the small dimensions required, the pattern must first be produced greatly enlarged. The original artwork, usually a drawing on a graph paper, undergoes positive/negative as well as mirror image reversal in every reduction and contact printing step. Therefore, the number of mask processing steps planned must be considered while preparing the artwork so that the artwork represents a positive or negative of the right polarity with respect to the final mask. The nature of the photoresist (positive or negative), too, has to be taken into account. The original artwork is reduced to final scale in one or more photographic reduction steps.

C.3.1.2. *Photoresist*

There are many organic compounds that undergo structural and chemical solubility changes when exposed to light, particularly in the UV region. However, practical photoresist systems, in addition to being UV-sensitive, must also possess the ability to form adherent and uniform coatings, which are not physically or chemically destroyed by the etching processes. The principal components of a photoresist solution are a polymer, a sensitizer, and the solvent. The polymers are characterized by unsaturated carbon bonds, which undergo cross-linking on exposure. Several photoresist systems are commercially available.

The photoresist can be applied to the substrate by spraying, spinning, dipping, flowing, or rolling. After application, the photoresist undergoes a prebaking to remove residual solvents, which, if left in the film, may weaken the adhesion of the photoresist to the substrate or inhibit cross-linking.

C.3.1.3. *Exposure*

The photoresist coated sample is exposed to UV light through the mask. Fine line patterns are exposed with point sources such as xenon flash lamps, which produce nearly collimated light if they are remote from the substrate. The exposure times are preset with test patterns. Both over- and underexposure are undesirable; the former produces cross-linking (in negative resists) in regions under the mask, resulting in line broadening and the latter causes only the resist near the surface to cross-link, which may lead to the entire pattern being washed off when the image is developed. With positive resists, an underexposed pattern leaves a film of insoluble polymer behind.

C.3.1.4. *Developing*

The photoresist images are developed by immersing and soaking in a suitable developer. Mild agitation is usually employed. A widely preferred method is spraying the developer onto the photoresist. The continued renewal of the solvent and the gentle brushing action of the spray lead to shorter development times and, hence, less swelling and better dimensional control.

It is usually necessary to postbake the developed resist images to evaporate residual solvents, enhance chemical stability of the polymer, and improve adhesion.

C.3.1.5. *Etching*

The reagents used to etch thin film patterns are the same etchants normally employed to dissolve the bulk metal. For thin film etching, the chemicals are applied in fairly dilute solutions to reduce the etch rate and avoid undercutting, i.e., attack of material under the polymer stencil mask. Table C.2 lists the commonly employed etchants for various metals. While aqueous solutions are most common, alcohols, like methanol, ethylene glycol, or glycerine are sometimes added to moderate the reaction.

Table C.2. Some Common Etchants for Various Materials Used in Solar Cells

Materials	Etchants
Metals	
Al	(1) 20% NaOH; 60–90 °C
	(2) 16–19 H_3PO_4, 1 HNO_3, 0–4 H_2O; 40 °C
	(3) 0.1 *M* $K_2B_4O_7$, 0.51 *M* KOH, 0.6 *M* $K_3[Fe(CN)_6]$; pH 13.6
Cu	(1) $FeCl_3$, 42° Baumé, 49 °C
	(2) 20–30% H_2SO_4, 10–20% CrO_3 or $K_2Cr_2O_7$; 49 °C
Au	(1) 4 g KI, 1 g I_2, 40 ml H_2O
	(2) 3 HCl, 1 HNO_3; 32–38 °C
	(3) 0.4 *M* $K_3[Fe(CN)_6]$, 0.2 *M* KCN, 0.1 *M* KOH
Ag	(1) 5–9 HNO_3, 1–5 H_2O; 39–49 °C
	(2) 11 g Fe $(NO_3)_2$, 9 ml H_2O; 44–49 °C
	(3) 4 CH_3OH, 1 NH_4OH, 1 H_2O_2
Ni	(1) 8 HNO_3, 5 CH_3 COOH, 2 H_2SO_4, H_2O
	(2) 1 HNO_3, 1 HCl, 3H_2O
	(3) 3 HNO_3, 1 H_2SO_4, 1 H_3PO_4 (98%), 5CH_3COOH; 85–95 °C
	(4) 10 H_2SO_4, 10 H_2O_2, H_3PO_4, 5–7 $NiSO_4$ (30%); 50 °C
Fe	(1) 3 HNO_3, 7 HCl, 30 H_2O; 60–70 °C
W	(1) 34 g KH_2PO_4, 13.4 g KOH, 33 g $K_3[Fe(CN)_6]$, H_2O to make 1 liter
	(2) 5% KOH, 5% $K_3[Fe(CN)_6]$, 1% surfactant; 23 °C, 2 A cm^{-2}; Pt cathode electrochemical etching
Mo	(1) 5 H_3PO_4, 3 HNO_3, 2H_2O
	(2) 1 H_2SO_4, 1 HNO_3, 1–5 H_2O; 25–54 °C
	(3) 38 H_3PO_4, 15 HNO_3, 30 CH_3COOH, 75 H_2O
Ta	(1) 9 NaOH or KOH (30%), 1 H_2O_2; 90 °C; H_2O
	(2) 1–2 HNO_3, 1 HF, 1–2 H_2O
	(3) 9 H_2SO_4, 1 HF, 35–45 °C; 2 A dm^{-2}, Pt or carbon cathode electrochemical etching

Table C.2 (continued)

Materials	Etchants
Ti	(1) 9 H_2O, 1 HF; 32 °C (2) 7 H_2O, 2 HNO_3, 1 HF; 32 °C (3) 180 ml C_2H_5OH, 20 ml n-butyl alcohol, 12 g $AlCl_3$, 56 g $ZnCl_2$; 30–50 V, 12 A dm^{-2}; stainless steel cathode electrochemical etching
Cr	(1) 1 Vol (18 NaOH, 2 ml H_2O), 3 vol (1 g $K_3[Fe(CN)_6]$), 3 ml H_2O
Pt	(1) 8 H_2O, 7 HCl, 1 HNO_3; 85 °C (2) Aqua regia
Pd	(1) 1 HCl, 10 HNO_3, 10 CH_3COOH (2) Aqua regia
Pb	(1) $FeCl_3$, 36–42° Baumé; 43–54 °C (2) 1 H_2O_2, 4 CH_3COOH (3) 1 HNO_3, 19 H_2O
Semiconductors	
Si	(1) HF, HNO_3, H_2O or CH_3COOH; 25 °C (2) NaOH (4%), add NaOCl (40%); 80 °C
Ge	(1) 50 wt% HF, 50 wt% H_2O_2; 25 °C (2) 5 H_2O, 1 H_2O_2; 25 °C (3) 9 HF, 1 HNO_3
GaAs	(1) 4 H_2SO_4, 1 H_2O_2, 1 H_2O; 50 °C (2) 1 NaOH (1 *M*), 1 H_2O_2 (0.76 *M*); 30 °C (3) 2 Br_2, 98 CH_3OH (4) 40 HCl, 4 H_2O_2, 1 H_2O; 20 °C
InP	(1) 99 CH_3OH, 1 Br_2 or 90 CH_3OH, 10 Br_2 (2) 1 HCl, 1 HNO_3
CdS	(1) 100 ml H_2O, 1 ml H_2SO_4, 0.08 g Cr_2O_3; 80 °C (2) 2:1 HCl:H_2O; ~42 °C
CdSe	(1) 30 HNO_3, 20 H_2SO_4, 10 CH_3COOH, 0.1 HCl; 40 °C
CdTe	(1) 3 HF, 2 H_2O_2, 1 H_2O (2) 2 HNO_3, 2 HCl, 1 H_2O (3) 20 ml H_2O, 10 ml HNO_3, 4 g $K_2Cr_2O_7$
ZnO	(1) Mineral acids or alkali or NH_4Cl solution
SnO_2	(1) HCl, Zn powder (2) 3–10 H_2O, 1 HCl electrolytic, 5–40 mA cm^{-2}
Cu_2S	(1) KCN
Cu_2Se	(1) KCN
$CuInSe_2$	(1) Aqua regia
Zn_3P_2	(1) 36% HCl; room temperature

(*Continued*)

Table C.2 (continued)

Materials	Etchants
Dielectrics	
SiO_2	(1) HF
	(2) 15 ml 49% HF, 10 ml 70% HNO_3, 300 ml H_2O
	(3) BHF
SiO	(1) HF + HNO_3
TiO_2	(1) HF
	(2) BHF
	(3) H_3PO_4 (Hot)
	(4) H_2SO_4
Ta_2O_5	(1) HF
	(2) BHF
	(3) H_3PO_4
	(4) H_2SO_4
Si_3N_4	(1) 48% HF
	(2) BHF
	(3) H_3PO_4
Nb_2O_5	(1) HF
	(2) BHF
	(3) H_3PO_4
	(4) H_2SO_4
	(5) NaOH (~30%)
ZnS	(1) HCl

APPENDIX D

Solar Cell Arrays

D.1. *Introduction*

An arrangement of solar cells, electrically connected into circuits, having the appearance of rows and columns is termed a solar cell array or simply a solar array or solar battery. The solar cell array constitutes the power generating component within the framework of a power system, as shown in Figure D.1a.

The solar cell array is constructed from several subsystems which include: (a) an optical subsystem, comprised of sunlight concentrators (in a concentrating system) and the array coverglass; (b) an electrical subsystem, consisting of solar cells and the associated wiring; (c) a mechanical subsystem, which includes the solar cell mechanical support; (d) an orientation and structural subsystem made up of the structural supports and sun-tracking mechanisms; (e) a status sensor subsystem, which encompasses performance monitoring devices and their circuits; (f) a thermal control subsystem represented by heat radiators, cooling fans, and thermal control coatings; and (g) an environmental protection subsystem, containing packaging and encapsulation material to minimize degradation of the solar cells from environmental effects. Figure D.1b illustrates the subsystems of an array in the form of a block diagram.

A solar cell converts, at best, about 20% of the incident solar energy into electrical power. A large fraction of the rest of the 80% incident power is wasted as heat. In a photovoltaic/thermal hybrid system, the heat energy is extracted by a thermal system, typically by a cooling fluid. The relatively low temperature of the fluid is insufficient to drive motors, but is useful for water heating and space cooling.

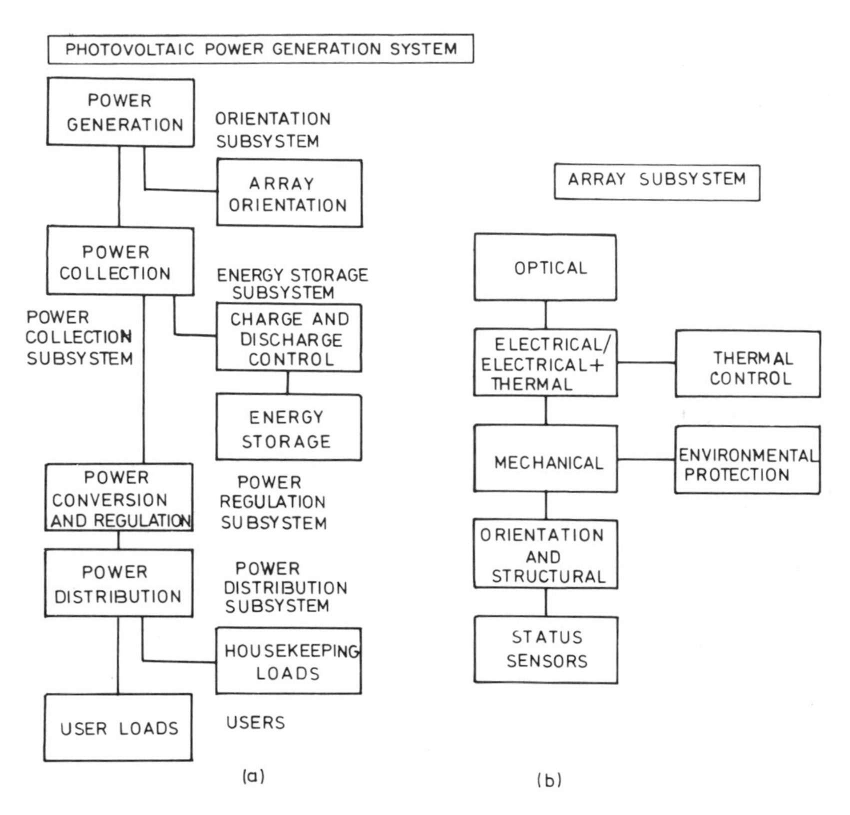

Figure D.1. (a) Block diagram of a power generation system containing an array as a subsystem. (b) Subsystems of an array (after Rauschenbach[1]).

D.2. *Types of Arrays*

Solar cell arrays are classified into two principal types: (i) terrestrial arrays designed for use within the earth's atmosphere, and (ii) space arrays designed to operate outside the earth's atmosphere. We shall restrict ourselves to a discussion of terrestrial arrays, which can be further categorized into two types: concentrator and nonconcentrator or flat plate arrays.

Flat plate arrays utilize the solar insolation as it falls naturally onto the solar cells. They may or may not be fully oriented toward the sun. Concentrator arrays utilize optical devices, primarily lenses or reflectors, to increase the sunlight intensity on the solar cells. Both flat plate and concentrator arrays may be body-mounted, rigidly held to the internal structure of a vehicle; or fixed, rigidly held to a terrestrial structure; or oriented, pointed in the direction of the sun. Sun-tracking is effected by orientation drive mechanisms called heliostats. In one-axis orientation, the arrays follow the sun by rotating about a single axis. In two-axis orientation, the arrays track the sun by rotating about two different axes. Obviously, perfect sun tracking is possible only in the latter case.

D.3. *Array Design and Performance*

The design of a solar cell array is determined by the performance required, which, in turn, is dictated by the specific application for which the array is designed. The design of an array involves several factors. An excellent treatment of the subject is contained in a recent treatise on solar cell arrays by Rauschenbach.[1] Our purpose in this chapter is to present the basic concepts and familiarize the reader with both the terminology and methodology of array design and performance. We have borrowed ideas liberally from Rauschenbach in the discussion that follows.

D.3.1. *Ratings*

We now define some terms in common usage that denote solar cell array peformance criteria:

1. Power output refers to the power (in W) available at the load circuit input terminals and is specified either as peak power or as average power during one day under certain conditions of illumination, solar cell temperature, and other factors. The power output may also be specified in terms of the current output at a particular voltage.

2. Energy output is the time-integrated value of power and indicates the amount of energy (in W-h) produced by the array during one day under specific conditions, as for power output.

3. Ampere-hour output, used when the array delivers electricity into energy storage batteries, gives the ampere-hour capacity.

4. Conversion efficiency is given either as power efficiency η_p or energy efficiency η_e by

$$\eta_p = \frac{\text{Power output from array}}{\text{Power input from sun}} \times 100\%$$

$$\eta_e = \frac{\text{Energy output from array}}{\text{Energy input from sun}} \times 100\%$$

5. Weight or mass, expressed in kilograms, indicates the quantity of material comprising a given array system.

6. Specific power can be in terms of power per unit array area (W/m^2), power per unit array mass (W/kg), or power per unit array cost (W/$).

7. Specific mass is defined as mass per unit area (kg/m^2), mass per unit power (kg/W), mass per unit energy (kg/W-hr), or area per unit mass (m^2/kg).

8. Specific cost is the cost per unit power ($/W), cost per unit energy ($/W-hr), cost per unit mass ($/kg), or cost per unit area ($/m^2).

D.3.2. *Operating Points*

The operating point of a resistively loaded solar cell is obtained from the intersection of the load line and the I–V characteristics of the cell. When the load is a combination of resistive and constant power loads, the combined load characteristic is obtained by summing the load's current at constant voltages, as shown in Figure D.2a. In this case two operating points are possible and the power system operation may be unstable, changing back and forth between the two operating points Q1 and Q2, as illustrated in Figure D.2b.

Let us consider a simple theree-component power system comprising an array, a battery, and a load, all connected in parallel such that the voltage V is common among them. The graphical solution of the problem is illustrated in Figure D.2c and d. When the battery is being charged [Figure D.2c], the available array current I_A is greater than the load current I_L and the battery charging current I_B is given by

$$|I_B| = I_A - I_L$$

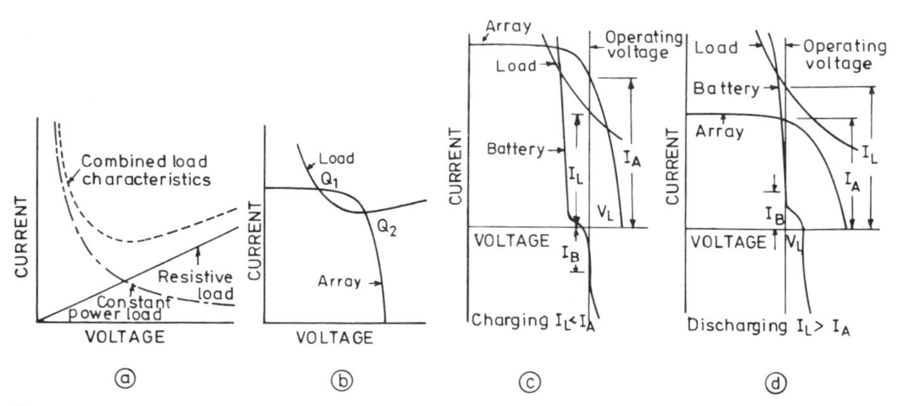

Figure D.2. (a) Combined load characteristics, (b) two possible operating points, (c) charging characteristics of a battery, and (d) discharging characteristics of a battery (after Rauschenbach[1]).

During discharge of the battery (Figure D.2d), I_A is less than the required load current I_L, thus requiring a current from the battery I_B of magnitude

$$|I_B| = I_L - I_A$$

D.3.3. *Array Models*

The power output from an individual solar cell is multiplied by the number of solar cells in an array. The cell's output current is multiplied by the number of cells connected in parallel and voltage output is multiplied by the number of cells connected in series. The cells connected in parallel form a submodule and the cells or submodules connected in series form a string.

We now define the following quantities:

N_s is the number of cells connected in series, N_p is the number of cells connected in parallel, N_t is the total number of cells in the array, V_c is the cell output voltage, V_a is the array output voltage, I_c is the cell output current capability, P_c is the cell output power capability, and P_a is the array output power capacity.

The following relationships define the array performance:

$$\begin{aligned} V_a &= N_s V_c \\ I_a &= N_p I_c \\ P_a &= N_t P_c = V_a I_a \\ P_c &= V_c I_c \quad \text{and} \\ N_t &= N_s N_p \end{aligned}$$

While analyzing solar cell arrays, we must take into account additional series resistance components arising from solar cell interconnections and array wiring and a voltage drop representing blocking diode losses. The current output of a cell at some standard conditions, say at light level Q, is given functionally by

$$\begin{aligned} I_c(v) &= I_L - I_0(v_0) \qquad v \geqslant 0 \\ &= I_L - vG(v) \qquad v < 0 \end{aligned} \tag{D.1}$$

where v is the terminal voltage, I_L is the terminal short-circuit current and $I_0(v_0)$ corresponds conceptually to the diode conduction current. $G(v)$ is a nonlinear conductive element in parallel with the output terminals and represents the cell reverse characteristics. At a different light intensity kQ, equation (D.1) may be written as

$$I_c(v) = kI_L - I_0(v_0 - \Delta v) \qquad v \geqslant 0$$
$$= kI_L - vG(v) \qquad v < 0 \tag{D.2a}$$

where

$$\Delta v = (1-k)I_L R_s \tag{D.2b}$$

with R_s the internal series resistance of the cell.

In Figure D.3a, Curve 2 shows a cell characteristic at intensity Q, while Curve 5 shows the characteristics of the same cell at zero intensity, i.e., at $k = 0$. Along the current axis, the curve has shifted by an amount $(1-k)I_L = I_L$ and along the voltage axis by an amount $\Delta v = (1-k)I_L R_s = I_L R_s$.

For a submodule composed of p equal cells connected in parallel, the terminal output current, $I_M(v)$, is given by

$$I_M(v) = p[kI_L - I_0(v_0 - \Delta v)] \qquad v \geqslant 0$$
$$= p[kI_L - vG(v)] \qquad v < 0 \tag{D.3}$$

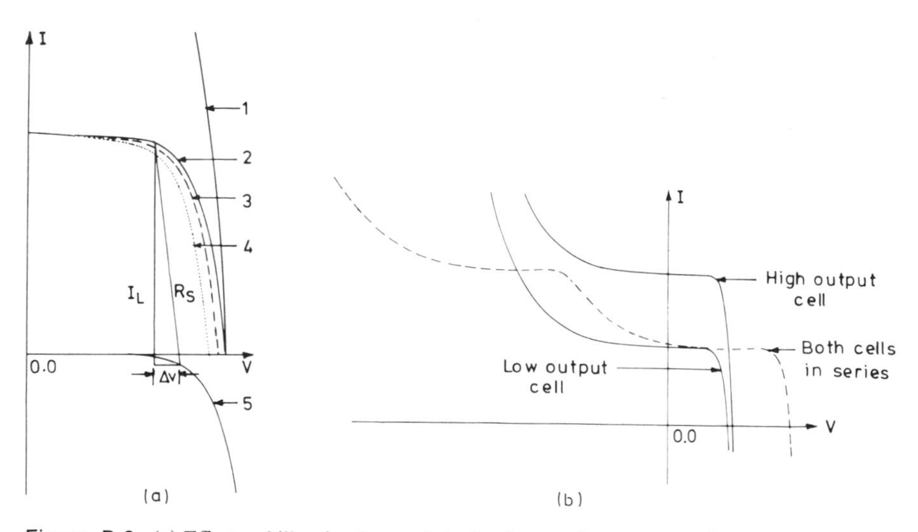

Figure D.3. (a) Effects of illumination and shadowing on the current–voltage characteristics of a solar cell and a submodule with two cells in parallel: (1) two illuminated cells in parallel, (2) one illuminated cell and approximate model for one illuminated and one dark cell in parallel, (3) accurate model for two cells in parallel, one illuminated and the other dark, (4) accurate model for two cells in parallel, illuminated at half intensity and approximate model for one shadowed cell in parallel with one illuminated cell. (b) I–V characteristics of two unequal cells connected in series (after Rauschenbach[1]).

To obtain the I-V characteristics of a string with s submodules in series, we first express equation (D.3) in terms of v and then sum at constant current values of I_s, namely

$$V_s(I_s) = \sum_{i=1}^{i=s} [v(I)_i]_{I_s} \tag{D.4}$$

Equation (D.4) may now be expressed in terms of I as

$$I_s(V_s) = p[kI_L - I_0(V_s - \Delta V)] \tag{D.5}$$

where

$$V_s = vs \quad \text{and} \quad \Delta V = (1-k)I_L s R_s$$

The array is comprised of several strings connected to a particular bus system. In the case where the array voltage V_A is forced by an energy storage battery, the string voltage V_s is also forced such that

$$V_s = V_A + V_D \tag{D.6}$$

and

$$I_s = I_D$$

where V_D is the isolation diode drop and I_s and I_D are the string and diode currents, respectively.

The current I_A of an array composed of m strings, each illuminated at different intensity, is given by

$$I_A(V_A) = \sum_{j=1}^{m} [I_s(V_A)_j]_{V_A}$$

$$= p \sum_{j=1}^{m} [k_j I_L - I_0(V_s - V_D - \Delta V)_j]_{V_A} \tag{D.7}$$

D.3.4. *Shadowing Effects*

Shadowing of a portion of a cell or a submodule results in a reduction in the total output owing to a reduction in the energy input to the cell and an increase in the internal energy losses in the non-illuminated portions of the cell. In the event of shadowing, the short-circuit current output of a partially shadowed cell is given by rI_L, where $r = A_i/A_t$, with A_t the total active area of the cell and A_i the illuminated active area of the cell. Hence, partial shadowing has the same effect on I_L as reduced light intensity on a non-shadowed cell. We must point out that the remainder of the cell characteristics does not follow the same relationship. Moreover, the above argument is valid only for moderate levels of illumination (up to two solar constants).

Consider a submodule containing p cells in parallel. Suppose rp cells are illuminated. The submodule current output is given by,[1] for $v \geqslant 0$,

$$I_M(v) = I_{\text{illum}} + I_{\text{dark}} \tag{D.8}$$

where

$$I_{\text{illum}} = rpkI_L - rpI_0(v_0 - \Delta v_1) \tag{D.8a}$$

$$\Delta v_1 = (1-k)I_L R_s \tag{D.8b}$$

and

$$I_{\text{dark}} = -(1-r)pI_0(v_0 - \Delta v_2) \tag{D.8c}$$

$$\Delta v_2 = I_L R_s \qquad (v < 0) \tag{D.8d}$$

The reverse characteristics ($v < 0$) are given by

$$I_M(v) = p[rkI_L - vG(v)] \tag{D.8e}$$

Curve 3 in Figure D.3a illustrates the effect of shadowing on a submodule consisting of $p = 2$ identical cells in parallel. The submodule is assumed to be partially shadowed with $r = 0.5$ such that one cell is illuminated and the other is in the dark. The effects of various shadowing conditions on the $I-V$ characteristics of a solar cell and a submodule are shown in Figure D.3a.

When two cells of unequal output are connected in series, the terminal voltage of this string, requiring $I_1 = I_2$, is obtained by summing the cell voltages at constant current values. This is illustrated in Figure D.3b. The lower output cell number 1 (shadowed cell) limits the output from the higher cell number 2. We emphasize here that the amount of limiting depends on the reverse characteristics of cell number 1. Thus, an analysis of cells connected in series must take into account the reverse breakdown characteristics as represented by the term $G(v)$ of equation (D.2a).

Assuming infinite breakdown voltages and zero reverse currents for each solar cell, we can write the characteristics of a partially shadowed string as[1]

$$I_s(V_s) = I(0) - J(V_s + \Delta V) \tag{D.9}$$

where $I(0) = rpkI_L$ and $J(V)$ for the string is analogous to $I_0(v_0)$ for the cell.

Sayed and Partain[2] have analyzed the effects of shading in general using the specific $I-V$ characteristics of a Cu_xS/CdS solar cell. Under identical shadowing conditions, the amount of power loss has been determined for various array configurations. The authors have treated the conditions leading to localized heating or hot spots and have enunciated criteria (including the effects of protective diodes) for their avoidance. General design rules have been formulated to obtain optimal solar array configurations.

For a more comprehensive and detailed account of shadowing effects, the reader should consult Rauschenbach[1] and Sayed and Partain.[2]

D.3.5. *Performance Prediction*

The electrical power output capability of an array at some specified operating conditions is the electrical performance of a solar cell array. The above must be distinguished from the actual output of the array, which depends upon the load demand. Performance prediction of an array, also termed performance analysis of output computation, involves the following: (i) solar cell electrical performance characterization, (ii) determination of the degradation factors related to solar cell array design and assembly, (iii) conversion of environmental considerations and criteria into solar cell operating temperatures, and (iv) computation of solar cell array power output capability.

Performance analysis is carried out in three steps: (i) gathering of input data, (ii) performing supporting analyses, and (iii) performing the array output analysis.

We have illustrated in a flow diagram (Figure D.4) the general analytical approach. The various terms in Figure D.4 are defined as follows: $F_{\tau a}$ — adhesive transmission, $F_{\tau c}$ — cover transmission, R_{sc} — cosine correction, F_d — light transmission loss factor owing to deposits and their darkening with time, α_s — absorptance, ε_H — emittance, q_c — array convective heat transfer, I_{sc} — short-circuit current, I_{mp} — maximum power current, V_{mp} — maximum power voltage, V_{oc} — open-circuit voltage, R_s — series resitance, β_I — temperature coefficient of current, β_V — temperature coefficient of voltage, F_c — cover loss/gain, F_A — assembly factor representing solar cell power output degradation owing to soldering, welding, etc., F_{TC} — solar cell array power output degradation from temperature cycling, V_D — blocking diode loss, V_W — wiring voltage loss, Z — number of zones, N_Z — number of panels in each zone, λ — panel spacing, σ — twist angle of panel, ϕ — tilt angle of panel, θ — sun angle, N_p — number of cells in parallel, and N_s — number of cells in series.

The techniques for determining the input data parameters are given elsewhere (Chapter 4) in this book.

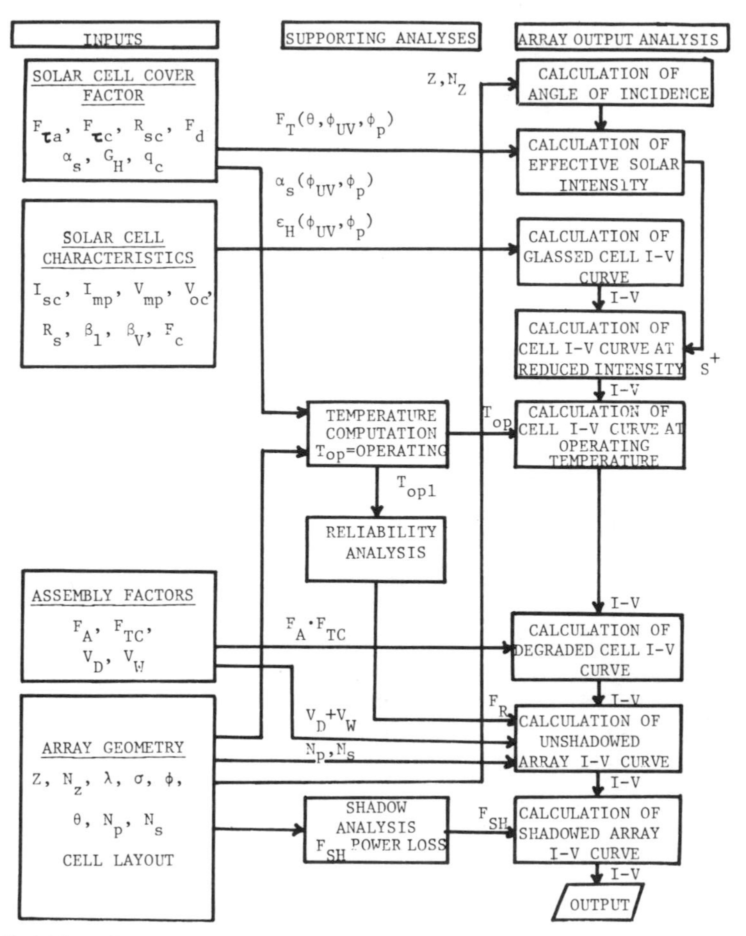

Figure D. 4. Flow diagram of array electrical performance analysis (after Rauschenbach[1]).

D.3.6. *Design Requirements*

While fabricating a solar cell array, the design requirements must be specified. These include: (i) Power output — average power, peak power, power profile per day, maximum power voltage profile, energy per day. (ii) Conditions under which power output is required — illumination (intensity, orientation, shadowing), temperature (operating range, ambient conditions), electrical losses (interconnections, wiring, blocking diodes, mismatch), reliability, mechanical (dimensions, mass, rigidity, strength), life, environmental (storage, ultraviolet radiation, handling, testing, installation, deposits, humidity, temperature cycling, weathering), and other miscellaneous information.

D.3.7. *Array Sizing*

There are two ways of calculating the required array area to satisfy the load requirements: the area method and the cell efficiency method.

1. Area method: The power output capability of an array, P_A, is given by

$$P_A = S \cdot \cos\Gamma \cdot \eta \cdot F \cdot A_A \tag{D.10}$$

where S is the solar intensity in $\mathrm{W\,m^{-2}}$, Γ is the angle between the sun line and the array normal, η is the solar cell efficiency, F is the sum total of all array design and degradation factors, and A_A is the array area. We can rewrite equation (D.10) as

$$A_A = (P_A/S)\cos\Gamma \cdot \eta \cdot F \tag{D.11}$$

2. Cell efficiency method: If N_t is the total number of solar cells in the array and A_c is the solar cell area, then

$$P_A = S \cdot \cos\Gamma \cdot \eta \cdot F \cdot N_t \cdot A_c \tag{D.12}$$

Rewriting equation (D.12), we get

$$N_t = \frac{P_A}{(S \cdot \cos\Gamma \cdot \eta \cdot F \cdot A_c)} \tag{D.13}$$

APPENDIX E

Concentrators

E.1. *Introduction*

One approach to decrease the array cost is to reduce the amount of currently expensive solar cell material required for a given power output by incorporating cheap sunlight concentrators into the array, to collect sunlight over large areas and focus onto small areas of solar cells, thus increasing the power output and reducing the dollar per watt figure for the array.

A concentrator photovoltaic system can be evaluated in three functional areas: (i) cell types, (ii) concentration methods, and (iii) tracking schemes. It must be realized that achieving a proper match among the three areas is critical for maximum power conversion performance of the system. Employment of concentrator systems brings in its wake several problems. The solar cells used in concentrators have the same general requirements as ordinary solar cells used in nonconcentrating systems, but to much more stringent specifications:

1. At 50 to 100 suns illumination (5 to 10 W of power incident per cm^2 of the cell surface), many of the bulk-material temperature and surface effects, which may not have an appreciable impact on cell performance at normal intensities, may become dominant factors in the overall performance. These factors include p/n junction geometry, lifetime of charge carriers, and surface recombination.

2. The top surface grid geometry and reflectivity very sensitively affect cell performance. A very finely spaced grid geometry has to be employed to ensure efficient collection of the large output current and reduce loss in fill factor owing to series resistance effects. In this context, the metallization on both the top and bottom surfaces and the interconnection contacts place a critical limit on output power, as the power now lost owing to contact resistance is greater by a factor of the concentration ratio (50–100, or more).

3. The increased waste heat per unit area of cell surface, resulting from the increased incident energy, necessitates cooling of the cells to limit cell operating temperatures to reasonable values. Passive cooling (using heat-sink fins) works on some designs, while active cooling with a liquid is required in other designs.

4. The solar cell geometry has to be matched to the concentrator type. Long strips composed of rectangular or square cells are appropriate for linear-focus concentrators. Small circular cells are required for point-focus concentrators.

Several novel structures for solar cells, suitable for operation under concentration, have been proposed and these have been discussed in Chapter 12. In a separate appendix we have reviewed briefly the considerations involved in grid design. In this appendix we shall summarize the various types of concentrators available and their salient characteristics. Before proceeding further, perhaps a word is necessary on the applicability of concentration in relation to thin film solar cells. It is true that at present concentrator photovoltaic systems do not employ thin film solar cells. This is primarily due to the low efficiencies of the present-day thin film devices. However, the renewed interest in thin film junctions has led to a better understanding of the devices, improvements in fabrication techniques and refinements in cell structure, with a concomitant steady improvement in cell performance. There is, thus, every cause for optimism and we expect that, in a few years, thin film solar cells will achieve levels of performance comparable to single crystal devices. Moreover, because of their complex geometry and structure, realization of high-efficiency multijunction/tandem cells would require thin film deposition techniques at some, if not most, stages of device fabrication.

E.2. *Definitions*

The most important property that characterizes a concentrator is the concentration ratio C, defined as the ratio of the area of the input beam to the area of the output beam. Thus, if a concentrator is modeled as a box with a plane entrance aperture of area A and a plane exit aperture of area A', then

$$C = A/A' \tag{E.1}$$

However, the actual concentration achieved is less than the geometric concentration C [defined by equation (E.1)] because of optical losses from reflection, scattering, absorption by the concentrator material, dust accumulation, and off-axis aberrations. The actual concentration C_a may be defined by the following two equations:

$$C_a = \eta_0 C \tag{E.2}$$

and

$$C_a = S_t / S_i \tag{E.3}$$

where η_0 is the optical efficiency of the system and S_i and S_t are the intensities incident on the entrance aperture and at the target plane, respectively. In practice, it is difficult to determine the optical losses and, hence, C_a accurately. Instead, we define the overall efficiency of the concentrator array system by

$$\eta_a = \frac{P_{\text{out}}}{S_a A_a} \tag{E.4}$$

where P_{out} is the electrical output from the array (or a cell), S_a is the illumination intensity at the entrance aperture and A_a is the entrance aperture area perpendicular to the principal optical axis of the system.

E.3. *Types of Concentrators*[1-7]

The various schemes employed for sunlight concentration include refraction, reflection, and wavelength conversion. Refraction is achieved by means of lenses which can be of the planoconvex, biconvex, or Fresnel type. Further, the lens surfaces may be spherical or aspherical. Reflection is accomplished by using mirrors.

Both refractive as well as reflective concentrators may be further classified as imaging or nonimaging, depending on whether the concentrator projects a reduced image of the sun onto the imaging plane in accordance with the laws of geometrical optics or simply collects the sun's rays onto a small area. An excellent treatment of nonimaging concentrators is given by Welford and Winston.[7] Depending on whether the focusing is at a point or along a line, the refractive and reflective concentrators, both imaging and nonimaging, can be further categorized as point-focusing (also known as axial, coaxial, or three-dimensional concentrators) or line-focusing (also termed troughs, linear, or two-dimensional concentrators) type. Some concentrators utilize several concentration stages and thus concentrators are sometimes classified by the number of stages. In Figure E.1 we give the schematic representation of the various types of concentrators.

In the wavelength conversion scheme, the solar spectrum is transformed and concentrated by means of incandescent bodies or by photoluminescent dyes[1] into a single wavelength, corresponding to the wavelength at which the solar cell exhibits the highest conversion efficiency.

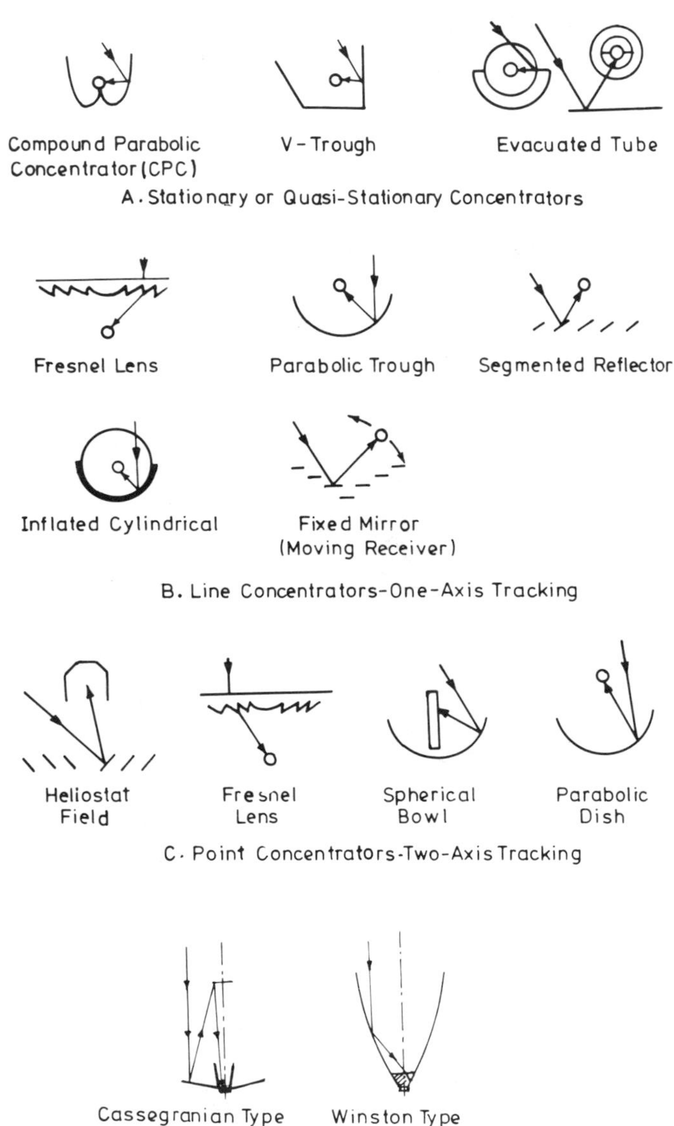

Figure E.1. Schematic diagrams of different types of concentrators.

E.4. *Tracking*

Most concentrator systems employ some form of tracking to collect the maximum amount of power from the sun each day and also to maintain uniformity of illumination over the cell/array. At the high concentrations envisaged, shading or partial shading of solar cells can affect the cell output disastrously. The concentration method and the concentration ratio desired dictate the accuracy with which the sun must be tracked, daily and seasonally. Systems with low concentration ratios (3–5) have wide acceptance angles, typically 6 to 20° compared with medium-concentration systems (~20–50). High-concentration systems (100–1000) have an acceptance angle of less than 1°. Obviously, two-axis tracking systems have the advantage of collecting more sunlight than one-axis trackers and allow the use of point-focusing concentrators. Thus, a two-axis tracking system becomes necessary where concentration ratios of more than 60 are required.

Several techniques are employed for tracking. Either clock drives or sun-sensing electronic circuits are used. Where the tracking accuracy is not very critical, as in low-concentration systems, tracking can be accomplished manually. Sun-sensing trackers utilize a pair of solar cells or light activated SCRs separated by a shadow band. An electronic circuit senses which cell is receiving more light and, subsequently, provides a signal to a relay or triac to run the motor and correct the orientation of the concentrator.

Mash and Ross[(2)] described an automatic two-axis sun-tracking concentrator, called the solar eyeball, which does not require motors or clockwork drives. Sun tracking is achieved using a solar energy powered pneumatic system. When the eyeball is misaligned, the sun's image falls on a heat exchanger in one of two air reservoirs adjacent to the cells, causing an expansion as a result of which a magnetized piston is forced against a fixed external magnetic field. The complete module rotates until the solar cells are brought back into focus.

In Table E.1 we have listed the various types of concentrators and their principal characteristics. The type of tracking required for each concentrator is also noted.

Finally, a particular concentrator photovoltaic design depends on the cells available, their cost and performance at various concentration ratios and temperatures, and the concentrator materials available. The design equation is given by

$$C_s = \frac{C_c - V_c + (C_p/C)}{\eta_c \eta_p} \tag{E.5}$$

where C_s is the cost of the system (in \$/peak kW$_e$), C_c is the cost of the

Table E.1. Characteristics of Some Typical Solar Concentrators

Type	Name	Method of concentration	Concentration ratio	Type of tracking required	Focal zone
Flat Reflectors	(1) Side mirrors	Reflective	1.5–3.0	None	Area
	(2) Fixed flat mirrors, movable focus	Reflective	20–50	One-axis	Line
	(3) Multiple heliostats	Reflective	100–1000	Two-axis	Point
Single curvature reflectors	(1) Truncated cones	Reflective	1.5–5	None and one-axis	Point
	(2) Compound parabolic concentrators	Reflective	3–10	None and one-axis	Point and line
	(3) Parabolic cylinder	Reflective	10–30	One-axis	Line
	(4) Reflecting linear Fresnel lens	Reflective	10–30	One-axis	Line
Double curvature reflectors	(1) Paraboloids	Reflective	50–1000	Two-axis	Point
	(2) Hemispheres	Reflective	25–500	Two-axis	Point
	(3) Reflecting circular Fresnel lens	Reflective	50–1000	Two-axis	Point
Refracting lenses	(1) Linear Fresnel lens	Refractive	3–50	None and one-axis	Line
	(2) Circular Fresnel lens	Refractive	50–1000	Two-axis	Point

concentrator ($/m^2 aperture), V_c is the value of the system as thermal collector ($/m^2), C_p is the cost of encapsulated, interconnected photovoltaics ($/m^2 cell area), C is the geometric concentration ratio, η_c is the optical efficiency of the concentrator system, and η_p is the photovoltaic efficiency at the operating temperature.

APPENDIX F

Degradation and Encapsulation of Solar Cells

F.1. *Introduction*

The degradation modes of a solar cell may be classified as either intrinsic or extrinsic, depending on whether the degradation is due to changes in the junction and its characteristics owing to internal changes in the properties of cell materials or the degradation is due to changes in the cell behavior brought about by external agencies such as oxygen, water vapors, and other ambient gases.

The primary mechanism of intrinsic degradation in a solar cell is an interdiffusion process involving the contacts and dopants in p/n homojunctions, and, in the case of heterojunctions and Schottky barriers, the major chemical components of the cell materials as well. Since the diffusion processes are thermally activated, accelerated life testing at elevated temperatures can provide values for the rate at which degradation in specific cell parameters will take place. A phenomenon closely related to diffusion is electromigration which takes place in some materials under the influence of an electric field in the solar cell.

The key extrinsic effect is the oxidation of semiconductor layers when no encapsulation or inadequate encapsulation is present. Other extrinsic mechanisms include deterioration of contact materials and AR coating and darkening of the encapsulation materials as a result of the effects of the UV content in the solar flux and/or weathering. In some cases, rapid temperature cycling can lead to grid/contact delamination.

F.2. *Degradation Processes*

Extensive accelerated life testing has been performed on Si and Cu_2S/CdS solar cells. Some of the major conclusions are presented in the following sections.

F.2.1. *Silicon*

Degradation mechanisms in single crystal Si solar cells are an example of the pure extrinsic type of degradation. On the other hand, SnO_2/n-Si SIS solar cells show both modes of degradation.[1] The major cause of intrinsic degradation in the SIS cells is the diffusion of top grid metal electrodes into SnO_2 and the presence of alkali atom impurities in SnO_2 that change the barrier height at the interface. A major source of extrinsic degradation in a-Si MIS solar cells is the injection of OH^- ions into the space-charge region.[1]

F.2.2. *Cu_2S/CdS*

The degradation of Cu_2S/CdS thin film solar cells has been studied extensively, partly because the cell exhibits a very broad variety of degradation processes and largely because of the R/D challenge of eliminating degradation in an otherwise viable and attractive device. Investigations in our laboratory have revealed that the degradation of both unencapsulated evaporated and sprayed cells can be divided into three regions, as shown in Figure F.1. The first region shows a very rapid decrease in the efficiency η, which drops to 5 to 6% within a few hours of the fabrication of the cell. This region is observed only in high-efficiency ($>8\%$) cells. The changes are primarily in J_{sc}, V_{oc} and FF remaining essentially unchanged. As the decrease is rapid, we believe that it is confined to the junction only. A possible mode could be the structural and topographical changes occurring in and around the junction region affecting the interface recombination processes and changing the built-in electric field. Another possible mechanism of such a rapid degradation could be the diffusion of Cu ions and formation of Cu_2S owing to the unreacted CuCl solution trapped in very fine intercolumnar deep pores which exist in the columnar evaporated CdS films. In the case of dense, small-grained, compact, sprayed CdS films, this mechanism is not expected to contribute significantly.

In the second region, a slow decrease in η takes place over a period of time that is determined by the nature of the materials constituting the cell and competing degradation mechanisms. Both the intrinsic and extrinsic modes are believed to operate in this region. There are four major processes of degradation in this region: diffusion, oxidation, electronic, and electrochemical.[2] The diffusion process having the intrinsic nature is probably an overlap between the first and second regions. Cu diffuses from the structurally unstable Cu_2S into the CdS, which results in a compensated region near the interface. As a consequence the series resistance increases, lowering FF, and the junction field is reduced, decreasing J_{sc}. The forward diffusion of the free Cu ion into CdS is assisted by the electric field and concentration gradient and occurs more rapidly along the diffusion pipes present at the columnar grain boundaries. The diffusion of Cu in CdS is via the formation of graded-composition Cu_xS–CdS. Thus, grain boundaries running down to the bottom contact act as regions of partial local shorts, reducing R_{sh}. The exposure of Cu_2S to humidity and/or oxidizing environment results in oxidation of a part of the Cu^+ (cuprous) ions to Cu^{2+} (cupric) and, hence, formation of nonstoichiometric Cu_2S, which is degenerate and

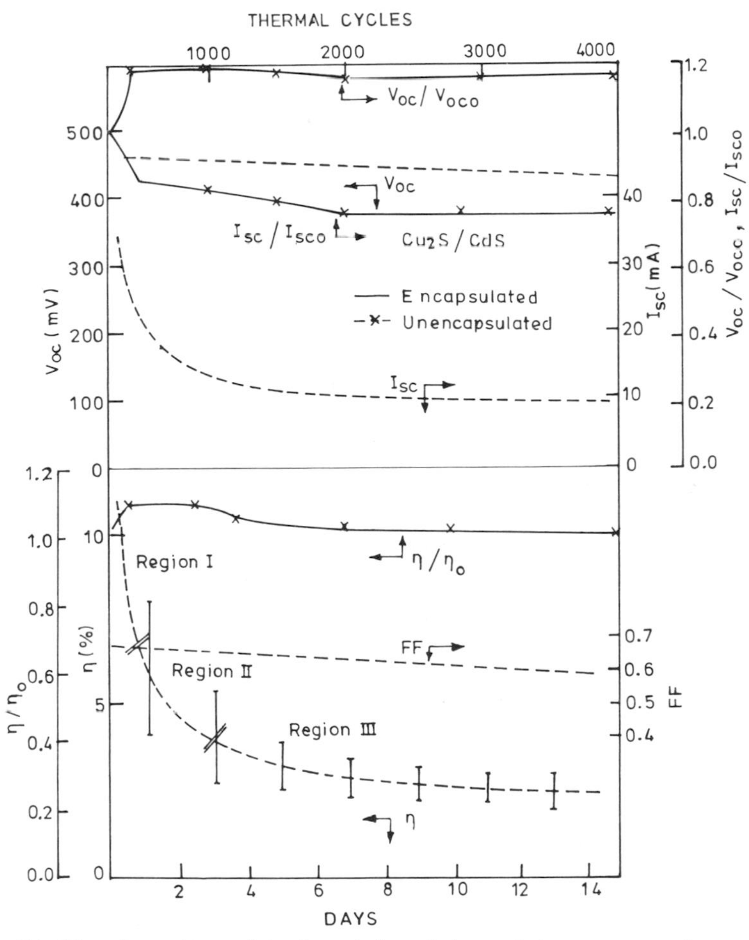

Figure F.1. Time dependence of the degradation of photovoltaic parameters of unencapsulated (-----) and encapsulated (———) Cu_2S/CdS thin film solar cells. The error bars on the efficiency curve of unencapsulated cells denote the spread in observed values. The regions I, II, III correspond to different degradation mechanisms. The subscript 0 for the encapsulated cells refers to the value immediately after encapsulation.

possesses a higher bandgap. These changes result in a decrease in the J_{sc} of the cell as the optical generation rate decreases. The oxidation process is dependent on temperature and environment and is accelerated by illumination with short wavelengths. This mechanism could occur even in cells encapsulated with a plastic which is permeable to gas and humidity.

In a cell under illumination or subject to an EMF, the I–V characteristics undergo a temporal evolution that is reversible in nature. In this electronic process, the electrons ejected from Cu_2S into CdS fill the traps at the interface and change the barrier height. The effect, seen as a lowering of V_{oc}, is closely linked to the presence of electron traps at the interface and their occupancy, as discussed in Chapter 7.

Electrochemical decomposition of Cu_2S subjected to an EMF larger than the decomposition EMF, E_D, occurs through an oxidation–reduction process liberating free Cu. The value of E_D and the J_{sc} of the cell are dependent on the stoichiometry of the initial Cu_xS. The liberated Cu starts to grow at the negative electrode, forming modules and filaments, under illumination and open-circuit conditions thus completely shorting the cell.

In the third region, a very slow decrease in η takes place with time. The mode of degradation is extrinsic and could be caused by the delamination of the base and front contacts and oxidation/corrosion of the contact metals. The final efficiency becomes almost constant between 3 and 4% after all the three stages of degradation of an unencapsulated cell.

The intrinsic degradation taking place in regions I and II could be controlled by the structural and electronic modification of the CdS and Cu_2S layers. This modification could be achieved in several ways. For CdS layers, incorporation of certain activators to promote crystal growth resulting in a decrease in the number of grain boundaries, modification of the surface topography and compactness, filling grain boundaries with certain additives to block the diffusion of Cu ions and controlling diffusion of Cu into the bulk of the CdS by an internal gradient doping electric field are some of the experimentally tried methods. A CdS:Al_2O_3 composite layer, obtained by spray pyrolysis, is an example of a useful material with a densely packed structure with Al_2O_3 in the grain boundaries that possesses a surface topography useful for making good junctions. Utilizing a two-layer structure of CdS/CdS:Al_2O_3, higher efficiency and better stability cells have been fabricated for a much smaller overall thickness. These cells are described in detail in Chapter 7.

The quality of Cu_xS, its initial stoichiometry and subsequent rate of decrease of x, can be controlled by selection of proper conversion conditions. In the chemiplating technique, almost stoichiometric Cu_2S is obtained by controlling the oxidation of Cu^+ ions in the conversion bath by adding certain reducing agents to the bath. The electroconversion technique is promising in this respect as it conveniently allows trace addition of

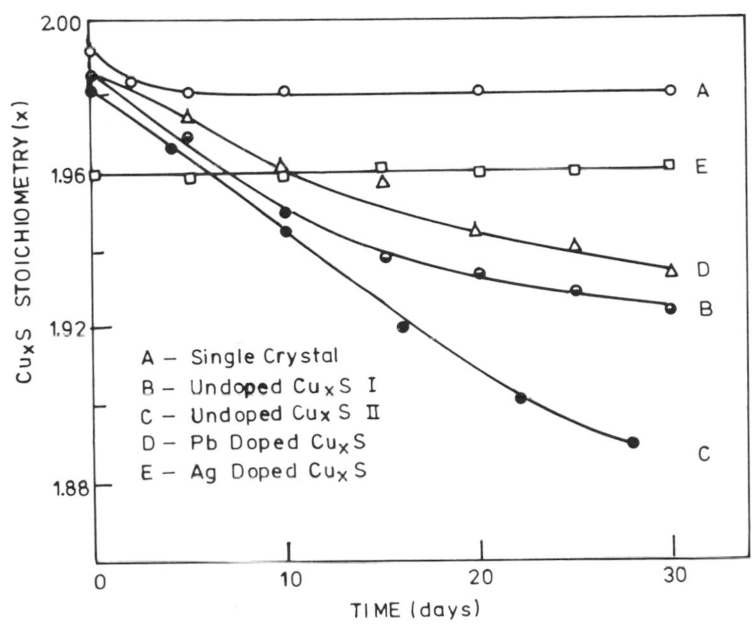

Figure F.2. Effect of doping on the temporal evolution of stoichiometry of Cu_xS films prepared by electroplating.

dopants like Ag and Pb, simply by suitable alteration of the plating bath composition. Figure F.2 compares the temporal stability of the stoichiometry of doped and undoped Cu_xS films and clearly illustrates the role of dopants in modifying the stoichiometry and its stability. Codeposition of an appropriate element in optimum amounts in this fashion could serve to form a barrier layer at the junction interface to retard Cu ion diffusion into CdS without altering the solar cell performance. It is, however, necessary to choose a proper dopant for the purpose.

F.3. *Encapsulation Procedures*

Adequate encapsulation or hermetic sealing can minimize or eliminate degradation, as illustrated in Figure F.1 for thin film Cu_2S/CdS solar cells. Several groups of workers have investigated different encapsulation processes, with varying degrees of success. Brief descriptions of the processes follow.

At I.I.T., Delhi, and Photon Power, USA, spray deposited Cu_2S/CdS back-wall solar cells are being stabilized against degradation by overcoating the active Cu_2S layer with a multilayer metal barrier with a thick top layer of Cu and/or Pb that acts as the top contact to which contact leads can be soldered. The stability of the thus modified device is considerably improved.

Shirland et al.[3] at Westinghouse have employed a SiO-7040 glass sputtered double layer as an integral cover over evaporated CdS cells. No lowering of J_{sc} and V_{oc} has been observed on testing after encapsulation and 3 weeks postannealing at room temperature.

Encapsulation, along with series connection of front-wall evaporated CdS cells, is done in a laminating press by the R/D group at Stuttgart, Germany.[4] The substrate glass, with cell layers deposited on it, and front cover glass, containing the grid laminated on an epoxy layer of a specific thickness, are pressed together at a pressure of 4 bars and a temperature of about 170 °C while simultaneously evacuating the space between the Cu_2S layer and the grid. The hot setting epoxy flows into the open spaces between the grid lines and makes good optical contact to the Cu_2S layer without flowing under the grid and insulating the grid from the Cu_2S layer. Outdoor tests of the encapsulated cells have shown a slight decrease in J_{sc} and an increase in V_{oc} initially. An overall decrease in efficiency of less than 5% of the initial value has been observed over a period of 600 days (Figure F.1).

Workers at Delaware[5] have tested thin film Cu_2S/CdS arrays, mounted on 25-mil galvanized steel and covered with 6.2-mm-thick plexiglass windows, over a period of 20 months under various humidity and temperature conditions while maintaining a continuous flow of N_2 through the hermetically sealed arrays. A life expectancy of as high as 11 years has been estimated for some cells.

Front-wall, gridded Cu_2S/CdS cells, fabricated on Ag coated kapton, have been encapsulated with kapton, FEP teflon, Aclar 33 C, Mylar, and polyethylene in a laminating press. Studies of the performance of the sealed cells indicated that FEP Teflon and Aclar 33 C were more suitable as encapsulants.[6] It was observed that all the cell parameters showed a relative increase in their value on aging at room temperature for a period of more than 1 day. Over a period of 100 days testing, the parameters returned to their gridded values.

From the preceding, it is apparent that the progress achieved to date in arresting the intrinsic and extrinsic degradation processes in Cu_2S/CdS thin film solar cells is encouraging, though limited. However, considerable work would be required to modify the compositional and structural factors in these cells to obtain stable cells with efficiencies greater than 10%.

A key requirement for a low-cost terrestrial cell is that the lifetime of the cell be about 20 years. The meaning of the lifetime has not yet been adequately defined in the photovoltaic terminology because of the lack of knowledge about the degradation mechanisms and their relationship to the externally applied physical agencies. Further, the degradation mechanisms of different types of cells could be quite different from each other. The traditional use of lifetime with a parameter falling to $1/e$ of its initial value may not be meaningful in solar cells. As a practical goal, time required for a 20% degradation in the output of a solar cell should be an adequate criterion for lifetime. If we expect a 20-year life per cell, it would be necessary to carry out accelerated tests on these cells in the laboratory under simulated conditions and correlate the performance of the cell under accelerated tests and the life of the cells under the actual working environment. These tests obviously need to be planned for each type of solar cell. With this in view, a detailed test plan has been proposed by workers at Battelle Columbus Laboratories for Cu_2S/CdS thin film cells, involving the use of at least 200 identical cells. The flow diagram of the tests is shown in Figure F.3.

F.4. *Properties of Encapsulants*

Covers for solar cell modules must be transparent to the solar spectrum, should provide hermetic sealing, and should not degrade under exposure to the UV component of the solar spectrum. The material characteristics of interest are weatherability, transparency, index of refraction, moisture permeability, fungal resistance, soil accumulation, tensile strength, coefficient of linear thermal expansion, impact resistance, abrasion resistance, insulation resistance, inflammability, and thermal conductivity.

The cover can be formed out of a transparent potting compound, hard/soft plastic, or an encapsulant. Glass or metal windows and substrates provide the greatest resistance to moisture permeation but an adequate edge sealing is essential for this configuration. Moisture permeability of plastics can be minimized by use of multiple layers but this reduces transmission to the solar cell. The cover could be a separate self-contained layer such as a glass or plastic sheet attached to the cell, or an integral layer deposited or formed in place on the cell. Transparent silicone resin pottant, a glass cover, and an epoxy frame assembly have been used to make self-contained covers for solar cells.[7] Problems of cover cracking from thermal stresses imposed by the frame are eliminated by a rubber belt used to cushion the module or by using thermally tempered glass. Modules fabricated with a thin glass cover plate have been exposed for 1 year without any degradation in performance of the cell.[4] Glass has also been used as a cover/substrate for sprayed Cu_2S/CdS terrestrial solar arrays. Tubular envelopes of glass or acrylic polymers have also been tried for encapsulating various solar cells.[7] Other techniques include the electrostatic bonding of cells to low expansion borosilicate glass without the use of organic adhesives and adhesive bonding of cells to window glass.

Table F.1. Properties of Polymers, Glasses, and Other Materials for Various Applications[a]

Application	Class of materials	Trade name (supplier)	Key properties					
			Light transmission	Weather-ability	Temperature range	Moisture resistance	Moisture barrier characteristics	Mechanical strength
Adhesives	Acrylic	Acryloid B-7 (RH) Cavalon 3100S (DP)	VG	VG	VG	VG	VG	VG
	Epoxy	Eccobond 45LV (EC) Epol Tek 310 (ET) Scotch-Weld 2216 B/A(M′)	G	G	VG	M	V	G
	Fluorocarbon	Teflon FEP (DP)	G	VG	VG	VG	VG	M
	Silicone	RTV-108 and RTV-118 (GE)	G–VG	VG	VG	VG	VG	M
Coatings	Acrylic	Eccocoat AC-8 (EC)	VG	VG	M	M	M	G
	Fluorocarbon	Kynar 202, Teflon FEP (DP)	G	VG	VG	VG	L	M
	Polyamide	Pyre ML (DP)	M	VG	VG	VG	M	VG
	Polyxylylene	Parylene C (UC)	G	VG	G	VG	U	U
	Silicone	DC 3140 (DC)	G–VG	VG	VG	VG	V	M
	Glasresin	Glass resin Type 650 (DI)	G	VG	VG	VG	V	M

Films	Acrylic	Korad A (RH)	VG	VG	M	M	M	G
	Fluorocarbon	Kynar (PC), Tedlar, Teflon FEP (DP)	G	VG	VG	VG	VG	M
	Polycarbonate	Lexan (UV-stabilized) (GE)	VG	M	VG	G	M	VG
	Polyester (TP)	Mylar (weatherable) (DP)	VG	M	VG	G	L	VG
Pottants	Epoxy	Epocast 212/9617 (FP) Stycast 1269A (EC)	G	G	G	M	V	G
	Silicone	RTV-615, RTV-619, RTV-655 (GE) Sylgard 184 (DC)	G–VG	VG	VG	VG	V	M
Sealants	Acrylic	MONO (TM)	VG	VG	M	M	M	G
	Butyl	Termco 440 (TM)	—	VG	VG	VG	VG	—
	EPR	Vistalon 404 (EC)	—	VG	VG	VG	VG	—
	Polysulfide	Lasto Meric (TM)	—	VG	VG	VG	VG	—
Sheet/Tubing	Acrylic Lucite (DP)	Plexiglass (RH)	VG	VG	M	M	M	G
	Modified acrylic	XT-375 (AC)	VG	VG	M	G	M	VG
	Polycarbonate	Lexan (GE) Tuffak (RH)	VG	M	VG	G	M	VG

[a] RH — Rohm and Hass Company; DP — E.I. duPont de Nemours, Inc.; EC — Emerson and Cuming, Inc.; ET — Epoxy Technology, Inc.; M′ — 3M Company; GE — General Electric Company; UC — Union Carbide Corporation; FP — Furane Plastics Company; TM — Tremco Manufacturing Company; EC — Exxon Chemical Company; AC — American Cyanamide Company; G — Good, VG — Very Good, M — Marginal, L — Low, V — Variable, U — Unknown, DC — Dow Corning Company; OI — Owens Illinois; PC — Pennwalt Corporation.

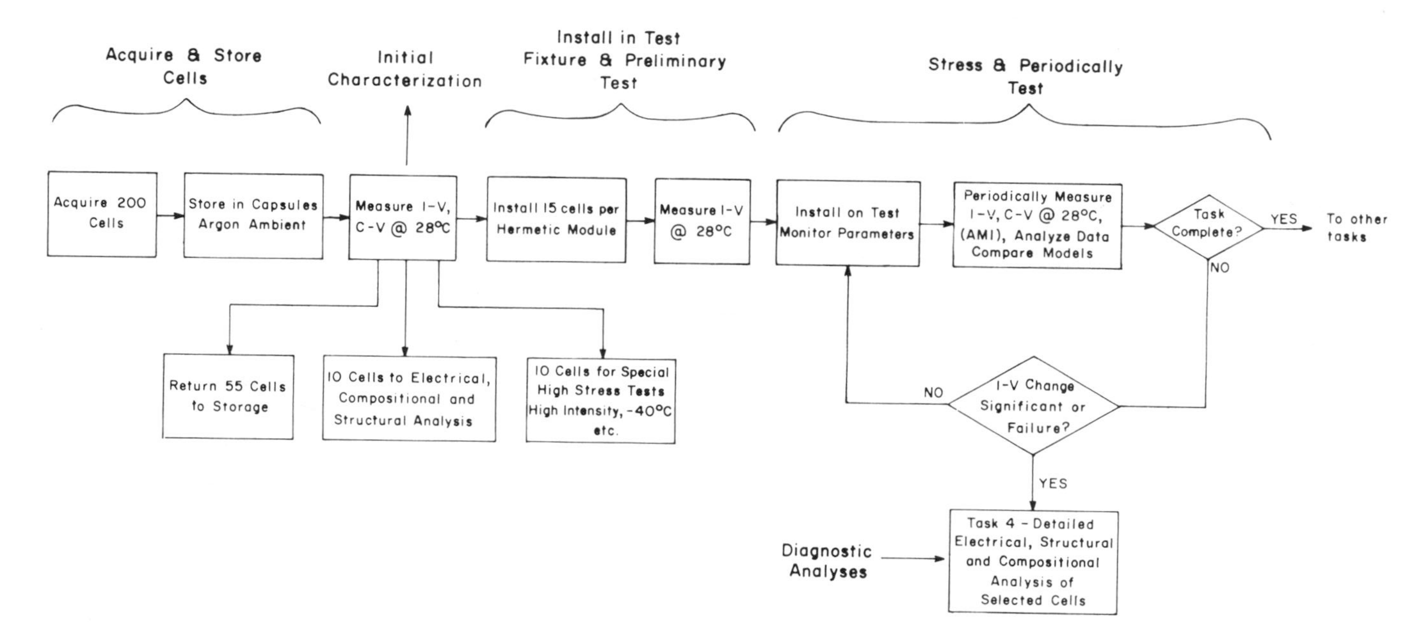

Figure F.3. Flow diagram of a proposed accelerated test program involving 200 identical Cu_2S/CdS thin film solar cells (courtesy Battelle Laboratories).

A variety of polymeric encapsulation materials like acrylics, epoxies, fluorocarbons, polycarbonates, polyesters, polyamides, polyxylylenes, silicones, and elastomeric sealants have also been tried. A brief summary of various encapsulants, their properties and performance, is given in Table F.1. The known moisture resistant plastics are teflon (PTFE) and PMMA. Polyethylene terephthalate and polycarbonate compounds have medium weather resistance, while polyethylene, PVC, cellulose, polystyrene, natural rubber, and nylon have low weather resistance. However, addition of antiozonants, UV stabilizers, etc. can significantly improve weather resistance characteristics.

The most useful classes of materials for windows are fluorocarbons, silicones, and acrylics. However, these materials are relatively expensive. Kel-F, PFA, CR-39, and various plexiglass formulations weather well, while Tedler, C-4 polycarbonate, and Tenite 479 weather poorly. Soil accumulation on hard surfaces, resulting in broadband transmission loss, is usually slight. Soft materials such as RTV-615, Sylgard 184, and Viton AHW accumulate large amounts of soil resulting in 20 to 60% transmission loss. Clearly, each material has its own characteristics which form the basis for its use for a particular solar cell.

In conclusion, our present knowledge about degradation processes, encapsulants, and encapsulation procedures for thin film solar cells is far from satisfactory and is undergoing a rapid evolutionary development of considerable promise.

References

References for Chapter 1

1. C. Marchetti, *Energy Systems — The Broader Context*, RM. 78–18, International Institute for Applied Systems Analysis, Laxenburg, Austria (1978).
2. Denis Hayes, *The Solar Energy Timetable*, Worldwatch Paper 19 (April 1978); *Energy: The Solar Prospect*, Worldwatch Paper 11 (March 1977).
3. H. Ehrenreich and J. Martin, *Physics Today* (September 1979), p. 25.
4. C.H. Henry, *J. Appl. Phys.*, **51**, 4494 (1980).
5. H.S. Rauschenbach, *Solar Cell Arrays*, Van Nostrand-Reinhold Company, New York (1980).
6. K.L. Chopra, *Thin Film Phenomena*, McGraw-Hill Book Company, New York (1969).
7. L.I. Maissel and R. Glang, *Handbook of Thin Film Technology*, McGraw-Hill Book Company, New York (1970).
8. L. Holland, *Vacuum Deposition of Thin Films*, John Wiley and Sons Inc., New York (1961).
9. J.L. Vossen and W. Kern (Eds.), *Thin Film Processes*, Academic Press, New York (1978).
10. D.E. Carlson and C.R. Wronski, *Amorphous Semiconductors* (Ed., M.H. Brodsky), Topics in Applied Physics, Vol. 36, Springer-Verlag, Heidelberg (1979), p. 287.
11. J.R. Szedon, F.A. Shirland, W.J. Biter, T.W. O'Keeffe, J.A. Stoll and S.J. Fonash, Cadmium Sulphide/Copper Sulphide Heterojunction Cell Research, Westinghouse Final Report DE-ACO3-77ET20429 (July 1979).
12. S.K. Deb and W.L. Wallace, *Proc. Soc. Photo-Optical Instrumentation Engineers*, **248**, 38 (1980).
13. F.A. Shirland and P. Rai Choudhury, Materials for Low Cost Solar Cells, Westinghouse Document No. 77-1F3-IETFS-P1 (1978).
14. T.L. Chu, S.S. Chu, G.A. Van der Leeden, C.J. Lin and J.R. Boyd, *Solid-State Electronics*, **21**, 781 (1978).
15. L.F. Buldhaupt, R.A. Mickelsen and W.S. Chen, *Proc. CdS/Cu_2S and CdS/Cu-Ternary Photovoltaic Cells Subcontractors In-Depth Review Meeting*, Washington, D.C. (1980), p. 1.
16. G.B. Reddy, V. Dutta, D.K. Pandya and K.L. Chopra, *Solar Energy Materials*, **5**, 187 (1981).
17. J.F. Jordan, *Proc. 11th IEEE Photovoltaic Specialists Conference*, Scottsdale (1975), p. 508.
18. R.D. Dupuis, P.D. Dapkus, R.D. Yingling and L.A. Mondy, *Appl. Phys. Lett.*, **31**, 201 (1977).

19. C.H. Seager, D.S. Ginley and J.D. Zook, *Appl. Phys. Lett.*, **36**, 831 (1980).
20. S.K. Ghandhi, J.M. Borrego, D. Reep, Y.S. Hsu and K.P. Pande, *Appl. Phys. Lett.*, **34**, 699 (1979).
21. Bodhraj, A.P. Thakoor, A. Banerjee, D.K. Pandya and K.L. Chopra, *Proc. National Solar Energy Congress*, Annamalai, India (1980), p. 322.
22. H. Kelly, Photovoltaic Power Systems: A Tour Through the Alternatives, *Science*, **199**, 634 (1978).
23. F. von Hippel and R.H. Williams, *Bulletin of the Atomic Scientists* (October 1977).
24. 1978 Mining Annual Review, *Mining Journal*, London (June 1978).
25. F.M. Vokes, *World Mineral Supplies, Assessment, and Perspective* (Eds., G.J.S. Gorett and M.H. Gorett), Elsevier, Amsterdam (1976), p. 65.
26. K.S. Parikh, *Energy*, The Macmillan Company (India) Limited (1976).
27. A.B. Meinel and M.P. Meinel, *Applied Solar Energy*, Addison-Wesley Publishing Company, California (1977), p. 51.

References for Chapter 2

1. J.P. McKelvey, *Solid-State and Semiconductor Physics*, Harper and Row, New York (1966).
2. C. Kittel, *Introduction to Solid State Physics*, John Wiley and Sons, New York (1956).
3. B.R. Nag, *Theory of Electrical Transport in Semiconductors*, Pergamon Press, Oxford (1972).
4. P.S. Kireev, *Semiconductor Physics*, Mir, Moscow (1975).
5. J.M. Ziman, *Principles of the Theory of Solids*, Cambridge University Press, Cambridge (1972).
6. W. Shockley, *Electrons and Holes in Semiconductors*, D. Van Nostrand Co. Inc., Princeton (1950).
7. T.S. Moss, G.J. Burell and B. Ellis, *Semiconductor Opto-Electronics*, Butterworths and Co. Ltd., London (1973).
8. N.F. Mott and E. A. Davies, *Electronic Processes in Non-Crystalline Materials*, Clarendon Press, Oxford (1971).
9. S.M. Sze, *Physics of Semiconductor Devices*, Wiley-Interscience, New York (1969).
10. K.L. Chopra, *Thin Film Phenomena*, McGraw Hill Book Co., New York (1969).
11. R.H. Bube, *Photoconductivity of Solids*, John Wiley and Sons Inc., New York (1967).
12. F. Herman, *Phys. Rev.*, **88**, 1210 (1952).
13. F. Herman, *Phys. Rev.*, **93**, 1214 (1954).
14. F. Herman, *Proc. IRE*, **43**, 1703 (1955).
15. F. Herman and J. Callaway, *Phys. Rev.*, **89**, 518 (1953).
16. J.S. Blakemore, *Elec. Commun.*, **29**, 131 (1952).
17. I.E. Tamm, *Z. Physik*, **76**, 849 (1932).
18. W. Shockley, *Phys. Rev.*, **56**, 317 (1939).
19. J. Kontecky, *J. Phys. Chem. Solids*, **14**, 233 (1960).
20. D. Pugh, *Phys. Rev. Lett.*, **12**, 390 (1964).
21. W. Shockley and G.L. Pearson, *Phys. Rev.*, **74**, 232 (1948).
22. A. Many, Y. Goldstein and N.B. Grover, *Semiconductor Surfaces*, North-Holland Publishing Company, Amsterdam (1971), p. 165.
23. R. Dalven, *Introduction to Applied Solid State Physics*, Plenum Press, New York (1980), p. 140.
24. W. Shockley and J.T. Last, *Phys. Rev.*, **107**, 392 (1957).

25. A.S. Grove, *Physics and Technology of Semiconductor Devices*, John Wiley, New York (1967), p. 267.
26. E.H. Putley, *The Hall Effect and Semiconductor Physics*, Dover Publications, Inc., New York (1968), p. 138.
27. R.L. Petritz, *Phys. Rev.*, **104**, 1508 (1956).
28. J. van den Broek, *Philips Research Report*, **22**, 367 (1967).
29. V. Snejdar, J. Jerhot, P. Hrebacka and J. Kohout, *Thin Solid Films*, **36**, 427 (1976).
30. J.W. Orton and M.J. Powell, *Prog. Rep. Phys.*, **43**, 1263 (1980); L. Kazmerski (Ed.), *Polycrystalline and Amorphous Thin Films and Devices*, Academic Press, New York (1980).
31. B. Abeles, P. Sheng, M.D. Coutts and Y. Arie, *Advances in Physics*, **24**, 407 (1975).
32. R. Landauer, *J. Appl. Phys.*, **23**, 779 (1952).
33. H. Fritzsche, *Amorphous and Liquid Semiconductors* (Ed., J. Tauc), Plenum Press, New York (1974).
34. M.H. Cohen, H. Fritzsche and S.R. Ovshinsky, *Phys. Rev. Lett.*, **22**, 1065 (1969).
35. E.A. Davis and N.F. Mott, *Philosophical Magazine*, **22**, 903 (1970).
36. M.H. Cohen, *J. Non-Crystalline Solids*, **4**, 391 (1970).
37. D. Emin, *Electronic and Structural Properties of Amorphous Semiconductors* (Eds., P.G. Le Comber and J. Mort), Academic Press, New York (1973).
38. N.F. Mott, *Electronic and Structural Properties of Amorphous Semiconductors* (Eds., P.G. Le Comber and J. Mort), Academic Press, New York (1973).
39. N.K. Hindley, *J. Non-Crystalline Solids*, **5**, 17 (1970).
40. L. Friedman, *J. Non-Crystalline Solids*, **6**, 329 (1971).
41. P. Nagels, *Amorphous Semiconductors* (Ed., M.H. Brodsky), Topics in Applied Physics, Vol. 36, Springer-Verlag, Heidelberg (1979).
42. N.F. Mott, *Philosophical Magazine*, **19**, 835 (1969).
43. J.E. Parrott, *Proc. Phys. Soc. London*, **85**, 1143 (1965).
44. A. Amith, *J. Phys. Chem. Solids*, **14**, 271 (1960).
45. J.G. Simmons, *J. Appl. Phys.*, **35**, 2472 (1964); **34**, 1793 (1963).
46. M. Born and E. Wolf, *Principles of Optics*, Pergamon Press, Oxford (1970).
47. O.S. Heavens, *Optical Properties of Thin Solid Films*, Dover Publications (1965).
48. H.A. MacLeod, *Thin Film Optical Filters*, Adam Hilger, London (1969).
49. R. Jacobsson, *Progress in Optics* (Ed., E. Wolf), North-Holland Publishing Co., Amsterdam (1965).
50. E. Shanthi, V. Dutta, A. Banerjee and K.L. Chopra, *J. Appl. Phys.*, **51**, 6243 (1980).
51. E.O. Kane, *Phys. Rev.*, **131**, 79 (1963).
52. B.I. Halperin and M. Lax, *Phys. Rev.*, **148**, 722 (1966).
53. W. Franz, *Z. Naturf.*, **139**, 484 (1958).
54. F. Urbach, *Phys. Rev.*, **92**, 1324 (1953).
55. O. Hunderi, *Phys. Rev. B*, **7**, 3419 (1973); *Thin Solid Films*, **37**, 275 (1976).
56. G. Mie, *Ann. Physik*, **25**, 377 (1908).
57. M. Kerker, *The Scattering of Light and Other Electromagnetic Radiations*, Academic Press, New York (1969).
58. C.G. Granqvist and O. Hunderi, *Phys. Rev. B*, **18**, 2897 (1978).
59. D. Polder and J.H. Van Santen, *Physica*, **12**, 257 (1946).
60. D.A.G. Bruggeman, *Ann. Phys.*, **24**, 636 (1935).
61. S. Koc, *Czech. J. Phys.*, **7**, 91 (1957).
62. J. Tauc and A. Abraham, *Czech. J. Phys.*, **9**, 95 (1959).
63. T.S. Moss, *J. Luminescence*, **7**, 359 (1973).
64. C.T. Sah, R.N. Noyce and W. Shockley, *Proc. IRE*, **45**, 1228 (1957).
65. R.N. Hall, *Phys. Rev.*, **87**, 387 (1952).
66. W. Shockley and W.T. Read, *Phys. Rev.*, **87**, 835 (1952).

67. A.F. Mayadas and M. Shatzkes, *Phys. Rev. B*, **1**, 1382 (1970).
68. C.R. Tellier and A.J. Tosser, *Thin Solid Films*, **52**, 53 (1978) and other references therein.

References for Chapter 3

1. S.M. Sze, *Physics of Semiconductor Devices*, John Wiley and Sons Inc., New York (1969).
2. A. Rothwarf and K.W. Boer, *Progress in Solid-State Chemistry*, **10**, 71 (1975).
3. W. Shockley, *Bell Syst. Tech. J.*, **28**, 435 (1949).
4. W. Shockley, *Electrons and Holes in Semiconductors*, Van Nostrand Co. Inc., New Jersey (1950).
5. H.J. Hovel, *Solar Cells, Semiconductors and Semimetals*, Vol. 11 (Eds., R.K. Willardson and A. C. Beer), Academic Press, New York (1975).
6. B.L. Sharma and R.K. Purohit, *Semiconductor Heterojunctions*, Pergamon Press Ltd., Oxford (1974).
7. A.G. Milnes and D.L. Feucht, *Heterojunctions and Metal-Semiconductor Junctions*, Academic Press, New York (1972).
8. A.M. Barnett and A. Rothwarf, *IEEE Trans. Electron Dev.*, **ED-27**, 615 (1980).
9. M.J. Adams and A. Nussbaum, *Solid-State Electronics*, **22**, 783 (1979).
10. W.A. Harrison, *J. Vac. Sci. Technol.*, **14**, 1016 (1977).
11. A.I. Gubanov, *Zh. Tekh. Fiz.*, **21**, 304 (1951).
12. A.I. Gubanov, *Zh. Eksper. Teor. Fiz.*, **21**, 79 (1951).
13. A.I. Gubanov, *Zh. Tekh. Fiz.*, **22**, 729 (1952).
14. R.L. Anderson, *IBM J. Res. Dev.*, **4**, 283 (1960).
15. R.L. Anderson, *Solid-State Electronics*, **5**, 341 (1962).
16. R.L. Anderson, *Proc. Int. Conf. on the Physics and Chemistry of Semiconductor Heterojunctions* (Ed.-in-chief, G. Szigeti), Vol. 2, Akademiai Kiado, Budapest (1971), p. 55.
17. S.S. Perlman and D.L. Feucht, *Solid-State Electronics*, **7**, 911 (1964).
18. U. Dolega, *Z. Natur.*, **18a**, 653 (1963).
19. C.J.M. Van Opdorp, Thesis, Technische Hogeschool, Eindhoven, Netherlands (1969).
20. R.H. Rediker, S. Stopek and J.H.R. Ward, *Solid-State Electronics*, **7**, 621 (1964).
21. P.C. Newman, *Electronic Letters*, **1**, 265 (1965).
22. A.R. Riben and D.L. Feucht, *Solid-State Electronics*, **9**, 1055 (1966).
23. A.R. Riben and D.L. Feucht, *Int. J. Electron.*, **20**, 583 (1966).
24. J.P. Donnelly and A.G. Milnes, *Proc. IEE*, **113**, 1468 (1966).
25. R.C. Kumar, *Int. J. Electron.*, **25**, 239 (1968).
26. W.G. Oldham and A.G. Milnes, *Solid-State Electronics*, **6**, 121 (1963).
27. W.G. Oldham and A.G. Milnes, *Solid-State Electronics*, **7**, 153 (1964).
28. C. Van Opdorp and H.K.J. Kanerva, *Solid-State Electronics*, **10**, 401 (1967).
29. L.J. Van Ruyven, J.M.P. Papenhuijen and A.C.J. Verhoven, *Solid-State Electronics*, **8**, 631 (1965).
30. A. Rothwarf, Final Report NSF/RANN/AER 72–03478 A03/FR/75 (1975), pp. 136–171.
31. H.A. Bethe, MIT Radiation Laboratory Report, 43–12 (1942).
32. W. Schottky, *Naturwiss.*, **26**, 843 (1938).
33. C. R. Crowell and S. M. Sze, *Solid-State Electronics*, **9**, 1035 (1966).
34. A.L. Fahrenbruch, Ph.D. Thesis, Stanford University (1973).
35. J. Lindmayer and A.G. Revesz, *Solid-State Electronics*, **14**, 647 (1971).
36. W.D. Gill, Ph.D. Thesis, Stanford University (1970).
37. P.F. Lindquist, Ph.D. Thesis, Stanford University (1970).

38. F.A. Shirland, *Adv. Energy Conversion*, **6**, 201 (1966).
39. F.A. Lindholm, J.G. Forsum and E.L. Burgess, *IEEE Trans. Electron Dev.*, **ED-26**, 165 (1979).
40. N.G. Tarr and D.L. Pulfrey, *Solid-State Electronics*, **22**, 265 (1979).
41. A. Rothwarf, *Proc. 13th IEEE Photovoltaic Specialists Conference*, Washington, D.C. (1978), p. 1312.
42. M.B. Prince, *J. Appl. Phys.*, **26**, 534 (1955).
43. J.J. Loferski, *J. Appl. Phys.*, **27**, 777 (1956).
44. J. J. Wysocki and P. Rappaport, *J. Appl. Phys.*, **31**, 571 (1960).
45. M. Wolf, *Proc. IRE*, **48**, 1246 (1960).
46. W. Shockley and H.J. Queisser, *J. Appl. Phys.*, **32**, 510 (1961).
47. A. deVos, *Energy Conversion*, **16**, 67 (1976).
48. A. deVos and H.J. Pauwels, *IEEE Trans. Electron Dev.*, **ED-24**, 388 (1977).
49. P. Moon, *J. Franklin Inst.*, **20**, 583 (1940).
50. D.L. Pulfrey, *Photovoltaic Power Generation*, Van Nostrand-Reinhold Company, New York (1978).
51. M. Wolf, *Proc. IEEE*, **51**, 674 (1963).
52. B. Ellis and T.S. Moss, *Solid-State Electronics*, **13**, 1 (1970).
53. J.G. Fossum, Sandia Laboratories, Energy Report, SLA-74-0273 (June 1974).
54. S.C. Tsaur, A.G. Milnes, R. Sahai and D.L. Feucht, Symposium on GaAs, Boulder, p. 156; Institute of Physics and Physical Society, London (1972).
55. W.H. Bullis and W.R. Runyan, *IEEE Trans. Electron Dev.*, **ED-14**, 75 (1967).
56. S. Kaye and G.P. Rolik, *IEEE Trans. Electron Dev.*, **ED-13**, 563 (1966).
57. R. van Overstraeten and W. Nuyts, *IEEE Trans. Electron Dev.*, **ED-16**, 632 (1969).
58. O. von Roos, *J. Appl. Phys.*, **49**, 3503 (1978).
59. M.P. Godlewski, C.R. Baraona and H.W. Brandhorst, Jr., *Proc. 10th IEEE Photovoltaic Specialists Conference*, Palo Alto (1973), p. 40.
60. S.J. Fonash, *J. Appl. Phys.*, **51**, 2115 (1980).
61. D.L. Feucht, *J. Vac. Sci. Technol.*, **14**, 57 (1977).
62. G. Vanhoutte and H. Pauwels, *Proc. 2nd European Commission Photovoltaic Solar Energy Conference* (Eds., R. Van Overstraeten and W. Palz), Berlin, West Germany (April 1979), p. 662.
63. H.J. Pauwels and G. Vanhoutte, *J. Phys. D:Appl. Phys.*, **11**, 649 (1978).
64. H.J. Pauwels, *Solid-State Electronics*, **22**, 988 (1979).
65. H.J. Pauwels, P. de Visschere and P. Reussens, *Solid-State Electronics*, **21**, 775 (1978).
66. K.W. Boer, *J. Appl. Phys.*, **50**, 5356 (1979).
67. R.H. Bube, F. Buch, A.L. Fahrenbruch, Y.Y. Ma and K.W. Mitchell, *IEEE Trans. Electron Dev.*, **ED-24**, 487 (1977).
68. R. Sahai and A.H. Milnes, *Solid-State Electronics*, **13**, 1289 (1970).
69. A.L. Fahrenbruch and J. Aranovich, in *Topics in Applied Physics*, Vol. 31, Heterojunction Phenomena and Interfacial Defects in Photovoltaic Converters, *Solar Energy Conversion* (Ed., B.O. Seraphin), Springer-Verlag, New York (1979), p. 257.
70. *Proc. Int. Workshop on Cadmium Sulphide Solar Cells and Other Abrupt Heterojunctions* (Eds., K.W. Boer and J.D. Meakin), Delaware (1975); NSF-RANN AER75-15858.
71. S.J. Fonash, *Solid-State Electronics*, **22**, 907 (1979).
72. S. Fonash and A. Ashok, *Proc. 14th IEEE Photovoltaic Specialists Conference*, San Diego (1980).
73. J.M. Woodall and H.J. Hovel, *J. Vac. Sci. Technol.*, **12**, 1000 (1975).
74. A. Rothwarf, Technical Report NSF/RANN/AER 72–03478 A04/TR76/1, University of Delaware (1976).
75. R. Hill, *Solid State and Electron Devices*, **2**, S 55 (1978).

76. J. R. Szedon, F.A. Shirland, W.J. Biter, J.A. Stoll, H.C. Dickey and T.W. O'Keeffe, Technical Progress Report No. 4, Contract EG-77-C-03-1577, Westinghouse R&D Center (1978).
77. H.W. Schock, *Proc. Workshop on II–VI Solar Cells and Similar Compounds*, Montpellier (1979), p. IX–1.
78. W. Palz, J. Besson, T. Nguyen Duy and J. Vedel, *Proc. 10th IEEE Photovoltaic Specialists Conference*, Palo Alto (1973), p. 69.
79. H.J. Hovel and J.M. Woodall, *Proc. 10th IEEE Photovoltaic Specialists Conference*, Palo Alto (1973), p. 25.
80. C. Feldman, N.A. Blum, H.K. Charles, Jr. and F.G. Satkiewicz, *J. Electronic Materials*, **7**, 309 (1978).
81. K.W. Boer and A. Rothwarf, *Annual Review of Materials Science*, **6**, 303 (1976).
82. H.J. Hovel and J.M. Woodall, *J. Electrochem. Soc.*, **29**, 1246 (1973).
83. K.W. Boer, *Int. NSF Workshop on Heterojunctions*, University of Delaware, Proc. NSF AER75 15858 (1975), p. 194.
84. H.C. Card and E.S. Yang, *IEEE Transac. Electron Dev.*, **ED-24**, 397 (1977).
85. S.I. Soclof and P.A. Iles, *Proc. 11th IEEE Photovoltaic Specialists Conference*, Scottsdale (1975), p. 56.
86. C. Lanza and H.J. Hovel, *IEEE Trans. Electron Dev.*, **ED-24**, 392 (1977).
87. A. Rothwarf, *Proc. 12th IEEE Photovoltaic Specialists Conference*, Baton Rouge (1977), p. 488.
88. L.M. Fraas, *J. Appl. Phys.*, **49**, 871 (1978).
89. L.L. Kazmerski, *Solid-State Electronics*, **21**, 1545 (1978).
90. L.L. Kazmerski, P. Sheldon and P.J. Ireland, *Thin Solid Films*, **58**, 95 (1979).
91. A. Rothwarf and A.M. Barnett, Technical Report IEC/PV/TR/76/2 (1976).
92. K. Rajkanan and J. Shewchun, *Solid-State Electronics*, **22**, 193 (1979).
93. P.J. Chen, S.C. Pao, A. Neugroschel and F.A. Lindholm, *IEEE Trans. Electron Dev.*, **ED-25,** 386 (1978).
94. M. Wolf and H. Rauschenbach, *Advanced Energy Conversion*, **3**, 455 (1963).
95. K. Lehovec and A. Fedotowsky, *Solid-State Electronics*, **20**, 725 (1977).
96. A. Flat and A.G. Milnes, *Solar Energy*, **25**, 283 (1980).
97. Ir. B. Jacobs, *Archiv für Elektronik und Übertragungstechnik*, **32**, 127 (1978).
98. J.T. Burill, D. Smith, K. Stirrup and W.J. King, *Proc. 6th IEEE Photovoltaic Specialists Conference*, Florida (1967), p. 81.
99. D. Redfield, *Appl. Phys. Lett.*, **26**, 647 (1974).
100. G.H. Hewig, F. Pfisterer, H.W. Schock and W.H. Bloss, *Proc. Workshop on the II–VI Solar Cells and Similar Compounds*, Montpellier (1979), p. VII–1.
101. B. Jacobs and G. DeMey, *Solid-State Electronics*, **21**, 1191 (1978).
102. G. DeMey, B. Jacobs and F. Fransen, *Electronics Letters*, **13**, 657 (1977).
103. R. Singh, M.A. Green and K. Rajkanan, *Solar Cells*, **3**, 95 (1981).
104. M.A. Green, Ph.D. Thesis, McMaster University (1974).
105. R. Singh, Ph.D. Thesis, McMaster University (1979).
106. M.A. Green, F.D. King and J. Shewchun, *Solid-State Electronics*, **17**, 551 (1974).
107. R. Singh and J. Shewchun, *Appl. Phys. Lett.*, **28**, 512 (1976).
108. J. Shewchun, J. DuBow, A. Myszkowski and R. Singh, *J. Appl. Phys.*, **49**, 855 (1978).
109. J. Shewchun, R. Singh and M.A. Green, *J. Appl. Phys.*, **48**, 765 (1977).
110. H. B. Michaelson, *J. Phys.*, **48**, 4729 (1977).
111. M.A. Green, *Appl. Phys. Lett.*, **33**, 179 (1978).
112. P. Richman, *MOS Field-Effect Transistor and Integrated Circuits*, Wiley, New York (1973), p. 160.
113. K. Rajkanan, R. Singh and J. Shewchun, *Solid-State Electronics*, **22**, 793 (1979).

114. M.A. Green, *J. Appl. Phys.*, **50**, 1116 (1979).
115. E. Bucher, *Appl. Phys.*, **17**, 1 (1978).
116. C.E. Norman and R.E. Thomas, *IEEE Trans. Elec. Dev.*, **ED-27**, 731 (1980).
117. R.B. Godfrey and M.A. Green, *IEEE Trans. Elec. Dev.*, **ED-27**, 737 (1980).

References for Chapter 4

1. Terrestrial Photovoltaic Measurement Procedures, Report No. ERDA/NASA/1022–77/16, NASA TM 73702 (June 1977).
2. H.W. Brandhorst, Jr. (Private communication).
3. M. Wolf and H. Rauschenbach, *Advanced Energy Conversion*, **3**, 455 (1963).
4. Development of a Thin Film Polycrystalline Solar Cell for Large Scale Terrestrial Use, University of Delaware, Report No. E (49–18)–2538 PR 77/1 (July 1977).
5. A. Banerjee, S.R. Das, A.P. Thakoor, H.S. Randhawa and K.L. Chopra, *Solid-State Electronics*, **22**, 495 (1979).
6. Research Directed to Stable High Efficiency CdS Solar Cells, University of Delaware, Report No. NSF/RANN/AER 72–03478 A04 (March 1976).
7. F.A. Shirland, *Advanced Energy Conversion*, **6**, 201 (1966).
8. J.L. Shay, S. Wagner, M. Bettini, K.J. Bachmann and E. Buehler, *IEEE Trans. Electron Dev.*, **ED-24**, 483 (1977).
9. S.M. Sze, *Physics of Semiconductor Devices*, Wiley Eastern Ltd., New Delhi (1979), p. 404.
10. S.R. Das, Ph.D. Thesis, Indian Institute of Technology, Delhi, India (1978).
11. L.C. Olsen and R.C. Bohara, *Proc. 11th IEEE Photovoltaic Specialists Conference*, Scottsdale (1975), p. 381.
12. E. Fabre, M. Mautref and A. Mircea, *Appl. Phys. Lett.*, **27**, 239 (1975).
13. C. Moses and D. Wasserman, *Solar Cells*, **1**, 218 (1980).
14. T. Warabisako, T. Saitoh, H. Itoh, N. Nakamura and T. Tokuyama, *Jpn. J. Appl. Phys.*, **17**, Supplement 17–1, 309 (1978).
15. C.T. Ho and J.D. Mathias, *Solid-State Electronics*, **24**, 115 (1981).
16. S.S. Chu, T.L. Chu and M.S. Lan, *J. Appl. Phys.*, **50**, 5805 (1979).
17. W.W. Anderson and A.D. Jonath, Final Report LMSC-D766341, Contract XJ-9-8033-1 (June 1980).
18. J.R. Szedon, F.A. Shirland, W.J. Biter, J.A. Stoll, H.C. Dickey and T.W. O'Keeffe, Technical Progress Report No. 4, Contract EG-77-C-03-1577 (November 1978).
19. T. Saitoh, T. Warabisako, H. Itoh, N. Nakumura, H. Tamura, S. Minagawa and T. Tokuyama, *Jpn. J. Appl. Phys.*, **16**, Supplement 16–1, 413 (1977).
20. M. Slapa, P.A. Tove and G. Boberg, *Nuclear Instruments and Methods*, **150**, 55 (1978).
21. P.A. Tove, U. Oscarsson, B. Drugge and W. Seibt, UPTEC Report 75 61 R (1975).
22. N.C. Wyeth, *Proc. Workshop on Cadmium Sulphide Solar Cells and Other Abrupt Heterojunctions* (Eds., K.W. Boer and J.D. Meakin), Delaware (1975), p. 575.
23. D.L. Staebler, Abstracts of the Amorphous Silicon/Materials Contractor's Review Meeting (May 29–June 1, 1979).
24. T.E. Furtak, D.C. Canfield and B.A. Parkinson, *J. Appl. Phys.*, **51**, 6018 (1980).
25. J.R. Szedon, F.A. Shirland, W.J. Biter, J.A. Stoll, H.C. Dickey and T.W. O'Keeffe, Westinghouse R and D Document No. 78-9F3-CADSO-R4 (1978).
26. A.K. Ghosh and T. Feng, *J. Appl. Phys.*, **44**, 2781 (1973).
27. T. Saitoh, S. Matsubara and S. Minagawa, *Jpn. J. Appl. Phys.*, **16**, 807 (1977).
28. L.L. Kazmerski, F.R. White, M.S. Ayyagiri and Y.J. Juang, *J. Vac. Sci. Technol.*, **14**, 65 (1977).

29. F.C. Jain and M.A. Melehy, *Appl. Phys. Lett.*, **27**, 36 (1975).
30. S. Martinuzzi and O. Mallem, *Phys. Stat. Sol.* (*a*), **16**, 339 (1973).
31. T.L. Chu, S.S. Chu, G.A. Van der Leeden, C.J. Lin and J.R. Boyd, *Solid-State Electronics*, **21**, 781 (1978).
32. M. Purushottam, S.R. Das, A.K. Mukerjee and K.L. Chopra (unpublished).
33. H. Hadley, *Proc. Workshop on Cadmium Sulphide Solar Cells and Other Abrupt Heterojunctions* (Eds., K.W. Boer and J.D. Meakin), Delaware (1975), p. 553.
34. A. Rohatgi, J.R. Davis, P. Rai-Choudhury, P.D. Blais and R.H. Hopkins, Study of Impurity Effects on Solar Cell by I–V Analysis, Westinghouse Research Report 78-IF5-SOLEP-R1 (May 1978).
35. C. Feldman, N.A. Blum, H.K. Charles, Jr. and F.G. Satkiewicz, *J. Electronic Materials*, **7**, 309 (1978).
36. S.R. Das, A. Banerjee and K.L. Chopra, *Solid-State Electronics*, **22**, 533 (1979).
37. M.J. Hampshire, *Phys. Stat. Sol.* (*a*), **1**, 57 (1970).
38. F. Pfisterer, *Proc. Workshop on the II–VI Solar Cells and Similar Compounds*, Montpellier (September 1979), p. VIII–1.
39. P.J. Chen, S.C. Pao, A. Neugroschel and F.A. Lindholm, *IEEE Trans. Electron Dev.*, **ED-25**, 386 (1978).
40. A. Neugroschel, P.J. Chen, S.C. Pao and F.A. Lindholm, *IEEE Trans. Electron Dev.*, **ED-25**, 485 (1978).
41. V.P. Singh, *Proc. 13th IEEE Photovoltaic Specialists Conference*, Washington (June 1978), p. 507.
42. A. Rothwarf, J. Phillips and N.C. Wyeth, *Proc. 13th IEEE Photovoltaic Specialists Conference*, Washington (June 1978), p. 399.
43. R.B. Hall and V.P. Singh, *J. Appl. Phys.*, **50**, 6406 (1979).
44. Development of a Thin Film Polycrystalline Solar Cell for Large Scale Terrestrial Use, University of Delaware, Report No. E(49–18)–2538 PR 76/2 (May 1977).
45. S. Martinuzzi, O. Mallem and T. Cabot, *Phys. Stat. Sol.* (*a*), **36**, 227 (1976).
46. H.R. Deppe, L. Schrader and R. Kassing, *Thin Solid Films*, **27**, 287 (1975).
47. M. Purushottam, S.R. Das, A.K. Mukerjee and K.L. Chopra, *Rev. Sci. Instr.*, **53**, 1749 (1982).
48. G.L. Miller, D.V. Lang and L.C. Kimerling, *Ann. Rev. Mater Sci.* (1977), p. 377.
49. A. Rohatgi, in: *Vacuum-Surfaces-Thin Films* (Eds., K.L. Chopra and T.C. Goel), Vanity Books, New Delhi (1981), p. 115.
50. P. Besomi and B. Wessels, *J. Appl. Phys.*, **51**, 4305 (1980).
51. D.V. Lang, *J. Appl. Phys.*, **45**, 3023 (1974).
52. M. Ozeki, J. Komeno, A. Shibatomi and S. Ohkawa, *J. Appl. Phys.*, **50**, 4808 (1979).
53. D.S. Day, M.Y. Tsai, B.G. Streetman and D.V. Lang, *J. Appl. Phys.*, **50**, 5093 (1979).
54. B.W. Wessels, *J. Appl. Phys.*, **48**, 1656 (1977).
55. C. Grill, G. Bastide, G. Sagnes and M. Rouzeyre, *J. Appl. Phys.*, **50**, 1375 (1979).
56. B.D. Cullity, *Elements of X-Ray Diffraction*, Addison-Wesley Publishing Company, Inc., Massachusetts (1967).
57. B.K. Gupta and O.P. Agnihotri, *Phil. Mag.*, **37**, 631 (1978).
58. K. Yamaguchi, H. Matsumoto, N. Nakayama and S. Ikegami, *Jpn. J. Appl. Phys.*, **15**, 1575 (1976).
59. R.R. Chamberlin and J.S. Skarman, *J. Electrochem. Soc.*, **113**, 86 (1966).
60. Y.Y. Ma and R.H. Bube, *J. Electrochem. Soc.*, **124**, 1430 (1977).
61. P.J. Grundy and G.A. Jones, *Electron Microscopy in the Study of Materials*, The Structures and Properties of Solids, Vol. 7 (Ed., B.R. Coles), Edward Arnold Ltd. (1976).
62. V.D. Vankar, S.R. Das, P. Nath and K.L. Chopra, *Phys. Stat. Sol.* (*a*), **45**, 665 (1978).
63. S.R. Das, V.D. Vankar, P. Nath and K.L. Chopra, *Thin Solid Films*, **51**, 257 (1978).

64. Final Report, Direct Solar Energy for Large Scale Terrestrial Use, Report No. NSF/RANN/AER 72–03478 A03/FR/75, University of Delaware (October 1975), p. 65.
65. T.S. TeVelde, *Energy Conversion*, **15**, 111 (1975).
66. J.H. Richardson, *Optical Microscopy for the Materials Sciences*, Marcel Dekker, Inc., New York (1971).
67. L.C. Burton, T. Hench, G. Storti and G. Haacke, *J. Electrochem. Soc.*, **123**, 1741 (1976).
68. T.L. Chu, J.C. Lien, H.C. Mollenkopf, S.C. Chu, K.W. Heizer, F.W. Voltmer and G.F. Wakefield, *Solar Energy*, **17**, 229 (1975).
69. F. Dutault, Ph.D. Thesis, L'Université de Haute Alsace et L'Université Louis Pasteur de Strasbourg (1979).
70. M.K. Mukerjee, F. Pfisterer, G.H. Hewig, H.W. Schock and W.H. Bloss, *J. Appl. Phys.*, **48**, 1538 (1977).
71. I.G. Greenfield and W.F. Tseng, Techniques for the Determination of the Phases in the Surface Layers of CdS/Cu_2S Cells, Report No. NSF/RANN/SE/GI34872/TR 73/15, University of Delaware (December 1973).
72. F.A. Shirland, Document No. 78-9F3-CADSO-P1, Westinghouse (September 1978).
73. J.M. Titchmarsh, H.R. Pettit and G.R. Booker, Structural and Electrical Studies of Semiconductor Materials and Devices, CVD Annual Report No. RU_4-9 (October 1973 — October 1974).
74. R.O. Müller, *Spectrochemical Analysis by X-Ray Fluorescence*, Plenum Press, New York (1972).
75. K. Yamaguchi, N. Nakayama, H. Matsumoto and S. Ikegami, *Jpn. J. Appl. Phys.*, **16**, 1203 (1977).
76. L.L. Kazmerski, R.B. Cooper, F.R. White and A.J. Merrill, *IEEE Trans. Electron Dev.*, (April 1977), p. 496.
77. J.R. Szedon, F.A. Shirland, W.J. Biter, T.W. O'Keeffe, J.A. Stoll and S.J. Fonash, Cadmium Sulphide/Copper Sulphide Heterojunction Cell Research, Final Report, Contract No. DE-AC03-77 ET 20429 (July 1979).
78. F.G. Satkiewicz and H.K. Charles, Jr., Sputter Ion Spectrometer Analysis of Copper Sulphide/Cadmium Sulphide Solar Cell Samples, Technical Memorandum APL/JHU TG 1284, Applied Physics Laboratory, The Johns Hopkins University (October 1975).
79. H.M. Windawi, *Proc. Workshop on Cadmium Sulphide Solar Cells and Other Abrupt Heterojunctions* (Eds., K.W. Boer and J.D. Meakin), Delaware (1975), p. 452.
80. E. Castel and J. Vedel, *Analysis*, **3**, 487 (1972).
81. L.F. Buldhaupt, R.A. Mickelsen, J.M. Stewart and W.S. Chen, Emerging Materials Systems for Solar Cell Applications — $Cu_{2-x}Se$, Final Report DE-AC04-79ET-23005, Boeing Aerospace Company (April 1980).
82. K.L. Chopra, *Thin Film Phenomena*, McGraw-Hill Book Company, New York (1969), pp. 732–741.
83. E. Shanthi, V. Dutta, A. Banerjee and K.L. Chopra, *J. Appl. Phys.*, **51**, 6243 (1980).
84. K. Rajkanan, *Mat. Res. Bull.*, **14**, 207 (1979).
85. Development of a Thin Film Polycrystalline Solar Cell for Large Scale Terrestrial Use, Quarterly Progress Report E(49–18)–2538, University of Delaware (July 1977).
86. H.K. Charles, Jr., R.J. King and A.P. Ariotedjo, *Solar Cells*, **1**, 327 (1979).
87. E.H. Putley, *The Hall Effect and Semiconductor Physics*, Dover Publications, Inc., New York (1960).
88. R.H. Bube, *Photoconductivity of Solids*, John Wiley and Sons, Inc., New York (1967).
89. T.S. Moss, G.J. Burrell and B. Ellis, *Semiconductor Opto-Electronics*, Butterworths and Co. Ltd., London (1973).
90. I.A. Baeu and E.G. Valyashko, *Sov. Phys. Solid State*, **7**, 2093 (1966).
91. E.O. Johnson, *J. Appl. Phys.*, **28**, 1349 (1957).

92. T.L. Chu and E.D. Stokes, *J. Electronic Materials*, **7**, 173 (1978).
93. M.A. Channon, J.R. Maltby, C.E. Reed and C.G. Scott, *J. Phys. D:Appl. Phys.*, **8**, L39 (1975).
94. J.E. Mahan, T.W. Ekstedt, R.I. Frank and R. Kaplow, *IEEE Trans. Electron Dev.*, **ED-26**, 733 (1979).
95. S.R. Dhariwal and N.K. Vasu, *IEEE Electron Dev. Lett.*, **EDL-2**, 53 (1981).
96. J.J. Oakes, I.G. Greenfield and L.D. Partain, *J. Appl. Phys.*, **48**, 2548 (1977).
97. H.W. Schock, *Proc. Workshop on the II–VI Solar Cells and Similar Compounds*, Montpellier (September 1979), p. IX–1.

References for Chapter 5

1. L. Holland, *Vacuum Deposition of Thin Films*, John Wiley and Sons, Inc., New York (1961).
2. K.L. Chopra, *Thin Film Phenomena*, McGraw-Hill Book Company, New York (1969).
3. L.I. Maissel and R. Glang (Eds.), *Handbook of Thin Film Technology*, McGraw-Hill Book Company, New York (1970).
4. J.L. Vossen and W. Kern (Eds.), *Thin Film Processes*, Academic Press, New York (1978).
5. G.V. Planer and L.S. Phillips, *Thick Film Circuits*, Butterworths and Company, London (1972).
6. *T.J. Coutts (Ed.), Active and Passive Thin Film Devices*, Academic Press, London (1978).
7. D.E. Carlson and C.R. Wronski, in: *Amorphous Semiconductors* (Ed., M.H. Brodsky), Topics in Applied Physics, Vol. 36, Springer-Verlag, New York (1979), p. 287.
8. W. Huber and A. Lopez-Otero, *Thin Solid Films*, **58**, 21 (1979).
9. L. Esaki and C.L. Chang, *Thin Solid Films*, **36**, 285 (1976).
10. D.R. Arthur, *J. Vac. Sci. Technol.*, **16**, 273 (1979).
11. M.W. Geis, D.C. Flanders and H.I. Smith, *Appl. Phys. Lett.*, **35**, 71 (1979).
12. H.I. Smith and D.C. Flanders, *Appl. Phys. Lett.*, **32**, 349 (1978).
13. K. Suzuki and M. Mizuhashi, *J. Vac. Soc. Jpn.*, **21**, 158 (1978).
14. J.A. Thornton and V.L. Hedgcoth, *J. Vac. Sci. Technol.*, **13**, 117 (1976).
15. J.A. Thornton, *Proc. CdS/Cu_2S and CdS/Cu-Ternary Photovoltaic Cells Sub-Contractors In-Depth Review Meeting*, Washington, D.C. (1980), p. 239.
16. J. Shewchun, *Proc. Photovoltaic Advanced Research and Development Meeting*, Solar Energy Research Institute, Colorado (1979), p. 254.
17. M. Foex, *Bull. Soc. Chim.* (France), **11**, 6 (1944).
18. R.R. Chamberlin, WPAFB Contract No. AF 331657–7919 (1962).
19. J.E. Hill and R.R. Chamberlin, U.S. Patent 3 148084 (1964).
20. R.R. Chamberlin and J.S. Skarman, *J. Electrochem. Soc.*, **113**, 86 (1966).
21. J.S. Skarman, *Solid State Electronics*, **8**, 17 (1965).
22. R.R. Chamberlin, *Chem. Bull.*, **15**, 698 (1966).
23. R.R. Chamberlin and J.S. Skarman, *Solid-State Electronics*, **9**, 819 (1966).
24. C.S. Wu, R.S. Fiegelson and R.H. Bube, *J. Appl. Phys.*, **43**, 756 (1972).
25. C.S. Wu and R.H. Bube, *J. Appl. Phys.*, **45**, 648 (1974).
26. Y.Y. Ma and R.H. Bube, *J. Electrochem. Soc.*, **124**, 1430 (1977).
27. R.H. Bube, F. Buch, A.L. Fahrenbruch, Y.Y. Ma and K.W. Mitchell, *IEEE Trans. Electron Dev.*, **ED-24**, 487 (1977).
28. R.S. Fiegelson, A.N. Diaye, S. Yin and R.H. Bube, *J. Appl. Phys.*, **48**, 3162 (1977).

29. F. Buch, A.L. Fahrenbruch and R.H. Bube, *J. Appl. Phys.*, **48**, 1596 (1978).
30. S.Y. Yin, A.L. Fahrenbruch and R.H. Bube, *J. Appl. Phys.*, **49**, 1294 (1978).
31. J. Bougnot, M. Perotin, J. Marucchi, M. Sirkis and M. Savelli, *Proc. 12th IEEE Photovoltaic Specialists Conference*, Baton Rouge (1976), p. 519.
32. M. Savelli, *Proc. Workshop on II–VI Solar Cells*, Montpellier (1979), p. I–1.
33. J. Bougnot, M. Savelli, J. Marucchi, M. Perotin, M. Marjin, O. Moris, C. Grill and R. Pommier, *Proc. Workshop on II–VI Solar Cells*, Montpellier (1979), p. II–1.
34. Oudea Coumar, Ph.D. Thesis, Université des Sciences et Techniques du Languedoc, Montpellier (1979).
35. P.K. Gogna, L.K. Malhotra and K.L. Chopra, *Research and Industry*, **22**, 74 (1977).
36. A. Banerjee, Prem Nath, V.D. Vankar and K.L. Chopra, *Phys. Stat. Sol.* (*a*), **46**, 723 (1978).
37. A. Banerjee, Prem Nath, S.R. Das and K.L. Chopra, *Proc. 7th Int. Solar Energy Congress*, New Delhi (1978); *Sun: Mankind's Future Source of Energy*, Pergamon, Oxford (1978), p. 675.
38. E. Shanthi, D.K. Pandya and K.L. Chopra, *Proc. 7th Int. Solar Energy Congress*, New Delhi (1978); *Sun: Mankind's Future Source of Energy*, Pergamon, Oxford (1978), p. 698.
39. A. Banerjee, Ph.D. Thesis, Indian Institute of Technology, New Delhi (1978).
40. A. Banerjee, S.R. Das, A.P. Takoor, H.S. Randhawa and K.L. Chopra, *Solid-State Electronics*, **22**, 495 (1979).
41. E. Shanthi, A. Banerjee, V. Dutta and K.L. Chopra, *Thin Solid Films*, **71**, 237 (1980).
42. E. Shanthi, V. Dutta, A. Banerjee and K.L. Chopra, *J. Appl. Phys.*, **51**, 6243 (1980).
43. K.L. Chopra, R.C. Kainthla, D.K. Pandya and A.P. Thakoor, *Physics of Thin Films*, Vol. 12, Academic Press, New York (1982).
44. C.M. Lampkin, *Prog. Cryst. Growth Characteristics*, **1**, 405 (1979).
45. F. Dutault, Ph.D. Thesis, L'Université de Haute Alsace et L'Université Louis Pasteur de Strasbourg (1979).
46. F. Dutault and J. Lahaye, *Proc. 2nd EC Photovoltaic Solar Energy Conference*, Berlin (Eds., R. Van Overstraeten and W. Palz), Reidel, Holland (1979), p. 898.
47. J.C. Manifacier, L. Szepessy, J.F. Bresse, M. Perotin and R. Stuck, *Mat. Res. Bull.*, **14**, 109 (1979).
48. Bodhraj, A.P. Thakoor, A. Banerjee, D.K. Pandya and K.L. Chopra, *Proc. Natl. Solar Energy Convention*, Annamalai University (1980), p. 322.
49. O.V. Varob'eva and E. S. Bessonova, *Steklo i Kapanika*, **21**, 9 (1964).
50. K. Chidamberam, L.K. Malhotra and K.L. Chopra, *Int. J. Energy Research*, **5**, 395 (1981).
51. J. Aranovich, A. Ortiz and R.H. Bube, *J. Vac. Sci. Tech.*, **16**, 994 (1979).
52. T.R. Vinerito, E.W. Rilu and L.H. Slack, *Am. Ceramic Soc. Bull.*, **54**, 217 (1955).
53. J. Kane, H.P. Schweizer and W. Kein, *J. Electrochem. Soc.*, **122**, 144 (1978).
54. V.F. Korzo and L.A. Ryabora, *Sov. Phys. Solid State*, **9**, 745 (1967).
55. V.F. Korzo and V.N. Chernayev, *Phys. Stat. Sol.* (*a*), **20**, 695 (1973).
56. J. Kane, H.P. Schweizer and W. Kern, *J. Electrochem. Soc.*, **29**, 155 (1975).
57. V.N. Semenov and Yu.E. Babenko, *Inorg. Mater.*, **14**, 193 (1978).
58. B.R. Mehta, A.P. Thakoor, D.K. Pandya and K.L. Chopra (to be published).
59. C.A. Vincent, *J. Electrochem. Soc.*, **119**, 515 (1972).
60. M. Van der Leij, Ph.D. thesis, Delf University (1979).
61. W.J. Deshotels, F. Augustine and A. Carlson, 2nd Quarterly Report, Contract NAS 7-203, Clevite Corp. (1963).
62. W.J. Deshotels, F Augustine, A. Carlson, J. Koening and M.P. Makowski, 3rd Quarterly Report, Contract NAS 7–203, Clevite Corp. (1963).
63. B.R. Pamplin, *Prog. Cryst. Growth Characteristics*, **1**, 395 (1979).

64. B.R. Pamplin and R.S. Fiegelson, *Mat. Res. Bull.*, **14**, 1 (1979); *Thin Solid Films*, **60**, 141 (1979).
65. E. Shanthi and K.L. Chopra (to be published).
66. R.E. Aitchison, *Australian J. Appl. Sci.*, **5**, 10 (1954).
67. A. Fischer, *Z. Natur.*, **9(a)**, 508 (1954).
68. H. Ludwig, *Silikattechnik*, **15**, 182 (1964).
69. P.W. Haayman, P.C. Van der Linden, D. Veeneman and G.H. Janssen, U.S. Patent 2772190 (1956).
70. I. Golovcenco, Gh.I. Rusu, V. Stefan and M. Rusu, *Iasi. Sect. Ib. Fiz.*, **11**, 77 (1965).
71. Philips Electrical Industries, Ltd., British Patent 732566 (1955).
72. J.W. McAuley, U.S. Patent 2692836 (1954).
73. Union des Verreries Mécaniques Belges, British Patent 892708.
74. R. Groth, *Phys. Stat. Sol.*, **14**, 69 (1966).
75. W.C. Lytle and A. E. Wagner, U.S. Patent 2740731 (1956).
76. M.S. Tarnopol, U.S. Patent 2694649 (1954).
77. R.S. Berg, R.D. Nasby and C. Lampkin, *J. Vac. Sci. Tech.*, **15**, 359 (1978).
78. A.P. Thakoor, B.R. Mehta, D.K. Pandya and K.L. Chopra, Int. Conf. Metallurgical Coatings, San Francisco (1981).
79. D.E. Bode, *Proc. Natl. Elec. Conf.*, **19**, 630 (1963).
80. D.E. Bode, T.H. Johnson and B.N. Maclian, *J. Appl. Opt.*, **4**, 327 (1965).
81. D.E. Bode, *Physics of Thin Films*, Vol. 3 (Eds., G. Hass and R.E. Thun), Academic Press, New York (1966), p. 275.
82. G.A. Kitaev, A.A. Uritskaya and S.G. Mokrushin, *Soviet J. Phys. Chem.*, **39**, 1101 (1965).
83. A.B. Lundin and G.A. Kitaev, *Inorg. Mater.*, **1**, 2107 (1965).
84. G.A. Kitaev, S.G. Mokrushin and A.A. Uritskaya, *Colloid J.*, **27**, 38 (1965).
85. G.M. Fofanov and G.A. Kitaev, *Sov. J. Inorg. Chem.*, **14**, 322 (1969).
86. G.A. Kitaev and T.S. Terekhova, *Sov. J. Inorg. Chem.*, **15**, 25 (1970).
87. G.A. Kitaev and T.P. Sokolova, *Sov. J. Inorg. Chem.*, **15**, 167 (1970).
88. N.C. Sharma, D.K. Pandya, H.K. Sehgal and K.L. Chopra, *Mater. Res. Bull.*, **11**, 1109 (1976).
89. N.C. Sharma, R.C. Kainthla, D.K. Pandya and K.L. Chopra, *Thin Solid Films*, **60**, 55 (1979).
90. R.C. Kainthla, D.K. Pandya and K.L. Chopra, *J. Electrochem. Soc.*, **127**, 77 (1980).
91. I. Kaur, D.K. Pandya and K.L. Chopra, *J. Electrochem. Soc.*, **127**, 943 (1980).
92. R.J. Cashman, *J. Opt. Soc. Am.*, **36**, 356 (1946).
93. D.O. Skovlin and R.A. Zingaro, *J. Electrochem. Soc.*, **111**, 42 (1964).
94. G.A. Kitaev, V.Ya Shcherbakova, V.I. Dvoinin and N.N. Belyaeva, *Zh. Prikl. Khimi*, **51**, 18 (1978).
95. R.C. Kainthla, Ph.D. Thesis, Indian Institute of Technology, Delhi (1980).
96. N.C. Sharma, D.K. Pandya, H.K. Sehgal and K.L. Chopra, *Thin Solid Films*, **42**, 383 (1977).
97. I. Kaur, Ph.D. Thesis, Indian Institute of Technology, New Delhi (1981).
98. N.C. Sharma, Ph.D. Thesis, Indian Institute of Technology, New Delhi (1978).
99. N.C. Sharma, D.K. Pandya, H.K. Sehgal and K.L. Chopra, *Thin Solid Films*, **59**, 157 (1979).
100. A. Vecht, *Physics of Thin Films*, Vol. 3 (Eds., G. Hass and R.E. Thun), Academic Press, New York (1966), p. 165.
101. S. Vojdani, A. Sharifnai and M. Doroudian, *Electron. Lett.*, **9**, 128 (1973).
102. N. Croitoru and S. Jakobson, *Thin Solid Films*, **56**, L5 (1979).
103. N. Nakayama, H. Matsumoto, K. Yamaguchi, S. Ikegami and Y. Hioki, *Jpn. J. Appl. Phys.*, **15**, 2281 (1976).

104. N. Nakayama, H. Matsumoto, A. Nakano, S. Ikegami, H. Uda and T. Yamashita, *Jpn. J. Appl. Phys.*, **19**, 703 (1980).
105. D.W. Hamer and J.V. Biggers, *Thick Film Hybrid Microcircuit Technology*, Wiley-Interscience, New York (1972).
106. J.W. Mellor, *A Comprehensive Treatise on Inorganic and Theoretical Chemistry*, Vol. 6, Longmans-Green, New York (1957).
107. C.F. Powell, J.H. Oxley and J.M. Blocher, Jr., (Eds.), *Vapor Deposition*, Wiley, New York (1966).
108. W.M. Feist, S.R. Stiele and D.W. Readey, *Physics of Thin Films*, Vol. 5 (Eds., G. Hass and R.E. Thun), Academic Press, New York (1969), pp. 237–322. Also references therein.
109. R.W. Christy, *J. Appl. Phys.*, **31**, 1680 (1960).
110. E.S. Wajda, B.W. Kippenham and W.H. White, *IBM J. Res. Develop.*, **4**, 288 (1960).
111. J.A. Amick, E.A. Roth and H. Gossenberger, *RCA Rev.*, **24**, 473 (1963).
112. G.A. Long and T. Stavish, *RCA Rev.*, **24**, 488 (1963).
113. M.V. Sullivan and G.H. Kolb, *J. Electrochem. Soc.*, **115**, 62C (1968).
114. W. Kern and V.S. Ban, *Thin Film Processes* (Eds., J.C. Vossen and W. Kern), Academic Press, New York (1978), p. 257. Also references therein.
115. T.L. Chu and K.N. Singh, *Solid-State Electronics*, **19**, 837 (1976).
116. G. Eriksson, *Acta Chem. Scand.*, **25**, 2651 (1971).
117. V.S. Ban and S.L. Gilbert, *J. Cryst. Growth*, **31**, 284 (1975).
118. F.C. Eversteijn, *Philips Res. Rep.*, **29**, 45 (1974).
119. F.R. Lever, *IBM J. Res. Develop.*, **8**, 460 (1964).
120. V.S. Ban, A.F. Gossenberger and J.J. Tietjen, *J. Appl. Phys.*, **43**, 2471 (1972).
121. D.W. Shaw, *J. Crystal Growth*, **31**, 130 (1975).
122. W. Kern and R.S. Rossler, *J. Vac. Sci. Technol.*, **14**, 1082 (1977).
123. J.D. Filby, S. Nielsen and G.J. Rich, *The Use of Thin Films in Physical Investigations* (Ed., J.C. Anderson), Academic Press, New York (1966), p. 233.
124. J.H. Oxley, *Vapor Deposition* (Eds., C.F. Powell, J.H. Oxley and J.M. Blocher, Jr.) Wiley, New York (1966), p. 493.
125. E. Sirtl, *J. Phys. Chem. Sol.*, **24**, 1285 (1967).
126. F.H. Nicoll, *J. Electrochem. Soc.*, **110**, 1165 (1962).
127. F. Bailly, G. Cohen-Solal and J. Mimila-Arroyo, *J. Electrochem. Soc.*, **126**, 1604 (1979).
128. J.G. May, *J. Electrochem. Soc.*, **112**, 710 (1965).
129. B.J. Curtis and H. Brunner, *J. Cryst. Growth*, **6**, 269 (1970).
130. J. Saraie, M. Akiyama and T. Tanaka, *Jpn. J. Appl. Phys.*, **11**, 1758 (1972).
131. G.E. Gottlich and F. Coro, *RCA Rev.*, **24**, 585 (1963).
132. R.B. Hall and J.D. Meakin, *Thin Solid Films*, **63**, 203 (1979).
133. D.A. Cusano, *Solid-State Electronics*, **6**, 217 (1963).
134. F. Pfisterer, Personal communication.
135. M. Savelli and J. Bougnot, in: *Solar Energy Conversion* (Ed., B.O. Seraphin), Topics in Applied Physics, Vol. 31, Springer-Verlag, New York (1979), p. 213.
136. H.W. Schock, G. Bilger, G.H. Hewig, F. Pfisterer and W.H. Bloss, *Proc. Int. Conf. Solar Electricity*, Toulouse (1976), p. 285.
137. S. Salkalachen, S. Jatar, A.C. Rastogi and V.G. Bhide, *Proc. National Solar Energy Convention*, Bombay (1978).
138. T.S. teVelde, *Energy Conversion*, **15**, 111 (1975).
139. S.R. Das, V.D. Vankar, Prem Nath and K.L. Chopra, *Thin Solid Films*, **51**, 257 (1978).
140. A.N. Casperd and R. Hill, *Proc. 1st EC Photovoltaic Solar Energy Conference*, Luxembourg (September 1977), p. 1131.
141. D.P. Saunders, U.S. Patent 3,032,484 (1962).

142. G.L. Schnable and J.G. Javes, U.S. Patent 3,017,332 (1962).
143. IBM Corporation, British Patent 1,052,856 (1965).
144. F.A. Lowenheim (Ed.), *Modern Electroplating*, John Wiley and Sons, New York (1974).
145. F.A. Lowenheim, *Electroplating*, McGraw-Hill Book Company (1978).
146. C.F. Coombs, Jr. (Ed.), *Printed Circuits Handbook*, McGraw-Hill Book Company, New York (1979).
147. G. Hodes, J. Manassen and D. Cahen, *Nature*, **261**, 403 (1976).
148. W.J. Danaher and L.E. Lyons, Nature, **271**, 139 (1978).
149. A.S. Baranski and W.R. Fawcett, *J. Electrochem. Soc.*, **127**, 766 (1980).
150. N. Nakayama, H. Matsumoto, A. Nakavo, S. Ikegami, H. Uda and T. Yamashita, *Jpn. J. Appl. Phys.*, **19**, 703 (1980).
151. S. Saksena, D.K. Pandya and K.L. Chopra, *Thin Solid Films*, **49**, 223 (1982).
152. C.J. Dell'Oca, D.L. Pulfrey and L. Young, *Physics of Thin Films* (Eds., M.H. Francombe and R.W. Hoffmann), Academic Press, New York (1971), p. 1.
153. D.S. Campbell, *Handbook of Thin Film Technology* (Eds., L.I. Maissel and R. Glang), McGraw-Hill Book Company, New York (1970), p. 5.1.
154. G.C. Wood, J.P. O'Sullivan and B. Vaszko, *J. Electrochem. Soc.*, **115**, 618 (1968).
155. J.L. Whitton, *J. Electrochem. Soc.*, **115**, 58 (1968).
156. S. Zwerdling and S. Sheff, *J. Electrochem. Soc.*, **107**, 338 (1960).
157. J.D.E. Beynon, G.G. Bloodworth and I.M. McLeod, *Solid-State Electronics*, **16**, 309 (1973).
158. P.F. Schmidt and W. Michel, *J. Electrochem. Soc.*, **104**, 230 (1957).
159. B. Miller and A. Heller, *Nature*, **262**, 680 (1976).
160. L. Young, *Anodic Oxide Films*, Academic Press, New York (1961).
161. D.R. Gabe, *Principles of Metal Surface Treatment and Protection*, Pergamon Press, New York (1978).
162. D.S. Campbell, Personal communication (1980).
163. H.C. Hamaker, *Trans. Faraday Soc.*, **36**, 279 (1948).
164. S.A. Troelstra, *Philips Tech. Rev.*, **12**, 293 (1951).
165. H. Kolemans and J.Th.G. Overbeck, *Discussions Faraday Soc.*, **18**, 52 (1954).
166. W.F. Pickard, *J. Electrochem. Soc.*, **115**, 105C (1968).
167. C.A. Hampel (Ed.), *Encyclopedia of Electrochemistry*, Reinhold Publishing Corporation, New York (1964), p. 544.
168. T.H. Oster, U.S. Patent, 3,200,052 (1965).
169. M. Benjamin and A.B. Osborn, *Trans. Faraday Soc.*, **36**, 287 (1940).
170. E.W. Williams, K. Jones, A.J. Griffiths, D.J. Roughley, J.M. Bell, J.H. Steven, M.J. Huson, M. Rhodes and T. Costich, *Proc. 2nd EC Photovoltaic Solar Energy Conference* (Eds., R. Van Overstraeten and W. Palz.), Berlin (April 1979), p. 874.
171. H.E. Graham, Jr., U.S. Patent, 2,800,447 (1957).
172. M.N. Fredenburg, U.S. Patent, 2,800,448 (1957).
173. S.O. Dorest, U.S. Patent 2,707,703 (1955).
174. A.L. Gray, U.S. Patent 2,530,366 (1950).
175. M. Feinleib, *Trans. Faraday Soc.*, **88**, 11 (1945).
176. M. Glusurit-Werke, Belgium Patent 643,520 (1964).
177. L.R. Dawson, *J. Appl. Phys.*, **48**, 2485 (1977).
178. A. Hadini and R. Thomas, *Thin Solid Films*, **81**, 247 (1981).
179. N. Tsuya and K. Arai, *Jpn. J. Appl. Phys.*, **18**, 207 (1979).
180. N. Tsuya and K. Arai, *Solid State Phys.*, **13**, 237 (1978).
181. Proceedings of the Photovoltaic Advanced Research and Development Meeting, Solar Energy Research Institute, Colorado (1979).

References for Chapter 6

1. Ch. Kühl, M. Druminski and K. Wittmaack, *Thin Solid Films*, **37**, 317 (1976).
2. A.V. Dvurechensky, N.N. Gerasimenko and L.P. Potapova, *Thin Solid Films*, **52**, 329 (1978).
3. J. Adamczewska and T. Budzynski, *Thin Solid Films*, **56**, 267 (1979).
4. W.A. Bryant, *Thin Solid Films*, **60**, 19 (1979).
5. T.L. Chu, S.S. Chu, C.L. Lin and R. Abderrassoul, *J. Appl. Phys.*, **50**, 919 (1979).
6. M. Hirose, M. Taniguchi and Y. Osaka, *J. Appl. Phys.*, **50**, 377 (1979).
7. T.L. Chu, *Proc. Symposium on the Material Science Aspects of Thin Film Systems for Solar Energy Conversion*, Tucson, Arizona (May 1974), p. 300.
8. T. Yoshihara, A. Yasuoka and H. Abe, *J. Electrochem. Soc.*, **127**, 1603 (1980).
9. T. Warabisako, T. Saitoh, H. Itoh, N. Nakamura and T. Tokuyama, *Jpn. J. Appl. Phys.*, **17**, Supplement 17–1, 309 (1978).
10. T.L. Chu, S.S. Chu, G.A. Van der Leeden, C.J. Lin and J.R. Boyd, *Solid-State Electronics*, **21**, 781 (1978).
11. T.I. Kamins, M.M. Mandurah and K.C. Saraswat, *J. Electrochem. Soc.*, **125**, 927 (1978).
12. A. Emmanuel and H.M. Pollock, *J. Electrochem. Soc.*, **120**, 1586 (1973).
13. T.L. Chu, J.C. Lien, H.C. Mollenkopf, S.C. Chu, K.W. Heizer, F.W. Voltmer and G.F. Wakefield, *Solar Energy*, **17**, 229 (1975).
14. T.L. Chu and K.N. Singh, *Solid-State Electronics*, **19**, 837 (1976).
15. T. Saitoh, T. Warabisako, H. Itoh, N. Nakamura, H. Tamura, S. Minagawa and T. Tokuyama, *Jpn. J. Appl. Phys.*, **16**, Supplement 16–1, 413 (1977).
16. T.L. Chu, S.S. Chu, C.L. Lin and R.M. Davis, *Abstracts of the Fourth Annual Photovoltaic Advanced Research and Development Conference*, Colorado (November 1980), pp. 173–176.
17. R.J.C. van Zolingen and A.H.M. Kipperman, *Thin Solid Films*, **58**, 89 (1979).
18. R.J.C. van Zolingen and A.H.M. Kipperman (personal communication).
19. P.H. Fang, L. Ephrath and W.B. Nowak, *Proc. Symposium on the Material Science Aspects of Thin Film Systems for Solar Energy Conversion*, Tucson, Arizona (May 1974), pp. 351–354.
20. C. Feldman, N.A. Blum, H.K. Charles, Jr. and F.G. Satkiewicz, *J. Electronic Mat.*, **7**, 309 (1978).
21. R. Harman, V. Tvarožek, O. Vaněk, M. Kempný and J. Liday, *Thin Solid Films*, **32**, 55 (1976).
22. H.J. Hinneberg, M. Weidner, G. Hecht and Chr. Weissmantel, *Thin Solid Films*, **33**, 29 (1976).
23. K. Haberle and E. Fröschle, *Thin Solid Films*, **61**, 105 (1979).
24. G. Baccarani, B. Ricco and G. Spadini, *J. Appl. Phys.*, **49**, 5565 (1978).
25. R.B. Maciolek, J.D. Heaps and J.D. Zook, *J. Electronic Mat.*, **8**, 31 (1979).
26. C.H. Seager, D.P. Ginley and J.D. Zook, *Appl. Phys. Lett.*, **36**, 831 (1980).
27. B.L. Grung, S.B. Schuldt, F.N. Schnut, J.D. Heaps and J.D. Zook, *Abstracts of the Fourth Annual Photovoltaic Advanced Research and Development Conference*, Colorado (November 1980), pp. 139–140.
28. K.R. Sarma, M.J. Rice and R.N. Legge, *Abstracts of the Fourth Annual Photovoltaic Advanced Research and Development Conference*, Colorado (November 1980), pp. 151–152.
29. K.R. Sarma and M.J. Rice, Jr., *IEEE Trans. Electron Dev.*, **ED-27**, 651 (1980).
30. W.R. Gass, R.E. Witkowski and I.E. Kanter, *Abstracts of the Fourth Annual Photovoltaic Advanced Research and Development Conference*, Colorado (November 1980), pp. 177–178.

31. J.F. Mahoney, J. Perel and T. Anestos, *Abstracts of the Fourth Annual Photovoltaic Advanced Research and Development Conference*, Colorado (November 1980), pp. 189–190.
32. A. Kubovy, J. Hamersky and B. Symersky, *Thin Solid Films*, **4**, 35 (1969).
33. D.S.H. Chan and A.E. Hill, *Thin Solid Films*, **35**, 337 (1976).
34. N.G. Dhere, N.R. Parikh and A. Ferreira, *Thin Solid Films*, **44**, 83 (1977).
35. M.A. Russak, J. Reichman, H. Witzke, S.K. Deb and S.N. Chen, *J. Electrochem. Soc.*, **127**, 725 (1980).
36. J. Hamersky, *Thin Solid Films*, **44**, 277 (1977); **38**, 101 (1976).
37. F.V. Shallcross, *Trans. Met. Soc. AIME*, **236**, 309 (1966).
38. H.W. Lehmann and R. Widner, *Thin Solid Films*, **33**, 301 (1976).
39. K. Tanaka, *Jpn. J. Appl. Phys.*, **9**, 1070 (1970).
40. R.W. Glew, *Thin Solid Films*, **46**, 59 (1977).
41. C.H.J. Liu and J.H. Wang, *Appl. Phys. Lett.*, **36**, 852 (1980).
42. B. Miller, A. Heller, M. Robbins, S. Menzes, K.C. Chang and J. Thomson, Jr., *J. Electrochem. Soc.*, **127**, 725 (1980).
43. G. Hodes, J. Manassen and D. Cahen, *Nature*, **261**, 403 (1976).
44. S. Chandra and R.K. Pandey, *Phys. Stat. Sol. (a)*, **59**, 787 (1980).
45. R.C. Kainthla, D.K. Pandya and K.L. Chopra, *J. Electrochem. Soc.*, **127**, 277 (1980).
46. R.C. Kainthla, Ph.D. Thesis, Indian Institute of Technology, Delhi (1980).
47. N.Kh. Abriksov, V.F. Bankina, L.V. Poretskaya, L.E. Shellimova and E.V. Skudnova, *Semiconducting II–VI, IV–VI, and V–VI Compounds*, Plenum Press, New York (1969).
48. N.G. Dhere and A. Goswami, *Indian J. Pure Appl. Phys.*, **7**, 398 (1969).
49. L. Däweritz and M. Dornics, *Phys. Stat. Sol. (a)*, **20**, K37 (1973).
50. R. Rentzsch and H. Berger, *Thin Solid Films*, **37**, 235 (1976).
51. F.H. Gejji and D.B. Holt, *J. Electrochem. Soc.*, **122**, 535 (1975).
52. D.B. Holt, *Thin Solid Films*, **24**, 1 (1974).
53. T.M. Ratcheva, Yu.D. Tchistiakov, G.A. Krasulin, A.V. Vanyukuv and D.H. Djoglev, *Phys. Stat. Sol.*, **16**, 315 (1973).
54. D.B. Holt, M.I. Abdalla, F.H. Gejji and D.M. Wilcox, *Thin Solid Films*, **37**, 91 (1976).
55. D.M. Wilcox and D.B. Holt, *Thin Solid Films*, **37**, 109 (1976).
56. R.S. Feigelson, A. N'Diaye, S. Yin and R.H. Bube, *J. Appl. Phys.*, **48**, 3162 (1977).
57. N.G. Dhere, N.R. Parikh and A. Ferreira, *Thin Solid Films*, **36**, 133 (1976).
58. J. Kutra, A-Sakalas, A. Zindulis and V. Zuk, *Thin Solid Films*, **55**, 421 (1978).
59. R.C. Kainthla, D.K. Pandya and K.L. Chopra, *Solid-State Electronics*, **25**, 73 (1982).
60. S.V. Svechnikov and E.B. Kaganovich, *Thin Solid Films*, **66**, 41 (1980).
61. R. Rentzsch and H. Berger, *Thin Solid Films*, **37**, 235 (1976).
62. R.C. Kainthla, D.K. Pandya and K.L. Chopra, *J. Electrochem. Soc.*, **129**, 99 (1982).
63. A.L. Fahrenbruch, *Proc. Symposium on the Materials Science Aspects of Thin Film Systems for Solar Energy Conversion*, Tucson (May 1974), p. 384.
64. G.K.M. Thutupalli and S.G. Tomlin, *J. Phys. D: Appl. Phys.*, **9**, 128 (1976).
65. G. Cohen-Solal, M. Barbe, D. Lincot, Y. Marfaing and R. Triboulet, *Proc. Workshop on the II–VI Solar Cells and Similar Compounds*, Montpellier (September 1979), p. XIII–1.
66. M. Barbe, J. Dixmier, G. Cohen-Solal, C. Sella and J.C. Martin, *Proc. Workshop on the II–VI Solar Cells and Similar Compounds*, Montpellier (September 1979), p. XVI–1.
67. W. Huber and A. Lopez-Otero, *Thin Solid Films*, **58**, 21 (1979).
68. T.L. Chu, S.S. Chu, Y. Pauleau, C.L. Jiang, E.D. Stokes, and R. Abderrassoul, *Abstracts of the Fourth Annual Photovoltaic Advanced Research and Development Conference*, Colorado (November 1980), p. 213.
69. D.A. Cusano, *Solid-State Electronics*, **6**, 217 (1963).
70. D. Lincot, J. Mimila-Arroyo, R. Triboulet, Y. Marfaing, G. Cohen-Solal, and M. Barbe,

Proc. 2nd EC Photovoltaic Solar Energy Conference (Eds., R. Van Overstraeten and W Palz), Berlin (April 1979), p. 424.
71. A.L. Fahrenbruch, F. Buch, K. Mitchell and R.H. Bube, *Proc. 11th IEEE Photovoltaic Specialists Conference*, Scottsdale (1975), p. 490.
72. J. Mimila-Arroyo, A. Bouazzi and G. Cohen-Solal, *Rev. de Phys. Appl.*, **12**, 423 (1977).
73. N. Nakayama, H. Matsumoto, K. Yamaguchi, S. Ikegami and Y. Hioki, *Jpn. J. Appl. Phys.*, **15**, 2281 (1976).
74. N. Nakayama, H. Matsumoto, A. Nakano, S. Ikegami, H. Uda and T. Yamashita, *Jpn. J. Appl. Phys.*, **19**, 703 (1980).
75. Laboratoire de Spectrométrie Physique, Université Scientifique et Médicale de Grenoble, *Proc. Workshop on II–VI Compound Thin Film Solar Cells*, Stuttgart (September 1978), p. 20–1.
76. M. Tsai and R.H. Bube, *J. Appl. Phys.*, **49**, 3397 (1978).
77. K.J. Bachmann, E. Buehler, J.L. Shay and S. Wagner, *Appl. Phys. Lett.*, **29**, 121 (1976).
78. J.L. Shay, S. Wagner, M. Bettini, K.J. Bachmann and E. Buehler, *IEEE Trans. Electron Dev.*, **ED-24**, 483 (1977).
79. T. Saitoh, S. Matsubara and S. Minagawa, *Jpn. J. Appl. Phys.*, **16**, 807 (1977).
80. L.L. Kazmerski, F.R. White, M.S. Ayyagiri, Y.J. Juang and R.P. Patterson, *J. Vac. Sci. Technol.*, **14**, 65 (1977).
81. K.R. Zanio, *Abstracts of the Fourth Annual Photovoltaic Advanced Research and Development Conference*, Colorado (November 1980), p. 227.
82. A. Catalano, V. Dalal, E.A. Fagen, R.B. Hall, J.V. Masi, J.D. Meakin, G. Warfield and A.M. Barnett, *Proc. 1st EC Photovoltaic Solar Energy Conference*, Luxembourg (1977), p. 644.
83. E.A. Fagen, *J. Appl. Phys.*, **50**, 6505 (1979).
84. R.H. Bube, A.L. Fahrenbruch, F.C. Wang and J. Ng, *Abstracts of the Fourth Annual Photovoltaic Advanced Research and Development Conference*, Colorado (November 1980), p. 217.
85. N.C. Wyeth and A. Catalano, *J. Appl. Phys.*, **50**, 1403 (1979).
86. K.P. Pande, D.H. Reep, S.K. Shastry, A.S. Weiner, J.M. Borrego and S.K. Ghandhi, *IEEE Trans. Electron Dev.*, **ED-27**, 635 (1980).
87. S.K. Ghandhi, J.M. Borrego, D. Reep, Y.S. Hsu and K.P. Pande, *Appl. Phys. Lett.*, **34**, 699 (1979).
88. S.M. Vernon, A.E. Blakeslee and H.J. Hovel, *J. Electrochem. Soc: Accelerated Brief Communications*, **126**, 703 (1979).
89. R.D. Dupuis, P.D. Dapkus, R.D. Yingling and L.A. Mondy, *Appl. Phys. Lett.*, **31**, 201 (1977).
90. S.S. Chu, T.L. Chu, Y.T. Lee, C.L. Jiang and A.B. Kuper, *Proc. 14th IEEE Photovoltaic Specialists Conference*, San Diego (1980), p. 1306.
91. S.S. Chu, T.L. Chu and M.S. Lan, *J. Appl. Phys.*, **50**, 5805 (1979).
92. S.S. Chu, T.L. Chu, H.T. Yang and K.H. Hong, *J. Electrochem. Soc.*, **125**, 1668 (1978).
93. S.S. Chu, T.L. Chu, C.L. Jiang, C.W. Loh, E.D. Stokes and J.M. Yu, *Abstracts of the Fourth Annual Photovoltaic Advanced Research and Development Conference*, Collorado (November 1980), p. 205.
94. S.S. Chu, T.L. Chu and Y.T. Lee, *IEEE Trans. Electron Dev.*, **ED-27**, 640 (1980).
95. R.J. Soukup, D.M. Mosher and A.K. Kulkarni, *Thin Solid Films*, **52**, 237 (1980).
96. L. Esaki and L.L. Chang, *Thin Solid Films*, **36**, 285 (1976).
97. M. Naganuma and K. Takahashi, *Thin Solid Films*, **32**, 42 (1976).
98. D.L. Partin, J.W. Chen, A.G. Milnes and L.F. Vassamillet, *J. Appl. Phys.*, **50**, 6845 (1979).
99. B.J. Baliga, R. Bhat and S.K. Ghandhi, *J. Appl. Phys.*, **46**, 3941 (1975).

100. N. Holonyak, Jr., R.M. Kolbas, E.A. Rezek, R. Chin, R.D. Dupuis and P.D. Dapkus, *J. Appl. Phys.*, **49**, 5392 (1978).
101. M.L. Timmons, S.M. Bedair, J.A. Hutchby, T.S. Colpitts, M. Simons and J.R. Hauser, *Proc. SERI Contractors Review Meeting*, Research Triangle Park (March 1981).
102. H. Morkoc, A.Y. Cho and C. Radice, Jr., *J. Appl. Phys.*, **51**, 4882 (1980).
103. G.B. Stringfellow and H. Künzel, *J. Appl. Phys.*, **51**, 3254 (1980).
104. H. Jäger and E. Seipp, *J. Appl. Phys.*, **49**, 3317 (1978).
105. B. Balland, R. Blondeau, L. Mayet, B. DeCremoux and P. Hirtz, *Thin Solid Films*, **65**, 275 (1980).
106. J.J.J. Jang, P.D. Dapkus, R.D. Dupuis and R.D. Yingling, *J. Appl. Phys.*, **51**, 3794 (1980).
107. N. Romeo, G. Sberveglieri and L. Terricone, *Thin Solid Films*, **55**, 413 (1978); *Thin Solid Films*, **43**, L15 (1977).
108. N.G. Dhere and N.R. Parikh, *Thin Solid Films*, **60**, 257 (1979).
109. P. Domens, M. Cadene, G.W. Cohen-Solal and S. Martinuzzi, *Proc. 1st EC Photovoltaic Solar Energy Conference*, Luxembourg (September 1977), p. 770.
110. T.H. Weng, *J. Electrochem. Soc.*, **126**, 1820 (1979).
111. V.D. Vankar, S.R. Das, P. Nath and K.L. Chopra, *Phys. Stat. Sol. (a)*, **45**, 665 (1978).
112. L.C. Burton, T.L. Hench and J.D. Meakin, *J. Appl. Phys.*, **50**, 6014 (1979).
113. A. Amith, *J. Vac. Sci. Technol.*, **15**, 353 (1978).
114. K. Mitchell, A.L. Fahrenbruch and R.H. Bube, *J. Vac. Sci. Technol.*, **12**, 909 (1975).
115. R.B. Hall, R.W. Birkmire, J.E. Phillips and J.D. Meakin, *Appl. Phys. Lett.*, **38**, 925 (1981).
116. S.R. Das, Ph.D. Thesis, Indian Institute of Technology, Delhi (1978).
117. G.H. Hewig and W.H. Bloss, *Thin Solid Films*, **45**, 1 (1977).
118. A.M. Casperd and R. Hill, *Proc. 1st EC Photovoltaic Solar Energy Conference*, Luxembourg (September 1977), p. 1131.
119. R.B. Hall, *Proc. Int. Workshop on Cadmium Sulphide Solar Cells and Other Abrupt Heterojunctions* (Eds., K.W. Boer and J.D. Meakin), Delaware (1975), p. 284.
120. L.D. Partain, J.J. Oakes and I.G. Greenfield, *Proc. Int. Workshop on Cadmium Sulphide Solar Cells and Other Abrupt Heterojunctions* (Eds., K.W. Boer and J.D. Meakin), Delaware (1975), p. 346.
121. W.F. Tseng, *Proc. Int. Workshop on Cadmium Sulphide Solar Cells and Other Abrupt Heterojunctions* (Eds., K.W. Boer and J.D. Meakin), Delaware (1975), p. 435.
122. R. Deppe and R. Kassing, *Proc. 7th Int. Vac. Congress and 3rd Int. Conf. Solid Surfaces*, Vienna (1977), p. 1927.
123. V.G. Bhide, S. Jatar and A.C. Rastogi, *Pramana*, **9**, 399 (1977).
124. E. Khwaja and S.G. Tomlin, *J. Phys. D: Appl. Phys.*, **8**, 581 (1975).
125. L.M. Fraas, W.P. Bleha and P. Braatz, *J. Appl. Phys.*, **46**, 491 (1975).
126. C. Wu and R.H. Bube, *J. Appl. Phys.*, **45**, 648 (1974).
127. L.C. Burton and T.L. Hench, *Appl. Phys. Lett.*, **29**, 612 (1976).
128. W.M. Kane, J.P. Spratt, L.W. Hershinger and I.H. Khan, *J. Electrochem. Soc.*, **113**, 136 (1966).
129. G. Storti and J. Culik, The Relationships Between Preparation Parameters, Operating Characteristics and Physical Processes in Cu_2S–CdS Thin Film Solar Cells, Technical Report IEC/PV/TR/76/6, University of Delaware (November 1976).
130. L.C. Burton, T. Hench, G. Storti and G. Haacke, *J. Electrochem. Soc.*, **123**, 1741 (1976).
131. F. Dutault, Ph.D. Thesis, L'Université de Haute Alsace et L'Université Louis Pasteur de Strasbourg (1979).
132. H.L. Kwok and W.C. Siu, *Thin Solid Films*, **61**, 249 (1979).

133. S. Martinuzzi, J. Oualid, F. Cabane-Brouty, A. Mostavan and J. Gervais, *Revue Phys. Appl.*, **14**, 237 (1979).
134. Oudeacoumar, Ph.D. Thesis, Université des Sciences et Techniques du Languedoc, Académie de Montpellier (1979).
135. R.S. Berg, R.D. Nasby and C. Lampkin, *J. Vac. Sci. Technol.*, **15**, 359 (1979).
136. A. Banerjee, Ph.D. Thesis, Indian Institute of Technology, Delhi (1978).
137. V.P. Singh and J.F. Jordan, *IEEE Electron Dev. Lett.*, **EDL-2**, 137 (1981).
138. H.P. Maruska and A.R. Young, *Abstracts of the Fourth Annual Photovoltaic Advanced Research and Development Conference*, Colorado (November, 1980), p. 135.
139. A. Banerjee, P. Nath, V.D. Vankar and K.L. Chopra, *Phys. Stat. Sol.* (*a*), **46**, 723 (1978).
140. B.K. Gupta and O.P. Agnihotri, *Phil. Mag.*, **37**, 631 (1978).
141. J. Bougnot, M. Perotin, J. Marucchi, M. Sirkis and M. Savelli, *Proc. 12th IEEE Photovoltaic Specialists Conference*, Baton Rouge (1976), p. 519.
142. F. Dutault and J. Lahaye, *Revue Phys. Appl.*, **15**, 253 (1980).
143. M.S. Alaee and M.D. Rouhani, *J. Electronic Mat.*, **8**, 289 (1979).
144. W.W. Anderson and A.D. Jonath, Cadmium Sulphide/Copper Sulphide Heterojunction Cell Research, Final Report LMSC-D-766341, Lockheed Palo Alto Research Laboratory (June 1980).
145. R. Hill, R. Harrison, G. Jenkins and R.L. Wilson, *Proc. Intl. Conf. Future Energy Concepts*, London (1979), p. 17.
146. S. Durand, *Thin Solid Films*, **44**, 43 (1977).
147. A. Piel and H. Murray, *Thin Solid Films*, **44**, 65 (1977).
148. R. Hill, *Solid-State and Electron Dev.*, **2**, S49 (1978).
149. D.C. Cameron, W. Duncan and W.M. Tsang, *Thin Solid Films*, **58**, 61 (1979).
150. K. Ito and T. Ohsawa, *Jpn. J. Appl. Phys.*, **16**, 11 (1977).
151. A. Yoshikawa and Y. Sakai, *J. Appl. Phys.*, **45**, 3521 (1974).
152. N. Croitoru and S. Jakobson, *Thin Solid Films*, **56**, L5 (1979).
153. Z. Porada and E. Schabowska, *Thin Solid Films*, **66**, L55 (1980).
154. I. Kaur, Ph.D. Thesis, Indian Institute of Technology, Delhi (1981).
155. I. Kaur, D.K. Pandya and K.L. Chopra, *J. Electrochem. Soc.*, **127**, 943 (1981).
156. G.A. Kitaev, V.Ya. Shcherbakova, V.I. Dvoinin and N.N. Belyaeva, *Zhur. Prikl. Khimii*, **51**, 18 (1978).
157. J. Vedel, M. Soubeyrand and E. Castel, *J. Electrochem. Soc.*, **124**, 177 (1977).
158. Y.Y. Ma and R.H. Bube, *J. Electrochem. Soc.*, **124**, 1430 (1977).
159. R.R. Chamberlin and J.S. Skarman, *J. Electrochem. Soc.*, **113**, 86 (1966).
160. A.P. Thakoor, B.R. Mehta, D.K. Pandya and K.L. Chopra, *Proc. Int. Conf. Metallurgical Coatings*, San Francisco (April 1981); Thin Solid Films, **83**, 231 (1981).
161. B.R. Mehta, A.P. Thakoor, D.K. Pandya and K.L. Chopra, to be published.
162. M. Takeuchi, Y. Sakagawa and H. Nagasaka, *Thin Solid Films*, **33**, 89 (1976).
163. M. Arienzo and J.J. Loferski, *Proc. 2nd EC Photovoltaic Solar Energy Conference* (Eds., R. Van Overstraeten and W. Palz), Berlin (April 1979), p. 361.
164. E.W. Williams, K. Jones, A.J. Griffiths, D.J. Roughley, J.M. Bell, J.H. Steven, M.J. Huson, M. Rhodes and T. Costich, *Proc. 2nd EC Photovoltaic Solar Energy Conference* (Eds., R. Van Overstraeten and W. Palz), Berlin (April 1979), p. 874.
165. A. Banerjee, S.R. Das, A.P. Thakoor, H.S. Randhawa and K.L. Chopra, *Solid-State Electronics*, **22**, 495 (1979).
166. P.K. Gogna, L.K. Malhotra and K.L. Chopra, *Research and Industry*, **22**, 74 (1977).
167. F.B. Micheletti and P. Mark, *Appl. Phys. Lett.*, **10**, 136 (1967).
168. C.S. Wu, R.S. Fiegelson and R.H. Bube, *J. Appl. Phys.*, **43**, 756 (1972).
169. M. Lichtensteiger, I. Lagnado and H.C. Gatos, *Appl. Phys. Lett.*, **15**, 418 (1969).

170. N.R. Pavaskar, C.A. Menezes and A.P.B. Sinha, *J. Electrochem. Soc.*, **124**, 743 (1977).
171. M. Cadene, M. Ginter-Lumbreras, S. Martinuzzi, J. Vedel and M. Soubeyrand, *Proc. Workshop on the II–VI Solar Cells and Similar Compounds*, Montpellier (September 1979), V–1.
172. R.B. Shafizade, I.V. Ivanova and M.M. Kazinets, *Thin Solid Films*, **35**, 169 (1976).
173. R.B. Shafizade, I.V. Ivanova and M.M. Kazinets, *Thin Solid Films*, **55**, 211 (1978).
174. L.F. Buldhaupt, R.A. Mickelsen, J.M. Stewart and W.S. Chen, Emerging Materials Systems for Solar Cell Applications-$Cu_{2-x}Se$, Boeing Aerospace Company, Final Report DE-AC04-79ET-23005 (April 1980).
175. M. Savelli and J. Bougnot, in: *Problems of the Cu_2S/CdS Cell, in Solar Energy Conversion* (Ed., B.O. Seraphin), Topics in Applied Physics, Vol. 31, Springer-Verlag, New York (1979), p. 213.
176. R. Hill, in *Thin Film Solar Cells, Active and Passive Thin Film Devices* (Ed., T.J. Coutts), Academic Press, London (1978), p. 487.
177. T. Ishikawa and S. Miyatani, *J. Phys. Soc. Jap.*, **42**, 159 (1977).
178. K. Okamoto and S. Kawai, *Jpn. J. Appl. Phys.*, **12**, 1132 (1973).
179. J. Besson, T. Nguyen Duy, A. Gauthier, W. Palz, C. Martin and J. Vedel *Proc. 11th IEEE Photovoltaic Specialists Conference*, Scottsdale (1975), p. 468.
180. F. Guastavino, Ph.D. Thesis, Université des Sciences et Techniques du Languedoc, Academie de Montpellier (1974).
181. J.R. Szedon, F.A. Shirland, W.J. Biter, T.W. O'Keeffe, J.A. Stoll and S.J. Fonash, Cadmium Sulphide/Copper Sulphide Heterojunction Cell Research, Final Report No. 78-9F3-CADSO-R5, Westinghouse R&D Center (1979).
182. H.E. Nastelin, J.M. Smith and A.L. Gombach, Final Report, Contract No. NAS3-13467, Clevite Corporation (June, 1971).
183. S.R. Das, V.D. Vankar, P. Nath and K.L. Chopra, *Thin Solid Films*, **51**, 257 (1978).
184. B. Rezig, S. Duchemin and F. Guastavino, *Solar Energy Mat.*, **2**, 53 (1979).
185. S. Couve, L. Gouskov, L. Szepessy, J. Vedel and E. Castel, *Thin Solid Films*, **15**, 223 (1973).
186. H.S. Randhawa, D.G. Brock, R.F. Bunshah, B.M. Basol and O.M. Stafsudd, *Abstracts of the Fourth Annual Photovoltaic Advanced Research and Development Conference*, Colorado (November 1980), p. 127.
187. N. Nakayama, *Jpn. J. Appl. Phys.*, **8**, 450 (1969).
188. S. Saxena, D.K. Pandya and K.L. Chopra, *Thin Solid Films* (commun.)
189. J.F. Jordan, *Proc. 11th IEEE Photovoltaic Specialists Conference*, Scottsdale (1975), p. 508.
190. J. Dielman, *Proc. Int. Workshop on Cadmium Sulphide Solar Cells and Other Abrupt Heterojunctions* (Eds., K.W. Boer and J.D. Meakin), Delaware (1975), p. 92.
191. J. L. Shay, S. Wagner and H.M. Kasper, *Appl. Phys. Lett.*, **27**, 89 (1975).
192. L.L. Kazmerski, R.B. Cooper, F.R. White and A.J. Merrill, *IEEE Trans. Electron. Dev.*, **ED–24**, 496 (1977).
193. A.F. Fray and P. Lloyd, *Thin Solid Films*, **58**, 29 (1979).
194. J. Piekoszewski, J.J. Loferski, R. Beaulieu, J. Beall, R. Roeslev and J. Shewchun, *Proc. 14th IEEE Photovoltaic Specialists Conference*, San Diego (1980), p. 980.
195. Y. Kokubun and M. Wada, *Jpn. J. Appl. Phys.*, **16**, 879 (1977).
196. L.L. Kazmerski, M.S. Ayyagiri, F.R. White and G.A. Sanborn, *J. Vac. Sci. Tech.*, **13**, 139 (1976).
197. R. Durny, A.E. Hill and R.D. Tomlinson, *Thin Solid Films*, **69**, 211 (1980).
198. W. Horig, H. Neumann and H. Sobota, *Thin Solid Films*, **48**, 67 (1978).
199. B. Pamplin and R.S. Fingleton, *Thin Solid Films*, **60**, 41 (1979).
200. F.R. White, A.H. Clarke, M.C. Graf and L.L. Kazmerski, *J. Appl. Phys.*, **50**, 544 (1979).

201. D.B. Fraser and H.D. Cook, *J. Electrochem. Soc.*, **119**, 1368 (1972).
202. G. Haacke, *J. Appl. Phys.*, **47**, 4086 (1976).
203. R.W. Keyes, *J. Appl. Phys.*, **30**, 454 (1959).
204. W.R. Sinclair, F.G. Peters, D.W. Stillinger and S.E. Koonee, *J. Electrochem. Soc.*, **112**, 1096 (1965).
205. E. Leja, T. Pisarkiewicz and A. Kolodzieg, *Thin Solid Films*, **67**, 45 (1980).
206. M. Hecq and E. Porteir, *Thin Solid Films*, **9**, 341 (1972).
207. E. Giani and R. Kelly, *J. Electrochem. Soc.*, **121**, 394 (1974).
208. J.A. Thornton and V.L. Hedgcoth, *J. Vac. Sci. Technol.*, **13**, 117 (1976).
209. N. Miyata, K. Miyaka and S. Nao, *Thin Solid Films*, **58**, 385 (1979).
210. G. Haacke, W.E. Mealmaker and L.A. Siegel, *Thin Solid Films*, **55**, 67 (1978).
211. J.L. Vossen, *RCA Review*, **33**, 289 (1971).
212. N. Miyata, K. Miyaka, K. Koga and T. Fukushima, *J. Electrochem. Soc.*, **127**, 918 (1980).
213. H.W. Lehmann and R. Widmer, *Thin Solid Films*, **27**, 359 (1975).
214. W.G. Haines and R.H. Bube, *J. Appl. Phys.*, **49**, 304 (1978).
215. J.L. Vossen, in: *Physics of Thin Films* (Eds., G. Hass, M.H. Francombe, and R.W. Hoffman), Vol. 9, Academic Press, New York (1976), p. 1.
216. J.C.C. Fan, *Appl. Phys. Lett.*, **34**, 515 (1979).
217. R.P. Howson, J.N. Avaratsiotis, M.I. Ridge and C.A. Bishop, *Appl. Phys. Lett.*, **35**, 161 (1979).
218. J. Aranovich, A. Ortiz and R.H. Bube, *J. Vac. Sci. Technol.*, **16**, 994 (1979).
219. E. Shanthi, A. Banerjee, V. Dutta and K.L. Chopra, *J. Appl. Phys.*, **53**, 1615 (1982).
220. E. Shanthi, A. Banerjee and K.L. Chopra, *Thin Solid Films*, **88**, 93 (1982).
221. E. Shanthi, V. Dutta, A. Banerjee and K.L. Chopra, *J. Appl. Phys.*, **51**, 6243 (1980).
222. J.C. Manifacier, L. Szepessy, J.F. Bresse, M. Perotin and R. Staek, *Mat. Res. Bull.*, **14**, 109 (1979).
223. E. Shanthi, Ph.D. Thesis, Indian Institute of Technology, Delhi (1981).
224. H. Kostlin, R. Jost and W. Lerns, *Phys. Stat. Sol.* (*a*), **29**, 87 (1975).
225. R. Groth, *Phys. Stat. Sol.*, **14**, 69 (1966).
226. B.J. Baliga and S.K. Gandhi, *J. Electrochem. Soc.*, **123**, 941 (1976).
227. Y.S. Hsu and S.K. Gandhi, *J. Electrochem. Soc.*, **126**, 1434 (1979).
228. A.K. Ghosh, C. Fishman and T. Feng, *J. Appl. Phys.*, **49**, 3490 (1978).
229. J.C. Manifacier, M.De Murica and J.P. Fillard, *Thin Solid Films*, **41**, 127 (1977).
230. M. Mizuhashi, *Thin Solid Films*, **70**, 91 (1980).
231. H. Watanabe, *Jpn. J. Appl. Phys.*, **9**, 1551 (1970).
232. F. Van der Maesen and C.H.M. Witmer, *Proc. 7th Int. Conf. on Physics of Semiconductors*, Paris (1964), Academic Press Inc., New York, p. 1211.
233. P. Nath and R.F. Bunshah, *Thin Solid Films*, **69**, 63 (1980).
234. J.H. Morgan and D.E. Brodie (Private communication).
235. K. Tanaka, A. Kunioka and Y. Sakai, *Jpn. J. Appl. Phys.*, **8**, 681 (1969).
236. H. Finkernath, *Z. Phys.*, **158**, 511 (1960).
237. V.K. Miloslavskii and A.I. Ranyuk, *Opt. Spectr.*, **11**, 289 (1961).
238. G. Helwig, *Z. Phys.*, **132**, 621 (1952).
239. M. Van der Leij, Ph.D. Thesis, Delf University (1979).
240. E. Shanthi, A. Banerjee, V. Dutta and K.L. Chopra, *Thin Solid Films*, **71**, 237 (1980).
241. J.R. Boshell and R. Waghorne, *Thin Solid Filsms*, **15**, 141 (1973).
242. H.K. Muller, *Phys. Stat. Sol.*, **27**, 723 (1968).
243. J.E. Morris, M.I. Ridge, C.A. Bishop and R.P. Howson, *J. Appl. Phys.*, **51**, 1847 (1980).
244. A.J. Steckl and G. Mohammed, *J. Appl. Phys.*, **51**, 3890 (1980).
245. A. Fischer, *Z. Natur. A*, **9**, 508 (1954).
246. R.L. Weiher and R.P. Ley, *J. Appl. Phys.*, **37**, 299 (1966).

247. Y. Ohhata, F. Shinoki and S. Yoshida, *Thin Solid Films*, **59**, 255 (1979).
248. A.J. Nozik, *Phys. Rev. B*, **6**, 453 (1972).
249. G. Haacke, *Ann. Rev. Mat. Sci.*, **7**, 73 (1977).
250. A.L. Fahrenbruch, J. Aranovich, F. Courreges, T. Chynoweth and R.H. Bube, *Proc. 13th IEEE Photovoltaic Specialists Conference*, Washington, D.C. (1978), p. 281.
251. D.E. Brodie, R. Singh, J.H. Morgan, J.D. Leslie, C.J. Moore and A.E. Dixon *Proc. 14th IEEE Photovoltaic Specialists Conference*, San Diego (1980), p. 468.
252. G. Haacke, *Appl. Phys. Lett.*, **28**, 622 (1976).
253. G. Haacke, H. Ando and W.E. Mealmaker, *J. Electrochem. Soc.*, **124**, 1923 (1977).
254. K.L. Chopra, *Thin Film Phenomena*, McGraw-Hill Book Company, New York (1969).
255. L.I. Maissel and R. Glang, *Handbook of Thin Film Technology*, McGraw-Hill Book Company, New York (1970).
256. T.J. Coutts, in: *Active and Passive Thin Film Devices* (Ed., T.J. Coutts), Academic Press, London (1978), p. 57.
257. F. Abeles, in: *Physics of Thin Films*, Vol. 6 (Eds., M.H. Francombe and R.W. Hoffman), Academic Press, New York (1971), p. 151.
258. D.C. Larson, in: *Physics of Thin Films*, Vol. 6 (Eds., M.H. Francombe and R.W. Hoffman), Academic Press, New York (1971), p. 81.
259. R. Suri, A.P. Thakoor and K.L. Chopra, *J. Appl. Phys.*, **46**, 2574 (1975).
260. S.I. Raider and R. Flitsch, *J. Vac. Sci. Technol.*, **13**, 58 (1976).
261. S.M. Sze, *J. Appl. Phys.*, **38**, 2951 (1967).
262. P. Gadenne, *Thin Solid Films*, **57**, 77 (1979).
263. F.E. Carlson, G.T. Howard, A.F. Turner and H.H. Schroeder, *J. Soc. Motion Picture Television Engr.*, **65**, 136 (1956).
264. E.M. Beesley, A. Makulec and H.H. Schroeder, *Illum. Engr.*, **58**, 380 (1963).
265. B.T. Barnes, *J. Opt. Soc. Am.*, **56**, 1546 (1966).
266. G. Hass and J.E. Waylonis, *J. Opt. Soc. Am.*, **50**, 1133 (1960).
267. L.G. Schulz and F.R. Tangherlini, *J. Opt. Soc. Am.*, **44**, 362 (1954).
268. L.G. Schulz, *J. Opt. Soc. Am.*, **44**, 557 (1954).
269. S. Major, V. Dutta and K.C. Chopra, 1982 (to be published).
270. A.P. Roth and D.F. Williams, *J. Appl. Phys.*, **52**, 6685 (1981).
271. J.B. Webb, D.F. Williams and M. Buchanan, *Appl. Phys. Lett.*, **39**, 640 (1981).

References for Chapter 7

1. P.A. Crossley, G.I. Noel and M. Wolf, Final Report NASW 1427 (1968).
2. A.G. Stanley, in: *Applied Solid State Science*, Vol. 5, Academic Press, Inc., New York (1975), p. 251.
3. R. Hill, in: *Active and Passive Thin Film Devices* (Ed., T.J. Coutts), Academic Press, London (1978), p. 487.
4. M. Savelli and J. Bougnot, in: *Solar Energy Conversion*, Topics in Applied Physics, Vol. 31 (Ed., B.O. Seraphin), Springer-Verlag, New York (1979).
5. *Proc. Int. Workshop CdS Solar Cells and Other Abrupt Heterojunctions* (Eds., K.W. Boer and J.D. Meakin), University of Delaware (1975).
6. A.M. Barnett, J.A. Bragagnolo, R.B. Hall, J.E. Phillips, and J.D. Meakin, *Proc. 13th IEEE Photovoltaic Specialists Conference*, Washington, D.C. (1978), p. 419.
7. R.B. Hall, R.W. Birkmire, J.E. Phillips and J.D. Meakin, *Appl. Phys. Lett.*, **38**, 925 (1981).
8. R.B. Hall and J.D. Meakin, *Thin Solid Films*, **63**, 203 (1979).

9. W.J. Deshotels, F. Augustine and A. Carlson, 2nd Quarterly Report, Contract NASW 203, Clevite Corporation (1963).
10. D.C. Reynolds, G. Leies, L.T. Antes and R.E. Marburger, *Phys. Rev.*, **96**, 533 (1954).
11. W. Arndt, G. Bilger, G.H. Hewig, F. Pfisterer, H.W. Schock, J. Worner and W.H. Bloss, *Proc. 2nd EC Photovoltaic Solar Energy Conference* (Eds., R. Van Overstraeten and W. Palz), Berlin (April 1979), p. 826.
12. W. Arndt, G. Bilger, W.H. Bloss, G.H. Hewig, F. Pfisterer, H.W. Schock and J. Worner, *Proc. Sym. Micro*-79, Banaras Hindu University, Varanasi (January 1979).
13. J.R. Szedon, F.A. Shirland, W.J. Biter, T.W. O'Keeffe, J.A. Stoll and S.J. Fonash, Cadmium Sulphide/Copper Sulphide Heterojunction Cell Research, Final Report, Contract DE-AC03-77ET 20429 (July 1979).
14. T.S. teVelde, *Energy Conversion*, **14**, 111 (1974).
15. S.R. Das, P. Nath, A. Banerjee and K.L. Chopra, *Solid State Commun.*, **21**, 49 (1977).
16. S.R. Das, V.D. Vankar, P. Nath and K.L. Chopra, *Thin Solid Films*, **51**, 257 (1978).
17. S.R. Das, P. Nath, A. Banerjee, V.D. Vankar and K.L. Chopra, *Proc. Int. Solar Energy Congress*, New Delhi (1978), p. 694.
18. P.K. Bhat, S.R. Das, D.K. Pandya and K.L. Chopra, *Solar Energy Materials*, **1**, 215 (1979).
19. A. Banerjee, S.R. Das, A.P. Thakoor, H.S. Randhawa and K.L. Chopra, *Solid-State Electronics*, **22**, 495 (1979).
20. S.R. Das, Ph.D. Thesis, Indian Institute of Technology, Delhi (1978).
21. A.N. Casperd and R. Hill, *Proc. 1st EC Photovoltaic Solar Energy Conference*, Luxembourg (September 1977), p. 1131.
22. S. Martinuzzi, F. Zapien-Nataren, D. Vassilevski, B. Bouchikhi, M. Cadene, M. Moutaki and M. Rolland (Personal communication).
23. N.K. Annamalai, *Proc. CdS/Cu_2S and CdS/Cu-Ternary Photovoltaic Cells Sub-Contractors In-Depth Review Meeting*, Washington, D.C. (1980), p. 259.
24. A. Banerjee, Ph.D. Thesis, Indian Institute of Technology, Delhi (1978).
25. A. Banerjee, P. Nath, S.R. Das and K.L. Chopra, *Proc. Int. Solar Energy Congress*, New Delhi (1978), p. 675.
26. Bodhraj, A.P. Thakoor, A. Banerjee, D.K. Pandya and K.L. Chopra, *Proc. Natl. Solar Energy Congress*, Annamalai (1980), p. 322.
27. S. Martinuzzi, J. Oualid, F. Cabane-Brouty, A. Mostavan and J. Gervais, *Rev. Phys. Appl.*, **14**, 237 (1979).
28. S. Martinuzzi, F. Cabane-Brouty, T. Cabot, A. Franco and J. Kalliontzis (Personal communication).
29. V.P. Singh, *Proc. 13th IEEE Photovoltaic Specialists Conference*, Washington, D.C. (1978), p. 507.
30. J.F. Jordan, *Proc. 11th IEEE Photovoltaic Specialists Conference*, Scottsdale (1975), p. 508.
31. W. Müller, H. Frey, K. Radler and K.H. Schuller, *Thin Solid Films*, **59**, 327 (1979).
32. R. Hill, R. Harrison, G. Jenkins and R.L. Wilson, *Proc. Int. Conf. Future Energy Concepts*, London (1979), p. 17.
33. J.A. Thornton, *Proc. CdS/Cu_2S and CdS/Cu-Ternary Photovoltaic Cells Sub-Contractors In-Depth Review Meeting*, Washington, D.C. (1980), p. 221.
34. W.W. Anderson, *Proc. CdS/Cu_2S and CdS/Cu-Ternary Photovoltaic Cells Sub-Contractors In-Depth Review Meeting*, Washington, D.C. (1980), p. 239.
35. N. Croitoru and S. Jakobson, *Thin Solid Films*, **56**, L5 (1979).
36. E.W. Williams, K. Jones, A.J. Griffiths, D.J. Roughley, J.M. Bell, J.H. Steven, M.J. Huson, M. Rhodes and T. Costich, *Proc. 2nd EC Photovoltaic Solar Energy Conference*, Berlin (April 1979), p. 874.

37. L.C. Burton, T. Hench, G. Storti and G. Haacke, *J. Electrochem. Soc.*, **123**, 1741 (1976).
38. S. Martinuzzi, J. Oualid, D. Sarti and J. Gervais, *Thin Solid Films*, **51**, 211 (1978).
39. S.R. Das and K.L. Chopra (unpublished).
40. L.C. Burton, T.L. Hench and J.D. Meakin, *J. Appl. Phys.*, **50**, 6014 (1979).
41. J. Bougnot, M. Savelli, J. Marucchi, M. Perotin, M. Marjan, O. Maris, C. Grill and R. Pommier, *Proc. Workshop on the II–VI Solar Cells and Similar Compounds*, Montpellier (September 1979), p. II–1.
42. A. Rothwarf in Direct Solar Energy Conversion for Large Scale Terrestrial Use, Final Report No. NSF/RANN/AER 72–03478 A02/FR/75, University of Delaware (October 1975).
43. G.H. Hewig, F. Pfisterer, H.W. Schock and W.H. Bloss, *Proc. Workshop on the II–VI Solar Cells and Similar Compounds*, Montpellier (September 1979), p. VIII–I.
44. J.R. Szedon, F.A. Shirland, W.J. Biter, J.A. Stoll, H.C. Dickey and T.W. O'Keeffe, Cadmium Sulphide/Copper Sulphide Heterojunction Cell Research, Technical Progress Report No. 4, Contract EG-77-C-03-1577 (November 1978).
45. F.A. Shirland, *J. Appl. Phys.*, **50**, 4714 (1979).
46. M.K. Mukherjee, F. Pfisterer, G.H. Hewig, H.W. Schock and W.H. Bloss, *J. Appl. Phys.*, **48**, 1538 (1977).
47. R.S. Berg, R.D. Nasby and C. Lampkin, *J. Vac. Sci. Technol.*, **15**, 359 (1978).
48. Oudeacoumar, Ph.D. Thesis, Université des Sciences et Techniques du Languedoc, Montpellier (1979).
49. A.P. Thakoor, Bodhraj, D.K. Pandya and K.L. Chopra, *Proc. Int. Conf. Metallurgical Coatings*, San Francisco (1981), *Thin Solid Films*, **83**, 231 (1981).
50. F. Dutault, Ph.D. Thesis, L'Université de Haute Alsace et L'Université Louis Pasteur de Strasbourg (1974).
51. B. Baron, A.W. Catalano and E.A. Fagen, *Proc. 13th IEEE Photovoltaic Specialists Conference*, Washington, D.C. (1978), p. 406.
52. W.F. Tseng and I.G. Greenfield, Direct Solar Energy Conversion for Large Scale Terrestrial Use, Report No. NSF/RANN/SE/GI 34872/TR74/1 (February 1974).
53. F. Pfisterer, H.W. Schock and J. Worner, *Proc. 3rd EC Photovoltaic Solar Energy Conference*, Cannes (1980).
54. F.G. Satkiewicz and H.K. Charles, Jr., Sputter Ion Mass Spectrometer Analysis of Copper Sulphide/Cadmium Sulphide Solar Cell Samples, Technical Memorandum No. APL/JHU TG 1284, Johns Hopkins University (October 1975).
55. L.C. Burton, D.W. Dwight and M.V. Zeller (Personal communication).
56. S. Martinuzzi, F. Zapien-Nataren and H. Amzil, *Proc. Workshop on the II–VI Solar Cells and Similar Compounds*, Montpellier (September 1979), p. III–1.
57. E.J. Hsieh, D. Miller, K.W. Vindelov and T.G. Brown, Proc. Workshop on Cadmium Sulphide Solar Cells and Other Abrupt Heterojunctions (Eds., K.W. Boer and J.D. Meakin), University of Delaware (1975), p. 301.
58. W.W. Anderson and A.D. Jonath, Cadmium Sulphide/Copper Sulphide Heterojunction Cell Research, Final Report LMSC 0766341, Contract XJ-9-8033-1, Lockheed Palo Alto Research Laboratory (June 1980).
59. Development of a Thin Film Polycrystalline Solar Cell for Large Scale Terrestrial Use, Quarterly Progress Report E(49–18)-2538, PR 77/1 University of Delaware (July 1977).
60. H. Hadley and J. Phillips, *Proc. Workshop on Cadmium Sulphide Solar Cells and Other Abrupt Heterojunctions* (Eds., K.W. Boer and J.D. Meakin), University of Delaware (1975), pp. 134, 533.
61. F.J. Bryant and R.W. Glew, *Energy Conversion*, **14**, 129 (1975).
62. F.A. Shirland, *Advanced Energy Conversion*, **6**, 201 (1966).

63. Research Directed to Stable High Efficiency CdS Solar Cells, Semi-Annual Progress Report NSF/RANN/AER 72–03478 A04, University of Delaware (March 1976).
64. Cadmium Sulphide/Copper Sulphide Heterojunction Cell Research, Final Report EG-77-C-03-1576, University of Delaware (May 1979).
65. A. Rothwarf, J. Phillips and N.C. Wyeth, *Proc. 13th IEEE Photovoltaic Specialists Conference*, Washington, D.C. (1978), p. 999.
66. M.J. Robertson and J. Woods, *Proc. 2nd EC Photovoltaic Solar Energy Conference* (Eds., R. Van Overstraeten and W. Palz), Berlin (April 1979), p. 309.
67. C. Moses and D. Wasserman, *Solar Cells*, **1**, 218 (1980).
68. F. Pfisterer, H.W. Schock and G.H. Hewig, *Proc. 2nd EC Photovoltaic Solar Energy Conference* (Eds., R. Van Overstraeten and W. Palz), Berlin (April 1979), p. 352.
69. H.W. Schock, *Proc. Workshop on the II–VI Solar Cells and Similar Compounds*, Montpellier (September 1979), p. IX–1.
70. L.D. Partain, J.J. Oakes and I.G. Greenfield, *Proc. Workshop on Cadmium Sulphide Solar Cells and Other Abrupt Heterojunctions* (Eds., K.W. Boer and J.D. Meakin), University of Dalaware (1975), p. 346.
71. J.J. Oakes, I.G. Greenfield and L.D. Partain, *Proc. 11th IEEE Photovoltaic Specialists Conference*, Scottsdale (1975), p. 454; *J. Appl. Phys.*, **48**, 2548 (1977).
72. W.D. Gill and R.H. Bube, *J. Appl. Phys.*, **41**, 1694 (1970).
73. N.C. Wyeth, *Proc. Workshop on Cadmium Sulphide Solar Cells and Other Abrupt Heterojunctions* (Eds., K.W. Boer and J.D. Meakin), University of Delaware (1975), p. 575.
74. S.R. Das, A. Banerjee and K.L. Chopra, *Solid State Electronics*, **22**, 533 (1979).
75. R.B. Hall and V.P. Singh, *J. Appl. Phys.*, **50**, 6406 (1979).
76. F. Pfisterer, *Proc. Workshop on the II–VI Solar Cells and Similar Compounds*, Montpellier (September 1979), p. VIII–1.
77. V.P. Singh, *Proc. Workshop on Cadmium Sulphide Solar Cells and Other Abrupt Heterojunctions* (Eds., K.W. Boer and J.D. Meakin), University of Delaware (1975), p. 560.
78. G. Storti and J. Culik, The Relationships Between Preparation Parameters, Operating Characteristics and Physical Processes in Cu_2S/CdS Thin Film Solar Cells, Technical Report IEC/PV/TR/76/6, University of Delaware (November 1976).
79. Development of a Thin Film Polycrystalline Solar Cell for Large Scale Terrestrial Use, Progress Report E(49–18)–2538, University of Delaware (January 1977).
80. S. Martinuzzi and O. Mallem, *Phys. Stat. Sol.* (*a*), **16**, 339 (1973).
81. R.B. Hall, *Proc. Workshop on Cadmium Sulphide Solar Cells and Other Abrupt Heterojunctions* (Eds., K.W. Boer and J.D. Meakin), University of Delaware (1975), p. 284.
82. F. Pfisterer, H.W. Schock and W.H. Bloss (Personal communication).
83. F. Pfisterer, G.H. Hewig and W.H. Bloss, *Proc. 11th IEEE Photovoltaic Specialists Conference*, Scottsdale (1975), p. 460.
84. Kh.T. Akramov, G.Ya. Ilmayov and T.M. Razylov, *Geliotekhnika*, **11**, 3 (1975).
85. J.J. Loferski, J. Shewchun, E.A. DeMeo, R. Arnott, E.E. Crisman, R. Beaulieu, H.L. Hwang and C.C. Wu, *Proc. 12th IEEE Photovoltaic Specialists Conference*, Baton Rouge (1976), p. 496.
86. B.G. Caswell and J. Woods, *Phys. Stat. Sol.* (*a*), **44**, K47 (1977).
87. W.E. Devaney, A.M. Barnett, G.M. Storti and J.D. Meakin, *IEEE Trans. Electron Dev.*, **ED-26**, 205 (1979).
88. J. Besson, T. Nguyen Duy, A. Gauthier, W. Palz, C. Martin and J. Vedel, *Proc. 11th IEEE Photovoltaic Specialists Conference*, Scottsdale (1975), p. 468.

89. W. Palz, J. Besson, T. Nguyen Duy and J. Vedel, *Proc. 10th IEEE Photovoltaic Specialists Conference*, Palo Alto (1973), p. 69.
90. J. Dielman, *Proc. Workshop on Cadmium Sulphide Solar Cells and Other Abrupt Heterojunctions* (Eds., K.W. Boer and J.D. Meakin), University of Delaware (1975), p. 92.
91. L.D. Partain and C.E. Birchenall, *Proc. Workshop on Cadmium Sulphide Solar Cells and Other Abrupt Heterojunctions* (Eds., K.W. Boer and J.D. Meakin), University of Delaware (1975), p. 355.
92. N.C. Wyeth and A.W. Catalano, *Proc. 12th IEEE Photovoltaic Specialists Conference*, Baton Rouge (1976), p. 471.
93. J. Marek, *Proc. 3rd EC Photovoltaic Solar Energy Conference*, Cannes (1980).
94. G. Ya. Umarov, Kh. T. Akramov, T.M. Razykov and A.T. Teshabaev, *Geliotekhnika*, **13**, 20 (1977).
95. L.C. Burton and T.L. Hench, *Appl. Phys. Lett.*, **29**, 612 (1976).
96. L.C. Burton, B. Baron, W. Devaney, T.L. Hench, S. Orenz and J.D. Meaking, Studies Related to $Zn_xCd_{1-x}S$ Solar Cells, Technical Report IEC/PV/TR/76/4, University of Delaware (November 1976).
97. V.P. Singh and J.F. Jordan, *IEEE Electron Dev. Lett.*, **EDL–2**, 137 (1981).
98. G.H. Hewig and W.H. Bloss (Personal communication).
99. B.G. Caswell, G.J. Russell and J. Woods, *J. Phys. D: Appl. Phys.*, **8**, 1889 (1975).
100. A. Rothwarf, Crystallite Size Considerations in Polycrystalline Solar Cells, Technical Report IEC/PV/TR/76/5, University of Delaware (November 1976).
101. A. Amith, *J. Vac. Sci. Technol.*, **15**, 353 (1978).
102. A. Amith, *J. Appl. Phys.*, **50**, 1160 (1979).
103. N. Nakayama, A. Gyobu and N. Morimoto, *Jpn. J. Appl. Phys.*, **10**, 1415 (1971).
104. J. Bougnot, F. Guastavino, G.M. Moussalli and M. Savelli, *Proc. Workshop on Cadmium Sulphide Solar Cells and Other Abrupt Heterojunctions* (Eds., K.W. Boer and J.D. Meakin), University of Delaware (1975), p. 337.
105. A. Vecht, Methods of Activating and Recrystallising Thin Films of II–VI Compounds, in: *Physics of Thin Films*, Vol. 3 (Eds., G. Hass and R.E. Thun), Academic Press, New York (1966), p. 165.
106. H.H. Woodbury, *J. Appl. Phys.*, **36**, 2287 (1965).
107. H. Luquet, L. Szepessy, J. Bougnot and M. Savelli, *Proc. 11th IEEE Photovoltaic Specialists Conference*, Scottsdale (1975), p. 445.

References for Chapter 8

1. P.A. Crossley, G.I. Noel and M. Wolf, Final Report NASW 1427 (1968).
2. T.L. Chu, S.S. Chu, C.L. Lin and R.M. Davis, *Abstracts of the Fourth Annual Photovoltaic Advanced Research and Development Conference*, Solar Energy Research Institute, Colorado (November 1980), p. 173.
3. K.R. Sarma, M.J. Rice and R.N. Legge, *Abstracts of the Fourth Annual Photovoltaic Advanced Research and Development Conference*, Solar Energy Research Institute, Colorado (November 1980), p. 151.
4. M. Wolf, *Proc. 8th IEEE Photovoltaic Specialists Conference*, Seattle (1970), p. 360.
5. J. Lindmayer and J.F. Allison, *COMSAT Tech. Rev.*, **3**, 1 (1973).
6. M.P. Godlewski, C.A. Baraona and H.W. Brandhorst, Jr., *Proc. 10th IEEE Photovoltaic Specialists Conference*, Palo Alto (1973), p. 40.

7. J. Mandelkorn, J.H. Lamneck and L.R. Scudder, *Proc. 10th IEEE Photovoltaic Specialists Conference*, Palo Alto (1973), p. 207.
8. J.G. Fossum, *IEEE Trans. Electron Dev.*, **ED-24**, 322 (1977).
9. C.P. Khattak and F. Schmid, *Proc. 13th IEEE Photovoltaic Specialists Conference*, Washington, D.C. (1978), p. 137.
10. R.G. Seidensticker, R.E. Kothmann, J.P. McHugh, C.S. Duncan, R.H. Hopkins, P.D. Blais, J.R. Davis and A. Rohatgi, *Proc. 13th IEEE Photovoltaic Specialists Conference*, Washington, D.C. (1978), p. 358.
11. B. Chalmer, H.E. LaBelle, Jr. and A.I. Mlavsky, *J. Crystal Growth*, **13–14**, 84 (1972).
12. B.H. Mackintosh, T. Swek, J.P. Kalejs, E.M. Sachs, S. Nagy and F.V. Wald, *Proc. 13th IEEE Photovoltaic Specialists Conference*, Washington, D.C. (1978), p. 376.
13. *Abstracts of the Fourth Annual Photovoltaic Advanced Research and Development Conference*, Solar Energy Research Institute, Colorado (November 1980), p. 256.
14. T.L. Chu, S.S. Chu, E.D. Stokes, C.L. Lin and R. Abderrassoul, *Proc. 13th IEEE Photovoltaic Specialists Conference*, Washington, D.C. (1978), p. 1106.
15. P.H. Robinson, R.V. D'Aiello, D. Richman and B.W. Faughan, *Proc. 13th IEEE Photovoltaic Specialists Conference*, Washington, D.C (1978), p. 1111.
16. T.L. Chu, S.S. Chu, G.A. Van der Leeden, C.J. Lin and J.R. Boyd, *Solid-State Electronics*, **21**, 781 (1978).
17. T. Warabisako, T. Saitoh, H. Itoh, N. Nakamura and T. Tokuyama, *Jpn. J. of Appl. Phys.*, **17** (Supplement 17–1), 309 (1978).
18. P.H. Robinson, R.V. D'Aiello and D. Richman, *Proc. 14th IEEE Photovoltaic Specialists Conference*, San Diego (1980), p. 54.
19. F. Schmid, C.P. Khattak and M. Basaran, *Abstracts of the Fourth Annual Photovoltaic Advanced Research and Development Conference*, Solar Energy Research Institute, Colorado (November 1980), p. 166.
20. T.L. Chu, J.C. Lien, H.C. Mollenkopf, S.C. Chu, K.W. Heizer, F.W. Voltmer and G.F. Wakefield, *Solar Energy*, **17**, 229 (1975).
21. T.L. Chu and K.N. Singh, *Solid-State Electronics*, **19**, 837 (1976).
22. T. Saitoh, T. Warabisako, H. Itoh, N. Nakamura, H. Tamura, S. Minagawa and T. Tokuyama, *Jpn. J. Appl. Phys.*, **16**, (Supplement 16–1), 413 (1977).
23. C. Feldman, N.A. Blum, H.K. Charles, Jr. and F. Satkiewicz, *J. Electronic Mat.*, **7**, 309 (1978).
24. R.B. Maciolek, J.D. Heaps and J.D. Zook, *J. Electronic Mat.*, **8**, 31 (1979).
25. W.A. Anderson, G. Rajeswaran, F. Kai and M. Jackson, *Abstracts of the Fourth Annual Photovoltaic Advanced Research and Development Conference*, Solar Energy Research Institute, Colorado (November 1980), p. 185.
26. E. Fabre and Y. Baudet (Communication); E. Fabre, M. Mautref and A. Mircea, *Appl. Phys. Lett.*, **27**, 239 (1975).
27. C.T. Ho, R.O. Bell and F.V. Wald, *Appl. Phys. Lett.*, **31**, 463 (1977).
28. C.T. Ho and J.D. Mathias, *Solid-State Electronics*, **24**, 115 (1981).
29. W.A. Anderson, G. Rajeswaran, K. Rajkanan and G. Hoeft, *IEEE Elec. Rev. Lett.*, **EDL-1**, 128 (1980).
30. C.H. Seager, D.S. Ginley and J.D. Zook, *Appl. Phys. Lett.*, **36**, 831 (1980).
31. S.I. Soclof and P.A. Iles, *Proc. 11th IEEE Photovoltaic Specialists Conference*, Scottsdale (1975), p. 56.
32. R.B. Hilborn, Jr. and J. Lin, *Proc. National Workshop on Low-Cost Polycrystalline Silicon Solar Cells*, Southern Methodist University, Dallas (May 1976), p. 246.
33. H.C. Card and E.S. Yang, *IEEE Trans. Electron. Dev.*, **ED–24**, 397 (1977).
34. H.J. Hovel, *Semiconductors and Semimetals*, Vol. 11; Solar Cells, Academic Press, New York (1975), p. 107.

35. T.L. Chu, H.C. Mollenkopf and S.S. Chu, *J. Electrochem. Soc.*, **123**, 106 (1976).
36. C.H. Seager and D.S. Ginley (unpublished).
37. H.C. Card and W. Hwang, *IEEE Trans. Electron Dev.*, **ED–27**, 700 (1980).
38. C.M. Wu, E.S. Yang, W. Hwang and H.C. Card, *IEEE Trans. Electron Dev.*, **ED–27**, 687 (1980).
39. D. Redfield, *Appl. Phys. Lett.*, **26**, 647 (1974).

References for Chapter 9

1. J.M. Woodall and H.J. Hovel, *J. Vac. Sci. Technol.*, **12**, 1000 (1975).
2. J.M. Woodall and H.J. Hovel, *Appl. Phys. Lett.*, **21**, 379 (1972); *Appl. Phys. Lett.*, **27**, 447 (1975).
3. Zh.I. Alferov, V.M. Andreev, M.B. Kagan, I.I. Protasov and V.G. Trofim, *Sov. Phys. Semicond.*, **4**, 2047 (1971).
4. L.W. James and R.L. Moon, *Appl. Phys. Lett.*, **26**, 467 (1975).
5. R.J. Stirn and Y.C.M. Yeh, *Appl. Phys. Lett.*, **27**, 95 (1975).
6. C.O. Bozler and J.C.C. Fan, *Appl. Phys. Lett.*, **31**, 629 (1977). J.C.C. Fan, C.O. Bozler and R.W. McClelland, *Proc. 15th IEEE Photovoltaic Specialists Conference*, Kissimmee (1981), p. 666.
7. S.S. Chu, T.L. Chu and Y.T. Lee, *IEEE Trans. Electron Dev.*, **ED–27**, 640 (1980).
8. S.S. Chu, T.L. Chu and M.S. Lan, *J. Appl. Phys.*, **50**, 5805 (1979).
9. S.S. Chu, T.L. Chu, H.T. Yang and K.H. Hong, *J. Electrochem. Soc.*, **125**, 1668 (1978).
10. S.S. Chu, T.L. Chu, Y.T. Lee, C.L. Jiang and A.B. Kuper, *Proc. 14th IEEE Photovoltaic Specialists Conference*, San Diego (January 1980), p. 1306.
11. R.D. Dupuis, P.D. Dapkus, R.D. Yingling and L.A. Mondy, *Appl. Phys. Lett.*, **31**, 201 (1977).
12. K.P. Pande, D.H. Reep, S.K. Shastry, A.J. Weiser, J.M. Borrego and S.K. Ghandhi, *IEEE Trans. Electron Dev.*, **ED–27**, 635 (1980).
13. S.M. Vernon, A.E. Blakeslee and H.J. Hovel, *J. Electrochem. Soc.*, **126**, 703 (1979).
14. S.K. Ghandhi, J.M. Borrego, D. Reep, Y.S. Hsu and K.P. Pande, *Appl. Phys. Lett.*, **34**, 699 (1979).
15. J. Mimila-Arroyo, A. Bouazzi and G. Cohen-Salal, *Revue Phys. Appl.*, **12**, 423 (1977).
16. G. Cohen-Solal, M. Barbe, D. Lincot, Y. Marfaing and R. Triboulet, *Proc. Workshop on the II–VI Solar Cells and Similar Compounds*, Montpellier (September 1979), p. XIII–I.
17. H. Jager, E. Seipp and B. Füssl, *Proc. Workshop on the II–VI Solar Cells and Similar Compounds*, Montpellier (September 1979), p. XVIII–1.
18. A.L. Fahrenbruch, J. Aranovich, F. Courreges, S.Y. Yin and R.H. Bube, *Proc. 2nd European Commission Photovoltaic Solar Energy Conference* (Eds., R. Van Overstraeten and W. Palz), Berlin (April 1979), p. 608.
19. K. Yamaguchi, N. Nakayama, H. Matsumoto and S. Ikegami, *Jpn. J. Appl. Phys.*, **16**, 1203 (1977).
20. K. Yamaguchi, N. Nakayama, H. Matsumoto, Y. Hioki and S. Ikegami, *Jpn. J. Appl. Phys.*, **14**, 1397 (1975).
21. J. Bernard, *Proc. Workshop on the II–VI Solar Cells and Similar Compounds*, Montpellier (September 1979), p. XV–1.
22. F.G. Courreges, A.L. Fahrenbruch and R.H. Bube, *J. Appl. Phys.*, **51**, 2175 (1980).
23. J. Calderer, H. Luquet and M. Savelli, *Proc. Workshop on the II–VI Solar Cells and Similar Compounds*, Montpellier (September 1979), p. XIV–1.
24. A.L. Fahrenbruch, F. Buch, K. Mitchell and R.H. Bube, *Proc. 11th IEEE Photovoltaic Specialists Conference*, Scottsdale (1975), p. 490.

25. R.H. Bube, F. Buch, A.L. Fahrenbruch, Y.Y. Ma and K.W. Mitchell, *IEEE Trans. Electron Dev.*, **ED–24**, 487 (1977).
26. N. Nakayama, H. Matsumoto, A. Nakano, S. Ikegami, H. Uda and T. Yamashita, *Jpn. J. Appl. Phys.*, **19**, 703 (1980).
27. N. Nakayama, H. Matsumoto, K. Yamaguchi, S. Ikegami and Y. Hioki, *Jpn. J. Appl. Phys.*, **15**, 228 (1976).
28. D.A. Cusano, *Solid-State Electronics*, **6**, 217 (1963).
29. G.P. Naumov and D.V. Nicholaena, *Sov. Phys. Solid State*, **3**, 2718 (1962).
30. J. Bernard, R. Lacon, C. Paparoditis and M. Rodot, *Revue Phys. Appl.*, **1**, 211 (1966).
31. E.W. Justi, G. Schneider and J. Secedynski, *Energy Conversion*, **13**, 53 (1973).
32. D. Bonnet and H. Rabenhorst, *Proc. Int. Colloquium on Solar Cells*, Toulouse (July 1970), p. 155; *Proc. 10th IEEE Photovoltaic Specialists Conference*, Palo Alto (1971), p. 129.
33. Yu.A. Vodakov, G.A. Lomakina, G.P. Naumov and Yu.A. Maslakovets, *Sov. Phys. Solid State*, **2**, 11 (1961).
34. H. Okimura, T. Matsumae and R. Makabe, *Thin Solid Films*, **71**, 53 (1980).
35. L.F. Buldhaupt, R.A. Mickelsen, J.M. Stewart and W.S. Chen, Emerging Materials Systems for Solar Cell Applications-$Cu_{2-x}Se$, Boeing Aerospace Company, Final Report DE-AC04-79ET-23005 (April 1980).
36. E. Rickus and D. Bonnet, *Proc. Workshop on the II–VI Solar Cells and Similar Compounds*, Montpellier (September 1979), p. XX–10.
37. F.J. García and M.S. Tomar, *Thin Solid Films*, **69**, 137 (1980).
38. E.A. Fagen, *J. Appl. Phys.*, **50**, 6505 (1979).
39. A. Catalano, V. Dalal, E.A. Fagen, R.B. Hall, J.V. Masi, J.D. Meakin, G. Warfield and A.M. Barnett, *Proc. 1st European Commission Photovoltaic Solar Energy Conference*, Luxembourg (1977), p. 644.
40. N.C. Wyeth and A. Catalano, *J. Appl. Phys.*, **50**, 1403 (1979).
41. *Abstracts of the Fourth Annual Photovoltaic Advanced Research and Development Conference*, Solar Energy Research Institute, Colorado (November 1980).
42. S. Wagner, J.L. Shay, K.J. Bachmann and E. Buehler, *Appl. Phys. Lett.*, **26**, 229 (1975).
43. J.L. Shay, S. Wagner, M. Bettini, K.J. Bachmann and E. Buehler, *IEEE Trans. Electron Dev.*, **ED–24**, 483 (1977).
44. K. Ito and T. Ohsawa, *Jpn. J. Appl. Phys.*, **14**, 1259 (1975).
45. L.L. Kazmerski, F.R. White, M.S. Ayyagiri, Y.J. Juang and R.P. Patterson, *J. Vac. Sci. Technol.*, **14**, 65 (1977).
46. T. Saitoh, S. Matsubara and S. Minagawa, *Jpn. J. Appl. Phys.*, **16**, 807 (1977).
47. K.J. Bachmann, E. Buehler, J.L. Shay and S. Wagner, *Appl. Phys. Lett.*, **29**, 121 (1976).
48. J.L. Shay, S. Wagner and H.M. Kasper, *Appl. Phys. Lett.*, **27**, 89 (1975).
49. S. Wagner, J.L. Shay, P. Migliorato and H.M. Kasper, *Appl Phys. Lett.*, **25**, 434 (1974).
50. Y. Kokubun and M. Wada, *Jpn. J. Appl. Phys.*, **16**, 879 (1977).
51. E. Elliott, R.D. Tomlinson, J. Parkes and M.J. Hampshire, *Thin Solid Films*, **20**, 525 (1974).
52. L.L. Kazmerski, M.S. Ayyagiri and G.A. Sanborn, *Bull. Am. Phys. Soc.*, **20**, 391 (1975).
53. L.L. Kazmerski, F.R. White and G.K. Morgan, *Appl. Phys. Lett.*, **29**, 268 (1976).
54. L.F. Buldhaupt, R.A. Mickelsen and W.S. Chen, *Proc. CdS/Cu_2S and CdS/Cu-Ternary Photovoltaic Cells Sub-Contractors In-Depth Review Meeting*, Washington, D.C. (September 1980), p. 1.
55. J.B. Mooney and C.W. Bates, *Proc. CdS/Cu_2S and CdS/Cu-Ternary Photovoltaic Cells Sub-Contractors In-Depth Review Meeting*, Washington, D.C. (September 1980), p. 17.
56. J.J. Loferski, *Proc. CdS/Cu_2S and CdS/Cu-Ternary Photovoltaic Cells Sub-Contractors In-Depth Review Meeting*, Washington, D.C. (September 1980), p. 37.

57. L.L. Kazmerski, R.B. Cooper, F.R. White and A.J. Merrill, *IEEE Trans. Electron Dev.*, **ED–24**, 496 (1977).
58. R.A. Mickelsen, Personal communication (July 1981).
59. L.C. Olsen and R.C. Bohara, *Proc. 11th IEEE Photovoltaic Specialists Conference*, Scottsdale (1975), p. 381.
60. P.H. Fang, *J. Appl. Phys.*, **45**, 4672 (1974).
61. A.K. Ghosh, D.L. Morel, T. Feng, R.F. Shaw and C.A. Rome, Jr., *J. Appl. Phys.*, **45**, 230 (1974).
62. A.K. Ghosh and T. Feng, *J. Appl. Phys.*, **44**, 2781 (1973).
63. C.W. Tang and A.C. Albrecht, *Nature*, **254**, 507 (1975).
64. R.O. Loutfy and J.H. Sharp, *J. Chem. Phys.* **71**, 1211 (1979).
65. D.L. Morel, A.K. Ghosh, T. Feng, E.L. Stogryn, P.E. Purwin, R.F. Shaw and C. Fishman, *Appl. Phys. Lett.*, **32**, 495 (1978).
66. M. Schoijet, *Solar Energy Materials*, **1**, 43 (1979).

References for Chapter 10

1. R.C. Chittick, J.H. Alexander and H.F. Sterling, *J. Electrochem. Soc.*, **116**, 77 (1969).
2. W.E. Spear, P.G. LeComber, S. Kinmond and M.H. Brodsky, *Appl. Phys. Lett.*, **28**, 105 (1976).
3. W.E. Spear and P.G. LeComber, *Solid State Commun.*, **17**, 1193 (1975).
4. D.E. Carlson, U.S. Patent No. 4,064,521 (1977).
5. D.E. Carlson and C.R. Wronski, *Appl. Phys. Lett.*, **28**, 671 (1976).
6. A. Madan, S.R. Ovshinsky and E. Benn, *Phil. Mag. B*, **40**, 259 (1979).
7. K. Rajkanan, R. Singh and W.A. Anderson, *Proc. 14th IEEE Photovoltaic Specialists Conference*, San Diego (1979), p. 439.
8. P. Viktorovitch and G. Moddel, *J. Appl. Phys.*, **51**, 4847 (1980).
9. J.A. McMillan and E.M. Peterson, *J. Appl. Phys.*, **50**, 5238 (1979).
10. P. John, I.M. Odeh, M.J.K. Thomas, M.J. Tricker, F. Riddoch and J.I.B. Wilson, *Phil. Mag. B*, **42**, 671 (1980).
11. M.H. Brodsky and P.A. Leary, *Proc. 8th Int. Conf. Amorphous and Liquid Semiconductors*, Cambridge (August 1979).
12. P. John, M. Odeh, M.J.K. Thomas, M.J. Tricker, J. McGill, A. Wallace and J.I.B. Wilson, *Proc. 8th Int. Conf. Amorphous and Liquid Semiconductors*, Cambridge (August 1979), Part I, p. 237.
13. R. Mosseri, C. Sella and J. Dixmier, *Phys. Stat. Sol. (a)*, **52**, 475 (1979).
14. A. Madan, J. McGill, W. Czubatyj, J. Yang and S.R. Ovshinsky, *Appl. Phys. Lett.*, **37**, 826 (1980).
15. J.I.B. Wilson, J. McGill and S. Kinmond, *Nature*, **272**, 152 (1978).
16. F. Riddoch, A. Wallace and J.I.B. Wilson, *Solar Cells*, **1**, 99 (1979).
17. J.I.B. Wilson and P. Robinson, *Solid-State Electronics*, **2**, 489 (1978).
18. S. Guha, K.L. Narasimhan, R.V. Navkhandewala and S.M. Pietruszko, *Appl. Phys. Lett.*, **37**, 572 (1980).
19. D.G. Ast and M.H. Brodsky, *Proc. 8th Int. Conf. Amorphous and Liquid Semiconductors*, Cambridge (August 1979); IBM Research Report No. RC 7598 (1979); *Physics of Semiconductors* (1978), p. 1159.
20. D.A. Anderson and W.E. Spear, *Phil. Mag.*, **36**, 695 (1977); **35**, 1 (1977).
21. B.Y. Tong, *Physics in Canada*, **36**, 26 (1980).

22. D.A. Anderson, G. Moddel, M.A. Paesler and W. Paul, *J. Vac. Sci. Technol.*, **16**, 906 (1979).
23. G.H. Bauer and G. Bilger (Personal communication).
24. R. Messier and I.S.T. Tsong, Black a-Si:H Sputtered Films for Photovoltaic Solar Cells, Final Technical Progress Report No. DOE/ET/23038–4, Contract No. DE-ACO3-79ET 23038 (1980).
25. *Abstracts of the Fourth Annual Photovoltaic Advanced Research and Development Conference*, Solar Energy Research Institute, Colorado (November 1980).
26. M.H. Brodsky and J.J. Cuomo, *IBM Technical Disclosure Bulletin*, **19**, 4802 (1977).
27. M.H. Brodsky, *Thin Solid Films*, **40**, L23 (1977).
28. M.H. Brodsky, M.A. Frisch, J.F. Ziegler and W.A. Lanford, *Appl. Phys. Lett.*, **30**, 561 (1977).
29. P.G. LeComber, D.I. Jones and W.E. Spear, *Phil. Mag.*, **35**, 1173 (1977).
30. D.E. Carlson, *IEEE Trans. Electron Dev.*, **ED–24**, 449 (1977).
31. C.R. Wronski, *IEEE Trans. Electron Dev.*, **ED–24**, 351 (1977).
32. T.D. Moustakas, D.A. Anderson and W. Paul, *Solid State Commun.*, **23**, 155 (1977).
33. T.D. Moustakas and W. Paul (Personal communication); *Phys. Rev. B*, **16**, 1564 (1977).
34. A.R. Moore, *Appl. Phys. Lett.*, **31**, 762 (1977).
35. Z.S. Jan and R.H. Bube, *J. Electronic Mat.*, **8**, 47 (1979).
36. P.G. LeComber and W.E. Spear, in: *Amorphous Semiconductors* (Ed., M.H. Brodsky), Topics in Applied Physics, Vol. 36, Springer-Verlag, Heidelberg (1979), p. 251.
37. D.E. Carlson and C.R. Wronski, in: *Amorphous Semiconductors* (Ed., M.H. Brodsky), Topics in Applied Physics, Vol. 36, Springer-Verlag, Heidelberg (1979), p. 287.
38. J.I.B. Wilson, J. McGill and D. Weaire, *Adv. Phys.*, **27**, 365 (1978).
39. D.E. Carlson, *Solar Energy Materials*, **3**, 503 (1980).
40. Abstracts of Amorphous Silicon/Materials Contractor's Review Meeting, RCA Laboratories, Princeton (May 29–June 1, 1979).
41. F. Riddoch and J.I.B. Wilson, *Solar Cells*, **2**, 141 (1980).
42. R.E. Stapleton, Amorphous Silicon Solar Cells, Report No. 79-1F5-THINS-M1 (June 1979).
43. R.A. Gibson, P.G. LeComber and W.E. Spear, *Appl. Phys.*, **21**, 307 (1980).
44. P. Nath, Ph.D. Thesis, Indian Institute of Technology, Delhi (1975).
45. K.L. Chopra and P. Nath, *Phys. Stat. Sol. (a)*, **33**, 333 (1976).
46. P.J. Zanzucchi, C.R. Wronski and D.E. Carlson, *J. Appl. Phys.*, **48**, 5227 (1977).
47. A. Triska, D. Dennison, H. Fritzsche, *Bull. Am. Phys. Soc.*, **20**, 392 (1975).
48. J.I. Pankove and D.E. Carlson, *Appl. Phys. Lett.*, **31**, 450 (1977).
49. G.J. Clark, C.W. White, D.D. Alfred, B.R. Appleton, C.W. Maggee and D.E. Carlson, *Appl. Phys. Lett.*, **31**, 582 (1977).
50. J.R. Pawlik and W. Paul, *Proc. 7th Int. Conf. Amorphous and Liquid Semiconductors*, Edinburgh (1977), p. 437.
51. D.A. Anderson, T.D. Moustakas and W. Paul, *Proc. 7th Int. Conf. Amorphous and Liquid Semiconductors*, Edinburgh (1977), p. 334.
52. M.A. Paesler and W. Paul, *Phil. Mag. B*, **41**, 393 (1980).
53. E.C. Freeman and W. Paul, *Phys. Rev. B*, **20**, (1978); *Phys. Rev. B*, **18**, 4288 (1978).
54. G.A.N. Connell and J.R. Pawlik, *Phys. Rev. B*, **13**, 787 (1976).
55. W. Paul, A.J. Lewis, G.A.N. Connell and J.D. Moustakas, *Solid State Commun.*, **20**, 969 (1976).
56. A.J. Lewis, G.A.N. Connell, W. Paul, J.R. Pawlik and R.J. Temkin, *Proc. Int. Conf. Tetrahedrally Bonded Amorphous Semiconductors* (Eds., M.H. Brodsky, S. Kirkpatrick and D. Weaire), American Institute of Physics, New York (1974), p. 27.
57. S.K. Bahl and S.M. Bhagat, *J. Non-Cryst. Solids*, **17**, 409 (1975).

58. F.H. Cocks, P.L. Jones, S.F. Cogan, L.J. Dimmey, S.R. Wright, A. Korhonnen, K.D. Moore, H. Park, A.J. Scharman, P.P. Thogersen and B.L. Zalph, *Abstracts of the Fourth Annual Photovoltaic Advanced Research and Development Conference*, Solar Energy Research Institute, Colorado (November 1980), p. 107.
59. J.C. Knights, *Phil. Mag.*, **34**, 663 (1976).
60. C.C. Tsai, H. Fritzsche, M.H. Tanielian, P.G. Gaczi, D.D. Persans and M.A. Vesaghi, *Proc. 7th Int. Conf. Amorphous and Liquid Semiconductors*, Edinburgh (1977), p. 339.
61. M.A. Paesler, D.A. Anderson, E.C. Freeman, G. Moddel and W. Paul, *Phys. Rev. Lett.*, **41**, 1492 (1978).
62. W.T. Pawlewicz, *J. Appl. Phys.*, **49**, 5595 (1978).
63. G. Turban, Y. Catherine and B. Grolleau, *Thin Solid Films*, **67**, 302 (1980).
64. I. Haller, *Bull. Am. Phys. Soc.*, **25**, 294 (1980).
65. M.H. Brodsky, M. Cardona and J.J. Cuomo, *Phys. Rev. B*, **16**, 3556 (1977).
66. B. von Roedern, L. Ley and M. Cardona, *Phys. Rev. Lett.*, **39**, 1576 (1977).
67. D. Weaire, *Contemporary Phys.*, **17**, 173 (1976).
68. A. Barna, P.B. Barna, G. Radnoczi, L. Toth and P. Thomas, *Phys. Stat. Sol. (a)*, **41**, 81 (1977).
69. J.C. Knights and R.A. Lujan, *Appl. Phys. Lett.*, **35**, 244 (1979).
70. J.C. Knights, G. Lucovsky and R.J. Nemanich, *J. Non-Cryst. Solids*, **32**, 393 (1979).
71. J.C. Knights, *J. Non-Cryst. Solids*, **35–36**, 159 (1980).
72. J.C. Knights, G. Lucovsky and R.J. Nemanich, *Phil. Mag. B*, **37**, 467 (1978).
73. G. Lucovsky, R.J. Nemanich and J.C. Knights, *Phys. Rev. B*, **19**, 2064 (1979).
74. J.C. Knights, T.M. Hayes and J.C. Mikkelsen, *Phys. Rev. Lett.*, **39**, 712 (1977).
75. R.A. Street, J.C. Knights and D.K. Biegelsen, *Phys. Rev. B*, **18**, 1880 (1978).
76. D.K. Biegelsen, J.C. Knights, R.A. Street, C.C. Tsang and R.M. White, *Phil. Mag. B*, **37**, 477 (1978).
77. J.C. Knights, *Solid State Commun.*, **21**, 983 (1977).
78. R. Tsu, M. Izu and S.R. Ovshinsky, *Bull. Am. Phys. Soc.*, **25**, 295 (1980).
79. L.R. Gilbert, R. Messier and R. Roy, *Thin Solid Films*, **54**, 151 (1978).
80. P. Swab, S.V. Krishnaswamy and R. Messier, *J. Vac. Sci. Technol.*, **17**, 362 (1980).
81. R. Messier, S.V. Krishnaswamy, L.R. Gilbert and P. Swab, *J. Appl. Phys.*, **51**, 1611 (1980).
82. D.E. Carlson, C.W. Magee and A.R. Triano, *J. Electrochem. Soc.*, **126**, 688 (1979).
83. V.A. Singh, C. Weigel, J.W. Corbett and L.M. Roth, *Phys. Stat. Sol. (b)*, **81**, 637 (1977).
84. M.H. Tanielian, H. Fritzsche and C.C. Tsai, *Bull. Am. Phys. Soc.*, **22**, 336 (1977).
85. W. Beyer and R. Fischer, *Appl. Phys. Lett.*, **31**, 850 (1977).
86. R.J. Loveland, W.E. Spear and A. Al-Sharboty, *J. Non-Cryst. Solids*, **13**, 55 (1973); **15**, 410 (1974).
87. W.E. Spear and P.G. LeComber, *Phil. Mag.*, **33**, 935 (1976).
88. D.L. Staebler and C.R. Wronski, *Extended Abstracts of the Fall Meeting of the Electrochemical Society*, Atlanta (October 1977), p. 805.
89. D.L. Staebler and C.R. Wronski, *Appl. Phys. Lett.*, **31**, 292 (1977).
90. C.R. Wronski and D.E. Carlson, *Proc. 7th Int. Conf. Amorphous and Liquid Semiconductors*, Edinburgh (1977), p. 452.
91. P.G. LeComber, A. Madan and W.E. Spear, *J. Non-Cryst. Solids*, **11**, 219 (1972).
92. W. Meyer and H. Neldel, *Z. Tech. Phys.*, **18**, 588 (1937).
93. J.G. Simmons, *Phys. Rev.*, **155**, 657 (1967).
94. P.G. LeComber and W.E. Spear, *Phys. Rev. Lett.*, **25**, 509 (1970).
95. W.E. Spear and P.G. LeComber, *J. Non-Cryst. Solids*, **8–10**, 727 (1972).
96. A. Madan, P.G. LeComber and W.E. Spear, *J. Non-Cryst. Solids*, **20**, 239 (1976).
97. D.E. Carlson and C.W. Magee, *Appl. Phys. Lett.*, **33**, 81 (1978).

98. D.E. Carlson and C.W. Magee, *Proc. 2nd EC Photovoltaic Solar Energy Conference*, Berlin (April 1979), p. 312.
99. J.J. Hanak and V. Korsun, *Proc. 13th IEEE Photovoltaic Specialists Conference*, Washington, D.C. (1978), p. 780.
100. D.E. Carlson, *Proc. 14th IEEE Photovoltaic Specialists Conference*, San Diego (1980), p. 1408.
101. D.E. Carlson and C.R. Wronski, *J. Electronic Mat.*, **6**, 95 (1977).
102. G.H. Dohler and H. Heyszenan, *Phys. Rev. B*, **12**, 641 (1975).
103. G.A. Swartz and R. Williams, *Proc. 14th IEEE Photovoltaic Specialists Conference*, San Diego (1980), p. 1224.
104. H.C. Card and E.H. Roderick, *J. Phys. D: Appl. Phys.*, **4**, 1589 (1971).
105. D.L. Staebler, *J. Non-Cryst. Solids*, **35–36**, 387 (1980).
106. R.A. Street and D.K. Biegelsen, *Solid State Commun.*, **33**, 1159 (1980).
107. R.S. Crandall, R. Williams and B.E. Tompkins, *J. Appl. Phys.*, **50**, 5506 (1979).
108. R. Crandall, *Appl. Phys. Lett.*, **36**, 607 (1980).
109. G.A. Swartz, See ref. 39.
110. T.S. Nashashibi, I.G. Austin and T.M. Searle, *Phil. Mag.*, **35**, 831 (1977).
111. Z.S. Jan, R.H. Bube and J.C. Knights, *J. Appl. Phys.*, **51**, 3278 (1980).
112. J.J. Hanak, B. Faughnan, V. Korsun and J.P. Pellicane, *Proc. 14th IEEE Photovoltaic Specialists Conference*, San Diego (1980), p. 1209.
113. R.H. Williams, R.R. Varma, W.E. Spear and P.G. LeComber, *J. Phys. C: Solid State*, **12**, L209 (1979).
114. W.E. Spear, *Adv. Phys.*, **26**, 312 (1977).
115. D.J. Morel, A.K. Ghosh, F. Feng, G.L. Stogryn, P.E. Purwin, R.S. Shaw and C. Fishman, *Appl. Phys. Lett.*, **32**, 495 (1978).
116. Y. Hamakawa, H. Okamoto and Y. Nitta, *Appl. Phys. Lett.*, **35**, 187 (1979).
117. D. Engemann, R. Fischer and J. Knecht, *Appl. Phys. Lett.*, **32**, 567 (1978).
118. J.C. Knights, R.A. Street and G. Lucokvsky, *J. Non-Cryst. Solids*, **35–36**, 391 (1980).
119. J.J. Hanak, V. Korsun and J.P. Pellicane, *Proc. 2nd EC Photovoltaic Solar Energy Conference*, Berlin (April 1979), p. 270.
120. K. Tanaka, S. Yamasaki, K. Nakagawa, A. Matsuda, H. Okushi, M. Matsumura and S. Iizima, *J. Non-Cryst. Solids*, **35–36**, 475 (1980).
121. M. Ohnishi, T. Fukatsu and S. Tsuda, *Proc. Electrochemical Society Meeting*, Montreal (May 9–14, 1982).
122. J.B. Webb and S.R. Das, *J. Appl. Phys.* (1983).

References for Chapter 11

1. M.D. Archer, *J. Appl. Electrochem.*, **5**, 17 (1975).
2. H. Gerischer, in: *Physical Chemistry*, Vol. IX A (Eds., H. Eyring, D. Henderson and W. Jost), Academic Press, New York (1970).
3. H. Gerischer, in: *Solar Photoelectrolysis with Semiconductor Electrodes*, Topics in Applied Physics, Vol. 31, Springer-Verlag, New York (1979), p. 115.
4. H. Gerischer, in: *Advances in Electrochemistry and Electrochemical Engineering*, Vol. 1 (Ed., P. Delahey), Interscience (1961), p. 139.
5. H. Gerischer, *Electroanalytical Chemistry and Interfacial Electrochemistry*, **58**, 263 (1975).
6. H. Gerischer, in: *Solar Power and Fuels* (Ed., J.R. Botton), Academic Press, New York (1977), p. 77.

7. M. Green, in: *Modern Aspects of Electrochemistry*, Vol. 2 (Ed., J. O'M. Bockris), Butterworths, London (1959), p. 343.
8. V.A. Myamlin and Yu.V. Pleskov, *Electrochemistry of Semiconductors*, Plenum Press, New York (1967).
9. K. Rajeshwar, P. Singh and J. DuBow, *Electrochim. Acta*, **23**, 1117 (1978).
10. M.J. Spaarnay, *The Electrical Double Layer*, Pergamon Press, Oxford (1972).
11. F. Lohmann, *Z. Naturf.*, **22a**, 843 (1967).
12. S.M. Sze, *Physics of Semiconductor Devices*, Wiley, New York (1969).
13. M.A. Butler, *J. Appl. Phys.*, **48**, 1914 (1977).
14. A.K. Ghosh and H.P. Maruska, *J. Electrochem. Soc.*, **124**, 1516 (1977).
15. J. Gobrecht and H. Gerischer, *Solar Energy Materials*, **2**, 131 (1979).
16. A.K. Ghosh, D.L. Morel, T. Feng, R.F. Shaw and C.A. Rowe, *J. Appl. Phys.*, **45**, 230 (1974).
17. A.B. Ellis, S.W. Kaiser and M.S. Wrighton, *J. Am. Chem. Soc.*, **98**, 6855 (1976).
18. H. Gerischer and W. Mindt, *Electrochim. Acta*, **13**, 1329 (1968).
19. A.J. Bard and M.S. Wrighton, in: *Semiconductor–Liquid Junction Solar Cells*, Vol. 77–3, Electrochemical Society, Princeton (1977), p. 195.
20. H. Gerischer, *J. Vac. Sci. Technol.*, **15**, 1422 (1978).
21. A. Fujishima and K. Honda, *Bull. Chem. Soc. Japan*, **44**, 1148 (1971); *Nature*, **238**, 37 (1972).
22. W.M. Latimer, *Oxidation Potentials*, Prentice Hall, New York (1952), p. 42.
23. J.M. Bolts and M.S. Wrighton, *J. Phys. Chem.*, **80**, 2641 (1976).
24. J. Manassen, D. Cahen, G. Hodes and A. Sofer, *Nature*, **263**, 97 (1976).
25. A.J. Nozik, *Nature*, **257**, 383 (1975).
26. A. Fujishima, K. Kohayakawa and K. Honda, *Bull. Chem. Soc. Japan*, **48**, 1041 (1975).
27. M.S. Wrighton, D.S. Ginley, P.T. Wolczanski, A.B. Ellis, D.L. Morse and A. Linz, *Proc. National Academy of Science (USA)*, **72**, 1518 (1975).
28. J. G. Mavroides, D.I. Tchernev, J.A. Kafalas and D.F. Kolesar, *Mat. Res. Bull.*, **10**, 1023 (1975).
29. W. Gissler, P.L. Lensi and S. Pizzini, *J. Appl. Electrochem.*, **6**, 9 (1976).
30. K.L. Hardee and A.J. Bard, *J. Electrochem. Soc.*, **122**, 739 (1975).
31. J. Kenney, D.H. Weinstein and G.M. Hass, *Nature*, **253**, 719 (1975).
32. H. Tamura, H. Yoneyama, C. Iwakura and T. Murai, *Bull. Chem. Soc. Japan*, **50**, 753 (1977).
33. H.P. Maruska and A.K. Ghosh, *Solar Energy Materials*, **1**, 237 (1979).
34. Y. Matsumoto, J. Kurimoto, Y. Amagasaki and E. Sato, *J. Electrochem. Soc.*, **127**, 2148 (1980).
35. J.H. Carey and B.G. Oliver, *Nature*, **259**, 554 (1976).
36. A.B. Bocarsly, J.M. Bolts, P.G. Cummins and M.S. Wrighton, *Appl. Phys. Lett.*, **31**, 568 (1977).
37. L.A. Harris, D.R. Cross and M.E. Gerstner, *J. Electrochem. Soc.*, **124**, 839 (1977).
38. L.A. Harris and R.H. Wilson, *J. Electrochem. Soc.*, **123**, 1010 (1976).
39. J.G. Mavroides, J.A. Kafalas, D.F. Kolesar, *Appl. Phys. Lett.*, **28**, 241 (1976).
40. M.S. Wrighton, A.B. Ellis, P.T. Wolczanski, D.L. Morse, H.B. Abrahamson and D.S. Ginley, *J. Am. Chem. Soc.*, **98**, 2774 (1976).
41. T. Watanabe, A. Fujishima and K. Honda, *Bull. Chem. Soc. Japan*, **49**, 355 (1976).
42. R.D. Nasby and R.K. Quinn, *Mat. Res. Bull.*, **11**, 985 (1976).
43. M. Okuda, K. Yoshida and N. Tanaka, *Jpn. J. Appl. Phys.*, **15**, 1599 (1976).
44. J.H. Kennedy and K.W. Frise, *J. Electrochem. Soc.*, **123**, 1683 (1976).
45. A.B. Ellis, E.W. Kaiser and M.S. Wrighton, *J. Phys. Chem.*, **80**, 1325 (1976).
46. P. Clecht, J. Martin, R. Oliver and C. Vallony, *C.R. Acad. Sci. Ser. C*, **282**, 887 (1976).

47. M.S. Wrighton, D.L. Morse, A.B. Ellis, D.S. Ginley and H.B. Abrahamson, *J. Am. Chem. Soc.*, **98**, 44 (1976).
48. K. Kim and H.A. Laitinen, *J. Electrochem. Soc.*, **122**, 53 (1975).
49. H. Gerischer, *Proc. 2nd EC Photovoltaic Solar Energy Conference* (Eds., R. Van Overstraeten and W. Palz), Berlin (April 1979), p. 408.
50. J. Augustynski, J. Hinden and C. Stalder, *J. Electrochem. Soc.*, **124**, 1063 (1977).
51. H.H. Kung, M.S. Jarrett, A.W. Sleight and A. Ferretti, *J. Appl. Phys.*, **48**, 2463 (1977).
52. M.A. Butler, R.D. Nasby and R.K. Quinn, *Solid State Commun.*, **19**, 1011 (1976).
53. G. Hodes, D. Cahen and J. Manassen, *Nature*, **260**, 312 (1976).
54. K.L. Hardee and A.J. Bard, *J. Electrochem. Soc.*, **124**, 215 (1977).
55. M.A. Butler, D.S. Ginley and M. Eibschutz, *J. Appl. Phys.*, **48**, 3070 (1977).
56. K.L. Hardee and A.J. Bard, *J. Electrochem. Soc.*, **123**, 1024 (1976).
57. R.K. Quinn, R.D. Nasby and R.J. Baughnan, *Mat. Res. Bull.*, **11**, 1011 (1976).
58. J.H. Kennedy and K.W. Freese, *J. Electrochem. Soc.*, **124**, 833 (1977).
59. A.J. Nozik, *Appl. Phys. Lett.*, **28**, 150 (1976).
60. H. Yoneyama, H. Sakamoto and H. Tamura, *Electrochim. Acta*, **20**, 341 (1975).
61. K. Ohashi, J. McCann and J.O'M. Bockris, *Int. J. Energy Research*, **1**, 259 (1977).
62. H. Gerischer and E. Meyer, *Z. Phys. Chem.*, **74**, 302 (1971).
63. R.A.L. Van den Berghe, W.P. Gomes and F. Cardon, *Z. Phys. Chem.*, **92**, 91 (1974).
64. K. Ohashi, J. McCann and J.O'M Bockris, *Nature*, **266**, 610 (1977).
65. K. Ohashi, K. Vosaki and J.O'M. Bockris, *Energy Research*, **1**, 25 (1977).
66. A. Yamamoto and S. Yano, *J. Electrochem. Soc.*, **122**, 260 (1975).
67. A.J. Nozik, *Appl. Phys. Lett.*, **29**, 150 (1976).
68. J. O'M. Bockris and K. Vosaki, *J. Electrochem. Soc.*, **124**, 1348 (1977).
69. P.A. Kohl, S.N. Frank and A.J. Bard, *J. Electrochem. Soc.*, **124**, 225 (1977).
70. S. Gourgand and D. Elliot, *J. Electrochem. Soc.*, **124**, 102 (1977).
71. M. Tomkiewicz and J.M. Woodall, *J. Electrochem. Soc.*, **124**, 1436 (1977).
72. H. Gerischer and J. Gobrecht, *Berlin Bunsen Ges. Phys. Chem.*, **80**, 327 (1976).
73. A.B. Ellis, S.W. Kaiser and M.S. Wrighton, *J. Am. Chem. Soc.*, **98**, 6855 (1976).
74. A. Heller, K.C. Chang and B. Miller, *J. Electrochem. Soc.*, **124**, 697 (1977).
75. B. Miller and A. Heller, *Nature*, **262**, 680 (1976).
76. C.C. Tsou and J.R. Cleveland, *J. Appl. Phys.*, **50**, 455 (1980).
77. M. Tsuiki, H. Minoura, T. Nakamura and Y. Veno, *J. Appl. Electrochem.*, **8**, 523 (1978).
78. A. Heller, G.P. Schwartz, R.G. Vadimsky, S. Menezes and B. Miller, *J. Electrochem. Soc.*, **125**, 1623 (1978).
79. A. Heller, K.C. Chang and B. Miller, *J. Am. Chem. Soc.*, **100**, 684 (1978).
80. S. Deb, W.L. Wallace and R. Noufi, *Abstracts of the Fourth Annual Photovoltaic Advanced Research and Development Conference*, Colorado (November 1980), p. 291
81. W.J. Danaher and L.E. Lyons, *Nature*, **271**, 139 (1978).
82. A. Heller and B. Miller, *Electrochim. Acta*, **25**, 29 (1980).
83. A.B. Ellis, S.W. Kaiser and M.S. Wrighton, *J. Am. Chem. Soc.*, **98**, 1635 (1976).
84. A.B. Ellis, S.W. Kaiser, J.M. Bolts and M.S. Wrighton, *J. Am. Chem. Soc.*, **99**, 2839 (1977).
85. A. Heller and B. Miller, in: *Interfacial Photoprocesses*: Energy Conversion and Synthesis (Ed., M.S. Wrighton), American Chemical Society (1980), p. 215.
86. B. Miller, A. Heller, M. Robbins, S. Menzes, K.C. Chang and J. Thomson, Jr., *J. Electrochem. Soc.*, **124**, 1019 (1977).
87. M.A. Russak, J. Reichmann, H. Witzke, S.K. Deb and S.N. Chen, *J. Electrochem. Soc.*, **127**, 725 (1980).
88. J.R. Owen, *Nature*, **267**, 504 (1977).
89. C.J. Liu and J.H. Wang, *Appl. Phys. Lett.*, **36**, 852 (1980).

90. L. Thompson, K. Rajeshwar, P. Singh, R.C. Kainthla and K.L. Chopra, *J. Electrochem. Soc.*, **128**, 1744 (1981).
91. R.C. Kainthla, Ph.D. Thesis, Indian Institute of Technology, Delhi (1980).
92. G. Hodes, J. Manassen and D. Cahen, *Nature*, **261**, 403 (1976).
93. S. Chandra and R.K. Pandey, *Phys. Stat. Sol.* (*a*), **59**, 787 (1980).
94. S. Deb, W.L. Wallace and R. Noufi, *Abstracts of the Fourth Annual Photovoltaic Advanced Research and Development Conference*, Colorado (November 1980), p. 296.
95. K.T.L. Desilva and D. Haneman, *J. Electrochem. Soc.*, **127**, 1554 (1980).
96. A. Heller, G.P. Schwartz, R.G. Vadimsky, S. Menezes and B. Miller, *J. Electrochem. Soc.*, **125**, 1156 (1978).
97. R. Noufi, D. Tench and L. Warren, *J. Am. Chem. Soc.*, **99**, 309 (1977).
98. G. Hodes, *Nature*, **285**, 29 (1980).
99. G. Hodes, D. Cahen, J. Manassen and M. David, *J. Electrochem. Soc.*, **127**, 2252 (1980).
100. G. Hodes, J. Manassen and D. Cahen, *J. Electrochem. Soc.*, **127**, 544 (1980).
101. B. Parkinson, A. Heller and B. Miller, *Appl. Phys. Lett.*, **33**, 521 (1978).
102. B.A. Parkinson, A. Heller and B. Miller, *J. Electrochem. Soc.*, **126**, 954 (1979).
103. K.C. Chang, A. Heller, B. Schwartz, S. Menezes and B. Miller, *Science*, **196**, 1097 (1977).
104. S. Menezes, A. Heller and B. Miller, *J. Electrochem. Soc.*, **127**, 1268 (1980).
105. B. Parkinson, A. Heller and B. Miller, *Proc. 13th IEEE Photovoltaic Specialists Conference*, Washington, D.C. (1978), p. 1253.
106. A.B. Ellis, J.M. Bolts, S.W. Kaiser and M.S. Wrighton, *J. Am. Chem. Soc.*, **99**, 2848 (1977).
107. A. Heller, B. Miller, S.S. Chu and Y.T. Lee, *J. Am. Chem. Soc.*, **101**, 7633 (1979).
108. H. Tributsch and J.C. Bennett, *J. Electroanalytical Chem.*, **81**, 97 (1977).
109. H. Tributsch, *Berlin Bunsen Ges. Phys. Chem.*, **81**, 361 (1977).
110. H. Tributsch, *Berlin Bunsen Ges. Phys. Chem.*, **82**, 169 (1978).
111. H. Tributsch, *J. Electrochem. Soc.*, **125**, 1086 (1978).
112. J. Gobrecht, H. Gerischer and H. Tributsch, *J. Electrochem. Soc.*, **125**, 2085 (1978).
113. F.F. Fan, H.S. White, B. Wheeler and A.J. Bard, *J. Electrochem. Soc.*, **127**, 518 (1980).
114. M. Robbins, K.J. Bachmann, V.G. Lambrecht, F.A. Thiel, J. Thomson, Jr., R.G. Vadimsky, S. Menezes, A. Heller and B. Miller, *J. Electrochem. Soc.*, **125**, 831 (1978).
115. R. Memming, *J. Electrochem. Soc.*, **125**, 117 (1978).
116. M.A. Butler, *J. Electrochem. Soc.*, **127**, 1273 (1980).
117. S.K. Deb and W.L. Wallace, *Proc. Society of Photo-Optical Instrumentation Engineers*, **248**, 38 (1980).
118. D.J. Miller and D. Haneman, *Solar Energy Materials*, **4**, 231 (1981).

References for Chapter 12

1. J.E. Parrott, *IEEE Trans. Electron Dev.*, **ED–21**, 89 (1974).
2. P.E. Gray, *IEEE Trans. Electron Dev.*, **ED–16**, 424 (1969).
3. S.M. Sze, *Physics of Semiconductor Devices*, John Wiley & Sons, New York (1969).
4. D.L. Pulfrey, *Photovoltaic Power Generation*, Van Nostrand-Reinhold Company, New York (1978).
5. S.R. Dhariwal, L.S. Kothari and S.C. Jain, *IEEE Trans. Electron Dev.*, **ED–23**, 504 (1976).
6. J.G. Fossum and E.L. Burgess, *Proc. 12th IEEE Photovoltaic Specialists Conference*, Baton Rouge (1976), p. 737

7. H.J. Hovel and J.M. Woodall, *Proc. 12th IEEE Photovoltaic Specialists Conference*, Baton Rouge (1976), p. 945.
8. L.W. James and R.L. Moon, *Proc. 11th IEEE Photovoltaic Specialists Conference*, Scottsdale (1975), p. 402.
9. J.A. Castle, *Proc. 12th IEEE Photovoltaic Specialists Conference*, Baton Rouge (1976), p. 751.
10. R.H. Dean, L.S. Napoli and S.G. Liu, *RCA Rev.*, **36**, 324 (1975).
11. H.J. Hovel, *Solar Cells, Semiconductors and Semimetals*, Vol. 11 (Eds., A.C. Beer and R.K. Willardson), Academic Press, New York (1975), p. 139.
12. P. Shah, *Solid-State Electronics*, **18**, 1099 (1975).
13. W.W. Lloyd, *Proc. 11th IEEE Photovoltaic Specialists Conference*, Scottsdale (1975), p. 349.
14. *Proc. 9th IEEE Photovoltaic Specialists Conference*, Silver Spring (1972).
15. R.J. Schwartz and M.D. Lammert, *Proc. IEEE Int. Electron Devices Meeting*, Washington, D.C. (1975), p. 353.
16. T.I. Chappell, *IEEE Trans. Electron Devices*, **ED–26**, 1091 (1979).
17. H.J. Hovel, R.J. Hodgson and J.M. Woodall, *Solar Energy Materials*, **2**, 19 (1979).
18. S. Sakai and M. Umeno, *J. Appl. Phys.*, **51**, 5018 (1980).
19. M. Ariezo and J.J. Loferski, *Proc. 13th IEEE Photovoltaic Specialists Conference*, Washington, D.C. (1978), p. 898.
20. S. Chiang, B.G. Carbajal and G.F. Wakefield, *13th IEEE Photovoltaic Specialists Conference*, Washington, DC (1975), p. 1290.
21. J.J. Loferski, *Proc. 12th IEEE Photovoltaic Specialists Conference*, Baton Rouge (1976), p. 957.
22. S.M. Bedair, S.B. Pathak and J.R. Hauser, *IEEE Trans. Electron Devices*, **ED–27**, 822 (1980).
23. M.F. Lamorte and D.H. Abbott, *IEEE Trans. Electron Devices*, **ED–27**, 831 (1980).
24. G.W. Masden and C.E. Backus, *Proc. 13th IEEE Photovoltaic Specialists Conference*, Washington, D.C. (1978), p. 853.
25. J.A. Cape, J.S. Harris, Jr. and R. Sahai, *Proc. 13th IEEE Photovoltaic Specialists Conference*, Washington D.C. (1978), p. 881.
26. R.L. Moon, L.W. James, H.A. Van der Plas, T.O. Yep, C.A. Antypas and U. Chai, *Proc. 13th IEEE Photovoltaic Specialists Conference*, Washington, D.C. (1978), p. 859.
27. A. Bennet and L.C. Olsen, *Proc. 13th IEEE Photovoltaic Specialists Conference*, Washington, D.C. (1978), p. 869.
28. E. Fanetti, C. Flores, G. Guarini, F. Paletta and D. Passoni, *Solar Cells*, **3**, 187 (1981).
29. H.A. Van der Plas, R.L. Moon, L.W. James, T.O. Yep and R.R. Fulks, *Proc. 2nd European Community Conference on Photovoltaic Solar Energy*, Berlin (1979), p. 507.
30. M.P. Vecchi, *Solar Energy*, **22**, 383 (1979).
31. L.M. Fraas and R.C. Knechtli, *Proc. 13th IEEE Photovoltaic Specialists Conference*, Washington, D.C. (1978), p. 885.
32. M.F. Lamorte and D. Abbott, *Solid-State Electronics*, **22**, 467 (1979).
33. S.M. Bedair, M.F. Lamorte and J.R. Hauser, *Appl. Phys. Lett.*, **34**, 38 (1979).
34. J.A. Hutchby, S.M. Bedair, A.D. Brooks, R.A. Connnor, M. Dubey, F.M. Stevens and M. Simons, SERI Contractors Review Meeting, Research Triangle Park, N.C. (March 1981).
35. M.L. Timmons, S.M. Bedair, J.A. Hutchby, T.S. Colpitts, M. Simons and J.R. Hauser, SERI Contractors Review Meeting, Research Triangle Park, N.C. (March 1981).
36. A.G. Milnes, *Proc. 13th IEEE Photovoltaic Specialists Conference*, Washington, D.C. (1978), p. 892.
37. C.H. Henry, *J. Appl. Phys.*, **51**, 4494 (1980).

References for Appendix A

1. *Solar Cells*, Ed., C.E. Backus, IEEE Press, New York (1976), pp. 1–9.
2. M.P. Thekaekara, *Data on Incident Solar Energy, The Energy Crisis and Energy from the Sun*, Institute of Environmental Sciences (1974).
3. Terrestrial Photovoltaic Measurement Procedures, NASA Lewis Research Center, Report No. ERDA/NASA/1022–77/16, NASA TM 73702 (June 1977).
4. Interim Solar Cell Testing Procedures for Terrestrial Applications, NASA Lewis Research Center, Report No. NASA TMX–71771 (1975).

References for Appendix B

1. H.A. MacLeod, *Thin Film Optical Filters*, Adam Hilger, London (1969).
2. H.A. MacLeod, in: *Active and Passive Thin Film Devices* (Ed., T.J. Coutts), Academic Press, London (1978).
3. O.S. Heavens, *Optics of Thin Films*, Dover, London (1965).
4. G. Seibert, Technical Note ESROTN-90 ESTEC (1969).
5. K. Kern and E. Tracy, *RCA Review*, **41**, 133 (1980).
6. Y.C.M. Yeh, F.P. Ernest and R.J. Stirn, *J. Appl. Phys.*, **47**, 4107 (1976).
7. B. Gandham, R. Hill, H.A. MacLeod and M. Bowden, *Solar Cells*, **1**, 3 (1979).
8. L. Chambouleyron and E. Saucedo, *Solar Energy Materials*, **1**, 299 (1979).
9. J.H. Apfel, *Proc. Sym. Material Science Aspects of Thin Film Systems for Solar Energy Conversion*, Tucson (May 1974), p. 276.
10. A. Rothwarf, J. Phillips and N.C. Wyeth, *Proc. 13th IEEE Photovoltaic Specialists Conference*, Washington, D.C. (1978).
11. P.H. Berning, *J. Opt. Soc. Am.*, **52**, 431 (1962).
12. P.B. Clapham and M.C. Hutley, *Nature*, **244**, 281 (1973).
13. G.H. Hewig, F. Pfisterer, H.W. Schock and W.H. Bloss, *Proc. Workshop on the II–VI Solar Cells and Similar Compounds*, Montpellier (September 1979), p. VII–1.

References for Appendix C

1. R.J. Handy, *Solid-State Electronics*, **10**, 765 (1967).
2. M. Wolf, *Proc. Inst. Radio Engrs.*, **48**, 1246 (1960).
3. C.R. Fang and J.R. Hauser, *Proc. 13th IEEE Photovoltaic Specialists Conference*, Washington (1978), p. 1306.
4. A. Flat and A.G. Milnes, *Solar Energy*, **25**, 283 (1980).
5. H.B. Serreze, *Proc. 13th IEEE Photovoltaic Specialists Conference*, Washington (1978), p. 609.
6. N.C. Wyeth, *Solid-State Electronics*, **20**, 629 (1977).
7. K.W. Heizer and T.L. Chu, *Solid-State Electronics*, **19**, 471 (1976).
8. N.C. Wyeth, Sheet Resistance Component of Series Resistance in a Solar Cell as a Function of Grid Geometry, Technical Report, IEC/PV/TR/77/2, University of Delaware (1977).

9. Progress Report E(49–18)–2538 PR76/1, University of Delaware (1977).
10. S.R. Das, Ph.D. Thesis, Indian Institute of Technology, Delhi (1978).
11. A.M. Barnett, J.D. Meakin and A. Rothwarf, Appendix C, Development of a Thin Film Polycrystalline Solar Cell for Large-Scale Terrestrial Use, Final Report E(49–18)–2538 FR77, University of Delaware (1977).
12. W. Arndt, G. Bilger, G.H. Hewig, F. Pfisterer, H.W. Schock, J. Worner, and W.H. Bloss, *Proc. 2nd European Commission Photovoltaic Solar Energy Conference*, Berlin (1979), p. 826.
13. T.F. Deucher, Thin Film CdS Solar Cell Fabrication Parameter Study, Report No. ARL 70–0099, Contract No. F33615–68–C–1182, Aerospace Research Laboratories (June 1970).
14. P. Rai Choudhary, Westinghouse, 1980 (Personal communication).
15. R. Glang and L.V. Gregor, in: *Handbook of Thin Film Technology* (Eds., L.I. Maissel and R.Glang), McGraw-Hill Book Company, New York (1970), p. 7–1.
16. W.S. De Forest, *Phoresist Materials and Processes*, McGraw-Hill Book Company, New York (1975).

References for Appendix D

1. H.S. Rauschenbach, *Solar Cell Arrays*, Van Nostrand-Reinhold Co., New York (1980).
2. M. Sayed and L. Partain, *Energy Conversion*, **14**, 61 (1975).

References for Appendix E

1. H.S. Rauschenbach, *Solar Cell Arrays*, Van Nostrand-Reinhold Company, New York (1980).
2. D.H. Mash and P.W. Ross, *Solid-State and Electron Devices*, **2**, 574 (1978).
3. C. Wyman, J. Castle and F. Kreith, *Solar Energy*, **24**, 517 (1980).
4. J. Furber, *Proc. Int. Solar Energy Congress*, New Delhi (1978); *Sun: Mankind's Future Source of Energy*, **2**, 720 (1978).
5. M.C. Merchant, *Sunworld*, **4**, 22 (1980).
6. D.L. Pulfrey, *Photovoltaic Power Generation*, Van Nostrand-Reinhold Company, New York (1978).
7. W.T. Welford and R. Winston, *The Optics of Nonimaging Concentrators*, Academic Press, New York (1978).

References for Appendix F

1. R. Singh, M.A. Green and K. Rajkanan, *Solar Cells*, **3**, 95 (1980).
2. J. Besson, T. Nguyen Duy, A. Gauthier, W. Palz, C. Martin and J. Vedel, *Proc. 11th IEEE Photovoltaic Specialists Conference*, Scottsdale (1975), p. 468.

3. F.A. Shirland, W.J. Biter, E.W. Greeneich, A.J. Simon and T.P. Brody, Westinghouse Research Labs., Final Report NSF/RANN/SE/AER74–14918 A01/FR/76 (February 1977).
4. W. Arndt, G. Bilger, G.H. Hewig, F. Pfisterer, H.W. Schock, J. Worner and W.H. Bloss, *Proc. 2nd European Commission Photovoltaic Solar Energy Conference*, Berlin (1979), p. 826.
5. H.M. Windawi, *Proc. 11th IEEE Photovoltaic Specialists Conference*, Scottsdale (1975), p. 464.
6. H.E. Nastelin, J.M. Smith and A.L. Gombach, Clevite Corporation, Final Report, Contract No. NAS3–13467 (June 1971).
7. H.S. Rauschenbach, *Solar Cell Arrays*, Van Nostrand-Reinhold Co., New York (1980).

Index